GRAVITATION
AND COSMOLOGY:
PRINCIPLES AND APPLICATIONS
OF THE GENERAL THEORY
OF RELATIVITY

GRAVITATION
AND COSMOLOGY:
PRINCIPLES AND APPLICATIONS
OF THE GENERAL THEORY
OF RELATIVITY

STEVEN WEINBERG
Massachusetts Institute of Technology

John Wiley & Sons, Inc. New York London Sydney Toronto

Library of Congress Cataloging in Publication Data:

Weinberg, Steven.
Gravitation and cosmology

Includes bibliographies.
1. General relativity (Physics). 2. Gravitation.
3. Cosmology.

QC6.W47   530.1´1   78-37175
ISBN 0-471-92567-5

Printed in the United States of America.

10 9 8 7 6 5 4

TO LOUISE

PREFACE

Now that this book is done I can look back, and identify two purposes which led me to begin writing, and which have guided the work to completion.

One good practical purpose was to bring together and assess the wealth of data provided over the last decade by new techniques in experimental physics and in optical, radio, radar, X-ray, and infrared astronomy. Of course, new data will keep coming in even as the book is being printed, and I cannot hope that this work will remain up to date forever. I do hope, however, that by giving a comprehensive picture of the experimental tests of general relativity and observational cosmology, I will help to prepare the reader (and myself) to understand the new data as they emerge. I have also tried to look a little way into the future, and to discuss what may be the next generation of experiments, especially those based on artificial satellites of the earth and sun.

There was another, more personal reason for my writing this book. In learning general relativity, and then in teaching it to classes at Berkeley and M.I.T., I became dissatisfied with what seemed to be the usual approach to the subject. I found that in most textbooks geometric ideas were given a starring role, so that a student who asked why the gravitational field is represented by a metric tensor, or why freely falling particles move on geodesics, or why the field equations are generally covariant would come away with an impression that this had something to do with the fact that space-time is a Riemannian manifold.

Of course, this *was* Einstein's point of view, and his preeminent genius necessarily shapes our understanding of the theory he created. However, I believe that the geometrical approach has driven a wedge between general relativity and the theory of elementary particles. As long as it could be hoped, as Einstein did hope, that matter would eventually be understood in geometrical terms, it made sense to give Riemannian geometry a primary role in describing the theory of gravitation. But now the passage of time has taught us not to expect that the strong, weak, and electromagnetic interactions can be understood in geometrical terms, and too great an emphasis on geometry can only obscure the deep connections between gravitation and the rest of physics.

In place of Riemannian geometry, I have based the discussion of general relativity on a principle derived from experiment: the Principle of the Equivalence of Gravitation and Inertia. It will be seen that geometric objects, such as the metric, the affine connection, and the curvature tensor, naturally find their way into a theory of gravitation based on the Principle of Equivalence and, of course, one winds up in the end with Einstein's general theory of relativity. However, I have tried here to put off the introduction of geometric concepts until they are needed, so that Riemannian geometry appears only as a mathematical tool for the exploitation of the Principle of Equivalence, and not as a fundamental basis for the theory of gravitation.

This approach naturally leads us to ask *why* gravitation should obey the Principle of Equivalence. In my opinion the answer is not to be found in the realm of classical physics, and certainly not in Riemannian geometry, but in the constraints imposed by the quantum theory of gravitation. It seems to be impossible to construct any Lorentz-invariant quantum theory of particles of mass zero and spin two, unless the corresponding classical field theory obeys the Principle of Equivalence. Thus the Principle of Equivalence appears as the best bridge between the theories of gravitation and of elementary particles. The quantum basis for the Principle of Equivalence is briefly touched upon here in a section on the quantum theory of gravitation, but it was not possible to go far into the quantum theory in this book.

The nongeometrical approach taken in this book has, to some extent, affected the choice of the topics to be covered. In particular, I have not discussed in detail the derivation and classification of complicated exact solutions of the Einstein field equations, because I did not feel that most of this material was needed for a fundamental understanding of the theory of gravitation, and hardly any of it seemed to be relevant to experiments that might be carried out in the foreseeable future. By this omission, I have left out much of the work done by professional general relativists over the past decade, but I have tried to provide an entrée to this work through references and bibliographies. I regret the omission here of a detailed discussion of the beautiful theorems of Penrose and Hawking on gravitational collapse; these theorems are briefly discussed in Sections 11.9 and 15.11, but an adequate discussion would have taken up too much time and space.

I have tried to give a comprehensive set of references to the experimental literature on general relativity and cosmology. I have also given references to detailed theoretical calculations whenever I have quoted their results. However, I have not tried to give complete references to all the theoretical material discussed in the book. Much of this material is now classical, and to search out the original references would be an exercise in the history of science for which I did not feel equipped. The mere absence of literature citations should not be interpreted as a claim that the work presented is original, but some of it is.

It is a pleasure to acknowledge the inestimable help I have received in writing this book. Students in my classes over the past seven years have, by their questions and comments, helped to free the calculations of errors and obscurities. I especially thank Jill Punsky for carefully checking many of the derivations. I have drawn very heavily on the knowledge of many colleagues, including Stanley Deser, Robert Dicke, George Field, Icko Iben, Jr., Arthur Miller, Philip Morrison, Martin Rees, Leonard Schiff, Maarten Schmidt, Joseph Weber, Rainier Weiss, and especially Irwin Shapiro. Finally, I am greatly indebted to Connie Friedman and Lillian Horton for typing and retyping the manuscript with inexhaustible skill and patience.

STEVEN WEINBERG

Cambridge, Massachusetts
April 1971

NOTATION

Latin indices i, j, k, l, and so on generally run over three spatial coordinate labels, usually, 1, 2, 3 or x, y, z.

Greek indices α, β, γ, δ, and so on generally run over the four space-time inertial coordinate labels 1, 2, 3, 0 or x, y, z, t.

Greek indices μ, ν, κ, λ, and so on generally run over the four coordinate labels in a general coordinate system.

Repeated indices are summed unless otherwise indicated.

The metric $\eta_{\alpha\beta}$ in an inertial coordinate system has diagonal elements $+1$, $+1$, $+1$, -1.

A dot over any quantity denotes the time derivative of that quantity.

Cartesian three-vectors are indicated by boldface type.

The speed of light is taken to be unity, except when c.g.s. units are indicated. Planck's constant is not taken to be unity.

CONTENTS

Sections marked with an asterisk are somewhat out of the book's main line of development and may be omitted in a first reading.

PART TWO
THE GENERAL THEORY OF RELATIVITY

3 THE PRINCIPLE OF EQUIVALENCE 67

1 Statement of the Principle 67

Equivalence of gravitation and inertia ☐ Analogy with metric geometry ☐ The weak and strong principles of equivalence

PART THREE
APPLICATIONS OF GENERAL RELATIVITY

I5 COSMOLOGY: THE STANDARD MODEL 469

1   Einstein's Equations 470

Robertson-Walker metric, affine connection, Ricci tensor ☐ First-order field equation ☐ Upper limit on the age of the universe ☐ Curvature and the future of the universe ☐ Mach's principle ☐ Newtonian cosmology

2   Density and Pressure of the Present Universe 475

Critical density ☐ Density of galactic mass ☐ Intergalactic mass inside and outside galactic clusters ☐ Radio, microwave, far-infrared, optical, X-ray, γ-ray, and cosmic ray densities ☐ Pressure

3   The Matter-Dominated Era 481

Time as a function of R ☐ Age of the universe ☐ Red shift versus luminosity distance and parallax distance ☐ Number counts ☐ Measurements of the age of the universe: uranium dating, globular clusters ☐ Particle and event horizons

4   Intergalactic Emission and Absorption Processes 491

Optical depth ☐ Stimulated emission ☐ The Einstein relation ☐ Isotropic background ☐ Resonant absorption ☐ Absorption trough ☐ Absorption and emission of 21-cm radiation ☐ Search for Lyman α absorption ☐ Isotropic X-ray background ☐ Thermal history of intergalactic hydrogen ☐ Thomson scattering ☐ Time delay by intergalactic plasma ☐ Extragalactic pulsars

5   The Cosmic Microwave Radiation Background 506

Black-body radiation ☐ Black-body temperature and antenna temperature ☐ Models with TR constant ☐ Specific photon entropy ☐ Hot models ☐ Estimates of black-body temperature in the cosmological theory of element synthesis ☐ Observation of the cosmic microwave radiation background ☐ Absorption by interstellar molecules ☐ Summary of measurements of black-body temperature ☐ Gray-body radiation and the Rayleigh-Jeans law ☐ Expected departures from the black-body spectrum ☐ Anisotropies of small and large angular scale ☐ Velocity of the solar system ☐ Homogenization of the universe ☐ Discrete source models ☐ Scattering of cosmic ray electrons, electrons in radio sources, cosmic ray photons, and protons

6   Thermal History of the Early Universe 528

Summary of the early history of the universe ☐ Time scale ☐ Thermal equilibrium ☐ Vanishing chemical potentials ☐ The lepton-photon era ☐ Conditions at $10^{12}°K$ ☐ Decoupling of neutrinos ☐ Neutrino temperature after electron-positron annihilation ☐ Time as a function of temperature ☐ Degenerate neutrinos ☐ Measurements of neutrino degeneracy

7 Helium Synthesis 545

Theories of nucleosynthesis ☐ Neutron-proton conversion rates ☐ Neutron abundance as a function of time ☐ Equilibrium abundances of complex nuclei ☐ The deuterium bottleneck ☐ Helium production at $10^9 °K$ ☐ Measurements of the cosmic helium abundance: stellar masses and luminosities, solar neutrino experiments, direct solar measurements, theory of globular clusters, stellar spectra, spectroscopy of interstellar matter, extragalactic measurements ☐ Modifications in the expected helium abundance: cool models, fast or slow models, neutrino interactions, degeneracy

8 The Formation of Galaxies 561

Jeans's theory ☐ Analogy with plasma waves ☐ Acoustic limit ☐ Jeans's mass ☐ Effect of black-body background radiation ☐ Phases of galactic growth ☐ Acoustic damping ☐ Critical mass ☐ Observation of protogalactic fluctuations as small-scale anisotropies in the microwave background

9 Newtonian Theory of Small Fluctuations 571

Unperturbed solutions ☐ First-order equations ☐ Plane-wave solutions ☐ Rotational modes ☐ Differential equation for compressional modes ☐ Zero-pressure solutions: growth from recombination to the present ☐ Zero-curvature solutions: stable and unstable modes

10 General-Relativistic Theory of Small Fluctuations 578

Dissipative terms in the energy-momentum tensor ☐ Unperturbed solutions ☐ Equivalent solutions ☐ Elimination of space-time and time-time components ☐ Perturbations in the affine connection, Ricci tensor, source tensor ☐ First-order Einstein equations and equations of motion ☐ Unphysical solutions ☐ Plane waves ☐ Radiative modes: absorption and instability of gravitational waves ☐ Rotational modes ☐ Compressional modes: four coupled equations, long wavelength limit, growth at early times

11 The Very Early Universe 588

Elementary and composite particle models ☐ Fossil quarks and gravitons ☐ Heat production by bulk viscosity ☐ Symmetric cosmologies ☐ Necessity of a past singularity ☐ Necessity of a future singularity ☐ Periodic cosmologies

Bibliography 597

References 599

COPYRIGHT ACKNOWLEDGEMENTS

PART ONE
PRELIMINARIES

"But the tale of history
forms a very strong bulwark
against the stream of time,
and to some extent checks
its irresistible flow, and, of
all things done in it, as
many as history has taken
over, it secures and binds
together, and does not allow
them to slip away into the
abyss of oblivion."
Anna Comnena, The Alexiad

I HISTORICAL INTRODUCTION

Physics is not a finished logical system. Rather, at any moment it spans a great confusion of ideas, some that survive like folk epics from the heroic periods of the past, and others that arise like utopian novels from our dim premonitions of a future grand synthesis. The author of a book on physics can impose order on this confusion by organizing his material in either of two ways: by recapitulating its history, or by following his own best guess as to the ultimate logical structure of physical law. Both methods are valuable; the great thing is not to confuse physics with history, or history with physics.

This book sets out the theory of gravitation according to what I think is its inner logic as a branch of physics, and not according to its historical development. It is certainly a historical fact that when Albert Einstein was working out general relativity, there was at hand a preexisting mathematical formalism, that of Riemannian geometry, that he could and did take over whole. However, this historical fact does not mean that the essence of general relativity necessarily consists in the application of Riemannian geometry to physical space and time. In my view, it is much more useful to regard general relativity above all as a theory of *gravitation*, whose connection with geometry arises from the peculiar empirical properties of gravitation, properties summarized by Einstein's Principle of the Equivalence of Gravitation and Inertia. For this reason, I have tried throughout this book to delay the introduction of geometrical objects, such as the metric, the affine connection, and the curvature, until the use of these objects could be motivated by considerations of physics. The order of chapters here thus bears very little resemblance to the order of history.

Nevertheless, because we must not allow the history of physics "to slip away into the abyss of oblivion," this first chapter presents a brief backward look at three great antecedents to general relativity—non-Euclidean geometry, the Newtonian theory of gravitation, and the principle of relativity. Their history is traced up to 1916, the year in which they were brought together by Einstein in the General Theory of Relativity.[1]

1   History of Non-Euclidean Geometry

Euclid showed in his *Elements*[2] how geometry could be deduced from a few definitions, axioms, and postulates. These assumptions for the most part dealt with the most fundamental properties of points, lines, and figures, and seem as self-evident to schoolboys in the twentieth century as they did to Hellenistic mathematicians in the third century B.C. However, one of Euclid's assumptions has always seemed a little less obvious than the others. The fifth postulate states

"If a straight line falling on two straight lines make the interior angles on the same side less than two right angles, the two straight lines if produced indefinitely meet on that side on which the angles are less than two right angles."

For two thousand years geometers tried to purify Euclid's system by proving that the fifth postulate is a logical consequence of his other assumptions. Today we know that this is impossible. Euclid was right, there is no logical inconsistency in a geometry without the fifth postulate, and if we want it we will have to put it in at the beginning rather than prove it at the end. However, the struggle to prove the fifth postulate is one of the great success stories in the history of mathematics, because it ultimately gave birth to modern non-Euclidean geometry.

The list of those who hoped to prove the fifth postulate as a theorem includes Ptolemy (d. 168), Proclos (410–485), Nasir al din al Tusi (thirteenth century), Levi ben Gerson (1288–1344), P. A. Cataldi (1548–1626), Giovanni Alfonso Borelli (1608–1679), Giordano Vitale (1633–1711), John Wallis (1616–1703), Geralamo Saccheri (1667–1733), Johann Heinrich Lambert (1728–1777), and Adrien Marie Legendre (1752–1833). Without exception, their efforts only succeeded in replacing the fifth postulate with some other equivalent postulate, which might or might not seem more self-evident, but which in any case could not be proved from Euclid's other postulates either. Thus, the Athenian neo-Platonist Proclos offered the substitute postulate: "If a straight line intersects one of two parallels, it will intersect the other also." (That is, if we define parallel lines as straight lines that do *not* intersect however far extended, then there can be at most one line that passes through any given point and is parallel to a given line.) John Wallis, Savillian Professor at Oxford, showed that Euclid's fifth postulate could be replaced with the equivalent statement "Given any figure there exists a figure, similar to it, of any size." And Legendre proved the equivalence of the fifth postulate with the statement "There is a triangle in which the sum of the three angles is equal to two right angles."[3]

The attempt to dispense with Euclid's fifth postulate began to take a different direction in the eighteenth century. In 1733 the Jesuit Geralamo Saccheri published a detailed study of what geometry would be like if the fifth postulate were false. He particularly examined the consequences of what he called the "hypothesis of the acute angle," that is, that "a straight line being given, there can be drawn a perpendicular to it and a line cutting it at an acute angle, which do not intersect each other."[3] However, Saccheri did not really think that this is possible; he still believed in the logical necessity of the fifth postulate, and explored non-Euclidean geometry only in the hope of eventually turning up a logical contradiction. Similar tentative explorations of non-Euclidean geometry were begun by Lambert and Legendre.

It seems to have been Carl Friedrich Gauss (1777–1855) who first had the courage to accept non-Euclidean geometry as a logical possibility. His gradual enlightenment is recorded in a series of letters[4] to W. Bólyai, Olbers, Schumacher, Gerling, Taurinus, and Bessel, extending from 1799 to 1844. In a letter dated 1824 he begged Taurinus to keep silent about the "heretical opinions" he had revealed. Gauss even went to the extent of surveying a triangle[40] in the Harz mountains formed by Inselberg, Brocken, and Hoher Hagen to see if the sum of its interior angles was 180°! (It was.) Then, in 1832, Gauss received a letter from his friend Wolfgang Bólyai, describing the non-Euclidean geometry developed by his son, Janos Bólyai (1802–1860), an Austrian army officer. He subsequently also learned that a professor in the Kazan, Nikolai Ivanovich Lobachevski (1793–1856), had obtained similar results in 1826.

Gauss, Bólyai, and Lobachevski had independently discovered what in modern terms is called the *two-dimensional space of constant negative curvature*. Such spaces are still very interesting; we shall see in the chapter on cosmography that the space in which we actually live may be a three-dimensional space of constant curvature. But to its discoverers the important thing about their new geometry was that it describes an infinite two-dimensional space in which all of Euclid's assumptions are satisfied—except the fifth postulate! In this it is unique, which perhaps explains why it was discovered more or less independently in Germany, Austria, and Russia. (The surface of a sphere also satisfies Euclidean geometry without the fifth postulate, but being finite it does not have room for parallel lines.) We shall see in Chapter 13, on symmetric spaces, that the two-dimensional space of constant negative curvature cannot be realized as a surface in ordinary three-dimensional Euclidean space, which is doubtless why it took two millennia to find it. And of course it also violates the alternative "common-sense" versions of Euclid's fifth postulate given by Proclos, Wallis, and Legendre—through a given point there can be drawn *infinitely* many lines parallel to any given line; *no* figures of different size are similar; and the sum of the angles of any triangle is *less* than 180°.

However, it still remained an open possibility that Euclid's fifth postulate could be derived from the others, for it was not at all obvious that the geometry of

Gauss, Bólyai, and Lobachevski did not contain a logical inconsistency. The usual way to "prove" that a system of mathematical postulates is self-consistent is to construct a model that satisfies the postulates out of some other system whose consistency is (for the moment) unquestioned. For both Euclidean and non-Euclidean geometry the "model" is provided by the theory of real numbers. Descartes' analytic geometry shows that if a point is identified with a pair of real numbers (x_1, x_2) and the distance between two points (x_1, x_2) and (X_1, X_2) is identified as $[(x_1 - X_1)^2 + (x_2 - X_2)^2]^{1/2}$, then all of Euclid's postulates can be proved as theorems about real numbers. In 1870 a similar analytic geometry[5] was constructed by Felix Klein (1849–1925) for the geometry of Gauss, Bólyai, and Lobachevski—a "point" is represented as a pair of real numbers x_1, x_2 with

$$x_1{}^2 + x_2{}^2 < 1 \tag{1.1.1}$$

and the distance $d(x, X)$ between two points x, X is defined by

$$\cosh\left[\frac{d(x, X)}{a}\right] = \frac{1 - x_1 X_1 - x_2 X_2}{(1 - x_1{}^2 - x_2{}^2)^{1/2}(1 - X_1{}^2 - X_2{}^2)^{1/2}} \tag{1.1.2}$$

where a is a fundamental length which sets the scale of the geometry. Note that this space *is* infinite, because $d(x, X) \to \infty$ as $X_1{}^2 + X_2{}^2$ approaches unity. With this definition of "point" and "distance" one can verify that this model satisfies all of Euclid's postulates except the fifth, and in fact obeys the geometry discovered by Gauss, Bólyai, and Lobachevski. Thus after two millennia the logical independence of Euclid's fifth postulate was at last established.

This was just the beginning of the development of non-Euclidean geometry. We saw that in order to discover the geometry of Gauss, Bólyai, and Lobachevski it was necessary to give up the idea that a curved surface could only be described in terms of its embedding in ordinary three-dimensional spaces. How then *can* we describe and classify curved spaces? To pick up our story we must go back to 1827 when Gauss published his *Disquisitiones generales circa superficies curvas*. Gauss for the first time distinguished the *inner* properties of a surface, that is, the geometry experienced by small flat bugs living in the surface, from its *outer* properties, that is, its embedding in a higher-dimensional space, and he realized that it is the inner properties of surfaces that are "most worthy of being diligently explored by geometers."

Gauss also realized that the essential inner property of any surface is the metric function $d(x, X)$, which gives the distance between x and X along the shortest path between them on the surface. For instance, a cone or a cylinder has the same local inner properties as a plane, since a plane can be rolled without stretching or tearing (i.e., without distorting metric relations) into a cone or a cylinder. On the other hand, all cartographers know that a sphere cannot be unrolled onto a plane surface without distortion, and thus its local inner properties are not the same as the plane's.

There is a simple example that has been used by Einstein, Wheeler, and others to illustrate how the inner properties of a surface can be discovered by exploring its metric. (See Figure 1.1.) Consider N points in a plane. We can use one point as an origin of coordinates and draw an x-axis through a second point, so that the

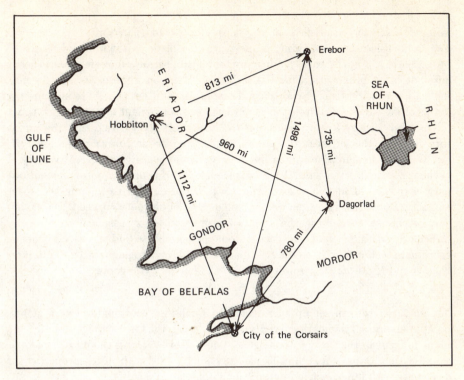

Figure 1.1   Problem: Is Middle Earth flat?

distances between the various points are described in terms of $(2N - 3)$ co-ordinates, that is, the x-coordinate of the second point and the x- and y-coordinates of the remaining $(N - 2)$ points. But there are $N(N - 1)/2$ different distances between the N points, and thus for large enough N these distances must be subject to M algebraic relations, where

$$M = \frac{N(N - 1)}{2} - (2N - 3) = \frac{(N - 2)(N - 3)}{2} \qquad (1.1.3)$$

For instance, in the simplest interesting case, $N = 4$, we can easily show that the distances d_{mn} between points m and n satisfy the single relation

$$
\begin{aligned}
0 = {} & d_{12}{}^4 d_{34}{}^2 + d_{13}{}^4 d_{24}{}^2 + d_{14}{}^4 d_{23}{}^2 + d_{23}{}^4 d_{14}{}^2 + d_{24}{}^4 d_{13}{}^2 + d_{34}{}^4 d_{12}{}^2 \\
& + d_{12}{}^2 d_{23}{}^2 d_{31}{}^2 + d_{12}{}^2 d_{24}{}^2 d_{41}{}^2 + d_{13}{}^2 d_{34}{}^2 d_{41}{}^2 + d_{23}{}^2 d_{34}{}^2 d_{42}{}^2 \\
& - d_{12}{}^2 d_{23}{}^2 d_{34}{}^2 - d_{13}{}^2 d_{32}{}^2 d_{24}{}^2 - d_{12}{}^2 d_{24}{}^2 d_{43}{}^2 - d_{14}{}^2 d_{42}{}^2 d_{23}{}^2 \\
& - d_{13}{}^2 d_{34}{}^2 d_{42}{}^2 - d_{14}{}^2 d_{43}{}^2 d_{32}{}^2 - d_{23}{}^2 d_{31}{}^2 d_{14}{}^2 - d_{21}{}^2 d_{13}{}^2 d_{34}{}^2 \\
& - d_{24}{}^2 d_{41}{}^2 d_{13}{}^2 - d_{21}{}^2 d_{14}{}^2 d_{43}{}^2 - d_{31}{}^2 d_{12}{}^2 d_{24}{}^2 - d_{32}{}^2 d_{21}{}^2 d_{14}{}^2
\end{aligned}
$$

$$(1.1.4)$$

This relation will be satisfied on any simply connected patch of a cylinder or a cone, which share the same inner properties as the plane, but it will *not* be satisfied by a table of airline distances among any four cities, because the earth's surface has different inner properties. There is a different relation appropriate to spherical surfaces, which *is* satisfied by airline mileage tables, and can be used to measure the radius of the earth. Of course, this is not the most convenient method and it is not the method used by Eratosthenes, but the important point here is that the curvature of the earth's surface can be determined from its local inner properties.

Were our imaginations given free rein, we could conceive of a great variety of peculiar metric functions $d(x, X)$. It was Gauss's great contribution to pick out one particular class of metric spaces, which was broad enough to include the space of Gauss, Bólyai, and Lobachevski as well as that of ordinary curved surfaces, but narrow enough to deserve the name of geometry. Gauss assumed that in any sufficiently small region of the space it would be possible to find a locally Euclidean coordinate system (ξ_1, ξ_2) so that the distance between two points with coordinates (ξ_1, ξ_2) and $(\xi_1 + d\xi_1, \xi_2 + d\xi_2)$ satisfies the law of Pythagoras,

$$ds^2 = d\xi_1{}^2 + d\xi_2{}^2 \tag{1.1.5}$$

For instance, we can set up such a locally Euclidean coordinate system at any point in an ordinary smooth curved surface by using the Cartesian coordinates of a plane tangent to the surface at the given point. However, this should not make us suppose that Gauss's assumption has anything to do with outer properties; it deals only with inner metric relations for infinitesimal neighborhoods.

If a surface is not Euclidean, it will not be possible to cover any *finite* part of it with a Euclidean coordinate system (ξ_1, ξ_2) satisfying the law of Pythagoras. Suppose that we use some other coordinate system (x_1, x_2) that *does* cover the space, and ask what form Gauss's assumption takes in these coordinates. It is easy to calculate that the distance ds between points (x_1, x_2) and $(x_1 + dx_1, x_2 + dx_2)$ is given by

$$ds^2 = g_{11}(x_1, x_2)\, dx_1{}^2 + 2g_{12}(x_1, x_2)\, dx_1\, dx_2 + g_{22}(x_1, x_2)\, dx_2{}^2 \tag{1.1.6}$$

where

$$g_{11} = \left(\frac{\partial \xi_1}{\partial x_1}\right)^2 + \left(\frac{\partial \xi_2}{\partial x_1}\right)^2$$

$$g_{12} = \left(\frac{\partial \xi_1}{\partial x_1}\right)\left(\frac{\partial \xi_1}{\partial x_2}\right) + \left(\frac{\partial \xi_2}{\partial x_1}\right)\left(\frac{\partial \xi_2}{\partial x_2}\right) \tag{1.1.7}$$

$$g_{22} = \left(\frac{\partial \xi_1}{\partial x_2}\right)^2 + \left(\frac{\partial \xi_2}{\partial x_2}\right)^2$$

This form for ds^2 is the hallmark of a *metric space*. [We shall see in Chapter 3 that this derivation can be reversed; given any space with ds given by (1.1.6), we can at any point choose *locally* Euclidean coordinates ξ_1, ξ_2 satisfying (1.1.5).] For the case of a sphere of radius a we can use spherical polar coordinates θ, φ, and the metric is

$$g_{\theta\theta} = a^2, \qquad g_{\theta\varphi} = 0, \qquad g_{\varphi\varphi} = a^2 \sin^2 \theta \tag{1.1.8}$$

It is the factor $\sin^2 \theta$ in $g_{\varphi\varphi}$ that gives a sphere different inner properties from a plane. In the geometry of Gauss, Bólyai, and Lobachevski, we can use the coordinates x_1, x_2 of Klein's model, and find from the posited formula for $d(x, X)$ that

$$g_{11} = \frac{a^2(1 - x_2{}^2)}{(1 - x_1{}^2 - x_2{}^2)^2} \qquad g_{12} = \frac{a^2 x_1 x_2}{(1 - x_1{}^2 - x_2{}^2)^2} \qquad g_{22} = \frac{a^2(1 - x_1{}^2)}{(1 - x_1{}^2 - x_2{}^2)^2}$$

$$(1.1.9)$$

The length of any path can be determined by integrating ds along the path.

The metric functions g_{ij} determine all inner properties of a metric space, but they also depend on how we choose the coordinate mesh. For instance, we can use polar coordinates r, θ to describe a plane surface, and find that the metric functions are

$$g_{rr} = 1 \qquad g_{r\theta} = 0 \qquad g_{\theta\theta} = r^2 \tag{1.1.10}$$

This does not *look* like a Euclidean space, but of course it is, as we can show formally by transforming to Cartesian coordinates $x = r \cos \theta$, $y = r \sin \theta$. More generally, a change of coordinates from (x_1, x_2) to (x_1', x_2') will change the metric functions g_{ij} to g_{ij}', where, for instance,

$$\begin{aligned}
g_{11}' &= \left(\frac{\partial \xi_1}{\partial x_1'}\right)^2 + \left(\frac{\partial \xi_2}{\partial x_1'}\right)^2 \\
&= \left(\frac{\partial \xi_1}{\partial x_1}\frac{\partial x_1}{\partial x_1'} + \frac{\partial \xi_1}{\partial x_2}\frac{\partial x_2}{\partial x_1'}\right)^2 + \left(\frac{\partial \xi_2}{\partial x_1}\frac{\partial x_1}{\partial x_1'} + \frac{\partial \xi_2}{\partial x_2}\frac{\partial x_2}{\partial x_1'}\right)^2 \\
&= g_{11}\left(\frac{\partial x_1}{\partial x_1'}\right)^2 + 2g_{12}\frac{\partial x_1}{\partial x_1'}\frac{\partial x_2}{\partial x_1'} + g_{22}\left(\frac{\partial x_2}{\partial x_1'}\right)^2
\end{aligned} \tag{1.1.11}$$

How then can we tell the inner properties of a space by looking at its metric coefficients? What we need is some function of the g_{ij} and their derivatives that depends only on the inner properties of the space and not, like the g_{ij}, also on the particular coordinate system chosen to describe the space.

Gauss found this function, and found it to be essentially unique; it is the so-called Gaussian curvature:

$$\begin{aligned}
K(x_1, x_2) = \frac{1}{2g}&\left[2\frac{\partial^2 g_{12}}{\partial x_1 \partial x_2} - \frac{\partial^2 g_{11}}{\partial x_2{}^2} - \frac{\partial^2 g_{22}}{\partial x_1{}^2}\right] \\
-\frac{g_{22}}{4g^2}&\left[\left(\frac{\partial g_{11}}{\partial x_1}\right)\left(2\frac{\partial g_{12}}{\partial x_2} - \frac{\partial g_{22}}{\partial x_1}\right) - \left(\frac{\partial g_{11}}{\partial x_2}\right)^2\right] \\
+\frac{g_{12}}{4g^2}&\left[\left(\frac{\partial g_{11}}{\partial x_1}\right)\left(\frac{\partial g_{22}}{\partial x_2}\right) - 2\left(\frac{\partial g_{11}}{\partial x_2}\right)\left(\frac{\partial g_{22}}{\partial x_1}\right)\right. \\
&\left.+ \left(2\frac{\partial g_{12}}{\partial x_1} - \frac{\partial g_{11}}{\partial x_2}\right)\left(2\frac{\partial g_{12}}{\partial x_2} - \frac{\partial g_{22}}{\partial x_1}\right)\right] \\
-\frac{g_{11}}{4g^2}&\left[\left(\frac{\partial g_{22}}{\partial x_2}\right)\left(2\frac{\partial g_{12}}{\partial x_1} - \frac{\partial g_{11}}{\partial x_2}\right) - \left(\frac{\partial g_{22}}{\partial x_1}\right)^2\right]
\end{aligned} \tag{1.1.12}$$

where g is the determinant

$$g(x_1, x_2) \equiv g_{11}g_{22} - g_{12}{}^2$$

(The reader should not quail at the awful appearance of this formula. After introducing a certain amount of mathematical formalism, we shall be able to derive and discuss the curvature in a far more compact and elegant notation, in Chapter 6.) By applying Eq. (1.1.12) to the metric functions (1.1.8) and (1.1.9), we find that the surface of a sphere is a space of constant positive curvature

$$K = \frac{1}{a^2} \qquad \text{(sphere)} \tag{1.1.13}$$

whereas the space of Gauss, Bólyai, and Lobachevski has constant negative curvature

$$K = -\frac{1}{a^2} \qquad \text{(G–B–L)} \tag{1.1.14}$$

(Incidentally, there is nothing very exotic about negative curvature; an ordinary saddle is negatively curved. It is the *constancy* of K that makes the geometry of Gauss, Bólyai, and Lobachevski unrealizable for ordinary curved surfaces. It is also obvious that only with K constant could the other postulates of Euclid be satisfied, because these other postulates describe an intrinsically homogeneous space, whereas if K varied from point to point then the inner properties of the space would vary with it.) Finally, if we apply our formula for K to the metric (1.1.10) that describes a plane in polar coordinates, then we find

$$K = 0 \qquad \text{(plane)} \tag{1.1.15}$$

as of course we must. Thus, however perverse we are in our choice of coordinate system, the inner properties of a space can still be revealed by the straightforward procedure of calculating K.

Having come so far, it was not long before mathematicians turned to the problem of describing the inner properties of curved spaces having three or more dimensions. It was not a trivial matter to expand the work of Gauss to more than two dimensions, because the inner properties of such spaces cannot be described by a single curvature function K. In D dimensions there will be $D(D + 1)/2$ independent metric functions g_{ij}, and our freedom to choose the D coordinates at will allows us to impose D arbitrary functional relations on the g_{ij}, leaving C functions that truly express the inner properties of the space, where

$$C = \frac{D(D + 1)}{2} - D = \frac{D(D - 1)}{2}$$

For $D = 2$, $C = 1$, as found by Gauss. For $D > 2$, $C > 1$, and the description of the geometry becomes much more complicated. This problem was completely solved in 1854 by Georg Friedrich Bernhard Riemann (1826–1866), who presented what we now call Riemannian geometry in his Göttingen inaugural lecture, *Über die Hypothesen, welche der Geometrie zu Grunde liegen*. Subsequent work by Christoffel, Ricci, Levi-Civita, Beltrami, and others developed Riemann's ideas into the beautiful mathematical structure described in our chapters on tensor analysis and curvature. However, it remained for Einstein to see the use physics could make of non-Euclidean geometry.

2   History of the Theory of Gravitation

At the end of the *Principia*, Isaac Newton (1642–1727) described gravitation as a cause that operates on the sun and planets "according to the quantity of solid matter which they contain and propagates on all sides to immense distances, decreasing always as the inverse square of the distances."[6] There are two parts to Newton's law, which were discovered in different ways, and which played different roles in the development of mechanics from Newton to Einstein.

It was of course Galileo Galilei (1564–1642) who discovered that bodies fall at a rate independent of their mass. His tools were an inclined plane to slow the fall, a water clock to measure its duration, and also a pendulum, to avoid rolling friction. These observations were later improved by Christaan Huygens (1629–1695). Newton could thus use his second law to conclude that the force exerted by gravitation is proportional to the mass of the body on which it acts; the third law then ensures that the force is also proportional to the mass of its source.

Newton was well aware that these conclusions might be only approximately true, and that the "inertial mass" entering in his second law might not be precisely the same as the "gravitational mass" appearing in the law of gravitation. If this were the case, we would have to write Newton's second law as

$$\mathbf{F} = m_i \mathbf{a} \tag{1.2.1}$$

and write the law of gravitation as

$$\mathbf{F} = m_g \mathbf{g} \tag{1.2.2}$$

where $\mathbf{g}$ is a field depending on position and other masses. The acceleration at a given point would be

$$\mathbf{a} = \left(\frac{m_g}{m_i}\right) \mathbf{g} \tag{1.2.3}$$

and would be different for bodies with different values for the ratio m_g/m_i; in particular pendulums of equal length would have periods proportional to $(m_i/m_g)^{1/2}$. Newton tested this possibility by experiments with pendulums of equal length but different composition, and found no difference in their periods. This result was later verified more accurately by Friedrich Wilhelm Bessel (1784–1846) in 1830. Then, in 1889, Roland von Eötvös[7] succeeded by a different method in showing that the ratio m_g/m_i does not differ from one substance to another by more than one part in 10^9. (See Figure 1.2.) Eötvös hung two weights A and B from the ends of a 40-cm beam suspended on a fine wire at its center. At equilibrium the beam would sag in such a way that

$$l_A(m_{gA}g - m_{iA}g_z') = l_B(m_{gB}g - m_{iB}g_z') \qquad (1.2.4)$$

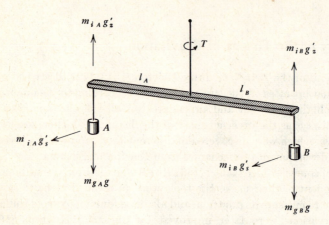

Figure 1.2 Schematic view of the Eötvös experiment.

where g is the earth's gravitational field, g_z' is the vertical component of the centripetal acceleration due to the earth's rotation, and l_A and l_B are the effective lever arms for the two weights. [Of course Eötvös chose weights and lever arms to be nearly equal, but the point of his method is that even if A is a little bigger than B, the beam will still sag just so as to make (1.2.4) correct.] At the latitude of Budapest the centripetal acceleration due to the earth's rotation also has an appreciable horizontal component g_s', giving to the balance a torque around the vertical axis equal to

$$T = l_A m_{iA} g_s' - l_B m_{iB} g_s'$$

Using the equilibrium condition to determine l_B, we have then

$$T = l_A m_{iA} g_s' \left[1 - \left(\frac{m_{gA}}{m_{iA}} g - g_z' \right) \left(\frac{m_{gB}}{m_{iB}} g - g_z' \right)^{-1} \right]$$

or, since g_z' is much less than g,

$$T = l_A g_s' m_{gA} \left[\frac{m_{iA}}{m_{gA}} - \frac{m_{iB}}{m_{gB}} \right]$$

Any inequality in the ratios m_i/m_g for the two weights would thus tend to twist the wire from which the balance was suspended. No twist was detected, and

Eötvös concluded from this that the difference of m_i/m_g for wood and platinum was less than 10^{-9}.

Einstein was very impressed with the observed equality of gravitational and inertial mass[8], and as we shall see, it served him as a signpost toward the Principle of Equivalence. (It also sets very stringent limits on any possible nongravitational forces that might exist. For instance, any new kind of electrostatic force in which the number of nucleons plays the role of charge would have to be much weaker than gravitation.[9]) In recent years a group under R. H. Dicke[10] at Princeton has improved on Eötvös' method, by using the gravitational field of the sun and the earth's centripetal acceleration toward the sun, rather than the rotation of the earth, to produce the torque on the balance. The advantage is that the angle between the direction of the sun and the balance arm changed with a 24-hr period, and so Dicke could filter out of his data any noise not at the diurnal frequency. In this way he concluded that "aluminum and gold fall toward the sun with the same acceleration, the accelerations differing from each other by at most one part in $10.$[11]" It has also been shown (with very much less precision) that neutrons fall with the same acceleration as ordinary matter,[11] and that the gravitational force on electrons in copper is the same as on free electrons.[12]

We now move on to the second part of Newton's law of gravitation, which says that the force decreases as the inverse square of the distance. This idea was not entirely original with Newton. Johannus Scotus Erigena (c. 800–c. 877) had guessed that heaviness and lightness vary with distance from the earth. This theory was taken up by Adelard of Bath (twelfth century), who realized that a stone dropped into a very deep well could fall no farther than the center of the earth. (Incidentally, Adelard also translated Euclid from Arabic into Latin, thus making it available to medieval Europe.) The first suggestion of an inverse-square law may have been made around 1640 by Ismael Bullialdus (1605–1694). However, it was certainly Newton who in 1665 or 1666 first deduced the inverse-square law from observations. He knew that the moon falls toward the earth a distance 0.0045 ft. each second, and he knew that the moon is 60 earth radii away from the center of the earth. Hence, if the gravitational force obeys an inverse-square law, then an apple in Lincolnshire (which is 1 earth radius away from the center of the earth) should fall in the first second 3600 times 0.0045 ft, or about 16 ft, in good agreement with the measured value. However, Newton did not publish this calculation for twenty years, because he did not know how to justify the fact that he had treated the earth as if its whole mass were concentrated at its center. Meanwhile, it became known to several members of the Royal Society, including Edmund Halley (1656–1742), Christopher Wren (1632–1723), and Robert Hooke (1635–1703), that Kepler's third law would imply an inverse-square law of force if the orbits of planets were circular. That is, if the squares of the periods, r^2/v^2, are proportional to the cubes of the radii r^3, then the centripetal acceleration v^2/r is proportional to $1/r^2$. However, the planets actually move on ellipses, not circles, and no one knew how to calculate their centripetal acceleration. Under Halley's instigation, Newton in 1684 proved that planets moving under the influence of an inverse-square-law force would indeed obey all the empirical laws of Johannes Kepler (1571–1630); that is, they would move on ellipses with the sun at a focus, they would sweep out equal areas in equal times, and the square of their periods would be proportional to the cube of their major axes. Finally, in 1685, Newton was able to complete his lunar calculation of 1665. These stupendous accomplishments were published on July 5, 1686, under the title *Philosophiae Naturalis Principia Mathematica*.[13]

In the following centuries Newton's law of gravitation met with a brilliant series of successes in explaining the motion of the moon and planets. Some irregularities in the orbit of Uranus remained unexplained until, in 1846, they were independently used by John Couch Adams (1819–1892) in England and Urbain Jean Joseph LeVerrier (1811–1877) in France to predict the existence and position of Neptune. The discovery of Neptune shortly thereafter was perhaps the most splendid verification of Newton's theory. The motion of the moon and Encke's comet (and, later, Halley's comet) still showed departures from Newtonian theory, but it was clear that nongravitational forces could be at work.

One problem remained. A year before his prediction of Neptune, LeVerrier had calculated that the observed precession of the perihelia of Mercury was 35″/century faster than what would be expected according to Newton's theory from the known perturbing fields of the other planets. This discrepancy was confirmed in 1882 by Simon Newcomb (1835–1909), who gave a value of 43″ for the excess centennial precession.[14] LeVerrier had thought that this excess was probably due to a group of small planets between Mercury and the sun, but after a careful search none were discovered. Newcomb then suggested that perhaps the matter responsible for the faint "zodiacal light" seen in the plane of the ecliptic of the solar system was also responsible for the excess precession of Mercury. However, his calculations showed that the amount of matter needed to account for the precession of Mercury would, if placed in the plane of the ecliptic, produce a rotation of the plane of the orbits (that is, a precession of the nodes) of both Mercury and Venus different from what had been observed. For this reason, Newcomb was led by 1895 "to drop these explorations as unsatisfactory, and to prefer provisionally the hypothesis that the Sun's gravitation is not exactly as the inverse square."[15]

Unfortunately this was not the last word. In 1896 H. H. Seeliger constructed an elaborate model of the zodiacal light, placing the matter responsible on ellipsoids close to the sun, which could account for the excess precession of Mercury without upsetting the agreement between theory and experiment for the rotation of the planes of the inner planets' orbits. Today we know that this model is totally wrong, and that there simply is not enough interplanetary matter to account for the observed excess precession of Mercury. However, Seeliger's hypothesis, together with the continued success of Newtonian theory elsewhere, convinced Newcomb that there was no need to alter the law of gravitation.[15]

I do not know whether Einstein was very much influenced, in creating general relativity, by the problem of the precession of Mercury's perihelia. However, there is no doubt that the first confirmation of his theory was that it predicted an excess precession of precisely 43″/century.

3   History of the Principle of Relativity

Newtonian mechanics defined a family of reference frames, the so-called *inertial frames*, within which the laws of nature take the form given in the *Principia*. For instance, the equations for a system of point particles interacting gravitationally are

$$m_N \frac{d^2 \mathbf{x}_N}{dt^2} = G \sum_M \frac{m_N m_M (\mathbf{x}_M - \mathbf{x}_N)}{|\mathbf{x}_M - \mathbf{x}_N|^3} \tag{1.3.1}$$

where m_N is the mass of the Nth particle and $\mathbf{x}_N$ is its Cartesian position vector at time t. It is a simple matter to check that these equations take the same form when written in terms of a new set of space-time coordinates:

$$\begin{aligned} \mathbf{x}' &= R\mathbf{x} + \mathbf{v}t + \mathbf{d} \\ t' &= t + \tau \end{aligned} \tag{1.3.2}$$

where $\mathbf{v}$, $\mathbf{d}$, and τ are any real constants, and R is any real orthogonal matrix. (If O and O' use the unprimed and primed coordinate system, respectively, then O' sees the O coordinate axes rotated by R, moving with velocity $\mathbf{v}$, displaced at $t = 0$ by $\mathbf{d}$, and O' sees the O clock running behind his own by a time τ.) The transformations (1.3.2) form a 10-parameter group (three Euler angles in R, plus three components each for $\mathbf{v}$ and $\mathbf{d}$, plus one τ) today called the *Galileo group*, and the invariance of the laws of motion under such transformations is today called Galilean invariance, or the *Principle of Galilean Relativity*.

What really impressed Newton about all this was that there are a great many more transformations that do *not* leave the equations of motion invariant. For instance, (1.3.1) does not retain its form if we transform into an accelerating or a rotating coordinate system, that is, if we let $\mathbf{v}$ or R depend on t. The equations of motion can hold in their usual form in only a limited class of coordinate systems, called *inertial frames*. What then determines which reference frames are inertial frames? Newton answered that there must exist an absolute space, and that the inertial frames were those at rest in absolute space, or in a state of uniform motion with respect to absolute space. In his words[16]:

> "Absolute space, in its own nature and with regard to anything external, always remains similar and unmovable. Relative space is some movable dimension or measure of absolute space, which our senses determine by its position with respect to other bodies, and is commonly taken for absolute space."

Newton also described several experiments that demonstrated what he interpreted as the effects of rotation with respect to absolute space. The most famous is the rotating bucket[17]:

"If a bucket, suspended by a long cord, is so often turned about that finally the cord is strongly twisted, then is filled with water, and held at rest together with the water; and afterwards by the action of a second force, it is suddenly set whirling about the contrary way, and continues, while the cord is untwisting itself, for some time in this motion; the surface of the water will at first be level, just as it was before the vessel began to move; but subsequently the vessel, by gradually communicating its motion to the water, will make it begin sensibly to rotate, and the water will recede little by little from the middle and rise up at the sides of the vessel; its surface assuming a concave form. (This experiment I have made myself.) . . . At first, when the *relative* motion of the water in the vessel was greatest, that motion produced no tendency whatever of recession from the axis, the water made no endeavor to move upwards towards the circumference, by rising at the sides of the vessel, but remained level, and for that reason its true circular motion had not yet begun. But afterwards, when the relative motion of the water had decreased, the rising of the water at the sides of the vessel indicated an endeavor to recede from the axis; and this endeavor reveals the real circular motion of the water, continually increasing till it had reached its greatest point, when relatively the water was at rest in the vessel. . . ."

Newton's conception of absolute space was rejected by his great opponent Gottfried Wilhelm von Leibniz (1646–1716), who argued that there is no philosophical need for any conception of space apart from the relations of material objects. The issue was debated in a famous series of letters[18] (1715–1716) between Leibniz and Newton's supporter, Samuel Clarke (1675–1729), and philosophers continued the argument, with Newton's position defended by Leonhard Euler (1707–1783) and Immanuel Kant (1724–1804) and attacked by Bishop George Berkeley (1685–1753) in his *Principles of Human Knowledge* (1710) and *Analyst* (1734). Of course none of this high-minded metaphysics led to any idea about how to develop a dynamical theory that might replace Newton's.

The first constructive attack on Newtonian absolute space was launched in the 1880's by the Austrian philosopher Ernst Mach (1836–1916). In his book *Die Mechanik in ihrer Entwicklung*[19] he remarks that

"Newton's experiment with the rotating vessel of water simply informs us, that the relative rotation of the water with respect to the sides of the vessel produces no noticeable centrifugal forces, but that such forces *are* produced by its relative motion with respect to the mass of the Earth and the other celestial bodies. No one is competent to say how the experiment would turn out if the sides of the vessel increased in thickness and mass until they were several leagues thick."

The hypothesis, that there is some influence of "the mass of the Earth and the other celestial bodies" which determines the inertial frames, is called *Mach's principle*.

There is a simple experiment that anyone can perform on a starry night, to clarify the issues raised by Mach's principle. First stand still, and let your arms hang loose at your sides. Observe that the stars are more or less unmoving, and that your arms hang more or less straight down. Then pirouette. The stars will seem to rotate around the zenith, and at the same time your arms will be drawn upward by centrifugal force. It would surely be a remarkable coincidence if the inertial frame, in which your arms hung freely, just happened to be the reference frame in which typical stars are at rest, unless there were some interaction between the stars and you that determined your inertial frame.

This argument can be made more precise. The surface of the earth is not exactly an inertial frame, and of course the rotation and revolution of the earth give the stars an apparent motion, but these effects can be eliminated by using the inertial frame defined by the solar system as a whole. In this inertial frame of reference the average observed rotation of the galaxies with respect to any axis through the sun is less than about 1 arc-sec/century![20]

We seem to be faced with an unavoidable choice: Either we admit that there is a Newtonian absolute space-time, which defines the inertial frames and with respect to which typical galaxies happen to be at rest, or we must believe with Mach that inertia is due to an interaction with the average mass of the universe. And if Mach is right, then the acceleration given a particle by a given force ought to depend not only on the presence of the fixed stars but also, very slightly, on the distribution of matter in the immediate vicinity of the particle. We shall see in Chapter 3 that Einstein's equivalence principle gives an answer to the problem of inertia that does not refer to a Newtonian absolute space and yet does not quite agree with the conclusions of Mach. The issue is not closed.

I have not yet mentioned special relativity because, despite its name, it really does not affect the antinomy between absolute and relative space. However, we shall have to formulate the equivalence principle in special-relativistic terms, so a detailed review of special relativity is presented in the next chapter; for the moment we only take a glance at its history.

The theory of electrodynamics presented in 1864 by James Clark Maxwell (1831–1879) clearly did not satisfy the principle of Galilean relativity. For one thing, Maxwell's equations predict that the speed of light in vacuum is a universal constant c, but if this is true in one coordinate system x^i, t, then it will not be true in the "moving" coordinate system x'^i, t' defined by the Galilean transformation (1.3.2). Maxwell himself thought that electromagnetic waves were carried by a medium,[21] the luminiferous ether, so that his equations would hold in only a limited class of Galilean inertial frames, that is, in those coordinate frames at rest with respect to the ether.

However, all attempts to measure the velocity of the earth with respect to the ether failed,[22] even though the earth has a velocity of 30 km/sec relative to the sun, and about 200 km/sec relative to the center of our galaxy. The most important experiment was that of Albert Abraham Michelson (1852–1931) and E. W. Morley,[23] which showed in 1887 that the velocity of light is the same, within 5 km/sec, for light traveling along the direction of the earth's orbital motion and transverse to it. The accuracy of this result has been recently improved to about 1 km/sec.[24]

The persistent failure of experimentalists to discover effects of the earth's motion through the ether led theorists, including George Francis Fitzgerald[25] (1851–1901), Hendrik Antoon Lorentz[26] (1853–1928), and Jules Henri Poincaré[27] (1854–1912) to suggest reasons why such "ether drift" effects should be in principle unobservable. (See Figure 1.3.) Poincaré in particular seems to have glimpsed the revolutionary implications that this would have for mechanics, and Whittaker[28] gives the credit for special relativity to Poincaré and Lorentz. Without entering this controversy,[29] it is safe to say that a comprehensive solution to the problems

CONSEIL DE PHYSIQUE SOLVAY
BRUXELLES 1911

Photo Couprie, Bruxelles

GOLDSCHMIDT PLANCK RUBENS LINDEMANN HASENOHRL
NERNST BRILLOUIN SOMMERFELD DE BROGLIE HOSTELET
SOLVAY KNUDSEN HERZEN JEANS RUTHERFORD
LORENTZ WARBURG WIEN EINSTEIN LANGEVIN
PERRIN Madame CURIE POINCARÉ KAMERLINGH ONNES

Figure 1.3   Founders of the Special Theory of Relativity, at the First Solvay Conference in 1911.

of relativity in electrodynamics and mechanics was first set out in detail in 1905 by Albert Einstein[30] (1879–1955).

Einstein proposed that the Galilean transformation (1.3.2) should be replaced with a different 10-parameter space-time transformation, called a *Lorentz transformation*, that does leave Maxwell's equations and the speed of light invariant. (It is not clear that Einstein was directly influenced by the Michelson-Morley experiment itself,[31] but he specifically refers to "the unsuccessful attempts to discover any motion of the earth relative to the 'light medium'" in his 1905 paper.[32]) The equations of Newtonian mechanics, such as Eq. (1.3.1), are not invariant under Lorentz transformations; therefore Einstein was led to modify the laws of motion so that they would be Lorentz-invariant. The new physics, consisting of Maxwell's electrodynamics and Einstein's mechanics, then satisfied a new principle of relativity, the Principle of Special Relativity, which says that all physical equations must be invariant under Lorentz transformations. These developments are discussed in detail in the next chapter.

The Lorentz group of transformations is not in any way larger than the Galileo group, and therefore the principle of relativity was not originated by the special theory of relativity, but rather *restored* by it. Before Maxwell, it might have been supposed that all of physics is invariant under the Galileo group. Maxwell's equations were not invariant under this group, and for half a century it appeared that only mechanics, not electrodynamics, obeys the principle of relativity. After Einstein, it was clear that the equations of both mechanics and electrodynamics are invariant, but with respect to Lorentz transformations, not Galileo transformations. The laws of physics in the form given them by Maxwell and Einstein could still only be true in a limited class of inertial reference frames, and the question of what determines these inertial frames was as mysterious after 1905 as in 1686.

It remained to construct a relativistic theory of gravitation. A crucial step toward this goal was taken in 1907, when Einstein introduced the Principle of Equivalence of Gravitation and Inertia,[33] and used it to calculate the red shift of light in a gravitational field. As we shall see in Chapter 3, this principle determines the effects of gravitation on arbitrary physical systems, but it does not determine the field equations for gravitation itself. Einstein tried to use the equivalence principle in 1911 to calculate the deflection of light in the sun's gravitational field,[34] but the structure of the field was not then correctly understood, and Einstein's answer was one-half the "correct" general-relativistic result, derived here in Chapter 8. A number of attempts were made in 1911–1912 by Einstein,[35] Abraham,[36] and Nordström[37] to construct relativistic field equations for a single scalar gravitational field, but Einstein soon became dissatisfied with all such theories, largely on aesthetic grounds. (The gravitational deflection of light by the sun had not yet been measured.) A collaboration with the mathematician Marcel Grossman led Einstein by 1913 to the view[38] that the gravitational field must be identified with the 10 components of the metric tensor of Riemannian space-time geometry. As discussed in Chapters 4 and 5, the Principle of Equivalence is incorporated into this formalism through the requirement that the physical equations be invariant under *general* coordinate transformations, not just Lorentz transformations, though I do not know to what extent this "General Principle of Relativity" took on in Einstein's mind a life of its own, apart from the Principle of Equivalence. During the next two years, Einstein presented to the Prussian Academy of Sciences a series of papers[39] in which he worked out the field equations for the metric tensor and calculated the gravitational deflection of light and the precession of the perihelia of Mercury. These magnificent achievements were finally summarized by Einstein in his 1916 paper,[1] titled "The Foundation of the General Theory of Relativity."

I BIBLIOGRAPHY

Not being an historian, I have been content to base this chapter on secondary sources, aside from the works of Newton, Mach, Maxwell, Newcomb, and Einstein quoted in the text. The authorities on whom I have drawn most heavily are listed below.

Non-Euclidean Geometry

☐ R. Bonola, *Non-Euclidean Geometry* (Dover Publications, New York, 1955).
☐ G. Sarton, *Ancient Science and Modern Civilization* (Yale University Press, New Haven, 1951), Chapter I.
☐ H. Weyl, *Space, Time, Matter*, 4th ed. (Dover Publications, New York, 1950), Chapter II.

Gravitation

☐ F. Cajori, historical and explanatory appendix to Isaac Newton's *Philosophiae Naturalis Principia Mathematica* (University of California Press, 1966).
☐ E. Guth, in *Relativity—Proceedings of the Relativity Conference in the Midwest*, ed. by M. Carmeli, S. I. Fickler, and L. Witten (Plenum Press, New York, 1970), p. 161.
☐ M. Jammer, *Concepts of Force* (Harper and Brothers, New York, 1962), Chapters IV–VII.
☐ E. Whittaker, *A History of the Theories of Aether and Electricity* (Thomas Nelson and Sons, Edinburgh, 1953), Vol. II, Chapter V.
☐ W. P. D. Wightman, *The Growth of Scientific Ideas* (Yale University Press, New Haven, 1951), Chapters VIII, X.

Relativity

☐ G. Holton, "On the Origins of the Special Theory of Relativity," *Am. J. Phys.*, **28**, 627 (1960).

☐ A. Koyré, *From the Closed World to the Infinite Universe* (Harper and Row, New York, 1958), Chapters VII, IX–XI.

☐ C. Møller, *The Theory of Relativity* (Oxford University Press, London, 1952), Chapter I.

☐ W. Pauli, *Theory of Relativity* (Pergamon Press, Oxford, 1958), Parts I, IV.50.

☐ E. Whittaker, *A History of Aether and Electricity* (Thomas Nelson and Sons, Edinburgh, 1953), Vol. I, Chapters VIII–X, XIII; Vol. II, Chapters II, V.

I REFERENCES

1. A. Einstein, Annalen der Phys., **49**, 769 (1916). For an English translation, see *The Principle of Relativity* (Methuen, 1923, reprinted by Dover Publications), p. 35.

2. The leading English edition is *Euclid's Elements*, translated with an introduction and commentary by T. L. Heath (rev. ed., Cambridge, 1926).

3. These quotations are taken from George Sarton, *Ancient Science and Modern Civilization* (University of Nebraska Press, 1954; reprinted by Harper and Brothers, New York, 1959), p. 26.

4. Quoted by R. Bonola, in *Non-Euclidean Geometry*, trans. by H. S. Carslaw (Dover Press, 1955), pp. 65–67.

5. F. Klein, Math. Ann., **4**, 573 (1871); **6**, 112 (1873); **37**, 544 (1890); quoted by H. Weyl, in *Space-Time-Matter*, trans. by H. L. Brose (Dover Press, 1952), p. 80. A Euclidean model for the Gauss-Bólyai-Lobachevski geometry was given in 1868 by E. Beltrami, *Saggio di interpretazione della geometria non-euclidea*, quoted by J. D. North in *The Measure of the Universe* (Oxford, 1965), p. 60.

6. Isaac Newton, *Philosophiae Naturalis Principia Mathematica*, trans. by Andrew Motte, revised and annotated by F. Cajori (University of California Press, 1966), p. 546.

7. R. v. Eötvös, Math. nat. Ber. Ungarn, **8**, 65 (1890); R. v. Eötvös, D. Pekár, and E. Fekete, Ann. Phys., **68**, 11 (1922). Also see J. Renner, Hung. Acad. Sci., Vol. 53, Part II (1935).

8. See, for example, A. Einstein, *The Meaning of Relativity* (2nd ed., Princeton, 1946), p. 56.

9. T. D. Lee and C. N. Yang, Phys. Rev., **98**, 1501 (1955).

10. R. H. Dicke, in *Relativity, Groups, and Topology*, ed. by C. DeWitt and B. S. DeWitt (Gordon and Breach, New York, 1964), p. 167; P. G. Roll, R. Krotkov, and R. H. Dicke, Ann. Phys. (N.Y.), **26**, 442 (1967).

11. J. W. T. Dobbs, J. A. Harvey, D. Paya, and H. Horstmann, Phys. Rev., **139,** B756 (1965).

12. F. C. Witteborn and W. M. Fairbank, Phys. Rev. Letters, **19,** 1049 (1967).

13. The most accessible edition is that of Florian Carjori, ref. 6.

14. S. Newcomb, Astronomical Papers of the American Ephemeris, **1,** 472 (1882).

15. S. Newcomb, article on "Mercury" in *The Encyclopaedia Britannica*, 11th ed., **XVIII,** 155 (1910–1911).

16. Ref. 6, p. 6 (a different translation is quoted here).

17. *Ibid.*, p. 10.

18. G. H. Alexander, *The Leibniz-Clarke Correspondence* (Manchester University Press, 1956). Excerpts are quoted by A. Koyré in *From the Closed World to the Infinite Universe* (Harper and Row, New York, 1958), Chapter XI. (See especially Leibniz's fifth letter.)

19. E. Mach, *The Science of Mechanics*, trans. by T. J. McCormack (2nd ed., Open Court Publishing Co., 1893).

20. L. I. Schiff, Rev. Mod. Phys., **36,** 510 (1964); G. M. Clemence, Rev. Mod. Phys., **19,** 361 (1947); **29,** 2 (1957).

21. James Clark Maxwell, article on "Ether" in *The Encyclopaedia Britannica*, 9th ed. (1875–1889); reprinted in *The Scientific Papers of James Clark Maxwell*, ed. by W. D. Niven (Dover Publications, 1965), p. 763. Also see Maxwell's *Treatise on Electricity and Magnetism*, Vol. II (Dover Publications, 1954), pp. 492–493.

22. For an account of these experiments, see C. Møller, *The Theory of Relativity* (Oxford Press, London, 1952), Chapter I.

23. A. A. Michelson and E. W. Morley, Am. J. Sci., **34,** 333 (1887); reprinted in *Relativity Theory: Its Origins and Impact on Modern Thought*, ed. by L. Pearce Williams (John Wiley and Sons, New York, 1968).

24. T. S. Jaseja, A. Javan, J. Murray, and C. H. Townes, Phys. Rev., **133,** A1221 (1964).

25. G. F. Fitzgerald, quoted by O. Lodge, Nature, **46,** 165 (1892). Also see O. Lodge, Phil. Trans. Roy. Soc., **184A** (1893).

26. H. A. Lorentz, Zittungsverslagen der Akad. van Wettenschappen, **1,** 74 (November 26, 1892); *Versuch einer Theorie der elektrischen und optische Erscheinungen in bewegten Körpern* (E. J. Brill, Leiden, 1895); Proc. Acad. Sci. Amsterdam (English version), **6,** 809 (1904). The third reference, and a translated excerpt from the second, are available in *The Principle of Relativity*, ref. 1.

27. J. H. Poincaré, *Rapports présentés au Congrès International de Physique réuni à Paris* (Gauthier-Villiers, Paris, 1900); speech at the St. Louis International Exposition in 1904, trans. by G. B. Halstead, *The Monist*, **15,** 1 (1905), reprinted in *Relativity Theory: Its Origins and Impact on Modern Thought*, ref. 23; *Rend. Circ. Mat. Palermo*, **21,** 129 (1906).

28. Sir Edmund Whittaker, *A History of The Theories of Aether and Electricity*, Vol. II (Thomas Nelson and Sons, London, 1953), Chapter I.

29. For a balanced view of this question, see G. Holton, Am. J. Phys., **28**, 627 (1960), reprinted in part in *Relativity Theory: Its Origins and Impact on Modern Thought*, ref. 23.

30. A. Einstein, Ann. Physik, **17**, 891 (1905); **18**, 639 (1905). Translations are given in *The Principle of Relativity*, ref. 1.

31. G. Holton, ref. 29, and Isis, **60**, 133 (1969).

32. See ref. 30, and also A. Grünbaum, in *Current Issues in the Philosophy of Science*, ed. by H. Feigl and G. Maxwell (Holt, Rinehart, and Winston, New York, 1961), reprinted in part in *Relativity Theory: Its Origins and Impact on Modern Thought*, ref. 23.

33. A. Einstein, Jahrb. Radioakt., **4**, 411 (1907); also see M. Planck, Sitzungsber. preuss. Akad. Wiss., June 13, 1907, p. 542; Ann. Phys. Leipzig, **26** (1908).

34. A. Einstein, Ann. Phys. Leipzig, **35**, 898 (1911). For an English translation, see *The Principle of Relativity*, ref. 1.

35. A Einstein, Ann. Phys. Leipzig, **38**, 355, 443 (1912).

36. M. Abraham, Lincei Atti, **20**, 678 (1911); Phys. Z., **13**, 1, 4, 176, 310, 311, 793 (1912); Nuovo Cimento, **4**, 459 (1912).

37. G. Nordström, Phys. Z., **13**, 1126 (1912); Ann. Phys. Leipzig, **40**, 856 (1913); **42**, 533 (1913); **43**, 1101 (1914); Phys. Z., **15**, 375 (1914); Ann. Acad. Sci. fenn., **57** (1914, 1915).

38. A. Einstein, Phys. Z., **14**, 1249 (1913); A. Einstein and M. Grossmann, Z. Math. Phys., **62**, 225 (1913); **63**, 215 (1914); A. Einstein, Vierteljahr Nat. Ges. Zürich, **58**, 284 (1913); Archives sci. phys. nat., **37**, 5 (1914); Phys. Z., **14**, 1249 (1913).

39. A. Einstein, Sitzungsber. preuss. Akad. Wiss., 1914, p. 1030; 1915, pp. 778, 799, 831, 844. Also see D. Hilbert, Nachschr. Ges. Wiss. Göttingen, November 20, 1915, p. 395.

40. This famous experiment may in fact be a myth. See A. I. Miller, Isis, to be published (1972).

"There are really four
dimensions, three which we
call the three planes of
Space, and a fourth, Time.
There is, however, a
tendency to draw an unreal
distinction between the
former three dimensions
and the latter, because it
happens that our conscious-
ness moves intermittently
in one direction along the
latter from the beginning
to the end of our lives."
"'That', said a very young
man, making spasmodic
efforts to relight his cigar
over the lamp; 'that . . .
very clear indeed.'" *H. G.
Wells, The Time Machine*

2 SPECIAL RELATIVITY

We now review Einstein's Special Theory of Relativity. This chapter, while
self-contained, is only a brief summary, and aims primarily at establishing our
notation and collecting some formulas that will be useful later. The reader who
needs a more extensive introduction to special relativity is advised to turn to one
of the books listed at the end of this chapter, and then return. The reader who feels
completely at home with the subject may find it desirable to move on immediately
to Chapter 3.

1 Lorentz Transformations

The Principle of Special Relativity states that the laws of nature are invariant
under a particular group of space-time coordinate transformations, called Lorentz
transformations. We saw at the end of Chapter 1 that Newton's laws of motion are
invariant under the Galilean coordinate transformations (1.3.2), but that Maxwell's

equations are not, and that Einstein resolved this conflict by replacing Galilean invariance with Lorentz invariance. I shall not continue this discussion in historical terms, but shall simply define the Lorentz transformations, and then show how Lorentz invariance guides our search for the laws of nature.

A Lorentz transformation is a transformation from one system of space-time coordinates x^α to another system x'^α, so that

$$x'^\alpha = \Lambda^\alpha{}_\beta x^\beta + a^\alpha \tag{2.1.1}$$

where a^α and $\Lambda^\alpha{}_\beta$ are constants, restricted by the conditions

$$\Lambda^\alpha{}_\gamma \Lambda^\beta{}_\delta \eta_{\alpha\beta} = \eta_{\gamma\delta} \tag{2.1.2}$$

with

$$\eta_{\alpha\beta} = \begin{cases} +1 & \alpha = \beta = 1, 2, \text{ or } 3 \\ -1 & \alpha = \beta = 0 \\ 0 & \alpha \neq \beta \end{cases} \tag{2.1.3}$$

In our notation α, β, γ, and so on, will always run over the four values 1, 2, 3, 0, with x^1, x^2, x^3 the Cartesian components of the position vector $\mathbf{x}$ and x^0 the time t. We shall use natural units in which the speed of light is unity, so all x^α have the dimension of length. Any index, like β in Eq. (2.1.1), that appears twice, once as a subscript and once as a superscript, is understood to be summed over unless otherwise noted; that is, Eq. (2.1.1) is an abbreviation for

$$x'^\alpha = \Lambda^\alpha{}_0 x^0 + \Lambda^\alpha{}_1 x^1 + \Lambda^\alpha{}_2 x^2 + \Lambda^\alpha{}_3 x^3 + a^\alpha$$

The fundamental property that distinguishes the Lorentz transformations is that they leave invariant the "proper time" $d\tau$, defined by

$$d\tau^2 \equiv dt^2 - d\mathbf{x}^2 = -\eta_{\alpha\beta}\, dx^\alpha\, dx^\beta \tag{2.1.4}$$

In a new coordinate system x'^α, the coordinate differentials are given by (2.1.1) as

$$dx'^\alpha = \Lambda^\alpha{}_\gamma\, dx^\gamma$$

so the new coordinate time will be

$$\begin{aligned} d\tau'^2 &= -\eta_{\alpha\beta}\, dx'^\alpha\, dx'^\beta \\ &= -\eta_{\alpha\beta}\Lambda^\alpha{}_\gamma \Lambda^\beta{}_\delta\, dx^\gamma\, dx^\delta \\ &= -\eta_{\gamma\delta}\, dx^\gamma\, dx^\delta \end{aligned}$$

and therefore

$$d\tau'^2 = d\tau^2 \tag{2.1.5}$$

It is this property that accounts for the observation by Michelson and Morley that the speed of light is the same in all inertial systems. A light wave front will have $|d\mathbf{x}/dt|$ equal to the speed of light, which in our units is unity; hence the propagation of light is described by the statement that

$$d\tau = 0 \tag{2.1.6}$$

Performing a Lorentz transformation does not change $d\tau$, so $d\tau'^2 = 0$, and therefore $|d\mathbf{x}'/dt'| = 1$; that is, the speed of light in the new coordinate system is still unity.

We can also show that the Lorentz transformations (2.1.1) are the *only* non-singular coordinate transformations $x \to x'$ that leave $d\tau^2$ invariant. (Nonsingular means that $x'(x)$ and $x(x')$ are well-behaved differentiable functions, so that the matrix $\partial x'^\alpha / \partial x^\beta$ has a well-defined inverse $\partial x^\beta / \partial x'^\alpha$.) A general coordinate transformation $x \to x'$ will change $d\tau$ into $d\tau'$, given by

$$d\tau'^2 = -\eta_{\alpha\beta} \, dx'^\alpha \, dx'^\beta$$

$$= -\eta_{\alpha\beta} \frac{\partial x'^\alpha}{\partial x^\gamma} \frac{\partial x'^\beta}{\partial x^\delta} \, dx^\gamma \, dx^\delta$$

If this is equal to $d\tau^2$ for all dx^γ, we must have

$$\eta_{\gamma\delta} = \eta_{\alpha\beta} \frac{\partial x'^\alpha}{\partial x^\gamma} \frac{\partial x'^\beta}{\partial x^\delta} \tag{2.1.7}$$

Differentiation with respect to x^ε gives

$$0 = \eta_{\alpha\beta} \frac{\partial^2 x'^\alpha}{\partial x^\gamma \, \partial x^\varepsilon} \frac{\partial x'^\beta}{\partial x^\delta} + \eta_{\alpha\beta} \frac{\partial x'^\alpha}{\partial x^\gamma} \frac{\partial^2 x'^\beta}{\partial x^\delta \, \partial x^\varepsilon}$$

To solve for the second derivatives, we add to this the same equation with γ and ε interchanged, and subtract the same with ε and δ interchanged; that is,

$$0 = \eta_{\alpha\beta} \left[\frac{\partial^2 x'^\alpha}{\partial x^\gamma \, \partial x^\varepsilon} \frac{\partial x'^\beta}{\partial x^\delta} + \frac{\partial^2 x'^\beta}{\partial x^\delta \, \partial x^\varepsilon} \frac{\partial x'^\alpha}{\partial x^\gamma} + \frac{\partial^2 x'^\alpha}{\partial x^\varepsilon \, \partial x^\gamma} \frac{\partial x'^\beta}{\partial x^\delta} + \frac{\partial^2 x'^\beta}{\partial x^\delta \, \partial x^\gamma} \frac{\partial x'^\alpha}{\partial x^\varepsilon} \right.$$

$$\left. - \frac{\partial^2 x'^\alpha}{\partial x^\gamma \, \partial x^\delta} \frac{\partial x'^\beta}{\partial x^\varepsilon} - \frac{\partial^2 x'^\beta}{\partial x^\varepsilon \, \partial x^\delta} \frac{\partial x'^\alpha}{\partial x^\gamma} \right]$$

The last term cancels the second, the penultimate cancels the fourth (because $\eta_{\alpha\beta} = \eta_{\beta\alpha}$), and the first equals the third, so we are left with

$$0 = 2\eta_{\alpha\beta} \frac{\partial^2 x'^\alpha}{\partial x^\gamma \, \partial x^\varepsilon} \frac{\partial x'^\beta}{\partial x^\delta}$$

But both $\eta_{\alpha\beta}$ and $\partial x'^\beta / \partial x^\delta$ are nonsingular matrices, so this immediately yields

$$0 = \frac{\partial^2 x'^\alpha}{\partial x^\gamma \, \partial x^\varepsilon} \tag{2.1.8}$$

The general solution of (2.1.8) is of course just the linear function (2.1.1), and by inserting (2.1.1) in (2.1.7) we see that $\Lambda^\alpha_{\ \beta}$ must be subject to the condition (2.1.2). This proof is an elementary example of the sort of thing we do in Chapter 13, on symmetric spaces. (Incidentally, if we had only assumed that the transformations

$x \to x'$ leave $d\tau$ invariant when $d\tau = 0$, that is, for a particle moving at the speed of light, then we would have found that these transformations are in general nonlinear, and form a 15-parameter group, the conformal group, which contains the Lorentz transformations as a subgroup. But the statement that a free particle moves at constant velocity would not be an invariant statement unless the velocity were that of light, and since there are massive particles in the world, we must reject the conformal group as a possible invariance of nature.)

The set of all Lorentz transformations of the form (2.1.1) is correctly called the *inhomogeneous Lorentz group*, or the *Poincaré group*. The subset with $a^\alpha = 0$ is called the *homogeneous Lorentz group*. Both the homogeneous and the inhomogeneous Lorentz groups have subgroups called the *proper* homogeneous and inhomogeneous Lorentz groups, defined by imposing on $\Lambda^\alpha{}_\beta$ the additional requirements

$$\Lambda^0{}_0 \geq 1; \qquad \text{Det } \Lambda = +1 \tag{2.1.9}$$

Note from (2.1.2) that

$$(\Lambda^0{}_0)^2 = 1 + \sum_{i=1,2,3} (\Lambda^i{}_0)^2 \geq 1 \tag{2.1.10}$$

and

$$(\text{Det } \Lambda)^2 = 1 \tag{2.1.11}$$

[Equation (2.1.10) follows upon setting $\gamma = \delta = 0$ in (2.1.2). Equation (2.1.11) is derived by writing Eq. (2.1.2) as a matrix equation $\eta = \Lambda^T \eta \Lambda$ and taking its determinant.] It follows that any $\Lambda^\alpha{}_\beta$ that can be converted to the identity $\delta^\alpha{}_\beta$ by a continuous variation of its parameters must be a proper Lorentz transformation, because it is impossible by a continuous change of parameters to jump from $\Lambda^0{}_0 \leq -1$ to $\Lambda^0{}_0 \geq +1$, or from Det $\Lambda = -1$ to Det $\Lambda = +1$, and the identity has $\Lambda^0{}_0 = +1$ and Det $\Lambda = +1$. The *improper* Lorentz transformations involve either space inversion (Det $\Lambda = -1$, $\Lambda^0{}_0 \geq 1$), which is now known not to be an exact symmetry of nature,[1] or time reversal (Det $\Lambda = -1$, $\Lambda^0{}_0 \leq -1$), which is strongly suspected to be not an exact symmetry of nature,[2] or their product. We are dealing almost exclusively with proper Lorentz transformations, and unless otherwise noted, any Lorentz transformation is assumed to satisfy Eq. (2.1.9).

The proper homogeneous Lorentz transformations have a further subgroup, consisting of the rotations, for which

$$\Lambda^i{}_j = R_{ij}, \qquad \Lambda^i{}_0 = \Lambda^0{}_i = 0, \qquad \Lambda^0{}_0 = 1$$

where R_{ij} is a unimodular orthogonal matrix (i.e., Det $R = 1$ and $R^T R = 1$) and the indices i, j run over the values 1, 2, 3. With regard to both rotations and the space-time translations $x^\alpha \to x^\alpha + a^\alpha$, there is no difference between the Lorentz group and the Galileo group discussed in Chapter 1. The difference arises only in those transformations, called *boosts*, that change the velocity of the coordinate frame. Suppose that one observer O sees a particle at rest, and a second observer O' sees it moving with velocity $\mathbf{v}$. From (2.1.1) we have

$$dx'^{\alpha} = \Lambda^{\alpha}{}_{\beta} \, dx^{\beta} \tag{2.1.12}$$

or, since $d\mathbf{x}$ vanishes,

$$dx'^{i} = \Lambda^{i}{}_{0} \, dt \qquad (i = 1, 2, 3) \tag{2.1.13}$$

$$dt' = \Lambda^{0}{}_{0} \, dt \tag{2.1.14}$$

Dividing $d\mathbf{x}'$ by dt' gives the velocity $\mathbf{v}$, so

$$\Lambda^{i}{}_{0} = v_{i} \Lambda^{0}{}_{0} \tag{2.1.15}$$

We can get a second relation between $\Lambda^{i}{}_{0}$ and $\Lambda^{0}{}_{0}$ by setting $\gamma = \delta = 0$ in Eq. (2.1.2):

$$-1 = \Lambda^{\alpha}{}_{0} \, \Lambda^{\beta}{}_{0} \eta_{\alpha\beta} = \sum_{i=1,2,3} (\Lambda^{i}{}_{0})^{2} - (\Lambda^{0}{}_{0})^{2} \tag{2.1.16}$$

The solution of Eqs. (2.1.15) and (2.1.16) is

$$\Lambda^{0}{}_{0} = \gamma \tag{2.1.17}$$

$$\Lambda^{i}{}_{0} = \gamma v_{i} \tag{2.1.18}$$

where

$$\gamma \equiv (1 - \mathbf{v}^{2})^{-1/2} \tag{2.1.19}$$

The other $\Lambda^{\alpha}{}_{\beta}$ are not uniquely determined, because if $\Lambda^{\alpha}{}_{\beta}$ carries a particle from rest to velocity $\mathbf{v}$, then so does $\Lambda^{\alpha}{}_{\gamma} \, R^{\gamma}{}_{\beta}$, where R is an arbitrary rotation. One convenient choice that satisfies Eq. (2.1.2) is

$$\Lambda^{i}{}_{j} = \delta_{ij} + v_{i} v_{j} \frac{(\gamma - 1)}{\mathbf{v}^{2}} \tag{2.1.20}$$

$$\Lambda^{0}{}_{j} = \gamma v_{j} \tag{2.1.21}$$

It can easily be seen that any proper homogeneous Lorentz transformation may be expressed as the product of a boost $\Lambda(\mathbf{v})$ times a rotation R.

2   Time Dilation

Although the Lorentz transformations were invented to account for the invariance of the speed of light, the change from Galilean relativity to special relativity had immediate kinematic consequences for material objects moving at speeds less than that of light. The simplest and most important is the time dilation of moving clocks. An observer looking at a clock at rest will see two ticks separated by a space-time interval $d\mathbf{x} = 0$, $dt = \Delta t$, where Δt is the nominal period between ticks intended by the manufacturer. He will calculate the proper time interval (2.1.4) as

$$d\tau \equiv (dt^2 - dx^2)^{1/2} = \Delta t$$

A second observer, who sees the same clock moving with velocity **v**, will observe that the two ticks are separated by a time interval dt' and also by a space interval $dx' = \mathbf{v}\, dt'$, and he will conclude that the proper time interval is

$$d\tau' \equiv (dt'^2 - dx'^2)^{1/2} = (1 - \mathbf{v}^2)^{1/2}\, dt'$$

But both observers are supposed to be using inertial coordinate systems, so their coordinate systems are related by a Lorentz transformation, and on comparing notes they must find that $d\tau = d\tau'$, in accordance with Eq. (2.1.5). It follows that the observer who sees the clock in motion will see it tick with a period

$$dt' = \Delta t(1 - \mathbf{v}^2)^{-1/2} \tag{2.2.1}$$

[For an alternate derivation, use Eqs. (2.1.14), (2.1.17), (2.1.19).] This relation is literally being verified every day by experiments that measure the mean lifetime of rapidly moving unstable particles from cosmic rays and accelerators. Such particles of course do not tick; instead (2.2.1) tells us here that a moving particle will have a mean life larger than it has at rest by a factor $(1 - \mathbf{v}^2)^{-1/2}$, in perfect agreement with the lifetime measurements made electronically or by measuring the free path length.

The time dilation (2.2.1) is not to be confused with the apparent time dilation or contraction known as the *Doppler effect*. If our "clock" is a moving source of light of frequency $\nu = 1/\Delta t$, then the time between emission of successive wave fronts (say, with a maximum value of some component of the electric field) is given by (2.2.1) as $dt' = \Delta t(1 - \mathbf{v}^2)^{-1/2}$. However, during this time the distance from the observer to the light source will have increased by an amount $v_r\, dt'$, where v_r is the component of **v** along the direction from observer to light source. Hence the period between *reception* of wave fronts will be

$$dt_0 = (1 + v_r)\, dt' = (1 + v_r)(1 - \mathbf{v}^2)^{-1/2}\, \Delta t$$

That is, the ratio of the frequency of the light actually measured by the observer to the frequency of the light source at rest is

$$\frac{\nu_{\text{obs}}}{\nu} = (1 + v_r)^{-1}(1 - \mathbf{v}^2)^{1/2} \tag{2.2.2}$$

If the light source is moving away, then $v_r > 0$, and this is necessarily a red shift. If the light source is moving transversely, then $v_r = 0$, and we have the pure time dilation red shift discussed above. If the light source is moving directly toward the observer, then $v_r = -v$, and (2.2.2) gives a violet shift by a factor

$$(1 + v)^{1/2}(1 - v)^{-1/2}$$

The transition from violet to red shift occurs for a source moving at an angle between straight toward the observer and at right angles to the line of sight.

3 Particle Dynamics

Let us suppose that a particle moves in a field of force at a velocity so high that Newtonian mechanics does not suffice to calculate its motion. Let us also suppose, as in the case of electrodynamics, that we know how to calculate the force **F** on our particle in any Lorentz frame in which, at a given moment, it is at rest. Then we could compute the motion of our particle by performing a Lorentz transformation to a frame in which the particle is at rest at some time t_0, computing the velocity $d\mathbf{v} = \mathbf{F}\, dt/m$ at the time $t_0 + dt$, performing another Lorentz transformation to bring the velocity to zero again, and so on. Fortunately, there is an easier way.

Let us define the *relativistic force* f^α acting on a particle with coordinates $x^\alpha(\tau)$ by

$$f^\alpha = m\,\frac{d^2 x^\alpha}{d\tau^2} \tag{2.3.1}$$

Clearly, if f^α were known, we could compute the motion of our particle. We shall relate f^α to the Newtonian force by noting two of its properties:

(A) If the particle is momentarily at rest, then the proper time interval $d\tau$ equals dt, so $f^\alpha = F^\alpha$, where F^i are the Cartesian components of the nonrelativistic force **F**, and

$$F^0 \equiv 0 \tag{2.3.2}$$

(B) Under a general Lorentz transformation (2.1.1), the coordinate differentials transform according to $dx'^\alpha = \Lambda^\alpha{}_\beta\, dx^\beta$, while $d\tau$ is invariant, so (2.3.1) tells us that f^α has the Lorentz transformation rule:

$$f'^\alpha = \Lambda^\alpha{}_\beta f^\beta \tag{2.3.3}$$

Any quantity such as dx^α or f^α that transforms according to Eq. (2.3.3) is called a *four-vector*.

Now suppose that our particle has velocity **v** at some moment t_0, and introduce a new coordinate system x'^α, defined by

$$x^\alpha = \Lambda^\alpha{}_\beta(\mathbf{v})x'^\beta$$

where $\Lambda(\mathbf{v})$ is the "boost" defined by Eqs. (2.1.17)–(2.1.21). Since $\Lambda(\mathbf{v})$ is constructed so as to carry a particle from rest to velocity **v**, and since our particle has velocity **v** at time t_0 in the coordinate system x^α, it must be at rest at this moment in the coordinate system x'^α. Hence, according to (A), the force four-vector in the coordinate system x'^{α} at time t_0 is equal to the nonrelativistic force F^{α}. And therefore, according to (B), the force in our original coordinate system is

$$f^{\alpha} = \Lambda^{\alpha}{}_{\beta}(\mathbf{v})F^{\beta} \tag{2.3.4}$$

or more explicitly, since $F^0 = 0$,

$$\mathbf{f} = \mathbf{F} + (\gamma - 1)\,\mathbf{v}\,\frac{(\mathbf{v} \cdot \mathbf{F})}{v^2} \tag{2.3.5}$$

$$f^0 = \gamma \mathbf{v} \cdot \mathbf{F} = \mathbf{v} \cdot \mathbf{f} \tag{2.3.6}$$

with $\mathbf{v}$ the instantaneous velocity.

Now that we know how to calculate f^{α}, we can use the differential equations (2.3.1) to calculate the four dependent variables $x^{\alpha}(\tau)$, and then eliminate τ to determine $\mathbf{x}(t)$. However, the initial values of $dx^{\alpha}/d\tau$ must be chosen so that $d\tau$ really is the proper time, that is, so that

$$-1 = \eta_{\alpha\beta}\frac{dx^{\alpha}}{d\tau}\frac{dx^{\beta}}{d\tau} \tag{2.3.7}$$

Note that (2.3.7) will be true for all τ if it is true at some initial τ, providing that its derivative vanishes, that is, providing that

$$0 = 2\eta_{\alpha\beta}f^{\alpha}\frac{dx^{\beta}}{d\tau} \tag{2.3.8}$$

That this *is* true can be seen either directly from (2.3.4), or more elegantly by noticing that the right-hand side is Lorentz-invariant:

$$\eta_{\alpha\beta}f'^{\alpha}\frac{dx'^{\beta}}{d\tau} = \eta_{\alpha\beta}\Lambda^{\alpha}{}_{\gamma}\,\Lambda^{\beta}{}_{\delta}f^{\gamma}\frac{dx^{\delta}}{d\tau}$$

$$= \eta_{\gamma\delta}f^{\gamma}\frac{dx^{\delta}}{d\tau}$$

and that it vanishes by virtue of (2.3.2) in a reference frame in which the particle is at rest.

4 Energy and Momentum

The relativistic form (2.3.1) of Newton's second law immediately suggests that we define an energy-momentum four-vector

$$p^{\alpha} \equiv m\frac{dx^{\alpha}}{d\tau} \tag{2.4.1}$$

and write the second law as

$$\frac{dp^\alpha}{d\tau} = f^\alpha \tag{2.4.2}$$

Recall that

$$d\tau \equiv (dt^2 - d\mathbf{x}^2)^{1/2} = (1 - \mathbf{v}^2)^{1/2}\, dt$$

where

$$\mathbf{v} \equiv \frac{d\mathbf{x}}{dt}$$

Then the space components of p^α form the momentum vector

$$\mathbf{p} = m\gamma\mathbf{v} \tag{2.4.3}$$

and its time component is the energy

$$p^0 \equiv E = m\gamma \tag{2.4.4}$$

where

$$\gamma \equiv \frac{dt}{d\tau} = (1 - \mathbf{v}^2)^{-1/2} \tag{2.4.5}$$

For small $\mathbf{v}$, these definitions give

$$\mathbf{p} = m\mathbf{v} + 0(\mathbf{v}^3) \tag{2.4.6}$$

$$E = m + \tfrac{1}{2}m\mathbf{v}^2 + 0(\mathbf{v}^4) \tag{2.4.7}$$

in agreement with the nonrelativistic formulas, except for the term m in E. (Recall that in our units 1 sec equals 3×10^{10} cm, so 1 g equals 9×10^{20} ergs.) Sometimes the factor $m\gamma$ is called the relativistic mass $\tilde{m}$, so that $\mathbf{p} = \tilde{m}\mathbf{v}$. I do not follow this custom here; for us, "mass" will always mean the constant m.

Why do we call $\mathbf{p}$ and E the relativistic momentum and energy? We can use these names for anything we like, but if the concepts of momentum and energy are to be useful they must be reserved for quantities that are *conserved*. The unique feature of our $\mathbf{p}$ and E is that, if one observer says that they are conserved in a reaction, then so will any other observer related to the first by a Lorentz transformation. Note that dx^α is a four-vector whereas m and $d\tau$ are invariants, so the p^α for any single particle is a four-vector; that is, it transforms under (2.1.1) like

$$p'^\alpha = \Lambda^\alpha{}_\beta\, p^\beta$$

Since Λ does not depend on anything but the Lorentz transformation being performed, it follows that in any reaction, the change of the sum of the p^α of all particles is also a four-vector:

$$\Delta \sum_n p'^\alpha_n = \Lambda^\alpha{}_\beta\, \Delta \sum_n p^\beta_n$$

(The sums run over all particles, and Δ denotes the difference between initial and final states.) The conservation of $\mathbf{p}$ and E in the original inertial frame tells us that $\Delta \sum_n p_n{}^\beta$ vanishes, so in any coordinate system related to the first by a Lorentz transformation they will still be conserved; that is, $\Delta \sum_n p_n'^\alpha$ will vanish.

(I shall not show here that $\mathbf{p}$ and E are the *only* functions of velocity whose conservation is Lorentz-invariant.[3] However, it is worth stressing that E must be conserved if $\mathbf{p}$ is. For suppose that momentum is conserved in two different coordinate systems related by a Lorentz transformation, that is,

$$\Delta \sum_n \mathbf{p}_n = 0 \qquad \Delta \sum_n \mathbf{p}_n' = 0$$

Since $\Delta \sum_n p_n{}^\alpha$ is a four-vector, we have

$$\Delta \sum_n p_n'^i = \Lambda^i{}_\beta \, \Delta \sum_n p_n{}^\beta$$

and using momentum conservation in both coordinate systems, this gives

$$0 = \Lambda^i{}_0 \, \Delta \sum_n p_n{}^0$$

But $\Lambda^i{}_0$ is not necessarily zero, so $p^0 = E$ is conserved.)

At zero velocity the energy E has the finite value m. For this reason we sometimes give the name "kinetic energy" to the quantity $E - m$, which for small $\mathbf{v}$ is approximately $\frac{1}{2}m\mathbf{v}^2$. If the total mass is conserved in a reaction (as in elastic scattering), then the kinetic energy is conserved, but if some mass is destroyed (as in radioactive decay or fusion or fission), then very large quantities of kinetic energy will be liberated, with consequences of well-known importance.

The velocity can be eliminated from Eqs. (2.4.3) and (2.4.4), yielding a relation between energy and momentum

$$E(\mathbf{p}) = (\mathbf{p}^2 + m^2)^{1/2} \tag{2.4.8}$$

This can also be derived by noting from (2.4.1) and the definition of $d\tau$ that

$$\eta_{\alpha\beta} p^\alpha p^\beta = -m^2 \tag{2.4.9}$$

For a photon or neutrino we must set $\mathbf{v}^2 = 1$ and $m = 0$, so (2.4.3) and (2.4.4) become indeterminate, but their ratio gives a relation useful for all particles

$$\frac{\mathbf{p}}{E} = \mathbf{v} \tag{2.4.10}$$

Note that for $m = 0$ Eq. (2.4.8) gives

$$E = |\mathbf{p}|$$

so $\mathbf{v}$ is a unit vector, as of course it must be for a massless particle.

5   Vectors and Tensors

Next we go on to electrodynamics and relativistic hydrodynamics, but it is convenient first to pause and outline a notation that makes the Lorentz transformation properties of physical quantities transparent. This notation will be extended in Chapter 4, on tensor analysis, to encompass general coordinate transformations, but in fact few changes will be needed.

We have already introduced the term "four-vector" for any quantity such as dx^α or f^α or p^α that undergoes the transformation

$$V^\alpha \to V^{\alpha\prime} = \Lambda^\alpha{}_\beta V^\beta \tag{2.5.1}$$

when the coordinate system is transformed by

$$x^\alpha \to x'^\alpha = \Lambda^\alpha{}_\beta x^\beta \tag{2.5.2}$$

More precisely, such a V^α should be called a *contravariant* four-vector, to distinguish it from a *covariant* four-vector, defined as a quantity U_α whose transformation rule is

$$U_\alpha \to U'_\alpha = \Lambda_\alpha{}^\beta U_\beta \tag{2.5.3}$$

where

$$\Lambda_\alpha{}^\beta \equiv \eta_{\alpha\gamma}\eta^{\beta\delta} \Lambda^\gamma{}_\delta \tag{2.5.4}$$

The matrix $\eta^{\beta\delta}$ introduced here is numerically the same as $\eta_{\beta\delta}$, that is,

$$\eta^{\beta\delta} = \eta_{\beta\delta} \tag{2.5.5}$$

but we write it with indices upstairs to conform with our summation convention. Note that

$$\eta^{\beta\delta}\eta_{\alpha\delta} = \delta^\beta{}_\alpha \equiv \begin{cases} +1 & \alpha = \beta \\ 0 & \alpha \neq \beta \end{cases} \tag{2.5.6}$$

so $\Lambda_\alpha{}^\beta$ is the inverse of the matrix $\Lambda^\beta{}_\alpha$, that is,

$$\Lambda_{\alpha'}{}^\gamma\Lambda^\alpha{}_\beta = \eta_{\alpha\delta}\eta^{\gamma\varepsilon}\Lambda^\delta{}_\varepsilon \Lambda^\alpha{}_\beta = \eta_{\varepsilon\beta}\eta^{\gamma\varepsilon} = \delta^\gamma{}_\beta \tag{2.5.7}$$

It follows that the scalar product of a contravariant with a covariant four-vector is invariant, that is,

$$U'_\alpha V'^\alpha = \Lambda_\alpha{}^\gamma \Lambda^\alpha{}_\beta U_\gamma V^\beta = U_\beta V^\beta \tag{2.5.8}$$

To every contravariant four-vector V^α there corresponds a covariant four-vector

$$V_\alpha \equiv \eta_{\alpha\beta}V^\beta \tag{2.5.9}$$

and to every covariant U_α there corresponds a contravariant

$$U^\alpha \equiv \eta^{\alpha\beta}U_\beta \tag{2.5.10}$$

Note that raising the index on V_α simply gives back V^α, and lowering the index on U^α simply gives back U_α,

$$\eta^{\alpha\beta} V_\beta = \eta^{\alpha\beta}\eta_{\beta\gamma} V^\gamma = V^\alpha$$
$$\eta_{\alpha\beta} U^\beta = \eta_{\alpha\beta}\eta^{\beta\gamma} U_\gamma = U_\alpha$$

Note also that (2.5.9) does yield a covariant, because

$$V'_\alpha = \eta_{\alpha\beta} V'^\beta = \eta_{\alpha\beta} \Lambda^\beta{}_\gamma V^\gamma = \eta_{\alpha\beta}\eta^{\gamma\delta} \Lambda^\beta{}_\gamma V_\delta$$
$$= \Lambda_\alpha{}^\delta V_\delta$$

in agreement with (2.5.3). Similarly, (2.5.10) does yield a contravariant.

Although any vector can be written in a contravariant or a covariant form, there are some vectors, such as dx^α, that appear more naturally contravariant and others that appear more naturally covariant. An example of the latter is the gradient $\partial/\partial x^\alpha$, which obeys the transformation rule

$$\frac{\partial}{\partial x^\alpha} \to \frac{\partial}{\partial x'^\alpha} = \frac{\partial x^\beta}{\partial x'^\alpha} \frac{\partial}{\partial x^\beta}$$

Multiplying (2.5.2) by $\Lambda_\alpha{}^\gamma$ gives

$$x^\gamma = \Lambda_\alpha{}^\gamma x'^\alpha$$

so

$$\frac{\partial x^\beta}{\partial x'^\alpha} = \Lambda_\alpha{}^\beta$$

and therefore the gradient is covariant:

$$\frac{\partial}{\partial x'^\alpha} = \Lambda_\alpha{}^\beta \frac{\partial}{\partial x^\beta} \tag{2.5.11}$$

One consequence is that the divergence of a contravariant vector $\partial V^\alpha/\partial x^\alpha$ is invariant. Another is that the scalar product of $\partial/\partial x^\alpha$ with itself, the d'Alembertian operator

$$\Box^2 = \eta^{\alpha\beta} \frac{\partial}{\partial x^\beta} \frac{\partial}{\partial x^\alpha} = \nabla^2 - \frac{\partial^2}{\partial t^2} \tag{2.5.12}$$

is also invariant.

Many physical quantities are not scalars or vectors, but more complicated objects called *tensors*. A tensor has several contravariant and/or covariant indices with corresponding Lorentz transformation properties, for example,

$$T^\gamma{}_{\alpha\beta} \to T'^\gamma{}_{\alpha\beta} = \Lambda^\gamma{}_\delta \Lambda_\alpha{}^\varepsilon \Lambda_\beta{}^\zeta T^\delta{}_{\varepsilon\zeta}$$

A contravariant or covariant vector can be regarded as a tensor with one index, and a scalar is a tensor with no indices. There are several ways of forming tensors out of other tensors:

(A) **Linear Combinations.** A linear combination of tensors with the same upper and lower indices is a tensor with these indices. For instance, if $R^\alpha{}_\beta$ and $S^\alpha{}_\beta$ are tensors, and a and b are scalars, and we define

$$T^\alpha{}_\beta \equiv aR^\alpha{}_\beta + bS^\alpha{}_\beta$$

then $T^\alpha{}_\beta$ is a tensor, that is,

$$\begin{aligned}
T'^\alpha{}_\beta &\equiv aR'^\alpha{}_\beta + bS'^\alpha{}_\beta \\
&= a\Lambda^\alpha{}_\gamma \Lambda_\beta{}^\delta R^\gamma{}_\delta + b\Lambda^\alpha{}_\gamma \Lambda_\beta{}^\delta S^\gamma{}_\delta \\
&= \Lambda^\alpha{}_\gamma \Lambda_\beta{}^\delta T^\gamma{}_\delta
\end{aligned}$$

(B) **Direct Products.** The product of the components of two tensors yields a tensor whose upper and lower indices consist of all the upper and lower indices of the two original tensors. For instance, if $A^\alpha{}_\beta$ and B^γ are tensors, and

$$T^\alpha{}_\beta{}^\gamma \equiv A^\alpha{}_\beta B^\gamma$$

then $T^\alpha{}_\beta{}^\gamma$ is a tensor, that is,

$$\begin{aligned}
T'^\alpha{}_\beta{}^\gamma &= A'^\alpha{}_\beta B'^\gamma \\
&= \Lambda^\alpha{}_\delta \Lambda_\beta{}^\varepsilon \Lambda^\gamma{}_\zeta T^\delta{}_\varepsilon{}^\zeta
\end{aligned}$$

(C) **Contraction.** Setting an upper and lower index equal and summing it over its values 0, 1, 2, 3, yields a tensor with these two indices absent. For instance, if $T^\alpha{}_\beta{}^{\gamma\delta}$ is a tensor and

$$T^{\alpha\gamma} \equiv T^\alpha{}_\beta{}^{\gamma\beta}$$

then $T^{\alpha\gamma}$ is a tensor, that is,

$$\begin{aligned}
T'^{\alpha\gamma} &\equiv T'^\alpha{}_\beta{}^{\gamma\beta} \\
&= \Lambda^\alpha{}_\delta \Lambda_\beta{}^\varepsilon \Lambda^\gamma{}_\zeta \Lambda^\beta{}_\kappa T^\delta{}_\varepsilon{}^{\zeta\kappa} \\
&= \Lambda^\alpha{}_\delta \Lambda^\gamma{}_\zeta \delta^\varepsilon{}_\kappa T^\delta{}_\varepsilon{}^{\zeta\kappa} \\
&= \Lambda^\alpha{}_\delta \Lambda^\gamma{}_\zeta T^{\delta\zeta}
\end{aligned}$$

(D) **Differentiation.** The derivative $\partial/\partial x^\alpha$ of any tensor is a tensor with one additional lower index α. For instance, if $T^{\beta\gamma}$ is a tensor and

$$T_\alpha{}^{\beta\gamma} \equiv \frac{\partial}{\partial x^\alpha} T^{\beta\gamma}$$

then $T_\alpha{}^{\beta\gamma}$ is a tensor, that is,

$$\begin{aligned}
T'_\alpha{}^{\beta\gamma} &\equiv \frac{\partial}{\partial x'^\alpha} T'^{\beta\gamma} \\[2mm]
&= \Lambda_\alpha{}^\delta \frac{\partial}{\partial x^\delta} \Lambda^\beta{}_\varepsilon \Lambda^\gamma{}_\zeta T^{\varepsilon\zeta} \\[2mm]
&= \Lambda_\alpha{}^\delta \Lambda^\beta{}_\varepsilon \Lambda^\gamma{}_\zeta T_\delta{}^{\varepsilon\zeta}
\end{aligned}$$

Note that the order of indices matters, even as between upper and lower indices. For instance, $T_\alpha{}^{\beta\gamma}$ may or may not be the same as $T^\beta{}_\alpha{}^\gamma$.

Aside from the scalars, there are three special tensors whose components are the same in all coordinate systems:

(i) **The Minkowski Tensor.** The definition of Lorentz transformations tells us immediately that $\eta_{\alpha\beta}$ is a covariant tensor,

$$\eta_{\alpha\beta} = \Lambda^\gamma{}_\alpha \Lambda^\delta{}_\beta \, \eta_{\gamma\delta}$$

Multiplying this equation by $\eta^{\alpha\varepsilon}\eta^{\beta\zeta}$ and using (2.5.6) and (2.5.4), we find that

$$\eta^{\varepsilon\zeta} = \eta^{\gamma\kappa}\eta^{\delta\lambda}\Lambda_\kappa{}^\varepsilon \Lambda_\lambda{}^\zeta \eta_{\gamma\delta}$$
$$= \eta^{\kappa\lambda}\Lambda_\kappa{}^\varepsilon \Lambda_\lambda{}^\zeta$$

so $\eta^{\alpha\beta}$ is a contravariant tensor. (Recall that $\eta_{\alpha\beta}$ and $\eta^{\alpha\beta}$ are numerically the same matrix, so this is a matrix that is both covariant and contravariant.) We can form a mixed tensor by lowering one index on $\eta^{\alpha\beta}$ or raising one index on $\eta_{\alpha\beta}$; this gives the Kronecker symbol

$$\delta^\alpha{}_\beta = \eta^{\alpha\gamma}\eta_{\gamma\beta}$$

That this is a tensor follows from rules (B) and (C) and the fact that $\eta^{\alpha\gamma}$ and $\eta_{\gamma\beta}$ are tensors.

(ii) **The Levi-Civita Tensor.** This is a quantity $\varepsilon^{\alpha\beta\gamma\delta}$ defined by

$$\varepsilon^{\alpha\beta\gamma\delta} = \begin{cases} +1 & \text{if } \alpha\beta\gamma\delta \text{ even permutation of } 0123 \\ -1 & \text{if } \alpha\beta\gamma\delta \text{ odd permutation of } 0123 \\ 0 & \text{otherwise} \end{cases} \qquad (2.5.13)$$

Note that

$$\Lambda^\alpha{}_\varepsilon \Lambda^\beta{}_\zeta \Lambda^\gamma{}_\kappa \Lambda^\delta{}_\lambda \, \varepsilon^{\varepsilon\zeta\kappa\lambda} \propto \varepsilon^{\alpha\beta\gamma\delta}$$

because the left-hand side must be odd under any single permutation of the indices $\alpha\beta\gamma\delta$. To find the constant of proportionality, set $\alpha\beta\gamma\delta = 0123$. The left-hand side is then simply the determinant of Λ, which for proper Lorentz transformations is unity. (See Section 2.1.) Thus the constant of proportionality is unity, that is,

$$\Lambda^\alpha{}_\varepsilon \Lambda^\beta{}_\zeta \Lambda^\gamma{}_\kappa \Lambda^\delta{}_\lambda \, \varepsilon^{\varepsilon\zeta\kappa\lambda} = \varepsilon^{\alpha\beta\gamma\delta} \qquad (2.5.14)$$

and therefore $\varepsilon^{\alpha\beta\gamma\delta}$ *is a tensor*.

(iii) **The Zero Tensor.** We can define a tensor with an arbitrary pattern of upper and lower indices by setting all its components equal to zero.

Since $\eta^{\alpha\beta}$ and $\eta_{\alpha\beta}$ are tensors, we can use them to raise or lower indices on an arbitrary tensor; rules (B) and (C) tell us that this gives a new tensor with one more upper or lower index and one less lower or upper index. For instance, if $T_{\alpha\beta\gamma}$ is a tensor, then so is

$$T_\alpha{}^\delta{}_\gamma \equiv \eta^{\delta\beta} T_{\alpha\beta\gamma}$$

In particular, we can lower some or all of the indices on the Levi-Civita tensor $\varepsilon^{\alpha\beta\gamma\delta}$. Lowering all the indices gives back the same numerical quantity except for a minus sign:

$$\varepsilon_{\alpha\beta\gamma\delta} = -\varepsilon^{\alpha\beta\gamma\delta} \tag{2.5.15}$$

The point of all this algebra is that it enables us to tell at a glance that an equation is Lorentz-invariant. The fundamental theorem is that *if two tensors, with the same upper and lower indices, are equal in one coordinate system, then they are equal in any other coordinate system related to the first by a Lorentz transformation.* For instance, if $T^\alpha{}_\beta = S^\alpha{}_\beta$, then

$$T'^\alpha{}_\beta = \Lambda^\alpha{}_\gamma \Lambda_\beta{}^\delta\, T^\gamma{}_\delta = \Lambda^\alpha{}_\gamma \Lambda_\beta{}^\delta\, S^\gamma{}_\delta$$
$$= S'^\alpha{}_\beta$$

In particular, *the statement that a tensor vanishes is Lorentz-invariant.*

The formalism outlined in this section is nothing but a description of the representations of the homogeneous Lorentz group. We shall explore these representations in greater generality in Section 2.12.

6 Currents and Densities

Suppose that we have a system of particles with position $\mathbf{x}_n(t)$ and charges e_n. The current and charge densities are usually defined by

$$\mathbf{J}(\mathbf{x}, t) \equiv \sum_n e_n \delta^3(\mathbf{x} - \mathbf{x}_n(t)) \frac{d\mathbf{x}_n(t)}{dt} \tag{2.6.1}$$

$$\varepsilon(\mathbf{x}, t) \equiv \sum_n e_n \delta^3(\mathbf{x} - \mathbf{x}_n(t)) \tag{2.6.2}$$

Here δ^3 is the Dirac delta function, defined by the statement that for any smooth function $f(x)$,

$$\int d^3x f(\mathbf{x}) \delta^3(\mathbf{x} - \mathbf{y}) = f(\mathbf{y})$$

We can unite $\mathbf{J}$ and ε into a four-vector J^α by setting

$$J^0 \equiv \varepsilon \tag{2.6.3}$$

that is

$$J^\alpha(x) \equiv \sum_n e_n \delta^3(\mathbf{x} - \mathbf{x}_n(t)) \frac{dx_n{}^\alpha(t)}{dt} \tag{2.6.4}$$

To show that this is a four-vector, define $x_n{}^0(t) = t$, and write (2.6.4) as

$$J^\alpha(x) = \int dt' \sum_n e_n \delta^4(x - x_n(t')) \frac{dx_n{}^\alpha(t')}{dt'}$$

The differentials dt' cancel, and hence can be replaced with an invariant $d\tau$:

$$J^\alpha(x) = \int d\tau \sum_n e_n \delta^4(x - x_n(\tau)) \frac{dx_n{}^\alpha(\tau)}{d\tau} \tag{2.6.5}$$

But $\delta^4(x - x_n(\tau))$ is a scalar (because Det $\Lambda = 1$) and $dx_n{}^\alpha$ is a four-vector, so J^α is a four-vector.

We also note that

$$\nabla \cdot \mathbf{J}(\mathbf{x}, t) = \sum_n e_n \frac{\partial}{\partial x^i} \delta^3(\mathbf{x} - \mathbf{x}_n(t)) \frac{dx_n{}^i(t)}{dt}$$

$$= -\sum_n e_n \frac{\partial}{\partial x_n{}^i} \delta^3(\mathbf{x} - \mathbf{x}_n(t)) \frac{dx_n{}^i(t)}{dt}$$

$$= -\sum_n e_n \frac{\partial}{\partial t} \delta^3(\mathbf{x} - \mathbf{x}_n(t))$$

$$= -\frac{\partial}{\partial t} \varepsilon(\mathbf{x}, t)$$

or, in four-dimensional language

$$\frac{\partial}{\partial x^\alpha} J^\alpha(x) = 0 \tag{2.6.6}$$

The Lorentz invariance of this statement is evident.

Whenever any current $J^\alpha(x)$ satisfies the invariant conservation law (2.6.6), we can form a total charge

$$Q \equiv \int d^3x J^0(x) \tag{2.6.7}$$

This quantity is time-independent, because (2.6.6) and Gauss's theorem give

$$\frac{dQ}{dt} = \int d^3x \frac{\partial}{\partial x^0} J^0(x) = -\int d^3x \, \nabla \cdot \mathbf{J}(x) = 0$$

If $J^\alpha(x)$ is a four-vector, then Q is not only constant but a scalar. To see this, write Q as

$$Q = \int d^4x J^\alpha(x) \partial_\alpha \theta(n_\beta x^\beta) \tag{2.6.8}$$

where θ is the step function

$$\theta(s) = \begin{cases} 1 & s > 0 \\ 0 & s < 0 \end{cases}$$

and n_λ is defined by

$$n_1 \equiv n_2 \equiv n_3 \equiv 0, \qquad n_0 \equiv +1$$

The effect of a Lorentz transformation on Q is then evidently simply to change n:

$$Q' = \int d^4x J^\alpha(x) \partial_\alpha \theta(n'_\beta x^\beta)$$

$$n'_\beta \equiv \Lambda^\gamma_{\ \beta} n_\gamma$$

and using (2.6.6), the change in Q is then

$$Q' - Q = \int d^4x \partial_\alpha [J^\alpha(x) \{\theta(n'_\beta x^\beta) - \theta(n_\beta x^\beta)\}]$$

The current $J^\alpha(x)$ can be presumed to vanish if $|\mathbf{x}| \rightarrow \infty$ with t fixed, whereas the function $\theta(n'_\beta x^\beta) - \theta(n_\beta x^\beta)$ vanishes if $|t| \rightarrow \infty$ with $\mathbf{x}$ fixed. Hence we can apply the four-dimensional Gauss theorem, and find $Q' - Q = 0$; that is, Q is a scalar. (For the current density J^0 defined by (2.6.2) the charge (2.6.7) is

$$Q = \sum_n e_n$$

which of course is a constant scalar; however, in dealing with the charge and current distributions of extended particles it is important to realize that (2.6.7) defines a time-independent scalar for *any* conserved four-vector J^α.)

7 Electrodynamics

Maxwell's equations for the electric and magnetic fields $\mathbf{E}$, $\mathbf{B}$ produced by a given charge density ε and current density $\mathbf{J}$ are

$$\mathbf{\nabla} \cdot \mathbf{E} = \varepsilon \qquad (2.7.1)$$

$$\mathbf{\nabla} \times \mathbf{B} = \frac{\partial \mathbf{E}}{\partial t} + \mathbf{J} \qquad (2.7.2)$$

$$\mathbf{\nabla} \cdot \mathbf{B} = 0 \qquad (2.7.3)$$

$$\mathbf{\nabla} \times \mathbf{E} = -\frac{\partial \mathbf{B}}{\partial t} \qquad (2.7.4)$$

To uncover the Lorentz transformation properties of **E** and **B**, we introduce a matrix $F^{\alpha\beta}$, defined by

$$F^{12} = B_3 \qquad F^{23} = B_1 \qquad F^{31} = B_2$$
$$F^{01} = E_1 \qquad F^{02} = E_2 \qquad F^{03} = E_3 \qquad (2.7.5)$$
$$F^{\alpha\beta} = -F^{\beta\alpha}$$

Then (2.7.1) and (2.7.2) can be written as

$$\frac{\partial}{\partial x^\alpha} F^{\alpha\beta} = -J^\beta \qquad (2.7.6)$$

(recall that $J^0 \equiv \varepsilon$) whereas (2.7.3) and (2.7.4) give

$$\varepsilon^{\alpha\beta\gamma\delta} \frac{\partial}{\partial x^\beta} F_{\gamma\delta} = 0 \qquad (2.7.7)$$

where $\varepsilon^{\alpha\beta\gamma\delta}$ is the Levi-Civita symbol defined in Section 2.5, and $F_{\gamma\delta}$ is the covariant defined as usual by

$$F_{\gamma\delta} \equiv \eta_{\gamma\alpha}\eta_{\delta\beta}F^{\alpha\beta}$$

Since J^α is a four-vector, we conclude that $F^{\alpha\beta}$ is a tensor,

$$F'^{\alpha\beta} = \Lambda^\alpha{}_\gamma \Lambda^\beta{}_\delta F^{\gamma\delta} \qquad (2.7.8)$$

because if $F^{\alpha\beta}$ is a solution of (2.7.6) and (2.7.7), then (2.7.8) will be a solution in a Lorentz-transformed coordinate system.

The electromagnetic force on a charged particle is

$$f^\alpha = e\eta_{\beta\gamma}F^{\alpha\beta}\frac{dx^\gamma}{d\tau} = eF^\alpha{}_\gamma\frac{dx^\gamma}{d\tau} \qquad (2.7.9)$$

That this is correct may be seen by repeating the arguments of Section 3. Equation (2.7.9) is correct in a reference system in which the particle is at rest because in this frame it gives $\mathbf{f} = e\mathbf{E}$, $f^0 = 0$, and it transforms like a four-vector, so it is correct for all velocities. Note incidentally that (2.7.9) and (2.4.2) give

$$\frac{d\mathbf{p}}{dt} = e[\mathbf{E} + \mathbf{v} \times \mathbf{B}]$$

so the formula for magnetic force follows as a consequence of special relativity.

There is a useful alternate form to the homogeneous equations (2.7.7):

$$\frac{\partial}{\partial x^\alpha} F_{\beta\gamma} + \frac{\partial}{\partial x^\beta} F_{\gamma\alpha} + \frac{\partial}{\partial x^\gamma} F_{\alpha\beta} = 0 \qquad (2.7.10)$$

Note that for α, β, γ all different, Eq. (2.7.10) is the same as (2.7.7); for instance, setting $\alpha = 0$ in Eq. (2.7.7) gives the same result as setting $\alpha\beta\gamma = 123$ in Eq.

(2.7.10). On the other hand, for two indices equal, Eq. (2.7.10) is an identity; for instance, if $\beta = \gamma$ then (2.7.10) reads

$$\frac{\partial}{\partial x^\beta} F_{\beta\alpha} + \frac{\partial}{\partial x^\beta} F_{\alpha\beta} = 0 \qquad \text{(not summed)}$$

and this is identically true because $F_{\alpha\beta} = -F_{\beta\alpha}$.

Equation (2.7.7) allows us to represent $F_{\gamma\delta}$ as a "curl" of a four-vector A_γ:

$$F_{\gamma\delta} = \frac{\partial}{\partial x^\gamma} A_\delta - \frac{\partial}{\partial x^\delta} A_\gamma \qquad (2.7.11)$$

(See section 4.11.)

We can change A_γ by a term $\partial_\gamma\varphi$ without affecting $F_{\gamma\delta}$, so A_γ may be defined so that

$$\partial^\alpha A_\alpha = 0 \qquad (2.7.12)$$

With (2.7.11) and (2.7.12), the rest of Maxwell's equations reduce to

$$\Box^2 A_\alpha = -J_\alpha \qquad (2.7.13)$$

8 Energy-Momentum Tensor

In Section 5 we introduced the density ε and current $\mathbf{J}$ of electric charge. We now give a similar definition for the density and current of the energy-momentum four-vector p^α. First consider a system of particles labeled n, with energy-momentum four-vectors $p_n^\alpha(t)$. The density of p^α is defined by

$$T^{\alpha 0}(\mathbf{x}t) \equiv \sum_n p_n^\alpha(t)\delta^3(\mathbf{x} - \mathbf{x}_n(t)) \qquad (2.8.1)$$

and its current is defined by

$$T^{\alpha i}(\mathbf{x}t) \equiv \sum_n p_n^\alpha(t) \frac{dx_n^i(t)}{dt} \delta^3(\mathbf{x} - \mathbf{x}_n(t)) \qquad (2.8.2)$$

These two definitions can be united into a single formula,

$$T^{\alpha\beta}(x) = \sum_n p_n^\alpha \frac{dx_n^\beta(t)}{dt} \delta^3(\mathbf{x} - \mathbf{x}_n(t)) \qquad (2.8.3)$$

where $x_n^0(t) \equiv t$. We note from (2.4.10) that

$$p_n^\beta = E_n \frac{dx_n^\beta}{dt}$$

so (2.8.3) can also be written as

$$T^{\alpha\beta}(x) = \sum_n \frac{p_n{}^\alpha p_n{}^\beta}{E_n} \delta^3(\mathbf{x} - \mathbf{x}_n(t)) \tag{2.8.4}$$

and we see that $T^{\alpha\beta}$ is *symmetric*:

$$T^{\alpha\beta}(x) = T^{\beta\alpha}(x) \tag{2.8.5}$$

We can also write (2.8.3) in analogy with (2.6.5) as

$$T^{\alpha\beta}(x) = \sum_n \int d\tau \, p_n{}^\alpha \frac{dx_n{}^\beta}{d\tau} \delta^4(x - x_n(\tau)) \tag{2.8.5a}$$

and we see that $T^{\alpha\beta}$ is a *tensor*, that is,

$$T'^{\alpha\beta} = \Lambda^\alpha{}_\gamma \Lambda^\beta{}_\delta \, T^{\gamma\delta}$$

under a Lorentz transformation (2.1.1).

The conservation law for $T^{\alpha\beta}$ will take a little more thought. Returning to (2.8.1) and (2.8.2) we see that

$$\frac{\partial}{\partial x^i} T^{\alpha i}(\mathbf{x}, t) = -\sum_n p_n{}^\alpha(t) \frac{dx_n{}^i(t)}{dt} \frac{\partial}{\partial x_n{}^i} \delta^3(\mathbf{x} - \mathbf{x}_n(t))$$

$$= -\sum_n p_n{}^\alpha(t) \frac{\partial}{\partial t} \delta^3(\mathbf{x} - \mathbf{x}_n(t))$$

$$= -\frac{\partial}{\partial t} T^{\alpha 0}(\mathbf{x}, t) + \sum_n \frac{dp_n{}^\alpha(t)}{dt} \delta^3(\dot{\mathbf{x}} - \mathbf{x}_n(t))$$

and so

$$\frac{\partial}{\partial x^\beta} T^{\alpha\beta} = G^\alpha \tag{2.8.6}$$

where G^α is the *density of force*:

$$G^\alpha(\mathbf{x}, t) \equiv \sum_n \delta^3(\mathbf{x} - \mathbf{x}_n(t)) \frac{dp_n{}^\alpha(t)}{dt} = \sum_n \delta^3(\mathbf{x} - \mathbf{x}_n(t)) \frac{d\tau}{dt} f_n{}^\alpha(t)$$

If the particles are free, then $p_n{}^\alpha$ is constant and $T^{\alpha\beta}$ is conserved, that is,

$$\frac{\partial}{\partial x^\beta} T^{\alpha\beta}(x) = 0 \tag{2.8.7}$$

The same is also true if the particles interact only during collisions that are strictly localized in space. In this case (2.8.6) gives

$$\frac{\partial}{\partial x^\beta} T^{\alpha\beta}(x) = \sum_c \delta^3(\mathbf{x} - \mathbf{x}_c(t)) \frac{d}{dt} \sum_{n \in c} p_n{}^\alpha(t)$$

where $\mathbf{x}_c(t)$ is the location of the cth collision going on at time t, and $n \in c$ means we sum only over the particles participating in the cth collision. But each collision conserves momentum, so $\sum_{n \in c} p_n{}^\alpha(t)$ must be time-independent, yielding the conservation equation (2.8.7).

The energy-momentum tensor (2.8.3) will not be conserved if the particles are subject to forces that act at a distance. For instance, consider a gas of charged particles, with charges e_n . Then (2.8.6), (2.4.1), and (2.7.9) give

$$\frac{\partial}{\partial x^\beta} T^{\alpha\beta}(x) = \sum_n e_n F^\alpha{}_\gamma(x) \frac{dx_n{}^\gamma}{dt} \delta^3(\mathbf{x} - \mathbf{x}_n(t))$$

and, using (2.6.4), this gives

$$\frac{\partial}{\partial x^\beta} T^{\alpha\beta}(x) = F^\alpha{}_\gamma(x) J^\gamma(x) \qquad (2.8.8)$$

Although this is not conserved, we can construct a conserved tensor by adding a purely electromagnetic term

$$T_{\text{em}}{}^{\alpha\beta} \equiv F^\alpha{}_\gamma F^{\beta\gamma} - \tfrac{1}{4}\eta^{\alpha\beta} F_{\gamma\delta} F^{\gamma\delta} \qquad (2.8.9)$$

That is, the electromagnetic energy and momentum densities are given by

$$T_{\text{em}}{}^{00} = \tfrac{1}{2}(\mathbf{E}^2 + \mathbf{B}^2) \qquad T_{\text{em}}{}^{i0} = (\mathbf{E} \times \mathbf{B})_i \qquad (2.8.10)$$

We note that

$$\frac{\partial}{\partial x^\beta} T_{\text{em}}{}^{\alpha\beta} = F^\alpha{}_\gamma \frac{\partial}{\partial x^\beta} F^{\beta\gamma} + F^{\beta\gamma} \frac{\partial}{\partial x^\beta} F^\alpha{}_\gamma - \tfrac{1}{2} F_{\gamma\delta} \frac{\partial}{\partial x_\alpha} F^{\gamma\delta}$$

[Here $\partial/\partial x_\alpha = \eta^{\alpha\beta}(\partial/\partial x^\beta)$.] With a little reshuffling of indices, this becomes

$$\frac{\partial}{\partial x^\beta} T_{\text{em}}{}^{\alpha\beta} = F^\alpha{}_\gamma \frac{\partial}{\partial x^\beta} F^{\beta\gamma} - \tfrac{1}{2} F_{\beta\gamma} \left(\frac{\partial}{\partial x_\alpha} F^{\beta\gamma} + \frac{\partial}{\partial x_\beta} F^{\gamma\alpha} + \frac{\partial}{\partial x_\gamma} F^{\alpha\beta} \right)$$

Using the Maxwell equations (2.7.6) and (2.7.10), we find

$$\frac{\partial}{\partial x^\beta} T_{\text{em}}{}^{\alpha\beta} = -F^\alpha{}_\gamma J^\gamma \qquad (2.8.11)$$

Comparing (2.8.8) with (2.8.11), we are led to redefine the energy-momentum tensor as

$$T^{\alpha\beta} = \sum_n p_n{}^\alpha \frac{dx_n{}^\beta}{dt} \delta^3(\mathbf{x} - \mathbf{x}_n(t)) + T_{\text{em}}{}^{\alpha\beta} \qquad (2.8.12)$$

This is again a symmetric tensor, and is now conserved

$$\partial_\alpha T^{\alpha\beta} = 0 \qquad (2.8.13)$$

We can continue to add more and more terms to $T^{\alpha\beta}$ to account for other fields and keep $T^{\alpha\beta}$ conserved. A systematic method for constructing these terms is presented in Chapter 12.

Just as the integral of the charge density J^0 is the total charge, the integral of the density $T^{\alpha 0}$ of p^α is the total p^α:

$$p^\alpha_{\text{total}} = \int d^3x T^{\alpha 0}(\mathbf{x}, t) \tag{2.8.14}$$

That this is a constant four-vector can be shown in the same way that we showed in Section 6 that the total charge (2.6.7) is a constant scalar.

9   Spin

One important use we can make of the energy-momentum tensor $T^{\alpha\beta}$ is to define angular momentum and spin. Consider first an isolated system, for which the *total* energy-momentum tensor $T^{\alpha\beta}$ is conserved

$$\frac{\partial}{\partial x^\gamma} T^{\beta\gamma} = 0$$

We can use T to construct another tensor,

$$M^{\gamma\alpha\beta} \equiv x^\alpha T^{\beta\gamma} - x^\beta T^{\alpha\gamma} \tag{2.9.1}$$

and because T is conserved and symmetric, M is also conserved:

$$\frac{\partial M^{\gamma\alpha\beta}}{\partial x^\gamma} = T^{\beta\alpha} - T^{\alpha\beta} = 0 \tag{2.9.2}$$

We can then form a *total angular momentum*

$$J^{\alpha\beta} = \int d^3x M^{0\alpha\beta} = -J^{\beta\alpha} \tag{2.9.3}$$

From (2.9.2) we see (by following the arguments of the last section) that $J^{\alpha\beta}$ is constant in time and is a tensor. We further note that

$$J^{ij} = \int d^3x(x^i T^{j0} - x^j T^{i0})$$

and since T^{j0} is the density of the jth component of momentum, we may regard J^{23}, J^{31}, and J^{12} as the 1-, 2-, and 3-components of the angular momentum. The other components of $J^{\alpha\beta}$ are

$$J^{0i} = tp^i - \int x^i T^{00} \, d^3x$$

These components have no clear physical significance, and in fact can be made to vanish if we fix the origin of coordinates to coincide with the "center of energy" at $t = 0$, that is, if at $t = 0$ the moment $\int x^i T^{00} d^3x$ vanishes.

Although a tensor with regard to the homogeneous Lorentz transformations $x^\alpha \to \Lambda^\alpha{}_\beta x^\beta$, the total angular momentum behaves peculiarly under the translation $x^\alpha \to x'^\alpha = x^\alpha + a^\alpha$. From (2.9.3) and (2.8.13) we find that

$$J^{\alpha\beta} \to J'^{\alpha\beta} = J^{\alpha\beta} + a^\alpha p^\beta - a^\beta p^\alpha \qquad (2.9.4)$$

This is of course because $J^{\alpha\beta}$ includes the orbital angular momentum, which is always defined with respect to some center of rotation. In order to isolate the *internal* part of $J^{\alpha\beta}$, it is convenient to define a *spin four-vector*

$$S_\alpha \equiv \tfrac{1}{2}\varepsilon_{\alpha\beta\gamma\delta} J^{\beta\gamma} U^\delta \qquad (2.9.5)$$

where $\varepsilon_{\alpha\beta\gamma\delta}$ is the completely antisymmetric tensor discussed in Section 5, and $U^\alpha \equiv p^\alpha/(-p_\beta p^\beta)^{1/2}$ is the four-vector velocity of the system. Because of the antisymmetry of $\varepsilon_{\alpha\beta\gamma\delta}$, the translation $x^\alpha \to x^\alpha + a^\alpha$, which changes $J^{\beta\gamma}$ according to the rule (2.9.4), does *not* change S_α. Furthermore, S_α is obviously a vector and is constant for a free particle

$$\frac{dS_\alpha}{dt} = 0 \qquad (2.9.6)$$

Finally we note that in the center-of-mass frame of the system $U^i = 0$ and $U^0 = 1$, so in this frame

$$S_1 = J^{23}, \qquad S_2 = J^{31}, \qquad S_3 = J^{12}, \qquad S_0 = 0 \qquad (2.9.7)$$

This justifies us in regarding S_α as the internal angular momentum of the system. Even when the velocity **U** is not zero, S_α really has only three independent components, because (2.9.5) gives

$$U^\alpha S_\alpha = 0 \qquad (2.9.8)$$

We use these properties of S_α later, when we discuss the precession of a gyroscope in free fall.

10 Relativistic Hydrodynamics

A great many macroscopic physical systems, including perhaps the universe itself, may be approximately regarded as *perfect fluids*. A perfect fluid is defined as having at each point a velocity **v**, such that an observer moving with this velocity sees the fluid around him as isotropic. This will be the case if the mean free path between collisions is small compared with the scale of lengths used by the observer. (For instance, a sound wave will propagate in air if its wavelength is large compared with the mean free path, but at very short wavelengths viscosity becomes important and the air stops acting like a perfect fluid.) We shall translate the above definition of a perfect fluid into a statement about the energy-momentum tensor.

First suppose that we are in a frame of reference (distinguished by a tilde) in which the fluid is at rest at some particular position and time. At this space-time point, the perfect fluid hypothesis tells us that the energy-momentum tensor takes the form characteristic of spherical symmetry:

$$\tilde{T}^{ij} = p\delta_{ij} \tag{2.10.1}$$

$$\tilde{T}^{i0} = T^{0i} = 0 \tag{2.10.2}$$

$$\tilde{T}^{00} = \rho \tag{2.10.3}$$

The coefficients p and ρ are called the *pressure* and the *proper energy density*, respectively. Now go into a reference frame at rest in the laboratory, and suppose that the fluid in this frame appears to be moving (at the given space-time point) with velocity $\mathbf{v}$. The connection between the comoving coordinates $\tilde{x}^\beta$ and the lab coordinates x^α is then

$$x^\alpha = \Lambda^\alpha{}_\beta(\mathbf{v})\tilde{x}^\beta$$

with $\Lambda^\alpha{}_\beta(\mathbf{v})$ the "boost" defined by Eqs. (2.1.17)–(2.1.21). But $T^{\alpha\beta}$ is a tensor, so in the lab frame it is

$$T^{\alpha\beta} = \Lambda^\alpha{}_\gamma(\mathbf{v})\Lambda^\beta{}_\delta(\mathbf{v})\tilde{T}^{\gamma\delta}$$

or explicitly

$$T^{ij} = p\delta_{ij} + (p + \rho)\frac{v_i v_j}{1 - \mathbf{v}^2} \tag{2.10.4}$$

$$T^{i0} = (p + \rho)\frac{v_i}{1 - \mathbf{v}^2} \tag{2.10.5}$$

$$T^{00} = \frac{(\rho + p\mathbf{v}^2)}{1 - \mathbf{v}^2} \tag{2.10.6}$$

To check that this *is* a tensor, we note that (2.10.4)–(2.10.6) can be integrated into a single equation:

$$T^{\alpha\beta} = p\eta^{\alpha\beta} + (p + \rho)U^\alpha U^\beta \tag{2.10.7}$$

where U^α is the velocity four-vector,

$$\mathbf{U} = \frac{d\mathbf{x}}{d\tau} = (1 - \mathbf{v}^2)^{-1/2}\,\mathbf{v}$$

$$U^0 = \frac{dt}{d\tau} = (1 - \mathbf{v}^2)^{-1/2} \tag{2.10.8}$$

normalized so that

$$U_\alpha U^\alpha = -1 \tag{2.10.9}$$

Indeed, Eq. (2.10.7) could have been derived very easily by noting that the quantity on the right-hand side is a tensor, which equals the tensor $T^{\alpha\beta}$ in a Lorentz frame moving with the fluid, and hence must equal $T^{\alpha\beta}$ in all Lorentz frames.

Apart from energy and momentum, a fluid will in general carry one or more conserved quantities, such as the charge, the number of baryons minus the number of antibaryons, or, at normal temperatures, the number of atoms. Let us consider one such conserved quantity, and refer to it for brevity as the "particle number." If n is the particle number density in a Lorentz frame that moves with the fluid at a given space-time point, then in this frame the particle current four-vector at this point is

$$\tilde{N}^i = 0 \qquad \tilde{N}^0 = n \qquad\qquad (2.10.10)$$

In any other Lorentz frame, in which the fluid at this point moves with velocity **v**, the particle current is related to (2.10.10) by the "boost" $\Lambda(\mathbf{v})$:

$$N^i = \Lambda^i{}_\beta(\mathbf{v})\tilde{N}^\beta = (1 - \mathbf{v}^2)^{-1/2} v^i n \qquad\qquad (2.10.11)$$

$$N^0 = \Lambda^0{}_\beta(\mathbf{v})\tilde{N}^\beta = (1 - \mathbf{v}^2)^{-1/2} n \qquad\qquad (2.10.12)$$

or, more concisely,

$$N^\alpha = nU^\alpha \qquad\qquad (2.10.13)$$

The motion of the fluid will be governed by the equations of conservation of energy and momentum,

$$0 = \frac{\partial T^{\alpha\beta}}{\partial x^\beta} = \frac{\partial p}{\partial x_\alpha} + \frac{\partial}{\partial x^\beta}\left[(\rho + p)U^\alpha U^\beta\right] \qquad\qquad (2.10.14)$$

and of the particle number:

$$0 = \frac{\partial N^\alpha}{\partial x^\alpha} = \frac{\partial}{\partial x^\alpha}\left(nU^\alpha\right) = \frac{\partial}{\partial t}\left(n(1 - \mathbf{v}^2)^{-1/2}\right) + \mathbf{\nabla} \cdot \left(n\mathbf{v}(1 - \mathbf{v}^2)^{-1/2}\right)$$
$$(2.10.15)$$

It is convenient to write (2.10.14) as separate three-vector and scalar equations. The three-vector equation is obtained by setting $\alpha = i$ in Eq. (2.10.14), writing $U^i = v^i U^0$, and then using Eq. (2.10.14) with $\alpha = 0$; this gives

$$\frac{\partial \mathbf{v}}{\partial t} + (\mathbf{v} \cdot \mathbf{\nabla})\mathbf{v} = -\frac{(1 - \mathbf{v}^2)}{(\rho + p)}\left[\mathbf{\nabla}p + \mathbf{v}\,\frac{\partial p}{\partial t}\right] \qquad\qquad (2.10.16)$$

The scalar equation is obtained by multiplying Eq. (2.10.14) by U_α; using the relation

$$0 = \frac{\partial}{\partial x^\beta}\left(U_\alpha U^\alpha\right) = 2U_\alpha \frac{\partial U^\alpha}{\partial x^\beta} \qquad\qquad (2.10.17)$$

we then have

$$0 = U_\alpha \frac{\partial T^{\alpha\beta}}{\partial x^\beta} = U^\beta \frac{\partial p}{\partial x^\beta} - \frac{\partial}{\partial x^\beta}[(p + \rho)U^\beta]$$

Using Eq. (2.10.15), we can write this as

$$0 = U^\beta \left[\frac{\partial p}{\partial x^\beta} - n\frac{\partial}{\partial x^\beta}\left(\frac{p+\rho}{n}\right)\right]$$

$$= -nU^\beta \left[p\frac{\partial}{\partial x^\beta}\left(\frac{1}{n}\right) + \frac{\partial}{\partial x^\beta}\left(\frac{\rho}{n}\right)\right] \tag{2.10.17a}$$

The second law of thermodynamics tells us that the pressure p, the energy density ρ, and the volume per particle $1/n$ may be expressed as functions of the temperature T and the entropy per particle σk, in such a way that

$$kT\, d\sigma = pd\left(\frac{1}{n}\right) + d\left(\frac{\rho}{n}\right) \tag{2.10.18}$$

(Boltzmann's constant k is introduced here to make σ dimensionless.) Our scalar equation (2.10.17a) can now be written

$$0 = U^\beta \frac{\partial \sigma}{\partial x^\beta} \propto \frac{\partial \sigma}{\partial t} + (\mathbf{v} \cdot \mathbf{\nabla})\sigma \tag{2.10.19}$$

The specific entropy σ is therefore constant in time at any point that moves along with the fluid. The fundamental equations of relativistic hydrodynamics are the "continuity equation" (2.10.15), the "Euler equations" (2.10.16), the "energy equation" (2.10.19), together with equations of state that give p and ρ in terms of n and σ.

In order to gain some insight into the possible equations of state, we may consider a fluid composed of structureless point particles that interact only in spatially localized collisions. As shown in Section 2.8, the energy-momentum tensor is

$$T^{\alpha\beta} = \sum_N \frac{p_N{}^\alpha p_N{}^\beta}{E_N} \delta^3(\mathbf{x} - \mathbf{x}_N) \tag{2.10.20}$$

[See Eq. (2.8.4).] In a comoving Lorentz frame, $T^{\alpha\beta}$ will have the isotropic form (2.10.1)–(2.10.3), so the pressure and energy density will be given in this frame by

$$p = \frac{1}{3}\sum_{i=1}^{3} T^{ii} = \frac{1}{3}\sum_N \frac{\mathbf{p}^2_N}{E_N} \delta^3(\mathbf{x} - \mathbf{x}_N) \tag{2.10.21}$$

$$\rho = T^{00} = \sum_N E_N \delta^3(\mathbf{x} - \mathbf{x}_N) \tag{2.10.22}$$

whereas the particle number density is, in analogy with (2.6.2),

$$n = \sum_N \delta^3(\mathbf{x} - \mathbf{x}_N) \tag{2.10.23}$$

It follows that in general

$$0 \leq p \leq \frac{\rho}{3} \tag{2.10.24}$$

For a cool, nonrelativistic gas, we can approximate

$$E_N \simeq m + \frac{\mathbf{p}^2_N}{2m}$$

so (2.10.22) gives

$$\rho \simeq nm + \tfrac{3}{2}p \tag{2.10.25}$$

For a hot, extremely relativistic gas, we have

$$E_N \simeq |\mathbf{p}_N| \gg m$$

so (2.10.22) gives

$$\rho \simeq 3p \gg nm \tag{2.10.26}$$

Both (2.10.25) and (2.10.26) can be incorporated into a single equation,

$$\rho - nm \simeq (\gamma - 1)^{-1}p \tag{2.10.27}$$

with

$$\gamma = \begin{cases} \tfrac{5}{3} & \text{nonrelativistic} \\ \tfrac{4}{3} & \text{extreme relativistic} \end{cases} \tag{2.10.28}$$

Equation (2.10.18) then gives

$$kT \, d\sigma = pd\left(\frac{1}{n}\right) + (\gamma - 1)^{-1} d\left(\frac{p}{n}\right) = \frac{n^{\gamma-1}}{\gamma - 1} d\left(\frac{p}{n^{\gamma}}\right) \tag{2.10.29}$$

Thus Eq. (2.10.19) takes the form

$$0 = \frac{\partial}{\partial t}\left(\frac{p}{n^{\gamma}}\right) + (\mathbf{v} \cdot \mathbf{V})\left(\frac{p}{n^{\gamma}}\right) \tag{2.10.30}$$

and (2.10.27) is to be used to express ρ in terms of n and p in Eq. (2.10.16). The proportionality expressed in Eq. (2.10.27) between internal energy and pressure actually holds, with various values of γ, over a class of fluids much wider than the simple gas of point particles discussed here. For all such fluids, the energy equation can be put in the form (2.10.30).

As an example, let us calculate the speed of sound in a static homogeneous relativistic fluid. In the unperturbed state, we have n, ρ, p, and σ constant in space and time, and $\mathbf{v} = 0$. The sound wave produces small changes n_1, ρ_1, p_1, and $\mathbf{v}_1$ in n, ρ, p, and $\mathbf{v}$, but according to (2.10.19), it leaves σ unchanged. To first order in small quantities, Eqs. (2.10.15) and (2.10.16) then read

$$\frac{\partial n_1}{\partial t} + n\mathbf{\nabla} \cdot \mathbf{v}_1 = 0$$

$$\frac{\partial \mathbf{v}_1}{\partial t} = -\frac{\mathbf{\nabla} p_1}{p + \rho}$$

But with $d\sigma = 0$, Eq. (2.10.18) gives

$$0 = -\frac{(p + \rho)}{n} n_1 + \rho_1$$

so that

$$\frac{\partial \mathbf{v}_1}{\partial t} = -\frac{v_s{}^2 \mathbf{\nabla} n_1}{n}$$

where

$$v^2{}_s \equiv \frac{p_1}{\rho_1} = \left(\frac{\partial p}{\partial \rho}\right)_{\sigma \text{ const}} \tag{2.10.31}$$

Combining the equations for n_1 and $\mathbf{v}_1$, we obtain a wave equation

$$0 = \left[\frac{\partial^2}{\partial t^2} - v_s{}^2 \mathbf{\nabla}^2\right] n_1$$

that shows that sound waves travel with the speed v_s, just as in a nonrelativistic fluid. The speed of sound is much less than the speed of light (i.e., unity) for a nonrelativistic fluid, but it increases with temperature, so it is worth checking whether v_s might exceed unity for a fluid of highly relativistic point particles, such as hydrogen above 10^{13} °K. In this case, (2.10.26) and (2.10.31) give a sound speed

$$v_s = \frac{1}{\sqrt{3}} \tag{2.10.32}$$

which is still safely less than unity. This conclusion would not be affected if electromagnetic forces were taken into account, because Eqs. (2.10.7) and (2.8.9) impose on the electromagnetic pressure p_{em} and energy density ρ_{em} the relation

$$0 = T_{em}{}^\alpha{}_\alpha = 3p_{em} - \rho_{em} \tag{2.10.33}$$

so the inclusion of p_{em} and ρ_{em} would not invalidate (2.10.26) or (2.10.32). It is an open question whether v_s remains less than unity when nonelectromagnetic forces are taken into account.[4]

11   Relativistic Imperfect Fluids*

The last section dealt with a perfect fluid, in which mean free paths and times are so short that perfect isotropy is maintained about any point moving with the fluid. In practice, one often has to deal with somewhat imperfect fluids, in which the pressure, density, or velocity vary appreciably over distances of the order of a mean free path, or over times of the order of a mean free time, or both. In such fluids, thermal equilibrium is not strictly maintained, and the fluid kinetic energy is dissipated as heat.

The correct treatment of dissipative effects for relativistic fluids raises certain delicate questions of principle, which do not arise in the nonrelativistic case. For this reason, and also because dissipation plays an increasingly important role in theories of the early universe (see Sections 15.8, 15.10, 15.11), it will be worth our while here to develop the outlines of the general theory of relativistic imperfect fluids.

We suppose that the presence of weak space-time gradients in an imperfect fluid has the effect of modifying the energy-momentum tensor and particle current vector by terms $\Delta T^{\alpha\beta}$ and ΔN^{α}, which are of first order in these gradients. Instead of (2.10.7) and (2.10.13), we then have

$$T^{\alpha\beta} = p\eta^{\alpha\beta} + (p + \rho)U^{\alpha}U^{\beta} + \Delta T^{\alpha\beta} \tag{2.11.1}$$

$$N^{\alpha} = nU^{\alpha} + \Delta N^{\alpha} \tag{2.11.2}$$

Once we allow such correction terms, the definitions of the pressure p, energy density ρ, particle density n, and fluid velocity U^{α} become somewhat ambiguous. The general practice is to define ρ and n as the total energy density and particle number density in a comoving frame:

$$T^{00} \equiv \rho \tag{2.11.3}$$

$$N^{0} \equiv n \tag{2.11.4}$$

a comoving frame being characterized by the condition that at a given point, the velocity four-vector is

$$U^{i} \equiv 0 \qquad U^{0} \equiv 1 \tag{2.11.5}$$

In addition, the pressure p is generally defined to be the same function of ρ and n [e.g., (2.10.27)] as in the case where all fluid gradients are negligible and dissipation is absent. Finally, it is necessary in a relativistic fluid to specify whether U^{α} is the velocity of energy transport or particle transport. In the approach of Landau and Lifshitz,[5] U^{α} is taken to be the velocity of energy transport, so that T^{i0} vanishes in a comoving frame. In the approach of Eckart,[6] U^{α} is taken to be the velocity of

---

* This section lies somewhat out of the book's main line of development, and may be omitted in a first reading.

particle transport, so that it is N^i that vanishes in a comoving frame. The two approaches are perfectly equivalent, but Eckart's seems to me to be slightly more convenient, and will be adopted here. With this definition of U^α, we then have in a comoving frame

$$N^i \equiv 0 \qquad (2.11.6)$$

A comparison of (2.11.3)–(2.11.6) with (2.11.1) and (2.11.2) shows that in a comoving frame, the dissipative terms $\Delta T^{\alpha\beta}$ and ΔN^α are subject to the constraints

$$\Delta T^{00} = \Delta N^0 = \Delta N^i = 0 \qquad (2.11.7)$$

and therefore, in a general Lorentz frame,

$$U^\alpha U^\beta \Delta T_{\alpha\beta} = 0 \qquad (2.11.8)$$

$$\Delta N^\alpha = 0 \qquad (2.11.9)$$

All effects of dissipation thus show up as contributions to $\Delta T^{\alpha\beta}$. Our task is now to construct the most general possible dissipative tensor $\Delta T^{\alpha\beta}$ allowed by Eq. (2.11.8) and by the second law of thermodynamics.

To this end, let us calculate the entropy produced by fluid motions. As in the last sections, we start by contracting the conservation law (2.8.7) with U_α:

$$0 = U_\alpha \frac{\partial}{\partial x^\beta} T^{\alpha\beta} \qquad (2.11.10)$$

By following the same reasoning that was used to derive (2.10.19) for a perfect fluid, one sees that in general

$$U_\alpha \frac{\partial}{\partial x^\beta} [p\eta^{\alpha\beta} + (p + \rho)U^\alpha U^\beta] = -kT \frac{\partial}{\partial x^\alpha} (n\sigma U^\alpha)$$

where T and σk are the temperature and entropy per particle, defined by Eq. (2.10.18). Hence (2.11.10) now reads

$$\frac{\partial}{\partial x^\alpha} (n\sigma U^\alpha) = \frac{1}{kT} U_\alpha \frac{\partial}{\partial x^\beta} \Delta T^{\alpha\beta}$$

or equivalently

$$\frac{\partial S^\alpha}{\partial x^\alpha} = -\frac{1}{T} \frac{\partial U_\alpha}{\partial x^\beta} \Delta T^{\alpha\beta} + \frac{1}{T^2} \frac{\partial T}{\partial x^\beta} U_\alpha \Delta T^{\alpha\beta} \qquad (2.11.11)$$

where

$$S^\alpha \equiv nk\sigma U^\alpha - T^{-1} U_\beta \Delta T^{\alpha\beta} \qquad (2.11.12)$$

The entropy density in a comoving frame is $nk\sigma = S^0$, so we may interpret S^α as the entropy current four-vector, and Eq. (2.11.11) thus gives the rate of entropy production per unit volume. The second law of thermodynamics then requires that $\Delta T^{\alpha\beta}$ be a linear combination of velocity and temperature gradients, such that the right-hand side of (2.11.11) is *positive* for all possible fluid configurations. Note that this is only possible because we have included the second term in Eq. (2.11.12); without this term, $\partial S^{\alpha}/\partial x^{\alpha}$ would not be simply quadratic in first derivatives, and hence could not be positive for all fluid configurations. Note also that $\Delta T^{\alpha\beta}$ is not allowed to involve gradients of p, ρ, n, and so on, because if it did then (2.11.11) would contain products of pressure or density gradients with velocity or temperature gradients, and, again, these products would not be positive for all fluid configurations.

It is convenient at this point to go over to a comoving frame, in which U^{α} has the form (2.11.5) at a given space-time point P. From (2.10.17), it follows that in this frame, all gradients of U^0 vanish at P. Setting U^i, $\partial U^0/\partial x^{\alpha}$, and ΔT^{00} equal to zero in Eq. (2.11.11), we find that in a Lorentz frame comoving at P, the rate of entropy production per unit volume at P is

$$\frac{\partial S^{\alpha}}{\partial x^{\alpha}} = -\left(\frac{1}{T}\,\dot{U}_i + \frac{1}{T^2}\frac{\partial T}{\partial x^i}\right)\Delta T^{i0} - \frac{1}{T}\frac{\partial U_i}{\partial x^j}\,\Delta T^{ij} \qquad (2.11.13)$$

In order for this to be positive for all possible fluid configurations, we must have

$$\Delta T^{i0} = -\chi\left(\frac{\partial T}{\partial x^i} + T\dot{U}_i\right) \qquad (2.11.14)$$

$$\Delta T^{ij} = -\eta\left(\frac{\partial U_i}{\partial x^j} + \frac{\partial U_j}{\partial x^i} - \frac{2}{3}\,\mathbf{\nabla}\cdot\mathbf{U}\delta_{ij}\right) - \zeta\mathbf{\nabla}\cdot\mathbf{U}\delta_{ij} \qquad (2.11.15)$$

with positive coefficients

$$\chi \geq 0, \qquad \eta \geq 0, \qquad \zeta \geq 0 \qquad (2.11.16)$$

so that (2.11.13) reads

$$\frac{\partial S^{\alpha}}{\partial x^{\alpha}} = \frac{\chi}{T^2}\,(\mathbf{\nabla}T + T\dot{\mathbf{U}})^2$$

$$+ \frac{\eta}{2T}\left(\frac{\partial U_i}{\partial x^j} + \frac{\partial U_j}{\partial x^i} - \frac{2}{3}\delta_{ij}\,\mathbf{\nabla}\cdot\mathbf{U}\right)\left(\frac{\partial U_i}{\partial x^j} + \frac{\partial U_j}{\partial x^i} - \frac{2}{3}\delta_{ij}\,\mathbf{\nabla}\cdot\mathbf{U}\right)$$

$$+ \frac{\zeta}{T}\,(\mathbf{\nabla}\cdot\mathbf{U})^2 \geq 0 \qquad (2.11.17)$$

Except for the relativistic correction $T\dot{\mathbf{U}}$ in (2.11.14), the form of (2.11.14) and (2.11.15) is the same as in the nonrelativistic theory of imperfect fluids,[5] and we therefore may identify χ, η, and ζ as the coefficients of *heat conduction, shear viscosity*, and *bulk viscosity*.

It now only remains to translate our results from the forms (2.11.5), (2.11.7), (2.11.14), (2.11.15), which are valid only in comoving frames, to forms valid in general Lorentz frames. Let us define a *shear tensor*,

$$W_{\alpha\beta} \equiv \frac{\partial U_\alpha}{\partial x^\beta} + \frac{\partial U_\beta}{\partial x^\alpha} - \frac{2}{3}\eta_{\alpha\beta}\frac{\partial U^\gamma}{\partial x^\gamma} \tag{2.11.18}$$

a *heat-flow vector*,

$$Q_\alpha \equiv \frac{\partial T}{\partial x^\alpha} + T\frac{\partial U_\alpha}{\partial x^\beta}U^\beta \tag{2.11.19}$$

and a projection tensor on the hyperplane normal to U^α:

$$H_{\alpha\beta} \equiv \eta_{\alpha\beta} + U_\alpha U_\beta \tag{2.11.20}$$

It is straightforward to check that in a comoving Lorentz frame, our formulas (2.11.7), (2.11.14), (2.11.15) for $\Delta T^{\alpha\beta}$ are satisfied by the tensor

$$\Delta T^{\alpha\beta} = -\eta H^{\alpha\gamma}H^{\beta\delta}W_{\gamma\delta}$$

$$-\chi(H^{\alpha\gamma}U^\beta + H^{\beta\gamma}U^\alpha)Q_\gamma - \zeta H^{\alpha\beta}\frac{\partial U^\gamma}{\partial x^\gamma} \tag{2.11.21}$$

Since this formula is Lorentz-invariant, and valid in a comoving Lorentz frame, it is valid in all Lorentz frames.

In general, the coefficients χT, η, and ζ might be expected on dimensional grounds to be of the order of the pressure, or the thermal energy density, times some sort of mean free time. However, there are important special cases[8] in which the bulk viscosity ζ is much smaller than η or χT. To see when this applies, note that (2.11.1) and (2.11.21) give the trace of the total energy-momentum tensor as

$$T^\alpha_{\ \alpha} = 3p - \rho - 3\zeta\frac{\partial U^\gamma}{\partial x^\gamma} \tag{2.11.22}$$

Suppose that we are dealing with a medium for which this trace can be expressed as a function of ρ and n alone:

$$T^\alpha_{\ \alpha} = f(\rho, n) \tag{2.11.23}$$

For instance, for the simple gas characterized by (2.10.20), this trace is

$$T^\alpha_{\ \alpha} = -\sum_N \frac{m^2}{E_N}\delta^3(\mathbf{x} - \mathbf{x}_N)$$

In the extreme relativistic case, we have $E_N \gg m$, so in this case (2.11.23) is satisfied, with

$$f(\rho, n) \simeq 0$$

In the nonrelativistic case, we have

$$\frac{1}{E_N} \simeq \frac{1}{m} - \left(\frac{E_N - m}{m^2}\right)$$

so in this case (2.11.23) is again satisfied, with

$$f(\rho, n) \simeq -mn + (\rho - mn)$$

In the absence of velocity gradients, Eqs. (2.11.22) and (2.11.23) would give a formula for the pressure

$$p = \tfrac{1}{3}[\rho + f(\rho, n)] \tag{2.11.24}$$

But we have agreed to define p in general as the same function of ρ and n as in the absence of dissipation, so (2.11.24) must hold even in the presence of velocity gradients, and therefore (2.11.22), (2.11.23), and (2.11.24) give

$$\zeta = 0 \tag{2.11.25}$$

It would be wrong, however, to conclude that ζ is generally negligible. As we have seen, the trace of the energy-momentum tensor for a simple gas is a function of ρ and n only in the extreme relativistic or extreme nonrelativistic limit; for kT of order m, $T^\alpha{}_\alpha$ cannot be expressed in the form (2.11.23), and the bulk viscosity is of the same order as the shear viscosity.[7] The bulk viscosity is also important[8] in a fluid that allows an easy exchange of energy between translational and internal degrees of freedom, as in the case of a gas of rough spheres.[9] Another case, of particular importance to cosmology, is that of a material medium with very short mean free times, interacting with radiation quanta with a finite mean free time τ. In this case, the coefficients of heat conduction, shear viscosity, and bulk viscosity are calculated to be[10]

$$\chi = \tfrac{4}{3}aT^3\tau \tag{2.11.26}$$

$$\eta = \tfrac{4}{15}aT^4\tau \tag{2.11.27}$$

$$\zeta = 4aT^4\tau\left[\frac{1}{3} - \left(\frac{\partial p}{\partial \rho}\right)_n\right]^2 \tag{2.11.28}$$

where a is the Stefan-Boltzmann constant, defined so that the radiation energy density is aT^4, and p and ρ are the total pressure and energy density of the matter and radiation. Note that, in general, χT, η, and ζ are comparable, but if the pressure and thermal energy are dominated by radiation, then $(\partial p/\partial \rho)_n \simeq \tfrac{1}{3}$ and, as expected, the bulk viscosity will be small.

12   Representations of the Lorentz Group*

The tensor formalism described in Section 2.5 is perfectly adequate for handling the problems of relativistic classical physics. However, there are certain formal advantages in looking at these Lorentz transformation rules in a more general way, from the perspective of the theory of the representations of the homogeneous Lorentz group. We shall see in Section 12.5, that this approach allows an elegant reformulation of the effects of gravitation on arbitrary physical systems. Also, it is only in this way that we can deal with fields of half-integer spin.

Under the general Lorentz transformation rule, a set of quantities ψ_n transform under a Lorentz transformation $\Lambda^\alpha{}_\beta$ into the new quantities:

$$\psi'_n = \sum_m [D(\Lambda)]_{nm}\psi_m \qquad (2.12.1)$$

In order for a Lorentz transformation Λ_1 followed by a Lorentz transformation Λ_2 to give the same result as the Lorentz transformation $\Lambda_1\Lambda_2$, it is necessary that the matrices $D(\Lambda)$ should furnish a *representation* of the Lorentz group, that is,

$$D(\Lambda_1)D(\Lambda_2) = D(\Lambda_1\Lambda_2) \qquad (2.12.2)$$

with matrix multiplication now understood. For instance, if ψ is a contravariant vector V^α, then $D(\Lambda)$ is simply

$$[D(\Lambda)]^\alpha{}_\beta = \Lambda^\alpha{}_\beta \qquad (2.12.3)$$

whereas for a covariant tensor $T_{\alpha\beta}$, the corresponding D-matrix is

$$[D(\Lambda)]_{\alpha\beta}{}^{\gamma\delta} = \Lambda_\alpha{}^\gamma \Lambda_\beta{}^\delta \qquad (2.12.4)$$

It is easy to check that (2.12.3) and (2.12.4) do satisfy the group multiplication rule (2.12.2). We can compile a catalogue of all possible Lorentz transformation rules by constructing the most general representation of the homogeneous Lorentz group.

In fact, the most general true representations of the homogeneous Lorentz group are provided by the tensor representations, such as (2.12.3) and (2.12.4), so we might expect that all quantities of physical interest should be tensors. However, there are additional representations of the *infinitesimal* Lorentz group, the *spinor representations*, that play an important role in relativistic quantum field theory. The infinitesimal Lorentz group consists of Lorentz transformations infinitesimally close to the identity, that is,

$$\Lambda^\alpha{}_\beta = \delta^\alpha{}_\beta + \omega^\alpha{}_\beta \qquad (2.12.5)$$
$$|\omega^\alpha{}_\beta| \ll 1$$

* This section lies somewhat out of the book's main line of development, and may be omitted in a first reading.

In order for this to satisfy the fundamental condition (2.1.2) for a Lorentz transformation, we must have

$$(\delta^{\alpha}{}_{\gamma} + \omega^{\alpha}{}_{\gamma})(\delta^{\beta}{}_{\delta} + \omega^{\beta}{}_{\delta})\eta_{\alpha\beta} = \eta_{\gamma\delta}$$

or, to first order in ω,

$$\omega_{\gamma\delta} = -\omega_{\delta\gamma} \tag{2.12.6}$$

with the indices on ω of course lowered with η:

$$\omega_{\gamma\delta} \equiv \eta_{\gamma\alpha}\omega^{\alpha}{}_{\delta}$$

For such a transformation, the matrix representation $D(\Lambda)$ must be infinitesimally close to the identity

$$D(1 + \omega) = 1 + \tfrac{1}{2}\omega^{\alpha\beta}\sigma_{\alpha\beta} \tag{2.12.7}$$

where $\sigma_{\alpha\beta}$ are a fixed set of matrices, which by virtue of (2.12.6) can always be chosen to be antisymmetric in α and β:

$$\sigma_{\alpha\beta} = -\sigma_{\beta\alpha} \tag{2.12.8}$$

For instance, for the tensor representations (2.12.3) and (2.12.4), we have

$$[\sigma_{\alpha\beta}]^{\gamma}{}_{\delta} = \delta_{\alpha}{}^{\gamma}\eta_{\beta\delta} - \delta_{\beta}{}^{\gamma}\eta_{\alpha\delta} \tag{2.12.9}$$

$$\begin{aligned}[\sigma_{\alpha\beta}]_{\gamma\delta}{}^{\varepsilon\zeta} &= \eta_{\alpha\gamma}\delta_{\beta}{}^{\varepsilon}\delta^{\zeta}{}_{\delta} - \eta_{\beta\gamma}\delta_{\alpha}{}^{\varepsilon}\delta^{\zeta}{}_{\delta} \\ &\quad + \eta_{\alpha\delta}\delta_{\beta}{}^{\zeta}\delta^{\varepsilon}{}_{\gamma} - \eta_{\beta\delta}\delta_{\alpha}{}^{\zeta}\delta^{\varepsilon}{}_{\gamma}\end{aligned} \tag{2.12.10}$$

The matrices $\sigma_{\alpha\beta}$ are not allowed to be just any set of constant matrices, but must be constrained so that $D(\Lambda)$ satisfies the group multiplication rule (2.12.2). It is convenient first to apply this rule to the product $\Lambda[1 + \omega]\Lambda^{-1}$:

$$D(\Lambda)D(1 + \omega)D(\Lambda^{-1}) = D(1 + \Lambda\omega\Lambda^{-1})$$

To zero order in ω, this simply says that $1 = 1$, whereas to first order, we must equate the coefficients of $\omega_{\alpha\beta}$ on both sides:

$$D(\Lambda)\sigma_{\alpha\beta}D(\Lambda^{-1}) = \sigma_{\gamma\delta}\,\Lambda^{\gamma}{}_{\alpha}\,\Lambda^{\delta}{}_{\beta} \tag{2.12.11}$$

If we now set $\Lambda = 1 + \omega$ (not necessarily with the same ω) and $\Lambda^{-1} = 1 - \omega$, then this will be satisfied to first order in ω provided that σ satisfies the commutation relations,

$$[\sigma_{\alpha\beta}, \sigma_{\gamma\delta}] = \eta_{\gamma\beta}\sigma_{\alpha\delta} - \eta_{\gamma\alpha}\sigma_{\beta\delta} + \eta_{\delta\beta}\sigma_{\gamma\alpha} - \eta_{\delta\alpha}\sigma_{\gamma\beta} \tag{2.12.12}$$

with square brackets denoting the usual matrix commutator

$$[u, v] \equiv uv - vu$$

The reader can easily check that the matrices (2.12.9) and (2.12.10) do satisfy Eq. (2.12.12). The problem of finding the general representations of the infinitesimal homogeneous Lorentz group is thus reduced to the problem of finding all matrices that satisfy the commutation relations (2.12.12).

These commutation relations can be put in a somewhat more familiar form by defining the matrices

$$
\begin{aligned}
a_1 &= \tfrac{1}{2}[-i\sigma_{23} + \sigma_{10}] & b_1 &= \tfrac{1}{2}[-i\sigma_{23} - \sigma_{10}] \\
a_2 &= \tfrac{1}{2}[-i\sigma_{31} + \sigma_{20}] & b_2 &= \tfrac{1}{2}[-i\sigma_{31} - \sigma_{20}] \\
a_3 &= \tfrac{1}{2}[-i\sigma_{12} + \sigma_{30}] & b_3 &= \tfrac{1}{2}[-i\sigma_{12} - \sigma_{30}]
\end{aligned}
\tag{2.12.13}
$$

Equation (2.12.12) then takes the form

$$
\mathbf{a} \times \mathbf{a} = i\mathbf{a}
\tag{2.12.14}
$$

$$
\mathbf{b} \times \mathbf{b} = i\mathbf{b}
\tag{2.12.15}
$$

$$
[a_i, b_j] = 0
\tag{2.12.16}
$$

Equations (2.12.14)–(2.12.16) are simply the commutation relations for a pair of independent angular momentum matrices. The rules for constructing such matrices can be found in any book on nonrelativistic quantum mechanics[11]: In the most general case **a** and **b** are a direct sum of "irreducible" components, each characterized by an integer or half-integer A or B, with

$$
\mathbf{a}^2 = A(A + 1) \qquad \mathbf{b}^2 = B(B + 1)
\tag{2.12.17}
$$

and with dimensionality $2A + 1$ or $2B + 1$. Thus the most general objects ψ_n, which transform linearly under infinitesimal homogeneous Lorentz transformations, can be decomposed into "irreducible" pieces, characterized by a pair of integers and/or half-integers (A, B), each piece having $(2A + 1)(2B + 1)$ components.

A straightforward calculation shows that the contravariant vector representation (2.12.9), as well as its covariant counterpart, has $A = B = \tfrac{1}{2}$. Any tensor representation, such as (2.12.10), can be regarded as a direct product of vector representations, so it consists only of irreducible components with $A + B$ an *integer*; for instance, the general second-rank tensor representation (2.12.10) consists of irreducible components with (A, B) equal to $(1, 1)$, $(1, 0)$, $(0, 1)$, and $(0, 0)$. The representations in which $A + B$ is a *half-integer* are quite distinct from the tensors, and are called *spinor representations*. The most familiar example is the Dirac electron field, which consists of components with (A, B) equal to $(\tfrac{1}{2}, 0)$ and $(0, \tfrac{1}{2})$.

The transformation property of any object under ordinary spatial rotations is determined by its behavior with respect to infinitesimal Lorentz transformations (2.12.5) for which $\omega_{i0} = 0$, and hence by the structure of the purely spatial components σ_{12}, σ_{23}, σ_{31} of $\sigma_{\alpha\beta}$. From these components, we can construct a matrix vector

$$\mathbf{s} = \mathbf{a} + \mathbf{b} = -i\{\sigma_{23}, \sigma_{31}, \sigma_{12}\} \qquad (2.12.18)$$

that according to (2.12.14)–(2.12.16) has the commutation relations of an angular momentum:

$$\mathbf{s} \times \mathbf{s} = i\mathbf{s} \qquad (2.12.19)$$

Any irreducible representation (A, B) of the homogeneous Lorentz group can be decomposed[11] into pieces with $\mathbf{s}^2$ equal to $s(s + 1)$, where s is an integer or half-integer between $|A - B|$ and $A + B$; each term describes excitations (e.g., particles) of spin s. It follows then from (2.12.18) that the tensor representations can describe only excitations with integer spin, whereas the spinor representations describe only excitations with half-integer spin.

Finite Lorentz transformations can be built up by multiplying together an infinite number of infinitesimal Lorentz transformations. In the same way, the tensor representations of the infinitesimal Lorentz group can be used to construct the tensor representations, such as (2.12.3) and (2.12.4), of the group of finite Lorentz transformations. However, if we try to construct spinor representations of the finite Lorentz transformations, we find that we can only get "representations up to a sign";[12] that is, the group multiplication law (2.12.2) will occasionally have a minus sign on the right-hand side. For instance, the product of two successive 180° rotations about a given axis does not give the unit matrix, but minus the unit matrix. The appearance of these minus signs means that a spinor field itself cannot be a physical observable, though even functions of spinor fields can be observables.

13 Temporal Order and Antiparticles*

One of the most striking features of the Lorentz transformations is that they do not leave invariant the order of events. For instance, suppose that in one reference frame an event at x_2 is observed to occur later than one at x_1, that is, $x_2{}^0 > x_1{}^0$. A second observer who sees the first observer moving with velocity $\mathbf{v}$ will see the events separated by a time difference

$$x_2'{}^0 - x_1'{}^0 = \Lambda^0{}_\alpha(\mathbf{v})(x_2{}^\alpha - x_1{}^\alpha)$$

where $\Lambda^\beta{}_\alpha(\mathbf{v})$ is the "boost" defined by (2.1.17)–(2.1.21). Applying (2.1.17) and (2.1.21) gives then

$$x_2'{}^0 - x_1'{}^0 = \gamma(x_2{}^0 - x_1{}^0) + \gamma\mathbf{v} \cdot (\mathbf{x}_2 - \mathbf{x}_1)$$

* This section lies somewhat out of the book's main line of development, and may be omitted in a first reading.

and this will be negative if

$$\mathbf{v} \cdot (\mathbf{x}_2 - \mathbf{x}_1) < -(x_2{}^0 - x_1{}^0) \qquad (2.13.1)$$

At first sight this might seem to raise the danger of a logical paradox. Suppose that the first observer sees a radioactive decay $A \to B + C$ at x_1, followed at x_2 by absorption of particle B, for example, $B + D \to E$. Does the second observer then see B absorbed at x_2 before it is emitted at x_1? The paradox disappears if we note that the speed $|\mathbf{v}|$ characterizing any Lorentz transformation $\Lambda(\mathbf{v})$ must be less than unity, so that (2.13.1) can be satisfied only if

$$|\mathbf{x}_2 - \mathbf{x}_1| > |x_2{}^0 - x_1{}^0| \qquad (2.13.2)$$

However, this is impossible, because particle B was assumed to travel from x_1 to x_2, and (2.13.2) would require its speed to be greater than unity, that is, than the speed of light. To put it another way, the temporal order of events at x_1 and x_2 is affected by Lorentz transformations only if $x_1 - x_2$ is *spacelike*, that is,

$$\eta_{\alpha\beta}(x_1 - x_2)^\alpha (x_1 - x_2)^\beta > 0$$

whereas a particle can travel from x_1 to x_2 only if $x_1 - x_2$ is *timelike*, that is,

$$\eta_{\alpha\beta}(x_1 - x_2)^\alpha (x_1 - x_2)^\beta < 0$$

Although the relativity of temporal order raises no problems for classical physics, it plays a profound role in quantum theories. The uncertainty principle tells us that when we specify that a particle is at position $\mathbf{x}_1$ at time t_1, we cannot also define its velocity precisely. In consequence there is a certain chance of a particle getting from x_1 to x_2 even if $x_1 - x_2$ is spacelike, that is, $|\mathbf{x}_1 - \mathbf{x}_2| > |x_1{}^0 - x_2{}^0|$. To be more precise, the probability of a particle reaching x_2 if it starts at x_1 is nonnegligible as long as

$$(\mathbf{x}_1 - \mathbf{x}_2)^2 - (x_1{}^0 - x_2{}^0)^2 \lesssim \frac{\hbar^2}{m^2}$$

where $\hbar$ is Planck's constant (divided by 2π) and m is the particle mass. (Such space-time intervals are very small even for elementary particle masses; for instance, if m is the mass of a proton then $\hbar/m = 2 \times 10^{-14}$ cm or in time units 6×10^{-25} sec. Recall that in our units 1 sec $= 3 \times 10^{10}$ cm.) We are thus faced again with our paradox; if one observer sees a particle emitted at x_1, and absorbed at x_2, and if $(\mathbf{x}_1 - \mathbf{x}_2)^2 - (x_1{}^0 - x_2{}^0)^2$ is positive (but less than $\hbar^2/m^2$), then a second observer may see the particle absorbed at $\mathbf{x}_2$ at a time t_2 *before* the time t_1 it is emitted at $\mathbf{x}_1$.

There is only one known way out of this paradox. The second observer must see a particle emitted at $\mathbf{x}_2$ and absorbed at $\mathbf{x}_1$. But in general the particle seen by the second observer will then necessarily be different from that seen by the first. For instance, if the first observer sees a proton turn into a neutron and a *positive* pi-meson at x_1 and then sees the pi-meson and some other neutron turn into a proton at x_2, then the second observer must see the neutron at x_2 turn into a proton and a particle of *negative* charge, which is then absorbed by a proton at x_1 that turns into a neutron. Since mass is a Lorentz invariant, the mass of the negative particle seen by the second observer will be equal to that of the positive pi-meson seen by the first observer. There is such a particle, called a negative pi-meson, and it does indeed have the same mass as the positive pi-meson. This reasoning leads us to the conclusion that for every type of charged particle there is an oppositely charged particle of equal mass, called its antiparticle. Note that this conclusion does not obtain in nonrelativistic quantum mechanics or in relativistic classical mechanics; it is only in relativistic quantum mechanics that antiparticles are a necessity.[13] And it is the existence of antiparticles that leads to the characteristic feature of relativistic quantum dynamics, that given enough energy we can create arbitrary numbers of particles and their antiparticles.

2 BIBLIOGRAPHY

Special Relativity

For a more comprehensive treatment of special relativity, see any one of the following books:

☐ J. L. Anderson, *Principles of Relativity Physics* (Academic Press, New York, 1967), Chapters 6–9.
☐ C. Møller, *The Theory of Relativity* (Oxford University Press, London, 1952), Chapters I–VII.
☐ W. Pauli, *Theory of Relativity*, trans. by G. Field (Pergamon Press, Oxford, 1958), Part I.
☐ W. Rindler, *Special Relativity* (2nd ed., Oliver and Boyd, Edinburgh, 1966).
☐ J. L. Synge, *Relativity: The Special Theory* (Interscience Publishers, New York, 1956).

Relativistic Hydrodynamics

☐ L. D. Landau and E. M. Lifshitz, *Fluid Mechanics*, trans. by J. B. Sykes and W. H. Reid (Pergamon Press, London, 1959), Chapter XV.

Representations of the Lorentz Group

☐ G. Ya. Lyubarskii, *The Application of Group Theory in Physics* (Pergamon Press, Oxford, 1960), Chapters XV, XVI.

2 REFERENCES

1. T. D. Lee and C. N. Yang, Phys. Rev., **104**, 254 (1956); C. S. Wu et al., Phys. Rev., **105**, 1413 (1957); R. Garwin, L. Lederman, and M. Weinrich, Phys. Rev., **105**, 1415 (1957); J. I. Friedman and V. L. Telegdi, Phys. Rev., **105**, 1681 (1957).

2. J. H. Christenson, J. W. Cronin, V. L. Fitch, and R. Turlay, Phys. Rev. Letters, **13**, 138 (1964).

3. See A. Einstein, Bull. Amer. Mat. Soc., April 1935, p. 223.

4. S. A. Bludman and M. A. Ruderman, Phys. Rev., **170,** 1176 (1968); **1**, 3243 (1970).

5. L. D. Landau and E. M. Lifshitz, *Fluid Mechanics*, trans. by J. B. Sykes and W. H. Reid (Pergamon Press, London, 1959), Section 127.

6. C. Eckart, Phys. Rev., **58**, 919 (1940).

7. J. L. Anderson, in *Relativity—Proceedings of the Relativity Conference in the Midwest*, ed. by M. Carmeli, S. I. Fickler, and L. Witten (Plenum Press, New York, 1969), p. 109; W. Israel and J. N. Vardalas, Nuovo Cimento Letters, **4**, 887 (1970).

8. L. Tisza, Phys. Rev., **61, 531** (1942).

9. See, for example, S. Chapman and T. G. Cowling, *The Mathematical Theory of Non-Uniform Gases* (2nd ed., Cambridge University Press, 1952), Note B and Chapter 11.

10. For χ, see C. W. Misner and D. H. Sharp, Phys. Letters, **15,** 279 (1965). For η, see C. W. Misner, Ap. J., **151,** 431 (1968). For ζ, χ, and η, see S. Weinberg, Ap. J., **168**, 175 (1971).

11. See, for example, L. I. Schiff, *Quantum Mechanics* (3rd ed., McGraw-Hill, New York, 1968), Section 27.

12. See, for example, E. P. Wigner, *Group Theory*, trans. by J. J. Griffin (Academic Press, New York, 1959), Chapter 15.

13. For a rigorous discussion of the necessity of antiparticles in relativistic quantum mechanics, see R. F. Streater and A. S. Wightman, *PCT, Spin & Statistics, and All That* (W. A. Benjamin, New York, 1964).

PART TWO
THE GENERAL THEORY
OF RELATIVITY

"Either the well was very deep, or she fell very slowly, for she had plenty of time as she went down to look about her, and to wonder what was going to happen next." *Lewis Carroll, Alice's Adventures in Wonderland*

3 THE PRINCIPLE OF EQUIVALENCE

The Principle of the Equivalence of Gravitation and Inertia tells us how an arbitrary physical system responds to an external gravitational field. We shall first see what this principle says, and then in the balance of this chapter we shall take a look at a few of its consequences. However, the appropriate mathematical technique for implementing the Principle of Equivalence is tensor analysis, and only after we complete the introduction to tensor analysis in the next chapter will we be able to make use of the full content of this principle.

1 Statement of the Principle

The Principle of Equivalence rests on the equality of gravitational and inertial mass, demonstrated by Galileo, Huygens, Newton, Bessel, and Eötvös. (See Section 1.2.) Einstein reflected that, as a consequence, no external static homogeneous gravitational field could be detected in a freely falling elevator, for the observers, their test bodies, and the elevator itself would respond to the field with the same acceleration. This can be easily proved for a system of particles N, moving with nonrelativistic velocities under the influence of forces $\mathbf{F}(\mathbf{x}_N - \mathbf{x}_M)$ (e.g., electrostatic or gravitational forces) and an external gravitational field $\mathbf{g}$. The equations of motion are

$$m_N \frac{d^2 \mathbf{x}_N}{dt^2} = m_N \mathbf{g} + \sum_M \mathbf{F}(\mathbf{x}_N - \mathbf{x}_M) \tag{3.1.1}$$

Suppose that we perform a non-Galilean space-time coordinate transformation

$$\mathbf{x}' = \mathbf{x} - \tfrac{1}{2}\mathbf{g}t^2 \qquad t' = t \tag{3.1.2}$$

Then $\mathbf{g}$ will be canceled by an inertial "force," and the equation of motion will become

$$m_N \frac{d^2 \mathbf{x}'_N}{dt'^2} = \sum_M \mathbf{F}(\mathbf{x}'_N - \mathbf{x}'_M) \tag{3.1.3}$$

Hence the original observer O who uses coordinates $\mathbf{x}t$, and his freely falling friend O' who uses $\mathbf{x}'t'$, will detect no difference in the laws of mechanics, except that O will say that he feels a gravitational field and O' will say that he does not. The equivalence principle says that this cancellation of gravitational by inertial force (and hence their equivalence) will obtain for *all* freely falling systems, whether or not they can be described by simple equations such as (3.1.1).

We are not yet ready to state the Principle of Equivalence in its final form, because the preceding remarks dealt only with a static homogeneous gravitational field. Had $\mathbf{g}$ depended on $\mathbf{x}$ or t, we would not have been able to eliminate it from the equations of motion by the acceleration (3.1.2). For example, the earth is in free fall about the sun, and for the most part we on earth do not feel the sun's gravitational field, but the slight inhomogeneity in this field (about 1 part in 6000 from noon to midnight) is enough to raise impressive tides in our oceans. Even the observers in Einstein's freely falling elevator would in principle be able to detect the earth's field, because objects in the elevator would be falling radially toward the center of the earth, and hence would approach each other as the elevator descended.

Although inertial forces do not exactly cancel gravitational forces for freely falling systems in an inhomogeneous or time-dependent gravitational field, we can still expect an approximate cancellation if we restrict our attention to such a small region of space and time that the field changes very little over the region. Therefore we formulate the equivalence principle as the statement that *at every space-time point in an arbitrary gravitational field it is possible to choose a "locally inertial coordinate system" such that, within a sufficiently small region of the point in question, the laws of nature take the same form as in unaccelerated Cartesian coordinate systems in the absence of gravitation.* There is a little vagueness here about what we mean by "the same form as in unaccelerated Cartesian coordinate systems," so to avoid any possible ambiguity we can specify that by this we mean the form given to the laws of nature by special relativity, for example, such equations as (2.3.1), (2.7.6), (2.7.7), (2.7.9), and (2.8.7). There is also a question of how small is "sufficiently small." Roughly speaking, we mean that the region must be small enough so that the gravitational field is sensibly constant throughout it, but we cannot be more precise until we learn how to represent the gravitational field mathematically. (See the end of Section 4.1.)

The attentive reader may have noticed a certain resemblance between the Principle of Equivalence and the axiom which Gauss took as the basis of non-Euclidean geometry. The Principle of Equivalence says that at any point in space-time we may erect a locally inertial coordinate system in which matter satisfies the laws of special relativity. We saw in Chapter 1 that Gauss assumed that at any point on a curved surface we may erect a locally Cartesian coordinate system in which distances obey the law of Pythagoras. Because of this deep analogy, we should expect the laws of gravitation to bear a strong resemblance to the formulas of Riemannian geometry. In particular, Gauss's assumption implies that all inner properties of a curved surface can be described in terms of derivatives $\partial \xi^\alpha / \partial x^\mu$ of the function $\xi^\alpha(x)$ that defines the transformation $x \to \xi$ from some general coordinate system x^μ covering the surface to the locally Cartesian system ξ^α, whereas the Principle of Equivalence tells us that all effects of a gravitational field can be described in terms of derivatives $\partial \xi^\alpha / \partial x^\mu$ of the function $\xi^\alpha(x)$ that defines the transformation from the "laboratory" coordinates x^μ to the locally inertial coordinates ξ^α. Furthermore, it was shown in Chapter 1 that the geometrically relevant functions of these derivatives are the quantities $g_{\mu\nu}$ defined by Eq. (1.1.7); we shall see in the following sections of this chapter that the gravitational field is described in just the same way.

Occasionally one finds references to a "weak Principle of Equivalence" and a "strong Principle of Equivalence." The strong Principle of Equivalence is just what I have already stated, with "laws of nature" meaning *all* the laws of nature. The weak principle is the same, but with "laws of nature" replaced by "laws of motion of freely falling particles." That is, the weak principle is nothing but a restatement of the observed equality of gravitational and inertial mass, whereas the strong principle is a generalization of these observations that governs the effects of gravitation on all physical systems.

The experiments of Eötvös, Dicke, and their predecessors (see Section 1.2) provide direct verification only of the weak Principle of Equivalence, but they provide some indirect evidence for the strong principle. The mass of different substances arises *in different proportions* from the masses of the neutrons and protons plus electrons of which they are composed, and from the strong and electromagnetic forces that bind these particles together, so the ratio of gravitational to inertial mass will be equal for all these substances only if it is equal for their constituents. Wapstra and Nijgh[1] have shown that the limits set by Eötvös on any possible inequality in the ratio of gravitational to inertial mass for glass, cork, antimonite, and brass imply that this ratio is equal for neutrons and protons plus electrons to 1 part in 6×10^5 and equal for neutrons and binding energies to 1 part in 1.2×10^4. To this accuracy, an observer in a freely falling coordinate system will detect no gravitational force on neutrons, hydrogen, or their binding energies. It would be difficult to conceive of a theory that satisfies this requirement and does not go all the way to the strong principle (that no gravitational effects of any sort can be felt in a locally inertial frame).

We might, however, distinguish two versions of the strong principle of equivalence, a "very strong principle," which applies to all phenomena, and a "medium-strong principle," which applies to all phenomena except gravitation itself. Certainly the experiments of Eötvös and Dicke are not accurate enough to say whether gravitational binding energies affect inertial and gravitational masses in the same way. This question might be settled by studying the motion of a small body in orbit about a large body that is itself in free fall in a gravitational field. For instance, the gravitational binding energy of the earth contributes a fraction -8.4×10^{-10} of its total mass, whereas the gravitational binding energy of an artificial satellite contributes a very much smaller fraction of its mass. Thus, if (to take an extreme case) the (negative) gravitational binding energy contributes fully to the inertial mass but not at all to the gravitational mass, then the ratio of gravitational to inertial mass of the satellite would be greater than that for the earth by a fraction 8.4×10^{-10}. The earth is in free fall, with the gravitational attraction of the sun balanced by the inertial force owing to the earth's revolution. The gravitational and inertial forces on the satellite owing to the presence of the sun and the earth's revolution are equal (neglecting for a moment the distance between the satellite and the earth's center of mass) to the gravitational and inertial forces on the earth times the ratio of gravitational or inertial masses, so these two forces are *not* in balance for the satellite, the gravitational force being greater than the inertial force by a fraction 8.4×10^{-10}. The acceleration owing to the sun's gravity is at the orbit of the earth about 6×10^{-4} of the acceleration owing to the earth's gravity at the surface of the earth, so we conclude that if the gravitational binding energy of the earth contributed fully to its inertial mass but not at all to its gravitational mass, then an artificial satellite in a low orbit about the earth would feel an effective attraction toward the sun equal to about 5.4×10^{-13} times its gravitational attraction toward the earth. This tiny effect would be entirely masked by a "tidal" force because the satellite is far from the center of mass of the earth, and there is no prospect of its being measured. This is a pity, because it is precisely the very strong assumption that the Principle of Equivalence applies to gravitational fields that will lead us in Chapter 5 to Einstein's field equations for gravitation.

2   Gravitational Forces

Consider a particle moving freely under the influence of purely gravitational forces. According to the Principle of Equivalence, there is a freely falling coordinate system ξ^α in which its equation of motion is that of a straight line in space-time, that is,

$$\frac{d^2\xi^\alpha}{d\tau^2} = 0 \qquad (3.2.1)$$

with $d\tau$ the proper time

$$d\tau^2 = -\eta_{\alpha\beta}\, d\xi^\alpha\, d\xi^\beta \qquad (3.2.2)$$

[Compare Eqs. (2.3.1) and (2.1.4).] Now suppose that we use any other coordinate system x^μ, which may be a Cartesian coordinate system at rest in the laboratory, but also may be curvilinear, accelerated, rotating, or what we will. The freely falling coordinates ξ^α are functions of the x^μ, and Eq. (3.2.1) becomes

$$0 = \frac{d}{d\tau}\left(\frac{\partial \xi^\alpha}{\partial x^\mu}\frac{dx^\mu}{d\tau}\right)$$

$$= \frac{\partial \xi^\alpha}{\partial x^\mu}\frac{d^2 x^\mu}{d\tau^2} + \frac{\partial^2 \xi^\alpha}{\partial x^\mu\,\partial x^\nu}\frac{dx^\mu}{d\tau}\frac{dx^\nu}{d\tau}$$

Multiply this by $\partial x^\lambda / \partial \xi^\alpha$, and use the familiar product rule

$$\frac{\partial \xi^\alpha}{\partial x^\mu}\frac{\partial x^\lambda}{\partial \xi^\alpha} = \delta^\lambda_\mu$$

This gives the equation of motion

$$0 = \frac{d^2 x^\lambda}{d\tau^2} + \Gamma^\lambda_{\mu\nu}\frac{dx^\mu}{d\tau}\frac{dx^\nu}{d\tau} \tag{3.2.3}$$

where $\Gamma^\lambda_{\mu\nu}$ is the *affine connection*, defined by

$$\Gamma^\lambda_{\mu\nu} \equiv \frac{\partial x^\lambda}{\partial \xi^\alpha}\frac{\partial^2 \xi^\alpha}{\partial x^\mu\,\partial x^\nu} \tag{3.2.4}$$

The proper time (3.2.2) may also be expressed in an arbitrary coordinate system,

$$d\tau^2 = -\eta_{\alpha\beta}\frac{\partial \xi^\alpha}{\partial x^\mu}\,dx^\mu\,\frac{\partial \xi^\beta}{\partial x^\nu}\,dx^\nu \tag{3.2.5}$$

or

$$d\tau^2 = -g_{\mu\nu}\,dx^\mu\,dx^\nu \tag{3.2.6}$$

where $g_{\mu\nu}$ is the *metric tensor*, defined by

$$g_{\mu\nu} \equiv \frac{\partial \xi^\alpha}{\partial x^\mu}\frac{\partial \xi^\beta}{\partial x^\nu}\,\eta_{\alpha\beta} \tag{3.2.7}$$

For a photon or a neutrino the equation of motion in a freely falling system is the same as (3.2.1), except that the independent variable cannot be taken as the proper time (3.2.2), because for massless particles the right-hand side of (3.2.2) vanishes. Instead of τ we can use $\sigma \equiv \xi^0$, so that (3.2.1) and (3.2.2) become

$$\frac{d^2 \xi^\alpha}{d\sigma^2} = 0$$

$$0 = -\eta_{\alpha\beta}\frac{d\xi^\alpha}{d\sigma}\frac{d\xi^\beta}{d\sigma}$$

Following the same reasoning as before, we find that the equation of motion in an arbitrary gravitational field and an arbitrary coordinate system is

$$\frac{d^2x^\mu}{d\sigma^2} + \Gamma^\mu_{\nu\lambda} \frac{dx^\nu}{d\sigma} \frac{dx^\lambda}{d\sigma} = 0 \tag{3.2.8}$$

$$0 = -g_{\mu\nu} \frac{dx^\mu}{d\sigma} \frac{dx^\nu}{d\sigma} \tag{3.2.9}$$

with $\Gamma^\mu_{\nu\lambda}$ and $g_{\mu\nu}$ given as before by (3.2.4) and (3.2.7).

Incidentally, in both (3.2.3) and (3.2.8) we do not need to know what τ and σ are in order to find the motion of our particle, for these equations when solved give $x^\mu(\tau)$ or $x^\mu(\sigma)$, and τ or σ can be eliminated to give $\mathbf{x}(t)$. The purpose of (3.2.6) is to tell us how to compute the proper time, whereas the purpose of (3.2.9) is to impose initial conditions appropriate to a massless particle. In particular, Eq. (3.2.9) tells us that the time dt for a photon to travel a distance $d\mathbf{x}$ is determined by the quadratic equation

$$0 = g_{00} \, dt^2 + 2g_{i0} \, dx^i \, dt + g_{ij} \, dx^i \, dx^j$$

with i and j summed over the values 1, 2, 3. The solution is

$$dt = \frac{1}{g_{00}} \left[-g_{i0} \, dx^i - \{(g_{i0}g_{j0} - g_{ij}g_{00}) \, dx^i \, dx^j\}^{1/2} \right] \tag{3.2.10}$$

and the time required for light to travel along any path may be calculated by integrating dt along the path.

The values of the metric tensor $g_{\mu\nu}$ and the affine connection $\Gamma^\lambda_{\mu\nu}$ at a point X in an arbitrary coordinate system x^μ provide enough information to determine the locally inertial coordinates $\xi^\alpha(x)$ in a neighborhood of X. First, we multiply Eq. (3.2.4) by $\partial\xi^\beta/\partial x^\lambda$ and use the product rule

$$\frac{\partial\xi^\beta}{\partial x^\lambda} \frac{\partial x^\lambda}{\partial\xi^\alpha} = \delta_\alpha^{\,\beta}$$

thereby obtaining the differential equations for ξ^α:

$$\frac{\partial^2\xi^\alpha}{\partial x^\mu \, \partial x^\nu} = \Gamma^\lambda_{\mu\nu} \frac{\partial\xi^\alpha}{\partial x^\lambda} \tag{3.2.11}$$

The solution is

$$\xi^\alpha(x) = a^\alpha + b^\alpha_{\;\mu}(x^\mu - X^\mu)$$
$$+ \tfrac{1}{2}b^\alpha_{\;\lambda}\Gamma^\lambda_{\mu\nu}(x^\mu - X^\mu)(x^\nu - X^\nu) + \cdots \tag{3.2.12}$$

where

$$a^\alpha = \xi^\alpha(X) \qquad b^\alpha_{\;\lambda} = \frac{\partial\xi^\alpha(X)}{\partial X^\lambda} \tag{3.2.13}$$

From Eq. (3.2.7) we also learn that

$$\eta_{\alpha\beta} b^\alpha{}_\mu b^\beta{}_\nu = g_{\mu\nu}(X) \tag{3.2.14}$$

Thus, given $\Gamma^\lambda_{\mu\nu}$ and $g_{\mu\nu}$ at X, the locally inertial coordinates ξ^α are determined to order $(x - X)^2$, except for the ambiguity in the constants a^α and $b^\alpha{}_\lambda$. The $b^\alpha{}_\lambda$ are determined by Eq. (3.2.13) up to a Lorentz transformation $b^\alpha{}_\mu \to \Lambda^\alpha{}_\beta b^\beta{}_\mu$, so the ambiguity in the solution for $\xi^\alpha(x)$ just reflects the fact that if ξ^α are locally inertial coordinates, then so are $\Lambda^\alpha{}_\beta \xi^\beta + c^\alpha$. Hence, since $\Gamma^\lambda_{\mu\nu}$ and $g_{\mu\nu}$ determine the locally inertial coordinates up to an inhomogeneous Lorentz transformation, and since the gravitational field can have no effects in a locally inertial coordinate system, we should not be surprised to find that all effects of gravitation are comprised in $\Gamma^\lambda_{\mu\nu}$ and $g_{\mu\nu}$. Note, however, that (3.2.12) satisfies (3.2.11) only at the point $x = X$; in order for it to be possible to solve (3.2.11) for all x, it is necessary for the derivatives of the affine connection to satisfy certain symmetry conditions, to be discussed in Chapter 5.

3 Relation between $g_{\mu\nu}$ and $\Gamma^\lambda_{\mu\nu}$

Our treatment of freely falling particles has shown that the field that determines the gravitational force is the "affine connection" $\Gamma^\lambda_{\mu\nu}$, whereas the proper time interval between two events with a given infinitesimal coordinate separation is determined by the "metric tensor" $g_{\mu\nu}$. We now show that $g_{\mu\nu}$ is also the gravitational potential; that is, its derivatives determine the field $\Gamma^\lambda_{\mu\nu}$.

We first recall the formula for the metric tensor, Eq. (3.2.7):

$$g_{\mu\nu} = \frac{\partial \xi^\alpha}{\partial x^\mu} \frac{\partial \xi^\beta}{\partial x^\nu} \eta_{\alpha\beta}$$

Differentiation with respect to x^λ gives

$$\frac{\partial g_{\mu\nu}}{\partial x^\lambda} = \frac{\partial^2 \xi^\alpha}{\partial x^\lambda \partial x^\mu} \frac{\partial \xi^\beta}{\partial x^\nu} \eta_{\alpha\beta} + \frac{\partial \xi^\alpha}{\partial x^\mu} \frac{\partial^2 \xi^\beta}{\partial x^\lambda \partial x^\nu} \eta_{\alpha\beta}$$

and recalling (3.2.11), we have

$$\frac{\partial g_{\mu\nu}}{\partial x^\lambda} = \Gamma^\rho_{\lambda\mu} \frac{\partial \xi^\alpha}{\partial x^\rho} \frac{\partial \xi^\beta}{\partial x^\nu} \eta_{\alpha\beta} + \Gamma^\rho_{\lambda\nu} \frac{\partial \xi^\alpha}{\partial x^\mu} \frac{\partial \xi^\beta}{\partial x^\rho} \eta_{\alpha\beta}$$

Using (3.2.7) again, we find

$$\frac{\partial g_{\mu\nu}}{\partial x^\lambda} = \Gamma^\rho_{\lambda\mu} g_{\rho\nu} + \Gamma^\rho_{\lambda\nu} g_{\rho\mu} \tag{3.3.1}$$

Before solving for Γ, it is necessary to point out a subtlety in the derivation of Eq. (3.3.1) that has been hidden by our too-compact notation. When we erect a locally inertial coordinate system $\xi^\alpha(x)$, we do so at a specific point X, and the coordinates that are locally inertial at X should be so labeled, as $\xi_X{}^\alpha(x)$. Thus Eqs. (3.2.7) and (3.2.11) should properly be written as

$$g_{\mu\nu}(X) = \left(\frac{\partial \xi_X^\alpha(x)}{\partial x^\mu} \frac{\partial \xi_X^\beta(x)}{\partial x^\nu} \eta_{\alpha\beta}\right)_{x=X} \tag{3.3.2}$$

$$\left(\frac{\partial^2 \xi_X^\alpha(x)}{\partial x^\mu \, \partial x^\nu}\right)_{x=X} = \Gamma^\lambda_{\mu\nu}(X) \left(\frac{\partial \xi_X^\alpha(x)}{\partial x^\lambda}\right)_{x=X} \tag{3.3.3}$$

When we differentiate (3.3.2) with respect to X^λ, we get two kinds of terms. The first kind arises because we set $x = X$; these contain just the second derivatives (3.3.3) and can be easily calculated as before. The second kind of term arises because $\xi_X^\alpha(x)$ carries a label X; these terms contain derivatives like

$$\left(\frac{\partial^2 \xi_X^\alpha(x)}{\partial X^\lambda \partial x^\mu}\right)_{x=X} \tag{3.3.4}$$

and do not seem to have anything to do with the metric or the affine connection. In order to deal with this second kind of term it is necessary to sharpen somewhat our interpretation of what is meant by "locally inertial" in the Principle of Equivalence. We shall see in Section 5 that first derivatives of the metric tensor may be measured by comparing the rates of identical clocks an infinitesimal space-time distance apart. Hence we shall interpret the Principle of Equivalence as meaning that *the locally inertial coordinates ξ_X^α that we construct at a given point X can be chosen so that the first derivatives of the metric tensor vanish at X*. In the coordinate system ξ_X^α the metric tensor at a point X' is given by (3.3.2) as

$$g_{\gamma\delta}^X(X') = \left(\frac{\partial \xi_{X'}^\alpha(x)}{\partial \xi_X^\gamma(x)} \frac{\partial \xi_{X'}^\beta(x)}{\partial \xi_X^\delta(x)} \eta_{\alpha\beta}\right)_{x=X'}$$

and our new interpretation of the Principle of Equivalence tells us that this quantity is stationary in X' at $X' = X$. In order to use this information, we introduce an arbitrary "laboratory" coordinate system x^μ, and write

$$g_{\mu\nu}(X') \equiv \left(\frac{\partial \xi_{X'}^\alpha(x)}{\partial x^\mu} \frac{\partial \xi_{X'}^\beta(x)}{\partial x^\nu} \eta_{\alpha\beta}\right)_{x=X'}$$

$$= g_{\gamma\delta}^X(X') \left(\frac{\partial \xi_X^\gamma(x)}{\partial x^\mu} \frac{\partial \xi_X^\delta(x)}{\partial x^\nu}\right)_{x=X'}$$

Differentiating with respect to X'^{λ} and setting $X' = X$ gives then (because $g^{X}_{\gamma\delta}(X')$ is stationary)

$$\frac{\partial g_{\mu\nu}(X)}{\partial X^{\lambda}} = g^{X}_{\gamma\delta}(X) \left(\frac{\partial}{\partial x^{\lambda}} \left\{ \frac{\partial \xi^{\gamma}_{X}(x)}{\partial x^{\mu}} \frac{\partial \xi^{\delta}_{X}(x)}{\partial x^{\nu}} \right\} \right)_{x=X}$$

$$= \eta_{\gamma\delta} \left(\frac{\partial^2 \xi^{\gamma}_{X}(x)}{\partial x^{\lambda} \partial x^{\mu}} \frac{\partial \xi^{\delta}_{X}(x)}{\partial x^{\nu}} + \frac{\partial \xi^{\gamma}_{X}(x)}{\partial x^{\mu}} \frac{\partial^2 \xi^{\delta}_{X}(x)}{\partial x^{\lambda} \partial x^{\nu}} \right)_{x=X}$$

No derivatives like (3.3.4) now appear, and we can use (3.3.2) and (3.3.3) as before to show that

$$\frac{\partial g_{\mu\nu}(X)}{\partial X^{\lambda}} = \Gamma^{\rho}_{\lambda\mu}(X)g_{\rho\nu}(X) + \Gamma^{\rho}_{\lambda\nu}(X)g_{\rho\mu}(X)$$

which is precisely Eq. (3.3.1).

Now we return to our previous compact notation, and solve for the affine connection. Add to Eq. (3.3.1) the same equation with μ and λ interchanged and subtract the same equation with ν and λ interchanged. We have then

$$\frac{\partial g_{\mu\nu}}{\partial x^{\lambda}} + \frac{\partial g_{\lambda\nu}}{\partial x^{\mu}} - \frac{\partial g_{\mu\lambda}}{\partial x^{\nu}} = g_{\kappa\nu}\Gamma^{\kappa}_{\lambda\mu} + g_{\kappa\mu}\Gamma^{\kappa}_{\lambda\nu}$$

$$+ g_{\kappa\nu}\Gamma^{\kappa}_{\mu\lambda} + g_{\kappa\lambda}\Gamma^{\kappa}_{\mu\nu}$$

$$- g_{\kappa\lambda}\Gamma^{\kappa}_{\nu\mu} - g_{\kappa\mu}\Gamma^{\kappa}_{\nu\lambda}$$

$$= 2g_{\kappa\nu}\Gamma^{\kappa}_{\lambda\mu} \qquad (3.3.5)$$

(Recall that $\Gamma^{\kappa}_{\mu\nu}$ and $g_{\mu\nu}$ are symmetric under interchange of μ and ν.) Define a matrix $g^{\nu\sigma}$ as the inverse of $g_{\nu\sigma}$, that is,

$$g^{\nu\sigma}g_{\kappa\nu} = \delta^{\sigma}_{\kappa} \qquad (3.3.6)$$

and multiply the above with $g^{\nu\sigma}$; this gives finally

$$\Gamma^{\sigma}_{\lambda\mu} = \frac{1}{2}g^{\nu\sigma} \left\{ \frac{\partial g_{\mu\nu}}{\partial x^{\lambda}} + \frac{\partial g_{\lambda\nu}}{\partial x^{\mu}} - \frac{\partial g_{\mu\lambda}}{\partial x^{\nu}} \right\} \qquad (3.3.7)$$

[It should be noted that (3.2.7) ensures that the metric tensor *does* have an inverse, given by

$$g^{\nu\sigma} \equiv g^{\sigma\nu} \equiv \eta^{\alpha\beta} \frac{\partial x^{\nu}}{\partial \xi^{\alpha}} \frac{\partial x^{\sigma}}{\partial \xi^{\beta}} \qquad (3.3.8)$$

for, using the familiar product rule

$$\frac{\partial x^{\nu}}{\partial \xi^{\alpha}} \frac{\partial \xi^{\gamma}}{\partial x^{\nu}} = \delta^{\gamma}_{\alpha}$$

we find

$$g^{v\sigma}g_{\kappa v} = \eta^{\alpha\beta}\frac{\partial x^v}{\partial \xi^\alpha}\frac{\partial x^\sigma}{\partial \xi^\beta}\eta_{\gamma\delta}\frac{\partial \xi^\gamma}{\partial x^\kappa}\frac{\partial \xi^\delta}{\partial x^v}$$

$$= \eta^{\alpha\beta}\frac{\partial x^\sigma}{\partial \xi^\beta}\eta_{\gamma\alpha}\frac{\partial \xi^\gamma}{\partial x^\kappa}$$

$$= \frac{\partial x^\sigma}{\partial \xi^\beta}\frac{\partial \xi^\beta}{\partial x^\kappa} = \delta^\sigma_\kappa$$

as required by (3.3.6).] Occasionally the right-hand side of Eq. (3.3.7) is called a *Christoffel symbol* and denoted

$$\begin{Bmatrix} \sigma \\ \lambda\mu \end{Bmatrix}.$$

One important consequence of the relation between the affine connection and the metric tensor is that the equation of motion of a freely falling particle automatically maintains the form of the proper time interval $d\tau$. Using (3.2.3) we may calculate that

$$\frac{d}{d\tau}\begin{Bmatrix} g_{\mu v}\frac{dx^\mu}{d\tau}\frac{dx^v}{d\tau} \end{Bmatrix} = \frac{\partial g_{\mu v}}{\partial x^\lambda}\frac{dx^\lambda}{d\tau}\frac{dx^\mu}{d\tau}\frac{dx^v}{d\tau}$$

$$+ g_{\mu v}\frac{d^2 x^\mu}{d\tau^2}\frac{dx^v}{d\tau} + g_{\mu v}\frac{dx^\mu}{d\tau}\frac{d^2 x^v}{d\tau^2}$$

$$= \left[\frac{\partial g_{\kappa\sigma}}{\partial x^\lambda} - g_{\mu\sigma}\Gamma^\mu_{\kappa\lambda} - g_{v\kappa}\Gamma^v_{\sigma\lambda}\right]\frac{dx^\kappa}{d\tau}\frac{dx^\sigma}{d\tau}\frac{dx^\lambda}{d\tau}$$

and (3.3.5) tells us that this vanishes, that is,

$$g_{\mu v}\frac{dx^\mu}{d\tau}\frac{dx^v}{d\tau} = -C \tag{3.3.9}$$

where C is a constant of the motion. Hence, once we choose initial conditions such that $d\tau^2$ is given by (3.2.6), we have $C = 1$, and (3.3.9) will ensure that (3.2.6) continues to hold along the particle's path. Similarly, for a massless particle the initial conditions give $C = 0$ (with τ replaced by some other parameter σ) and the equations of motion will keep $g_{\mu v}\, dx^\mu\, dx^v$ zero along the path.

An additional consequence of the relation (3.3.5) is that we are enabled to formulate the law of motion of freely falling bodies as a variational principle. Let us introduce an arbitrary parameter p to describe the path, and write the proper time elapsed when the particle falls from point A to B as

$$T_{BA} = \int_A^B \frac{d\tau}{dp}\, dp = \int_A^B \left\{ -g_{\mu v}\frac{dx^\mu}{dp}\frac{dx^v}{dp} \right\}^{1/2} dp$$

Now vary the path from $x^\mu(p)$ to $x^\mu(p) + \delta x^\mu(p)$, keeping fixed the endpoints, that is, setting $\delta x^\mu = 0$ at p_A and p_B. The change in T_{BA} is

$$\delta T_{BA} = \frac{1}{2} \int_A^B \left\{ -g_{\mu\nu} \frac{dx^\mu}{dp} \frac{dx^\nu}{dp} \right\}^{-1/2} \left\{ -\frac{\partial g_{\mu\nu}}{\partial x^\lambda} \delta x^\lambda \frac{dx^\mu}{dp} \frac{dx^\nu}{dp} - 2g_{\mu\nu} \frac{d\delta x^\mu}{dp} \frac{dx^\nu}{dp} \right\} dp$$

The first factor within the integrand is just $dp/d\tau$, so the integral can be rewritten as

$$\delta T_{BA} = -\int_A^B \left\{ \frac{1}{2} \frac{\partial g_{\mu\nu}}{\partial x^\lambda} \delta x^\lambda \frac{dx^\mu}{d\tau} \frac{dx^\nu}{d\tau} + g_{\mu\nu} \frac{d\delta x^\mu}{d\tau} \frac{dx^\nu}{d\tau} \right\} d\tau$$

We now integrate by parts, neglecting the endpoint contributions because δx^μ vanishes at A and B. This gives

$$\delta T_{BA} = -\int_A^B \left\{ \frac{1}{2} \frac{\partial g_{\mu\nu}}{\partial x^\lambda} \frac{dx^\mu}{d\tau} \frac{dx^\nu}{d\tau} - \frac{\partial g_{\lambda\nu}}{\partial x^\sigma} \frac{dx^\sigma}{d\tau} \frac{dx^\nu}{d\tau} - g_{\lambda\nu} \frac{d^2 x^\nu}{d\tau^2} \right\} \delta x^\lambda \, d\tau$$

Inserting Eq. (3.3.5) and recalling that $\Gamma_{\mu\nu}^\lambda$ is symmetric in its lower indices, we find

$$\delta T_{BA} = -\int_A^B \left\{ \frac{d^2 x^\nu}{d\tau^2} + \Gamma_{\mu\sigma}^\nu \frac{dx^\mu}{d\tau} \frac{dx^\sigma}{d\tau} \right\} g_{\lambda\nu} \, \delta x^\lambda \, d\tau \qquad (3.3.10)$$

Hence the space-time path taken by a particle that obeys the equations (3.2.3) for free fall will be such that the proper time elapsed is in extremum (and usually a minimum), that is,

$$\delta T_{BA} = 0$$

We may therefore express the equations of motion (3.2.3) geometrically, by saying that a particle in free fall through the curved space-time called a gravitational field will move on the shortest (or longest) possible path between two points, "length" being measured by the proper time. Such paths are called *geodesics*. For instance, we can think of the sun as distorting space-time just as a heavy weight distorts a rubber sheet, and can consider a comet's path as being bent toward the sun to keep the path as "short" as possible. However, this geometrical analogy is an *a posteriori* consequence of the equations of motion derived from the equivalence principle, and plays no necessary role in our considerations.

4   The Newtonian Limit

To make contact with Newton's theory, let us consider the case of a particle moving slowly in a weak stationary gravitational field. If the particle is sufficiently slow, we may neglect $d\mathbf{x}/d\tau$ with respect to $dt/d\tau$, and write (3.2.3) as

$$\frac{d^2 x^\mu}{d\tau^2} + \Gamma_{00}^\mu \left(\frac{dt}{d\tau} \right)^2 = 0$$

Since the field is stationary, all time derivatives of $g_{\mu\nu}$ vanish, and therefore

$$\Gamma^{\mu}_{00} = -\tfrac{1}{2} g^{\mu\nu} \frac{\partial g_{00}}{\partial x^{\nu}}$$

Finally, since the field is weak, we may adopt a nearly Cartesian coordinate system in which

$$g_{\alpha\beta} = \eta_{\alpha\beta} + h_{\alpha\beta} \qquad |h_{\alpha\beta}| \ll 1 \tag{3.4.1}$$

so to first order in $h_{\alpha\beta}$,

$$\Gamma^{\alpha}_{00} = -\tfrac{1}{2} \eta^{\alpha\beta} \frac{\partial h_{00}}{\partial x^{\beta}}$$

Using this affine connection in the equations of motion then gives

$$\frac{d^2 \mathbf{x}}{d\tau^2} = \frac{1}{2} \left(\frac{dt}{d\tau}\right)^2 \nabla h_{00}$$

$$\frac{d^2 t}{d\tau^2} = 0$$

The solution of the second equation is that $dt/d\tau$ equals a constant (as could also be seen by computing $d\tau$ with $h_{\alpha\beta}$ neglected), so dividing the equation for $d^2\mathbf{x}/d\tau^2$ by $(dt/d\tau)^2$, we find

$$\frac{d^2 \mathbf{x}}{dt^2} = \tfrac{1}{2} \nabla h_{00} \tag{3.4.2}$$

The corresponding Newtonian result is

$$\frac{d^2 \mathbf{x}}{dt^2} = -\nabla \phi \tag{3.4.3}$$

where ϕ is the gravitational potential, which at a distance r from the center of a spherical body of mass M takes the form

$$\phi = -\frac{GM}{r} \tag{3.4.4}$$

Comparing (3.4.2) with (3.4.3), we conclude that

$$h_{00} = -2\phi + \text{constant}$$

Furthermore, the coordinate system must become Minkowskian at great distances, so h_{00} vanishes at infinity, and if we define ϕ to vanish at infinity [as in (3.4.4)], we find that the constant here is zero, so $h_{00} = -2\phi$, and returning to the metric (3.4.1),

$$g_{00} = -(1 + 2\phi) \tag{3.4.5}$$

The gravitational potential ϕ is of the order of 10^{-39} at the surface of a proton, 10^{-9} at the surface of the earth, 10^{-6} at the surface of the sun, and 10^{-4} at the surface of a white dwarf star, so evidently the distortion in $g_{\mu\nu}$ produced by gravitation is generally very slight. (In c.g.s. units ϕ has the dimensions of a squared velocity; in our units ϕ is the c.g.s. value divided by the square of the c.g.s. speed of light.)

5 Time Dilation

Consider a clock in an arbitrary gravitational field, moving with arbitrary velocity, not necessarily in free fall. The equivalence principle tells us that its rate is unaffected by the gravitational field if we observe the clock from a locally inertial coordinate system ξ^α, so according to Section 2.2, the space-time interval $d\xi^\alpha$ between ticks is governed in this system by

$$\Delta t = (-\eta_{\alpha\beta} \, d\xi^\alpha \, d\xi^\beta)^{1/2}$$

where Δt is the period between ticks when the clock is at rest in the absence of gravitation. Hence in any arbitrary coordinate system the space-time interval between ticks will be governed by

$$\Delta t = \left(-\eta_{\alpha\beta} \frac{\partial \xi^\alpha}{\partial x^\mu} \, dx^\mu \frac{\partial \xi^\beta}{\partial x^\nu} \, dx^\nu\right)^{1/2}$$

or, introducing the metric tensor (3.2.7),

$$\Delta t = (-g_{\mu\nu} \, dx^\mu \, dx^\nu)^{1/2}$$

If the clock has velocity dx^μ/dt, then the time interval dt between ticks will be given by

$$\frac{dt}{\Delta t} = \left(-g_{\mu\nu} \frac{dx^\mu}{dt} \frac{dx^\nu}{dt}\right)^{-1/2} \tag{3.5.1}$$

In particular, if the clock is at rest this becomes

$$\frac{dt}{\Delta t} = (-g_{00})^{-1/2} \tag{3.5.2}$$

We cannot observe the time dilation factors appearing in (3.5.1) and (3.5.2) by merely measuring the time interval dt between ticks and comparing with the value Δt specified by the manufacturer, because the gravitational field affects our time standards in exactly the same way as it affects the clock being studied. That is, if our standard clock says that a certain physical process takes 1 sec at rest in the absence of gravitation, then it will also tell us that it takes 1 sec in the presence of gravitation, both standard clock and process being affected by the field in the same way. However, we can compare the time dilation factors at two different points in a field. For instance, suppose that at point 1 we observe the light coming from a particular atomic transition at point 2. If points 1 and 2 are at rest in a stationary gravitational field, then the time taken for a wave crest to travel from 2 to 1 will be a constant, given by the integral of (3.2.10) over the path, and therefore the time between the arrival at point 1 of successive crests will equal the time dt_2 between their departure at point 2, given by (3.5.2) as

$$dt_2 = \Delta t(-g_{00}(x_2))^{-1/2}$$

If the same atomic transition occurs at point 1, then the time between crests of the light waves to be observed at point 1 will be

$$dt_1 = \Delta t(-g_{00}(x_1))^{-1/2}$$

Hence, for a given atomic transition, the ratio of the frequency (observed at point 1) of the light from point 2 to that of the light from point 1 will be

$$\frac{v_2}{v_1} = \left(\frac{g_{00}(x_2)}{g_{00}(x_1)}\right)^{1/2} \tag{3.5.3}$$

In the weak field limit $g_{00} \simeq -1 - 2\phi$ and $\phi \ll 1$, so $v_2/v_1 = 1 + \Delta v/v$, where

$$\frac{\Delta v}{v} = \phi(x_2) - \phi(x_1) \tag{3.5.4}$$

(For a *uniform* gravitational field, this result could be derived directly from the Principle of Equivalence, without introducing a metric or affine connection.)

Let us apply Eq. (3.5.4) to the case of light from the sun's surface observed on the earth. The sun's gravitational potential can be calculated as

$$\phi_\odot = \frac{-GM_\odot}{R_\odot}$$

where $M_\odot$ and $R_\odot$ are the sun's mass and radius,

$$M_\odot = 1.97 \times 10^{33} \text{ g}$$
$$R_\odot = 0.695 \times 10^6 \text{ km}$$

and G is the gravitational constant

$$G = 6.67 \times 10^{-8} \text{ erg cm/gm}^2 = 7.41 \times 10^{-29} \text{ cm/gm} \tag{3.5.5}$$

(Here we have used our convention that $c = 1$ to set 1 sec $= 3 \times 10^{10}$ cm; in c.g.s. units the quantity 7.41×10^{-29} cm/gm would have to be called G/c^2.) We find that the potential on the surface of the sun is

$$\phi_\odot = -2.12 \times 10^{-6}$$

The gravitational potential of the earth is negligible in comparison, so ideally the frequency of light from the sun should be shifted to the red by 2.12 parts per million as compared with light emitted by terrestrial atoms.

The difficulty in measuring the solar gravitational red shift can be appreciated if we reflect that a motion of the source by a velocity v along the earth–sun direction will produce an additional Doppler frequency shift $\Delta v/v = v$ [recall Eq. (2.2.2)], so the gravitational red shift can be masked by a velocity $v = 2 \times 10^{-6}$, or in c.g.s. units, $v = 0.6$ km/sec. It is not the rotation of the earth or the sun that bothers us; these are known effects which can easily be taken into account. Thermal effects are more serious; at a temperature of 3000°K the thermal velocity of typical light elements (C, N, O) is about 2 km/sec, giving a Doppler broadening about three times larger than the expected gravitational red shift. However, thermal motions only broaden lines, not shift them, so they too can be lived with. The really bothersome problems arise from unknown Doppler shifts owing to the convection of gases in the solar atmosphere. In fact, the frequency shift is observed to vary from place to place on the solar disk, and is occasionally even toward the blue! The convection tends to be vertical, so we can minimize the Doppler shifts it produces by looking at the limb of the sun, where the motion is mostly at right angles to the line of sight. Until recently the best result that could be achieved in this way was that the solar gravitational red shift is of the order of 2 parts per million.[2] In the last few years improved observational techniques[3] have yielded a much better value of the red shift, equal to 1.05 ± 0.05 times the predicted value. However, it is too early to say that this result closes the story, at least until it can be corroborated.

The red shifts are much larger for white dwarf stars like Sirius B and 40 Eridani B. Such stars have masses typically of the order of one solar mass, and radii of the order of 1/10 to 1/100 the sun's radius, so the red shift of spectral lines from their surface is about 10 to 100 times greater than for the sun, or roughly 1 part in 10^4 to 10^5. Although this alleviates problems arising from convective Doppler shifts or temperature or pressure, a new difficulty enters here: It is difficult to determine the value of the gravitational potential ϕ with which to compare the measured value of $\Delta v/v$. If we know the mass of a white dwarf star, we can deduce a rough value for its radius and surface gravitational potential from astrophysical theory,[4] but the only white dwarf stars whose mass can be measured are members of binary star pairs. For instance the mass of Sirius B is determined by calculating the total mass of Sirius A and B from their separation and period, and then subtracting the mass of Sirius A calculated from stellar theory. However, the scattering of light from Sirius A by the atmosphere of Sirius B makes the gravitational red shift on Sirius B very difficult to measure.[13] On the other hand, 40 Eridani B is sufficiently far from 40 Eridani A so that scattering of light is no problem, and the mass of 40 Eridani B can be determined separately from that of A by locating their center of mass in addition to measuring their period and separation. However, because 40 Eridani B and A are so far apart, their period of revolution is very long, and there has not yet been time to determine the mass of B very accurately. The best predicted value of the surface gravitational potential is $\phi = -(5.7 \pm 1) \times 10^{-5}$, in good agreement with the observed[5] red shift $\Delta v/v = -(7 \pm 1) \times 10^{-5}$. Taking account of Stark shifts in the spectrum of 40 Eridani B appears to improve the agreement.[5a]

The empirical evidence for the red shift predicted by the Principle of Equivalence was much improved in 1960, by a terrestrial experiment performed by Pound and Rebka.[6] They allowed a γ-ray emitted by a 14.4 keV, 0.1 μsec transition in Fe[57] to fall 22.6 m, and observed its resonant absorption by an Fe[57] target. (Normally *resonant* absorption is impossible for such a narrow γ-ray line, because the recoil of the emitting nucleus lowers the γ-ray energy below the nuclear energy difference, whereas to produce the inverse transition in the target nucleus, which also recoils, a little more energy than the nuclear energy difference is needed. This experiment was made possible by the Mossbauer effect,[7] in which the recoil momentum on emission and absorption is taken up by the whole crystal, so that essentially no energy is lost to recoil on emission or absorption.) The difference in the gravitational potential from top to bottom is

$$\Delta\phi = \phi_{\text{top}} - \phi_{\text{bottom}} = -\frac{(980 \text{ cm/sec}^2)(2260 \text{ cm})}{(3 \times 10^{10} \text{ cm/sec})^2}$$

$$= -2.46 \times 10^{-15}$$

and if the equivalence principle is correct we would expect the photon arriving at the target to be shifted upward in frequency by an amount $\Delta v/v = -\Delta\phi$, lowering the counting rate by a factor

$$C = \frac{\Gamma^2}{\Delta v^2 + \Gamma^2}$$

where Γ is the full width of the γ-ray line at half-maximum. (Note that Γ appears here rather than $\Gamma/2$, because we have to fold together an emission coefficient proportional to $[(v + \Delta v)^2 + (\Gamma/2)^2]^{-1}$ with an absorption coefficient proportional to $[v^2 + (\Gamma/2)^2]^{-1}$). But in this transition the fractional width was $\Gamma/v = 1.13 \times 10^{-12}$, which is larger than the predicted $\Delta v/v$ by a factor of 460, so the reduction in the counting rate was by only 1 part in 2.1×10^5! This would seem to make the experiment impossible, and indeed Pound and Rebka had originally thought that they might have to let the γ-ray fall several kilometers in order to get a frequency shift Δv comparable with Γ, but happily they thought of a trick which let them measure very small frequency shifts. Their idea was to move the γ-ray source up and down with velocity $v_0 \cos \omega t$, where ω is some arbitrary fixed frequency (10 – 50 cps) and v_0 is also arbitrary, but much greater than $-\Delta\phi$, that is, much greater than 7.4×10^{-5} cm/sec. To the gravitational violet shift Δv_G there is then added a larger Doppler shift $\Delta v_D/v = -v_0 \cos \omega t$ (see Section 2.2), so the counting rate is reduced by a time-dependent factor

$$C(t) = \frac{\Gamma^2}{(\Delta v_G + \Delta v_D)^2 + \Gamma^2} = \frac{\left(\dfrac{\Gamma}{v}\right)^2}{\left(\dfrac{\Delta v_G}{v} - v_0 \cos \omega t\right)^2 + \left(\dfrac{\Gamma}{v}\right)^2}$$

$$\simeq \frac{\left(\dfrac{\Gamma}{v}\right)^2}{v_0{}^2 \cos^2 \omega t + \left(\dfrac{\Gamma}{v}\right)^2} \left\{ 1 + \frac{2\,\dfrac{\Delta v_G}{v}\,v_0 \cos \omega t}{v_0{}^2 \cos^2 \omega t + \left(\dfrac{\Gamma}{v}\right)^2} \right\}$$

and Δv_G can be picked out by looking for a term linear in $\cos \omega t$, for instance, by measuring the asymmetry between the number of counts registered when the source is going up (for example, $\cos \omega t > 1/\sqrt{2}$) and down ($\cos \omega t < -1/\sqrt{2}$). In this way Pound and Rebka obtained a value for $\Delta v_G/v$ about four times larger than the expected value 2.46×10^{-15}. This discrepancy was actually an intrinsic frequency shift owing to the difference between the source and target crystals (including differences in their temperature) and was removed by subtracting the asymmetry in γ-ray counts when the source is below the target from the asymmetry when the target is below the source. The final result for the gravitational frequency shift was $\Delta v/v = (2.57 \pm 0.26) \times 10^{-15}$, in excellent agreement with the predicted value 2.46×10^{-15}. The agreement between theory and experiment has since[8] been improved to about 1 percent.

There have also been proposals[8a] to measure the gravitational red shift of light from an artificial satellite. At a point directly below perigee there is no first-order Doppler shift because the time for the light to reach us from the satellite is momentarily constant. In this case the frequency shift of the emitted light must be determined from (3.5.1), whereas the frequency shift of our laboratory time standards may be calculated from (3.5.2), if we ignore the rotation of the earth. It follows that the frequency v_s of a given atomic line from the satellite will be related to the frequency v_e of the same line on earth by

$$\frac{v_s}{v_e} = \frac{\left(-g_{\mu\nu}\dfrac{dx^\mu}{dt}\dfrac{dx^\nu}{dt}\right)_s^{1/2}}{(-g_{00})_\oplus^{1/2}} \tag{3.5.6}$$

The velocity v_s of the satellite is given by

$$v_s^2 = -\phi_s = \frac{GM_\oplus}{R_\oplus + H}$$

where H is the altitude of the satellite and $M_\oplus$ and $R_\oplus$ are the mass and radius of the earth,

$$M_\oplus = 5.983 \times 10^{27} \text{ g}$$

$$R_\oplus = 6.371 \times 10^8 \text{ cm}$$

In the weak field approximation we have

$$\left(-g_{\mu\nu} \frac{dx^\mu}{dt} \frac{dx^\nu}{dt}\right)_s \simeq -(g_{00})_s - v_s^2 = 1 + 2\phi_s - v_s^2$$

$$\simeq 1 - \frac{3GM_\oplus}{R_\oplus + H}$$

and

$$(-g_{00})_\oplus \simeq 1 + 2\phi_\oplus \simeq 1 - \frac{2GM_\oplus}{R_\oplus}$$

so to this order Eq. (3.5.6) gives a frequency ratio $v_s/v_e = 1 + \Delta v/v$, where

$$\frac{\Delta v}{v} = -\frac{3}{2} \frac{GM_\oplus}{R_\oplus + H} + \frac{GM_\oplus}{R_\oplus}$$

$$\simeq -3.47 \times 10^{-10} \left\{ \frac{3R_\oplus}{R_\oplus + H} - 2 \right\}$$

We see that at low altitudes there is a purely special-relativistic red shift (see Section 2.2), to which is added at higher altitudes a general-relativistic violet shift, yielding a net red shift for $H < R_\oplus/2$ and a net violet shift for $H > R_\oplus/2$.

Incidentally, the gravitational red shift of light rising from a lower to a higher gravitational potential can to some extent be understood as a consequence of quantum theory, energy conservation, and the "weak" Principle of Equivalence. When a photon is produced at point 1 by some heavy nonrelativistic apparatus, an observer in a locally inertial coordinate system moving with the apparatus will see its internal energy and hence its inertial mass change by an amount related to the photon frequency v_1 he observes, that is, by

$$\Delta m_1 = -h v_1$$

where $h = 6.625 \times 10^{-27}$ erg sec is Planck's constant. Suppose that the photon is then absorbed at point 2 by a second heavy apparatus; an observer in a freely falling system will see the apparatus change in inertial mass by an amount related to the photon frequency v_2 he observes, that is, by

$$\Delta m_2 = h v_2$$

However, the total internal plus gravitational potential energy of the two pieces of apparatus must be the same before and after these events, so

$$0 = \Delta m_1 + \phi_1 \, \Delta m_1 + \Delta m_2 + \phi_2 \, \Delta m_2$$

and therefore

$$\frac{v_2}{v_1} = \frac{1 + \phi_1}{1 + \phi_2} \simeq 1 + \phi_1 - \phi_2$$

in agreement with our previous result. (Also, it makes no difference whether the photon frequencies are measured in locally inertial systems, because the gravitational field in any other frame will affect the rate of the observer's standard clock in the same way as it affects the v's.) This result can be interpreted as saying that a photon in a gravitational field has "kinetic energy" hv and "potential energy" $hv\phi$, their sum remaining constant. However, I have insisted on including a non-relativistic emitter and absorber in the above calculation, because the concept of gravitational potential energy for a photon is otherwise without foundation.

This derivation rests on the Principle of Equivalence in *two* respects: It assumes that the change in gravitational mass of the apparatus equals the change in its inertial mass and hence its internal energy; and it also assumes that in a freely falling frame the relation between photon energy and frequency is unaffected by the presence of gravitational fields. Hence even if we suppose that the Eötvös-Dicke experiments could improve to an unlimited accuracy, and that gravitational mass were found to equal inertial mass exactly, still there would be some point in verifying the gravitational red shift of spectral lines, as an independent test of the Principle of Equivalence.

6 Signs of the Times

The relation between the Minkowski metric $\eta_{\alpha\beta}$ and the metric tensor $g_{\mu\nu}$ of the theory of gravitation may be expressed in a matrix notation

$$g = D^T \eta D \qquad (3.6.1)$$

where g is for the purposes of this section a 4×4 matrix (not a determinant) whose elements are the $g_{\mu\nu}$, η is a matrix whose elements are the $\eta_{\alpha\beta}$, and D is the matrix

$$D_{\alpha\mu} \equiv \frac{\partial \xi^\alpha}{\partial x^\mu} \qquad (3.6.2)$$

with D^T its transpose

$$D^T_{\mu\alpha} \equiv D_{\alpha\mu}$$

It has been tacitly assumed as part of the Principle of Equivalence that the transformation from laboratory coordinates x^μ to locally inertial coordinates ξ^α is nonsingular; that is, ξ^α is a differentiable function of x^μ and x^μ is a differentiable function of ξ^α. It follows that there exists a matrix

$$D_{\mu\alpha}^{-1} \equiv \frac{\partial x^\mu}{\partial \xi^\alpha} \tag{3.6.3}$$

which is reciprocal to D, that is,

$$(D^{-1}D)_{\mu\nu} = \frac{\partial x^\mu}{\partial \xi^\alpha} \frac{\partial \xi^\alpha}{\partial x^\nu} = \delta_\nu^\mu$$

so that D must have nonvanishing determinant

$$\text{Det } D \neq 0 \tag{3.6.4}$$

A transformation of the form (3.6.1) with D having a nonvanishing determinant is called a *congruence*.

The fact that $g_{\mu\nu}$ is related to $\eta_{\alpha\beta}$ by the congruence (3.6.1) does not mean that the eigenvalues of $g_{\mu\nu}$ are the same as those of $\eta_{\alpha\beta}$, as would be the case if this were a similarity transformation. (Indeed, there are *no* invariant functions of the components of the metric tensor, although there are invariant functions of the $g_{\mu\nu}$ *and* their derivatives, as shown in Chapter 6.) However, there is a theorem known as *Sylvester's law of inertia*[9] that states that the *numbers* of eigenvalues that are respectively positive, negative, or zero do not change under such a congruence. We conclude then that the metric tensor $g_{\mu\nu}$ must like $\eta_{\alpha\beta}$ have three positive eigenvalues, one negative eigenvalue, and no zero eigenvalues. It is this property of the metric that distinguishes our familiar $(3 + 1)$-dimensional space-time from 4-dimensional space, or $(2 + 2)$-dimensional space-time, or worse.

7   Relativity and Anisotropy of Inertia

We have already seen in Section 1.3 that Newton and Mach came to different conclusions about the origin of inertia. Newton believed that inertial forces, such as centrifugal force, must arise from acceleration with respect to "absolute space," whereas Mach argued that they were more likely caused by acceleration with respect to the mass of the celestial bodies. The distinction is not one of metaphysics but of physics, for if Mach were right then a large mass could produce small changes in the inertial forces observed in its vicinity, whereas if Newton were right then no such effect could occur.

Einstein considered himself a follower of Mach, but in fact the answer given by the equivalence principle to the problem of inertia lies somewhere between that of Newton and Mach. The inertial frames, that is, the "freely falling coordinate systems," are indeed determined by the local gravitational field, which arises from all the matter of the universe, far and near. However, once in an inertial frame, the laws of motion [such as Eq. (2.3.1)] are completely unaffected by the presence of nearby masses, either gravitationally or in any other way. For instance, the mass of the sun determines the motion of the freely falling earth, but once we fix our coordinate frame to the earth we cannot detect the gravitational field of the sun, as shown with great accuracy by the experiment of Dicke. (See Section 1.2. Actually, the fact that the earth is not an infinitesimal neighborhood means that we can detect the sun's field through tidal effects, as already discussed in Section 3.1.) The celestial bodies play a role here because the gravitational field equations for $g_{\mu\nu}$ need boundary conditions at infinity, and these are provided by the requirement that at great distances from the sun $g_{\mu\nu}$ merge with the cosmic gravitational field produced by all the mass in the universe. We are not yet ready to go into the details of the field equations and cosmology, but we can anticipate that the gravitational field determined by the mass of the sun and these cosmic boundary conditions is such that planetary orbits far from the sun do not precess with respect to the typical stars, in agreement with observation. (See Section 15.1.)

These points are so important that they are worth repeating. In the absence of nearby matter, the inertial frames are determined by the mean cosmic gravitational field, which is in turn determined by the mean mass density of the stars, so it is not surprising that their inertial frames are at rest, or in a state of uniform non-rotating motion, with respect to the typical stars. When a large mass like the sun is brought close, it changes the inertial frames so that they accelerate toward the mass, but the laws of motion in these freely falling frames are still those of special relativity, and show no effects of the surrounding mass distribution. In this sense, the equivalence principle and Mach's principle are in direct opposition.

The issue between Mach and Einstein can be drawn by asking whether in fact the presence of large nearby masses does affect the laws of motion, other than by determining the inertial frames? Cocconi and Salpeter pointed out[10] that there is a large mass near us, the Milky Way galaxy, and that Mach's principle would suggest slight differences in inertial mass when a particle is accelerated toward or away from the galactic center. This was checked experimentally by Hughes, Robinson, and Beltran-Lopez,[11] and in a similar experiment, by Drever.[12] (See Figure 3.1.) Hughes et al. observed the resonant absorption of photons by a Li^7 nucleus in a 4700 Gauss magnetic field. The ground state has spin 3/2, so it splits in a magnetic field into four energy levels, which should be equally spaced if the laws of nuclear physics are rotationally invariant. In this case the three transitions among neighboring states should have the same energy and the photon absorption coefficient should show a single sharp peak at this energy. However, if inertia were anisotropic then the four magnetic substates would not be exactly equally spaced, and there would be not one but three closely spaced resonance lines. Hughes et al. found that no such splitting greater than the line width of 5.3×10^{-21} MeV occurred over a 12-hr period, during which the rotation of the earth carried the magnetic field from 22° toward the galactic center to 104° away from the galactic center. If we think of the Li^7 nucleus as a single proton with angular momentum

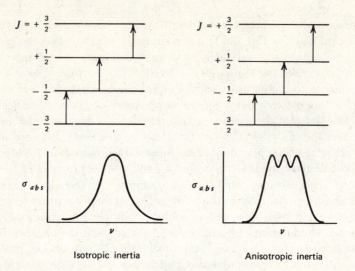

Isotropic inertia   *Anisotropic inertia*

Figure 3.1   The Li^7 absorption spectrum as a test of the isotropy of inertia (frequency differences and line splitting greatly exaggerated).

3/2 bound by a central potential to the other nucleons, then the anisotropy Δm in the proton mass must be such that

$$\Delta \left(\frac{p^2}{2m} \right) \simeq \frac{\Delta m}{m} \left(\frac{p^2}{2m} \right) \leq 5.3 \times 10^{-21} \text{ MeV}$$

where $p^2/2m$ is the proton kinetic energy. Since $p^2/2m$ is greater than $\frac{1}{2}$ MeV, we can conclude that the anisotropy in inertial mass is subject to the inequality

$$\frac{\Delta m}{m} \lesssim 10^{-20}$$

At least in this regard, the evidence strongly favors the equivalence principle rather than Mach's principle.

3 BIBLIOGRAPHY

Assorted Books on General Relativity

□ R. Adler, M. Bazin, M. Schiffer, *Introduction to General Relativity* (McGraw-Hill, New York, 1965).

☐ J. L. Anderson, *Principles of Relativity Physics* (Academic Press, New York, 1967).

☐ P. G. Bergmann, *Introduction to the Theory of Relativity* (Prentice-Hall, Englewood Cliffs, N.J., 1942).

☐ A. S. Eddington, *The Mathematical Theory of Relativity* (Cambridge University Press, Cambridge, 1960).

☐ A. Einstein, *The Meaning of Relativity* (Princeton University Press, Princeton, N. J., 1946).

☐ V. Fock, *The Theory of Space, Time, and Gravitation*, trans. by N. N. Kemmer (2nd rev. ed., Macmillan, New York, 1964).

☐ C. Møller, *The Theory of Relativity* (Clarendon Press, Oxford, 1952).

☐ W. Pauli, *Theory of Relativity*, trans. by G. Field (Pergamon Press, Oxford, 1958).

☐ E. Schroedinger, *Space-Time Structure* (Cambridge University Press, Cambridge, 1950).

☐ J. L. Synge, *Relativity: The General Theory* (Interscience Publishers, New York, 1960).

☐ H. Weyl, *Space-Time-Matter*, trans. by H. L. Brose (Dover Publications, New York, 1952).

On the experimental tests of the Principle of Equivalence, see the bibliography to Chapter 8, especially the article by Dicke.

3 REFERENCES

1. A. H. Wapstra and G. J. Nijgh, Physica, **21**, 796 (1955).

2. M. G. Adam, Mon. Nat. Roy. Astron. Soc., **119,** 460 (1959). For a review and earlier references, see B. Bertotti, D. Brill, and R. Krotkov, in *Gravitation*, ed. by L. Witten (Wiley, New York, 1962), pp. 23–27.

3. J. Brault, Bull. Am. Phys. Soc., **8**, 28 (1963). Also see J. E. Blamont and F. Roddier, Phys. Rev. Letters, **7**, 437 (1961).

4. M. Schwarzschild, *Structure and Evolution of the Stars* (Princeton University Press, Princeton, N. J., 1958), Chapter VII.

5. D. M. Popper, Astrophys. J., **120**, 316 (1954). For other white dwarfs, see J. L. Greenstein and V. Trimble, Ap. J., **149**, 283 (1967).

5a. W. L. Wiese and D. E. Kelleher, Ap. J., **166**, L59 (1971).

6. R. V. Pound and G. A. Rebka, Phys. Rev. Letters, **4**, 337 (1960); for their original proposal, see Phys. Rev. Letters, **3**, 439 (1959).

7. R. L. Mössbauer, Z. Physik, **151**, 124 (1958); Naturwissenschaften, **45,** 538 (1958); Z. Naturforsch, **14a**, 211 (1959).

8. R. V. Pound and J. L. Snider, Phys. Rev. Letters, **13**, 539 (1964).

8*a*. D. Kleppner, N. F. Ramsey, and R. F. C. Vessot, Astrophys. Space Sci., **6**, 13 (1970).

9. See, for example, H. W. Turnbull and A. C. Aitken, *An Introduction to the Theory of Canonical Matrices* (Dover Publications, New York, 1961), p. 89.

10. G. Cocconi and E. E. Salpeter, Phys. Rev. Letters, **4**, 176 (1960). However, see R. H. Dicke, Phys. Rev. Letters, **7**, 359 (1961).

11. V. W. Hughes, H. G. Robinson, and V. Beltran-Lopez, Phys. Rev. Letters, **4**, 342 (1960).

12. R. W. P. Drever, Phil. Mag., **6**, 683 (1961).

13. Spectroscopic studies of Sirius B have very recently yielded an estimate $(2.8 \pm 0.1) \times 10^{-4}$ for the dimensionless surface gravitational potential, and a value $(3.0 \pm 0.5) \times 10^{-4}$ for the fractional red shift; see J. L. Greenstein, J. B. Oke, and H. L. Shipman, Ap. J. **169**, 563 (1971).

4 TENSOR ANALYSIS

We have already noticed that the Principle of Equivalence of Gravitation and Inertia establishes a deep analogy between non-Euclidean geometry and the theory of gravitation. This chapter is devoted to an outline of the language common to both, that of tensor analysis.

1 The Principle of General Covariance

In the last chapter we demonstrated one way of using the Principle of Equivalence to assess the effects of gravitation on physical systems: We wrote down the equations that hold, for general gravitational fields, in locally inertial coordinate systems (i.e., the equations of special relativity, such as $d^2\xi^\alpha/d\tau^2 = 0$) and then performed a coordinate transformation to find the corresponding equations in the laboratory coordinate system. We could continue to follow this approach, but it would lead us into very tedious calculations when we come to the field equations for electromagnetism and gravitation. Instead, we shall follow a different method, one that is of precisely the same physical content, but is much more elegant in appearance and convenient in execution. This method is based on an alternative version of the Principle of Equivalence, known as the *Principle of General Covariance*. It states that a physical equation holds in a general gravitational field, if two conditions are met:

1. The equation holds in the absence of gravitation; that is, it agrees with the laws of special relativity when the metric tensor $g_{\alpha\beta}$ equals the Minkowski tensor $\eta_{\alpha\beta}$ and when the affine connection $\Gamma^{\alpha}_{\beta\gamma}$ vanishes.

2. The equation is generally covariant; that is, it preserves its form under a general coordinate transformation $x \to x'$.

To see that the Principle of General Covariance follows from the Principle of Equivalence, let us suppose that we are in an arbitrary gravitational field, and consider any equation that satisfies the two above conditions. From (2), we learn that the equation will be true in all coordinate systems if it is true in any one coordinate system. But at any given point there is a class of coordinate systems, the locally inertial systems, in which the effects of gravitation are absent. Condition (1) then tells us that our equation holds in these systems, and hence in all other coordinate systems.

It should be stressed that general covariance by itself is empty of physical content.[1] Any equation can be *made* generally covariant by writing it in any one coordinate system, and then working out what it looks like in other arbitrary coordinate systems. Indeed, from childhood we have become familiar with the appearance of physical equations in non-Cartesian systems, such as polar coordinates, and in noninertial systems, such as rotating coordinates. The significance of the Principle of General Covariance lies in its statement about the effects of gravitation, that a physical equation by virtue of its general covariance will be true in a gravitational field if it is true in the absence of gravitation.

The meaning of general covariance can be brought forward by comparing it with Lorentz invariance. Just as any equation can be made generally covariant, so any equation can be made Lorentz-invariant, by writing it in one coordinate system and then working out what it looks like after a Lorentz transformation. However, if we do this with a nonrelativistic equation like Newton's second law, we find after making it Lorentz-invariant that a new quantity has entered the equation, which of course is the velocity of the coordinate frame with respect to the original reference frame. The requirement that this velocity *not* appear in the transformed equation is what we call the Principle of Special Relativity, or "Lorentz invariance" for short, and this requirement places very powerful restrictions on the original equation. Similarly, when we make an equation generally covariant, new ingredients will enter, that is, the metric tensor $g_{\mu\nu}$ and the affine connection $\Gamma^{\lambda}_{\mu\nu}$. The difference is that we do not require that these quantities drop out at the end, and hence we do not obtain any restrictions on the equation we start with; rather, we exploit the presence of $g_{\mu\nu}$ and $\Gamma^{\lambda}_{\mu\nu}$ to represent gravitational fields. To put this briefly: The Principle of General Covariance is *not* an invariance principle, like the Principle of Galilean or Special Relativity, but is instead a statement about the effects of gravitation, and about nothing else. In particular, general covariance does not imply Lorentz invariance—there are generally covariant theories of gravitation that allow the construction of inertial frames at any point in a gravitational field, but that satisfy Galilean relativity rather than special relativity in these frames.[1a]

Any physical principle, such as general covariance, which takes the form of an invariance principle but whose content is actually limited to a restriction on the interactions of one particular field, is called a *dynamic symmetry*.[2] There are other dynamic symmetries of importance in physics, such as local gauge invariance, which governs the interactions of the electromagnetic field, and chiral symmetry,[3] which governs the interactions of the pi-meson field. We shall return to the analogy between general relativity and electrodynamics several times in following chapters.

The Principle of General Covariance can only be applied on a scale that is small compared with the space-time distances typical of the gravitational field, for it is only on this small scale that we are assured by the Principle of Equivalence of being able to construct a coordinate system in which the effects of gravitation are absent. For instance, the radius of the moon is not so very much smaller than the earth-moon separation, so we cannot accurately calculate the motion of the moon by finding generally covariant equations that reduce to the correct equations for a freely moving moon in the absence of gravitation. We can, however, treat the moon as a ball of rock and calculate its motion by applying the Principle of General Covariance to determine the gravitational force on each infinitesimal element of the lunar mass.

There are in general many generally covariant equations that reduce to a given special-relativistic equation in the absence of gravitation. However, because we only apply the Principle of General Covariance on a small scale compared with the scale of the gravitational field, we usually expect that it is only $g_{\mu\nu}$ and its first derivatives that enter our generally covariant equations. With this understanding we shall see in this and the next chapter that the Principle of General Covariance makes an unambiguous statement about the effects of gravitational fields on any system, or part of a system, that is sufficiently small.

2   Vectors and Tensors

In order to construct physical equations that are invariant under general coordinate transformation, we must know how the quantities described by the equations behave under these transformations. For some quantities, those defined directly in terms of coordinate differentials, the transformation properties may be determined by a straightforward calculation. For other quantities, such as the electromagnetic fields, the transformation properties are partially a matter of definition. However, there is a tendency for all quantities of physical interest to transform in a reasonably simple way, for otherwise it would be difficult to put them together to form invariant equations. In this section we describe one class of objects whose transformation properties are particularly simple, giving examples (where we can) from quantities defined directly in terms of the coordinate system.

The simplest of all transformation rules is that of the *scalars*, which simply do not change under general coordinate transformations. The obvious example is a pure number, like 137 or π or zero. Another example is the proper time $d\tau$, given by Eq. (3.2.6); in fact, we shall see below that the metric tensor $g_{\mu\nu}$ is defined to transform in just such a way as to keep $d\tau^2$ invariant.

The next simplest transformation rule is that of a *contravariant vector*, V^μ, which under a coordinate transformation $x^\mu \to x'^\mu$ transforms into

$$V'^\mu = V^\nu \frac{\partial x'^\mu}{\partial x^\nu} \tag{4.2.1}$$

For instance, the rules of partial differentiation give

$$dx'^\mu = \frac{\partial x'^\mu}{\partial x^\nu} dx^\nu \tag{4.2.2}$$

so the coordinate differential is a contravariant vector. A very closely related transformation rule is that of a *covariant vector* U_μ, which under a coordinate transformation $x^\mu \to x'^\mu$ transforms into

$$U'_\mu = \frac{\partial x^\nu}{\partial x'^\mu} U_\nu \tag{4.2.3}$$

For instance, if ϕ is a scalar field, then $\partial\phi/\partial x^\mu$ is a covariant vector, because in a transformed coordinate system the gradient is

$$\frac{\partial \phi}{\partial x'^\mu} = \frac{\partial x^\nu}{\partial x'_\mu} \frac{\partial \phi}{\partial x^\nu} \tag{4.2.4}$$

in agreement with (4.2.3).

From the contravariant and covariant vectors we can immediately generalize to the *tensors*. A tensor with upper indices $\mu, \nu, \ldots$ and lower indices $\kappa, \lambda, \ldots$ transforms like the product of contravariant vectors $U^\mu W^\nu \cdots$ and covariant vectors $V_\kappa Y_\lambda \cdots$. For instance, under a coordinate transformation $x \to x'$ a tensor $T^\mu{}_\nu{}^\lambda$ will transform into

$$T'^\mu{}_\nu{}^\lambda = \frac{\partial x'^\mu}{\partial x^\kappa} \frac{\partial x^\rho}{\partial x'^\nu} \frac{\partial x'^\lambda}{\partial x^\sigma} T^\kappa{}_\rho{}^\sigma \tag{4.2.5}$$

If all indices are upstairs the tensor is called contravariant; if all indices are downstairs the tensor is called covariant; otherwise it is called mixed. The most important example is the metric tensor, defined in Section 3.2 for a general coordinate system x^μ by

$$g_{\mu\nu} \equiv \eta_{\alpha\beta} \frac{\partial \xi^\alpha}{\partial x^\mu} \frac{\partial \xi^\beta}{\partial x^\nu}$$

where ξ^α is a locally inertial coordinate system. In a different coordinate system x'^μ the metric tensor is

$$g'_{\mu\nu} = \eta_{\alpha\beta} \frac{\partial \xi^\alpha}{\partial x'^\mu} \frac{\partial \xi^\beta}{\partial x'^\nu}$$

$$= \eta_{\alpha\beta} \frac{\partial \xi^\alpha}{\partial x^\rho} \frac{\partial x^\rho}{\partial x'^\mu} \frac{\partial \xi^\beta}{\partial x^\sigma} \frac{\partial x^\sigma}{\partial x'^\nu}$$

and therefore

$$g'_{\mu\nu} = g_{\rho\sigma} \frac{\partial x^\rho}{\partial x'^\mu} \frac{\partial x^\sigma}{\partial x'^\nu} \tag{4.2.6}$$

We see that $g_{\mu\nu}$ is indeed a covariant tensor. Its inverse is a contravariant tensor, for if we define $g^{\lambda\mu}$ so that

$$g^{\lambda\mu} g_{\mu\nu} = \delta^\lambda_\nu$$

we shall have

$$\frac{\partial x'^\lambda}{\partial x^\rho} \frac{\partial x'^\mu}{\partial x^\sigma} g^{\rho\sigma} g'_{\mu\nu} = \frac{\partial x'^\lambda}{\partial x^\rho} \frac{\partial x'^\mu}{\partial x^\sigma} g^{\rho\sigma} \frac{\partial x^\kappa}{\partial x'^\mu} \frac{\partial x^\eta}{\partial x'^\nu} g_{\kappa\eta}$$

$$= \frac{\partial x'^\lambda}{\partial x^\rho} g^{\rho\kappa} \frac{\partial x^\eta}{\partial x'^\nu} g_{\kappa\eta} = \frac{\partial x'^\lambda}{\partial x^\rho} \frac{\partial x^\rho}{\partial x'^\nu} = \delta^\lambda_\nu$$

and therefore

$$\frac{\partial x'^\lambda}{\partial x^\rho} \frac{\partial x'^\mu}{\partial x^\sigma} g^{\rho\sigma} = g'^{\lambda\mu} \tag{4.2.7}$$

as required for a contravariant tensor. Finally, the Kronecker symbol δ^μ_ν is a mixed tensor, because

$$\delta^\mu_\nu \frac{\partial x'^\rho}{\partial x^\mu} \frac{\partial x^\nu}{\partial x'^\sigma} = \frac{\partial x'^\rho}{\partial x^\mu} \frac{\partial x^\mu}{\partial x'^\sigma} = \delta^\rho_\sigma \tag{4.2.8}$$

Aside from the scalars and zero, δ^μ_ν (together with its direct products) is the only tensor whose components are the same in all coordinate systems.

A vector is just a tensor with one index and a scalar is a tensor with no indices, so it will not generally be necessary to give the scalars and vectors special treatment in the following. However, the reader should be warned that not everything is a tensor; in particular, the affine connection $\Gamma^\nu_{\mu\lambda}$, despite its appearance, is *not* a tensor.

We can now recognize one very large class of invariant equations: Any equation will be invariant under general coordinate transformations if it states the equality of two tensors with the same upper and lower indices. For instance, if $A^\mu{}_\nu{}^\lambda$ and $B^\mu{}_\nu{}^\lambda$ are two tensors with the transformation rule (4.2.5), and if in the x^μ coordinate system $A^\mu{}_\nu{}^\lambda = B^\mu{}_\nu{}^\lambda$, then obviously in the x'^μ coordinate system

$A'^{\mu}{}_{\nu}{}^{\lambda} = B'^{\mu}{}_{\nu}{}^{\lambda}$. In particular, since zero is any kind of tensor we want, a statement that a given tensor vanishes is invariant under general coordinate transformations. In contrast, a statement that is not an equality between tensors of the same kind (for instance, $T^{\mu\nu} = 5$ or $V^{\mu} = U_{\mu}$) may be numerically true in a limited class of coordinate systems, but not in all coordinate systems.

3   Tensor Algebra

The next step in our program of constructing equations invariant under general coordinate transformations is to learn how to put tensors together to form other tensors. This is accomplished through a few simple algebraic operations:

(A) **Linear Combinations.** A linear combination of tensors with the same upper and lower indices is a tensor with these indices. For instance, let $A^{\mu}{}_{\nu}$ and $B^{\mu}{}_{\nu}$ be mixed tensors, and let

$$T^{\mu}{}_{\nu} \equiv aA^{\mu}{}_{\nu} + bB^{\mu}{}_{\nu}$$

where a and b are scalars; then $T^{\mu}{}_{\nu}$ is a tensor, because

$$T'^{\mu}{}_{\nu} \equiv aA'^{\mu}{}_{\nu} + bB'^{\mu}{}_{\nu}$$

$$= a\, \frac{\partial x'^{\mu}}{\partial x^{\rho}} \frac{\partial x^{\sigma}}{\partial x'^{\nu}} A^{\rho}{}_{\sigma} + b\, \frac{\partial x'^{\mu}}{\partial x^{\rho}} \frac{\partial x^{\sigma}}{\partial x'^{\nu}} B^{\rho}{}_{\sigma}$$

$$= \frac{\partial x'^{\mu}}{\partial x^{\rho}} \frac{\partial x^{\sigma}}{\partial x'^{\nu}} T^{\rho}{}_{\sigma}$$

(B) **Direct Products.** The product of the components of two tensors yields a tensor whose upper and lower indices consist of all the upper and lower indices of the two original tensors. For instance, if $A^{\mu}{}_{\nu}$ and B^{ρ} are tensors, and

$$T^{\mu}{}_{\nu}{}^{\rho} \equiv A^{\mu}{}_{\nu}B^{\rho}$$

then $T^{\mu}{}_{\nu}{}^{\rho}$ is a tensor, that is,

$$T'^{\mu}{}_{\nu}{}^{\rho} \equiv A'^{\mu}{}_{\nu}B'^{\rho}$$

$$= \frac{\partial x'^{\mu}}{\partial x^{\lambda}} \frac{\partial x^{\kappa}}{\partial x'^{\nu}} A^{\lambda}{}_{\kappa} \frac{\partial x'^{\rho}}{\partial x^{\sigma}} B^{\sigma}$$

$$= \frac{\partial x'^{\mu}}{\partial x^{\lambda}} \frac{\partial x^{\kappa}}{\partial x'^{\nu}} \frac{\partial x'^{\rho}}{\partial x^{\sigma}} T^{\lambda}{}_{\kappa}{}^{\sigma}$$

(C) **Contraction.** Setting an upper and lower index equal and summing it over its four values yields a new tensor with these two indices absent. For instance, if $T^{\mu}{}_{\nu}{}^{\rho\sigma}$ is a tensor and

$$T^{\mu\rho} \equiv T^{\mu}{}_{\nu}{}^{\rho\nu}$$

then $T'^{\mu\rho}$ is a tensor, that is,

$$
\begin{aligned}
T'^{\mu\rho} &= T'^{\mu}{}_{\nu}{}^{\rho\nu} \\[6pt]
&= \frac{\partial x'^{\mu}}{\partial x^{\kappa}}\frac{\partial x^{\lambda}}{\partial x'^{\nu}}\frac{\partial x'^{\rho}}{\partial x^{\eta}}\frac{\partial x'^{\nu}}{\partial x^{\tau}}\, T^{\kappa}{}_{\lambda}{}^{\eta\tau} \\[6pt]
&= \frac{\partial x'^{\mu}}{\partial x^{\kappa}}\frac{\partial x'^{\rho}}{\partial x^{\eta}}\, T^{\kappa}{}_{\lambda}{}^{\eta\lambda} \\[6pt]
&= \frac{\partial x'^{\mu}}{\partial x^{\kappa}}\frac{\partial x'^{\rho}}{\partial x^{\eta}}\, T^{\kappa\eta}
\end{aligned}
$$

These three operations can of course be combined in various ways. One particularly important combined operation results in the *raising and lowering* of indices. If we take the direct product of a contravariant or mixed tensor T with the metric tensor $g_{\mu\nu}$, and contract the index μ with one of the contravariant indices of T, we get a new tensor in which this contravariant index is replaced by a covariant index ν. For instance, if $T^{\mu\rho}{}_{\sigma}$ is a tensor, and we define

$$
S_{\nu}{}^{\rho}{}_{\sigma} \equiv g_{\mu\nu} T^{\mu\rho}{}_{\sigma}
$$

then by rules (B) and (C), $S_{\nu}{}^{\rho}{}_{\sigma}$ will be a tensor. Similarly, if we take the direct product of a covariant or mixed tensor T with the inverse metric tensor $g^{\mu\nu}$, and contract the index μ with one of the covariant indices of T, we get a new tensor in which this covariant index is replaced by a contravariant index ν. For instance, if $S_{\mu}{}^{\rho}{}_{\sigma}$ is a tensor, and we define

$$
R^{\nu\rho}{}_{\sigma} \equiv g^{\mu\nu} S_{\mu}{}^{\rho}{}_{\sigma}
$$

then $R^{\nu\rho}{}_{\sigma}$ is also a tensor. Note that lowering an index and then raising it again gives back the original tensor; for instance, in the examples cited above, we lowered an index on T to get S and then raised it again to get R, so $R = T$, because

$$
\begin{aligned}
R^{\nu\rho}{}_{\sigma} &\equiv g^{\mu\nu} S_{\mu}{}^{\rho}{}_{\sigma} \equiv g^{\mu\nu} g_{\mu\lambda} T^{\lambda\rho}{}_{\sigma} \\[6pt]
&= \delta^{\nu}{}_{\lambda} T^{\lambda\rho}{}_{\sigma} = T^{\nu\rho}{}_{\sigma}
\end{aligned}
$$

By raising and lowering indices we can write a tensor with N indices in 2^{N} different ways. Since these are all physically equivalent, it is customary to use the same symbol for all 2^{N} tensors, distinguishing them only by their index locations.

For the sake of completeness, it should be mentioned that the tensor obtained by raising one index on the metric tensor $g_{\mu\nu}$ or by lowering one index on the inverse metric tensor $g^{\mu\nu}$, is precisely the Kronecker tensor, because

$$
g^{\mu\lambda} g_{\lambda\nu} = \delta^{\mu}{}_{\nu}
$$

Also, raising both indices on $g_{\mu\nu}$ gives the inverse tensor

$$g^{\lambda\mu}g^{\kappa\nu}g_{\mu\nu} = g^{\lambda\mu}\delta^{\kappa}{}_{\mu} = g^{\lambda\kappa}$$

and lowering both indices on $g^{\lambda\kappa}$ gives the metric tensor $g_{\mu\nu}$.

The reader will have noticed that this discussion of tensor algebra is precisely the same as the corresponding discussion in the chapter on special relativity (see Section 2.5) with one important exception: I have not yet mentioned differentiation. This is because the derivative of a tensor is in general not a tensor. In Section 6 we shall see that there is a kind of differentiation, called covariant differentiation, that provides one more way of constructing tensors from other tensors.

4   Tensor Densities

Despite the ubiquity of tensors, there is nothing sacred about the tensor transformation law. One very important example of a nontensor is the determinant of the metric tensor

$$g \equiv -\,\text{Det}\, g_{\mu\nu} \qquad\qquad (4.4.1)$$

The transformation rule for the metric tensor can be regarded as a matrix equation

$$g'_{\mu\nu} = \frac{\partial x^{\rho}}{\partial x'^{\mu}}\, g_{\rho\sigma}\, \frac{\partial x^{\sigma}}{\partial x'^{\nu}}$$

and taking the determinant, we find that

$$g' = \left|\frac{\partial x}{\partial x'}\right|^{2} g \qquad\qquad (4.4.2)$$

where $|\partial x/\partial x'|$ is the *Jacobian* of the transformation $x' \to x$; that is, it is the determinant of the matrix $\partial x^{\rho}/\partial x'^{\mu}$. A quantity such as g, which transforms like a scalar except for extra factors of the Jacobian, is called a *scalar density*, and similarly a quantity that transforms as a tensor except for extra factors of the Jacobian determinant is called a *tensor density*. The number of factors of the determinant $|\partial x'/\partial x|$ is called the *weight* of the density; for instance, we see from (4.4.2) that g is a density of weight -2, because

$$\left|\frac{\partial x}{\partial x'}\right| = \left|\frac{\partial x'}{\partial x}\right|^{-1} \qquad\qquad (4.4.3)$$

as can be seen by taking the determinant of the equation

$$\frac{\partial x^{\mu}}{\partial x'^{\lambda}}\, \frac{\partial x'^{\lambda}}{\partial x^{\nu}} = \delta^{\mu}{}_{\nu}$$

Any tensor density of weight W can be expressed as an ordinary tensor times a factor $g^{-W/2}$. For instance, a tensor density $\mathscr{I}^\mu{}_\nu$ of rank W has the transformation rule

$$\mathscr{I}'^\mu{}_\nu = \left|\frac{\partial x'}{\partial x}\right|^W \frac{\partial x'^\mu}{\partial x^\lambda}\frac{\partial x^\kappa}{\partial x'^\nu}\mathscr{I}^\lambda{}_\kappa \tag{4.4.4}$$

and using (4.4.2), we see that

$$g'^{W/2}\mathscr{I}'^\mu{}_\nu = \frac{\partial x'^\mu}{\partial x^\lambda}\frac{\partial x^\kappa}{\partial x'^\nu}g^{W/2}\mathscr{I}^\lambda{}_\kappa \tag{4.4.5}$$

The importance of tensor densities arises from the fundamental theorem of integral calculus,[4] that under a general coordinate transformation $x \to x'$, the volume element d^4x becomes

$$d^4x' = \left\|\left|\frac{\partial x'}{\partial x}\right|\right\| d^4x \tag{4.4.6}$$

Hence the product of d^4x with a tensor density of weight -1 transforms like an ordinary tensor. In particular, $\sqrt{g}\, d^4x$ is an invariant volume element.

There is one tensor density whose components are the same in all coordinate systems; it is the Levi-Civita tensor density $\varepsilon^{\mu\nu\lambda\kappa}$. To define this quantity in a general coordinate system, we must arbitrarily order the coordinate indices in a reference sequence, for example, x, y, z, t or r, θ, φ, t, and so on. Then $\varepsilon^{\mu\nu\lambda\kappa}$ is defined by

$$\varepsilon^{\mu\nu\lambda\kappa} = \begin{cases} +1 & \mu\nu\lambda\kappa \text{ even permutation of reference sequence} \\ -1 & \mu\nu\lambda\kappa \text{ odd permutation of reference sequence} \\ 0 & \text{some indices equal} \end{cases} \tag{4.4.7}$$

To see that this is a tensor density, consider the quantity

$$\frac{\partial x'^\rho}{\partial x^\mu}\frac{\partial x'^\sigma}{\partial x^\nu}\frac{\partial x'^\eta}{\partial x^\lambda}\frac{\partial x'^\xi}{\partial x^\kappa}\varepsilon^{\mu\nu\lambda\kappa} \tag{4.4.8}$$

We note that this is completely antisymmetric in the indices ρ, σ, η, ξ and therefore proportional to $\varepsilon^{\rho\sigma\eta\xi}$. To determine the proportionality constant, let $\rho\sigma\eta\xi$ take the values of the reference sequence; then (4.4.8) is just the determinant $|\partial x'/\partial x|$, and so

$$\frac{\partial x'^\rho}{\partial x^\mu}\frac{\partial x'^\sigma}{\partial x^\nu}\frac{\partial x'^\eta}{\partial x^\lambda}\frac{\partial x'^\xi}{\partial x^\kappa}\varepsilon^{\mu\nu\lambda\kappa} = \left|\frac{\partial x'}{\partial x}\right|\varepsilon^{\rho\sigma\eta\xi} \tag{4.4.9}$$

Thus $\varepsilon^{\mu\nu\lambda\kappa}$ is a tensor density of weight -1. We can form an ordinary contravariant tensor by multiplying $\varepsilon^{\mu\nu\lambda\kappa}$ by $g^{-1/2}$. We can also form a covariant density by lowering indices in the usual way, that is,

$$\varepsilon_{\rho\sigma\eta\xi} \equiv g_{\rho\mu}g_{\sigma\nu}g_{\eta\lambda}g_{\xi\kappa}\varepsilon^{\mu\nu\lambda\kappa} \tag{4.4.10}$$

This is antisymmetric in its indices, and therefore proportional to $\varepsilon^{\rho\sigma\eta\xi}$. By setting $\rho\sigma\eta\xi$ equal to the reference sequence, we find that the proportionality constant must be $-g$, so

$$\varepsilon_{\rho\sigma\eta\xi} = -g\varepsilon^{\rho\sigma\eta\xi} \tag{4.4.11}$$

The reader may easily verify that $\varepsilon_{\rho\sigma\eta\xi}$ is a covariant tensor density of weight -1.

The rules of tensor algebra may be easily extended to encompass tensor densities.

(A) The linear combination of two tensor densities of the *same weight* W is a tensor density of weight W.

(B) The direct product of two tensor densities of weight W_1, W_2 yields a tensor density of weight $W_1 + W_2$.

(C) The contraction of indices on a tensor density of weight W yields a tensor density of weight W. From (B) and (C) it follows that raising and lowering indices does not change the weight of a tensor density.

5   Transformation of the Affine Connection

Apart from the rather trivial example of the tensor densities, there appears throughout the laws of physics one other very important nontensor, the affine connection. We recall its definition,

$$\Gamma^\lambda_{\mu\nu} = \frac{\partial x^\lambda}{\partial \xi^\alpha} \frac{\partial^2 \xi^\alpha}{\partial x^\mu \, \partial x^\nu} \tag{4.5.1}$$

where $\xi^\alpha(x)$ is the locally inertial coordinate system. Passing from x^μ to a different system x'^μ, we find that

$$\begin{aligned}
\Gamma'^\lambda_{\mu\nu} &\equiv \frac{\partial x'^\lambda}{\partial \xi^\alpha} \frac{\partial^2 \xi^\alpha}{\partial x'^\mu \, \partial x'^\nu} \\
&= \frac{\partial x'^\lambda}{\partial x^\rho} \frac{\partial x^\rho}{\partial \xi^\alpha} \frac{\partial}{\partial x'^\mu} \left(\frac{\partial x^\sigma}{\partial x'^\nu} \frac{\partial \xi^\alpha}{\partial x^\sigma} \right) \\
&= \frac{\partial x'^\lambda}{\partial x^\rho} \frac{\partial x^\rho}{\partial \xi^\alpha} \left[\frac{\partial x^\sigma}{\partial x'^\nu} \frac{\partial x^\tau}{\partial x'^\mu} \frac{\partial^2 \xi^\alpha}{\partial x^\tau \, \partial x^\sigma} + \frac{\partial^2 x^\sigma}{\partial x'^\mu \, \partial x'^\nu} \frac{\partial \xi^\alpha}{\partial x^\sigma} \right]
\end{aligned}$$

and referring back to Eq. (4.5.1), this is

$$\Gamma'^\lambda_{\mu\nu} = \frac{\partial x'^\lambda}{\partial x^\rho} \frac{\partial x^\tau}{\partial x'^\mu} \frac{\partial x^\sigma}{\partial x'^\nu} \Gamma^\rho_{\tau\sigma} + \frac{\partial x'^\lambda}{\partial x^\rho} \frac{\partial^2 x^\rho}{\partial x'^\mu \, \partial x'^\nu} \tag{4.5.2}$$

The first term on the right is what we would expect if $\Gamma^\lambda_{\mu\nu}$ were a tensor; the second term is inhomogeneous, and makes it a nontensor.

Tensor analysis provides a very simple way of establishing the relation between $\Gamma^\lambda_{\mu\nu}$ and $g_{\mu\nu}$. Note that

$$\frac{\partial}{\partial x'^\kappa} g'_{\mu\nu} = \frac{\partial}{\partial x'^\kappa} \left(g_{\rho\sigma} \frac{\partial x^\rho}{\partial x'^\mu} \frac{\partial x^\sigma}{\partial x'^\nu} \right)$$

$$= \frac{\partial g_{\rho\sigma}}{\partial x^\tau} \frac{\partial x^\tau}{\partial x'^\kappa} \frac{\partial x^\rho}{\partial x'^\mu} \frac{\partial x^\sigma}{\partial x'^\nu} + g_{\rho\sigma} \frac{\partial^2 x^\rho}{\partial x'^\kappa \partial x'^\mu} \frac{\partial x^\sigma}{\partial x'^\nu}$$

$$+ g_{\rho\sigma} \frac{\partial^2 x^\rho}{\partial x'^\kappa \partial x'^\nu} \frac{\partial x^\sigma}{\partial x'^\mu}$$

so

$$\frac{\partial}{\partial x'^\mu} g'_{\kappa\nu} + \frac{\partial}{\partial x'^\nu} g'_{\kappa\mu} - \frac{\partial}{\partial x'^\kappa} g'_{\mu\nu} = \frac{\partial x^\tau}{\partial x'^\kappa} \frac{\partial x^\rho}{\partial x'^\mu} \frac{\partial x^\sigma}{\partial x'^\nu} \left(\frac{\partial g_{\sigma\tau}}{\partial x^\rho} + \frac{\partial g_{\rho\tau}}{\partial x^\sigma} - \frac{\partial g_{\rho\sigma}}{\partial x^\tau} \right)$$

$$+ 2g_{\rho\sigma} \frac{\partial^2 x^\rho}{\partial x'^\mu \partial x'^\nu} \frac{\partial x^\sigma}{\partial x'^\kappa}$$

It follows that

$$\begin{Bmatrix} \lambda \\ \mu\nu \end{Bmatrix}' = \frac{\partial x'^\lambda}{\partial x^\rho} \frac{\partial x^\tau}{\partial x'^\mu} \frac{\partial x^\sigma}{\partial x'^\nu} \begin{Bmatrix} \rho \\ \tau\sigma \end{Bmatrix} + \frac{\partial x'^\lambda}{\partial x^\rho} \frac{\partial^2 x^\rho}{\partial x'^\mu \partial x'^\nu} \tag{4.5.3}$$

where

$$\begin{Bmatrix} \lambda \\ \mu\nu \end{Bmatrix} \equiv \tfrac{1}{2} g^{\lambda\kappa} \left[\frac{\partial g_{\kappa\nu}}{\partial x^\mu} + \frac{\partial g_{\kappa\mu}}{\partial x^\nu} - \frac{\partial g_{\mu\nu}}{\partial x^\kappa} \right] \tag{4.5.4}$$

Subtracting (4.5.3) from (4.5.2), we see that $\Gamma^\lambda_{\mu\nu}$ minus $\begin{Bmatrix} \lambda \\ \mu\nu \end{Bmatrix}$ is a *tensor*,

$$\left[\Gamma^\lambda_{\mu\nu} - \begin{Bmatrix} \lambda \\ \mu\nu \end{Bmatrix} \right]' = \frac{\partial x'^\lambda}{\partial x^\rho} \frac{\partial x^\tau}{\partial x'^\mu} \frac{\partial x^\sigma}{\partial x'^\nu} \left[\Gamma^\rho_{\tau\sigma} - \begin{Bmatrix} \rho \\ \tau\sigma \end{Bmatrix} \right] \tag{4.5.5}$$

The equivalence principle tells us that there is a special coordinate system ξ_X in which, at a given point X, the effects of gravitation are absent. In this system there can be no gravitational force on free particles, so $\Gamma^\lambda_{\mu\nu}$ vanishes, and there can be no gravitational red shift between infinitesimally separated points, so the first derivatives of $g_{\mu\nu}$ vanish. Since $\Gamma^\rho_{\tau\sigma} - \begin{Bmatrix} \rho \\ \tau\sigma \end{Bmatrix}$ vanishes in a locally inertial coordinate system, and since it is a tensor, it must vanish in all coordinate systems, that is,

$$\Gamma^\lambda_{\mu\nu} = \begin{Bmatrix} \lambda \\ \mu\nu \end{Bmatrix} \tag{4.5.6}$$

It is useful to have at hand an alternative formula for the inhomogeneous term in the transformation rule of $\Gamma^\lambda_{\mu\nu}$. Differentiate the identity

$$\frac{\partial x'^\lambda}{\partial x^\rho} \frac{\partial x^\rho}{\partial x'^\nu} = \delta^\lambda_\nu$$

with respect to x'^{μ}; we find immediately that

$$\frac{\partial x'^{\lambda}}{\partial x^{\rho}} \frac{\partial^2 x^{\rho}}{\partial x'^{\mu} \partial x'^{\nu}} = - \frac{\partial x^{\rho}}{\partial x'^{\nu}} \frac{\partial x^{\sigma}}{\partial x'^{\mu}} \frac{\partial^2 x'^{\lambda}}{\partial x^{\rho} \partial x^{\sigma}} \tag{4.5.7}$$

We can therefore write (4.5.2) as

$$\Gamma'^{\lambda}_{\mu\nu} = \frac{\partial x'^{\lambda}}{\partial x^{\rho}} \frac{\partial x^{\tau}}{\partial x'^{\mu}} \frac{\partial x^{\sigma}}{\partial x'^{\nu}} \Gamma^{\rho}_{\tau\sigma} - \frac{\partial x^{\rho}}{\partial x'^{\nu}} \frac{\partial x^{\sigma}}{\partial x'^{\mu}} \frac{\partial^2 x'^{\lambda}}{\partial x^{\rho} \partial x^{\sigma}} \tag{4.5.8}$$

This is just what we would have found by first performing the inverse transformation $x' \to x$, and then solving for $\Gamma'^{\lambda}_{\mu\nu}$.

We are now in a position to use the Principle of General Covariance to give an alternative proof that a freely falling particle obeys the equation of motion

$$\frac{d^2 x^{\mu}}{d\tau^2} + \Gamma^{\mu}_{\nu\lambda} \frac{dx^{\nu}}{d\tau} \frac{dx^{\lambda}}{d\tau} = 0 \tag{4.5.9}$$

where

$$d\tau^2 = -g_{\mu\nu} \, dx^{\mu} \, dx^{\nu} \tag{4.5.10}$$

First, note that Eqs. (4.5.9) and (4.5.10) are true in the absence of gravitation, because setting $\Gamma^{\mu}_{\nu\lambda}$ equal to zero and $g_{\mu\nu}$ equal to $\eta_{\mu\nu}$ gives

$$\frac{d^2 x^{\mu}}{d\tau^2} = 0 \qquad d\tau^2 = -\eta_{\mu\nu} \, dx^{\mu} \, dx^{\nu}$$

and these are the correct equations for a free particle in special relativity. Second, note that (4.5.9) and (4.5.10) are invariant under a general coordinate transformation, for

$$\frac{d^2 x'^{\mu}}{d\tau^2} = \frac{d}{d\tau}\left(\frac{\partial x'^{\mu}}{\partial x^{\nu}} \frac{dx^{\nu}}{d\tau}\right) = \frac{\partial x'^{\mu}}{\partial x^{\nu}} \frac{d^2 x^{\nu}}{d\tau^2} + \frac{\partial^2 x'^{\mu}}{\partial x^{\nu} \partial x^{\lambda}} \frac{dx^{\lambda}}{d\tau} \frac{dx^{\nu}}{d\tau}$$

whereas (4.5.8) gives

$$\Gamma'^{\mu}_{\sigma\tau} \frac{dx'^{\sigma}}{d\tau} \frac{dx'^{\tau}}{d\tau} = \frac{\partial x'^{\mu}}{\partial x^{\nu}} \Gamma^{\nu}_{\lambda\rho} \frac{dx^{\lambda}}{d\tau} \frac{dx^{\rho}}{d\tau} - \frac{\partial^2 x'^{\mu}}{\partial x^{\nu} \partial x^{\lambda}} \frac{dx^{\lambda}}{d\tau} \frac{dx^{\nu}}{d\tau}$$

Adding these two equations, we find that the left-hand side of Eq. (4.5.9) is a vector, that is,

$$\frac{d^2 x'^{\mu}}{d\tau^2} + \Gamma'^{\mu}_{\nu\lambda} \frac{dx'^{\nu}}{d\tau} \frac{dx'^{\lambda}}{d\tau} = \frac{\partial x'^{\mu}}{\partial x^{\kappa}} \left(\frac{d^2 x^{\kappa}}{d\tau^2} + \Gamma^{\kappa}_{\sigma\rho} \frac{dx^{\sigma}}{d\tau} \frac{dx^{\rho}}{d\tau} \right) \tag{4.5.11}$$

Thus Eq. (4.5.9), as well as (4.5.10), is manifestly covariant. The Principle of General Covariance then tells us that (4.5.9) and (4.5.10) are true in general gravitational fields, because, to repeat the reasoning of Section 1, they are true in all coordinate systems if true in any one system, and they *are* true in the locally inertial coordinate systems.

6 Covariant Differentiation

We have already remarked that differentiation of a tensor does not generally yield another tensor. For instance, consider a contravariant vector V^μ, whose transformation law is

$$V'^\mu = \frac{\partial x'^\mu}{\partial x^\nu} V^\nu$$

Differentiating with respect to x'^λ gives

$$\frac{\partial V'^\mu}{\partial x'^\lambda} = \frac{\partial x'^\mu}{\partial x^\nu} \frac{\partial x^\rho}{\partial x'^\lambda} \frac{\partial V^\nu}{\partial x^\rho} + \frac{\partial^2 x'^\mu}{\partial x^\nu \partial x^\rho} \frac{\partial x^\rho}{\partial x'^\lambda} V^\nu \qquad (4.6.1)$$

The first term on the right is what we would expect if $\partial V^\mu / \partial x^\lambda$ were a tensor; the second term is what destroys the tensor behavior.

Although $\partial V^\mu / \partial x^\lambda$ is not a tensor, we can use it to construct a tensor. Using Eq. (4.5.8), we see that

$$\Gamma'^\mu_{\lambda\kappa} V'^\kappa = \left[\frac{\partial x'^\mu}{\partial x^\nu} \frac{\partial x^\rho}{\partial x'^\lambda} \frac{\partial x^\sigma}{\partial x'^\kappa} \Gamma^\nu_{\rho\sigma} - \frac{\partial^2 x'^\mu}{\partial x^\rho \partial x^\sigma} \frac{\partial x^\rho}{\partial x'^\lambda} \frac{\partial x^\sigma}{\partial x'^\kappa} \right] \frac{\partial x'^\kappa}{\partial x^\eta} V^\eta$$

$$= \frac{\partial x'^\mu}{\partial x^\nu} \frac{\partial x^\rho}{\partial x'^\lambda} \Gamma^\nu_{\rho\sigma} V^\sigma - \frac{\partial^2 x'^\mu}{\partial x^\rho \partial x^\sigma} \frac{\partial x^\rho}{\partial x'^\lambda} V^\sigma \qquad (4.6.2)$$

Adding (4.6.1) and (4.6.2), we find that the inhomogeneous terms cancel, yielding

$$\frac{\partial V'^\mu}{\partial x'^\lambda} + \Gamma'^\mu_{\lambda\kappa} V'^\kappa = \frac{\partial x'^\mu}{\partial x^\nu} \frac{\partial x^\rho}{\partial x'^\lambda} \left(\frac{\partial V^\nu}{\partial x^\rho} + \Gamma^\nu_{\rho\sigma} V^\sigma \right) \qquad (4.6.3)$$

Thus we are led to define a *covariant derivative*

$$V^\mu_{;\lambda} \equiv \frac{\partial V^\mu}{\partial x^\lambda} + \Gamma^\mu_{\lambda\kappa} V^\kappa \qquad (4.6.4)$$

and (4.6.3) tells us that $V^\mu_{;\lambda}$ is a tensor:

$$V'^\mu_{;\lambda} = \frac{\partial x'^\mu}{\partial x^\nu} \frac{\partial x^\rho}{\partial x'^\lambda} V^\nu_{;\rho}$$

We can also define a covariant derivative for a covariant vector V_μ. We recall its transformation rule

$$V'_\mu = \frac{\partial x^\rho}{\partial x'^\mu} V_\rho$$

Differentiate with respect to x'^ν:

$$\frac{\partial V'_\mu}{\partial x'^\nu} = \frac{\partial x^\rho}{\partial x'^\mu} \frac{\partial x^\sigma}{\partial x'^\nu} \frac{\partial V_\rho}{\partial x^\sigma} + \frac{\partial^2 x^\rho}{\partial x'^\mu \partial x'^\nu} V_\rho \qquad (4.6.5)$$

From (4.5.2) we have

$$
\Gamma'^{\lambda}_{\mu\nu} V'_{\lambda} = \left[\frac{\partial x'^{\lambda}}{\partial x^{\tau}} \frac{\partial x^{\rho}}{\partial x'^{\mu}} \frac{\partial x^{\sigma}}{\partial x'^{\nu}} \Gamma^{\tau}_{\rho\sigma} + \frac{\partial x'^{\lambda}}{\partial x^{\tau}} \frac{\partial^2 x^{\tau}}{\partial x'^{\mu} \partial x'^{\nu}} \right] \frac{\partial x^{\kappa}}{\partial x'^{\lambda}} V_{\kappa}
$$

$$
= \frac{\partial x^{\rho}}{\partial x'^{\mu}} \frac{\partial x^{\sigma}}{\partial x'^{\nu}} \Gamma^{\kappa}_{\rho\sigma} V_{\kappa} + \frac{\partial^2 x^{\kappa}}{\partial x'^{\mu} \partial x'^{\nu}} V_{\kappa} \tag{4.6.6}
$$

The inhomogeneous terms will cancel if we subtract (4.6.6) from (4.6.5):

$$
\frac{\partial V'_{\mu}}{\partial x'^{\nu}} - \Gamma'^{\lambda}_{\mu\nu} V'_{\lambda} = \frac{\partial x^{\rho}}{\partial x'^{\mu}} \frac{\partial x^{\sigma}}{\partial x'^{\nu}} \left(\frac{\partial V_{\rho}}{\partial x^{\sigma}} - \Gamma^{\kappa}_{\rho\sigma} V_{\kappa} \right) \tag{4.6.7}
$$

We therefore define a covariant derivative of a covariant vector

$$
V_{\mu;\nu} = \frac{\partial V_{\mu}}{\partial x^{\nu}} - \Gamma^{\lambda}_{\mu\nu} V_{\lambda} \tag{4.6.8}
$$

and Eq. (4.6.7) tells us that $V_{\mu;\nu}$ is a tensor:

$$
V'_{\mu;\nu} = \frac{\partial x^{\rho}}{\partial x'^{\mu}} \frac{\partial x^{\sigma}}{\partial x'^{\nu}} V_{\rho;\sigma} \tag{4.6.9}
$$

It is obvious how these definitions are to be extended to a general tensor. The covariant derivative with respect to x^{ρ} of a tensor $T:::$ equals $\partial T:::/\partial x^{\rho}$, plus for each contravariant index μ a term given by $\Gamma^{\mu}_{\nu\rho}$ times T with μ replaced with ν, minus for each covariant index λ a term $\Gamma^{\kappa}_{\lambda\rho}$ times T with λ replaced with κ. For instance,

$$
T^{\mu\sigma}{}_{\lambda;\rho} = \frac{\partial}{\partial x^{\rho}} T^{\mu\sigma}{}_{\lambda} + \Gamma^{\mu}_{\rho\nu} T^{\nu\sigma}{}_{\lambda} + \Gamma^{\sigma}_{\rho\nu} T^{\mu\nu}{}_{\lambda} - \Gamma^{\kappa}_{\lambda\rho} T^{\mu\sigma}{}_{\kappa} \tag{4.6.10}
$$

The reader may easily verify that this *is* a tensor.

We can also extend the idea of covariant differentiation to tensor densities. The easiest way to do this is to recall that if $\mathscr{I}$ is a tensor density of weight W, then $g^{W/2}\mathscr{I}$ is an ordinary tensor. Its covariant derivative is also a tensor, and multiplying by $g^{-W/2}$ gives back a tensor density of weight W. Hence the covariant derivative of a tensor density of weight W is defined by

$$
\mathscr{I}:::_{;\rho} \equiv g^{-W/2} (g^{W/2} \mathscr{I}:::)_{;\rho} \tag{4.6.11}
$$

and we need not check that this *is* a tensor density of weight W. The effect is that the covariant derivative with respect to x^{ρ} of a tensor density $\mathscr{I}$ of weight W is constructed just as if it were an ordinary tensor, except that we add an extra term $(W/2g)\mathscr{I}:::(\partial g/\partial x^{\rho})$. For instance,

$$
\mathscr{I}^{\mu}{}_{\lambda;\rho} \equiv \frac{\partial}{\partial x^{\rho}} \mathscr{I}^{\mu}{}_{\lambda} + \Gamma^{\mu}_{\rho\nu} \mathscr{I}^{\nu}{}_{\lambda} - \Gamma^{\kappa}_{\lambda\rho} \mathscr{I}^{\mu}{}_{\kappa} + \frac{W}{2g} \frac{\partial g}{\partial x^{\rho}} \mathscr{I}^{\mu}{}_{\lambda} \tag{4.6.12}
$$

The combination of covariant differentiation with the algebraic operations described in Section 3 gives results similar to those for ordinary differentiation. In particular:

(A) The covariant derivative of a linear combination of tensors (with constant coefficients) is the same linear combination of the covariant derivatives. For instance if α and β are constants, then

$$(\alpha A^\mu{}_\nu + \beta B^\mu{}_\nu)_{;\lambda} = \alpha A^\mu{}_{\nu;\lambda} + \beta B^\mu{}_{\nu;\lambda} \tag{4.6.13}$$

(B) The covariant derivative of a direct product of tensors obeys the Leibniz rule. For instance,

$$(A^\mu{}_\nu B^\lambda)_{;\rho} = A^\mu{}_{\nu;\rho} B^\lambda + A^\mu{}_\nu B^\lambda{}_{;\rho} \tag{4.6.14}$$

(C) The covariant derivative of a contracted tensor is the contraction of the covariant derivative. For instance, setting $\sigma = \lambda$ in Eq. (4.6.10) gives

$$T^{\mu\lambda}{}_{\lambda;\rho} = \frac{\partial}{\partial x^\rho} T^{\mu\lambda}{}_\lambda + \Gamma^\mu{}_{\rho\nu} T^{\nu\lambda}{}_\lambda \tag{4.6.15}$$

the last two terms canceling.

We also note that the covariant derivative of the metric tensor is zero, because it vanishes in locally inertial coordinates where $\Gamma^\mu{}_{\nu\lambda}$ and $\partial g_{\mu\nu}/\partial x^\lambda$ vanish, and a tensor, zero in one coordinate system, is zero in all systems. The same result can be obtained more directly by noting that

$$g_{\mu\nu;\lambda} = \frac{\partial g_{\mu\nu}}{\partial x^\lambda} - \Gamma^\rho{}_{\lambda\mu} g_{\rho\nu} - \Gamma^\rho{}_{\lambda\nu} g_{\rho\mu}$$

Eq. (3.3.1) tells us that this vanishes:

$$g_{\mu\nu;\lambda} = 0 \tag{4.6.16}$$

(This argument can be reversed to provide yet another derivation of the relation between $g_{\mu\nu}$ and $\Gamma^\lambda{}_{\mu\nu}$.) We can also show in the same way that the covariant derivatives of the other forms of the metric tensor also vanish, that is,

$$g^{\mu\nu}{}_{;\lambda} = 0 \tag{4.6.17}$$

$$\delta^\mu{}_{\nu;\lambda} = 0 \tag{4.6.18}$$

From (4.6.16)–(4.6.18) it follows that the operations of covariant differentiation and raising and lowering indices commute; for example,

$$(g^{\mu\nu} V_\nu)_{;\lambda} = g^{\mu\nu} V_{\nu;\lambda} \tag{4.6.19}$$

The importance of covariant differentiation arises from *two* of its properties: It converts tensors to other tensors, and it reduces to ordinary differentiation in the absence of gravitation, that is, when $\Gamma^\mu{}_{\nu\lambda} = 0$. These properties suggest the following algorithm for assessing the effects of gravitation on physical systems:

Write the appropriate special-relativistic equations that hold in the absence of gravitation, replace $\eta_{\mu\nu}$ with $g_{\mu\nu}$, and replace all derivatives with covariant derivatives. The resulting equations will be generally covariant and true in the absence of gravitation, and therefore, according to the Principle of General Covariance, they will be true in the presence of gravitational fields, provided always that we work on a space-time scale sufficiently small compared with the scale of the gravitational field.

7   Gradient, Curl, and Divergence

There are some special cases where the covariant derivative takes a particularly simple form. Simplest of all is, of course, the covariant derivative of a scalar, which is just the ordinary gradient

$$S_{;\mu} = \frac{\partial S}{\partial x^{\mu}} \tag{4.7.1}$$

Another simple special case is the covariant curl. Recall that

$$V_{\mu;\nu} \equiv \frac{\partial V_{\mu}}{\partial x^{\nu}} - \Gamma^{\lambda}_{\mu\nu} V_{\lambda}$$

Since $\Gamma^{\lambda}_{\mu\nu}$ is symmetric in μ and ν, the covariant curl is just the ordinary curl

$$V_{\mu;\nu} - V_{\nu;\mu} = \frac{\partial V_{\mu}}{\partial x^{\nu}} - \frac{\partial V_{\nu}}{\partial x^{\mu}} \tag{4.7.2}$$

Another special case that will take a little more work is the covariant divergence of a contravariant vector

$$V^{\mu}_{;\mu} \equiv \frac{\partial V^{\mu}}{\partial x^{\mu}} + \Gamma^{\mu}_{\mu\lambda} V^{\lambda} \tag{4.7.3}$$

We note that $\Gamma^{\mu}_{\mu\lambda}$ is given by

$$\Gamma^{\mu}_{\mu\lambda} = \tfrac{1}{2} g^{\mu\rho} \left\{ \frac{\partial g_{\rho\mu}}{\partial x^{\lambda}} + \frac{\partial g_{\rho\lambda}}{\partial x^{\mu}} - \frac{\partial g_{\mu\lambda}}{\partial x^{\rho}} \right\}$$

$$= \tfrac{1}{2} g^{\mu\rho} \frac{\partial g_{\rho\mu}}{\partial x^{\lambda}} \tag{4.7.4}$$

We may evaluate this easily if we recall that for an arbitrary matrix M,

$$\text{Tr} \left\{ M^{-1}(x) \frac{\partial}{\partial x^{\lambda}} M(x) \right\} = \frac{\partial}{\partial x^{\lambda}} \ln \text{Det } M(x) \tag{4.7.5}$$

where Det denotes the determinant and Tr the trace, that is, the sum of the diagonal elements. To prove (4.7.5), consider the variation in ln Det M owing to a variation δx^λ in x^λ:

$$\delta \ln \text{Det } M \equiv \ln \text{Det } (M + \delta M) - \ln \text{Det } M$$

$$= \ln \frac{\text{Det } (M + \delta M)}{\text{Det } M}$$

$$= \ln \text{Det } M^{-1}(M + \delta M)$$

$$= \ln \text{Det } (1 + M^{-1}\, \delta M)$$

$$\to \ln (1 + \text{Tr } M^{-1}\, \delta M)$$

$$\to \text{Tr } M^{-1}\, \delta M$$

Taking the coefficient of δx^λ in both sides gives Eq. (4.7.5). Applying (4.7.5) to the case where M is the matrix $g_{\rho\mu}$, we find from (4.7.4) that

$$\Gamma^\mu_{\mu\lambda} = \frac{1}{2} \frac{\partial}{\partial x^\lambda} \ln g = \frac{1}{\sqrt{g}} \frac{\partial}{\partial x^\lambda} \sqrt{g} \tag{4.7.6}$$

With (4.7.3), we find that the covariant divergence is precisely

$$V^\mu_{;\mu} = \frac{1}{\sqrt{g}} \frac{\partial}{\partial x^\mu} \sqrt{g}\, V^\mu \tag{4.7.7}$$

One immediate consequence is a covariant form of Gauss's theorem: If V^μ vanishes at infinity then

$$\int d^4x \sqrt{g}\, V^\mu_{;\mu} = 0 \tag{4.7.8}$$

Note the appearance here of a factor $\sqrt{g}$ that makes $d^4x \sqrt{g}$ invariant.

We can also use (4.7.6) to simplify the formula for the covariant divergence of a tensor. For instance,

$$T^{\mu\nu}_{;\mu} \equiv \frac{\partial T^{\mu\nu}}{\partial x^\mu} + \Gamma^\mu_{\mu\lambda}T^{\lambda\nu} + \Gamma^\nu_{\mu\lambda}T^{\mu\lambda}$$

and, applying (4.7.6), we find that

$$T^{\mu\nu}_{;\mu} = \frac{1}{\sqrt{g}} \frac{\partial}{\partial x^\mu} (\sqrt{g}\, T^{\mu\nu}) + \Gamma^\nu_{\mu\lambda}T^{\mu\lambda} \tag{4.7.9}$$

In particular, if $T^{\mu\lambda} = -T^{\lambda\mu}$, then the last term drops out, so

$$A^{\mu\nu}_{;\mu} = \frac{1}{\sqrt{g}} \frac{\partial}{\partial x^\mu} (\sqrt{g}\, A^{\mu\nu}) \qquad \text{for } A^{\mu\nu} \text{ antisymmetric} \tag{4.7.10}$$

There is one other special case of some importance. For a covariant tensor $A_{\mu\nu}$ the covariant derivative is

$$A_{\mu\nu;\lambda} \equiv \frac{\partial A_{\mu\nu}}{\partial x^{\lambda}} - \Gamma^{\rho}_{\mu\lambda}A_{\rho\nu} - \Gamma^{\rho}_{\nu\lambda}A_{\mu\rho}$$

Suppose that $A_{\mu\nu}$ is antisymmetric; that is,

$$A_{\mu\nu} = -A_{\nu\mu}$$

If we twice add to $A_{\mu\nu;\lambda}$ the same tensor with indices cyclically permuted, we find by virtue of the symmetry of $\Gamma^{\rho}_{\mu\lambda}$ and the antisymmetry of $A_{\rho\nu}$ that all Γ-terms cancel, yielding

$$A_{\mu\nu;\lambda} + A_{\lambda\mu;\nu} + A_{\nu\lambda;\mu} = \frac{\partial A_{\mu\nu}}{\partial x^{\lambda}} + \frac{\partial A_{\lambda\mu}}{\partial x^{\nu}} + \frac{\partial A_{\nu\lambda}}{\partial x^{\mu}} \qquad \text{for } A \text{ antisymmetric}$$

(4.7.11)

8 Vector Analysis in Orthogonal Coordinates*

The reader may be wondering what the tensor analysis formalism outlined in this chapter has to do with the familiar formulas for gradient, curl, and divergence in the classical curvilinear coordinate systems. These are three-dimensional coordinate systems characterized by the condition that g_{ij} is diagonal, that is,

$$g_{ij} = h_i^2 \delta_{ij} \qquad (i, j = 1, 2, 3) \tag{4.8.1}$$

where h_i is some function of the coordinates.[5] (The summation convention is suspended for the duration of this section.) The inverse metric tensor is then

$$g^{ij} = h_i^{-2}\delta_{ij} \tag{4.8.2}$$

The invariant proper length is now

$$ds^2 \equiv \sum_{i,j} g_{ij}\, dx^i\, dx^j = h_1^{\,2}(dx^1)^2 + h_2^{\,2}(dx^2)^2 + h_3^{\,2}(dx^3)^2 \tag{4.8.3}$$

and the invariant volume element is

$$dV \equiv (\text{Det } g)^{1/2}\, dx^1\, dx^2\, dx^3 = h_1 h_2 h_3\, dx^1\, dx^2\, dx^3 \tag{4.8.4}$$

---

* This section lies somewhat out of the book's main line of development, and may be omitted in a first reading.

What are usually called the components of a vector **V** in elementary treatments are not the covariant components V_i or the contravariant components V^i, but the "ordinary" components $\bar{V}_i$:

$$\bar{V}_i \equiv h_i V^i = h_i^{-1} V_i \tag{4.8.5}$$

The scalar product of two vectors is then very simple:

$$\mathbf{V} \cdot \mathbf{U} \equiv \sum_{ij} g_{ij} V^i U^j = \bar{V}_1 \bar{U}_1 + \bar{V}_2 \bar{U}_2 + \bar{V}_3 \bar{U}_3 \tag{4.8.6}$$

[This, of course, is the motivation for the definition (4.8.5).] However, the gradient of a scalar is now a little more complicated:

$$\mathbf{V}_i S \equiv \overline{S_{;i}} = h_i^{-1} \frac{\partial S}{\partial x^i} \tag{4.8.7}$$

The curl of a vector V is likewise defined by taking the "ordinary" components of a vector

$$(\mathbf{V} \times \mathbf{V})_i \equiv h_i \sum_{jk} (\text{Det } g)^{-1/2} \varepsilon^{ijk} V_{j;k}$$

$$= h_i \sum_{jk} (h_1 h_2 h_3)^{-1} \varepsilon^{ijk} \frac{\partial}{\partial x^j} h_k \bar{V}_k \tag{4.8.8}$$

(We have used (4.7.2), since ε^{ijk} is antisymmetric in j and k.) For instance, the first component of the curl is

$$(\mathbf{V} \times \mathbf{V})_1 = \frac{1}{h_2 h_3} \left(\frac{\partial}{\partial x^2} h_3 \bar{V}_3 - \frac{\partial}{\partial x^3} h_2 \bar{V}_2 \right) \tag{4.8.9}$$

The divergence of a vector V is nothing but the covariant divergence (4.7.7):

$$\mathbf{V} \cdot \mathbf{V} \equiv \sum_i V^i_{;i} = (\text{Det } g)^{-1/2} \sum_i \frac{\partial}{\partial x^i} (\text{Det } g)^{1/2} V^i$$

$$= (h_1 h_2 h_3)^{-1} \left(\frac{\partial}{\partial x^1} h_2 h_3 \bar{V}_1 + \frac{\partial}{\partial x^2} h_1 h_3 \bar{V}_2 + \frac{\partial}{\partial x^3} h_1 h_2 \bar{V}_3 \right) \tag{4.8.10}$$

The Laplacian of a scalar S is the divergence of its gradient

$$\mathbf{V}^2 S \equiv \sum_{ij} (g^{ij} S_{;i})_{;j} \tag{4.8.11}$$

or combining (4.8.10) with (4.8.7),

$$\mathbf{V}^2 S \equiv (h_1 h_2 h_3)^{-1} \left[\frac{\partial}{\partial x^1} \frac{h_2 h_3}{h_1} \frac{\partial S}{\partial x^1} + \frac{\partial}{\partial x^2} \frac{h_1 h_3}{h_2} \frac{\partial S}{\partial x^2} + \frac{\partial}{\partial x^3} \frac{h_1 h_2}{h_3} \frac{\partial S}{\partial x^3} \right] \tag{4.8.12}$$

The reader may easily check that the usual formulas for gradient, curl, divergence and Laplacian are obtained if the h_i take the forms appropriate for spherical or cylindrical coordinates.

9   Covariant Differentiation Along a Curve

This chapter has dealt until now with tensor fields defined over all space-time; we now consider tensors $T(\tau)$ defined only over a curve $x^\mu(\tau)$. Obvious examples come to mind, such as the momentum $P^\mu(\tau)$ or spin $S_\mu(\tau)$ of a single particle. For such tensors it would of course be meaningless to talk about covariant differentiation with respect to x^μ, but we can define a covariant derivative with respect to the invariant τ that parameterizes the curve.

Consider first a contravariant vector $A^\mu(\tau)$, with transformation rule

$$A'^{\mu}(\tau) = \frac{\partial x'^{\mu}}{\partial x^{\nu}} A^{\nu}(\tau) \tag{4.9.1}$$

It should be noted that the partial derivative $\partial x'^\mu / \partial x^\nu$ is to be evaluated at $x^\nu = x^\nu(\tau)$, so that it depends on τ. Therefore, differentiating with respect to τ, we find two terms,

$$\frac{dA'^{\mu}(\tau)}{d\tau} = \frac{\partial x'^{\mu}}{\partial x^{\nu}} \frac{dA^{\nu}(\tau)}{d\tau} + \frac{\partial^2 x'^{\mu}}{\partial x^{\nu}\, \partial x^{\lambda}} \frac{dx^{\lambda}}{d\tau} A^{\nu}(\tau) \tag{4.9.2}$$

The second derivative $\partial^2 x'^\mu / \partial x^\nu\, \partial x^\lambda$ is the same as that responsible for the inhomogeneous term in the transformation formula (4.5.8) for the affine connection, so we are led to define a covariant derivative along the curve $x^\mu(\tau)$ by

$$\frac{DA^{\mu}}{D\tau} \equiv \frac{dA^{\mu}}{d\tau} + \Gamma^{\mu}_{\nu\lambda} \frac{dx^{\lambda}}{d\tau} A^{\nu} \tag{4.9.3}$$

Eqs. (4.5.8), (4.9.1), and (4.9.2) then show that this is a vector:

$$\frac{DA'^{\mu}}{D\tau} = \frac{\partial x'^{\mu}}{\partial x^{\nu}} \frac{DA^{\nu}}{D\tau} \tag{4.9.4}$$

The similarity between (4.9.3) and the formula (4.6.4) for the covariant derivative of a vector field is evident.

The same considerations lead us to define the covariant derivative along a curve $x^\mu(\tau)$ of a covariant vector $B_\mu(\tau)$ by

$$\frac{DB_{\mu}}{D\tau} = \frac{dB_{\mu}}{d\tau} - \Gamma^{\lambda}_{\mu\nu} \frac{dx^{\nu}}{d\tau} B_{\lambda} \tag{4.9.5}$$

and with (4.5.2) we can easily show that this is a vector:

$$\frac{DB'_{\mu}}{D\tau} = \frac{\partial x^{\nu}}{\partial x'^{\mu}} \frac{DB_{\nu}}{D\tau} \tag{4.9.6}$$

In the same way, the covariant derivative along a curve $x^\mu(\tau)$ of a general tensor $T(\tau)$ is defined by adding to $dT/d\tau$ a term such as that in (4.9.3) for each upper index, and subtracting a term such as that in (4.9.5) for each lower index. For instance,

$$\frac{DT^{\mu}{}_{\nu}}{D\tau} \equiv \frac{dT^{\mu}{}_{\nu}}{d\tau} + \Gamma^{\mu}_{\lambda\rho} \frac{dx^{\lambda}}{d\tau} T^{\rho}{}_{\nu} - \Gamma^{\sigma}_{\lambda\nu} \frac{dx^{\lambda}}{d\tau} T^{\mu}{}_{\sigma} \qquad (4.9.7)$$

and

$$\frac{DT'^{\mu}{}_{\nu}}{D\tau} = \frac{\partial x'^{\mu}}{\partial x^{\rho}} \frac{\partial x^{\sigma}}{\partial x'^{\nu}} \frac{DT^{\rho}{}_{\sigma}}{D\tau} \qquad (4.9.8)$$

The properties of covariant differentiation outlined in Sections 6 through 8 can be easily extended to covariant derivatives along a curve.

It should be mentioned that the covariant derivative of a tensor *field* along a curve may be determined from its ordinary covariant derivative; for instance, if $T^{\mu}{}_{\nu}$ is a tensor field then (4.9.6) gives

$$\frac{DT^{\mu}{}_{\nu}}{D\tau} = T^{\mu}{}_{\nu;\lambda} \frac{dx^{\lambda}}{d\tau} \qquad (4.9.9)$$

However, we shall see in Chapter 6 that tensors defined along curves cannot always be promoted to tensor fields, and for these the derivative $D/D\tau$ is the only covariant derivative available.

It is often the case that a vector $A^{\mu}(\tau)$ carried along a curve by a particle does not change at τ if viewed from a reference frame $\xi_{x(\tau)}$ that is locally inertial at $x(\tau)$. (This is true for a particle's momentum and spin if it is subject to purely gravitational forces; see Section 5.1.) In this frame the affine connection as well as $dA^{\mu}/d\tau$ vanishes, so

$$\frac{DA^{\mu}}{D\tau} = 0 \qquad (4.9.10)$$

This being a covariant statement, and true at $x(\tau)$ in the locally inertial system $\xi_{x(\tau)}$, it is therefore true in all coordinate systems. The vector A^{μ} is then subject to the first-order differential equations

$$\frac{dA^{\mu}}{d\tau} = -\Gamma^{\mu}_{\nu\lambda} \frac{dx^{\lambda}}{d\tau} A^{\nu} \qquad (4.9.11)$$

that define A^{μ} for all τ, given A^{μ} at some initial τ. A vector $A^{\mu}(\tau)$ defined in this way along a curve $x^{\mu}(\tau)$ is said to be defined by *parallel transport*. Any tensor can be defined along a curve by parallel transport by requiring its covariant derivative along the curve to vanish.

10 The Electromagnetic Analogy*

I emphasized in Section 1 of this chapter that general covariance is not an ordinary symmetry principle like Lorentz invariance, but is rather a dynamical

---

* This section lies somewhat out of the book's main line of development, and may be omitted in a first reading.

principle that governs the effect of gravitational fields. As such, it bears a strong resemblance to another "dynamic symmetry," local gauge invariance, which governs the effects of electromagnetic fields. Local gauge invariance says that the differential equations satisfied by a set of charged fields $\psi(x)$ and the electromagnetic potential $A_\alpha(x)$ retain the same form when these fields are subjected to the transformations[6]

$$\psi(x) \rightarrow \psi(x)e^{ie\varphi(x)} \tag{4.10.1}$$

$$A_\alpha(x) \rightarrow A_\alpha(x) + \frac{\partial}{\partial x^\alpha}\,\varphi(x) \tag{4.10.2}$$

where e is the charge of the particle represented by ψ and $\varphi(x)$ is an arbitrary function of the space-time coordinates x^α. How are we to construct gauge-invariant equations? Note that derivatives of a charged field ψ do not behave under gauge transformations like ψ, but rather

$$\frac{\partial}{\partial x^\alpha}\,\psi(x) \rightarrow \frac{\partial}{\partial x^\alpha}\,[\psi(x)e^{ie\varphi(x)}]$$

$$= e^{ie\varphi(x)}\left[\frac{\partial\psi(x)}{\partial x^\alpha} + ie\psi(x)\frac{\partial\varphi(x)}{\partial x^\alpha}\right]$$

just as derivatives of tensors do not behave like tensors under general coordinate transformations. It follows that an equation such as

$$(\Box^2 - m^2)\psi(x) = 0 \qquad \text{where } \Box^2 \equiv \eta^{\alpha\beta}\frac{\partial}{\partial x^\alpha}\frac{\partial}{\partial x^\beta}$$

is *not* gauge-invariant, just as it is not generally covariant. Also note that the electromagnetic potential $A_\mu(x)$ obeys an inhomogeneous gauge transformation law, just as the affine connection obeys the inhomogeneous transformation law (4.5.2) for general coordinate transformations. In tensor analysis we put together derivatives of tensors and the affine connection to form "covariant derivatives" that transform like tensors. In electrodynamics we put together derivatives of fields and the vector potential to form "gauge-covariant derivatives"

$$\mathcal{D}_\alpha\psi(x) \equiv \left[\frac{\partial}{\partial x^\alpha} - ieA_\alpha(x)\right]\psi(x) \tag{4.10.3}$$

that transform like the fields themselves,

$$\mathcal{D}_\alpha\psi(x) \rightarrow [\mathcal{D}_\alpha\psi(x)]e^{ie\varphi(x)} \tag{4.10.4}$$

An equation that is invariant under gauge transformations with φ constant (such invariance is simply tantamount to charge conservation) will be invariant under the general gauge transformation (4.10.1)–(4.10.2) provided that it is constructed only out of fields $\psi(x)$ and their gauge-covariant derivatives $\mathcal{D}_\alpha\psi(x)$, just as an equation that is invariant under Lorentz transformations will be invariant under general coordinate transformations provided that it is constructed out of tensors and their covariant derivatives. For instance, we can write a gauge-invariant equation that might represent the effect of electromagnetism on a charged scalar field $\psi(x)$ as

$$[\eta^{\alpha\beta}\mathscr{D}_\alpha\mathscr{D}_\beta + m^2]\psi(x) = 0 \tag{4.10.5}$$

or, in more detail,

$$\left(\Box^2 - 2ieA^\alpha\frac{\partial}{\partial x^\alpha} - ie\frac{\partial A^\alpha}{\partial x^\alpha} - e^2 A^\alpha A_\alpha + m^2\right)\psi(x) = 0$$

One important property of such theories is that they admit the construction of a conserved gauge-invariant current; in this example we can define

$$J_\alpha(x) \equiv -ie\{\psi^\dagger(x)\mathscr{D}_\alpha\psi(x) - \psi(x)[\mathscr{D}_\alpha\psi(x)]^\dagger\}$$

(A dagger denotes complex conjugation, or in quantum theories the Hermitian adjoint.) That this is gauge-invariant is obvious; to see that it is conserved, we write

$$\frac{\partial}{\partial x^\alpha}J^\alpha(x) = -ie\left\{\frac{\partial\psi^\dagger(x)}{\partial x^\alpha}\left(\frac{\partial\psi(x)}{\partial x_\alpha} - ieA^\alpha(x)\psi(x)\right) - \frac{\partial\psi(x)}{\partial x^\alpha}\left(\frac{\partial\psi^\dagger(x)}{\partial x^\alpha} + ieA^\alpha(x)\psi^\dagger(x)\right)\right.$$

$$\left. + \psi^\dagger(x)(\mathscr{D}^\alpha + ieA^\alpha(x))\mathscr{D}_\alpha\psi(x) - \psi(x)[(\mathscr{D}^\alpha + ieA^\alpha(x))\mathscr{D}_\alpha\psi(x)]^\dagger\right\}$$

$$= \psi^\dagger(x)\mathscr{D}^\alpha\mathscr{D}_\alpha\psi(x) - \psi(x)[\mathscr{D}^\alpha\mathscr{D}_\alpha\psi(x)]^\dagger$$

and using (4.10.5) this gives

$$\frac{\partial}{\partial x^\alpha}J^\alpha(x) = 0$$

We can thus use this current in the right-hand side of Maxwell's equations (2.7.6), and these equations will then be gauge-invariant also. We see in Chapter 7 that the field equations for gravitation are constructed in an analogous manner.

    The analogy between the gauge invariance of electrodynamics and the general covariance of general relativity can be extended to a similar dynamic symmetry, called chirality,[3] that governs the interactions of pi-mesons. A proper explanation of this point would fill another book.

11 p-Forms and Exterior Derivatives*

    Antisymmetric tensors and their antisymmetrized derivatives possess certain remarkably simple and useful properties, some of which we have already en-

---

* This section lies somewhat out of the book's main line of development, and may be omitted in a first reading.

countered in Section 4.7. In order to deal with these properties in a unified way, mathematicians have developed a general formalism, known as the *theory of differential forms*.[7] Unfortunately, the rather abstract and compact notation associated with this formalism has in recent years seriously impeded communication between pure mathematicians and physicists. This section presents some of the fundamental results of the theory of differential forms, but in the tensor notation familiar to physicists, rather than the recondite notation favored by mathematicians.

A covariant tensor of rank p, which is antisymmetric under exchange of any pair of indices, will be called a *p-form*. In n dimensions, the number of algebraically independent components of a p-form is just the binomial coefficient

$$\binom{n}{p} \equiv \frac{n!}{p!\,(n-p)!} \tag{4.11.1}$$

For instance, a scalar field is a 0-form, a covariant vector field is a 1-form, and an antisymmetric covariant tensor with two indices is a 2-form.

Linear combinations of p-forms are p-forms. However, the direct product $s_{\mu\nu}\ldots t_{\rho\sigma}\ldots$ of a p-form $s_{\mu\nu}\ldots$ and a q-form $t_{\rho\sigma}\ldots$ is *not* a $(p+q)$-form, because it is not completely antisymmetric. We can form a $(p+q)$-form $s \wedge t$ by antisymmetrizing the direct product:

$$(s \wedge t)_{\mu_1\cdots\mu_{p+q}} \equiv \text{Antisym}\,\{s_{\mu_1\cdots\mu_p} t_{\mu_{p+1}\cdots\mu_{p+q}}\} \tag{4.11.2}$$

where, in general, "Antisym" denotes an average over all permutations Π of the indices,

$$\text{Antisym}\,\{u_{\mu_1\mu_2\cdots\mu_m}\} \equiv \frac{1}{m!} \sum_{\Pi} \delta_{\Pi} u_{\mu_{\Pi 1}\mu_{\Pi 2}\cdots\mu_{\Pi m}} \tag{4.11.3}$$

with a sign-factor δ_{Π} which is $+1$ or -1 according to whether Π consists of an even or odd number of permutations of individual index pairs:

$$\delta_{\Pi} \equiv \begin{cases} +1 & \Pi \text{ even} \\ -1 & \Pi \text{ odd} \end{cases} \tag{4.11.4}$$

The antisymmetrized direct product (4.11.2) is known as the *exterior product*. For instance, the exterior product of a 0-form s and a 1-form t_μ is simply the ordinary product

$$(s \wedge t)_\mu \equiv st_\mu$$

whereas the exterior product of a 1-form s_μ and a 1-form t_ν is the 2-form

$$(s \wedge t)_{\mu\nu} \equiv \tfrac{1}{2}(s_\mu t_\nu - s_\nu t_\mu)$$

The reader can easily verify that the exterior product is *associative*,

$$(s \wedge t) \wedge u = s \wedge (t \wedge u) \tag{4.11.5}$$

and *bilinear*,

$$(\alpha_1 s_1 + \alpha_2 s_2) \wedge t = \alpha_1(s_1 \wedge t) + \alpha_2(s_2 \wedge t)$$
$$s \wedge (\alpha_1 t_1 + \alpha_2 t_2) = \alpha_1(s \wedge t_1) + \alpha_2(s \wedge t_2) \tag{4.11.6}$$

(Here α_1 and α_2 are scalars.) However, the exterior product is not commutative; rather, if s is a p-form and t is a q-form, then

$$(s \wedge t) = (-1)^{pq}(t \wedge s) \tag{4.11.7}$$

This is a good spot to pause, and mention that in books on the mathematical theory of differential forms,[7] a p-form t is generally not represented by the tensor components $t_{\mu\nu}\ldots$, but rather by the "differential form"

$$\omega \equiv t_{\mu\nu}\ldots(dx^\mu \wedge dx^\nu \wedge \cdots)$$

The symbol dx^μ here denotes a quantity that transforms like a coordinate differential, that is, as a contravariant vector, but whose products, unlike ordinary products of coordinate differentials, are associative and *anticommutative*:

$$(dx^\mu \wedge dx^\nu) \wedge dx^\lambda = dx^\mu \wedge (dx^\nu \wedge dx^\lambda)$$
$$dx^\mu \wedge dx^\nu = -dx^\nu \wedge dx^\mu$$

The product $\omega_1 \wedge \omega_2$ of differential forms ω_1 and ω_2 has tensor coefficients $t_{\mu\nu}\ldots$ that are just given by the exterior product of the tensor coefficients of ω_1 and ω_2. The associativity and commutativity rules (4.11.5) and (4.11.7) of the exterior product then follow trivially from the associative and anticommutative properties of the product $dx^\mu \wedge dx^\nu$. As already indicated, we shall not use this language here; for us, a p-form will be simply an antisymmetrical tensor, not the corresponding differential form.

The point of developing the theory of p-forms separately from the rest of tensor analysis emerges when we study their derivatives. The partial derivative operator $\partial/\partial x^\mu$ is a covariant vector, or in other words a 1-form, so given any p-form t we can define a $(p+1)$-form Dt, known as the *exterior derivative* of t, by simply taking the exterior product of $\partial/\partial x$ with t:

$$Dt \equiv \frac{\partial}{\partial x} \wedge t \tag{4.11.8}$$

or, in more detail,

$$(Dt)_{\mu_1 \cdots \mu_{p+1}} \equiv \text{Antisym} \left\{ \frac{\partial}{\partial x^{\mu_1}} t_{\mu_2 \cdots \mu_{p+1}} \right\} \tag{4.11.9}$$

For instance, the exterior derivative of a 0-form t is simply the ordinary gradient

$$(Dt)_\mu = \frac{\partial t}{\partial x^\mu}$$

whereas the exterior derivative of a 1-form t_μ is just the "curl"

$$(Dt)_{\mu\nu} \equiv \frac{1}{2}\left(\frac{\partial t_\nu}{\partial x^\mu} - \frac{\partial t_\mu}{\partial x^\nu}\right)$$

In *three* dimensions, the exterior derivative of a 2-form t_{ij} may be expressed as an ordinary divergence:

$$(Dt)_{ijk} = \tfrac{1}{3}\varepsilon_{ijk}\left(\frac{\partial t_{23}}{\partial x^1} + \frac{\partial t_{31}}{\partial x^2} + \frac{\partial t_{12}}{\partial x^3}\right)$$

The first remarkable property of the exterior derivative is that, acting on a tensor p-form, it gives a *tensor* $(p + 1)$-form. The easiest way to see this is just to note that the partial derivatives used to define the exterior derivative can be replaced with covariant derivatives,

$$(Dt)_{\mu_1 \cdots \mu_{p+1}} = \text{Antisym}\,(t_{\mu_2 \cdots \mu_{p+1};\mu_1}) \tag{4.11.10}$$

because the contribution of the affine connections that appear in the covariant derivatives vanish upon antisymmetrization. Our previously derived results (4.7.1), (4.7.2), and (4.7.11) are special cases of Eq. (4.11.10) for $p = 0$, $p = 1$, and $p = 2$.

From the associativity and commutativity rules (4.11.5) and (4.11.7) for the exterior derivative, we can easily derive a simple formula for the exterior derivative of the exterior product of a p-form s and a q-form t:

$$D(s \wedge t) = Ds \wedge t + (-1)^{pq}Dt \wedge s$$
$$= Ds \wedge t + (-1)^p s \wedge Dt \tag{4.11.11}$$

From the same rules, it also follows that repeated exterior derivatives vanish,

$$D^2 t \equiv \frac{\partial}{\partial x} \wedge \left(\frac{\partial}{\partial x} \wedge t\right) = \left(\frac{\partial}{\partial x} \wedge \frac{\partial}{\partial x}\right) \wedge t = 0 \tag{4.11.12}$$

This latter result is known as *Poincaré's lemma*. Among the special cases of this lemma are two well-known results of three-dimensional vector analysis, that a gradient has zero curl and a curl has zero divergence.

The question naturally arises whether the converse to Poincaré's lemma is also valid. That is, if s is a $(p + 1)$-form for which

$$Ds = 0 \tag{4.11.13}$$

then can we express s as

$$s = Dt \tag{4.11.14}$$

for some p-form t? The answer is yes, provided that the region $\mathscr{R}$ over which (4.11.13) holds, and over which we want (4.11.14) to hold, can be deformed to a point. In general, we say that a region $\mathscr{R}$ can be deformed to a point y^μ if every point x^μ in $\mathscr{R}$ can be connected with the point y^μ by a path $X^\mu(\lambda; x)$ lying entirely in $\mathscr{R}$; here λ is a real parameter that can be taken to run from 0 to 1, with

$$X^\mu(0; x) = y^\mu \qquad X^\mu(1; x) = x^\mu$$

It is straightforward to verify that if (4.11.13) holds over such a region $\mathscr{R}$, then (4.11.14) will be satisfied throughout this region by the p-form

$$t_{\mu_1 \cdots \mu_p}(x) = (p + 1) \int_0^1 \frac{\partial X^\nu(\lambda; x)}{\partial \lambda} \frac{\partial X^{\nu_1}(\lambda; x)}{\partial x^{\mu_1}} \cdots \frac{\partial X^{\nu_p}(\lambda; x)}{\partial x^{\mu_p}} s_{\nu \nu_1 \cdots \nu_p}(X(\lambda; x)) \, d\lambda$$

(4.11.15)

The well-known results of three-dimensional vector analysis, that a vector may be expressed as a gradient if it has zero curl, or as a curl if it has zero divergence, may be regarded as special cases of this theorem for $p = 0$ and $p = 1$, respectively. Maxwell's equations provide an example in four dimensions: The field-strength tensor $F_{\alpha\beta}$ is a 2-form that according to Eq. (2.7.10) has vanishing exterior derivative, so that it can be expressed as the exterior derivative of a 1-form, conventionally denoted $-2A_\alpha$,

$$F_{\alpha\beta} = \frac{\partial A_\beta}{\partial x^\alpha} - \frac{\partial A_\alpha}{\partial x^\beta}$$

as in Eq. (2.7.11). In general, the p-form t satisfying Eq. (4.11.14) is not unique; given one such t, the most general p-form satisfying (4.11.14) is of the form

$$t' = t + Du \qquad (4.11.16)$$

where u is an arbitrary $(p - 1)$-form. For instance, if A_α is one vector potential whose curl is $F_{\alpha\beta}$, then the most general such vector potential is given by the "gauge transformation"

$$A'_\alpha = A_\alpha + \frac{\partial \Phi}{\partial x^\alpha}$$

where Φ is an arbitrary 0-form, that is, an arbitrary scalar.

Just as the exterior derivative provides a natural generalization of the familiar gradient, curl, and divergence, so also it is possible to construct a scalar integral of p-forms over manifolds of dimensionality p, which provides a natural generalization of the familiar volume integrals of scalar densities and surface integrals of normal components of vector densities. A manifold $\mathscr{M}$ of dimensionality p in an n-dimensional space is simply a region within which the n coordinates x^μ may be expressed in a smooth one-to-one way as functions of p parameters u^i:

$$x^\mu = x^\mu(u^1, u^2, \ldots, u^p) \qquad (4.11.17)$$

Actually, it is often impossible to cover the whole of a manifold with a *single* set of u-coordinates; in the general case, it is necessary to introduce different sets of u-

coordinates in different overlapping patches of the manifold, with the proviso that in the overlap between one patch with coordinates u^i and another patch with coordinates $\bar{u}^i$, the u^i can be expressed in a smooth one-to-one way as functions of the $\bar{u}^i$, and vice versa. We shall actually be concerned here with what are called *orientable manifolds*, for which the coordinates in each patch can be chosen so that in the overlap regions all determinants $|\partial u/\partial \bar{u}|$ are positive-definite. For instance, the surface of a sphere is orientable. For simplicity of notation, the discussion below does not take account of these complications, but it should be kept in mind that more than one set of u-coordinates may be needed to cover the manifold. With this understanding, the integral of a p-form t over a manifold $\mathcal{M}$ of dimensionality p is defined as the repeated integral

$$\int_{\mathcal{M}} t \, dV_p \equiv \int t_{\mu_1 \cdots \mu_p} \frac{\partial x^{\mu_1}}{\partial u^1} \cdots \frac{\partial x^{\mu_p}}{\partial u^p} \, du^1 \cdots du^p \tag{4.11.18}$$

with limits of integration set by the boundaries of the manifold.

This integral is obviously a scalar with respect to transformations of the x^μ coordinates used to define the p-form. It is also necessary to consider how the integral behaves if we decide to describe the manifold with a new set of parameters $\bar{u}^1 \cdots \bar{u}^p$ instead of $u^1 \cdots u^p$. Taking account of the antisymmetry of t, it is easy to see that in this case the integrand changes by a factor, the determinant $|\partial u/\partial \bar{u}|$, whereas the p-dimensional volume element changes by a positive factor $||\partial \bar{u}/\partial u||$. Thus the whole integral either is unchanged or changes by a minus sign, according to whether the determinant $|\partial u/\partial \bar{u}|$ is positive or negative. (We tacitly assume that the transformation $u^i \to \bar{u}^i$ is nonsingular, so that this determinant cannot vanish, and therefore keeps the same sign throughout $\mathcal{M}$.) This result shows, incidentally, that when several u-coordinate systems are needed to cover the manifold, the integral of a p-form over the overlap between two patches described by coordinates u^i and $\bar{u}^i$ can be evaluated using either coordinate system, provided that the determinant $|\partial u/\partial \bar{u}|$ is positive; it is for this reason that we have to restrict our attention to manifolds that are orientable.

The simplest example of an integral of the form (4.11.18) is provided by the special case where p is equal to the dimensionality n of the x^μ coordinate space. Here the x^μ coordinates themselves may be used as the u-coordinates, so that (4.11.18) becomes in this case

$$\int_{\mathcal{M}} t \, dV_n = \int t_{12 \cdots n} \, dx^1 \, dx^2 \cdots dx^n$$

Note that the integrand $t_{12 \cdots n}$, in addition to being one component of a tensor, is also a scalar density of weight -1, because we can write it as

$$t_{12 \cdots n} = \frac{1}{n!} \, \varepsilon^{\mu_1 \cdots \mu_n} t_{\mu_1 \cdots \mu_n}$$

and $\varepsilon^{\mu\nu\cdots}$ is a tensor density of weight -1. (See Section 4.4.) In the next simplest case, where $p = n - 1$, we can write (4.11.18) in the familiar form

$$\int_{\mathcal{M}} t \, dV_{n-1} = \int t^\mu \, dS_\mu$$

where t^μ is a vector density, defined by

$$t_{\mu_1\cdots\mu_p} \equiv \varepsilon_{\mu_1\cdots\mu_p\mu} t^\mu$$

and dS_μ is a surface element oriented normal to the manifold:

$$dS_\mu \equiv \varepsilon_{\mu_1\cdots\mu_p\mu} \frac{\partial x^{\mu_1}}{\partial u^1} \cdots \frac{\partial x^{\mu_p}}{\partial u^p} \, du^1 \cdots du^p$$

With this general definition of the integrals of p-forms, it is possible to prove that the integral of the exterior derivative of a p-form over a manifold of dimensionality $(p + 1)$ is simply the integral of the p-form itself over the p-dimensional boundary of the manifold:[7]

$$\int_{\mathcal{M}} Dt \, dV_{p+1} = \int_{\text{boundary of } \mathcal{M}} t \, dV_p \qquad (4.11.19)$$

(We shall not go into the problem of defining the orientation of the boundary, which is needed in order to specify the sign of the right-hand side). Stokes's theorem and Gauss's theorem are simply the special cases of this general formula for $n = 3$, $p = 1$ and $n = 3$, $p = 2$, respectively.

4 REFERENCES

1. E. Kretschmann, Ann. Phys. Leipzig, **53**, 575 (1917).
1a. K. O. Friedrichs, Math. Ann., **98**, 566 (1928).
2. E. P. Wigner, *Symmetries and Reflections* (Indiana University Press, Bloomington, Ind., 1967).
3. For this approach to chiral symmetry, see, for example, S. Weinberg in *Lectures on Elementary Particles and Quantum Field Theory* (M.I.T. Press, Cambridge, Mass., 1970), p. 283.

4. See, for example, L. M. Graves, *The Theory of Functions of A Real Variable* (McGraw-Hill, New York, 1956).

5. See, for example, J. A. Stratton, *Electromagnetic Theory* (McGraw-Hill, New York, 1941), Sections 1.14–1.18.

6. See, for example, L. I. Schiff, *Quantum Mechanics* (3rd ed., McGraw-Hill, New York, 1968), p. 399.

7. For an extremely readable text on the theory of differential forms, see H. Flanders, *Differential Forms* (Academic Press, New York, 1963).

Some guide the course of wand'ring orbs on high,
Or roll the planets through the boundless sky.
Some less refined, beneath the moon's pale light
Pursue the stars that shoot athwart the night,
Or suck the mists in grosser air below,
Or dip their pinions in the painted bow
Or blow fierce tempests on the wintry main,
Or o'er the glebe distill the kindly rain.
Alexander Pope, The Rape of the Lock

5 EFFECTS OF GRAVITATION

We now return to physics, and apply what we have learned in the last chapter to determine the effects of gravitation on the equations of mechanics and electrodynamics. The technique to be used is that afforded by the Principle of General Covariance: We must first write the equations as they hold in special relativity, then decide how each quantity in the equations is to transform under general coordinate transformations, and then replace $\eta_{\mu\nu}$ with $g_{\mu\nu}$ and all derivatives with covariant derivatives. The resulting equations will be generally covariant and true in the absence of gravitation, and hence true in arbitrary gravitational fields, provided that the system in question is small enough compared with the scale of the fields.

1 Particle Mechanics

A particle not under the influence of any force will in special relativity have constant four-velocity U^α and constant spin S_α, that is,

$$\frac{dU^\alpha}{d\tau} = 0 \qquad \left(U^\alpha \equiv \frac{d\xi^\alpha}{d\tau} \right) \tag{5.1.1}$$

$$\frac{dS_\alpha}{d\tau} = 0 \tag{5.1.2}$$

Recall that S_α is defined in the rest frame of the particle to have the components $\{S, 0\}$, so that in an arbitrary Lorentz frame it satisfies the further relation

$$S_\alpha U^\alpha = 0 \qquad (5.1.3)$$

In order to make these equations generally covariant, we define vectors U^μ and S_μ in a general coordinate system x^μ by

$$U^\mu \equiv \frac{\partial x^\mu}{\partial \xi^\alpha} U_f{}^\alpha = \frac{dx^\mu}{d\tau} \qquad (5.1.4)$$

$$S_\mu \equiv \frac{\partial \xi^\alpha}{\partial x^\mu} S_{f\alpha} \qquad (5.1.5)$$

where $U_f{}^\alpha$ and $S_{f\alpha}$ are components of U and S in the freely falling coordinate system ξ^α. Although U^μ and S_μ are vectors, $dU^\mu/d\tau$ and $dS_\mu/d\tau$ are not, but we saw in Section 4.9 that there can be defined vector derivatives $DU^\mu/D\tau$ and $DS_\mu/D\tau$, which reduce when $\Gamma^\mu_{\nu\tau} = 0$ to the ordinary derivatives $dU^\mu/d\tau$ and $dS_\mu/d\tau$. The correct equations for the particle position and spin are dictated by the Principle of General Covariance to be

$$\frac{DU^\mu}{D\tau} = 0 \qquad \frac{DS_\mu}{D\tau} = 0 \qquad (5.1.6)$$

or, in more detail,

$$\frac{dU^\mu}{d\tau} + \Gamma^\mu_{\nu\lambda} U^\nu U^\lambda = 0 \qquad (5.1.7)$$

$$\frac{dS_\mu}{d\tau} - \Gamma^\lambda_{\mu\nu} U^\nu S_\lambda = 0 \qquad (5.1.8)$$

In addition, (5.1.3) becomes now

$$S_\mu U^\mu = 0 \qquad (5.1.9)$$

To repeat the reasoning of Section 4.1, these equations are true in the presence of gravitational fields because they are generally covariant and true in the absence of gravitation, reducing when $\Gamma^\mu_{\nu\lambda}$ vanishes to Eqs. (5.1.1)–(5.1.3). That is, the Principle of Equivalence tells us that there are locally inertial coordinate systems in which (5.1.6)–(5.1.9) are valid (provided always that our particle is sufficiently small), and general covariance then ensures that these equations hold in the laboratory reference frame.

We recognize in Eqs. (5.1.7) and (5.1.8) the differential equations for parallel transport of the vectors U^μ and S_μ. Since $U^\mu \equiv dx^\mu/d\tau$, Eq. (5.1.7) is nothing but the familiar equation for free fall, derived previously by differentiating (5.1.4) with respect to τ and using (5.1.1); it should be evident that a good deal of work is saved by using general covariance instead of our previous direct approach. Equation (5.1.8) describes the precession of gyroscopes in free fall, and is further discussed in Chapter 9. For the present, we note that $S_\mu S^\mu$ is constant, because the ordinary derivative of a scalar is the same as its covariant derivative

$$\frac{d}{d\tau}(S_\mu S^\mu) = \frac{D}{D\tau}(S_\mu S^\mu) = 0 \qquad (5.1.10)$$

If the particle is not in free fall, then $DU^\mu/D\tau$ does not vanish, and instead of (5.1.7) we have

$$\frac{DU^\mu}{D\tau} \equiv \frac{f^\mu}{m} \qquad (5.1.11)$$

where m is the particle mass and f^μ is a contravariant force vector. This can also be written as

$$m\frac{d^2 x^\mu}{d\tau^2} = f^\mu - m\Gamma^\mu_{\nu\lambda}\frac{dx^\nu}{d\tau}\frac{dx^\lambda}{d\tau}$$

The term containing $m\Gamma^\mu_{\nu\lambda}$ evidently plays the role of a gravitational force. We can always calculate f^μ if we know its value f^α_f in freely falling frames ξ^α, for the requirement that f^μ behave as a vector gives it uniquely as

$$f^\mu = \frac{\partial x^\mu}{\partial \xi^\alpha}f^\alpha_f \qquad (5.1.12)$$

The electromagnetic force will be determined in the next section.

It sometimes happens that a particle is acted on by a force f^μ without experiencing any torque. In this event, an observer in a locally inertial coordinate frame that is momentarily at rest with respect to the particle will see no precession of the spin axis; that is, dS/dt will vanish. But in this particular coordinate system dx/dt also vanishes, so we may write the condition of zero torque as a Lorentz-invariant statement

$$\frac{dS^\alpha}{d\tau} \propto U^\alpha$$

and this will be valid in any locally inertial coordinate system, comoving or not. Now, what is the constant of proportionality? Let us set

$$\frac{dS^\alpha}{d\tau} = \Phi U^\alpha$$

We recall that S_α is defined so that

$$S_\alpha U^\alpha = 0$$

and therefore

$$0 = \frac{d}{d\tau}(S_\alpha U^\alpha) = \Phi U_\alpha U^\alpha + S_\alpha \frac{dU^\alpha}{d\tau}$$

so

$$\Phi = S_\alpha \frac{dU^\alpha}{d\tau} = S_\alpha \frac{f^\alpha}{m}$$

The spin vector therefore suffers a change given by

$$\frac{dS^\alpha}{d\tau} = \left(S_\beta \frac{f^\beta}{m} \right) U^\alpha \tag{5.1.13}$$

This phenomenon is known as the *Thomas precession*.[1] If we now turn on a gravitational field, Eq. (5.1.13) and the Principle of General Covariance tell us that the spin precesses according to the rule

$$\frac{DS^\mu}{D\tau} = \left(S_\nu \frac{f^\nu}{m} \right) U^\mu = S_\nu \frac{DU^\nu}{D\tau} U^\mu \tag{5.1.14}$$

A vector obeying this differential equation is said to be defined by *Fermi transport*;[2] parallel transport is the special case for $f^\mu = 0$.

2   Electrodynamics

We recall that in the absence of gravitational fields the Maxwell equations of electrodynamics can be written as

$$\frac{\partial}{\partial x^\alpha} F^{\alpha\beta} = -J^\beta \tag{5.2.1}$$

$$\frac{\partial}{\partial x_\alpha} F_{\beta\gamma} + \frac{\partial}{\partial x_\beta} F_{\gamma\alpha} + \frac{\partial}{\partial x_\gamma} F_{\alpha\beta} = 0 \tag{5.2.2}$$

where J^β is the current four-vector $\{\mathbf{J}, \varepsilon\}$, and $F^{\alpha\beta}$ is the field-strength tensor, $F^{12} = B_3$, $F^{01} = E_1$, and so on. (See Section 2.7.) Suppose that we *define* $F^{\mu\nu}$ and J^μ in general coordinates by the requirements that they reduce to $F^{\alpha\beta}$ and J^β in locally inertial Minkowskian coordinates, and that they behave as tensors under general coordinate transformations. (That is, if $\tilde{F}^{\alpha\beta}$ and $\tilde{J}^\alpha$ are the values measured in a locally inertial frame, then $F^{\mu\nu} \equiv (\partial x^\mu / \partial \xi^\alpha)(\partial x^\nu / \partial \xi^\beta)\tilde{F}^{\alpha\beta}$ and $J^\mu \equiv (\partial x^\mu / \partial \xi^\alpha)\tilde{J}^\alpha$.) We can then *make* (5.2.1) and (5.2.2) generally covariant by replacing all derivatives by covariant derivatives:

$$F^{\mu\nu}_{\ \ ;\mu} = -J^\nu \tag{5.2.3}$$

$$F_{\mu\nu;\lambda} + F_{\lambda\mu;\nu} + F_{\nu\lambda;\mu} = 0 \tag{5.2.4}$$

it being now understood that indices are to be raised and lowered with $g_{\mu\lambda}$ instead of $\eta_{\alpha\gamma}$, that is,

$$F_{\lambda\kappa} \equiv g_{\lambda\mu} g_{\kappa\nu} F^{\mu\nu} \tag{5.2.5}$$

Since $F^{\mu\nu}$ and $F_{\mu\nu}$ are antisymmetric, we may use (4.7.10) and (4.7.11) to rewrite the Maxwell equations as

$$\frac{\partial}{\partial x^{\mu}} \sqrt{g}\, F^{\mu\nu} = -\sqrt{g}\, J^{\nu} \tag{5.2.6}$$

$$\frac{\partial}{\partial x^{\lambda}} F_{\mu\nu} + \frac{\partial}{\partial x^{\nu}} F_{\lambda\mu} + \frac{\partial}{\partial x^{\mu}} F_{\nu\lambda} = 0 \tag{5.2.7}$$

Equations (5.2.3)–(5.2.7) are true in the absence of gravitation and generally covariant and hence, according to the Principle of General Covariance, also true in arbitrary gravitational fields.

The electromagnetic force on a particle of charge e is given in the absence of gravitation by Eq. (2.7.9):

$$f^{\alpha} = eF^{\alpha}{}_{\beta} \frac{dx^{\beta}}{d\tau} \tag{5.2.8}$$

We immediately conclude that in general coordinates the electromagnetic force in an arbitrary gravitational field is

$$f^{\mu} = eF^{\mu}{}_{\nu} \frac{dx^{\nu}}{d\tau} \tag{5.2.9}$$

where of course

$$F^{\mu}{}_{\nu} \equiv g_{\nu\lambda} F^{\mu\lambda} \tag{5.2.10}$$

Once again, we are using the Principle of General Covariance; Eq. (5.2.9) obviously reduces to (5.2.8) in locally inertial Minkowskian coordinates, and it is generally covariant, because f^{μ} is a vector (Section 5.1), $dx^{\nu}/d\tau$ is a vector, and $F^{\mu}{}_{\nu}$ is *defined* as a tensor; therefore (5.2.9) is true.

It is instructive to evaluate the current vector J^{ν}. In special relativity it is

$$J^{\alpha} = \sum_{n} e_{n} \int \delta^{4}(x - x_{n})\, dx_{n}^{\alpha} \tag{5.2.11}$$

the integral being taken along the trajectory of the nth particle. [See Eq. (2.6.5).] The four-dimensional delta function in a general coordinate system is defined by

$$\int d^{4}x\, \Phi(x) \delta^{4}(x - y) = \Phi(y) \tag{5.2.12}$$

Since $g^{1/2}\, d^{4}x$ is a scalar, $g^{-1/2}\delta^{4}(x - y)$ must be a scalar, which of course reduces to the ordinary delta function in special relativity, where $g = 1$. (In some works it is this scalar that is defined as the delta function.) Thus the contravariant vector that reduces to J^{α} in the absence of gravitation is

$$J^{\mu}(x) = g^{-1/2}(x) \sum_{n} e_{n} \int \delta^{4}(x - x_{n})\, dx_{n}^{\mu} \tag{5.2.13}$$

Note that the conservation law $\partial J^\alpha/\partial x^\alpha = 0$ of special relativity becomes in general relativity $J^\mu{}_{;\mu} = 0$ or, using (4.7.7),

$$\frac{\partial}{\partial x^\mu}\left(g^{1/2}J^\mu\right) = 0 \tag{5.2.14}$$

The factor $g^{-1/2}$ in (5.2.13) is just what is needed to cancel the $g^{1/2}$ in (5.2.14), so that (5.2.14) still just expresses the constancy of the e_n.

3   Energy-Momentum Tensor

The density and current of energy and momentum were united in Section 2.8 into a symmetric tensor $T^{\alpha\beta}$ satisfying the conservation equations

$$\frac{\partial T^{\alpha\beta}}{\partial x^\alpha} = G^\beta \tag{5.3.1}$$

where G^β is the density of the external force f^β acting on the system. (For an isolated system, $G^\beta = 0$.) Define $T^{\mu\nu}$ and G^ν as contravariant tensors that reduce to the special relativistic $T^{\alpha\beta}$ and G^β in the absence of gravitation. Then the generally covariant equation that agrees with (5.3.1) in locally inertial systems is

$$T^{\mu\nu}{}_{;\mu} = G^\nu \tag{5.3.2}$$

or, using (4.7.9),

$$\frac{1}{\sqrt{g}}\frac{\partial}{\partial x^\mu}\left(\sqrt{g}\,T^{\mu\nu}\right) = G^\nu - \Gamma^\nu_{\mu\lambda}T^{\mu\lambda} \tag{5.3.3}$$

The factor $\sqrt{g}$ is familiar from electrodynamics, and arises from the fact that the invariant volume is $\sqrt{g}\,d^4x$. In contrast, the second term on the right represents a gravitational *force* density. Just as we would expect, this force depends on the system on which it acts only through the energy-momentum tensor.

For a system of point particles the special-relativistic energy-momentum tensor is given in Section 2.8 as

$$T^{\alpha\beta} = \sum_n m_n \int \frac{dx_n{}^\alpha}{d\tau}\,dx_n{}^\beta \delta^4(x - x_n) \tag{5.3.4}$$

the integral again being taken along the particle trajectory. Following precisely the same reasoning as we did for J^α in the last section, we conclude that the contravariant tensor that agrees with (5.3.4) in the absence of gravitation is

$$T^{\mu\nu} = g^{-1/2}\sum_n m_n \int \frac{dx_n{}^\mu}{d\tau}\,dx_n{}^\nu \delta^4(x - x_n) \tag{5.3.5}$$

For an electromagnetic field $F^{\alpha\beta}$ the special-relativistic energy-momentum tensor was calculated in Section 2.8 as

$$T^{\alpha\beta} = F^{\alpha}{}_{\gamma}F^{\beta\gamma} - \tfrac{1}{4}\eta^{\alpha\beta}F_{\gamma\delta}F^{\gamma\delta} \qquad (5.3.6)$$

It takes no effort to see that the contravariant tensor that agrees with (5.3.6) in the absence of gravitation is

$$T^{\mu\nu} = F^{\mu}{}_{\lambda}F^{\nu\lambda} - \tfrac{1}{4}g^{\mu\nu}F_{\lambda\kappa}F^{\lambda\kappa} \qquad (5.3.7)$$

For a system consisting of particles and radiation, the energy-momentum tensor is the sum of (5.3.5) and (5.3.7).

Returning for a moment to the purely material energy-momentum tensor (5.3.5), we easily compute that

$$\int T^{\mu 0}g^{1/2}\,d^3x = \sum_n m_n \frac{dx_n{}^{\mu}}{d\tau}$$

the sum running over all particles in the volume of integration. This suggests that $T^{\mu 0}g^{1/2}$ is to be regarded in general as the spatial density of energy and momentum. In particular, we are tempted to define the energy, momentum, and angular momentum for an arbitrary system by

$$P^{\mu} \equiv \int T^{\mu 0}g^{1/2}\,d^3x \qquad (5.3.8)$$

$$J^{\mu\nu} \equiv \int (x^{\mu}T^{\nu 0} - x^{\nu}T^{\mu 0})g^{1/2}\,d^3x \qquad (5.3.9)$$

However, these quantities are *not* contravariant tensors and are *not* conserved, because $T^{\mu\nu}g^{1/2}$ is not conserved, that is, $\partial(T^{\mu\nu}g^{1/2})/\partial x^{\nu}$ does not vanish, owing to the exchange of energy and momentum between matter and gravitation.

4   Hydrodynamics and Hydrostatics

In the absence of gravitation, the energy-momentum tensor of a perfect fluid is that given by (2.10.7):

$$T^{\alpha\beta} = p\eta^{\alpha\beta} + (p + \rho)U^{\alpha}U^{\beta} \qquad (5.4.1)$$

where U^{α} is the fluid four-velocity, $U^0 = (1 - \mathbf{v}^2)^{-1/2}$, $\mathbf{U} = \mathbf{v}U^0$. The contravariant tensor that reduces to (5.4.1) in the absence of gravitation is

$$T^{\mu\nu} = pg^{\mu\nu} + (p + \rho)U^{\mu}U^{\nu} \qquad (5.4.2)$$

where U^{μ} is the local value of $dx^{\mu}/d\tau$ for a comoving fluid element. Note that p and ρ are always defined as the pressure and energy density measured by an observer in a locally inertial frame that happens to be moving with the fluid at the instant of measurement, and are therefore scalars. The conditions of energy-momentum conservation give the hydrodynamic equations

$$0 = T^{\mu\nu}{}_{;\nu} = \frac{\partial p}{\partial x^\nu} g^{\mu\nu} + g^{-1/2} \frac{\partial}{\partial x^\nu} g^{1/2}(p + \rho)U^\mu U^\nu$$

$$+ \Gamma^\mu_{\nu\lambda}(p + \rho)U^\nu U^\lambda \qquad (5.4.3)$$

The last term represents the gravitational force on the system. Note also that since $\eta_{\alpha\beta}U^\alpha U^\beta = -1$ in the absence of gravitation, we must in the presence of gravitation have

$$g_{\mu\nu}U^\mu U^\nu = -1 \qquad (5.4.4)$$

Consider as an example the case of a fluid in hydrostatic equilibrium. Since it is not moving, (5.4.4) gives

$$U^0 = (-g_{00})^{-1/2} \qquad U^\lambda = 0 \quad \text{for } \lambda \neq 0$$

Furthermore, all temporal derivatives of $g_{\mu\nu}$, p, or ρ vanish. In particular,

$$\Gamma^\mu_{00} = -\tfrac{1}{2}g^{\mu\nu}\frac{\partial g_{00}}{\partial x^\nu}$$

and

$$\frac{\partial}{\partial x^\nu}[(p + \rho)U^\mu U^\nu] = 0$$

Multiplying (5.4.3) by $g_{\mu\lambda}$ gives then

$$-\frac{\partial p}{\partial x^\lambda} = (p + \rho)\frac{\partial}{\partial x^\lambda}\ln(-g_{00})^{1/2} \qquad (5.4.5)$$

This is trivial for $\lambda = 0$, whereas for λ spacelike it is nothing but the ordinary nonrelativistic equation of hydrostatic equilibrium, except that $p + \rho$ appears instead of the mass density, and $\ln(-g_{00})^{1/2}$ appears instead of the gravitational potential. This equation is soluble if p is given as a function of ρ. We then find that

$$\int \frac{dp(\rho)}{p(\rho) + \rho} = -\ln\sqrt{-g_{00}} + \text{constant} \qquad (5.4.6)$$

For instance, if $p(\rho)$ is given by a power law:

$$p(\rho) \propto \rho^N \qquad (5.4.7)$$

then (5.4.6) gives for $N \neq 1$

$$\frac{\rho + p}{\rho} \propto (-g_{00})^{(1-N)/2N} \qquad (5.4.8)$$

and for $N = 1$

$$\rho \propto (-g_{00})^{-(p+\rho)/2p} \tag{5.4.9}$$

This, incidentally, shows that gravitation can never produce hydrostatic equilibrium in a finite highly relativistic fluid with $p = \rho/3$, for then (5.4.9) gives

$$\rho \propto (-g_{00})^{-2} \tag{5.4.10}$$

Since ρ must vanish outside the fluid, g_{00} would have to become singular at its surface.

5 REFERENCES

1. L. H. Thomas, Nature, **117**, 514 (1926).
2. E. Fermi, Atti. R. Accad. Rend. Cl. Sc. Fis. Mat. Nat., **31**, 21 (1922).

"When I behold, upon the
night's starr'd face, huge
cloudy symbols . . ."
*John Keats, When I Have
Fears That I May Cease
to Be*

6 CURVATURE

We are going to work out the gravitational field equations by applying the Principle of Equivalence to gravitation itself. As in the last chapter, it is most convenient to apply this principle by looking for field equations that are generally covariant and that reduce to the proper form for weak fields. Thus we must address ourselves to the question: What tensors can be formed from the metric tensor and its derivatives? In this chapter we treat this as a purely mathematical problem, as in fact it was treated by Gauss and Riemann; the information we compile here will then be used in the next chapter to guide us in our search for the field equations of gravitation.

1 Definition of the Curvature Tensor

We want to construct a tensor out of the metric tensor and its derivatives. If we use only $g_{\mu\nu}$ and its first derivatives, then no new tensor can be constructed, for at any point we can find a coordinate system in which the first derivatives of the metric tensor vanish, so in this coordinate system the desired tensor must be equal to one of those that can be constructed out of the metric tensor *alone*, (e.g., $g_{\mu\nu}$ or $g^{\mu\nu}$ or $\varepsilon^{\mu\nu\lambda\eta}/\sqrt{g}$, and so on), and since this is an equality between tensors it must be true in all coordinate systems.

The next simplest possibility is to construct a tensor out of the metric tensor and its first and second derivatives. To do this, let us recall the transformation rule for the affine connection:

$$\Gamma^\lambda_{\mu\nu} = \frac{\partial x^\lambda}{\partial x'^\tau} \frac{\partial x'^\rho}{\partial x^\mu} \frac{\partial x'^\sigma}{\partial x^\nu} \Gamma'^\tau_{\rho\sigma} + \frac{\partial x^\lambda}{\partial x'^\tau} \frac{\partial^2 x'^\tau}{\partial x^\mu \partial x^\nu} \tag{6.1.1}$$

(This is Eq. (4.5.2), with primed and unprimed coordinates interchanged.) It is the inhomogeneous term on the right that keeps $\Gamma^\lambda_{\mu\nu}$ from being a tensor, so let us isolate this term:

$$\frac{\partial^2 x'^\tau}{\partial x^\mu \partial x^\nu} = \frac{\partial x'^\tau}{\partial x^\lambda} \Gamma^\lambda_{\mu\nu} - \frac{\partial x'^\rho}{\partial x^\mu} \frac{\partial x'^\sigma}{\partial x^\nu} \Gamma'^\tau_{\rho\sigma} \tag{6.1.2}$$

To get rid of the left-hand side, we use the commutativity of partial differentiation. Differentiation with respect to x^κ gives

$$\frac{\partial^3 x'^\tau}{\partial x^\kappa \partial x^\mu \partial x^\nu} = \Gamma^\lambda_{\mu\nu} \left(\frac{\partial x'^\tau}{\partial x^\eta} \Gamma^\eta_{\kappa\lambda} - \frac{\partial x'^\rho}{\partial x^\kappa} \frac{\partial x'^\sigma}{\partial x^\lambda} \Gamma'^\tau_{\rho\sigma} \right)$$

$$- \Gamma'^\tau_{\rho\sigma} \frac{\partial x'^\rho}{\partial x^\mu} \left(\frac{\partial x'^\sigma}{\partial x^\eta} \Gamma^\eta_{\kappa\nu} - \frac{\partial x'^\eta}{\partial x^\kappa} \frac{\partial x'^\xi}{\partial x^\nu} \Gamma'^\sigma_{\eta\xi} \right)$$

$$- \Gamma'^\tau_{\rho\sigma} \frac{\partial x'^\sigma}{\partial x^\nu} \left(\frac{\partial x'^\rho}{\partial x^\eta} \Gamma^\eta_{\kappa\mu} - \frac{\partial x'^\eta}{\partial x^\kappa} \frac{\partial x'^\xi}{\partial x^\mu} \Gamma'^\rho_{\eta\xi} \right)$$

$$+ \frac{\partial x'^\tau}{\partial x^\lambda} \frac{\partial \Gamma^\lambda_{\mu\nu}}{\partial x^\kappa} - \frac{\partial x'^\rho}{\partial x^\mu} \frac{\partial x'^\sigma}{\partial x^\nu} \frac{\partial x'^\eta}{\partial x^\kappa} \frac{\partial \Gamma'^\tau_{\rho\sigma}}{\partial x'^\eta}$$

or, collecting similar terms and juggling indices a bit,

$$\frac{\partial^3 x'^\tau}{\partial x^\kappa \partial x^\mu \partial x^\nu} = \frac{\partial x'^\tau}{\partial x^\lambda} \left(\frac{\partial \Gamma^\lambda_{\mu\nu}}{\partial x^\kappa} + \Gamma^\eta_{\mu\nu} \Gamma^\lambda_{\kappa\eta} \right)$$

$$- \frac{\partial x'^\rho}{\partial x^\mu} \frac{\partial x'^\sigma}{\partial x^\nu} \frac{\partial x'^\eta}{\partial x^\kappa} \left(\frac{\partial \Gamma'^\tau_{\rho\sigma}}{\partial x'^\eta} - \Gamma'^\tau_{\rho\lambda} \Gamma'^\lambda_{\eta\sigma} - \Gamma'^\tau_{\lambda\sigma} \Gamma'^\lambda_{\eta\rho} \right)$$

$$- \Gamma'^\tau_{\rho\sigma} \frac{\partial x'^\sigma}{\partial x^\lambda} \left(\Gamma^\lambda_{\mu\nu} \frac{\partial x'^\rho}{\partial x^\kappa} + \Gamma^\lambda_{\kappa\nu} \frac{\partial x'^\rho}{\partial x^\mu} + \Gamma^\lambda_{\kappa\mu} \frac{\partial x'^\rho}{\partial x^\nu} \right) \tag{6.1.3}$$

Now subtracting the same equation with ν and κ interchanged, we find that all terms involving products of Γ with Γ' drop out, leaving

$$0 = \frac{\partial x'^\tau}{\partial x^\lambda} \left(\frac{\partial \Gamma^\lambda_{\mu\nu}}{\partial x^\kappa} - \frac{\partial \Gamma^\lambda_{\mu\kappa}}{\partial x^\nu} + \Gamma^\eta_{\mu\nu} \Gamma^\lambda_{\kappa\eta} - \Gamma^\eta_{\mu\kappa} \Gamma^\lambda_{\nu\eta} \right)$$

$$- \frac{\partial x'^\rho}{\partial x^\mu} \frac{\partial x'^\sigma}{\partial x^\nu} \frac{\partial x'^\eta}{\partial x^\kappa} \left(\frac{\partial \Gamma'^\tau_{\rho\sigma}}{\partial x'^\eta} - \frac{\partial \Gamma'^\tau_{\rho\eta}}{\partial x'^\sigma} - \Gamma'^\tau_{\lambda\sigma} \Gamma'^\lambda_{\eta\rho} + \Gamma'^\tau_{\lambda\eta} \Gamma'^\lambda_{\sigma\rho} \right)$$

This may be written as a transformation rule,

$$R'^\tau_{\rho\sigma\eta} = \frac{\partial x'^\tau}{\partial x^\lambda} \frac{\partial x^\mu}{\partial x'^\rho} \frac{\partial x^\nu}{\partial x'^\sigma} \frac{\partial x^\kappa}{\partial x'^\eta} R^\lambda_{\mu\nu\kappa} \tag{6.1.4}$$

where

$$R^\lambda_{\ \mu\nu\kappa} \equiv \frac{\partial \Gamma^\lambda_{\mu\nu}}{\partial x^\kappa} - \frac{\partial \Gamma^\lambda_{\mu\kappa}}{\partial x^\nu} + \Gamma^\eta_{\mu\nu}\Gamma^\lambda_{\kappa\eta} - \Gamma^\eta_{\mu\kappa}\Gamma^\lambda_{\nu\eta} \tag{6.1.5}$$

Equation (6.1.4) says that $R^\lambda_{\ \mu\nu\kappa}$ is a tensor; it is called the *Riemann-Christoffel curvature tensor*.

The existence of the tensor $R^\lambda_{\ \mu\nu\kappa}$ raises once again the question of whether or not the Principle of Equivalence or the Principle of General Covariance uniquely determines the effects of gravitation on arbitrary physical systems. For instance, let us ask whether the correct equation of motion for a freely falling particle of spin S_μ might be of the form

$$0 = \frac{d^2 x^\lambda}{d\tau^2} + \Gamma^\lambda_{\mu\nu}\frac{dx^\mu}{d\tau}\frac{dx^\nu}{d\tau} + f R^\lambda_{\ \mu\nu\kappa}\frac{dx^\mu}{d\tau}\frac{dx^\nu}{d\tau} S^\kappa \tag{6.1.6}$$

(with f an unknown scalar) instead of the familiar form

$$0 = \frac{d^2 x^\lambda}{d\tau^2} + \Gamma^\lambda_{\mu\nu}\frac{dx^\mu}{d\tau}\frac{dx^\nu}{d\tau} \tag{6.1.7}$$

Both Eqs. (6.1.6) and (6.1.7) are generally covariant, and both reduce in the absence of gravitation to the correct special-relativistic equation $dU^\alpha/d\tau = 0$. How then can we tell whether (6.1.6) or (6.1.7) is correct?

The answer again is one of *scale*. Suppose that our particle has a characteristic linear dimension d, and that the gravitational field has a characteristic space-time dimension D. The Riemann-Christoffel tensor has one more derivative of the metric than the affine connection, so the ratio of the third term in (6.1.6) to the second term is proportional to $1/D$; dimensional considerations then require that this ratio be roughly of order d/D. Thus, barring special circumstances that might make one term or the other anomalously large or small, we can regard the last term in (6.1.6) as being negligible if our particle is very much smaller than the characteristic dimensions of the gravitational field, and (6.1.7) is the correct equation of motion. Of course, if our particle is *not* much smaller than the gravitational field scale (as in the case of the moon moving in the gravitational field of the earth), then the Principle of Equivalence or the Principle of General Covariance must be applied to the infinitesimal elements of which the particle is composed, although (6.1.6) or (6.1.7) might give a fair phenomenological representation of the motion of the whole particle.

2 Uniqueness of the Curvature Tensor

We next prove that $R^\lambda_{\ \mu\nu\kappa}$ is the *only* tensor that can be constructed from the metric tensor and its first and second derivatives, and is linear in the second derivatives.

For this purpose it proves extremely convenient to fix our attention on a particular point X, and adopt a locally inertial coordinate system, in which at this point the affine connection $\Gamma^\lambda_{\mu\nu}$ vanishes. Furthermore, we consider only the limited class of coordinate transformations that leave the affine connection zero; according to Eq. (6.1.1), these are simply the transformations $x \to x'$ with

$$\left(\frac{\partial^2 x'^\tau}{\partial x^\mu \, \partial x^\nu}\right)_{x=X} = 0 \tag{6.2.1}$$

Any quantity that transforms as a tensor under general coordinate transformations will have to transform as a tensor under this limited class of transformations, and this requirement proves sufficiently strong for our purposes.

Since the affine connection vanishes at X, all first derivatives of the metric tensor vanish at X [see Eq. (3.3.5)] and our desired new tensor must be just a linear combination of the second derivatives of the metric tensor, or equivalently, of the first derivatives of the affine connection. We see from Eq. (6.1.3) that for $\Gamma^\lambda_{\mu\nu}$ and $\Gamma'^\tau_{\rho\sigma}$ zero, the derivatives of the affine connection obey the transformation rule

$$\frac{\partial \Gamma'^\tau_{\rho\sigma}}{\partial x'^\eta} = \frac{\partial x^\mu}{\partial x'^\rho} \frac{\partial x^\nu}{\partial x'^\sigma} \frac{\partial x^\kappa}{\partial x'^\eta} \frac{\partial x'^\tau}{\partial x^\lambda} \frac{\partial \Gamma^\lambda_{\mu\nu}}{\partial x^\kappa}$$

$$- \frac{\partial x^\mu}{\partial x'^\rho} \frac{\partial x^\nu}{\partial x'^\sigma} \frac{\partial x^\kappa}{\partial x'^\eta} \frac{\partial^3 x'^\tau}{\partial x^\kappa \, \partial x^\mu \, \partial x^\nu} \qquad \text{at } x = X \tag{6.2.2}$$

What linear combination of $\partial\Gamma/\partial x$ can we take that will behave like a tensor? Clearly, it must be such as to eliminate the inhomogeneous term in this transformation rule. However, at any given point X the inhomogeneous term is a completely arbitrary function of its indices ρ, σ, η, subject only to the condition that it be symmetric in these indices. Hence the only way of taking a linear combination of $\partial\Gamma/\partial x$ that will transform as a tensor under all transformations $x \to x'$ satisfying (6.2.1) is to antisymmetrize in κ and ν (or equivalently in κ and μ), so that (6.2.2) becomes

$$T'^\tau_{\rho\sigma\eta} = \frac{\partial x^\mu}{\partial x'^\rho} \frac{\partial x^\nu}{\partial x'^\sigma} \frac{\partial x^\kappa}{\partial x'^\eta} \frac{\partial x'^\tau}{\partial x^\lambda} \, T^\lambda_{\mu\nu\kappa} \qquad \text{at } x = X$$

where at $x = X$

$$T^\lambda_{\mu\nu\kappa} \equiv \frac{\partial \Gamma^\lambda_{\mu\nu}}{\partial x^\kappa} - \frac{\partial \Gamma^\lambda_{\mu\kappa}}{\partial x^\nu} \tag{6.2.3}$$

That is, the desired tensor must be a $T^\lambda_{\mu\nu\kappa}$ given by (6.2.3) when Γ vanishes. But when $\Gamma = 0$ the Riemann-Christoffel tensor satisfies (6.2.3), so in locally inertial systems $T^\lambda_{\mu\nu\kappa} = R^\lambda_{\mu\nu\kappa}$. But this is an equality between tensors, so since it is true in one class of coordinate systems it is true in all coordinate systems; that is, the only tensor T of the form desired is just $R^\lambda_{\mu\nu\kappa}$.

Of course other tensors can be formed by using the metric tensor itself to form linear combinations of the $R^\lambda{}_{\mu\nu\kappa}$. The most prominent are the contracted forms, the *Ricci tensor*,

$$R_{\mu\kappa} \equiv R^\lambda{}_{\mu\lambda\kappa} \tag{6.2.4}$$

and the *curvature scalar*,

$$R = g^{\mu\kappa} R_{\mu\kappa} \tag{6.2.5}$$

3 Round Trips by Parallel Transport

Both for its own sake and as preparation for the next section, we now take up the question whether a vector S_μ, when carried along a closed curve C by the equation of parallel transport (see Sections 4.9 and 5.1)

$$\frac{dS_\mu}{d\tau} = \Gamma^\lambda_{\mu\nu} \frac{dx^\nu}{d\tau} S_\lambda \tag{6.3.1}$$

will come back to its original value after one complete circuit of the curve.

We can answer this question by applying the method used in the familiar proof of Stokes's theorem. Consider the curve C as the edge of some two-dimensional surface A, and divide A into small cells bounded by little closed curves C_N. The change in S_μ when parallel-transported around C can be written as the sum of the changes in S_μ when transported around each of these little curves,

$$\Delta S_\mu = \sum_N \Delta_N S_\mu \tag{6.3.2}$$

because the change in S_μ around any one interior cell is canceled by the changes around adjacent cells, leaving only the contribution from the outer cell edges that make up C. Therefore we must ask only whether S_μ changes when parallel-transported around a small closed curve. If the curve is small enough, we can expand $\Gamma^\lambda_{\mu\nu}(x)$ around some point $X \equiv x(\tau_0)$ on the curve

$$\Gamma^\lambda_{\mu\nu}(x) = \Gamma^\lambda_{\mu\nu}(X) + (x^\rho - X^\rho)\frac{\partial}{\partial X^\rho}\Gamma^\lambda_{\mu\nu}(X) + \cdots \tag{6.3.3}$$

Then (6.3.1) gives to first order in $x^\mu - X^\mu$

$$S_\mu(\tau) = S_\mu(\tau_0) + \Gamma^\lambda_{\mu\nu}(X)(x^\nu(\tau) - X^\nu)S_\lambda(\tau_0) + \cdots \tag{6.3.4}$$

and by using (6.3.3) and (6.3.4) in (6.3.1) we obtain an equation valid to second order,

$$S_\mu(\tau) \simeq S_\mu(\tau_0) + \int_{\tau_0}^\tau \left[\Gamma^\lambda_{\mu\nu}(X) + (x^\rho(\tau) - X^\rho)\frac{\partial}{\partial X^\rho}\Gamma^\lambda_{\mu\nu}(X) + \cdots\right]$$

$$\times [S_\lambda(\tau_0) + S_\sigma(\tau_0)\Gamma^\sigma_{\lambda\rho}(X)(x^\rho(\tau) - X^\rho) + \cdots]\frac{dx^\nu(\tau)}{d\tau}\,d\tau$$

or discarding terms of third or higher order in $x - X$,

$$S_\mu(\tau) \simeq S_\mu(\tau_0) + \Gamma^\lambda_{\mu\nu}(X)S_\lambda(\tau_0) \int_{\tau_0}^{\tau} \frac{dx^\nu}{d\tau} d\tau$$

$$+ \left\{ \frac{\partial}{\partial X^\rho} \Gamma^\sigma_{\mu\nu}(X) + \Gamma^\sigma_{\lambda\rho}(X)\Gamma^\lambda_{\mu\nu}(X) \right\} S_\sigma(\tau_0) \int_{\tau_0}^{\tau} (x^\rho - X^\rho) \frac{dx^\nu}{d\tau} d\tau$$

If $x^\mu(\tau)$ returns to its original value X^μ at some $\tau = \tau_1$, then obviously

$$\int_{\tau_0}^{\tau_1} \frac{dx^\nu}{d\tau} d\tau = 0$$

so the change in S_μ when parallel-transported around the small closed curve $x^\mu(\tau)$ is of second order:

$$\Delta S_\mu \equiv S_\mu(\tau_1) - S_\mu(\tau_0)$$

$$= \left\{ \frac{\partial}{\partial X^\rho} \Gamma^\sigma_{\mu\nu}(X) + \Gamma^\lambda_{\mu\nu}(X)\Gamma^\sigma_{\lambda\rho}(X) \right\} S_\sigma(\tau_0) \oint x^\rho \, dx^\nu \qquad (6.3.5)$$

where

$$\oint x^\rho \, dx^\nu = \int_{\tau_0}^{\tau_1} x^\rho \frac{dx^\nu}{d\tau} d\tau$$

This integral does not generally vanish; for instance, if our curve is a small parallelogram with edges δa^μ, δb^μ, it equals

$$\oint x^\rho \, dx^\nu = \delta a^\rho \, \delta b^\nu - \delta a^\nu \, \delta b^\rho$$

However, it is always antisymmetric in ρ and ν, as can be seen by partial integration:

$$\oint x^\rho \, dx^\nu = \int_{\tau_0}^{\tau_1} \frac{d}{d\tau} (x^\rho x^\nu) \, d\tau - \int_{\tau_0}^{\tau_1} x^\nu \frac{dx^\rho}{d\tau} d\tau = - \oint x^\nu \, dx^\rho \qquad (6.3.6)$$

Thus the coefficient of this integral in (6.3.5) can be replaced by its antisymmetric part, which is just half the curvature tensor (6.1.5), so

$$\Delta S_\mu = \tfrac{1}{2} R^\sigma_{\mu\nu\rho} S_\sigma \oint x^\rho \, dx^\nu \qquad (6.3.7)$$

Our conclusion is that an arbitrary vector S_μ will not change when parallel-transported around an arbitrary small closed curve at X, if and only if $R^\sigma_{\mu\nu\rho}$ vanishes at X. We have already remarked that the change in S_μ when parallel-transported about a finite closed curve C may be computed by breaking up the area A bounded by C into small cells and then adding up the changes in S_μ when parallel-transported around the edges of these cells; hence, if $R^\sigma_{\mu\nu\rho}$ vanishes throughout A, then an arbitrary vector S_μ will not change when parallel-transported around C.

Now suppose that $R^{\sigma}{}_{\mu\nu\rho}$ does vanish. Consider a closed curve consisting of two segments A and B joining points x^{μ} and X^{μ}. The change in a vector S_{μ} when parallel-transported from x to X along A must be canceled by the change in S_{μ} when parallel-transported along B from X to x, that is,

$$\Delta^{A}_{X \to x} S_{\mu} + \Delta^{B}_{x \to x} S_{\mu} = 0$$

But the change in S_{μ} when parallel-transported from x to X along B is minus the change when parallel-transported from X to x along B:

$$\Delta^{B}_{x \to X} S_{\mu} = -\Delta^{B}_{X \to x} S_{\mu}$$

and therefore

$$\Delta^{A}_{X \to x} S_{\mu} = \Delta^{B}_{X \to x} S_{\mu} \tag{6.3.8}$$

That is, we get the same value of S_{μ} by parallel transportation from X to x, irrespective of which curve we follow. (For instance, if two gyroscopes are placed in different intersecting orbits about the earth, and have the same orientation when they pass close to each other at X^{μ}, then any difference in their orientations when they next pass close to each other at x^{μ} will be a measure of some average of the curvature produced by the earth's gravitational field.)

It follows that, given S_{μ} at X, we may determine a *field* $S_{\mu}(x)$, defined throughout the space-time region where $R^{\sigma}{}_{\mu\nu\rho}$ vanishes, by parallel transport from X to x; Eq. (6.3.8) ensures that the $S_{\mu}(x)$ so defined will depend only on x, and not on the path from X to x. For this field the derivative along any curve $x(\tau)$ is

$$\frac{dS_{\mu}}{d\tau} = \frac{\partial S_{\mu}}{\partial x^{\nu}} \frac{dx^{\nu}(\tau)}{d\tau}$$

and since the direction of $dx^{\nu}(\tau)/d\tau$ is arbitrary, Eq. (6.3.1) becomes

$$\frac{\partial S_{\mu}}{\partial x^{\nu}} = \Gamma^{\lambda}_{\mu\nu} S_{\lambda} \tag{6.3.9}$$

or, in other words,

$$S_{\mu;\nu} = 0 \tag{6.3.10}$$

Hence, if the curvature tensor vanishes, we may always construct solutions of Eq. (6.3.9), with any given value of $S_{\mu}(X)$, by parallel transport of S_{μ} from X to x. Conversely, if there exists any covariant vector field with vanishing covariant derivatives, then (6.3.1) will certainly be satisfied, and since parallel transport cannot change a field when carried about any closed curve, we conclude from (6.3.7) that

$$R^{\sigma}{}_{\mu\nu\rho} S_{\sigma} = 0 \tag{6.3.11}$$

throughout the region where S_{σ} satisfies (6.3.10).

(These conclusions could have been obtained by using well-known results from the theory of partial differential equations[1] instead of the method of parallel displacement. In this approach Eq. (6.3.11) appears as the necessary and sufficient condition that Eq. (6.3.9) can be solved by a power series expansion in $x^\mu - X^\mu$.)

4   Gravitation versus Curvilinear Coordinates

Suppose that we are presented with a metric tensor $g_{\mu\nu}(x)$ that is not just a constant. How can we tell if the space is really permeated by a gravitational field, or if $g_{\mu\nu}$ merely represents the metric $\eta_{\alpha\beta}$ of special relativity written in curvilinear coordinates? In other words, how can we tell whether there is a set of Minkowskian coordinates $\xi^\alpha(x)$ that everywhere satisfy the conditions

$$\eta^{\alpha\beta} = g^{\mu\nu} \frac{\partial \xi^\alpha(x)}{\partial x^\mu} \frac{\partial \xi^\beta(x)}{\partial x^\nu} \tag{6.4.1}$$

Note that the equivalence principle only says that at every point X we can find locally inertial coordinates $\xi_X(x)$ that satisfy (6.4.1) in an infinitesimal neighborhood of X; what we are asking now is whether we can find one set of coordinates $\xi^\alpha(x)$ that satisfy Eq. (6.4.1) everywhere. For example, given the metric coefficients

$$g_{rr} = 1, \qquad g_{\theta\theta} = r^2, \qquad g_{\varphi\varphi} = r^2 \sin^2 \theta, \qquad g_{tt} = -1 \tag{6.4.2}$$

we know that there is a set of $\xi^\alpha s$ satisfying (6.4.1), that is,

$$\xi^1 = r \sin \theta \cos \varphi, \qquad \xi^2 = r \sin \theta \sin \varphi, \qquad \xi^3 = r \cos \theta, \qquad \xi^4 = t \tag{6.4.3}$$

but how could we have told that (6.4.2) was really equivalent to the Minkowski metric $\eta_{\alpha\beta}$, if we weren't clever enough to have recognized it as simply $\eta_{\alpha\beta}$ in spherical polar coordinates? Or, on the other hand, if we change g_{rr} in (6.4.2) to an arbitrary function of r, how can we tell that this really represents a gravitational field, that is, how can we tell that Eqs. (6.4.1) now have no solution?

The answer is contained in the following theorem: The necessary and sufficient conditions for a metric $g_{\mu\nu}(x)$ to be equivalent to the Minkowski metric $\eta_{\alpha\beta}$ [in the sense that there is a transformation $x \to \xi$ satisfying (6.4.1)] are, first, that the curvature tensor calculated from $g_{\mu\nu}$ must everywhere vanish,

$$R^\lambda_{\ \mu\nu\kappa} = 0 \tag{6.4.4}$$

and, second, that at some point X the matrix $g^{\mu\nu}(X)$ has three positive and one negative eigenvalues.

The *necessity* of these two conditions is obvious. Suppose that we can find a coordinate system $\xi^\alpha(x)$ satisfying (6.4.1). In this coordinate system the metric is $\eta_{\alpha\beta}$, all components of the affine connection vanish, and hence the Riemann tensor $R^{\alpha}{}_{\beta\gamma\delta}$ vanishes. But the vanishing of a tensor is an invariant statement, so $R^{\lambda}{}_{\mu\nu\kappa}$ must have vanished in the original x^{μ} coordinate system. Also, we have already noted in Section 3.6 that a "congruence" like Eq. (6.4.1) requires $\eta_{\alpha\beta}$ and $g^{\mu\nu}$ to have the same numbers of positive, negative, and zero eigenvalues everywhere.

To prove the sufficiency of Eq. (6.4.4) for the existence of a system of "everywhere inertial" coordinates $\xi^{\alpha}(x)$ satisfying (6.4.1), we shall actually construct the $\xi^{\alpha}(x)$. First we note that at any point X we may find a matrix $d^{\alpha}{}_{\mu}$ for which

$$\eta^{\alpha\beta} = g^{\mu\nu}(X) \, d^{\alpha}{}_{\mu} \, d^{\beta}{}_{\nu} \tag{6.4.5}$$

(For, since $g^{\mu\nu}(X)$ is a symmetric matrix, we can find an orthogonal matrix $O^{\alpha}{}_{\mu}$ for which the matrix OgO^{T} is diagonal, that is, for which

$$O^{\alpha}{}_{\mu}g^{\mu\nu}O^{\beta}{}_{\nu} = D^{\alpha\beta}$$

$$D^{\alpha\beta} = \begin{cases} D^{\alpha} & \alpha = \beta \\ 0 & \alpha \neq \beta \end{cases}$$

We are assuming that three of the eigenvalues D^{α} are positive and one negative, and can always label the rows of $O^{\alpha}{}_{\mu}$ so that it is D^{1}, D^{2}, and D^{3} that are positive and D^{0} that is negative. Then to satisfy (6.4.5) we need only choose $d^{i}{}_{\mu} = D^{i}{}_{\mu}/\sqrt{D^{i}}$ for $i = 1, 2, 3$, and $d^{0}{}_{\mu} = D^{0}{}_{\mu}/\sqrt{-D^{0}}$.) Next, we define quantities $D^{\alpha}{}_{\mu}(x)$ by the differential equations

$$\frac{\partial D^{\alpha}{}_{\mu}}{\partial x^{\nu}} = \Gamma^{\lambda}{}_{\mu\nu}D^{\alpha}{}_{\lambda} \tag{6.4.6}$$

with the initial condition

$$D^{\alpha}{}_{\mu} = d^{\alpha}{}_{\mu} \qquad \text{at } x = X \tag{6.4.7}$$

We showed in the last section that such equations can always be solved, providing that $R^{\lambda}{}_{\mu\nu\kappa}$ vanishes. (The quantities $D^{\alpha}{}_{\mu}$ are to be thought of as four covariant vectors $D^{0}{}_{\mu}$, $D^{1}{}_{\mu}$, $D^{2}{}_{\mu}$, $D^{3}{}_{\mu}$ rather than as a single tensor.) Since $\partial D^{\alpha}{}_{\mu}/\partial x^{\nu}$ is symmetric in μ and ν, we can write the vectors $D^{\alpha}{}_{\mu}$ as gradients of scalars, which we define to be the locally inertial coordinates $\xi^{\alpha}(x)$:

$$\frac{\partial \xi^{\alpha}}{\partial x^{\mu}} = D^{\alpha}{}_{\mu} \tag{6.4.8}$$

with initial values $\xi^{\alpha}(X)$ some arbitrary constants. To see that these ξ coordinates do satisfy (6.4.1), note first that

$$\frac{\partial}{\partial x^{\rho}} (g^{\mu\nu}D^{\alpha}{}_{\mu}D^{\beta}{}_{\nu}) = 0 \tag{6.4.9}$$

This can be verified by direct calculation or, more simply, by noting that (6.4.6) just says that $D^\alpha{}_{\mu;\rho}$ vanishes, and $g^{\mu\nu}{}_{;\rho}$ also vanishes, so $(g^{\mu\nu}D^\alpha{}_\mu D^\beta{}_\nu)_{;\rho}$ vanishes; but since $g^{\mu\nu}D^\alpha{}_\mu D^\beta{}_\nu$ is a scalar, this means that its ordinary derivative vanishes. But (6.4.7) and (6.4.5) show that $g^{\mu\nu}D^\alpha{}_\mu D^\beta{}_\nu$ equals $\eta^{\alpha\beta}$ at $x = X$, so since it is a constant this holds everywhere:

$$\eta^{\alpha\beta} = g^{\mu\nu}D^\alpha{}_\mu D^\beta{}_\nu \qquad \text{(all } x\text{)} \tag{6.4.10}$$

Equation (6.4.1) follows immediately from (6.4.8) and (6.4.10).

5   Commutation of Covariant Derivatives

There is another way to see that the tensor $R^\lambda{}_{\mu\nu\kappa}$ expresses the presence or absence of a true gravitational field. Consider the second covariant derivative of a covariant vector V_λ:

$$V_{\mu;\nu;\kappa} = \frac{\partial}{\partial x^\kappa} V_{\mu;\nu} - \Gamma^\lambda_{\nu\kappa} V_{\mu;\lambda} - \Gamma^\lambda_{\mu\kappa} V_{\lambda;\nu}$$

$$= \frac{\partial^2 V_\mu}{\partial x^\nu \, \partial x^\kappa} - \frac{\partial V_\lambda}{\partial x^\kappa} \Gamma^\lambda_{\mu\nu} - V_\lambda \frac{\partial}{\partial x^\kappa} \Gamma^\lambda_{\mu\nu}$$

$$- \Gamma^\lambda_{\nu\kappa} \frac{\partial V_\mu}{\partial x^\lambda} + \Gamma^\lambda_{\nu\kappa} \Gamma^\sigma_{\mu\lambda} V_\sigma$$

$$- \Gamma^\lambda_{\mu\kappa} \frac{\partial V_\lambda}{\partial x^\nu} + \Gamma^\lambda_{\mu\kappa} \Gamma^\sigma_{\lambda\nu} V_\sigma$$

The terms involving first and second derivatives of V_μ are symmetric in ν and κ, but the terms involving V_μ itself contain an antisymmetric part,

$$V_{\mu;\nu;\kappa} - V_{\mu;\kappa;\nu} = -V_\sigma R^\sigma{}_{\mu\nu\kappa} \tag{6.5.1}$$

In the same way, we could show that

$$V^\lambda{}_{;\nu;\kappa} - V^\lambda{}_{;\kappa;\nu} = V^\sigma R^\lambda{}_{\sigma\nu\kappa} \tag{6.5.2}$$

Similar formulas hold for any tensor; for instance,

$$T^\lambda{}_{\mu;\nu;\kappa} - T^\lambda{}_{\mu;\kappa;\nu} = T^\sigma{}_\mu R^\lambda{}_{\sigma\nu\kappa} - T^\lambda{}_\sigma R^\sigma{}_{\mu\nu\kappa} \tag{6.5.3}$$

Thus, if the curvature tensor vanishes, then covariant derivatives *commute*, as would be expected for a coordinate system that can be transformed into a Minkowski coordinate system.

6   Algebraic Properties of $R_{\lambda\mu\nu\kappa}$

The algebraic properties of the curvature tensor are greatly clarified if we consider, instead of $R^{\lambda}{}_{\mu\nu\kappa}$, its fully covariant form

$$R_{\lambda\mu\nu\kappa} \equiv g_{\lambda\sigma} R^{\sigma}{}_{\mu\nu\kappa} \tag{6.6.1}$$

Using (6.1.5) and (3.3.7), this is

$$R_{\lambda\mu\nu\kappa} = \tfrac{1}{2} g_{\lambda\sigma} \frac{\partial}{\partial x^{\kappa}} g^{\sigma\rho} \left\{ \frac{\partial g_{\rho\mu}}{\partial x^{\nu}} + \frac{\partial g_{\rho\nu}}{\partial x^{\mu}} - \frac{\partial g_{\mu\nu}}{\partial x^{\rho}} \right\}$$

$$- \tfrac{1}{2} g_{\lambda\sigma} \frac{\partial}{\partial x^{\nu}} g^{\sigma\rho} \left\{ \frac{\partial g_{\rho\mu}}{\partial x^{\kappa}} + \frac{\partial g_{\rho\kappa}}{\partial x^{\mu}} - \frac{\partial g_{\mu\kappa}}{\partial x^{\rho}} \right\}$$

$$+ g_{\lambda\sigma} \{ \Gamma^{\eta}{}_{\mu\nu}\Gamma^{\sigma}{}_{\kappa\eta} - \Gamma^{\eta}{}_{\mu\kappa}\Gamma^{\sigma}{}_{\nu\eta} \}$$

We use the relation

$$g_{\lambda\sigma} \frac{\partial}{\partial x^{\kappa}} g^{\sigma\rho} = -g^{\sigma\rho} \frac{\partial}{\partial x^{\kappa}} g_{\lambda\sigma}$$

$$= -g^{\sigma\rho} (\Gamma^{\eta}{}_{\kappa\lambda} g_{\eta\sigma} + \Gamma^{\eta}{}_{\kappa\sigma} g_{\eta\lambda})$$

and obtain

$$R_{\lambda\mu\nu\kappa} = \frac{1}{2} \left[\frac{\partial^{2} g_{\lambda\nu}}{\partial x^{\kappa}\, \partial x^{\mu}} - \frac{\partial^{2} g_{\mu\nu}}{\partial x^{\kappa}\, \partial x^{\lambda}} - \frac{\partial^{2} g_{\lambda\kappa}}{\partial x^{\nu}\, \partial x^{\mu}} + \frac{\partial^{2} g_{\mu\kappa}}{\partial x^{\nu}\, \partial x^{\lambda}} \right]$$

$$- [\Gamma^{\eta}{}_{\kappa\lambda} g_{\eta\sigma} + \Gamma^{\eta}{}_{\kappa\sigma} g_{\eta\lambda}] \Gamma^{\sigma}{}_{\mu\nu}$$

$$+ [\Gamma^{\eta}{}_{\nu\lambda} g_{\eta\sigma} + \Gamma^{\eta}{}_{\nu\sigma} g_{\eta\lambda}] \Gamma^{\sigma}{}_{\mu\kappa}$$

$$+ g_{\lambda\sigma} [\Gamma^{\eta}{}_{\mu\nu}\Gamma^{\sigma}{}_{\kappa\eta} - \Gamma^{\eta}{}_{\mu\kappa}\Gamma^{\sigma}{}_{\nu\eta}]$$

Most of the $\Gamma\Gamma$ terms cancel, leaving us with

$$R_{\lambda\mu\nu\kappa} = \frac{1}{2} \left[\frac{\partial^{2} g_{\lambda\nu}}{\partial x^{\kappa}\, \partial x^{\mu}} - \frac{\partial^{2} g_{\mu\nu}}{\partial x^{\kappa}\, \partial x^{\lambda}} - \frac{\partial^{2} g_{\lambda\kappa}}{\partial x^{\nu}\, \partial x^{\mu}} + \frac{\partial^{2} g_{\mu\kappa}}{\partial x^{\nu}\, \partial x^{\lambda}} \right]$$

$$+ g_{\eta\sigma} [\Gamma^{\eta}{}_{\nu\lambda}\Gamma^{\sigma}{}_{\mu\kappa} - \Gamma^{\eta}{}_{\kappa\lambda}\Gamma^{\sigma}{}_{\mu\nu}] \tag{6.6.2}$$

From (6.6.2) we may read off the algebraic properties of the curvature tensor:

(A) Symmetry:

$$R_{\lambda\mu\nu\kappa} = R_{\nu\kappa\lambda\mu} \tag{6.6.3}$$

(B) Antisymmetry:

$$R_{\lambda\mu\nu\kappa} = -R_{\mu\lambda\nu\kappa} = -R_{\lambda\mu\kappa\nu} = +R_{\mu\lambda\kappa\nu} \tag{6.6.4}$$

(C) Cyclicity:

$$R_{\lambda\mu\nu\kappa} + R_{\lambda\kappa\mu\nu} + R_{\lambda\nu\kappa\mu} = 0 \tag{6.6.5}$$

We have already mentioned that $R_{\lambda\mu\nu\kappa}$ may be contracted to give the Ricci tensor

$$R_{\mu\kappa} = g^{\lambda\nu}R_{\lambda\mu\nu\kappa} \tag{6.6.6}$$

The symmetry property (A) shows that the Ricci tensor is symmetric,

$$R_{\mu\kappa} = R_{\kappa\mu} \tag{6.6.7}$$

and the antisymmetry property (B) tells us that $R_{\mu\kappa}$ is essentially the only second-rank tensor that can be formed from $R_{\lambda\mu\nu\kappa}$, since multiplying (6.6.4) with $g^{\lambda\nu}$, $g^{\lambda\mu}$, and $g^{\nu\kappa}$ gives

$$R_{\mu\kappa} = -g^{\lambda\nu}R_{\mu\lambda\nu\kappa} = -g^{\lambda\nu}R_{\lambda\mu\kappa\nu} = +g^{\lambda\nu}R_{\mu\lambda\kappa\nu}$$

$$g^{\lambda\mu}R_{\lambda\mu\nu\kappa} = g^{\nu\kappa}R_{\lambda\mu\nu\kappa} = 0$$

From the antisymmetry property (B) we also see that there is essentially only one way of contracting $R_{\lambda\mu\nu\kappa}$ to construct a scalar:

$$R \equiv g^{\lambda\nu}g^{\mu\kappa}R_{\lambda\mu\nu\kappa} = -g^{\lambda\nu}g^{\mu\kappa}R_{\mu\lambda\nu\kappa}$$

$$0 = g^{\lambda\mu}g^{\nu\kappa}R_{\lambda\mu\nu\kappa}$$

Finally, (C) eliminates the one other scalar that might have been formed in four dimensions, that is,

$$\frac{1}{\sqrt{g}}\varepsilon^{\lambda\mu\nu\kappa}R_{\lambda\mu\nu\kappa} = 0$$

7   Description of Curvature in N Dimensions*

For the moment let us consider a general space of N dimensions. To count the number of algebraically independent components of $R_{\lambda\mu\nu\kappa}$, it is convenient to adopt what may be called the *Petrov notation*,[2] and think of $R_{\lambda\mu\nu\kappa}$ as a matrix $R_{(\lambda\mu)(\nu\kappa)}$ with "indices" $(\lambda\mu)$ and $(\nu\kappa)$. From (6.6.4) we see that each "index" takes a number of independent values equal to the number of independent elements of an antisymmetric matrix in N dimensions, or $\frac{1}{2}N(N-1)$. From (6.6.3) we see that $R_{(\lambda\mu)(\nu\kappa)}$ is symmetric in these "indices," so (6.6.3) and (6.6.4) alone would leave $R_{\lambda\mu\nu\kappa}$ with a number of independent components equal to the number of independent elements of a symmetric matrix in $\frac{1}{2}N(N-1)$ dimensions, or

$$\tfrac{1}{2}[\tfrac{1}{2}N(N-1)][\tfrac{1}{2}N(N-1)+1] = \tfrac{1}{8}N(N-1)(N^2-N+2)$$

---

* This section lies somewhat out of the book's main line of development, and may be omitted in a first reading.

Equations (6.6.3) and (6.6.4) also make the cyclic sum $R_{\lambda\mu\nu\kappa} + R_{\lambda\kappa\mu\nu} + R_{\lambda\nu\kappa\mu}$ completely antisymmetric, so Eq. (6.6.5) adds $N(N-1)(N-2)(N-3)/4!$ further constraints, leaving $R_{\lambda\mu\nu\kappa}$ with a number of independent components equal to

$$C_N = \tfrac{1}{8}N(N-1)(N^2-N+2) - \tfrac{1}{24}N(N-1)(N-2)(N-3)$$

or combining terms

$$C_N = \tfrac{1}{12}N^2(N^2-1) \tag{6.7.1}$$

In one dimension the curvature tensor R_{1111} always vanishes, as can be seen from (6.6.4) or (6.6.5) or from the fact that (6.7.1) gives $C_1 = 0$ independent components. It may strike the reader as odd that a curved line should have zero curvature, but this just emphasizes that $R_{\lambda\mu\nu\kappa}$ reflects only the inner properties of the space, not how it is embedded in a higher dimensional space. Indeed, we note that the transformation rule for the metric tensor in one dimension is

$$g'_{11} = \left(\frac{dx}{dx'}\right)^2 g_{11}$$

so that g'_{11} can be made equal to ± 1 everywhere by choosing

$$x' = \int dx \sqrt{\pm g_{11}}$$

In two dimensions (6.7.1) gives $R_{\lambda\mu\nu\kappa}$ only *one* independent component, which can be taken as R_{1212}; the other components are related to R_{1212} by Eq. (6.6.4):

$$R_{1212} = -R_{2112} = -R_{1221} = R_{2121}$$

$$R_{1111} = R_{1122} = R_{2211} = R_{2222} = 0$$

These formulas can be summarized more elegantly by

$$R_{\lambda\mu\nu\kappa} = (g_{\lambda\nu}g_{\mu\kappa} - g_{\lambda\kappa}g_{\mu\nu})\frac{R_{1212}}{g}$$

where g is the determinant $g_{11}g_{22} - g_{12}^2$. Contracting λ with ν gives the Ricci tensor

$$R_{\mu\kappa} = g_{\mu\kappa}\frac{R_{1212}}{g} \tag{6.7.2}$$

and contracting μ and κ gives the curvature scalar

$$R = \frac{2R_{1212}}{g} \tag{6.7.3}$$

so the curvature tensor is

$$R_{\lambda\mu\nu\kappa} = \tfrac{1}{2}R(g_{\lambda\nu}g_{\mu\kappa} - g_{\lambda\kappa}g_{\mu\nu}) \tag{6.7.4}$$

The *Gaussian curvature K* discussed in the first section of this book is defined by

$$K \equiv -\frac{R}{2} = -\frac{R_{1212}}{g} \tag{6.7.5}$$

(The factor $-\frac{1}{2}$ is of purely historical interest.) Equation (1.1.12) follows from (6.6.2) and (6.7.5).

In three dimensions (6.7.1) gives the curvature tensor $C_3 = 6$ independent components. This is also the number of independent components of the Ricci tensor $R_{\mu\kappa}$ in three dimensions, so we may anticipate that here $R_{\lambda\mu\nu\kappa}$ may be expressed in terms of $R_{\mu\kappa}$ alone. By using the covariance, symmetry, and contraction properties of $R_{\lambda\mu\nu\kappa}$, we may further guess that this relation is

$$R_{\lambda\mu\nu\kappa} = g_{\lambda\nu}R_{\mu\kappa} - g_{\lambda\kappa}R_{\mu\nu} - g_{\mu\nu}R_{\lambda\kappa} + g_{\mu\kappa}R_{\lambda\nu}$$
$$- \tfrac{1}{2}(g_{\lambda\nu}g_{\mu\kappa} - g_{\lambda\kappa}g_{\mu\nu})R \tag{6.7.6}$$

To prove that (6.7.6) is correct, let us adopt a coordinate system such that $g_{\mu\nu}$ vanishes for $\mu \neq \nu$ at some point X. (This can be managed by choosing $\partial x'^\mu/\partial x^\lambda$ at X as the orthogonal matrix that diagonalizes $g_{\mu\nu}$ at X.) In this system we have at X

$$R_{12} = g^{33}R_{1323}$$

so

$$R_{1323} = g_{33}R_{12}$$

in agreement with (6.7.6). Furthermore,

$$R_{11} = g^{22}R_{1212} + g^{33}R_{1313}$$
$$R_{22} = g^{33}R_{2323} + g^{11}R_{2121}$$

so

$$g_{22}R_{11} + g_{11}R_{22} = 2R_{1212} + g^{33}(g_{22}R_{1313} + g_{11}R_{2323})$$
$$= R_{1212} + g_{11}g_{22}(g^{11}g^{22}R_{1212} + g^{11}g^{33}R_{1313}$$
$$+ g^{22}g^{33}R_{2323})$$

or

$$R_{1212} = g_{22}R_{11} + g_{11}R_{22} - \tfrac{1}{2}g_{11}g_{22}R$$

again in agreement with (6.7.6). The other independent components of $R_{\lambda\mu\nu\kappa}$ are R_{1223}, R_{1213}, R_{2323}, and R_{3131}, which can be obtained from R_{1323} and R_{1212} by permuting the values 1, 2, 3; so (6.7.6) holds for these components as well. Since (6.7.6) thus holds in a coordinate system that is orthogonal at X, and is manifestly covariant, it holds in general.

It is only in four or more dimensions that the full Riemann-Christoffel tensor $R_{\lambda\mu\nu\kappa}$ is needed to describe the curvature of a space. For instance, in four dimensions (6.7.1) gives the curvature tensor $C_4 = 20$ independent components, whereas

$R_{\mu\kappa}$ has only 10 independent components, so $R_{\lambda\mu\nu\kappa}$ has 10 components beyond those which can be expressed in terms of $R_{\mu\kappa}$.

The $\frac{1}{12}N^2(N^2 - 1)$ components of $R_{\lambda\mu\nu\kappa}$ describe the curvature of a general N-dimensional space, but they do not do so in an invariant manner, for the values of these components depend not only on the intrinsic properties of the space but also on the particular coordinate system chosen. The invariant characterization of a curved space must be in terms of *scalars* constructed from $R_{\lambda\mu\nu\kappa}$ and $g_{\mu\nu}$. Let us count how many such scalars there are. The N^2 quantities $\partial x'^{\mu}/\partial x^{\nu}$ can for a general coordinate transformation $x \to x'$ be made anything we like at a given point X. Hence the $\frac{1}{12}N^2(N^2 - 1)$ independent components of $R_{\lambda\mu\nu\kappa}$ and the $\frac{1}{2}N(N + 1)$ independent components of $g_{\mu\nu}$ at this point may by general coordinate transformation be subjected to N^2 algebraic conditions; the number of scalars that can be constructed from $R_{\lambda\mu\nu\kappa}$ and $g_{\mu\nu}$ is therefore

$$\tfrac{1}{12}N^2(N^2 - 1) + \tfrac{1}{2}N(N + 1) - N^2 = \tfrac{1}{12}N(N - 1)(N - 2)(N + 3)$$

(6.7.7)

The case $N = 2$ is an exception to this argument, because in two dimensions there is a one-parameter subgroup of coordinate transformations that has no effect on $g_{\mu\nu}$ and $R_{\lambda\mu\nu\kappa}$; the correct number of invariants here is not zero but *one*, that is, the curvature scalar R itself. This exception does not occur for higher dimensional spaces, so (6.7.7) holds for $N \geq 3$. For $N = 3$, Eq. (6.7.7) tells us that there are *three* curvature scalars, which can conveniently be chosen as the three roots of the secular equation

$$\mathrm{Det}\,(R_{\mu\nu} - \lambda g_{\mu\nu}) = 0$$

or equivalently as the three quantities

$$R \qquad R_{\mu\nu}R^{\mu\nu} \qquad \frac{\mathrm{Det}\,R}{\mathrm{Det}\,g}$$

For $N = 4$, Eq. (6.7.7) tells us that there are *fourteen* curvature scalars. To enumerate them (and for other purposes as well) it is convenient to decompose $R_{\lambda\mu\nu\kappa}$ into terms that depend only on the Ricci tensor $R_{\mu\nu}$ plus a term $C_{\lambda\mu\nu\kappa}$ that has no nontrivial contractions. In $N \geq 3$ dimensions this decomposition is

$$R_{\lambda\mu\nu\kappa} \equiv \frac{1}{N - 2}\,(g_{\lambda\nu}R_{\mu\kappa} - g_{\lambda\kappa}R_{\mu\nu} - g_{\mu\nu}R_{\lambda\kappa} + g_{\mu\kappa}R_{\lambda\nu})$$

$$- \frac{R}{(N - 1)(N - 2)}\,(g_{\lambda\nu}g_{\mu\kappa} - g_{\lambda\kappa}g_{\mu\nu}) + C_{\lambda\mu\nu\kappa}$$

The tensor $C_{\lambda\mu\nu\kappa}$ is called the *Weyl tensor*[3] or the *conformal tensor*. (The latter name is used because the necessary and sufficient condition for the existence of a coordinate system in which $g_{\mu\nu}$ is proportional to a constant matrix throughout the space, is that $C_{\lambda\mu\nu\kappa}$ vanish everywhere.[4]) This tensor has the same algebraic properties as $R_{\lambda\mu\nu\kappa}$, and in addition it satisfies the $\frac{1}{2}N(N + 1)$ conditions

$$C^\lambda{}_{\mu\lambda\kappa} = 0$$

so the number of its linearly independent components is

$$\tfrac{1}{12}N^2(N^2 - 1) - \tfrac{1}{2}N(N + 1) = \tfrac{1}{12}N(N + 1)(N + 2)(N - 3)$$

[Equation (6.7.6) simply says that $C_{\lambda\mu\nu\kappa} = 0$ for $N = 3$.] Barring degeneracies, the curvature invariants can be described as consisting of all the components of the Weyl tensor for that unique choice of coordinate axes that makes $R_{\mu\nu}$ and $g_{\mu\nu}$ diagonal, the elements of $g_{\mu\nu}$ being $+1$'s, -1's, and 0's, plus the N eigenvalues of $R_{\mu\nu}$. However, this enumeration breaks down when some of the eigenvalues of $R_{\mu\nu}$ are degenerate. A particularly interesting case is that for $R_{\mu\nu} = 0$, which we shall see in the next chapter describes physical gravitational fields in empty space. In this case the curvature invariants for $N = 4$ are the 10 vanishing components of $R_{\mu\nu}$ (the vanishing of a tensor is an invariant statement) plus the four quantities

$$C^{\lambda\mu\nu\kappa}C_{\lambda\mu\nu\kappa} \qquad\qquad \frac{\varepsilon^{\lambda\mu}{}_{\rho\sigma}C^{\rho\sigma\nu\kappa}C_{\lambda\mu\nu\kappa}}{\sqrt{g}}$$

$$C_{\lambda\mu\nu\kappa}C^{\nu\kappa\rho\sigma}C_{\rho\sigma}{}^{\lambda\mu} \qquad\qquad \frac{C_{\lambda\mu\nu\kappa}C^{\nu\kappa\rho\sigma}\varepsilon_{\rho\sigma}{}^{\tau\xi}C_{\tau\xi}{}^{\lambda\mu}}{\sqrt{g}}$$

Petrov[2] has given an equivalent description of the four nonvanishing curvature invariants as roots of a secular equation, and has classified various algebraic types of Weyl tensor according to the degeneracies of these roots.

Finally, it should be emphasized that (6.7.7) gives the number of *algebraically* independent curvature invariants. There are in general differential relations among these invariants, and the number of functionally independent curvature invariants is less than (6.7.7).

8   The Bianchi Identities

The curvature tensor obeys important differential identities, in addition to the algebraic identities discussed in Section 6. These can be most easily derived at a given point x, by adopting a locally inertial coordinate system in which $\Gamma^\lambda_{\mu\nu}$ (but not its derivatives) vanish at x. Then at x, Eq. (6.6.1) gives

$$R_{\lambda\mu\nu\kappa;\eta} = \frac{1}{2}\frac{\partial}{\partial x^\eta}\left(\frac{\partial^2 g_{\lambda\nu}}{\partial x^\kappa \, \partial x^\mu} - \frac{\partial^2 g_{\mu\nu}}{\partial x^\kappa \, \partial x^\lambda} - \frac{\partial^2 g_{\lambda\kappa}}{\partial x^\mu \, \partial x^\nu} + \frac{\partial^2 g_{\mu\kappa}}{\partial x^\nu \, \partial x^\lambda}\right)$$

all other terms being at least of first order in Γ. By permuting ν, κ, and η cyclically we obtain the *Bianchi identities*

$$R_{\lambda\mu\nu\kappa;\eta} + R_{\lambda\mu\eta\nu;\kappa} + R_{\lambda\mu\kappa\eta;\nu} = 0 \qquad\qquad (6.8.1)$$

These equations are manifestly generally covariant, so since they hold in locally inertial systems they hold in general. (They can also, of course, be checked by direct calculation.)

We shall be particularly concerned with the contracted form of (6.8.1). Recalling that the covariant derivatives of $g^{\lambda\nu}$ vanish, we find on contraction of λ with ν that

$$R_{\mu\kappa;\eta} - R_{\mu\eta;\kappa} + R^{\nu}{}_{\mu\kappa\eta;\nu} = 0 \qquad (6.8.2)$$

Contracting again gives

$$R_{;\eta} - R^{\mu}{}_{\eta;\mu} - R^{\nu}{}_{\eta;\nu} = 0$$

or

$$(R^{\mu}{}_{\eta} - \tfrac{1}{2}\delta^{\mu}{}_{\eta}R)_{;\mu} = 0 \qquad (6.8.3)$$

An equivalent but more familiar form is

$$(R^{\mu\nu} - \tfrac{1}{2}g^{\mu\nu}R)_{;\mu} = 0 \qquad (6.8.4)$$

9 The Geometric Analogy*

We have seen in this chapter that the nonvanishing of the tensor $R_{\lambda\mu\nu\kappa}$ is the true expression of the presence of a gravitational field. We also saw in Chapter 1 that Gauss was led to introduce the Gaussian curvature $K = -R/2$ as the true measure of the departure of a two-dimensional geometry from that of Euclid, and that Riemann subsequently introduced the curvature tensor $R_{\lambda\mu\nu\kappa}$ to generalize the concept of curvature to three or more dimensions. It is therefore not surprising that Einstein and his successors have regarded the effects of a gravitational field as producing a change in the geometry of space and time. At one time it was even hoped that the rest of physics could be brought into a geometric formulation, but this hope has met with disappointment, and the geometric interpretation of the theory of gravitation has dwindled to a mere analogy, which lingers in our language in terms like "metric," "affine connection," and "curvature," but is not otherwise very useful. The important thing is to be able to make predictions about images on the astronomers' photographic plates, frequencies of spectral lines, and so on, and it simply doesn't matter whether we ascribe these predictions to the physical effect of gravitational fields on the motion of planets and photons or to a curvature of space and time. (The reader should be warned that these views are heterodox and would meet with objections from many general relativists.)

Despite the preceding remarks, it is worth mentioning without proof just what the tensor $R_{\lambda\mu\nu\kappa}$ has to do with the curvature of a Riemannian space. Given a point X in a space of an arbitrary number of dimensions, and given two vectors

---

* This section lies somewhat out of the book's main line of development, and may be omitted in a first reading.

a^μ, b^μ defined at X, we may construct through X a family of "geodesic curves" $x^\mu = x^\mu(\tau, \alpha, \beta)$ defined by

$$\frac{d^2 x^\mu}{d\tau^2} + \Gamma^\mu_{\nu\lambda} \frac{dx^\nu}{d\tau} \frac{dx^\lambda}{d\tau} = 0$$

$$\left(\frac{dx^\mu}{d\tau}\right)_{x=X} = \alpha a^\mu + \beta b^\mu$$

the numbers α, β being allowed to run over all real values. These curves fill out a two-dimensional surface $S(a, b)$ through X, and the Gaussian curvature of this surface at X is[5]

$$K(a, b) = \frac{R_{\lambda\mu\nu\kappa} a^\lambda b^\mu a^\nu b^\kappa}{(g_{\lambda\kappa} g_{\mu\nu} - g_{\lambda\nu} g_{\mu\kappa}) a^\lambda b^\mu a^\nu b^\kappa} \tag{6.9.1}$$

From Eq. (6.7.4) we see that in two dimensions $K(a, b)$ is independent of a and b and is just $-R/2$.

10 Geodesic Deviation*

The introduction of the curvature tensor was motivated here by the need to construct suitable field equations for the gravitational field. However, the curvature tensor is also useful in expressing the effects of gravitation on physical systems.

For instance, consider a pair of nearby freely falling particles that travel on trajectories $x^\mu(\tau)$ and $x^\mu(\tau) + \delta x^\mu(\tau)$. The equations of motion are

$$0 = \frac{d^2 x^\mu}{d\tau^2} + \Gamma^\mu_{\nu\lambda}(x) \frac{dx^\nu}{d\tau} \frac{dx^\lambda}{d\tau}$$

$$0 = \frac{d^2}{d\tau^2} [x^\mu + \delta x^\mu] + \Gamma^\mu_{\nu\lambda}(x + \delta x) \frac{d}{d\tau} [x^\nu + \delta x^\nu] \frac{d}{d\tau} [x^\lambda + \delta x^\lambda]$$

Evaluating the difference between these equations to first order in δx^μ gives

$$0 = \frac{d^2 \, \delta x^\mu}{d\tau^2} + \frac{\partial \Gamma^\mu_{\nu\lambda}}{\partial x^\rho} \delta x^\rho \frac{dx^\nu}{d\tau} \frac{dx^\lambda}{d\tau}$$

$$+ 2\Gamma^\mu_{\nu\lambda} \frac{dx^\nu}{d\tau} \frac{d \, \delta x^\lambda}{d\tau}$$

---

* This section lies somewhat out of the book's main line of development, and may be omitted in a first reading.

or, in terms of covariant derivatives along the curve $x^\mu(\tau)$ (see Section 4.9),

$$\frac{D^2}{D\tau^2}\,\delta x^\lambda = R^\lambda{}_{\nu\mu\rho}\,\delta x^\mu\,\frac{dx^\nu}{d\tau}\frac{dx^\rho}{d\tau} \qquad (6.10.1)$$

Although a freely falling particle appears to be at rest in a coordinate frame falling with the particle, a pair of nearby freely falling particles will exhibit a relative motion that can reveal the presence of a gravitational field to an observer that falls with them. This is of course not a violation of the Principle of Equivalence, because the effect of the right-hand side of (6.10.1) becomes negligible when the separation between particles is much less than the characteristic dimensions of the field.

6 BIBLIOGRAPHY

☐ L. P. Eisenhart, *Riemannian Geometry* (Princeton University Press, Princeton, N. J., 1926).

☐ J. A. Schouten, *Ricci-Calculus* (Springer-Verlag, Berlin, 1954).

Also, see Bibliography, Chapter 3.

6 REFERENCES

1. See, for example, L. P. Eisenhart, *Continuous Groups of Transformations* (Dover Publications, New York, 1961), p. 1.
2. A. Z. Petrov, Uch. zap. Kazan Gos. Univ., **114**, No. 8, 55 (1954) [trans. no. 29, Jet Propulsion Laboratory, Pasadena, Cal. 1963]; *Einstein Spaces*, trans. by R. F. Kelleher (Pergamon Press, Oxford, 1969), Chapter 3.
3. H. Weyl, Mat. Z., **2**, 384 (1918).
4. See, for example, L. P. Eisenhart, *Riemannian Geometry* (Princeton University Press, Princeton, N. J., 1926), Section 28.
5. Reference 4, Section 25.

7 EINSTEIN'S FIELD EQUATIONS

Chapters 3 through 5 have provided us with one-half of a complete theory of gravitation, that is, with a mathematical description of gravitational fields that dictates their effects on arbitrary physical systems. In this chapter we move on to the second half of general relativity, that is, to the differential equations that determine the gravitational fields themselves.

1 Derivation of the Field Equations

The field equations for gravitation are inevitably going to be more complicated than those for electromagnetism. Maxwell's equations are linear because the electromagnetic field does not itself carry charge, whereas gravitational fields do carry energy and momentum (see Section 5.3) and must therefore contribute to their own source. That is, the gravitational field equations will have to be nonlinear partial differential equations, the nonlinearity representing the effect of gravitation on itself.

In dealing with these nonlinear effects we are guided once again by the Principle of Equivalence. At any point X in an arbitrarily strong gravitational field, we can define a locally inertial coordinate system such that

$$g_{\alpha\beta}(X) = \eta_{\alpha\beta} \tag{7.1.1}$$

$$\left(\frac{\partial g_{\alpha\beta}(x)}{\partial x^\gamma}\right)_{x=X} = 0 \tag{7.1.2}$$

Hence for x near X, the metric tensor $g_{\alpha\beta}$ can differ from $\eta_{\alpha\beta}$ only by terms quadratic in $x - X$. In this coordinate system the gravitational field is weak near X, and we can hope to describe the field by *linear* partial differential equations. And once we know what these weak-field equations are, we can find the general field equations by reversing the coordinate transformation that made the field weak.

Unfortunately, we have very little empirical information about the weak-field equations. This is not for any fundamental reason, but rather because gravitational radiation is so weakly generated and absorbed by matter, that it has not yet certainly been detected. However, although forgivable, our ignorance does prevent us from proceeding as directly as we did in previous chapters, and some guesswork will be unavoidable.

First let us recall that in a weak static field produced by a nonrelativistic mass density ρ, the time-time component of the metric tensor is approximately given by

$$g_{00} \simeq -(1 + 2\phi)$$

[See Eq. (3.4.5).] Here ϕ is the Newtonian potential, determined by Poisson's equation

$$\mathbf{\nabla}^2 \phi = 4\pi G \rho$$

where G is Newton's constant, equal to 6.670×10^{-8} in c.g.s. units. Furthermore, the energy density T_{00} for nonrelativistic matter is just equal to its mass density

$$T_{00} \simeq \rho$$

Combining the above, we have then

$$\mathbf{\nabla}^2 g_{00} = -8\pi G T_{00} \tag{7.1.3}$$

This field equation is only supposed to hold for weak static fields generated by nonrelativistic matter, and is not even Lorentz invariant as it stands. However, (7.1.3) leads us to *guess* that the weak-field equations for a general distribution $T_{\alpha\beta}$ of energy and momentum take the form

$$G_{\alpha\beta} = -8\pi G T_{\alpha\beta} \tag{7.1.4}$$

where $G_{\alpha\beta}$ is a linear combination of the metric and its first and second derivatives. It follows then from the Principle of Equivalence that the equations which govern gravitational fields of arbitrary strength must take the form

$$G_{\mu\nu} = -8\pi G T_{\mu\nu} \tag{7.1.5}$$

where $G_{\mu\nu}$ is a tensor which reduces to $G_{\alpha\beta}$ for weak fields.

In general, there will be a variety of tensors $G_{\mu\nu}$ that can be formed from the metric tensor and its derivatives, and that reduce in the weak-field limit to a given $G_{\alpha\beta}$. Let us imagine $G_{\mu\nu}$ to be expanded in a sum of products of derivatives of the metric, and classify each term according to the total number N of derivatives of metric components. (For example, a term with $N = 3$ could be linear in third derivatives of the metric, or a product of a first derivative with a second derivative, or a product of three first derivatives.) The whole of $G_{\mu\nu}$ must have the dimensions of a second derivative, so each term of type $N \neq 2$ appears multiplied with a constant having the dimensions of length to the power $N - 2$; such terms will become negligible for gravitational fields of sufficiently large or small space-time scale if $N > 2$ or $N < 2$, respectively. In order to remove the ambiguity in $G_{\mu\nu}$, we shall assume that *the gravitational field equations are uniform in scale*, so that only terms with $N = 2$ are allowed.

Let us review what we know about the left-hand-side of the field equation (7.1.5):

(A) By definition, $G_{\mu\nu}$ is a tensor.

(B) By assumption, $G_{\mu\nu}$ consists only of terms with $N = 2$ derivatives of the metric; that is, $G_{\mu\nu}$ contains only terms that are either linear in the second derivatives or quadratic in the first derivatives of the metric.

(C) Since $T_{\mu\nu}$ is symmetric, so is $G_{\mu\nu}$.

(D) Since $T_{\mu\nu}$ is conserved (in the sense of covariant differentiation) so is $G_{\mu\nu}$:

$$G^{\mu}{}_{\nu;\mu} = 0 \qquad (7.1.6)$$

(E) For a weak stationary field produced by nonrelativistic matter the 00 component of (7.1.5) must reduce to (7.1.3), so in this limit

$$G_{00} \simeq \nabla^2 g_{00} \qquad (7.1.7)$$

These properties are all we will need to find $G_{\mu\nu}$.

We saw in Section 6.2 that the most general way of constructing a field satisfying (A) and (B) is by contraction of the curvature tensor $R^{\lambda}{}_{\mu\nu\kappa}$. The antisymmetry property of $R_{\lambda\mu\nu\kappa}$ discussed in Section 6.6 shows that there are only two tensors that can be formed by contracting $R_{\lambda\mu\nu\kappa}$; that is, the Ricci tensor $R_{\mu\kappa} \equiv R^{\lambda}{}_{\mu\lambda\kappa}$, and the curvature scalar $R = R^{\mu}{}_{\mu}$. Hence (A) and (B) require $G_{\mu\nu}$ to take the form

$$G_{\mu\nu} = C_1 R_{\mu\nu} + C_2 g_{\mu\nu} R \qquad (7.1.8)$$

where C_1 and C_2 are constants. This is automatically symmetric [see Eq. (6.6.7)], so (C) tells us nothing new. Using the Bianchi identity (6.8.3) gives the covariant divergence of $G_{\mu\nu}$ as

$$G^{\mu}{}_{\nu;\mu} = \left(\frac{C_1}{2} + C_2\right) R_{;\nu}$$

so (D) allows two possibilities: either $C_2 = -C_1/2$, or $R_{;\nu}$ vanishes everywhere. We can reject the second possibility, because (7.1.8) and (7.1.5) give

$$G^{\mu}{}_{\mu} = (C_1 + 4C_2)R = -8\pi G T^{\mu}{}_{\mu}$$

Thus if $R_{;\nu} \equiv \partial R/\partial x^\nu$ vanishes, then so must $\partial T^\mu{}_\mu/\partial x^\nu$, and this is not the case in the presence of inhomogeneous nonrelativistic matter. We conclude then that $C_2 = -C_1/2$, so (7.1.8) becomes

$$G_{\mu\nu} = C_1(R_{\mu\nu} - \tfrac{1}{2}g_{\mu\nu}R) \qquad (7.1.9)$$

Finally, we use the property (E) to fix the constant C_1. A nonrelativistic system always has $|T_{ij}| \ll |T_{00}|$, so we are concerned here with a case where $|G_{ij}| \ll |G_{00}|$, or using (7.1.9),

$$R_{ij} \simeq \tfrac{1}{2}g_{ij}R$$

Furthermore, we deal here with a weak field, so $g_{\alpha\beta} \simeq \eta_{\alpha\beta}$. The curvature scalar is therefore given by

$$R \simeq R_{kk} - R_{00} \simeq \tfrac{3}{2}R - R_{00}$$

or

$$R \simeq 2R_{00} \qquad (7.1.10)$$

Using (7.1.10) and (7.1.1) in (7.1.9), we find

$$G_{00} \simeq 2C_1 R_{00} \qquad (7.1.11)$$

To calculate R_{00} for a weak field we may use the linear part of $R_{\lambda\mu\nu\kappa}$, given by Eq. (6.6.2) as

$$R_{\lambda\mu\nu\kappa} = \frac{1}{2}\left[\frac{\partial^2 g_{\lambda\nu}}{\partial x^\kappa \, \partial x^\mu} - \frac{\partial^2 g_{\mu\nu}}{\partial x^\kappa \, \partial x^\lambda} - \frac{\partial^2 g_{\lambda\kappa}}{\partial x^\nu \, \partial x^\mu} + \frac{\partial^2 g_{\mu\kappa}}{\partial x^\nu \, \partial x^\lambda}\right]$$

When the field is static all time derivatives vanish, and the components we need become

$$R_{0000} \simeq 0 \qquad R_{i0j0} \simeq \frac{1}{2}\frac{\partial^2 g_{00}}{\partial x^i \, \partial x^j}$$

Hence (7.1.11) gives

$$G_{00} \simeq 2C_1(R_{i0i0} - R_{0000}) \simeq C_1 \nabla^2 g_{00}$$

and comparing this with (7.1.7), we find that (E) is satisfied if and only if $C_1 = 1$. Setting $C_1 = 1$ in (7.1.9) completes our calculation of $G_{\mu\nu}$:

$$G_{\mu\nu} = R_{\mu\nu} - \tfrac{1}{2}g_{\mu\nu}R \qquad (7.1.12)$$

With (7.1.5), this gives the *Einstein field equations*

$$R_{\mu\nu} - \tfrac{1}{2}g_{\mu\nu}R = -8\pi G T_{\mu\nu} \qquad (7.1.13)$$

An alternative form is sometimes useful. Contracting (7.1.13) with $g^{\mu\nu}$ gives

$$R - 2R = -8\pi G T^\mu{}_\mu$$

or

$$R = 8\pi G T^\mu{}_\mu \tag{7.1.14}$$

and using this in (7.1.13), we have

$$R_{\mu\nu} = -8\pi G(T_{\mu\nu} - \tfrac{1}{2}g_{\mu\nu}T^\lambda{}_\lambda) \tag{7.1.15}$$

Of course we can also go from (7.1.15) back to (7.1.14) and (7.1.13), so (7.1.13) and (7.1.15) should be regarded as entirely equivalent forms of the Einstein field equations.

In a vacuum $T_{\mu\nu}$ vanishes, so from (7.1.15) we see that the Einstein field equations in *empty space* are just

$$R_{\mu\nu} = 0 \tag{7.1.16}$$

In a space-time of two or three dimensions this would imply the vanishing of the full curvature tensor $R_{\lambda\mu\nu\kappa}$, and the consequent absence of a gravitational field. (See Section 6.4.) It is only in four or more dimensions that true gravitational fields can exist in empty space.

We might be willing to relax assumption (B), and allow $G_{\mu\nu}$ to contain terms with *fewer* than two derivatives of the metric. The freedom to use first derivatives does not allow any new terms in $G_{\mu\nu}$ (see Section 6.1), but if we can use the metric tensor itself, then one new term is possible, equal to $g_{\mu\nu}$ times a constant λ. The field equations would then read

$$R_{\mu\nu} - \tfrac{1}{2}g_{\mu\nu}R - \lambda g_{\mu\nu} = -8\pi G T_{\mu\nu}$$

The term $\lambda g_{\mu\nu}$ was originally introduced by Einstein[1] for cosmological reasons (which have since disappeared); for this reason, λ is called the *cosmological constant*. This term satisfies the requirements (A), (C), and (D), but does not satisfy (E), so λ must be very small so as not to interfere with the successes of Newton's theory of gravitation. Except in Chapter 16, I am assuming throughout this book that $\lambda = 0$.

2   Another Derivation*

The derivation of Einstein's equations in the last section made heavy use of the assumption that the left-hand side $G_{\mu\nu}$ is a tensor depending solely on the metric and its first and second derivatives. We might consider using a more general

---

* This section lies somewhat out of the book's main line of development, and may be omitted in a first reading.

tensor, which involves elements unrelated to the metric tensor or its derivatives, such as

$$\left(\frac{\partial x^\mu}{\partial \xi^\alpha_X(x)} \, \frac{\partial^3 \xi^\alpha_X(x)}{\partial x^\nu \, \partial x^\lambda \, \partial x^\rho} \right)_{x=X} \tag{7.2.1}$$

where $\xi^\alpha_X(x)$ are the coordinates locally inertial at X. (Reference to the precise definitions (3.3.2) and (3.3.3) of the metric and affine connection will show that (7.2.1) is not related to their derivatives.) Such a tensor could be constructed by writing

$$G_{\mu\nu} \equiv \left(\frac{\partial \xi^\alpha_X(x)}{\partial x^\mu} \, \frac{\partial \xi^\beta_X(x)}{\partial x^\nu} \, G^X_{\alpha\beta}(x) \right)_{x=X} \tag{7.2.2}$$

where $G^X_{\alpha\beta}$ is the most general possible linear combination of second derivatives of the metric tensor in the ξ_X coordinate system allowed by Lorentz covariance and symmetry, that is,

$$G_{\alpha\beta} = a_1 \Box^2 g_{\alpha\beta} + a_2 \left(\frac{\partial^2 g_\beta{}^\gamma}{\partial \xi^\alpha \, \partial \xi^\gamma} + \frac{\partial^2 g_\alpha{}^\gamma}{\partial \xi^\beta \, \partial \xi^\gamma} \right) + a_3 \eta_{\alpha\beta} \frac{\partial^2 g^{\gamma\delta}}{\partial \xi^\gamma \, \partial \xi^\delta}$$

$$+ b_1 \frac{\partial^2 g^\gamma{}_\gamma}{\partial \xi^\alpha \, \partial \xi^\beta} + b_2 \eta_{\alpha\beta} \Box^2 g^\gamma{}_\gamma \tag{7.2.3}$$

with a_1, a_2, a_3, b_1, b_2 five arbitrary dimensionless constants. [We have dropped the label X. Also, all indices are raised and lowered with the Minkowski tensors $\eta^{\alpha\beta}$ and $\eta_{\alpha\beta}$, and $\Box^2$ is the d'Alembertian $\Box^2 \equiv \eta^{\alpha\beta}(\partial/\partial \xi^\alpha)(\partial/\partial \xi^\beta)$.] For perfectly general values of the five constants a_1, a_2, a_3, b_1, b_2 this $G_{\mu\nu}$ would indeed depend on foreign elements such as (7.2.1). However, it is remarkable that by making use of energy-momentum conservation, and the validity of Newton's theory for weak static fields produced by nonrelativistic matter, we can put such stringent requirements on the constants $a_1, \ldots, b_2$ that the terms involving (7.2.1) drop out, and we get Einstein's theory.

In a weak field the requirement of energy and momentum conservation yields the ordinary conservation law $\partial T^\alpha{}_\beta / \partial \xi^\alpha = 0$, and therefore the assumed field equations $G_{\alpha\beta} = -8\pi G T_{\alpha\beta}$ require that

$$0 = \frac{\partial}{\partial \xi^\alpha} \, G^\alpha{}_\beta = (a_1 + a_2) \Box^2 \frac{\partial}{\partial \xi^\alpha} \, g^\alpha{}_\beta + (a_2 + a_3) \left(\frac{\partial}{\partial \xi^\beta} \frac{\partial^2 g^{\gamma\delta}}{\partial \xi^\gamma \, \partial \xi^\delta} \right)$$

$$+ (b_1 + b_2) \Box^2 \frac{\partial}{\partial \xi^\beta} \, g^\gamma{}_\gamma$$

Hence $a_1 + a_2$, $a_2 + a_3$, and $b_1 + b_2$ must all vanish, giving

$$G_{\alpha\beta} = a_1 \left\{ \Box^2 g_{\alpha\beta} - \frac{\partial^2 g_\beta{}^\gamma}{\partial \xi^\alpha \, \partial \xi^\gamma} - \frac{\partial^2 g_\alpha{}^\gamma}{\partial \xi^\beta \, \partial \xi^\gamma} + \eta_{\alpha\beta} \frac{\partial^2 g^{\gamma\delta}}{\partial \xi^\gamma \, \partial \xi^\delta} \right\}$$

$$+ b_1 \left\{ \frac{\partial^2 g^\gamma{}_\gamma}{\partial \xi^\alpha \, \partial \xi^\beta} - \eta_{\alpha\beta} \Box^2 g^\gamma{}_\gamma \right\} \tag{7.2.4}$$

To determine a_1 and b_1 we pass to the Newtonian limit. For a static field, (7.2.4) gives

$$G_{ii} + G_{00} = a_1 \nabla^2 (g_{ii} + g_{00}) - b_1 \nabla^2 (g_{ii} - g_{00})$$

(Repeated Latin indices are summed over the values 1, 2, 3.) For a nonrelativistic material system $|T_{ij}|$ is much less than $|T_{00}|$, so we obtain the field equation

$$(a_1 + b_1)\nabla^2 g_{00} + (a_1 - b_1)\nabla^2 g_{ii} = -8\pi G T_{00} \qquad (7.2.5)$$

We want the field equations in this limit to *imply* Newton's law,

$$\nabla^2 g_{00} = -8\pi G T_{00}$$

but (7.2.5) is the only one of the field equations to involve only g_{00} and/or g_{ii}, so we must require that $a_1 = b_1 = \frac{1}{2}$. The left-hand side of the weak-field equations is then

$$G_{\alpha\beta} = \frac{1}{2}\left\{ \Box^2 g_{\alpha\beta} - \frac{\partial^2 g_\beta{}^\gamma}{\partial\xi^\alpha\,\partial\xi^\gamma} - \frac{\partial^2 g_\alpha{}^\gamma}{\partial\xi^\beta\,\partial\xi^\gamma} + \frac{\partial^2 g^\gamma{}_\gamma}{\partial\xi^\alpha\,\partial\xi^\beta} \right\}$$
$$+ \tfrac{1}{2}\eta_{\alpha\beta}\left\{ \frac{\partial^2 g^{\gamma\delta}}{\partial\xi^\gamma\,\partial\xi^\delta} - \Box^2 g^\gamma{}_\gamma \right\} \qquad (7.2.6)$$

But Eq. (6.6.2) shows that for a weak field the Ricci tensor is

$$R_{\alpha\beta} = \frac{1}{2}\left\{ \Box^2 g_{\alpha\beta} - \frac{\partial^2 g_\beta{}^\gamma}{\partial\xi^\alpha\,\partial\xi^\gamma} - \frac{\partial^2 g_\alpha{}^\gamma}{\partial\xi^\beta\,\partial\xi^\gamma} + \frac{\partial^2 g^\gamma{}_\gamma}{\partial\xi^\alpha\,\partial\xi^\beta} \right\}$$

so (7.2.5) gives the field equation as

$$G_{\alpha\beta} = R_{\alpha\beta} - \tfrac{1}{2}\eta_{\alpha\beta}R = -8\pi G T_{\alpha\beta} \qquad (7.2.7)$$

The Principle of Equivalence then immediately yields the Einstein equations for a general field,

$$R_{\mu\nu} - \tfrac{1}{2}g_{\mu\nu}R = -8\pi G T_{\mu\nu} \qquad (7.2.8)$$

for (7.2.8) is generally covariant and reduces in locally inertial coordinate systems to (7.2.6). Thus, if we want a more general equation than Einstein's, which reduces in the weak-field limit to a second-order equation with (7.2.4) on the left-hand side, then we must pay the price of allowing new elements such as (7.2.1) to enter, *and* we must give up the possibility of deriving Newton's theory as a limiting case.

3   The Brans-Dicke Theory

Long-range forces are known to be transmitted by the gravitational field $g_{\mu\nu}$ and by the electromagnetic potential A_μ. It is natural then to suspect that other long-range forces may be produced by scalar fields. Such theories have been suggested since before general relativity; this section describes the latest and possibly the best motivated theory in which a scalar field shares the stage with gravitation, that of Brans and Dicke.[2]

The starting point for Brans and Dicke is the idea of Mach, that the phenomenon of inertia ought to arise from accelerations with respect to the general mass distribution of the universe. (See Section 1.3.) Thus the inertial masses of the various elementary particles ought not to be fundamental constants, but should rather represent the particles' interaction with some cosmic field. But the absolute scale of the elementary particle masses (as opposed to their ratios, which presumably have nothing to do with cosmic fields) can be measured only by measuring gravitational accelerations Gm/r^2, so an equivalent conclusion is that the gravitational constant G ought to be related to the average value of a scalar field ϕ, which is coupled to the mass density of the universe.

The simplest generally covariant field equation for such a scalar field would be

$$\square^2\phi = 4\pi\lambda T_{M\mu}^{\mu} \tag{7.3.1}$$

where $\square^2\phi = \phi_{;\rho}^{\ ,\rho}$ is now the invariant d'Alembertian, λ is a coupling constant, and $T_M^{\mu\nu}$ is the energy-momentum tensor of the matter (i.e., everything but gravitation and the ϕ-field) of the universe. We can make a rough estimate of the average value of ϕ by computing the central potential of a gas sphere with the cosmic mass density $\rho \sim 10^{-29}$ g cm^{-3} and radius equal to the apparent radius of the universe $R \sim 10^{28}$ cm. (See Chapter 14.) This gives an average value

$$\langle\phi\rangle \sim \lambda\rho R^2 \sim \lambda \times 10^{27} \text{ g cm}^{-1} \tag{7.3.2}$$

Note that 10^{27} g cm^{-1} is reasonably close to the constant $1/G = 1.35 \times 10^{28}$ g cm^{-1}; hence we normalize ϕ so that

$$\langle\phi\rangle \simeq \frac{1}{G} \tag{7.3.3}$$

and (7.3.2) then shows that λ is a dimensionless number of order unity. These considerations led Brans and Dicke to suggest that the correct field equations for gravitation are obtained by replacing G with $1/\phi$ and including an energy-momentum tensor $T_\phi^{\mu\nu}$ for the ϕ-field in the source of the gravitational field:

$$R^{\mu\nu} - \tfrac{1}{2}g^{\mu\nu}R = -\frac{8\pi}{\phi}\left[T_M^{\mu\nu} + T_\phi^{\mu\nu}\right] \tag{7.3.4}$$

We do not, however, wish to give up the successes of the Principle of Equivalence, such as the equality of gravitational and inertial mass, and the gravitational time dilation. Brans and Dicke therefore require that it is only $g_{\mu\nu}$, and not ϕ, that enters in the equations of motion of particles and photons. Therefore the equation describing the interchange of energy between matter and gravitation is the same as in Einstein's theory:

$$T_{M\ \nu;\mu}^{\ \mu} \equiv \frac{\partial T_M^{\mu}{}_{\nu}}{\partial x^{\mu}} + \Gamma_{\mu\rho}^{\mu} T_M^{\rho}{}_{\nu} - \Gamma_{\mu\nu}^{\rho} T_M^{\mu}{}_{\rho} = 0 \tag{7.3.5}$$

The Bianchi identities tell us that the left-hand side of Eq. (7.3.4) has vanishing covariant divergence, so by multiplying (7.3.4) by ϕ and taking the covariant divergence, we find

$$(R^\mu_{\ v} - \tfrac{1}{2}\delta^\mu_{\ v}R)\phi_{;\mu} = -8\pi T_\phi{}^\mu_{\ v;\mu} \tag{7.3.6}$$

This requirement proves sufficient to determine $T_\phi{}^\mu_{\ v}$. The most general symmetric tensor that can be built up from terms each of which involves two derivatives of one or two ϕ fields, and ϕ itself, is

$$
\begin{aligned}
T_\phi{}^\mu_{\ v} = {}& A(\phi)\phi_;{}^\mu\phi_{;v} + B(\phi)\delta^\mu_{\ v}\phi_{;\rho}\phi_;{}^\rho \\
& + C(\phi)\phi_;{}^\mu_{\ ;v} + \delta^\mu_{\ v}D(\phi)\square^2\phi
\end{aligned}
\tag{7.3.7}
$$

A straightforward calculation gives

$$
\begin{aligned}
T_\phi{}^\mu_{\ v;\mu} = {}& [A'(\phi) + B'(\phi)]\phi_;{}^\mu\phi_{;v}\phi_{;\mu} \\
& + [A(\phi) + D'(\phi)]\phi_{;v}\square^2\phi \\
& + [A(\phi) + 2B(\phi) + C'(\phi)]\phi_;{}^\mu_{\ ;v}\phi_{;\mu} \\
& + D(\phi)(\square^2\phi)_{;v} + C(\phi)\square^2(\phi_{;v})
\end{aligned}
\tag{7.3.8}
$$

(A prime here means the derivative with respect to ϕ.) The first term of Eq. (7.3.6) is determined by Eq. (6.5.2) as

$$\phi_{;\sigma}R^\sigma_{\ v} = \phi_;{}^\mu_{\ ;\mu;v} - \phi_{;v};{}^\mu_{\ ;\mu} = (\square^2\phi)_{;v} - \square^2(\phi_{;v}) \tag{7.3.9}$$

Also, by taking the trace of Eq. (7.3.4) and using (7.3.1), we find

$$R = \frac{8\pi}{\phi}\left[\frac{1}{4\pi\lambda}\square^2\phi + (A(\phi) + 4B(\phi))\phi_;{}^\mu\phi_{;\mu} + (C(\phi) + 4D(\phi))\square^2\phi\right]$$

so the left-hand side of (7.3.6) is

$$
\begin{aligned}
(R^\mu_{\ v} &- \tfrac{1}{2}\delta^\mu_{\ v}R)\phi_{;\mu} \\
={}& (\square^2\phi)_{;v} - \square^2(\phi_{;v}) \\
& - \frac{4\pi}{\phi}\phi_{;v}\left[\left(\frac{1}{4\pi\lambda} + C(\phi) + 4D(\phi)\right)\square^2\phi + (A(\phi) + 4B(\phi))\phi_;{}^\mu\phi_{;\mu}\right]
\end{aligned}
\tag{7.3.10}
$$

By comparing the coefficients of $(\square^2\phi)_{;v}$, $\square^2(\phi_{;v})$, $\phi_{;v}\square^2\phi$, $\phi_;{}^\mu\phi_{;\mu}\phi_{;v}$, and $\phi_;{}^\mu_{\ ;v}\phi_{;\mu}$ in Eqs. (7.3.8) and (7.3.10), we find that Eq. (7.3.6) requires

$$1 = -8\pi D(\phi)$$

$$-1 = -8\pi C(\phi)$$

$$-\frac{4\pi}{\phi}\left(\frac{1}{4\pi\lambda} + C(\phi) + 4D(\phi)\right) = -8\pi(A(\phi) + D'(\phi))$$

$$-\frac{4\pi}{\phi}(A(\phi) + 4B(\phi)) = -8\pi(A'(\phi) + B'(\phi))$$

$$0 = A(\phi) + 2B(\phi) + C'(\phi)$$

The unique solution is

$$A(\phi) = \frac{\omega}{8\pi\phi} \qquad B(\phi) = -\frac{\omega}{16\pi\phi}$$

$$C(\phi) = \frac{1}{8\pi} \qquad D(\phi) = -\frac{1}{8\pi} \tag{7.3.11}$$

where ω is a convenient dimensionless constant given by

$$\omega = \frac{1}{\lambda} - \frac{3}{2}$$

or

$$\lambda = \frac{2}{3 + 2\omega} \tag{7.3.12}$$

The field equations (7.3.1) and (7.3.4) of the Brans-Dicke theory now read

$$\square^2 \phi = \frac{8\pi}{3 + 2\omega} T_M{}^\mu{}_\mu \tag{7.3.13}$$

$$R_{\mu\nu} - \tfrac{1}{2} g_{\mu\nu} R = -\frac{8\pi}{\phi} T_{M\mu\nu} - \frac{\omega}{\phi^2} (\phi_{;\mu}\phi_{;\nu} - \tfrac{1}{2} g_{\mu\nu}\phi_{;\rho}\phi_;{}^\rho)$$

$$- \frac{1}{\phi} (\phi_{;\mu;\nu} - g_{\mu\nu} \square^2 \phi) \tag{7.3.14}$$

Our previous estimate indicated that λ is of order unity, so we expect that ω is of order unity. If ω is much larger than unity, then (7.3.13) gives $\square^2\phi = 0(1/\omega)$, and therefore

$$\phi = \langle\phi\rangle + 0\left(\frac{1}{\omega}\right) = \frac{1}{G} + 0\left(\frac{1}{\omega}\right) \tag{7.3.15}$$

Using this in (7.3.14) gives then

$$R_{\mu\nu} - \tfrac{1}{2} g_{\mu\nu} R = -8\pi G T_{M\mu\nu} + 0\left(\frac{1}{\omega}\right)$$

Thus the Brans-Dicke theory goes over to the Einstein theory in the limit $\omega \to \infty$.

It must be stressed that the role of the scalar field in the Brans-Dicke theory is confined to its effect on the gravitational field equations. Once $g_{\mu\nu}$ is calculated, the effects of gravitation on arbitrary physical systems are to be determined exactly as described in Chapters 3 through 5.

Throughout most of this book it will be assumed that there is no scalar field ϕ that contributes to long-range interactions. However, from time to time we return to the Brans-Dicke theory in order to see what changes it would make in the predictions of general relativity.

4   Coordinate Conditions

The symmetric tensor $G_{\mu\nu}$ has 10 independent components, so Einstein's field equations (7.1.13) comprise 10 algebraically independent equations. The unknown metric tensor also has 10 algebraically independent components, and at first sight one would think that the Einstein equations (with appropriate boundary conditions) would suffice to determine the $g_{\mu\nu}$ uniquely. However, this is not so. Although algebraically independent, the 10 $G_{\mu\nu}$ are related by four differential identities, the Bianchi identities [see Eq. (6.8.3)]:

$$G^{\mu}{}_{\nu;\mu} = 0$$

Thus there are not 10 functionally independent equations, but only $10 - 4 = 6$, leaving us with four degrees of freedom in the 10 unknowns $g_{\mu\nu}$. These degrees of freedom correspond to the fact that if $g_{\mu\nu}$ is a solution of Einstein's equation, then so is $g'_{\mu\nu}$, where $g'_{\mu\nu}$ is determined from $g_{\mu\nu}$ by a general coordinate transformation $x \to x'$. Such a coordinate transformation involves four arbitrary functions $x'^{\mu}(x)$, giving to the solutions of (7.1.13) just four degrees of freedom.

The failure of Einstein's equations to determine $g_{\mu\nu}$ uniquely is closely analogous with the failure of Maxwell's equation to determine the vector potential A_{μ} uniquely. When written in terms of the vector potential, Maxwell's equations read

$$\square^2 A_{\alpha} - \frac{\partial^2}{\partial x^{\alpha} \, \partial x^{\beta}} \, A^{\beta} = -J_{\alpha} \tag{7.4.1}$$

[See Eq. (2.7.6) and (2.7.11).] There are four equations for the four unknowns, but they do not determine A_{α} uniquely, because the left-hand sides of these equations are related by a differential identity analogous to the Bianchi identities:

$$\frac{\partial}{\partial x^{\alpha}} \left\{ \square^2 A^{\alpha} - \frac{\partial^2}{\partial x_{\alpha} \, \partial x^{\beta}} \, A^{\beta} \right\} \equiv 0$$

Thus the number of functionally independent equations is really only $4 - 1 = 3$, and there is one degree of freedom in the solution for the four A_{α}. This degree of freedom of course corresponds to gauge invariance; given any solution A_{α}, we can find another solution $A'_{\alpha} \equiv A_{\alpha} + \partial \Lambda / \partial x^{\alpha}$, with Λ arbitrary.

The ambiguity in the solutions of Maxwell's and Einstein's equations can be removed by main force. In the case of Maxwell's equation we do this by choosing a particular gauge. For instance, given any solution A_{α}, we can always construct a solution A'_{α} such that

$$\partial_{\alpha} A'^{\alpha} = 0 \tag{7.4.2}$$

by setting

$$A'_{\alpha} \equiv A_{\alpha} + \frac{\partial \Phi}{\partial x^{\alpha}}$$

where Φ is defined by

$$\Box^2\Phi = -\frac{\partial A^\alpha}{\partial x^\alpha}$$

Such a solution is said to be in the *Lorentz gauge*. The condition (7.4.2) when added to the *three* independent equations (7.4.1) completes a system of four equations that, with appropriate boundary conditions, will generally determine the four A_α uniquely. In the same way, we can eliminate the ambiguity in the metric tensor by adopting some particular coordinate system. The choice of a coordinate system can be expressed in four *coordinate conditions*, which, when added to the six independent Einstein equations, determine an unambiguous solution.

One particularly convenient choice of a coordinate system is represented by the *harmonic coordinate conditions*

$$\Gamma^\lambda \equiv g^{\mu\nu}\Gamma^\lambda_{\mu\nu} = 0 \tag{7.4.3}$$

To see that it is always possible to choose a coordinate system in which this holds, we recall the transformation equations of the affine connection

$$\Gamma'^\lambda_{\mu\nu} = \frac{\partial x'^\lambda}{\partial x^\rho}\frac{\partial x^\tau}{\partial x'^\mu}\frac{\partial x^\sigma}{\partial x'^\nu}\Gamma^\rho_{\tau\sigma} - \frac{\partial x^\rho}{\partial x'^\nu}\frac{\partial x^\sigma}{\partial x'^\mu}\frac{\partial^2 x'^\lambda}{\partial x^\rho\,\partial x^\sigma}$$

[See Eq. (4.5.8).] Contracting this with $g'^{\mu\nu}$, we find

$$\Gamma'^\lambda = \frac{\partial x'^\lambda}{\partial x^\rho}\Gamma^\rho - g^{\rho\sigma}\frac{\partial^2 x'^\lambda}{\partial x^\rho\,\partial x^\sigma} \tag{7.4.4}$$

Hence if Γ^ρ does not vanish, we can always define a new coordinate system x'^λ by solving the second-order partial differential equations

$$g^{\rho\sigma}\frac{\partial^2 x'^\lambda}{\partial x^\rho\,\partial x^\sigma} = \frac{\partial x'^\lambda}{\partial x^\rho}\Gamma^\rho$$

and Eq. (7.4.4) then gives $\Gamma'^\lambda = 0$ in the x'-system.

The four conditions (7.4.3) are of course not generally covariant, since their purpose is to remove the ambiguity in the metric tensor owing to the general covariance of the Einstein equations. Although we cannot write them as covariant equations, they can be put in a somewhat more elegant form by expressing the affine connection in terms of the metric tensor:

$$\Gamma^\lambda = \tfrac{1}{2}g^{\mu\nu}g^{\lambda\kappa}\left\{\frac{\partial g_{\kappa\mu}}{\partial x^\nu} + \frac{\partial g_{\kappa\nu}}{\partial x^\mu} - \frac{\partial g_{\mu\nu}}{\partial x^\kappa}\right\}$$

We recall that

$$g^{\lambda\kappa}\frac{\partial g_{\kappa\mu}}{\partial x^\nu} = -g_{\kappa\mu}\frac{\partial g^{\lambda\kappa}}{\partial x^\nu}$$

$$\tfrac{1}{2}g^{\mu\nu}\frac{\partial g_{\mu\nu}}{\partial x^\kappa} = g^{-1/2}\frac{\partial}{\partial x^\kappa}g^{1/2}$$

[See Eq. (4.7.5).] This then gives

$$\Gamma^\lambda = -g^{-1/2} \frac{\partial}{\partial x^\kappa} (g^{1/2} g^{\lambda\kappa}) \qquad (7.4.5)$$

and the harmonic coordinate conditions read

$$\frac{\partial}{\partial x^\kappa} (\sqrt{g}\, g^{\lambda\kappa}) = 0 \qquad (7.4.6)$$

We are now in a position to explain the term "harmonic coordinates." A function ϕ is said to be harmonic if $\Box^2\phi$ vanishes, where $\Box^2$ is the invariant d'Alembertian, defined by

$$\Box^2\phi \equiv (g^{\lambda\kappa}\phi_{;\lambda})_{;\kappa} \qquad (7.4.7)$$

Using (4.7.1), (4.7.7), and (7.4.5), this is

$$\Box^2\phi = g^{\lambda\kappa} \frac{\partial^2\phi}{\partial x^\lambda\, \partial x^\kappa} - \Gamma^\lambda \frac{\partial\phi}{\partial x^\lambda} \qquad (7.4.8)$$

If $\Gamma^\lambda = 0$ then the coordinates are themselves harmonic functions,

$$\Box^2 x^\mu = 0 \qquad (7.4.9)$$

thus justifying our application of the adjective "harmonic" to such coordinate systems.

In the absence of gravitational fields, the obvious harmonic coordinate system is that of Minkowski, in which $g^{\lambda\kappa} = \eta^{\lambda\kappa}$ and $g = 1$, so that (7.4.6) is satisfied trivially. In the presence of weak gravitational fields the harmonic coordinate systems may be pictured as nearly Minkowskian. Another related advantage of the harmonic coordinate condition is that, as shown in Chapters 9 and 10, its use produces a very great simplification in the weak-field equations, similar to the simplification brought to Maxwell's equations by use of the Lorentz gauge.

5 The Cauchy Problem

We can gain further insight into the mathematical content of Einstein's equations by applying them to the traditional initial value problem of Cauchy. Suppose that we are given $g_{\mu\nu}$ and $\partial g_{\mu\nu}/\partial x^0$ everywhere on the "plane" $x^0 = t$. If we could extract from the field equations a formula for $\partial^2 g_{\mu\nu}/\partial(x^0)^2$ everywhere at $x^0 = t$, we could then compute $g_{\mu\nu}$ and $\partial g_{\mu\nu}/\partial x^0$ at a time $x^0 = t + \delta t$, and by continuing this process $g_{\mu\nu}$ could be computed for all x^i and x^0.

At first sight this looks feasible, because we need 10 second derivatives, and there are 10 field equations. But let us look more closely at the left-hand side

$G^{\mu\nu} \equiv R^{\mu\nu} - \frac{1}{2}g^{\mu\nu}R$ of the field equations. The Bianchi identities (6.8.4) tell us that

$$\frac{\partial}{\partial x^0} G^{\mu 0} \equiv - \frac{\partial}{\partial x^i} G^{\mu i} - \Gamma^{\mu}_{\nu\lambda}G^{\lambda\nu} - \Gamma^{\nu}_{\nu\lambda}G^{\mu\lambda}$$

The right-hand side contains no time derivatives higher than $\partial^2/\partial(x^0)^2$, so neither does the left-hand side, and therefore $G^{\mu 0}$ contains no time derivatives higher than $\partial/\partial x^0$. Thus we cannot learn anything about the time evolution of the gravitational field from the four equations

$$G^{\mu 0} = -8\pi G T^{\mu 0} \tag{7.5.1}$$

Rather, these equations must be imposed as constraints on the initial data, that is, on $g_{\mu\nu}$ and $\partial g_{\mu\nu}/\partial x^0$ at $x^0 = t$.

This leaves as "dynamical" equations only the other six Einstein equations

$$G^{ij} = -8\pi G T^{ij} \tag{7.5.2}$$

When we solve these equations for the 10 second derivatives $\partial^2 g_{\mu\nu}/\partial(x^0)^2$, we must encounter a fourfold ambiguity, which of course we could not have hoped to escape since it is always possible to make coordinate transformations that leave $g_{\mu\nu}$ and $\partial g_{\mu\nu}/\partial x^0$ unchanged at $x^0 = t$ but that do alter $g_{\mu\nu}$ everywhere else. To be more specific, what we find is that (7.5.2) determines the six $\partial^2 g^{ij}/\partial(x^0)^2$, but leaves the other four derivatives $\partial^2 g^{\mu 0}/\partial(x^0)^2$ indeterminate. This ambiguity can be removed by imposing four coordinate conditions that fix the coordinate system. For instance, if we adopt the harmonic coordinate condition discussed in the last section, the second time derivative of $\sqrt{g}\, g^{\mu 0}$ can be determined by differentiating (7.4.6) with respect to time:

$$\frac{\partial^2}{\partial(x^0)^2} (\sqrt{g}\, g^{\mu 0}) = - \frac{\partial^2}{\partial x^0\, \partial x^i} \sqrt{g}\, g^{\mu i} \tag{7.5.3}$$

and the 10 equations (7.5.2) and (7.5.3) suffice to determine the second time derivatives of all $g_{\mu\nu}$.

When the initial value problem is solved in this way, the constraints (7.5.1) on the initial data need only be imposed once. The Bianchi identities and the conservation of energy and momentum tell us that whether or not the Einstein field equations are satisfied, we must have

$$(G^{\mu\nu} + 8\pi G T^{\mu\nu})_{;\nu} = 0$$

Let us apply this at $x^0 = t$. Imposing the initial data constraints (7.5.1), and determining the second derivatives from (7.5.2), the quantity in brackets will vanish everywhere at $x^0 = t$, so this gives

$$\frac{\partial}{\partial x^0} (G^{\mu 0} + 8\pi G T^{\mu 0}) = 0 \qquad \text{at } x^0 = t$$

and the fields computed at $x^0 = t + dt$ will therefore automatically also satisfy the constraints (7.5.1). Thus this method of solving the initial value problem is one that can be programmed for an automatic computer, once we find an initial metric at $x^0 = t$ that satisfies the constraints (7.5.1).

6 Energy, Momentum, and Angular Momentum of Gravitation

The physical significance of the Einstein equations can be clarified by writing them in an entirely equivalent form that, because not *manifestly* covariant, reveals their relation to the wave equations of elementary particle physics. Let us adopt a coordinate system that is quasi-Minkowskian, in the sense that the metric $g_{\mu\nu}$ approaches the Minkowski metric $\eta_{\mu\nu}$ at great distances from the finite material system under study. (This is the case in harmonic coordinate systems, and others as well.) We then write

$$g_{\mu\nu} = \eta_{\mu\nu} + h_{\mu\nu} \tag{7.6.1}$$

so that $h_{\mu\nu}$ vanishes at infinity. (However, $h_{\mu\nu}$ is *not* assumed to be small everywhere.) The part of the Ricci tensor linear in $h_{\mu\nu}$ is then

$$R^{(1)}{}_{\mu\kappa} \equiv \frac{1}{2}\left(\frac{\partial^2 h^\lambda{}_\lambda}{\partial x^\mu \, \partial x^\kappa} - \frac{\partial^2 h^\lambda{}_\mu}{\partial x^\lambda \, \partial x^\kappa} - \frac{\partial^2 h^\kappa{}_\kappa}{\partial x^\lambda \, \partial x^\mu} + \frac{\partial^2 h_{\mu\kappa}}{\partial x^\lambda \, \partial x_\lambda} \right) \tag{7.6.2}$$

[See Eq. (6.6.2). We are adopting the convenient convention that indices on $h_{\mu\nu}$, $R^{(1)}_{\mu\nu}$, and $\partial/\partial x^\lambda$ are raised and lowered with η's, for example, $h^\lambda{}_\lambda \equiv \eta^{\lambda\nu}h_{\lambda\nu}$ and $\partial/\partial x_\lambda \equiv \eta^{\lambda\nu}\,\partial/\partial x^\nu$, whereas indices on true tensors such as $R_{\mu\kappa}$ are raised and lowered with g's as usual.] The exact Einstein equations can then be written as

$$R^{(1)}{}_{\mu\kappa} - \tfrac{1}{2}\eta_{\mu\kappa}R^{(1)\lambda}{}_\lambda = -8\pi G[T_{\mu\kappa} + t_{\mu\kappa}] \tag{7.6.3}$$

where

$$t_{\mu\kappa} \equiv \frac{1}{8\pi G}\, [R_{\mu\kappa} - \tfrac{1}{2}g_{\mu\kappa}R^\lambda{}_\lambda - R^{(1)}{}_{\mu\kappa} + \tfrac{1}{2}\eta_{\mu\kappa}R^{(1)\lambda}{}_\lambda] \tag{7.6.4}$$

Equation (7.6.3) has just the form we should expect for the wave equation of a field of spin 2 (see Section 10.2) but with the peculiarity that its "source" $T_{\mu\kappa} + t_{\mu\kappa}$ depends explicitly on the field $h_{\mu\nu}$. We interpret this feature by saying that the field $h_{\mu\nu}$ is generated by the total densities and fluxes of energy and momentum, and $t_{\mu\kappa}$ is simply the energy-momentum "tensor" of the gravitational field itself. That is, we interpret the quantity

$$\tau^{\nu\lambda} \equiv \eta^{\nu\mu}\eta^{\lambda\kappa}[T_{\mu\kappa} + t_{\mu\kappa}] \tag{7.6.5}$$

as the total energy-momentum "tensor" of matter and gravitation. There are several properties of $\tau^{\nu\lambda}$ that support this interpretation:

(A) The quantities $R^{(1)}{}_{\mu\kappa}$ obey the linearized Bianchi identities:

$$\frac{\partial}{\partial x^v}[R^{(1)v\lambda} - \tfrac{1}{2}\eta^{v\lambda}R^{(1)\mu}{}_\mu] \equiv 0 \tag{7.6.6}$$

It therefore follows from the field equations (7.6.3) that $\tau^{v\lambda}$ is locally conserved:

$$\frac{\partial}{\partial x^v}\tau^{v\lambda} = 0 \tag{7.6.7}$$

Note that although $T^{v\lambda}$ obeys the covariant conservation law $T^{v\lambda}{}_{;v} = 0$, which really describes the *exchange* of energy between matter and gravitation, the quantity $\tau^{v\lambda}$ is conserved in the ordinary sense. In particular, for any finite system of volume V bounded by a surface S, Eq. (7.6.7) tells us that

$$\frac{d}{dt}\int_V \tau^{0\lambda}\,d^3x = -\int_S \tau^{i\lambda}n_i\,dS \tag{7.6.8}$$

where $\mathbf{n}$ is the unit outward normal to the surface. Hence we may interpret

$$P^\lambda \equiv \int_V \tau^{0\lambda}\,d^3x \tag{7.6.9}$$

as the total energy-momentum "vector" of the system, including matter, electro-magnetism, *and* gravitation; $\tau^{i\lambda}$ is the corresponding flux.

(B) Besides being conserved, $\tau^{v\lambda}$ is also symmetric,

$$\tau^{v\lambda} = \tau^{\lambda v} \tag{7.6.10}$$

and therefore

$$\frac{\partial}{\partial x^\mu}M^{\mu v\lambda} = 0 \tag{7.6.11}$$

where

$$M^{\mu v\lambda} \equiv \tau^{\mu\lambda}x^v - \tau^{\mu v}x^\lambda \tag{7.6.12}$$

We can thus interpret $M^{0v\lambda}$ and $M^{iv\lambda}$ as the density and flux of a total angular momentum

$$J^{v\lambda} \equiv \int d^3x M^{0v\lambda} = -J^{\lambda v} \tag{7.6.13}$$

that is constant if $M^{iv\lambda}$ vanishes on the surface of the volume of integration.

(C) We can compute $t_{\mu\kappa}$ as a power series in h, and find that the first term is *quadratic*:

$$t_{\mu\kappa} = \frac{1}{8\pi G}[-\tfrac{1}{2}h_{\mu\kappa}R^{(1)\lambda}{}_\lambda + \tfrac{1}{2}\eta_{\mu\kappa}h^{\rho\sigma}R^{(1)}{}_{\rho\sigma} + R^{(2)}{}_{\mu\kappa} - \tfrac{1}{2}\eta_{\mu\kappa}\eta^{\rho\sigma}R^{(2)}{}_{\rho\sigma}] + \bigcirc(h^3) \tag{7.6.14}$$

where $R^{(2)}{}_{\mu\kappa}$ is the second-order part of the Ricci tensor, given by (6.6.2) as

$$R^{(2)}{}_{\mu\kappa} = -\tfrac{1}{2}h^{\lambda\nu}\left[\frac{\partial^2 h_{\lambda\nu}}{\partial x^\kappa\,\partial x^\mu} - \frac{\partial^2 h_{\mu\nu}}{\partial x^\kappa\,\partial x^\lambda} - \frac{\partial^2 h_{\lambda\kappa}}{\partial x^\nu\,\partial x^\mu} + \frac{\partial^2 h_{\mu\kappa}}{\partial x^\nu\,\partial x^\lambda}\right]$$

$$+ \frac{1}{4}\left[2\frac{\partial h^\nu{}_\sigma}{\partial x^\nu} - \frac{\partial h^\nu{}_\nu}{\partial x^\sigma}\right]\left[\frac{\partial h^\sigma{}_\mu}{\partial x^\kappa} + \frac{\partial h^\sigma{}_\kappa}{\partial x^\mu} - \frac{\partial h_{\mu\kappa}}{\partial x_\sigma}\right]$$

$$- \frac{1}{4}\left[\frac{\partial h_{\sigma\kappa}}{\partial x^\lambda} + \frac{\partial h_{\sigma\lambda}}{\partial x^\kappa} - \frac{\partial h_{\lambda\kappa}}{\partial x^\sigma}\right]\left[\frac{\partial h^\sigma{}_\mu}{\partial x_\lambda} + \frac{\partial h^{\sigma\lambda}}{\partial x^\mu} - \frac{\partial h^\lambda{}_\mu}{\partial x_\sigma}\right] \qquad (7.6.15)$$

The example of electrodynamics would have led us to expect the energy-momentum "tensor" of gravitation to start with a term quadratic in $h_{\mu\nu}$. [Compare Eq. (2.8.9).] The presence in $t_{\mu\kappa}$ of terms of third and higher order simply means that the gravitational interaction of the gravitational field with itself also contributes to the total energy and momentum. Of course, when the gravitational field is weak, $h_{\mu\nu}$ is small, so our inclusion of $t_{\lambda\nu}$ in (7.6.5) (and our use of η to raise indices) does not seriously change our picture of the energy-momentum content of physical systems.

(D) Though not generally covariant, $t_{\mu\kappa}$, $\tau^{\nu\lambda}$, and $M^{\mu\nu\lambda}$ are at least Lorentz-covariant. Thus for a closed system P^λ and $J^{\nu\lambda}$ are not only constant, but also Lorentz-covariant. (See Section 2.6.)

(E) We chose at the beginning of this section to work in a coordinate system in which $h_{\mu\nu}$ vanishes at infinity. Far away from the finite material system that produces the gravitational field, $T_{\mu\kappa}$ is zero and $t_{\mu\kappa}$ is of order h^2, so the source term on the right-hand side of the field equations (7.6.3) is effectively confined to a finite region. This suggests that in a large variety of physical problems $h_{\mu\nu}$ will behave at great distances as do the potentials in electrostatics or Newtonian gravitational theory, that is, for $r \to \infty$,

$$h_{\mu\nu} = O\!\left(\frac{1}{r}\right) \qquad \frac{\partial h_{\mu\nu}}{\partial x^\lambda} = O\!\left(\frac{1}{r^2}\right) \qquad \frac{\partial^2 h_{\mu\nu}}{\partial x^\lambda\,\partial x^\rho} = O\!\left(\frac{1}{r^3}\right) \qquad (7.6.16)$$

In this case, (7.6.14) shows that

$$t_{\mu\kappa} = O\!\left(\frac{1}{r^4}\right) \qquad (7.6.17)$$

so the integral $\int \tau^{0\lambda}\,d^3x$ that gives the total energy and momentum *converges*. This is why it was so important to identify the coordinate system as quasi-Minkowskian; if $g_{\mu\nu}$ approached the metric of spherical polar coordinates at infinity, then our definitions (7.6.1) and (7.6.4) would have led to a gravitational energy density concentrated at infinity! (Note though that (7.6.16) and (7.6.17) are not always valid. If the system is eternally radiating gravitational waves (see Chapter 10), then $h_{\mu\nu}$ oscillates so that $\partial h_{\mu\nu}/\partial x^\lambda$ and $\partial^2 h_{\mu\nu}/\partial x^\lambda\,\partial x^\rho$ are of the same order as $h_{\mu\nu}$, giving an infinite total energy, which is what we would expect for gravitational radiation filling all space. In this case not even $h_{\mu\nu}$ behaves like $1/r$.)[2a]

(F) By its construction, $\tau^{\nu\lambda}$ is clearly the energy-momentum "tensor" we determine when we measure the gravitational field produced by any system. Indeed, there are many possible definitions of the energy-momentum "tensor" of gravitation that share most of the good properties of our $t_{\mu\kappa}$ (these definitions are usually based on the action principle; see Chapter 12), but $t_{\mu\kappa}$ is specially picked out by its role in (7.6.3) as part of the source of $h_{\mu\nu}$.

(G) Although calculation of $t_{\mu\kappa}$ in specific physical problems can be a nuisance, it is fortunately possible to avoid this calculation if all we want is the total energy and momentum of the system. The left-hand side of the field equations (7.6.3) can be written as

$$R^{(1)\nu\lambda} - \tfrac{1}{2}\eta^{\nu\lambda}R^{(1)\mu}{}_{\mu} = \frac{\partial}{\partial x^{\rho}}Q^{\rho\nu\lambda} \tag{7.6.18}$$

where

$$Q^{\rho\nu\lambda} \equiv \frac{1}{2}\left\{ \frac{\partial h^{\mu}{}_{\mu}}{\partial x_{\nu}}\eta^{\rho\lambda} - \frac{\partial h^{\mu}{}_{\mu}}{\partial x_{\rho}}\eta^{\nu\lambda} - \frac{\partial h^{\mu\nu}}{\partial x^{\mu}}\eta^{\rho\lambda} + \frac{\partial h^{\mu\rho}}{\partial x^{\mu}}\eta^{\nu\lambda} + \frac{\partial h^{\nu\lambda}}{\partial x_{\rho}} - \frac{\partial h^{\rho\lambda}}{\partial x_{\nu}} \right\} \tag{7.6.19}$$

Note that $Q^{\rho\nu\lambda}$ is antisymmetric in its first two indices,

$$Q^{\rho\nu\lambda} = -Q^{\nu\rho\lambda} \tag{7.6.20}$$

from which follows the differential identity (7.6.6). By using the field equations (7.6.3) in conjunction with (7.6.18) we find for the total energy-momentum "vector" (7.6.9) the value

$$P^{\lambda} = -\frac{1}{8\pi G}\int_{V}\frac{\partial Q^{\rho 0\lambda}}{\partial x^{\rho}}\,d^{3}x = -\frac{1}{8\pi G}\int_{V}\frac{\partial Q^{i0\lambda}}{\partial x^{i}}\,d^{3}x$$

and using Gauss's theorem gives

$$P^{\lambda} = -\frac{1}{8\pi G}\int Q^{i0\lambda}n_{i}r^{2}\,d\Omega \tag{7.6.21}$$

the integral being taken over a large sphere of radius r, with **n** the outward normal and $d\Omega$ the differential solid angle; that is,

$$r \equiv (x_{i}x_{i})^{1/2} \qquad n_{i} \equiv \frac{x_{i}}{r} \qquad d\Omega = \sin\theta\,d\theta\,d\varphi$$

(Repeated Latin indices are summed over 1, 2, 3.) In greater detail, the total energy and momentum are given by (7.6.19) and (7.6.21) as

$$P^{j} = -\frac{1}{16\pi G}\int\int\left\{ -\frac{\partial h_{kk}}{\partial t}\delta_{ij} + \frac{\partial h_{k0}}{\partial x^{k}}\delta_{ij} - \frac{\partial h_{j0}}{\partial x^{i}} + \frac{\partial h_{ij}}{\partial t} \right\}n_{i}r^{2}\,d\Omega \tag{7.6.22}$$

$$P^{0} = -\frac{1}{16\pi G}\int\int\left\{ \frac{\partial h_{jj}}{\partial x^{i}} - \frac{\partial h_{ij}}{\partial x^{j}} \right\}n_{i}r^{2}\,d\Omega \tag{7.6.23}$$

By the same reasoning, the total angular momentum "tensor" (7.6.13) is

$$J^{\nu\lambda} = \int d^3x (x^\nu \tau^{0\lambda} - x^\lambda \tau^{0\nu})$$

$$= \frac{-1}{8\pi G} \int d^3x \left(x^\nu \frac{\partial Q^{i0\lambda}}{\partial x^i} - x^\lambda \frac{\partial Q^{i0\nu}}{\partial x^i} \right)$$

As remarked in Section 2.9, the physically interesting components of $J^{\nu\lambda}$ are the three independent space-space components:

$$J_1 \equiv J^{23} \qquad J_2 \equiv J^{31} \qquad J_3 \equiv J^{12}$$

Using Gauss's theorem again, these components are given by

$$J^{jk} = -\frac{1}{16\pi G} \int \left\{ -x_j \frac{\partial h_{0k}}{\partial x^i} + x_k \frac{\partial h_{0j}}{\partial x^i} \right.$$

$$\left. + x_j \frac{\partial h_{ki}}{\partial t} - x_k \frac{\partial h_{ji}}{\partial t} + h_{0k}\delta_{ij} - h_{0j}\delta_{ik} \right\} n_i r^2 \, d\Omega$$

$$(7.6.24)$$

Thus, in order to calculate the total momentum, energy, and angular momentum of an arbitrary finite system, it is only necessary to know the asymptotic behavior of $h_{\mu\nu}$ at great distances.

(H) It has been shown that P^0 is always *positive*, and takes the value zero only for matter-free empty space.[3]

(I) Although $\tau^{\nu\lambda}$ is not a tensor and P^λ is not a vector, the total energy and momenta have the important property of being invariant under any coordinate transformation that reduces at infinity to the identity. Such a transformation will be of the form

$$x^\mu \rightarrow x'^\mu = x^\mu + \varepsilon^\mu(x)$$

where $\varepsilon^\mu(x)$ vanishes as $r \rightarrow \infty$, although $\varepsilon^\mu(x)$ need not be small at finite distances. The metric tensor in the new coordinate system is

$$g'^{\mu\nu} = g^{\rho\sigma} \left(\delta^\mu_{\ \rho} + \frac{\partial \varepsilon^\mu}{\partial x^\rho} \right) \left(\delta^\nu_{\ \sigma} + \frac{\partial \varepsilon^\nu}{\partial x^\sigma} \right)$$

For $r \rightarrow \infty$ both ε^μ and $h_{\mu\nu}$ are small, so we can calculate $g'^{\mu\nu}$ to first order in ε^μ and $h_{\mu\nu}$ by setting $g^{\rho\sigma} \simeq \eta^{\rho\sigma} - h^{\rho\sigma}$ and expanding; this gives

$$g'^{\mu\nu} \simeq \eta^{\mu\nu} - h'^{\mu\nu}$$

where

$$h'^{\mu\nu} = h^{\mu\nu} - \frac{\partial \varepsilon^\mu}{\partial x_\nu} - \frac{\partial \varepsilon^\nu}{\partial x_\mu}$$

The change in the quantity (7.6.19) produced by this coordinate transformation is then given for $r \to \infty$ by

$$\Delta Q^{\rho\nu\lambda} = \frac{1}{2} \left\{ -\frac{\partial^2 \varepsilon^\mu}{\partial x^\mu \, \partial x_\nu} \eta^{\rho\lambda} + \frac{\partial^2 \varepsilon^\mu}{\partial x^\mu \, \partial x_\rho} \eta^{\nu\lambda} + \Box^2 \varepsilon^\nu \eta^{\rho\lambda} \right.$$

$$\left. - \Box^2 \varepsilon^\rho \eta^{\nu\lambda} - \frac{\partial^2 \varepsilon^\nu}{\partial x_\rho \, \partial x_\lambda} + \frac{\partial^2 \varepsilon^\rho}{\partial x_\nu \, \partial x_\lambda} \right\}$$

or

$$\Delta Q^{\rho\nu\lambda} = \frac{\partial}{\partial x^\sigma} D^{\sigma\rho\nu\lambda}$$

where

$$D^{\sigma\rho\nu\lambda} \equiv \frac{1}{2} \left\{ -\frac{\partial \varepsilon^\sigma}{\partial x_\nu} \eta^{\rho\lambda} + \frac{\partial \varepsilon^\sigma}{\partial x_\rho} \eta^{\nu\lambda} + \frac{\partial \varepsilon^\nu}{\partial x_\sigma} \eta^{\rho\lambda} - \frac{\partial \varepsilon^\rho}{\partial x_\sigma} \eta^{\nu\lambda} - \frac{\partial \varepsilon^\nu}{\partial x_\rho} \eta^{\sigma\lambda} + \frac{\partial \varepsilon^\rho}{\partial x_\nu} \eta^{\sigma\lambda} \right\}$$

We note that D is totally antisymmetric in its first three indices

$$D^{\sigma\rho\nu\lambda} = -D^{\rho\sigma\nu\lambda} = -D^{\sigma\nu\rho\lambda} = -D^{\nu\rho\sigma\lambda}$$

and therefore the change in the surface integral takes the form

$$\Delta P^\lambda = -\frac{1}{8\pi G} \int \left(\frac{\partial D^{\sigma i 0\lambda}}{\partial x^\sigma} \right) n_i r^2 \, d\Omega$$

$$= -\frac{1}{8\pi G} \int \left(\frac{\partial D^{j i 0\lambda}}{\partial x^j} \right) n_i r^2 \, d\Omega$$

or, using Gauss's theorem again,

$$\Delta P^\lambda = -\frac{1}{8\pi G} \int \left(\frac{\partial^2 D^{j i 0\lambda}}{\partial x^i \, \partial x^j} \right) d^3x = 0 \tag{7.6.25}$$

We may note as a corollary that P^λ transforms as a four-vector under any transformation that leaves the metric $\eta_{\mu\nu}$ at infinity unchanged, because any such transformation can be expressed as the product of a Lorentz transformation $x^\mu \to \Lambda^\mu{}_\nu x^\nu + a^\mu$, under which P^λ transforms as a four-vector (see (D) above), times a transformation that approaches the identity at infinity and hence does not change P^λ.

(J) If the matter in our system is divided into *distant* subsystems S_n, the gravitational field can be approximated by writing $h_{\mu\nu}$ as the sum of the $h^n_{\mu\nu}$'s that would be produced by each subsystem acting alone. (Interference terms between these different $h^n_{\mu\nu}$'s may be neglected in $t_{\mu\kappa}$, because any place where one $h^n_{\mu\nu}$ is large, all others are small.) It follows then from the calculation of P^λ in (E) above that the total energy and momentum are equal to the sum of the values $P_n{}^\lambda$ for each subsystem alone.

The energy-momentum "vector" P^λ defined by (7.6.9) is conserved, is a Lorentz four-vector, and is additive. What more could we ask? Any four quantities with these properties are uniquely determined to be the usual momentum and energy (as can be shown formally by applying the conservation laws to a collision in which distant subsystems come together, interact, and then go off to infinity again[4]).

The arguments of this section can be turned around to provide yet another derivation[5] of Einstein's field equations. Suppose that we set out to construct equations for a long-range field of spin 2. General group-theoretic considerations require them to take the form[6]

$$R^{(1)}{}_{\mu\kappa} - \tfrac{1}{2}\eta_{\mu\kappa}R^{(1)\lambda}{}_\lambda = \Theta_{\mu\kappa} \qquad (7.6.26)$$

with $\Theta_{\mu\kappa}$ some source function, which because of the identities (7.6.6) must be conserved

$$\frac{\partial}{\partial x_\mu}\Theta_{\mu\kappa} = 0 \qquad (7.6.27)$$

It will not do to set $\Theta_{\mu\kappa}$ proportional to the energy-momentum tensor $T_{\mu\kappa}$ of matter alone, because matter can interchange energy and momentum with gravitation, and therefore $T_{\mu\kappa}$ does not satisfy (7.6.27). We *must* include in $\Theta_{\mu\kappa}$ terms involving h itself, and when these terms are calculated by imposing the condition (7.6.27), we find that the field equation (7.6.26) must be simply (7.6.3), which is equivalent to Einstein's theory. We are thus led back to the remark at the beginning of this chapter, that the major difference between the electromagnetic and gravitational fields is that the source of the electromagnetic potential A^α is a conserved current J^α that does not involve A^α because the electromagnetic field is not itself charged, whereas the source of the gravitational field $h_{\mu\kappa}$ is a conserved "tensor" $\tau^{\mu\kappa}$ that *must* involve $h_{\mu\kappa}$ because the gravitational field does carry energy and momentum.

7 BIBLIOGRAPHY

☐ Y. Bruhat, "The Cauchy Problem," in *Gravitation: An Introduction to Current Research*, ed. by L. Witten (Wiley, New York, 1962), p. 130.

☐ A. Lichnerowicz, *Relativistic Hydrodynamics and Magnetohydrodynamics* (W. A. Benjamin, New York, 1967), Chapter 1.

☐ A. Trautman, "Conservation Laws in General Relativity," in *Gravitation: An Introduction to Current Research, op cit.*, p. 169.

Also, see Bibliography, Chapter 3.

7 REFERENCES

1. A Einstein, Sitz. Preuss. Akad. Wiss., 142, 1917. For an English translation, see *The Principle of Relativity* (Methuen, 1923, reprinted by Dover Publications), p. 35.
2. C. H. Brans and R. H. Dicke, Phys. Rev., **124**, 925 (1961). For an equivalent formulation, see R. H. Dicke, Phys. Rev., **125**, 2163 (1962).
2a. R. Arnowitt, S. Deser, and C. Misner, quoted by C. Misner, *Proceedings of the Conference on the Theory of Gravitation* (Gautier-Villars, Paris, 1964), p. 189.
3. D. R. Brill and S. Deser, Ann. Phys. (N.Y.), **50**, 542 (1968); S. Deser, Nuovo Cimento, **55B**, 593 (1968); D. Brill and S. Deser, Phys. Rev. Letters, **20**, 8 (1968). D. Brill, S. Deser, and L. Faddeev, Phys. Lett., **26A**, 538 (1968).
4. A. Einstein, Bull. Am. Mat. Soc., April, 1935, p. 223.
5. S. N. Gupta, Proc. Phys. Soc., **A65**, 161, 608 (1952); Phys. Rev., **96**, 1683 (1954); Rev. Mod. Phys., **29**, 334 (1957). W. Thirring, Ann. Phys. (N.Y.), **16**, 96 (1961). S. Deser, Gen. Rel. and Grav, **1**, 9 (1970).
6. See, for example, S. Weinberg, Phys. Rev., **138**, 988 (1965).

PART THREE
APPLICATIONS OF
GENERAL RELATIVITY

8 CLASSIC TESTS OF EINSTEIN'S THEORY

Einstein suggested three tests of general relativity:
  (A) The gravitational red shift of spectral lines.
  (B) The deflection of light by the sun.
  (C) The precession of the perihelia of the orbits of the inner planets.

Since then, one other test has been carried out:
  (D) The time delay of radar echoes passing the sun.

And another soon will be:
  (E) The precession of a gyroscope in orbit around the earth.

All five tests are carried out in empty space and in gravitational fields that are to a good approximation static and [except for (E)] spherically symmetric, so our first task will be to solve the Einstein vacuum field equations under the simplifying assumptions of isotropy and time independence. The results will then be used to treat tests (B) through (D). We have already seen in Chapter 3 that (A) tests only the Principle of Equivalence, so it need not be considered further here, whereas (E) involves anisotropic effects owing to the rotation of the earth, and will be discussed in Chapter 9.

1 The General Static Isotropic Metric

For the moment we put aside Einstein's equations, and consider what is the most general metric tensor that can represent a static isotropic gravitational field.

By "static and isotropic" we mean that it must be possible to find a set of "quasi-Minkowskian" coordinates x^1, x^2, x^3, $x^0 \equiv t$, such that the invariant proper time $d\tau^2 \equiv -g_{\mu\nu}\,dx^\mu\,dx^\nu$ does not depend on t, and depends on $\mathbf{x}$ and $d\mathbf{x}$ only through the rotational invariants $d\mathbf{x}^2$, $\mathbf{x} \cdot d\mathbf{x}$, and $\mathbf{x}^2$. The most general proper time interval is then

$$d\tau^2 = F(r)\,dt^2 - 2E(r)\,dt\,\mathbf{x} \cdot d\mathbf{x}$$
$$- D(r)(\mathbf{x} \cdot d\mathbf{x})^2 - C(r)\,d\mathbf{x}^2 \tag{8.1.1}$$

where F, E, D, and C are unknown functions of

$$r \equiv (\mathbf{x} \cdot \mathbf{x})^{1/2}$$

(Scalar products of three-vectors are throughout this chapter defined as usual, e.g., $\mathbf{x} \cdot d\mathbf{x} = x^1\,dx^1 + x^2\,dx^2 + x^3\,dx^3$, etc.) A deeper derivation of Eq. (8.1.1) will be given in Chapter 13; for the present we can regard (8.1.1) as a definition of what we mean by a static isotropic metric, or alternatively as an ansatz that allows us to find some solutions of the field equations.

It is convenient to replace $\mathbf{x}$ with spherical polar coordinates r, θ, φ, defined as usual by

$$x^1 = r \sin\theta \cos\varphi \qquad x^2 = r \sin\theta \sin\varphi \qquad x^3 = r \cos\theta$$

The proper time interval (8.1.1) then becomes

$$d\tau^2 = F(r)\,dt^2 - 2rE(r)\,dt\,dr$$
$$- r^2 D(r)\,dr^2 - C(r)\,(dr^2 + r^2\,d\theta^2 + r^2 \sin^2\theta\,d\varphi^2) \tag{8.1.2}$$

We are free to reset our clocks by defining a new time coordinate

$$t' \equiv t + \Phi(r)$$

with Φ an arbitrary function of r. This allows us to eliminate the off-diagonal element g_{tr} by setting

$$\frac{d\Phi}{dr} = -\frac{rE(r)}{F(r)}$$

The proper time (8.1.2) then becomes

$$d\tau^2 = F(r)\,dt'^2 - G(r)\,dr^2 - C(r)\,(dr^2 + r^2\,d\theta^2 + r^2 \sin^2\theta\,d\varphi^2) \tag{8.1.3}$$

where

$$G(r) \equiv r^2 \left(D(r) + \frac{E^2(r)}{F(r)} \right)$$

We are also free to redefine the radius r, and thereby impose one further relation on the functions F, G, and C. For instance, suppose that we define

$$r'^2 \equiv C(r)r^2$$

Then the proper time (8.1.3) takes what is called the *standard form*

$$d\tau^2 = B(r') \, dt'^2 - A(r') \, dr'^2 - r'^2 \, (d\theta^2 + \sin^2 \theta \, d\varphi^2) \qquad (8.1.4)$$

where

$$B(r') \equiv F(r)$$

$$A(r') \equiv \left(1 + \frac{G(r)}{C(r)}\right)\left(1 + \frac{r}{2C(r)}\frac{dC(r)}{dr}\right)^{-2}$$

Alternatively, we could define

$$r'' = \exp \int \left(1 + \frac{G(r)}{C(r)}\right)^{1/2} \frac{dr}{r}$$

and (8.1.3) would then appear in what is called the *isotropic form*,

$$d\tau^2 = H(r'') \, dt'^2 - J(r'') \, (dr''^2 + r''^2 \, d\theta^2 + r''^2 \sin^2 \theta \, d\varphi^2) \qquad (8.1.5)$$

where

$$H(r'') \equiv F(r)$$

$$J(r'') \equiv \frac{C(r)r^2}{r''^2}$$

We shall do most of our work with a metric of the "standard" form:

$$d\tau^2 = B(r) \, dt^2 - A(r) \, dr^2 - r^2 \, (d\theta^2 + \sin^2 \theta \, d\varphi^2) \qquad (8.1.6)$$

(We drop primes on r and t from now on.) The metric tensor has the nonvanishing components

$$g_{rr} = A(r) \qquad g_{\theta\theta} = r^2 \qquad g_{\varphi\varphi} = r^2 \sin^2 \theta \qquad g_{tt} = -B(r) \qquad (8.1.7)$$

with functions $A(r)$ and $B(r)$ that are to be determined by solving the field equations. Since $g_{\mu\nu}$ is diagonal, it is easy to write down all the nonvanishing components of its inverse:

$$g^{rr} = A^{-1}(r) \qquad g^{\theta\theta} = r^{-2} \qquad g^{\varphi\varphi} = r^{-2} (\sin \theta)^{-2} \qquad g^{tt} = -B^{-1}(r)$$

$$(8.1.8)$$

Furthermore, the determinant of the metric tensor is $-g$, where

$$g = r^4 A(r) B(r) \sin^2 \theta \qquad (8.1.9)$$

so the invariant volume element is

$$\sqrt{g} \, dr \, d\theta \, d\varphi = r^2 \sqrt{A(r)B(r)} \, \sin \theta \, dr \, d\theta \, d\varphi \qquad (8.1.10)$$

The affine connection can be computed from the usual formula:

$$\Gamma^{\lambda}_{\mu\nu} = \tfrac{1}{2} g^{\lambda\rho} \left(\frac{\partial g_{\rho\mu}}{\partial x^{\nu}} + \frac{\partial g_{\rho\nu}}{\partial x^{\mu}} - \frac{\partial g_{\mu\nu}}{\partial x^{\rho}} \right)$$

Its only nonvanishing components are

$$\Gamma^{r}_{rr} = \frac{1}{2A(r)} \frac{dA(r)}{dr} \qquad \Gamma^{r}_{\theta\theta} = -\frac{r}{A(r)}$$

$$\Gamma^{r}_{\varphi\varphi} = -\frac{r \sin^2\theta}{A(r)} \qquad \Gamma^{r}_{tt} = \frac{1}{2A(r)} \frac{dB(r)}{dr}$$

$$\Gamma^{\theta}_{r\theta} = \Gamma^{\theta}_{\theta r} = \frac{1}{r} \qquad \Gamma^{\theta}_{\varphi\varphi} = -\sin\theta\cos\theta$$

$$\Gamma^{\varphi}_{\varphi r} = \Gamma^{\varphi}_{r\varphi} = \frac{1}{r} \qquad \Gamma^{\varphi}_{\varphi\theta} = \Gamma^{\varphi}_{\theta\varphi} = \cot\theta$$

$$\Gamma^{t}_{tr} = \Gamma^{t}_{rt} = \frac{1}{2B(r)} \frac{dB(r)}{dr} \qquad (8.1.11)$$

We also need the Ricci tensor. It is given by (6.2.4) and (6.1.5) as

$$R_{\mu\kappa} = \frac{\partial \Gamma^{\lambda}_{\mu\lambda}}{\partial x^{\kappa}} - \frac{\partial \Gamma^{\lambda}_{\mu\kappa}}{\partial x^{\lambda}} + \Gamma^{\eta}_{\mu\lambda}\Gamma^{\lambda}_{\kappa\eta} - \Gamma^{\eta}_{\mu\kappa}\Gamma^{\lambda}_{\lambda\eta} \qquad (8.1.12)$$

(Note that despite its appearance, the first term is symmetric in μ and κ, because (4.7.6) gives $\Gamma^{\lambda}_{\mu\lambda}$ equal to $\frac{1}{2}\partial \ln g/\partial x^{\mu}$.) Inserting in (8.1.12) the components of the affine connection given by (8.1.11), we find

$$R_{rr} = \frac{B''(r)}{2B(r)} - \frac{1}{4}\left(\frac{B'(r)}{B(r)}\right)\left(\frac{A'(r)}{A(r)} + \frac{B'(r)}{B(r)}\right) - \frac{1}{r}\left(\frac{A'(r)}{A(r)}\right) \qquad (8.1.13)$$

$$R_{\theta\theta} = -1 + \frac{r}{2A(r)}\left(-\frac{A'(r)}{A(r)} + \frac{B'(r)}{B(r)}\right) + \frac{1}{A(r)}$$

$$R_{\varphi\varphi} = \sin^2\theta R_{\theta\theta}$$

$$R_{tt} = -\frac{B''(r)}{2A(r)} + \frac{1}{4}\left(\frac{B'(r)}{A(r)}\right)\left(\frac{A'(r)}{A(r)} + \frac{B'(r)}{B(r)}\right) - \frac{1}{r}\left(\frac{B'(r)}{A(r)}\right)$$

$$R_{\mu\nu} = 0 \qquad \text{for } \mu \neq \nu$$

(A prime now means differentiation with respect to r.) The results that $R_{r\theta}$, $R_{r\varphi}$, $R_{t\theta}$, $R_{t\varphi}$, and $R_{\theta\varphi}$ vanish, and that $R_{\varphi\varphi} = \sin^2\theta R_{\theta\theta}$, are merely consequences of the rotational invariance of the metric, whereas the result that R_{rt} vanishes is because we have set our clocks so that the metric is invariant under the time-reversal transformation $t \rightarrow -t$.

Neither the standard nor the isotropic coordinates are harmonic, but we can easily use the results (8.1.7) and (8.1.11) for the metric and affine connection in standard coordinates to construct harmonic coordinates X_1, X_2, X_3, t. We set

$$X_1 = R(r) \sin \theta \cos \varphi \qquad X_2 = R(r) \sin \theta \sin \varphi \qquad X_3 = R(r) \cos \theta$$

$$(8.1.14)$$

A straightforward calculation gives then

$$\Box^2 X_i \equiv g^{\mu\nu} \left[\frac{\partial^2 X_i}{\partial x^\mu \, \partial x^\nu} - \Gamma^\lambda_{\mu\nu} \frac{\partial X_i}{\partial x^\lambda} \right]$$

$$= \left(\frac{X_i}{AR} \right) \left[\left(\frac{B'}{2B} + \frac{2}{r} - \frac{A'}{2A} \right) R' + R'' - \frac{2A}{r^2} R \right]$$

Also, the standard time coordinate t satisfies

$$\Box^2 t = 0$$

Thus the coordinates X_1, X_2, X_3, t are harmonic if $R(r)$ satisfies the differential equation

$$\frac{d}{dr} \left(r^2 B^{1/2} A^{-1/2} \frac{dR}{dr} \right) - 2A^{1/2} B^{1/2} R = 0 \qquad (8.1.15)$$

In these harmonic coordinates the proper time (8.1.6) becomes

$$d\tau^2 = B \, dt^2 - \frac{r^2}{R^2} d\mathbf{X}^2 - \left[\frac{A}{R^2 R'^2} - \frac{r^2}{R^4} \right] (\mathbf{X} \cdot d\mathbf{X})^2 \qquad (8.1.16)$$

2   The Schwarzschild Solution

We now apply the Einstein field equations to the general static isotropic metric. We use the standard form discussed in the last section, that is,

$$d\tau^2 = B(r) \, dt^2 - A(r) \, dr^2 - r^2 \, d\theta^2 - r^2 \sin^2 \theta \, d\varphi^2 \qquad (8.2.1)$$

The field equations for empty space are

$$R_{\mu\nu} = 0 \qquad (8.2.2)$$

The components of the Ricci tensor are given for this metric by Eq. (8.1.13). We see that it will suffice to set R_{rr}, $R_{\theta\theta}$, and R_{tt} equal to zero. We also see that

$$\frac{R_{rr}}{A} + \frac{R_{tt}}{B} = -\frac{1}{rA} \left(\frac{A'}{A} + \frac{B'}{B} \right) \qquad (8.2.3)$$

so (8.2.2) requires that $B'/B = -A'/A$, or

$$A(r)B(r) = \text{constant} \tag{8.2.4}$$

Furthermore, we impose on A and B the boundary condition that for $r \to \infty$ the metric tensor must approach the Minkowski tensor in spherical coordinates, that is,

$$\lim_{r \to \infty} A(r) = \lim_{r \to \infty} B(r) = 1 \tag{8.2.5}$$

From (8.2.4) and (8.2.5) we have then

$$A(r) = \frac{1}{B(r)} \tag{8.2.6}$$

Since (8.2.3) now vanishes, it remains to make R_{rr} and $R_{\theta\theta}$ vanish. Using (8.2.6) in (8.1.13), we find

$$R_{\theta\theta} = -1 + B'(r)r + B(r) \tag{8.2.7}$$

$$R_{rr} = \frac{B''(r)}{2B(r)} + \frac{B'(r)}{rB(r)} = \frac{R'_{\theta\theta}(r)}{2rB(r)} \tag{8.2.8}$$

so it is sufficient to set $R_{\theta\theta}$ equal to zero, that is,

$$\frac{d}{dr}(rB(r)) = rB'(r) + B(r) = 1$$

The solution is

$$rB(r) = r + \text{constant} \tag{8.2.9}$$

To fix the constant of integration we recall that at great distances from a central mass M, the component $g_{tt} \equiv -B$ must approach $-1 - 2\phi$, where ϕ is the Newtonian potential $-MG/r$. (See Section 3.4.) Hence the constant of integration is $-2MG$, and our final solution is

$$B(r) = \left[1 - \frac{2MG}{r}\right] \tag{8.2.10}$$

$$A(r) = \left[1 - \frac{2MG}{r}\right]^{-1} \tag{8.2.11}$$

The full metric is given by

$$d\tau^2 = \left[1 - \frac{2MG}{r}\right]dt^2 - \left[1 - \frac{2MG}{r}\right]^{-1} dr^2 - r^2\,d\theta - r^2 \sin^2\theta\,d\varphi^2$$

$$\tag{8.2.12}$$

This solution was found by K. Schwarzschild in 1916.

The Schwarzschild solution is expressed in Eq. (8.2.12) in its "**standard**" form. We can also express it in the equivalent "isotropic" form, by introducing a new radius variable

$$\rho \equiv \tfrac{1}{2}\left[r - MG + (r^2 - 2MGr)^{1/2}\right] \tag{8.2.13}$$

or

$$r = \rho\left(1 + \frac{MG}{2\rho}\right)^2$$

Substituting this in Eq. (8.2.12) gives

$$d\tau^2 = \frac{(1 - MG/2\rho)^2}{(1 + MG/2\rho)^2}\,dt^2 - \left(1 + \frac{MG}{2\rho}\right)^4 (d\rho^2 + \rho^2\,d\theta^2 + \rho^2\sin^2\theta\,d\varphi^2) \tag{8.2.14}$$

We can also construct harmonic coordinates

$$X_1 = R\sin\theta\cos\varphi; \qquad X_2 = R\sin\theta\sin\varphi; \qquad X_3 = R\cos\theta; \qquad t$$

by using for R a solution of the differential equation (8.1.15), which here becomes

$$\frac{d}{dr}\left(r^2\left[1 - \frac{2MG}{r}\right]\frac{dR}{dr}\right) - 2R = 0$$

One convenient solution is

$$R = r - MG$$

The metric is then given by Eq. (8.1.16):

$$d\tau^{2\prime} = \left(\frac{1 - MG/R}{1 + MG/R}\right)dt^2 - \left(1 + \frac{MG}{R}\right)^2 d\mathbf{X}^2 - \left(\frac{1 + MG/R}{1 - MG/R}\right)\frac{M^2G^2}{R^4}(\mathbf{X}\cdot d\mathbf{X})^2 \tag{8.2.15}$$

with $R^2 \equiv \mathbf{X}^2$ now understood.

We identified the integration constant M with the mass of the sun by comparison with Newton's theory. In fact, we can show that M is precisely equal to the total energy P^0 of the sun *and its gravitational field*. Let us write the standard form of the metric in quasi-Minkowskian coordinates, by defining

$$x^1 \equiv r\sin\theta\cos\varphi, \qquad x^2 = r\sin\theta\sin\varphi, \qquad x^3 = r\cos\theta$$

Then Eq. (8.2.12) becomes

$$d\tau^2 = \left[1 - \frac{2MG}{r}\right]dt^2 - \left\{\left[1 - \frac{2MG}{r}\right]^{-1} - 1\right\}r^{-2}(\mathbf{x}\cdot d\mathbf{x})^2 - d\mathbf{x}^2$$

Since $g_{\mu\nu}$ is time independent and g_{i0} vanishes, it follows from (7.6.22) that the total momentum P^i of the system vanishes, which of course it must do since the system is static and isotropic. To calculate the total energy, we need the asymptotic behavior of the spatial part of the metric; as $r \to \infty$,

$$h_{ij} \equiv g_{ij} - \delta_{ij} \to \frac{2MG}{r} n_i n_j + O\left(\frac{1}{r^2}\right)$$

where $n_i \equiv x^i/r$. To calculate the integral (7.6.21) we use the relations

$$\frac{\partial r}{\partial x^i} = n_i$$

$$\frac{\partial n_i}{\partial x^j} = \frac{\delta_{ij} - n_i n_j}{r}$$

and find

$$\frac{\partial h_{jj}}{\partial x^i} - \frac{\partial h_{ij}}{\partial x^j} \to -\frac{4MG}{r^2} n_i + O\left(\frac{1}{r^3}\right)$$

so Eq. (7.6.23) gives the total energy of matter and gravitation here as

$$P^0 = M \tag{8.2.16}$$

The reader may check that the same result would be given by the isotropic or harmonic forms of the Schwarzschild solution. Finally, Eq. (7.6.24) gives for the total angular momentum here the expected value zero.

3 Other Metrics

The general kinematic framework provided by the Principle of Equivalence rests on a much firmer foundation than do Einstein's field equations. Indeed, in Chapters 3 through 5 we were led almost inevitably from the equality of gravitational and inertial mass to the full formalism of tensor analysis and general covariance, whereas in contrast the derivation of Einstein's equations in Chapter 7 contained a strong element of guesswork, and in any case there might exist a long-range scalar field, like that of Brans and Dicke, that would alter the field equations. It is therefore very useful to test general relativity by assuming that the usual rules for the motion of particles and photons in a given metric field $g_{\mu\nu}$ still apply, but that the metric may be different from that calculated from the Einstein equations.

In any case we would expect the metric produced by a static spherically symmetric body like the sun to be expressible in the "standard," "isotropic," and "harmonic" forms given in Section 8.1, and we would further expect that the metric coefficients [e.g., $A(r)$ and $B(r)$] could be expanded as power series in the small parameter MG/r. Such an expansion was given by Eddington and Robertson[1] for the metric in its isotropic form:

$$d\tau^2 = \left(1 - 2\alpha \frac{MG}{\rho} + 2\beta \frac{M^2G^2}{\rho^2} + \cdots\right) dt^2$$

$$- \left(1 + 2\gamma \frac{MG}{\rho} + \cdots\right) (d\rho^2 + \rho^2\, d\theta^2 + \rho^2 \sin^2\theta\, d\varphi^2) \qquad (8.3.1)$$

where α, β, and γ are unknown dimensionless parameters. (The reason for carrying this expansion to order M^2G^2/ρ^2 in g_{00} and only to order MG/ρ in g_{ij} is that in applications to celestial mechanics g_{ij} will always get multiplied with an extra factor $v^2 \sim MG/\rho$.) Comparing with the isotropic form (8.2.14) of the Schwarzschild solution, we see that the predictions of the Einstein field equations can be neatly summarized as

$$\alpha = \beta = \gamma = 1 \qquad (8.3.2)$$

In contrast, the Brans-Dicke theory discussed in Section 7.3 gives a metric (see Section 9.9) that can be expressed as in (8.3.1), with

$$\alpha = \beta = 1, \qquad \gamma = \frac{\omega + 1}{\omega + 2} \qquad (8.3.3)$$

where ω is the unknown dimensionless parameter of this theory. In order to decide whether Einstein, or Brans and Dicke, or someone else, has the right field equations, what must be done is to measure α, β, and γ.

We shall generally be doing our calculations with the metric in its "standard" form, so it will prove convenient to convert the Robertson expansion (8.3.1) to this form by defining

$$r \equiv \rho \left(1 + \gamma \frac{MG}{\rho} + \cdots\right) \qquad (8.3.4)$$

or

$$\rho = r \left(1 - \gamma \frac{MG}{r} + \cdots\right)$$

A simple calculation gives

$$d\tau^2 = \left(1 - 2\alpha \frac{MG}{r} + 2(\beta - \alpha\gamma) \frac{M^2G^2}{r^2} + \cdots\right) dt^2$$

$$- \left(1 + 2\gamma \frac{MG}{r} + \cdots\right) dr^2 - r^2\, d\theta^2 - r^2 \sin^2\theta\, d\varphi^2$$

$$(8.3.5)$$

Also, we can construct harmonic coordinates $\mathbf{X}$, t by using for $\mathbf{X}$

$$X_1 = R \sin\theta \cos\varphi, \quad X_2 = R \sin\theta \sin\varphi, \quad X_3 = R \cos\theta$$

with R satisfying the differential equation (8.1.15):

$$0 = \frac{d}{dr} r^2 \left(1 - (\alpha + \gamma)\frac{MG}{r} + \cdots\right)\frac{dR}{dr} - 2\left(1 - (\alpha - \gamma)\frac{MG}{r} + \cdots\right) R$$

The solution is

$$R = \left(1 + \frac{(\alpha - 3\gamma)MG}{2r} + \cdots\right) r \tag{8.3.6}$$

and (8.1.16) gives the metric (with $R^2 \equiv \mathbf{X}^2$):

$$d\tau^2 = \left[1 - 2\alpha\frac{MG}{R} + (\alpha\gamma - \alpha^2 + 2\beta)\frac{M^2G^2}{R^2} + \cdots\right] dt^2$$

$$- \left[1 + \frac{(3\gamma - \alpha)MG}{R} + \cdots\right] d\mathbf{X}^2$$

$$- \frac{[(\alpha - \gamma)MG/R + \cdots](\mathbf{X} \cdot d\mathbf{X})^2}{R^2} \tag{8.3.7}$$

Comparing (8.3.5) and (8.3.7) with the corresponding exact solutions (8.2.12) and (8.2.15) shows again that Einstein's theory gives $\alpha = \beta = \gamma = 1$.

The prediction that $\alpha = 1$ really just follows from the empirical definition of the mass M. Note that Eq. (8.3.1) would give a slowly moving particle far from the origin a centripetal acceleration equal to

$$-g = -\Gamma_{tt}^r = \frac{1}{2}\frac{\partial g_{tt}}{\partial r} = -\frac{\alpha MG}{r^2} \quad \text{(for } MG/r \ll 1 \text{ and } v^2 \ll 1\text{)}$$

whereas in fact the masses of the sun and planets are *measured* by setting $g = MG/r^2$; hence we must absorb α into M or, in other words, we must choose $\alpha = 1$. Only if it were possible to determine M by some independent nongravitational measurement would it make sense to ask whether in fact α is exactly unity.

With $\alpha = 1$, the metric functions given by (8.3.5) are

$$B(r) = 1 - \frac{2MG}{r} + 2(\beta - \gamma)\frac{M^2G^2}{r^2} + \cdots \tag{8.3.8}$$

$$A(r) = 1 + 2\gamma\frac{MG}{r} + \cdots \tag{8.3.9}$$

As shown in Chapter 3, the gravitational red shift experiment only measures the term $-2MG/r$ in $B(r)$, and hence can only verify the Principle of Equivalence. We shall see that, of the other tests of general relativity listed at the beginning of this chapter, (B) and (D) can only test whether $\gamma \simeq 1$, whereas (C), the precession of perihelia, verifies that $2\gamma - \beta \simeq 1$. (To the extent that we ignore the rotation of the earth, (E) also only tests whether $\gamma \simeq 1$.)

4 General Equations of Motion

We now consider the motion of a freely falling material particle or photon in a static isotropic gravitational field. First let us consider the most general such metric in the standard form derived in Section 1, that is,

$$d\tau^2 = B(r)\, dt^2 - A(r)\, dr^2 - r^2\, d\theta^2 - r^2 \sin^2 \theta\, d\varphi^2 \tag{8.4.1}$$

The equations of free fall are

$$\frac{d^2 x^\mu}{dp^2} + \Gamma^\mu_{\nu\lambda} \frac{dx^\nu}{dp} \frac{dx^\lambda}{dp} = 0 \tag{8.4.2}$$

where p is a parameter describing the trajectory. In general $d\tau$ is proportional to dp, so for a material particle we could normalize p so that $p = \tau$. However, for a photon the proportionality constant $d\tau/dp$ vanishes, and since we wish to treat photons as well as massive particles, we shall find it convenient to reserve the right to fix the normalization of p independently from that of τ.

Using the nonvanishing components of the affine connection given by Eq. (8.1.11), we find from (8.4.2) that

$$0 = \frac{d^2 r}{dp^2} + \frac{A'(r)}{2A(r)} \left(\frac{dr}{dp}\right)^2 - \frac{r}{A(r)} \left(\frac{d\theta}{dp}\right)^2 - r \frac{\sin^2 \theta}{A(r)} \left(\frac{d\varphi}{dp}\right)^2 + \frac{B'(r)}{2A(r)} \left(\frac{dt}{dp}\right)^2 \tag{8.4.3}$$

$$0 = \frac{d^2\theta}{dp^2} + \frac{2}{r} \frac{d\theta}{dp} \frac{dr}{dp} - \sin\theta\cos\theta \left(\frac{d\varphi}{dp}\right)^2 \tag{8.4.4}$$

$$0 = \frac{d^2\varphi}{dp^2} + \frac{2}{r} \frac{d\varphi}{dp} \frac{dr}{dp} + 2\cot\theta \frac{d\varphi}{dp} \frac{d\theta}{dp} \tag{8.4.5}$$

$$0 = \frac{d^2 t}{dp^2} + \frac{B'(r)}{B(r)} \frac{dt}{dp} \frac{dr}{dp} \tag{8.4.6}$$

(A prime denotes d/dr.) We solve these equations by looking for constants of the motion.

Since the field is isotropic, we may consider the orbit of our particle to be confined to the equatorial plane, that is,

$$\theta = \frac{\pi}{2} \tag{8.4.7}$$

Then (8.4.4) is immediately satisfied, and we can forget about θ as a dynamical variable. Dividing (8.4.5) and (8.4.6) by $d\varphi/dp$ and dt/dp, respectively, we next find

$$\frac{d}{dp}\left\{\ln\frac{d\varphi}{dp} + \ln r^2\right\} = 0 \tag{8.4.8}$$

$$\frac{d}{dp}\left\{\ln\frac{dt}{dp} + \ln B\right\} = 0 \tag{8.4.9}$$

This yields two constants of the motion. One of them will be absorbed immediately into the definition of p; we choose to normalize p so that the solution of (8.4.9) is

$$\frac{dt}{dp} = \frac{1}{B(r)} \tag{8.4.10}$$

Since $B(r)$ is close to unity, p is nearly equal to the coordinate time t. The other constant is obtained from (8.4.8), and plays the role of an angular momentum per unit mass

$$r^2\frac{d\varphi}{dp} = J \text{ (constant)} \tag{8.4.11}$$

Inserting (8.4.7), (8.4.10), and (8.4.11) in (8.4.3) gives the remaining equation of motion as

$$0 = \frac{d^2r}{dp^2} + \frac{A'(r)}{2A(r)}\left(\frac{dr}{dp}\right)^2 - \frac{J^2}{r^3A(r)} + \frac{B'(r)}{2A(r)B^2(r)} \tag{8.4.12}$$

By multiplying this equation with $2A(r)\,dr/dp$, we may write it as

$$\frac{d}{dp}\left\{A(r)\left(\frac{dr}{dp}\right)^2 + \frac{J^2}{r^2} - \frac{1}{B(r)}\right\} = 0$$

and our last constant of the motion is therefore

$$A(r)\left(\frac{dr}{dp}\right)^2 + \frac{J^2}{r^2} - \frac{1}{B(r)} = -E \text{ (constant)} \tag{8.4.13}$$

The proper time τ may now be determined from (8.4.1), (8.4.7), (8.4.10), (8.4.11), and (8.4.13); we find that

$$d\tau^2 = E\,dp^2 \tag{8.4.14}$$

in accordance with our earlier remark that (8.4.2) forces $d\tau/dp$ to be constant. We see that E must take the values

$$E > 0 \quad \text{for material particles} \tag{8.4.15}$$

$$E = 0 \quad \text{for photons} \tag{8.4.16}$$

Also, $A(r)$ is in practice always positive, so (8.4.13) tells us that our particle can reach radius r only if

$$\frac{J^2}{r^2} + E \leq \frac{1}{B(r)} \tag{8.4.17}$$

The parameter p may be eliminated everywhere by using (8.4.10) in (8.4.11), (8.4.13), and (8.4.14); we then have

$$r^2 \frac{d\varphi}{dt} = JB(r) \tag{8.4.18}$$

$$\frac{A(r)}{B^2(r)} \left(\frac{dr}{dt} \right)^2 + \frac{J^2}{r^2} - \frac{1}{B(r)} = -E \tag{8.4.19}$$

$$d\tau^2 = EB^2(r) \, dt^2 \tag{8.4.20}$$

For a slowly moving particle in a weak field J^2/r^2, $(dr/dt)^2$, $A - 1$, and $B - 1 \simeq 2\phi$ will all be small, and to first order in these quantities the above equations of motion become

$$r^2 \frac{d\varphi}{dt} \simeq J$$

$$\frac{1}{2} \left(\frac{dr}{dt} \right)^2 + \frac{J^2}{2r^2} + \phi \simeq \frac{1 - E}{2}$$

These are the same equations as would hold in Newton's theory, with $(1 - E)/2$ playing the role of an energy per unit mass.

To see how the exact equations of motion work in a simple case, consider a particle in a circular orbit at radius R. Since dr/dt vanishes, Eq. (8.4.19) gives

$$\frac{J^2}{R^2} - \frac{1}{B(R)} + E = 0 \tag{8.4.21}$$

Also, for equilibrium at this radius, the derivative at R of the left-hand side must also vanish, so

$$-\frac{2J^2}{R^3} + \frac{B'(R)}{B^2(R)} = 0 \tag{8.4.22}$$

(If we regard a circle as the limit of an ellipse with perihelia $R - \delta$ and aphelia $R + \delta$, then (8.4.19) shows that $J^2/r^2 - 1/B(r) + E$ must vanish at $r = R \pm \delta$, and this gives (8.4.21) and (8.4.22) in the limit $\delta \to 0$.) From (8.4.21) and (8.4.22) we find

$$E = \frac{1}{B(R)} \left(1 - \frac{RB'(R)}{2B(R)} \right) \tag{8.4.23}$$

$$J^2 = \frac{B'(R)R^3}{2B^2(R)} \tag{8.4.24}$$

Using (8.4.24) in (8.4.18) gives the rate of revolution as

$$\frac{d\varphi}{dt} = \left(\frac{B'(R)}{2R}\right)^{1/2}$$

(8.4.25)

whereas (8.4.23) and (8.4.20) give the proper time as

$$\frac{d\tau}{dt} = \sqrt{B(R) - \tfrac{1}{2}RB'(R)}$$

(8.4.26)

By using the Robertson expansion (8.3.8), we find

$$\frac{d\varphi}{dt} = \left(\frac{MG}{R^3}\right)^{1/2}\left[1 - \frac{(\beta - \gamma)MG}{R} + \cdots\right]$$

(8.4.27)

$$\frac{d\tau}{dt} = \left[1 - \frac{3MG}{R} + \cdots\right]$$

(8.4.28)

In most applications of general relativity we are more interested in the shape of orbits, that is, in r as a function of φ, than in their time history. The orbit shape can be obtained directly by eliminating dp from (8.4.11) and (8.4.13); this gives

$$\frac{A(r)}{r^4}\left(\frac{dr}{d\varphi}\right)^2 + \frac{1}{r^2} - \frac{1}{J^2 B(r)} = -\frac{E}{J^2}$$

(8.4.29)

The solution may then be determined by a quadrature:

$$\varphi = \pm \int \frac{A^{1/2}(r)\,dr}{r^2\left(\dfrac{1}{J^2 B(r)} - \dfrac{E}{J^2} - \dfrac{1}{r^2}\right)^{1/2}}$$

(8.4.30)

5   Unbound Orbits: Deflection of Light by the Sun

Consider a particle or photon approaching the sun from very great distances. (See Figure 8.1.) At infinity the metric becomes Minkowskian, that is, $A(\infty) = B(\infty) = 1$, and we expect motion on a straight line at constant velocity V, that is,

$$b \simeq r \sin(\varphi - \varphi_\infty) \simeq r(\varphi - \varphi_\infty)$$

$$-V \simeq \frac{d}{dt}(r\cos(\varphi - \varphi_\infty)) \simeq \frac{dr}{dt}$$

where b is the "impact parameter" and φ_∞ is the incident direction. Inserting these in (8.4.18) and (8.4.19), we see that they do satisfy the equations of motion at infinity, where $A = B = 1$, and that the constants of the motion are

$$J = bV^2$$

(8.5.1)

$$E = 1 - V^2$$

(8.5.2)

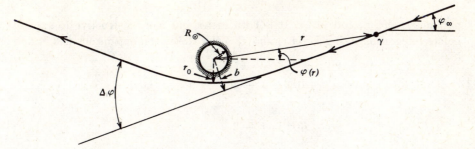

Figure 8.1 Quantities referred to in the calculation of the deflection of light by the sun. (Deflection greatly exaggerated.)

(Of course a photon has $V = 1$, and as we have already seen, this gives $E = 0$.) It is often more convenient to express J in terms of the distance r_0 of closest approach to the sun, rather than the impact parameter b. At r_0, $dr/d\varphi$ vanishes, so (8.4.29) and (8.5.2) give

$$J = r_0 \left(\frac{1}{B(r_0)} - 1 + V^2 \right)^{1/2} \tag{8.5.3}$$

The orbit is then described by (8.4.30), that is,

$$\varphi(r) = \varphi_\infty + \int_r^\infty \frac{A^{1/2}(r)\, dr}{r^2 \left(\frac{1}{r_0{}^2} \left[\frac{1}{B(r)} - 1 + V^2 \right] \left[\frac{1}{B(r_0)} - 1 + V^2 \right]^{-1} - \frac{1}{r^2} \right)^{1/2}} \tag{8.5.4}$$

The total change in φ as r decreases from infinity to its minimum value r_0 and then increases again to infinity is just twice its change from ∞ to r_0, that is, $2|\varphi(r_0) - \varphi_\infty|$. If the trajectory were a straight line, this would equal just π; hence the *deflection* of the orbit from a straight line is

$$\Delta\varphi = 2|\varphi(r_0) - \varphi_\infty| - \pi \tag{8.5.5}$$

If this is positive, then the angle φ changes by more than $180°$, that is, the trajectory is bent *toward* the sun; if $\Delta\varphi$ is negative then the trajectory is bent away from the sun.

For a photon $V^2 = 1$, and (8.5.4) gives

$$\varphi(r) - \varphi_\infty = \int_r^\infty A^{1/2}(r) \left[\left(\frac{r}{r_0} \right)^2 \left(\frac{B(r_0)}{B(r)} \right) - 1 \right]^{-1/2} \frac{dr}{r} \tag{8.5.6}$$

If we used the exact values of $A(r)$ and $B(r)$ given by the Schwarzschild solution (8.2.10), (8.2.11), then we would obtain $\varphi(r)$ and $\Delta\varphi$ as elliptic integrals of the usual sort, which could only be evaluated numerically by expanding in the small parameters MG/r_0 and MG/r. It is both easier and more instructive to expand before integrating, using for $A(r)$ and $B(r)$ the Robertson expansions (8.3.8) and (8.3.9):

$$A(r) = 1 + 2\gamma \frac{MG}{r} + \cdots$$

$$B(r) = 1 - 2 \frac{MG}{r} + \cdots$$

The argument of the second square root in (8.5.6) is then

$$\left(\frac{r}{r_0}\right)^2 \left(\frac{B(r_0)}{B(r)}\right) - 1 = \left(\frac{r}{r_0}\right)^2 \left[1 + 2MG\left(\frac{1}{r} - \frac{1}{r_0}\right) + \cdots \right] - 1$$

$$= \left[\left(\frac{r}{r_0}\right)^2 - 1\right]\left[1 - \frac{2MGr}{r_0(r + r_0)} + \cdots \right]$$

so (8.5.6) gives

$$\varphi(r) - \varphi_\infty = \int_r^\infty \frac{dr}{r\left[\left(\frac{r}{r_0}\right)^2 - 1\right]^{1/2}}\left[1 + \frac{\gamma MG}{r} + \frac{MGr}{r_0(r + r_0)} + \cdots\right]$$

The integral is elementary, and gives

$$\varphi(r) - \varphi_\infty = \sin^{-1}\left(\frac{r_0}{r}\right) + \frac{MG}{r_0}\left(1 + \gamma - \gamma\sqrt{1 - \left(\frac{r_0}{r}\right)^2} - \sqrt{\frac{r - r_0}{r + r_0}}\right) + \cdots$$

$$(8.5.7)$$

Hence to first order in MG/r_0, the deflection (8.5.5) is

$$\Delta\varphi = \frac{4MG}{r_0}\left(\frac{1 + \gamma}{2}\right) \tag{8.5.8}$$

(To this order, we could just as well replace r_0 here with the impact parameter b.)

For a light ray deflected by the sun we must use $M = M_\odot = 1.97 \times 10^{33}g$, that is, $MG = M_\odot G = 1.475$ km, and the minimum value of r_0 is $R_\odot = 6.95 \times 10^5$ km, so (8.5.8) gives here

$$\Delta\varphi = \left(\frac{R_\odot}{r_0}\right)\theta_\odot \tag{8.5.9}$$

where

$$\theta_\odot \equiv \frac{4M_\odot G}{R_\odot}\left(\frac{1 + \gamma}{2}\right) = 1.75''\left(\frac{1 + \gamma}{2}\right) \tag{8.5.10}$$

Furthermore, general relativity gives $\gamma = 1$, so it predicts a deflection toward the sun, with $\theta_\odot = 1.75''$. (For light just grazing Jupiter the deflection is only $0.02''$, so there seems little hope of observing the deflection of light by any other body than the sun.) In the Brans-Dicke theory (8.5.10) and (8.3.3) give a deflection constant

$$\theta_{\odot} = \frac{4 M_{\odot} G}{R_{\odot}} \left(\frac{2\omega + 3}{2\omega + 4} \right) \tag{8.5.11}$$

Whenever we obtain a prediction from general relativity the question always arises (or should arise) whether the result obtained really refers to an objective physical measurement or whether it has folded into it arbitrary subjective elements dependent on our choice of coordinate system. In the case at hand, we should ask ourselves what the predicted change in φ really has to do with the positions of stellar images on photographic plates. Fortunately, the answer is here quite simple, for this is really a *scattering* experiment. The light ray comes in from a very great distance, is deflected as it passes close to the sun, and is detected on earth, more than 200 solar radii away from the sun. At the points of origin and detection the metric is sensibly Minkowskian, and at these distances there is no question about the meaning of φ; it is the azimuthal angle in a system of coordinates within which light rays define lines that are essentially straight. Hence we can relate $\Delta\varphi$ to the shift of stellar images on photographic plates by the ordinary rules of geometric optics. (We are here neglecting effects of the gravitational field of the earth itself, because this field is on the earth's surface more than 10^3 times weaker than that of the sun on the sun's surface.) We would have to be a good deal more careful about the operational significance of our φ if we had to predict the deflection of light by the sun as seen from an observatory deep within the sun's gravitational field, as, for instance, from an orbiting satellite a few solar radii away from the sun.

Another conceptual difficulty that may arise here has to do with our treatment of the photon as a quantum of light moving as would any other particle that happened to have velocity close to unity, that is, to c. Actually, no use is being made of quantum mechanics. The wavelength of light is so small compared with the scale of the solar gravitational field (i.e., 10^{-5} cm as compared with 10^{10} cm) that at any point in this field we can erect a locally inertial coordinate system that covers a huge (say, 10^{15}) number of wavelengths. The Principle of Equivalence tells us that in such a coordinate system light behaves as it does in gravitation-free empty space, and since the wavelength is so small, this means that diffraction is negligible and each element of a wave front moves in a straight line at unit velocity. This statement, when rewritten in the noninertial coordinate systems of astronomy, is nothing but our equation of motion (8.4.2). (This argument, incidentally, shows why the deflection of light cannot depend on its polarization.)

Now let us see how Einstein's prediction (8.5.9) compares with observation. The deflection angle $\Delta\varphi$ is classically measured by comparing the apparent positions of stars that happen to lie near the solar disk during an eclipse, when their light comes close to the sun and yet may be detected, with their positions at night six months earlier, when these stars lie on opposite sides of the earth from the sun, and their light does not pass close to the sun on its way to us. Subtracting φ

(six months earlier) from φ (eclipse) then, in principle, should give $\Delta\varphi$. However, there is an unavoidable change in the *scale* of the photographs over a six-month interval, owing partly to small changes in the temperature and in the mechanical configuration of the telescope and camera over so long a time. A change in the scale of the photograph would give an apparent deflection of any star toward or away from the sun by an angle proportional to the distance r_0 at which its light passes the sun; hence what is done in practice is to compare observations with a theoretical curve

$$\Delta\varphi = \theta_\odot \left(\frac{R_\odot}{r_0}\right) + S\left(\frac{r_0}{R_\odot}\right) \tag{8.5.12}$$

where S is the unknown scale constant (often called α) and θ is an angle to be compared with the theoretical value 1.75″. There are other effects that could contribute to $\Delta\varphi$, such as refraction of the starlight in the solar corona or as it enters the colder air in the moon's shadow, but none of these is believed to play an important role.

Observations cannot be carried closer to the sun's disk than $r_0 \approx 2R_\odot$, but they can still be used to determine $\theta_\odot$ by fitting the observed $\Delta\varphi$ values to the theoretical curve (8.5.12). The difficulty with this program is just that $\Delta\varphi$ is very difficult to measure accurately in the brief time available during an eclipse. In 1919 eclipse expeditions were sent to two small islands, Sobral, off the northeast coast of Brazil, and Principe, in the Gulf of Guinea. About a dozen stars in all were studied, and yielded values[2] 1.98 ± 0.12″ and 1.61 ± 0.31″, in substantial agreement with Einstein's prediction $\theta_\odot = 1.75″$. It was perhaps this dramatic result more than any other success that brought general relativity to the attention of the general public in the 1920's.

Since 1919 there have been measurements on about 380 stars observed during the eclipses of 1922, 1929, 1936, 1947, and 1952, which we summarize in Table 8.1 (taken from the summary of von Klüber[3]). The values obtained for $\theta_\odot$ vary from 1.3″ to 2.7″, but mostly lie between 1.7 and 2″. The most recent of these results is $\Delta\varphi = 1.70 \pm 0.10″$, in very good agreement with Einstein's prediction, but it is not clear that the systematic error here is really smaller than for previous observations. From all this we can conclude that there definitely is a deflection of light greater than the value $\theta_\odot = 0.875″$ that would be predicted for $\gamma = 0$ (i.e., $A(r) = 1$), but as to its precise value we can say little more than that $\theta_\odot$ is somewhere between 1.6 and 2.2″; that is, γ is between about 0.9 and 1.3. It may become possible to improve the accuracy of this determination in the near future by using photoelectric techniques to monitor star positions without waiting for an eclipse.

Recent developments in radio astronomy[4] have made it possible to measure the deflection of radio signals by the sun with potentially far greater accuracy than is possible in optical astronomy. The angular accuracy of optical observations is limited by inhomogeneities in the earth's atmosphere to about 0.1″, whereas a radio interferometer with wavelength λ and baseline D can in principle measure

Table 8.1. Measurements of the Deflection of Light by the Sun.[3] The fourth column gives the minimum and maximum values for the distance of closest approach of the light ray to the sun's center for the various stars studied. The fifth column gives the deduced value for the deflection of a light ray that just grazes the sun's surface.

Eclipse	Site	Number of Stars	$r_0/R_\odot$	$\theta_\odot$ (sec)	Ref.
May 29, 1919	Sobral	7	2–6	1.98 ± 0.16	a
	Principe	5	2–6	1.61 ± 0.40	a
September 21, 1922	Australia	11–14	2–10	1.77 ± 0.40	b
	Australia	18	2–10	1.42 to 2.16	c
	Australia	62–85	2.1–14.5	1.72 ± 0.15	d
	Australia	145	2.1–42	1.82 ± 0.20	e
May 9, 1929	Sumatra	17–18	1.5–7.5	2.24 ± 0.10	f
June 19, 1936	U.S.S.R.	16–29	2–7.2	2.73 ± 0.31	g
	Japan	8	4–7	1.28 to 2.13	h
May 20, 1947	Brazil	51	3.3–10.2	2.01 ± 0.27	i
February 25, 1952	Sudan	9–11	2.1–8.6	1.70 ± 0.10	j

a F. W. Dyson, A. S. Eddington, and C. Davidson, Phil. Trans. Roy. Soc., **220A**, 291 (1920); Mem. Roy. Astron. Soc., **62**, 291 (1920).

b G. F. Dodwell and C. R. Davidson, Mon. Nat. Roy. Astron. Soc., **84**, 150 (1924).

c C. A. Chant and R. K. Young, Publ. Dominion Astron. Obs., **2**, 275 (1924).

d W. W. Campbell and R. Trumpler, Lick Observ. Bull., **11**, 41 (1923); Publ. Astron. Soc. Pacific, **35**, 158 (1923).

e W. W. Campbell and R. Trumpler, Lick Observ. Bull., **13**, 130 (1928).

f E. F. Freundlich, H. v. Klüber, and A. v. Brunn, Ab. Preuss. Akad. Wiss., No. 1, 1931; Z. Astrophys., **3**, 171 (1931).

g A. A. Mikhailov, C. R. Acad. Sci. USSR (N. S.), **29**, 189 (1940).

h T. Matukuma, A. Onuki, S. Yosida, and Y. Iwana, Jap. J. Astron. and Geophys., **18**, 51 (1940).

i G. van Biesbroeck, Astron. J., **55**, 49, 247 (1949).

j G. van Biesbroeck, Astron. J., **58**, 87 (1953).

angles with an accuracy of order $\lambda/2\pi D$ radians; this is $0.1''$ for $\lambda = 3$ cm and $D = 10$ km, and proportionately less for longer baselines.

One complication that bothers the astronomer more at radio than at optical frequencies is the refraction of rays in the solar corona. At X-band frequencies (8000–12500 MHz) the refraction is quite small, and can be eliminated by excluding data taken when the radio signal passes within about 2 solar radii of the sun's surface. However, at S-band frequencies (2000–4000 MHz) it is necessary to analyze the data in terms of a model, in which part of the deflection arises from general relativity, and the rest is produced by the corona. The parameters describing the solar corona can in principle be measured by this technique (using several frequencies) at the same time as the general-relativistic parameter, but the electron densities in the corona change with time, and it appears that the only really satisfactory method for dealing with coronal refraction is to use a radio frequency in the X-band, or above.

Each October, the quasi-stellar source 3C279 is occulted by the sun, and a number of radio astronomy groups have taken this opportunity to observe the changes in the angle (about 9.5°) between 3C279 and the quasi-stellar source 3C273, during the period just before and just after occultation. The results are given in Table 8.2. Again, we see general relativity confirmed, but not yet well enough to distinguish between the theories of Einstein and Brans and Dicke. However, the data taken with very long baselines, such as the "Goldstack" baseline of 3900 km, contain enough information in principle to measure angular positions to about 0.001″. It is hoped that the analysis of this data will eventually lead to a really precise determination of γ.

Table 8.2. Interferometric measurements of the deflection of radio waves from the source 3C279 by the sun. The data are analyzed in terms of the deflection $\theta_\odot$ that would be produced by a radio signal just grazing the sun.

Facility	Radar Frequency (MHz)	Baseline (km)	Period	$\theta_\odot$ (sec)	Ref.
Owens Valley	9602	1.0662	9/30–10/15, 1969	1.77 ± 0.20	a
Goldstone	2388	21.566	10/2–10/10, 1969	$1.82 \begin{array}{c} +\ 0.24 \\ -\ 0.17 \end{array}$	b
Goldstone/ Haystack	7840	3899.92	9/30–10/15, 1969	1.80 ± 0.2	c
NRAO	2695 & 8085	2.7	10/2–10/12, 1970	1.57 ± 0.08	d
	2697 & 4993.8	1.41	10/8, 1970	1.87 ± 0.3	e

a G. A. Seielstad, R. A. Sramek, and K. W. Weller, Phys. Rev. Letters, **24**, 1373 (1970).
b D. O. Muhleman, R. D. Ekers, and E. B. Fomalont, Phys. Rev. Letters, **24**, 1377 (1970).
c I. I. Shapiro, private communication.
d R. A. Sramek, Ap. J., **167**, L55 (1971).
e J. M. Hill, Mon. Not. Roy. Astron. Soc., **153**, 7P (1971).

6 Bound Orbits: Precession of Perihelia

Now consider a test particle bound in an orbit around the sun. (See Figure 8.2.) At perihelia and aphelia, r reaches its minimum and maximum values r_- and r_+, and at both points $dr/d\varphi$ vanishes, so (8.4.29) gives

$$\frac{1}{r_\pm{}^2} - \frac{1}{J^2 B(r_\pm)} = -\frac{E}{J^2}$$

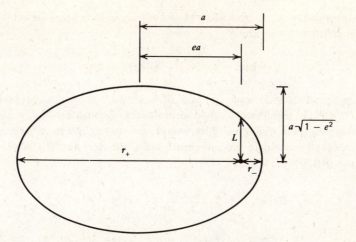

Figure 8.2 Elements of an ellipse referred to in the calculation of the precession of planetary orbits. (The ellipse here has the same eccentricity as the orbit of Icarus.)

From these two equations we can derive values for the two constants of the motion:

$$E = \frac{\dfrac{r_+^{\,2}}{B(r_+)} - \dfrac{r_-^{\,2}}{B(r_-)}}{r_+^{\,2} - r_-^{\,2}} \tag{8.6.1}$$

$$J^2 = \frac{\dfrac{1}{B(r_+)} - \dfrac{1}{B(r_-)}}{\dfrac{1}{r_+^{\,2}} - \dfrac{1}{r_-^{\,2}}} \tag{8.6.2}$$

The angle swept out by the position vector as r increases from r_- is given by Eq. (8.4.30) as

$$\varphi(r) = \varphi(r_-) + \int_{r_-}^{r} A^{1/2}(r) \left[\frac{1}{J^2 B(r)} - \frac{E}{J^2} - \frac{1}{r^2} \right]^{-1/2} \frac{dr}{r^2}$$

or, using (8.6.1) and (8.6.2),

$$\varphi(r) - \varphi(r_-)$$

$$= \int_{r_-}^{r} \left[\frac{r_-^{\,2}(B^{-1}(r) - B^{-1}(r_-)) - r_+^{\,2}(B^{-1}(r) - B^{-1}(r_+))}{r_+^{\,2} r_-^{\,2}(B^{-1}(r_+) - B^{-1}(r_-))} - \frac{1}{r^2} \right]^{-1/2}$$

$$\times \, A^{1/2}(r) r^{-2} \, dr \tag{8.6.3}$$

The change in φ as r decreases from r_+ to r_- is the same as the change in φ as r increases from r_- to r_+, so the total change in φ per revolution is $2|\varphi(r_+) - \varphi(r_-)|$.

This would equal 2π if the orbit was a closed ellipse, so in general the orbit precesses in each revolution by an angle

$$\Delta\varphi = 2|\varphi(r_+) - \varphi(r_-)| - 2\pi \qquad (8.6.4)$$

Using the exact values of $A(r)$ and $B(r)$ given by the Schwarzschild solution (8.2.10), (8.2.11) in (8.6.3) would yield formulas for $\varphi(r)$ and $\Delta\varphi$ as elliptic integrals, and to evaluate them numerically we would have to expand in MG/r and $MG/r_\pm$. Instead we shall expand in the integrand, using for $A(r)$ and $B(r)$ the Robertson expansions (8.3.8), (8.3.9):

$$A(r) = 1 + 2\gamma\,\frac{MG}{r} + \cdots \qquad (8.6.5)$$

$$B(r) = 1 - \frac{2MG}{r} + \frac{2(\beta - \gamma)M^2G^2}{r^2} + \cdots$$

Note that there is a complete cancellation in (8.6.3) of the leading term in $B(r)$ but not of that in $A(r)$, so to calculate φ and $\Delta\varphi$ to first order in $Mg/r_\pm$ we need $B(r)$ to *second* order in MG/r, whereas $A(r)$ will be needed only to first order.

It saves a great deal of work if we realize that by using the expansion

$$B^{-1}(r) \simeq 1 + \frac{2MG}{r} + \frac{2(2 - \beta + \gamma)M^2G^2}{r^2}$$

we make the argument of the first square root in (8.6.3) a quadratic function of $1/r$. Furthermore, it vanishes at $r = r_\pm$, so

$$\frac{r_-^2(B^{-1}(r) - B^{-1}(r_-)) - r_+^2(B^{-1}(r) - B^{-1}(r_+))}{r_+^2 r_-^2(B^{-1}(r_+) - B^{-1}(r_-))}$$

$$- \frac{1}{r^2} = C\left(\frac{1}{r_-} - \frac{1}{r}\right)\left(\frac{1}{r} - \frac{1}{r_+}\right) \qquad (8.6.6)$$

The constant C can be determined by letting $r \to \infty$:

$$C = \frac{r_+^2(1 - B^{-1}(r_+)) - r_-^2(1 - B^{-1}(r_-))}{r_+ r_-(B^{-1}(r_+) - B^{-1}(r_-))}$$

or factoring out of numerator and denominator a common factor $2(r_- - r_+)MG$:

$$C \simeq 1 - (2 - \beta + \gamma)MG\left(\frac{1}{r_+} + \frac{1}{r_-}\right) \qquad (8.6.7)$$

Using (8.6.5)–(8.6.7) in (8.6.3) gives then

$$\varphi(r) - \varphi(r_-) \simeq \left[1 + \tfrac{1}{2}(2 - \beta + \gamma)MG\left(\frac{1}{r_+} + \frac{1}{r_-}\right)\right]$$

$$\times \int_{r_-}^{r} \frac{\left[1 + \dfrac{\gamma MG}{r}\right]dr}{r^2\left[\left(\dfrac{1}{r_-} - \dfrac{1}{r}\right)\left(\dfrac{1}{r} - \dfrac{1}{r_+}\right)\right]^{1/2}}$$

The integral is made trivial by introducing a new variable ψ:

$$\frac{1}{r} \equiv \frac{1}{2}\left(\frac{1}{r_+} + \frac{1}{r_-}\right) + \frac{1}{2}\left(\frac{1}{r_+} - \frac{1}{r_-}\right)\sin\psi \qquad (8.6.8)$$

We then find

$$\varphi(r) - \varphi(r_-) = \left[1 + \tfrac{1}{2}(2 - \beta + 2\gamma)MG\left(\frac{1}{r_+} + \frac{1}{r_-}\right)\right]\left[\psi + \frac{\pi}{2}\right]$$

$$- \tfrac{1}{2}\gamma MG\left(\frac{1}{r_+} - \frac{1}{r_-}\right)\cos\psi \qquad (8.6.9)$$

At aphelion $\psi = \pi/2$, so (8.6.4) and (8.6.9) give the precession per revolution as

$$\Delta\varphi = \left(\frac{6\pi MG}{L}\right)\left(\frac{2 - \beta + 2\gamma}{3}\right) \text{ (radians/revolution)} \qquad (8.6.10)$$

where L is the dimension of the ellipse called the *semilatus rectum*

$$\frac{1}{L} \equiv \frac{1}{2}\left(\frac{1}{r_+} + \frac{1}{r_-}\right)$$

The elements of planetary orbits usually found in tables are the semimajor axis a and eccentricity e, defined by

$$r_\pm = (1 \pm e)a$$

Hence we can determine L from a and e by using the formula

$$L = (1 - e^2)a$$

Einstein's field equations yield $\beta = \gamma = 1$, so they predict a precession

$$\Delta\varphi = 6\pi\frac{MG}{L} \text{ radians/revolution} \qquad (8.6.11)$$

This is positive, meaning that the whole orbit should precess in the same direction as the motion of the test particle. In the Brans-Dicke theory (8.6.10) and (8.3.3) give

$$\Delta\varphi = \left(\frac{6\pi MG}{L}\right)\left(\frac{3\omega + 4}{3\omega + 6}\right) \qquad (8.6.12)$$

Once again, we should ask ourselves what the predicted value of $\Delta\varphi$ means. This is not a scattering experiment like the deflection of light by the sun; here we are dealing with an object that never gets out to infinity where the metric is Minkowskian. Any observation of the test particle's motion by optical or radar astronomers will make use of light rays that are themselves affected by the gravitational field, and if careful corrections are not made for the deflection of light, the astronomer's reported $\varphi(r)$ values will, at any given radius r, contain errors of the order of MG/L. [See Eq. (8.5.8).] However, in practice these fine points do not really matter, because the precession is *cumulative*. Equation (8.6.10) shows that after N revolutions the perihelia will have advanced by an angle of order NMG/L, so if $N \gg 1$ it is unnecessary to worry about an error in φ of order MG/L. Indeed, Eq. (8.6.11) tells us that the perihelion will return to its original azimuth after $L/3MG \gg 1$ revolutions, a prediction that clearly has nothing to do with how we define r or φ.

For Mercury we must take $L = 55.3 \times 10^6$ km, and of course $MG = 1.475$ km, so Eq. (8.6.11) gives $\Delta\varphi = 0.1038''$ per revolution. Since Mercury makes 415 revolutions per century, the prediction of general relativity is that

$$\Delta\varphi = 43.03'' \text{ per century } (\mercury)$$

Fortunately there are accurate observations of Mercury going back to 1765. These data were reanalyzed by Clemence[5] in 1943; he finds $\Delta\varphi = 43.11 \pm 0.45''$ per century, essentially confirming Newcomb's earlier value (see Section 1.2), and in excellent agreement with general relativity. Taken at face value, this agreement shows that the correction factor in Eq. (8.6.11) is

$$\left(\frac{2 - \beta + 2\gamma}{3} \right) = 1.00 \pm 0.01$$

This is by far the most important experimental verification of general relativity, both by virtue of its high accuracy, and because it alone is sensitive to the coefficient β appearing in the second-order term in g_{tt}.

Table 8.3. Comparison of Theoretical and Observed Centennial Precessions of Planetary Orbits.[6]

Planet	a (10^6 km)	e	$\dfrac{6\pi MG}{L}$	Revolutions Century	$\Delta\varphi$ (seconds/century) Gen. Rel.	$\Delta\varphi$ (seconds/century) Observed
Mercury ($\mercury$)	57.91	0.2056	$0.1038''$	415	43.03	43.11 ± 0.45
Venus ($\venus$)	108.21	0.0068	$0.058''$	149	8.6	8.4 ± 4.8
Earth ($\oplus$)	149.60	0.0167	$0.038''$	100	3.8	5.0 ± 1.2
Icarus	161.0	0.827	$0.115''$	89	10.3	9.8 ± 0.8

Results[6] for Venus, the earth, and Icarus are listed together with those for Mercury in Table 8.3. Evidently the accuracy available from the major planets degrades rapidly as we move away from the sun, both because the smaller eccentricities make observation of the perihelia more uncertain, and because as L increases, the precession per revolution and the revolutions per century both decrease. Icarus was only discovered in 1949, but is in some respects the most useful object to study because its small size, its close approach to the earth, and the large eccentricity of its orbit allow its precession to be determined with high accuracy. Suggestions have been made to put an artificial satellite into an eccentric orbit close to the sun; for instance, a satellite with $L = 10R_\odot$ would have a centennial precession equal to $8250''$! The trouble here is that a small object would be subject to nongravitational perturbations such as radiation pressure, solar wind, micrometeorites, which of course have a negligible effect on Mercury and Icarus.

There are two caveats that should be kept in mind in assessing the agreement of the observed advance of perihelia with the prediction of general relativity. First, there are many known perturbations that contribute to the precession of planetary orbits. In particular, Newtonian theory would give Mercury a precession

$$\Delta\varphi_N = 5557.62 \pm 0.20'' \, (\mathⅢ☿)$$

of which about $5025''$ is due to the rotation of the earth-based astronomical coordinate system, and about $532''$ is due to gravitational perturbations calculated by Newtonian perturbation theory from the motion of the other planets, chiefly Venus, earth, and Jupiter. The precession actually observed is

$$\Delta\varphi_{\text{OBS}} = 5600.73 \pm 0.41'' \, (☿)$$

and the value $\Delta\varphi = 43.11 \pm 0.45''$ quoted above for the "observed" anomalous precession is obtained by subtracting the Newtonian precession from what is observed, that is,

$$\Delta\varphi = \Delta\varphi_{\text{OBS}} - \Delta\varphi_N \qquad (8.6.13)$$

One may ask how we know that this is the right quantity to compare with the general-relativistic result of $43.03''$ per century. That is, how do we know that the total precession is correctly given by adding the Newtonian value $\Delta\varphi_N$, calculated while forgetting all effects of general relativity, to the Einstein value $\Delta\varphi_{\text{GR}}$, calculated forgetting all effects of planetary perturbations? To some extent this question can be answered by noting that the general-relativistic corrections to $\Delta\varphi_N$ would be of order MG/L times $\Delta\varphi_N$, or only about $10^{-4''}$ per century. For a fuller answer we shall have to wait until the discussion in the next chapter of the post-Newtonian approximation. But even granting that (8.6.13) is in principle correct, we should be aware that a very small systematic error in either $\Delta\varphi_N$ or $\Delta\varphi_{\text{OBS}}$ could completely destroy the agreement between theory and observation.

The second warning is that very small *unknown* effects may possibly be contributing to the observed precession of perihelia an amount comparable with that expected from general relativity. Indeed, we saw in Chapter 1 that Newcomb had by 1911 abandoned his earlier suggestion of a small departure from the inverse-square law, because the observed anomalous precession of 43″ per century could also be explained within Newtonian mechanics as due to the gravitational field produced by the matter which causes the "zodiacal light." (Today we know that there is not enough matter between Mercury and the sun to produce any appreciable precession.) It is also possible that the sun is slightly oblate,[7] in which case its Newtonian potential would have an r^{-3} term, giving the planets an anomalous precession per revolution decreasing as the inverse square of their distance from the sun. Table 8.2 shows that in fact the observed anomalous precession per revolution decreases roughly as $1/r$, not $1/r^2$, in agreement with the prediction of general relativity. Even more important, a large solar oblateness would produce an anomalous precession of the planes of the inner planets' orbits that is not observed.[8] These two arguments together rule out the possibility of explaining *all* of the observed anomalous precession as owing to solar oblateness, but this explanation might account for up to 20% of the observed effect. To test this hypothesis Dicke and Goldenberg[9] scanned the solar disk photoelectrically during the period June 1 to September 23, 1966. They concluded that the sun's polar diameter is shorter than its equatorial diameter by 5.0 ± 0.7 parts in 10^5. Taken at face value, this would give the perihelia of Mercury an extra precession of 3.4″ per century, so that only 39.6″ per century would be left as the relativistic effect, an 8% disagreement with Einstein's prediction of 43.03″ per century. The Brans-Dicke theory can account for an excess centennial precession of 39.6″ if we take $\omega = 6.4$. However, there are several reasons for hesitating before we give up general relativity:

(A) To account for the solar oblateness the solar interior would have to be rotating about once in one or two days, very much faster than the observed rotation rate (i.e., once in 25 days) of the sun's surface. This difference in rotation rates could perhaps be explained[10] as due to a magnetic torque induced by the solar wind, which retards the rotation of the surface, but it is not clear that this configuration would be dynamically stable.[11]

(B) Two very elaborate series of measurements[12] made with the Göttingen heliometer during the period 1891–1902 gave for the difference between the sun's equator and polar diameters the values 0.36 ± 0.78 parts in 10^5 and -0.10 ± 0.47 parts in 10^5, respectively, in agreement with each other and with perfect sphericity, but in disagreement with the result $+5.0 \pm 0.7$ parts in 10^5 of Dicke and Goldenberg. The Göttingen results were also supported by subsequent heliometer measurements. To quote Ashbrook:[12]

"What are we to make of all this? In view of the astronomical evidence, can the sun's polar diameter be 0.1 seconds shorter than its equatorial, as Drs. Dicke and Goldenberg think? Was there some unsuspected subtle systematic error in the Princeton experiment? Or was there some unrecognized effect in all the heliometer series?"

To do Dicke and Goldenberg justice, it should be noted that the axis of the oblate ellipsoid they observe follows the axis of the sun's rotation in its apparent annual oscillation, which certainly suggests that they are seeing something real.

(C) Even if the visible solar surface is oblate, what does this really tell us about the shape of its mass distribution and about the sun's gravitational field? Dicke[7] argues that the observed solar surface coincides with the gravitational equipotential surface, but this conclusion rests on a good deal of astrophysical theory and could be wrong.

(D) Finally, if Dicke and Goldenberg are right, then the 1% agreement between Einstein's prediction and the observed anomalous precession is a mere coincidence.

7   Radar Echo Delay

The classic tests of general relativity discussed in previous sections deal only with the shape of the trajectories of photons and planets. In recent years the development of high-speed electronics and high-power radar has opened to us the possibility of measuring motion as a function of *time* with the accuracy needed to test Einstein's equations. In particular, I. I. Shapiro has proposed[13] and, together with a group at the Lincoln Laboratory, has carried out measurements[14,15] of the time required for radar signals to travel to the inner planets and be reflected back to earth.

As a first step toward understanding the significance of these measurements, let us calculate the time required for a radar signal to go from one point with $r = r_1, \theta = \pi/2, \varphi = \varphi_1$, to a second point $r = r_2, \theta = \pi/2, \varphi = \varphi_2$. The equation governing the time history of orbits is Eq. (8.4.19):

$$\frac{A(r)}{B^2(r)} \left(\frac{dr}{dt}\right)^2 + \frac{J^2}{r^2} - \frac{1}{B(r)} = -E$$

We are dealing here with a light ray, so $E = 0$. Furthermore, $(dr/dt)^2$ must vanish at the distance $r = r_0$ of closest approach to the sun, so

$$J^2 = \frac{r_0{}^2}{B(r_0)}$$

The equation of motion of a photon is then

$$\frac{A(r)}{B^2(r)} \left(\frac{dr}{dt}\right)^2 + \left(\frac{r_0}{r}\right)^2 B^{-1}(r_0) - B^{-1}(r) = 0 \qquad (8.7.1)$$

From (8.7.1) we see that the time required for light to go from r_0 to r or from r to r_0 is

$$t(r, r_0) = \int_{r_0}^{r} \left(\frac{A(r)/B(r)}{\left[1 - \frac{B(r)}{B(r_0)} \left(\frac{r_0}{r} \right)^2 \right]} \right)^{1/2} dr \qquad (8.7.2)$$

and of course the total time required for light to go from point 1 to point 2 is (for $|\varphi_1 - \varphi_2| > \pi/2$)

$$t_{12} = t(r_1, r_0) + t(r_2, r_0) \qquad (8.7.3)$$

In order to evaluate the integral (8.7.2) we once again use in the integrand the Robertson expansion of Section 3:

$$A(r) \simeq 1 + \frac{2\gamma MG}{r} \qquad B(r) \simeq 1 - \frac{2MG}{r}$$

We then have

$$1 - \frac{B(r)}{B(r_0)} \left(\frac{r_0}{r} \right)^2 \simeq 1 - \left[1 + 2MG \left(\frac{1}{r_0} - \frac{1}{r} \right) \right] \left(\frac{r_0}{r} \right)^2$$

$$\simeq \left(1 - \frac{r_0^2}{r^2} \right) \left[1 - \frac{2MGr_0}{r(r + r_0)} \right]$$

so (8.7.2) gives, to first order in MG/r and MG/r_0,

$$t(r, r_0) \simeq \int_{r_0}^{r} \left(1 - \frac{r_0^2}{r^2} \right)^{-1/2} \left[1 + \frac{(1 + \gamma)MG}{r} + \frac{MGr_0}{r(r + r_0)} \right] dr$$

The integral is now elementary, and we find that the time required for light to go from r_0 to r is

$$t(r, r_0) \simeq \sqrt{r^2 - r_0^2} + (1 + \gamma)MG \ln \left(\frac{r + \sqrt{r^2 - r_0^2}}{r_0} \right)$$

$$+ MG \left(\frac{r - r_0}{r + r_0} \right)^{1/2} \qquad (8.7.4)$$

The leading term $\sqrt{r^2 - r_0^2}$ is what we should expect if light traveled in straight lines at unit velocity. The other terms evidently produce a general-relativistic *delay* in the time it takes a radar signal to travel to Mercury and back. (Note that a delay is opposite to what would have been expected from our experience with slowly moving bodies like comets.) This excess delay is a maximum when Mercury is at superior conjunction and the radar signal just grazes the sun; in this case r_0 is about equal to the radius of the sun, $r_0 \simeq R_\odot$, and is much smaller than the distances $r_\oplus$ and r_φ of the earth and Mercury from the sun, so the maximum round-trip excess time delay is given by (8.7.3) and (8.7.4) as

$$(\Delta t)_{max} \equiv 2[t(r_\oplus, R_\odot) + t(r_\varphi, R_\odot) - \sqrt{r_\oplus{}^2 - R_\odot{}^2} - \sqrt{r_\varphi{}^2 - R_\odot{}^2}]$$

$$\simeq 4M_\odot G \left\{ 1 + \left(\frac{1 + \gamma}{2}\right) \ln\left(\frac{4r_\varphi r_\oplus}{R_\odot{}^2}\right) \right\}$$

$$\simeq 5.9 \text{ km} \left\{ 1 + 11.2 \left(\frac{1 + \gamma}{2}\right) \right\} \tag{8.7.5}$$

If Einstein's field equations are correct, then $\gamma = 1$, and the maximum excess time delay will be

$$(\Delta t)_{max} \simeq 72 \text{ km} = 240 \ \mu\text{sec} \tag{8.7.6}$$

There is no difficulty in telling time to within a microsecond over the time interval of order 20 min during which the radar signal goes to Mercury and returns. Nevertheless this experiment presents extraordinary difficulties in execution and interpretation.

One trouble is that the radar signal is not reflected from just one "specular point" on Mercury's surface, but rather comes from a good-sized area, and is therefore spread out in arrival time by several hundred microseconds. Shapiro's group deals with this problem by a technique called "delay-Döppler mapping," that is, by measuring the distribution of the return signal power in frequency as well as arrival time. Each element of the reflecting surface has a characteristic velocity relative to the radar antenna (owing to the rotation and orbital motion of both the earth and Mercury) and therefore reflects the radar signal with a characteristic Döppler shift in frequency. Thus, if the reflecting properties of the surface are known, it is possible to deduce the arrival time of the echo from the point on Mercury's surface closest to the earth by analyzing the observed distribution of the echo in arrival time and frequency. (The reflecting properties of the surface are determined by studying the echo when Mercury is near inferior conjunction, where the signal-to-noise ratio is greatest and general relativity has no appreciable effect on the radar travel time.)

A more fundamental difficulty is that in order to compute an excess time delay to within 10 μsec, for instance, we have to know the time that the radar signal would have taken in the absence of the sun's gravitation to that accuracy; that is, we have to know the distance

$$(r_\oplus{}^2 - r_0{}^2)^{1/2} + (r_\varphi{}^2 - r_0{}^2)^{1/2}$$

to within 1.5 km! Here $r_\oplus$, r_φ, and r_0 are the distances (in "standard" coordinates) from the center of the sun to the radar antenna on earth, to the point on Mercury's surface closest to the earth, and to the point of closest approach of the radar signal to the sun. However, optical astronomy alone certainly does not provide us with the locations of the centers of Mercury or the earth, or Mercury's radius, to anything like the needed accuracy. Indeed, this accuracy is so demanding that it is necessary to specify whether one is dealing with standard, isotropic, or harmonic coordinates; needless to say, the U.S. Naval Observatory does not usually draw such fine distinctions! Shapiro's group deals with this problem by using general relativity theory itself to calculate $r_\oplus(t)$, $r_\yen(t)$, and $r_0(t)$ in terms of a large set of unknown parameters, including β, γ, $M_\odot G$, the equatorial radius of Mercury, and the positions and velocities of Mercury and the earth at some initial time. These parameters are then determined by fitting the observed radar travel times to Mercury and back with the theoretical formulas (8.7.3) and (8.7.4).

The first run, using the 7840 MHz Haystack radar at Lincoln Laboratory during the superior conjunctions of Mercury of April 28 to May 20, 1967, and August 15 to September 10, 1967, gave good agreement between theory and observation.[14] To put it quantitatively, if the radar delay times are computed using Eqs. (8.7.3) and (8.7.4) with γ left arbitrary, then the best fit is found for $\gamma = 0.8 \pm 0.4$. (For purely technical reasons, β was taken as unity in the preliminary analysis.) Further observations at Haystack and improvements in data analysis have since improved this result to[15]

$$\gamma = 1.03 \pm 0.1 \qquad (8.7.7)$$

(See Figure 8.3.) In addition, Shapiro has reanalyzed[16] over 400,000 older optical observations of the sun, moon, and planets in conjunction with the new radar

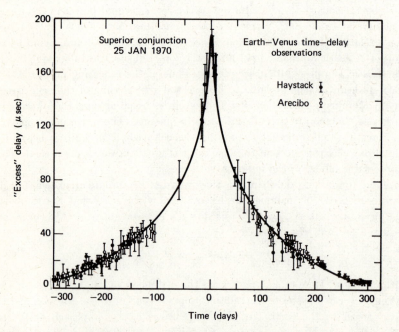

Figure 8.3 Comparison of theory with observation for the time delay of radar echoes from Venus.[15] (Courtesy of I. I. Shapiro.)

data, and finds that the quadrupole term in the sun's gravitational potential has the value $J_2 = (-0.8 \pm 2.5) \times 10^{-5}$, with J_2 defined by the Legendre expansion:

$$\phi_\odot = -\frac{GM_\odot}{r}\left\{1 - \sum_{l=2}^{\infty} J_l \left(\frac{R_\odot}{r}\right)^l P_l(\cos\theta)\right\}$$

For comparison, the solar oblateness found by Dicke and Goldenberg corresponds to the quadrupole term $J_2 = (2.7 \pm 0.5) \times 10^{-5}$. If J_2 is constrained to vanish, then Shapiro's analysis gives values for the extra precessions of the perihelia of the orbits of Mercury and Mars that are respectively 0.99 ± 0.01 and 1.07 ± 0.1 times the values predicted by general relativity.

Shapiro has also proposed[17] the measurement of time delays in the arrival of radio pulses from pulsars. The pulsar CP0952 approaches to within $5°$ of the sun, and at such times radio pulses would be delayed by about 50 μsec.

Recently a group[18] at the Jet Propulsion Laboratory measured the time delays of radar signals, sent from the earth to transponders aboard the artificial satellites Mariner 6 and 7 and thence back to earth, during the period March–June 1970 when these satellites were near superior conjunction. The best data were taken on April 28, 1970, when the radar signals passed within three solar radii of the sun's center. Analysis of these data gives a time delay within 5% of that predicted by general relativity. Unfortunately, the radar frequencies used were in the S-band, about 2300 MHz, so the solar corona played a troublesome role here. (See Section 8.5.) In addition, the Mariner satellites are small enough to be appreciably affected by nongravitational forces, chiefly solar radiation pressure, gas leakage, and thrust imbalances in the attitude control system.

The sensitivity of the radar echo arrival times to fine details of orbital motion makes the calculation of "theoretical" arrival times an enormously difficult task, which cannot be given the simple analytic treatment appropriate to this book. There is, however, some insight to be gained by looking at a simple calculation in a situation so highly idealized that we can deal with it here. Consider as a reflector a point planet labeled "1" in a circular orbit about the sun of radius r_1, and let the radar antenna be placed on a planet "2" that lies in the plane of planet 1's orbit ($\theta = \pi/2$), but is so far from the sun that its position can be taken as fixed, with $r_2 \gg r_1$ and $\phi_2 = 0$. (The change in ϕ_2 during the radar signal travel time vanishes as $r_2^{-1/2}$). A radar signal emitted from planet 2 at time t will reach planet 1 at a time t_1 given (for $|\phi_1| > \pi/2$) by

$$t_1 = t + t(r_1, r_0) + t(r_2, r_0)$$

or using (8.7.4) with $r_2 \to \infty$:

$$t_1 = t + T + (r_1{}^2 - r_0{}^2)^{1/2} + MG\left(\frac{r_1 - r_0}{r_1 + r_0}\right)^{1/2}$$

$$+ (1 + \gamma)MG \ln\left(\frac{[r_1 + (r_1{}^2 - r_0{}^2)^{1/2}]r_1}{r_0{}^2}\right) \qquad (8.7.8)$$

where T is a large constant:

$$T \equiv r_2 + MG + (1 + \gamma)MG \ln\left(\frac{r_2}{r_1}\right) \tag{8.7.9}$$

The azimuthal angle of the planet at this time is given by Eq. (8.4.27):

$$\varphi_1 = \varphi(0) + \omega t_1 \tag{8.7.10}$$

$$\omega \simeq \left(\frac{MG}{r_1{}^3}\right)^{1/2}\left(1 - \frac{(\beta - \gamma)MG}{r_1}\right) \tag{8.7.11}$$

Finally, r_0 is calculated by setting φ_1 equal to the value determined from Eq. (8.5.7),

$$\varphi_1 = [\varphi(r_0) - \varphi(r_1)] + [\varphi(r_0) - \varphi(\infty)]$$

$$= \pi - \sin^{-1}\left(\frac{r_0}{r_1}\right) + \left(\frac{MG}{r_0}\right)\left[1 + \gamma + \gamma\left(1 - \frac{r_0{}^2}{r_1{}^2}\right)^{1/2} + \left(\frac{r_1 - r_0}{r_1 + r_0}\right)^{1/2}\right]$$

or to first order in MG:

$$r_0 \simeq r_1 \sin \varphi_1 - MG \cot \varphi_1 \left[1 + \gamma - \gamma \cos \varphi_1 + \left(\frac{1 - \sin \varphi_1}{1 + \sin \varphi_1}\right)^{1/2}\right] \tag{8.7.12}$$

Using (8.7.10)–(8.7.12) in (8.7.8) gives the relation between the times t and t_1 of radar signal emission and reflection:

$$t_1 = t + T - a \cos(\omega t_1 + \varphi(0)) - b\{1 - \ln[1 + \cos(\omega t_1 + \varphi(0))]\} \tag{8.7.13}$$

where

$$a \equiv r_1 - \gamma MG \tag{8.7.14}$$

$$b \equiv (1 + \gamma)MG \tag{8.7.15}$$

Equation (8.7.13) can be solved for $t_1(t)$, and the arrival time of the echo back at the antenna can then be determined as

$$t_2(t) = t + 2[t_1(t) - t] = 2t_1(t) - t \tag{8.7.16}$$

By comparing this theoretical prediction with the observed arrival times of the radar echo, we can, in principle, determine the five parameters

$$T \quad a \quad b \quad \omega \quad \varphi(0)$$

But these five parameters depend on the six unknowns r_1, r_2, MG, γ, β, and $\varphi(0)$, so even if our measurements and Eqs. (8.7.13)–(8.7.16) were perfectly accurate, we still could not determine both β and γ. The best we can do is to eliminate r_1

and MG from the formulas (8.7.11), (8.7.14), and (8.7.15) for ω, a, and b, obtaining thereby a formula for γ:

$$1 + \gamma = ba^{-3}\omega^{-2}\left[1 + 0\left(\frac{MG}{a}\right)\right] \qquad (8.7.17)$$

Note that in this case β cannot, even in principle, be determined from radar time delay measurements.[19] It would be possible to determine both β and γ from observations of radar echoes from *two* reflecting planets in circular orbits, since in this case there are ten observable parameters and only eight unknowns. More important, it is possible to measure β by radar observations of Mercury alone, because its orbit is sufficiently eccentric for its precession seriously to affect radar arrival times.

8 The Schwarzschild Singularity*

The reader will probably have noticed that the Schwarzschild solution (8.2.12) becomes singular at $r = 2MG$. This radius corresponds to $\rho = MG/2$ and $R = MG$, so we see that this singularity also occurs when the metric is expressed in its isotropic form (8.2.14) or in its harmonic form (8.2.15). The radius $2GM$ at which the singularity occurs in standard coordinates is called the *Schwarzschild radius* of the mass M.

It should immediately be stressed that there is no Schwarzschild singularity in the gravitational field of any known object in the universe. The Schwarzschild singularity appeared in the solution of Einstein's vacuum equations $R_{\mu\nu} = 0$, and is therefore irrelevant if the radius $2GM$ lies within the massive body, where we must use the full Einstein equation (7.1.13). For the sun the Schwarzschild radius $2GM_\odot$ is 2.95 km, deep in the solar interior, and we shall see in Chapter 11 that the solution of the full Einstein equation inside a stable star exhibits no Schwarzschild singularity, or any other singularity. For a proton the Schwarzschild radius is 10^{-50} cm, and this is 37 orders of magnitude smaller than the characteristic proton radius of about 10^{-13} cm! In Chapter 11 we discuss the possibility that a very massive body might collapse to a radius smaller than its Schwarzschild radius, but with this one hypothetical exception the Schwarzschild singularity does not seem to have much relevance to the real world.

It is nonetheless instructive to imagine a body so small and massive that the radius $2GM$ lies outside it, in empty space. The Schwarzschild solution then holds down to this radius and actually displays a singularity. But is this singularity real? We can readily calculate the four nonvanishing curvature invariants described in Section 6.7, and find that they are all perfectly well behaved at the

---

* This section lies somewhat out of the book's main line of development, and may be omitted in a first reading.

Schwarzschild radius, although they do become singular at the origin. This suggests that the apparent Schwarzschild singularity may be only an artifact of the coordinate systems we have used. (If any one of the curvature invariants had been singular at the Schwarzschild radius, then this singularity would of course have been present in all coordinate systems.) Only a few years ago a coordinate system was found that allows us to avoid talking about a Schwarzschild singularity, if we are willing to allow the world an unusual topology.[20] To exhibit this reinterpretation of the Schwarzschild singularity, we introduce a new set of coordinates r', θ, φ, t', defined by

$$r'^2 - t'^2 \equiv T^2 \left(\frac{r}{2GM} - 1 \right) \exp \left(\frac{r}{2GM} \right) \tag{8.8.1}$$

$$\frac{2r't'}{r'^2 + t'^2} \equiv \tanh \left(\frac{t}{2MG} \right) \tag{8.8.2}$$

where T is an arbitrary constant. The Schwarzschild solution (8.2.12) then becomes

$$d\tau^2 = \left(\frac{32G^3M^3}{rT^2} \right) \exp \left(\frac{-r}{2GM} \right) (dt'^2 - dr'^2) - r^2 \, d\theta^2 - r^2 \sin^2 \theta \, d\varphi^2 \tag{8.8.3}$$

where r is now to be understood as a function of $r'^2 - t'^2$ defined by Eq. (8.8.1). The metric is nonsingular as long as r^2 is well defined and positive-definite, that is, as long as

$$r'^2 > t'^2 - T^2$$

Hence, during the time interval $0 < t' < T$, the metric is a perfectly smooth finite function of r' for all real r'. Indeed, even $g_{\theta\theta}$ and $g_{\varphi\varphi}$ do not vanish when $r' = 0$, so that when we approach the origin $r' = 0$ there is nothing to keep us from continuing right through to negative r'! The space described by (8.3.3) is therefore singularity-free, but consists of two identical sheets $r' > 0$ and $r' < 0$, joined in a smooth way by a branch point at $r' = 0$. When t' reaches the time T the two sheets detach from each other, and thereafter the metric has a real singularity at $r' = \pm \sqrt{t'^2 - T^2}$, that is, at $r = 0$. However, even so, the metric has no singularity at the radius $r' = t'$ that corresponds to the Schwarzschild radius $r = 2GM$.

To repeat, this discussion of the Schwarzschild singularity does not apply to any gravitational field actually known to exist anywhere in the universe. Indeed, it does not even apply to gravitational collapse (see Section 11.9) because for $t' < T$ space is empty for all r'. However, like Aesop's fables, it is useful because it points to a moral, that what appears in one coordinate system to be a singularity may in another coordinate system have quite a different interpretation.

8 BIBLIOGRAPHY

☐ V. B. Braginskii and V. N. Rudenko, "Relativistic Gravitational Experiments," Usp. Fiz. Nauk, **100**, 395 (1970) [trans. Soviet Physics Uspekhi, **13**, 165 (1970)].

☐ B. Bertotti, D. Brill, and R. Krotkov, "Experiments in Gravitation," in *Gravitation: An Introduction to Current Research*, ed. by L. Witten (Wiley, New York, 1962), p. 1.

☐ R. H. Dicke, "Experimental Relativity," in *Relativity, Groups, and Topology*, ed. by C. DeWitt and B. DeWitt (Gordon and Breach Science Publishers, New York, 1964), p. 163.

☐ F. J. Dyson, "Experimental Tests of General Relativity," in *Relativity Theory and Astrophysics 1. Relativity and Cosmology*, ed. by J. Ehlers (American Mathematical Society, Providence, R. I., 1967), p. 117.

☐ L. I. Schiff, "Comparison of Theory and Observation in General Relativity," in *Relativity Theory and Astrophysics 1. Relativity and Cosmology*, op. cit., p. 105.

☐ K. S. Thorne and C. M. Will, "High Precision Tests of General Relativity," Comments Astrophys. and Space Phys., **2**, 35 (1970).

8 REFERENCES

1. H. P. Robertson in *Space Age Astronomy*, ed. by A. J. Deutsch and W. B. Klemperer (Academic Press, New York, 1962), p. 228. For an earlier version of this expansion, see A. S. Eddington, *The Mathematical Theory of Relativity* (2nd ed., Cambridge University Press, 1924), p. 105.

2. F. W. Dyson, A. S. Eddington, and C. Davidson, Phil. Trans. Roy. Soc. (London), **220A**, 291 (1920); Mem. Roy. Astron. Soc., **62**, 291 (1920).

3. H. von Klüber in *Vistas in Astronomy*, ed. by A. Beer (Pergamon Press, New York, 1960), Vol. 3, p. 47. Reanalyses of some of this data are listed by B. Bertotti, D. Brill, and R. Krotkov, in *Gravitation: An Introduction to Current Research*, ed. by L. Witten (Wiley, New York, 1962), p. 1. Also see R. J. Trumpler, Helvetia Physica Acta Suppl., **IV**, 106 (1956); A. A. Mikhailov, Astron. Zh., **33**, 912 (1956).

4. I. I. Shapiro, Science, **157**, 806 (1967).

5. G. M. Clemence, Astron. Papers Am. Ephemeris, **11**, part 1 (1943); Rev. Mod. Phys., **19**, 361 (1947).

6. For the inner planets, see G. M. Clemence, ref. 5; R. L. Duncombe, Astron. J., **61**, 174 (1956); Astron. Papers Am. Ephemeris, **16**, part 1 (1958); R. L. Duncombe and G. M. Clemence, Astron. J., **63**, 456 (1958). For Icarus, see I. I. Shapiro, W. B. Smith, M. E. Ash, and S. Herrick, Astron.

J. **76,** 588 (1971); also I. I. Shapiro, M. E. Ash, and W. B. Smith, Phys. Rev. Letters, **20,** 1517 (1968); J. H. Lieske and G. Null, Astron. J., **74,** 297 (1969).

7. R. H. Dicke, Nature, **202,** 432 (1964); I. W. Roxburgh, Icarus, **3,** 92 (1964).

8. I. I. Shapiro, Icarus, **4,** 549 (1965).

9. R. H. Dicke and H. M. Goldenberg, Phys. Rev. Letters, **18,** 313 (1967).

10. J. C. Brandt, Ap. J., **144,** 1221 (1966).

11. P. Goldreich and G. Schubert, Ap. J., **154,** 1005 (1969). R. H. Dicke, Ap. J., **159,** 1 (1970). R. H. Dicke, to be published.

12. The heliometer results are summarized by J. Ashbrook, Sky and Telescope, **34,** 229 (1967).

13. I. I. Shapiro, Phys. Rev. Letters, **13,** 789 (1964).

14. I. I. Shapiro, G. H. Pettengill, M. E. Ash, M. L. Stone, W. B. Smith, R. P. Ingalls, and R. A. Brockelman, Phys. Rev. Letters, **20,** 1265 (1968).

15. I. I. Shapiro, M. E. Ash, R. P. Ingalls, W. B. Smith, D. B. Campbell, R. B. Dyce, R. F. Jurgens, and G. H. Pettengill, Phys. Rev. Letters, **26,** 1132 (1971).

16. I. I. Shapiro, report at the Third "Cambridge" Conference on Relativity, June 8, 1970 (unpublished).

17. I. I. Shapiro, Science, **162,** 352 (1968).

18. J. D. Anderson, report at the Third "Cambridge" Conference on General Relativity, June 8, 1970 (unpublished).

19. For an interesting exchange on this point, see D. K. Ross and L. I. Schiff, Phys. Rev., **141,** 1215 (1966); I. I. Shapiro, Phys. Rev., **145,** 1005 (1966).

20. M. D. Kruskal, Phys. Rev., **119,** 1743 (1960); see also C. Fronsdal, Phys. Rev., **116,** 778 (1959).

9 POST-NEWTONIAN CELESTIAL MECHANICS

The Einstein field equations are nonlinear, and therefore cannot in general be solved exactly. It is true that, by imposing the symmetry requirements of time independence and spatial isotropy, we were able to find one useful exact solution, the Schwarzschild metric, but we cannot actually make use of the full content of this solution, because in fact the solar system is *not* static and isotropic. Indeed, the Newtonian effects of the planets' gravitational fields are an order of magnitude greater than the first corrections due to general relativity, and completely swamp the higher corrections that are in principle provided by the exact Schwarzschild solution.

What we need then is not to find more exact solutions, but rather to develop some systematic approximation method that will not rely on any assumed symmetry properties of the system. There are two such methods that have been particularly useful; they are called the *post-Newtonian approximation*[1] and the *weak-field approximation*. The first is adapted to a system of slowly moving particles bound together by gravitational forces, such as the solar system, and is the subject of this chapter. The second method treats the fields in a lower order of approximation but does not assume that the matter moves nonrelativistically; it is therefore suited to handle the subject of gravitational radiation, and will be discussed in the next chapter. There obviously is an area of overlap between the two approximation methods, that is, for slowly moving particles moving in very weak fields, but it is best to keep them separate because of their separate applications.

The post-Newtonian approximation was historically derived[1] as a by-product of the study of the *problem of motion*: Do the equations of motion of massive particles follow from the gravitational field equations alone? According to the point of view adopted in this book, the equations of motion in general relativity should be derived from the equations of motion in special relativity and the Principle of Equivalence. Therefore, in this chapter the post-Newtonian approximation is discussed for its own sake, not as part of the problem of motion.

1   The Post-Newtonian Approximation

Consider a system of particles that, like the sun and planets, are bound together by their mutual gravitational attraction. Let $\bar{M}$, $\bar{r}$, and $\bar{v}$ be typical values of the masses, separations, and velocities of these particles. It is a familiar result of Newtonian mechanics that the typical kinetic energy $\frac{1}{2}\bar{M}\bar{v}^2$ will be roughly of the same order of magnitude as the typical potential energy $G\bar{M}^2/\bar{r}$, so

$$\bar{v}^2 \sim \frac{G\bar{M}}{\bar{r}} \tag{9.1.1}$$

(For instance, a test particle in a circular orbit of radius r about a central mass M will have velocity v given in Newtonian mechanics by the exact formula $v^2 = GM/r$.) The post-Newtonian approximation may be described as a method for obtaining the motions of the system to one higher power of the small parameters $G\bar{M}/\bar{r}$ and $\bar{v}^2$ than given by Newtonian mechanics. It is sometimes referred to as an expansion in inverse powers of the speed of light, but since in our units this speed is unity we prefer to say that our expansion parameter is $\bar{v}^2$, or equivalently, $G\bar{M}/\bar{r}$.

We must begin by asking what we need. The equations of motion of the particles are

$$\frac{d^2x^\mu}{d\tau^2} + \Gamma^\mu{}_{\nu\lambda}\frac{dx^\nu}{d\tau}\frac{dx^\lambda}{d\tau} = 0$$

From this we may compute the accelerations as

$$\frac{d^2x^i}{dt^2} = \left(\frac{dt}{d\tau}\right)^{-1}\frac{d}{d\tau}\left[\left(\frac{dt}{d\tau}\right)^{-1}\frac{dx^i}{d\tau}\right]$$

$$= \left(\frac{dt}{d\tau}\right)^{-2}\frac{d^2x^i}{d\tau^2} - \left(\frac{dt}{d\tau}\right)^{-3}\frac{d^2t}{d\tau^2}\frac{dx^i}{d\tau}$$

$$= -\Gamma^i{}_{\nu\lambda}\frac{dx^\nu}{dt}\frac{dx^\lambda}{dt} + \Gamma^0{}_{\nu\lambda}\frac{dx^\nu}{dt}\frac{dx^\lambda}{dt}\frac{dx^i}{dt}$$

This may be written in more detail as

$$\frac{d^2x^i}{dt^2} = -\Gamma^i{}_{00} - 2\Gamma^i{}_{0j}\frac{dx^j}{dt} - \Gamma^i{}_{jk}\frac{dx^j}{dt}\frac{dx^k}{dt}$$

$$+ \left[\Gamma^0{}_{00} + 2\Gamma^0{}_{0j}\frac{dx^j}{dt} + \Gamma^0{}_{jk}\frac{dx^j}{dt}\frac{dx^k}{dt}\right]\frac{dx^i}{dt} \tag{9.1.2}$$

In the Newtonian approximation discussed in Section 3.4 we treated all velocities as vanishingly small and kept only terms of first order in the difference between $g_{\mu\nu}$ and the Minkowski tensor $\eta_{\mu\nu}$, and we found

$$\frac{d^2x^i}{dt^2} \simeq -\Gamma^i{}_{00} \simeq \frac{1}{2}\frac{\partial g_{00}}{\partial x^i}$$

But $g_{00} - 1$ is of order $G\bar{M}/\bar{r}$, so the Newtonian approximation gives d^2x^i/dt^2 to order $G\bar{M}/\bar{r}^2$, that is, to order $\bar{v}^2/r$. *Therefore our objective in using the post-Newtonian approximation will be to compute d^2x^i/dt^2 to order $\bar{v}^4/\bar{r}$.* Inspection of (9.1.2) shows that we shall need the various components of the affine connection to the following orders:

$$\text{We need } \Gamma^i{}_{00} \text{ to order } \frac{\bar{v}^4}{\bar{r}}$$

$$\text{We need } \Gamma^i{}_{0j} \text{ to order } \frac{\bar{v}^3}{\bar{r}}$$

$$\text{We need } \Gamma^i{}_{jk} \text{ to order } \frac{\bar{v}^2}{\bar{r}}$$

$$\text{We need } \Gamma^0{}_{00} \text{ to order } \frac{\bar{v}^3}{\bar{r}}$$

$$\text{We need } \Gamma^0{}_{0j} \text{ to order } \frac{\bar{v}^2}{\bar{r}}$$

$$\text{We need } \Gamma^0{}_{jk} \text{ to order } \frac{\bar{v}}{\bar{r}} \tag{9.1.3}$$

From our experience with the Schwarzschild solution, we expect that it should be possible to find a coordinate system in which the metric tensor is nearly equal to the Minkowski tensor $\eta_{\mu\nu}$, the corrections being expandable in powers of $\bar{M}G/\bar{r} \sim \bar{v}^2$. In particular, we expect

$$g_{00} = -1 + \overset{2}{g}_{00} + \overset{4}{g}_{00} + \cdots \tag{9.1.4}$$

$$g_{ij} = \delta_{ij} + \overset{2}{g}_{ij} + \overset{4}{g}_{ij} + \cdots \tag{9.1.5}$$

$$g_{i0} = \overset{3}{g}_{i0} + \overset{5}{g}_{i0} + \cdots \tag{9.1.6}$$

the symbol $\overset{N}{g}_{\mu\nu}$ denoting the term in $g_{\mu\nu}$ of order $\bar{v}^N$. Odd powers of $\bar{v}$ occur in g_{i0} because g_{i0} must change sign under the time-reversal transformation $t \to -t$. The real justification for these expansions will come below when we show that they lead to a consistent solution of the Einstein field equations.

The inverse of the metric tensor is defined by the equations

$$g^{i\mu}g_{0\mu} = g^{i0}g_{00} + g^{ij}g_{j0} = 0 \qquad (9.1.7)$$

$$g^{0\mu}g_{0\mu} = g^{00}g_{00} + g^{0i}g_{0i} = 1 \qquad (9.1.8)$$

$$g^{i\mu}g_{j\mu} = g^{i0}g_{j0} + g^{ik}g_{jk} = \delta_{ij} \qquad (9.1.9)$$

We expect that

$$g^{00} = -1 + \overset{2}{g}{}^{00} + \overset{4}{g}{}^{00} + \cdots \qquad (9.1.10)$$

$$g^{ij} = \delta_{ij} + \overset{2}{g}{}^{ij} + \overset{4}{g}{}^{ij} + \cdots \qquad (9.1.11)$$

$$g^{i0} = \overset{3}{g}{}^{i0} + \overset{5}{g}{}^{i0} + \cdots \qquad (9.1.12)$$

and inserting these expansions into the defining equations (9.1.7)–(9.1.9), we find

$$\overset{2}{g}{}^{00} = -\overset{2}{g}_{00} \qquad \overset{2}{g}{}^{ij} = -\overset{2}{g}_{ij} \qquad \overset{3}{g}{}^{i0} = \overset{3}{g}_{i0} \qquad \text{etc.} \qquad (9.1.13)$$

The affine connection may now be obtained from the familiar formula

$$\Gamma^{\mu}{}_{\nu\lambda} = \tfrac{1}{2}g^{\mu\rho}\left\{\frac{\partial g_{\rho\nu}}{\partial x^{\lambda}} + \frac{\partial g_{\rho\lambda}}{\partial x^{\nu}} - \frac{\partial g_{\nu\lambda}}{\partial x^{\rho}}\right\}$$

In computing $\Gamma^{\mu}{}_{\nu\lambda}$ we must take into account the fact that the scales of distance and time in our systems are set by $\bar{r}$ and $\bar{r}/\bar{v}$, respectively, so the space and time derivatives should be regarded as being of order

$$\frac{\partial}{\partial x^i} \sim \frac{1}{\bar{r}} \qquad \frac{\partial}{\partial t} \sim \frac{\bar{v}}{\bar{r}}$$

Using our estimates (9.1.4)–(9.1.6) and (9.1.10)–(9.1.13) we find that the components $\Gamma^{i}{}_{00}$, $\Gamma^{i}{}_{jk}$, and $\Gamma^{0}{}_{0i}$ have the expansions

$$\Gamma^{\mu}{}_{\nu\lambda} = \overset{2}{\Gamma}{}^{\mu}{}_{\nu\lambda} + \overset{4}{\Gamma}{}^{\mu}{}_{\nu\lambda} + \cdots \qquad (\text{for } \Gamma^{i}{}_{00}, \Gamma^{i}{}_{jk}, \Gamma^{0}{}_{0i}) \qquad (9.1.14)$$

while the components $\Gamma^{i}{}_{0j}$, $\Gamma^{0}{}_{00}$, and $\Gamma^{0}{}_{ij}$ have the expansions

$$\Gamma^{\mu}{}_{\nu\lambda} = \overset{3}{\Gamma}{}^{\mu}{}_{\nu\lambda} + \overset{5}{\Gamma}{}^{\mu}{}_{\nu\lambda} + \cdots \qquad (\text{for } \Gamma^{i}{}_{0j}, \Gamma^{0}{}_{00}, \Gamma^{0}{}_{ij}) \qquad (9.1.15)$$

the symbol $\overset{N}{\Gamma^\mu{}_{\nu\lambda}}$ denoting the term in $\Gamma^\mu{}_{\nu\lambda}$ of order $\bar{v}^N/\bar{r}$. The components called for by (9.1.3) are explicitly

$$\overset{2}{\Gamma^i{}_{00}} = -\frac{1}{2}\frac{\partial \overset{2}{g_{00}}}{\partial x^i} \tag{9.1.16}$$

$$\overset{4}{\Gamma^i{}_{00}} = -\frac{1}{2}\frac{\partial \overset{4}{g_{00}}}{\partial x^i} + \frac{\partial \overset{3}{g_{i0}}}{\partial t} + \tfrac{1}{2}\overset{2}{g_{ij}}\frac{\partial \overset{2}{g_{00}}}{\partial x^j} \tag{9.1.17}$$

$$\overset{3}{\Gamma^i{}_{0j}} = \frac{1}{2}\left[\frac{\partial \overset{3}{g_{i0}}}{\partial x^j} + \frac{\partial \overset{2}{g_{ij}}}{\partial t} - \frac{\partial \overset{3}{g_{j0}}}{\partial x^i}\right] \tag{9.1.18}$$

$$\overset{2}{\Gamma^i{}_{jk}} = \frac{1}{2}\left[\frac{\partial \overset{2}{g_{ij}}}{\partial x^k} + \frac{\partial \overset{2}{g_{ik}}}{\partial x^j} - \frac{\partial \overset{2}{g_{jk}}}{\partial x^i}\right] \tag{9.1.19}$$

$$\overset{3}{\Gamma^0{}_{00}} = -\frac{1}{2}\frac{\partial \overset{2}{g_{00}}}{\partial t} \tag{9.1.20}$$

$$\overset{2}{\Gamma^0{}_{0i}} = -\frac{1}{2}\frac{\partial \overset{2}{g_{00}}}{\partial x^i} \tag{9.1.21}$$

$$\overset{1}{\Gamma^0{}_{ij}} = 0 \tag{9.1.22}$$

Evidently, we shall have to know the components g_{ij} to order $\bar{v}^2$, g_{i0} to order $\bar{v}^3$, and g_{00} to order $\bar{v}^4$. This should be contrasted with the Newtonian approximation, in which we needed g_{00} to order $\bar{v}^2$ and g_{i0} and g_{ij} only to zeroth order.

To calculate the Ricci tensor we shall use Eq. (6.1.5):

$$R_{\mu\kappa} \equiv R^\lambda{}_{\mu\lambda\kappa} = \frac{\partial \Gamma^\lambda{}_{\mu\lambda}}{\partial x^\kappa} - \frac{\partial \Gamma^\lambda{}_{\mu\kappa}}{\partial x^\lambda} + \Gamma^\eta{}_{\mu\lambda}\Gamma^\lambda{}_{\kappa\eta} - \Gamma^\eta{}_{\mu\kappa}\Gamma^\lambda{}_{\eta\lambda}$$

From (9.1.14) and the expansions (9.1.15) and (9.1.16) we find that the components of $R_{\mu\kappa}$ have the expansions:

$$R_{00} = \overset{2}{R_{00}} + \overset{4}{R_{00}} + \cdots \tag{9.1.23}$$

$$R_{i0} = \overset{3}{R_{i0}} + \overset{5}{R_{i0}} + \cdots \tag{9.1.24}$$

$$R_{ij} = \overset{2}{R_{ij}} + \overset{4}{R_{ij}} + \cdots \tag{9.1.25}$$

where $\overset{N}{R}_{\mu\nu}$ denotes the term in $R_{\mu\nu}$ of order $\bar{v}^N/\bar{r}^2$. The terms we can calculate from the "known" terms in the affine connection are

$$\overset{2}{R}_{00} = -\frac{\partial \overset{2}{\Gamma}{}^i_{00}}{\partial x^i} \tag{9.1.26}$$

$$\overset{4}{R}_{00} = \frac{\partial \overset{3}{\Gamma}{}^i_{0i}}{\partial t} - \frac{\partial \overset{4}{\Gamma}{}^i_{00}}{\partial x^i} + \overset{2}{\Gamma}{}^0_{0i}\overset{2}{\Gamma}{}^i_{00} - \overset{2}{\Gamma}{}^i_{00}\overset{2}{\Gamma}{}^j_{ij} \tag{9.1.27}$$

$$\overset{3}{R}_{i0} = \frac{\partial \overset{2}{\Gamma}{}^j_{ij}}{\partial t} - \frac{\partial \overset{3}{\Gamma}{}^j_{0i}}{\partial x^j} \tag{9.1.28}$$

$$\overset{2}{R}_{ij} = \frac{\partial \overset{2}{\Gamma}{}^0_{i0}}{\partial x^j} + \frac{\partial \overset{2}{\Gamma}{}^k_{ik}}{\partial x^j} - \frac{\partial \overset{2}{\Gamma}{}^k_{ij}}{\partial x^k} \tag{9.1.29}$$

Using (9.1.16)–(9.1.21), these give

$$\overset{2}{R}_{00} = \tfrac{1}{2}\nabla^2 \overset{2}{g}_{00} \tag{9.1.30}$$

$$\overset{4}{R}_{00} = \frac{1}{2}\frac{\partial^2 \overset{2}{g}_{ii}}{\partial t^2} - \frac{\partial^2 \overset{3}{g}_{i0}}{\partial x^i \partial t} + \tfrac{1}{2}\nabla^2 \overset{4}{g}_{00} - \tfrac{1}{2}\overset{2}{g}_{ij}\frac{\partial^2 \overset{2}{g}_{00}}{\partial x^i \partial x^j}$$
$$-\frac{1}{2}\left(\frac{\partial \overset{2}{g}_{ij}}{\partial x^j}\right)\left(\frac{\partial \overset{2}{g}_{00}}{\partial x^i}\right) + \frac{1}{4}\left(\frac{\partial \overset{2}{g}_{00}}{\partial x^i}\right)\left(\frac{\partial \overset{2}{g}_{00}}{\partial x^i}\right) + \frac{1}{4}\left(\frac{\partial \overset{2}{g}_{00}}{\partial x^i}\right)\left(\frac{\partial \overset{2}{g}_{jj}}{\partial x^i}\right) \tag{9.1.31}$$

$$\overset{3}{R}_{i0} = \frac{1}{2}\frac{\partial^2 \overset{2}{g}_{jj}}{\partial x^i \partial t} - \frac{1}{2}\frac{\partial^2 \overset{3}{g}_{j0}}{\partial x^i \partial x^j} - \frac{1}{2}\frac{\partial^2 \overset{2}{g}_{ij}}{\partial x^j \partial t} + \tfrac{1}{2}\nabla^2 \overset{3}{g}_{i0} \tag{9.1.32}$$

$$\overset{2}{R}_{ij} = -\frac{1}{2}\frac{\partial^2 \overset{2}{g}_{00}}{\partial x^i \partial x^j} + \frac{1}{2}\frac{\partial^2 \overset{2}{g}_{kk}}{\partial x^i \partial x^j} - \frac{1}{2}\frac{\partial^2 \overset{2}{g}_{ik}}{\partial x^k \partial x^j} - \frac{1}{2}\frac{\partial^2 \overset{2}{g}_{kj}}{\partial x^k \partial x^i} + \tfrac{1}{2}\nabla^2 \overset{2}{g}_{ij} \tag{9.1.33}$$

A tremendous simplification can be achieved at this point by choosing a suitable coordinate system. We showed in Section 7.4 that it is always possible to define the x^μ so that they obey the harmonic coordinate conditions

$$g^{\mu\nu}\Gamma^\lambda_{\mu\nu} = 0 \tag{9.1.34}$$

Using (9.1.10)–(9.1.13) and (9.1.16)–(9.1.21), we find that the vanishing of the third-order term in $g^{\mu\nu}\Gamma^0_{\mu\nu}$ gives

$$0 = \frac{1}{2}\frac{\partial \overset{2}{g}_{00}}{\partial t} - \frac{\partial \overset{3}{g}_{0i}}{\partial x^i} + \frac{1}{2}\frac{\partial \overset{2}{g}_{ii}}{\partial t} \tag{9.1.35}$$

while the vanishing of the second-order term in $g^{\mu\nu}\Gamma^i_{\ \mu\nu}$ gives

$$0 = \frac{1}{2}\frac{\partial \overset{2}{g}_{00}}{\partial x^i} + \frac{\partial \overset{2}{g}_{ij}}{\partial x^j} - \frac{1}{2}\frac{\partial \overset{2}{g}_{jj}}{\partial x^i} \tag{9.1.36}$$

It follows that

$$\frac{1}{2}\frac{\partial^2 \overset{2}{g}_{ii}}{\partial t^2} - \frac{\partial^2 \overset{3}{g}_{io}}{\partial x^i\,\partial t} + \frac{1}{2}\frac{\partial^2 \overset{2}{g}_{00}}{\partial t^2} = 0$$

$$\frac{\partial^2 \overset{2}{g}_{ii}}{\partial t\,\partial x^j} - \frac{\partial^2 \overset{3}{g}_{io}}{\partial x^i\,\partial x^j} - \frac{\partial^2 \overset{2}{g}_{ij}}{\partial x^i\,\partial t} = 0$$

$$\frac{\partial^2 \overset{2}{g}_{ij}}{\partial x^k\,\partial x^j} + \frac{\partial^2 \overset{2}{g}_{kj}}{\partial x^j\,\partial x^i} - \frac{\partial^2 \overset{2}{g}_{jj}}{\partial x^i\,\partial x^k} + \frac{\partial^2 \overset{2}{g}_{00}}{\partial x^i\,\partial x^k} = 0$$

so (9.1.30)–(9.1.33) now give simplified formulas for the Ricci tensor:

$$\overset{2}{R}_{00} = \tfrac{1}{2}\mathbf{V}^2 \overset{2}{g}_{00} \tag{9.1.37}$$

$$\overset{4}{R}_{00} = \tfrac{1}{2}\mathbf{V}^2 \overset{4}{g}_{00} - \frac{1}{2}\frac{\partial^2 \overset{2}{g}_{00}}{\partial t^2} - \tfrac{1}{2}\overset{2}{g}_{ij}\frac{\partial^2 \overset{2}{g}_{00}}{\partial x^i\,\partial x^j} + \tfrac{1}{2}(\mathbf{V}^2 \overset{2}{g}_{00})^2 \tag{9.1.38}$$

$$\overset{3}{R}_{0i} = \tfrac{1}{2}\mathbf{V}^2 \overset{3}{g}_{i0} \tag{9.1.39}$$

$$\overset{2}{R}_{ij} = \tfrac{1}{2}\mathbf{V}^2 \overset{2}{g}_{ij} \tag{9.1.40}$$

We are now ready to make use of the Einstein field equations, which we take in the form

$$R_{\mu\nu} = -8\pi G(T_{\mu\nu} - \tfrac{1}{2}g_{\mu\nu}T^\lambda_{\ \lambda}) \tag{9.1.41}$$

From their interpretation as the energy density, momentum density, and momentum flux, we expect that T^{00}, T^{i0}, and T^{ij} will have the expansions

$$T^{00} = \overset{0}{T}{}^{00} + \overset{2}{T}{}^{00} + \cdots \tag{9.1.42}$$

$$T^{i0} = \overset{1}{T}{}^{i0} + \overset{3}{T}{}^{i0} + \cdots \tag{9.1.43}$$

$$T^{ij} = \overset{2}{T}{}^{ij} + \overset{4}{T}{}^{ij} + \cdots \tag{9.1.44}$$

where $\overset{N}{T}{}^{\mu\nu}$ denotes the term in $T^{\mu\nu}$ of order $(\bar{M}/\bar{r}^3)\bar{v}^N$. (In particular $\overset{0}{T}{}^{00}$ is the density of rest-mass, while $\overset{2}{T}{}^{00}$ is the nonrelativistic part of the energy density.) What we need is

$$S_{\mu\nu} = T_{\mu\nu} - \tfrac{1}{2}g_{\mu\nu}T^\lambda_{\ \lambda} \tag{9.1.45}$$

But $G\bar{M}/\bar{r}$ is of order $\bar{v}^2$, so (9.1.4)–(9.1.6) and (9.1.42)–(9.1.44) give

$$S_{00} = \overset{0}{S}_{00} + \overset{2}{S}_{00} + \cdots \tag{9.1.46}$$

$$S_{i0} = \overset{1}{S}_{i0} + \overset{3}{S}_{i0} + \cdots \tag{9.1.47}$$

$$S_{ij} = \overset{0}{S}_{ij} + \overset{2}{S}_{ij} + \cdots \tag{9.1.48}$$

where $\overset{N}{S}_{\mu\nu}$ denotes the term in $S_{\mu\nu}$ of order $\bar{M}\bar{v}^N/\bar{r}^3$. In particular

$$\overset{0}{S}_{00} = \tfrac{1}{2}\overset{0}{T}{}^{00} \tag{9.1.49}$$

$$\overset{2}{S}_{00} = \tfrac{1}{2}[\overset{2}{T}{}^{00} - 2\overset{2}{g}_{00}\overset{0}{T}{}^{00} + \overset{2}{T}{}^{ii}] \tag{9.1.50}$$

$$\overset{1}{S}_{i0} = -\overset{1}{T}{}^{0i} \tag{9.1.51}$$

$$\overset{0}{S}_{ij} = +\tfrac{1}{2}\delta_{ij}\overset{0}{T}{}^{00} \tag{9.1.52}$$

Using (9.1.37)–(9.1.40) and (9.1.46)–(9.1.52) in the field equations (9.1.41), we find that the field equations in harmonic coordinates are indeed consistent with the expansions we have been using, and give

$$\nabla^2 \overset{2}{g}_{00} = -8\pi G \overset{0}{T}{}^{00} \tag{9.1.53}$$

$$\nabla^2 \overset{4}{g}_{00} = \frac{\partial^2 \overset{2}{g}_{00}}{\partial t^2} + \overset{2}{g}_{ij}\frac{\partial^2 \overset{2}{g}_{00}}{\partial x^i \, \partial x^j} - \left(\frac{\partial \overset{2}{g}_{00}}{\partial x^i}\right)\left(\frac{\partial \overset{2}{g}_{00}}{\partial x^i}\right)$$
$$- 8\pi G[\overset{2}{T}{}^{00} - 2\overset{2}{g}_{00}\overset{0}{T}{}^{00} + \overset{2}{T}{}^{ii}] \tag{9.1.54}$$

$$\nabla^2 \overset{3}{g}_{i0} = +16\pi G \overset{1}{T}{}^{i0} \tag{9.1.55}$$

$$\nabla^2 \overset{2}{g}_{ij} = -8\pi G \, \delta_{ij}\overset{0}{T}{}^{00} \tag{9.1.56}$$

From (9.1.53) we find as expected

$$\overset{2}{g}_{00} = -2\phi \tag{9.1.57}$$

where ϕ is the *Newtonian potential*, defined by Poisson's equation

$$\nabla^2 \phi = 4\pi G \overset{0}{T}{}^{00} \tag{9.1.58}$$

Also $\overset{2}{g}_{00}$ must vanish at infinity, so the solution is

$$\phi(\mathbf{x}, t) = -G \int d^3x' \, \frac{\overset{0}{T}{}^{00}(\mathbf{x}', t)}{|\mathbf{x} - \mathbf{x}'|} \tag{9.1.59}$$

From (9.1.56) we find that the solution for $\overset{2}{g}_{ij}$ that vanishes at infinity is

$$\overset{2}{g}_{ij} = -2\delta_{ij}\phi \tag{9.1.60}$$

On the other hand, $\overset{3}{g}_{i0}$ is a new vector potential $\boldsymbol{\zeta}$:

$$\overset{3}{g}_{i0} \equiv \zeta_i \tag{9.1.61}$$

and the solution of (9.1.55) that vanishes at infinity is

$$\zeta_i(\mathbf{x}, t) = -4G \int \frac{\overset{1}{T}{}^{i0}(\mathbf{x}'t)}{|\mathbf{x} - \mathbf{x}'|} d^3x' \tag{9.1.62}$$

Finally, we may simplify (9.1.54) by using (9.1.57), (9.1.58), and the identity

$$\frac{\partial\phi}{\partial x^i}\frac{\partial\phi}{\partial x^i} \equiv \frac{1}{2}\nabla^2\phi^2 - \phi\nabla^2\phi$$

The result is

$$\overset{4}{g}_{00} = -2\phi^2 - 2\psi \tag{9.1.63}$$

where ψ is a second potential

$$\nabla^2\psi = \frac{\partial^2\phi}{\partial t^2} + 4\pi G[\overset{2}{T}{}^{00} + \overset{2}{T}{}^{ii}] \tag{9.1.64}$$

Again, $\overset{4}{g}_{00}$ must vanish at infinity, so the solution is

$$\psi(\mathbf{x}, t) = -\int \frac{d^3x'}{|\mathbf{x} - \mathbf{x}'|}\left[\frac{1}{4\pi}\frac{\partial^2\phi(\mathbf{x}', t)}{\partial t^2} + G\overset{2}{T}{}^{00}(\mathbf{x}', t) + G\overset{2}{T}{}^{ii}(\mathbf{x}', t)\right] \tag{9.1.65}$$

The coordinate condition (9.1.35) imposes on ϕ and $\boldsymbol{\zeta}$ the further relation

$$4\frac{\partial\phi}{\partial t} + \nabla\cdot\boldsymbol{\zeta} = 0 \tag{9.1.66}$$

while the other coordinate condition (9.1.36) is now automatically satisfied. We shall see in Section 3 that (9.1.66) is also satisfied by our solutions, by virtue of the conservation conditions obeyed by $T^{\mu\nu}$.

Using (9.1.57), (9.1.60), (9.1.61), and (9.1.63) in (9.1.16)–(9.1.22) gives the desired components of the affine connection

$$\overset{2}{\Gamma}{}^i_{00} = \frac{\partial \phi}{\partial x^i} \tag{9.1.67}$$

$$\overset{4}{\Gamma}{}^i_{00} = \frac{\partial}{\partial x^i}(2\phi^2 + \psi) + \frac{\partial \zeta_i}{\partial t} \tag{9.1.68}$$

$$\overset{3}{\Gamma}{}^i_{0j} = \frac{1}{2}\left(\frac{\partial \zeta_i}{\partial x^j} - \frac{\partial \zeta_j}{\partial x^i}\right) - \delta_{ij}\frac{\partial \phi}{\partial t} \tag{9.1.69}$$

$$\overset{2}{\Gamma}{}^i_{jk} = -\delta_{ij}\frac{\partial \phi}{\partial x^k} - \delta_{ik}\frac{\partial \phi}{\partial x^j} + \delta_{jk}\frac{\partial \phi}{\partial x^i} \tag{9.1.70}$$

$$\overset{3}{\Gamma}{}^0_{00} = \frac{\partial \phi}{\partial t} \tag{9.1.71}$$

$$\overset{2}{\Gamma}{}^0_{0i} = \frac{\partial \phi}{\partial x^i} \tag{9.1.72}$$

As a bonus, we can now also calculate three additional terms in the affine connection that will play a role in post-Newtonian hydrodynamics:

$$\overset{3}{\Gamma}{}^0_{ij} = -\frac{1}{2}\left(\frac{\partial \zeta_i}{\partial x^j} + \frac{\partial \zeta_j}{\partial x^i}\right) - \delta_{ij}\frac{\partial \phi}{\partial t} \tag{9.1.73}$$

$$\overset{4}{\Gamma}{}^0_{i0} = \frac{\partial \psi}{\partial x^i} \tag{9.1.74}$$

$$\overset{5}{\Gamma}{}^0_{00} = \frac{\partial \psi}{\partial t} + \zeta \cdot \nabla\phi \tag{9.1.75}$$

2   Particle and Photon Dynamics

Before continuing our calculation of the post-Newtonian metric, we shall take a quick look back at the problem with which we started, that of computing the acceleration of a freely falling particle to order $\bar{v}^4/\bar{r}$. (Detailed applications of the post-Newtonian method are given in Sections 9.5–9.9.) Inserting the terms (9.1.67)–(9.1.72) of the affine connection into (9.1.2) immediately gives the equation of motion:

$$\frac{d\mathbf{v}}{dt} = -\nabla(\phi + 2\phi^2 + \psi) - \frac{\partial \zeta}{\partial t} + \mathbf{v}\times(\nabla\times\zeta)$$

$$+ 3\mathbf{v}\frac{\partial \phi}{\partial t} + 4\mathbf{v}(\mathbf{v}\cdot\nabla)\phi - \mathbf{v}^2\nabla\phi \tag{9.2.1}$$

where $v^i \equiv dx^i/dt$.

In addition, we need to know how to convert the harmonic coordinate time t into the proper time τ measured on a freely falling body of velocity $\mathbf{v}$. By definition

$$\left(\frac{d\tau}{dt}\right)^2 = -g_{00} - 2g_{i0}v^i - g_{ij}v^iv^j$$

To order $\bar{v}^4$, this is

$$\left(\frac{d\tau}{dt}\right)^2 = 1 - [\overset{2}{\mathbf{v}^2 + g_{00}}] - [\overset{4}{g_{00}} + 2\overset{3}{g_{i0}}v^i + \overset{2}{g_{ij}}v^iv^j]$$

or using (9.1.57), (9.1.60), (9.1.61), and (9.1.63):

$$\left(\frac{d\tau}{dt}\right)^2 = 1 + [2\phi - \mathbf{v}^2] + 2[\phi^2 + \psi - \boldsymbol{\zeta}\cdot\mathbf{v} + \phi\mathbf{v}^2]$$

The brackets enclose terms of order $\bar{v}^2$ and $\bar{v}^4$. By using the power series expansion of $\sqrt{1 + x}$, we find to order $\bar{v}^4$:

$$\frac{d\tau}{dt} = 1 + \phi - \tfrac{1}{2}\mathbf{v}^2 - \tfrac{1}{8}(2\phi - \mathbf{v}^2)^2 + \phi^2 + \psi - \boldsymbol{\zeta}\cdot\mathbf{v} + \phi\mathbf{v}^2$$

or

$$\frac{d\tau}{dt} = 1 - L \tag{9.2.2}$$

where

$$L = \tfrac{1}{2}\mathbf{v}^2 - \phi - \tfrac{1}{2}\phi^2 - \tfrac{3}{2}\phi\mathbf{v}^2 + \tfrac{1}{8}(\mathbf{v}^2)^2 - \psi + \boldsymbol{\zeta}\cdot\mathbf{v} \tag{9.2.3}$$

Since $\int (d\tau/dt)\, dt$ is stationary, we can regard L as the *Lagrangian* for a single particle, and we can derive the equations of motion from the Lagrange equation

$$\frac{d}{dt}\frac{\partial L}{\partial v^i} = \frac{\partial L}{\partial x^i} \tag{9.2.4}$$

(Acting on ϕ or $\boldsymbol{\zeta}$, d/dt is to be taken as $\partial/\partial t + \mathbf{v}\cdot\boldsymbol{\nabla}$.) The reader may readily check that (9.2.4) agrees with Eq. (9.2.1).

The post-Newtonian fields can also be used to calculate the acceleration of a photon in a gravitational field to order $\bar{v}^2$. (Here $\bar{v}$ is of course not the photon speed; it is the typical velocity of the material particles of the system.) Since the velocity $u_i \equiv dx^i/dt$ of the photon is of order unity, Eq. (9.1.2) now gives its acceleration as

$$\frac{du_i}{dt} = -\overset{2}{\Gamma}{}^i{}_{00} - \overset{2}{\Gamma}{}^i{}_{jk}u_ju_k + 2u_i\overset{2}{\Gamma}{}^0{}_{0j}u_j + \mathrm{O}(\bar{v}^3)$$

Using (9.1.67), (9.1.70), and (9.1.72), this gives

$$\frac{d\mathbf{u}}{dt} = -(1 + \mathbf{u}^2)\boldsymbol{\nabla}\phi + 4\mathbf{u}(\mathbf{u}\cdot\boldsymbol{\nabla}\phi) + \mathrm{O}(\bar{v}^3)$$

We also note that the photon speed is given by the condition

$$0 = -g_{\mu\nu} \frac{dx^\mu}{dt} \frac{dx^\nu}{dt} = 1 - \mathbf{u}^2 + 2(1 + \mathbf{u}^2)\phi + \bigcirc(\bar{v}^3)$$

or

$$|\mathbf{u}| = 1 + 2\phi + \bigcirc(\bar{v}^3) \tag{9.2.5}$$

Hence to the required accuracy we can replace $\mathbf{u}^2$ by unity in the photon acceleration:

$$\frac{d\mathbf{u}}{dt} = -2\nabla\phi + 4\mathbf{u}(\mathbf{u} \cdot \nabla\phi) + \bigcirc(\bar{v}^3) \tag{9.2.6}$$

It is somewhat more convenient to write this as an equation for the unit direction vector $\hat{\mathbf{u}} \equiv \mathbf{u}/|\mathbf{u}|$:

$$\frac{d\hat{\mathbf{u}}}{dt} = \hat{\mathbf{u}} \times (\hat{\mathbf{u}} \times \nabla\phi) + \bigcirc(\bar{v}^3) \tag{9.2.7}$$

3 · The Energy-Momentum Tensor

In order to complete the computational program outlined in Section 1, we must show how to calculate the energy-momentum tensor $T^{\mu\nu}$ that serves as the source of the gravitational fields. We shall first consider how the conservation laws of energy and momentum appear in the post-Newtonian approximation. In general, the conservation laws read $T^{\mu\nu}{}_{;\mu} = 0$, or in more detail

$$\frac{\partial}{\partial x^\mu} T^{\mu\nu} = -\Gamma^\nu{}_{\mu\lambda} T^{\mu\lambda} - \Gamma^\mu{}_{\mu\lambda} T^{\lambda\nu} \tag{9.3.1}$$

The term of order $M\bar{v}/\bar{r}^4$ with $\nu = 0$ gives

$$\frac{\partial}{\partial t} \overset{0}{T}{}^{00} + \frac{\partial}{\partial x^i} \overset{1}{T}{}^{i0} = 0 \tag{9.3.2}$$

since all Γ's are at least of order $\bar{v}^2/\bar{r}$. This may be regarded as the law of conservation of *mass*; it should not surprise us to find mass conserved in the post-Newtonian approximation, for a large rate of conversion of mass into energy would produce temperatures at which the particles of the system moved relativistically, in conflict with the assumption that $\bar{v} \ll 1$. Apart from its intrinsic importance, Eq. (9.3.2) is here indispensable to us, because it is needed for consistency of the harmonic coordinate conditions. From (9.1.53) and (9.1.55), we see that (9.3.2) implies

$$0 = \nabla^2 \left(-2 \frac{\partial \overset{2}{g}_{00}}{\partial t} + \frac{\partial \overset{3}{g}_{0i}}{\partial x^i} \right) = \nabla^2 \left(4 \frac{\partial\phi}{\partial t} + \nabla \cdot \boldsymbol{\zeta} \right)$$

Since ϕ and ζ must vanish at infinity, we conclude that

$$4\frac{\partial\phi}{\partial t} + \nabla\cdot\zeta = 0$$

thus verifying the coordinate condition (9.1.66).

Returning to Eq. (9.3.1), we find that the $\nu = i$ term of order $M\bar{v}^2/\bar{r}^4$ gives

$$\frac{\partial}{\partial t}\overset{1}{T}{}^{0i} + \frac{\partial}{\partial x^i}\overset{2}{T}{}^{ij} = -\overset{2}{\Gamma}{}^i{}_{00}\overset{0}{T}{}^{00}$$

or, using (9.1.67),

$$\frac{\partial}{\partial t}\overset{1}{T}{}^{0i} + \frac{\partial}{\partial x^i}\overset{2}{T}{}^{ij} = -\frac{\partial\phi}{\partial x^i}\overset{0}{T}{}^{00} \tag{9.3.3}$$

Since T^{ij} is the flux of momentum, we recognize in this the conservation of momentum; note that the right-hand side is just the Newtonian gravitational force density, equal to the mass density $\overset{0}{T}{}^{00}$ times $-\nabla\phi$.

There are no other conservation laws which involve *only* the terms in $T^{\mu\nu}$ needed to calculate the fields in the post-Newtonian approximation, that is, only $\overset{0}{T}{}^{00}$, $\overset{2}{T}{}^{00}$, $\overset{1}{T}{}^{i0}$, and $\overset{2}{T}{}^{ij}$. Further, we note that the two conservation laws (9.3.2) and (9.3.3) involve $g_{\mu\nu}$ only through ϕ, which can be calculated in the Newtonian approximation. Hence, the procedure to be followed is essentially *iterative*. We must first solve the Newtonian equations of motion, use the solution (plus the equations of state) to determine the terms $\overset{0}{T}{}^{00}$, $\overset{2}{T}{}^{00}$, $\overset{1}{T}{}^{i0}$, and $\overset{2}{T}{}^{ij}$, compute the post-Newtonian fields ψ and ζ, recompute the motions of the particles, and so on. It can be shown[1] that this procedure can be kept going; that is, to compute the fields in the Nth approximation we need to know terms in $T^{\mu\nu}$ that satisfy conservation laws that involve the fields only in the $(N-1)$th approximation. We shall content ourselves here with writing the conservation laws that govern terms in $T^{\mu\nu}$ of the next highest order than those appearing in (9.3.2) and (9.3.3). The $\nu = 0$ term of (9.3.1) of order $\overline{M}\bar{v}^3/\bar{r}^4$ and the $\nu = i$ term of (9.3.1) of order $\overline{M}\bar{v}^4/\bar{r}^4$ give

$$\frac{\partial}{\partial t}\overset{2}{T}{}^{00} + \frac{\partial}{\partial x^i}\overset{3}{T}{}^{i0} = -(2\overset{3}{\Gamma}{}^0{}_{00} + \overset{3}{\Gamma}{}^i{}_{i0})\overset{0}{T}{}^{00} - (3\overset{2}{\Gamma}{}^0{}_{0i} + \overset{2}{\Gamma}{}^j{}_{ji})\overset{1}{T}{}^{0i}$$

$$\frac{\partial}{\partial t}\overset{3}{T}{}^{i0} + \frac{\partial}{\partial x^j}\overset{4}{T}{}^{ij} = -\overset{4}{\Gamma}{}^i{}_{00}\overset{0}{T}{}^{00} - \overset{2}{\Gamma}{}^i{}_{00}\overset{2}{T}{}^{00}$$

$$- (2\overset{3}{\Gamma}{}^i{}_{0j} + \delta_{ij}\overset{3}{\Gamma}{}^0{}_{00} + \delta_{ij}\overset{3}{\Gamma}{}^k{}_{0k})\overset{1}{T}{}^{0j}$$

$$- (\overset{2}{\Gamma}{}^i{}_{jk} + \overset{2}{\Gamma}{}^0{}_{0j}\delta_{ik} + \overset{2}{\Gamma}{}^l{}_{lj}\delta_{ik})\overset{2}{T}{}^{jk}$$

or using (9.1.67)–(9.1.72)

$$\frac{\partial}{\partial t} \overset{2}{T}{}^{00} + \frac{\partial}{\partial x^i} \overset{3}{T}{}^{i0} = \overset{0}{T}{}^{00} \frac{\partial \phi}{\partial t} \tag{9.3.4}$$

$$\frac{\partial}{\partial t} \overset{3}{T}{}^{0i} + \frac{\partial}{\partial x^j} \overset{4}{T}{}^{ij} = -\overset{0}{T}{}^{00} \left[\frac{\partial}{\partial x^i}(2\phi^2 + \psi) + \frac{\partial \zeta_i}{\partial t} \right] - \overset{2}{T}{}^{00} \frac{\partial \phi}{\partial x^i}$$

$$- \overset{1}{T}{}^{0j} \left[\frac{\partial \zeta_i}{\partial x^j} - \frac{\partial \zeta_j}{\partial x^i} - 4\,\delta_{ij}\, \frac{\partial \phi}{\partial t} \right]$$

$$- \overset{2}{T}{}^{jk} \left[\delta_{jk} \frac{\partial \phi}{\partial x^i} - 4\delta_{ik} \frac{\partial \phi}{\partial x^j} \right] \tag{9.3.5}$$

As promised, the source terms $\overset{2}{T}{}^{00}$, $\overset{3}{T}{}^{i0}$, and $\overset{4}{T}{}^{ij}$, which would be needed to calculate the *post*-post-Newtonian fields, obey conservation laws that involve the metric only to the post-Newtonian order.

We still need a model with which to calculate the energy-momentum tensor. The simplest such model is that of an assembly of freely falling particles that interact only gravitationally and, perhaps, in localized collisions. From Eq. (5.3.5) we have

$$T^{\mu\nu}(\mathbf{x}, t) = g^{-1/2}(\mathbf{x}, t) \sum_n m_n \frac{dx_n^\mu(t)}{dt} \frac{dx_n^\nu(t)}{dt} \left(\frac{d\tau_n}{dt} \right)^{-1} \delta^3(\mathbf{x} - \mathbf{x}_n(t)) \tag{9.3.6}$$

where m_n, $x_n^\mu(t)$, and τ_n are the mass, space-time position, and proper time of the nth particle, and $-g$ is the determinant of $g_{\mu\nu}$. An elementary calculation using Eq. (4.7.5) gives

$$g = 1 + \overset{2}{g} + \overset{4}{g} + \cdots$$

where $\overset{N}{g}$ is of order $\bar{v}^N$, and in particular

$$\overset{2}{g} = \eta^{\mu\nu} \overset{2}{g}_{\mu\nu} = -\overset{2}{g}_{00} + \overset{2}{g}_{ii} = -4\phi \tag{9.3.7}$$

Using (9.3.7) and (9.2.3) in (9.3.6), we find

$$\overset{0}{T}{}^{00} = \sum_n m_n \delta^3(\mathbf{x} - \mathbf{x}_n) \tag{9.3.8}$$

$$\overset{2}{T}{}^{00} = \sum_n m_n (\phi + \tfrac{1}{2}\mathbf{v}_n^2) \delta^3(\mathbf{x} - \mathbf{x}_n) \tag{9.3.9}$$

$$\overset{1}{T}{}^{i0} = \sum_n m_n v_n^i \delta^3(\mathbf{x} - \mathbf{x}_n) \tag{9.3.10}$$

$$\overset{2}{T}{}^{ij} = \sum_n m_n v_n^i v_n^j \delta^3(\mathbf{x} - \mathbf{x}_n) \tag{9.3.11}$$

where $\mathbf{v}_n \equiv d\mathbf{x}_n/dt$. To impose the conservation laws, we must recall that

$$\frac{\partial}{\partial t} \delta^3(\mathbf{x} - \mathbf{x}_n(t)) = v_n{}^i \frac{\partial}{\partial x_n^i} \delta^3(\mathbf{x} - \mathbf{x}_n(t)) = -\mathbf{v}_n \cdot \nabla \delta^3(\mathbf{x} - \mathbf{x}_n(t))$$

so

$$\frac{\partial}{\partial t} \overset{0}{T}{}^{00} + \frac{\partial}{\partial x^i} \overset{1}{T}{}^{i0} = 0$$

$$\frac{\partial}{\partial t} \overset{1}{T}{}^{0i} + \frac{\partial}{\partial x^i} \overset{2}{T}{}^{ij} = \sum_n m_n \frac{dv_n{}^i}{dt} \delta^3(\mathbf{x} - \mathbf{x}_n)$$

We see that the mass-conservation equation (9.3.2) is automatically satisfied, while the equation (9.3.3) of momentum conservation is satisfied if and only if each particle obeys the Newtonian equation of motion:

$$\frac{d\mathbf{v}_n}{dt} = -\nabla\phi(x_n) \tag{9.3.12}$$

The program for calculating the motion of a set of gravitating point particles is therefore

(A) Solve the Newtonian problem; that is, solve Eqs. (9.3.12) and (9.1.58) for $\phi(x)$ and $\mathbf{x}_n(t)$. (This is the only step that is not always straightforward.)

(B) Use the results of (A) and Eqs. (9.3.8)–(9.3.11) to compute the terms $\overset{0}{T}{}^{00}, \overset{2}{T}{}^{00}, \overset{1}{T}{}^{i0}, \overset{2}{T}{}^{ij}$ of the energy-momentum tensor.

(C) Use the results of (A) and (B), and Eqs. (9.1.62) and (9.1.65), to compute the post-Newtonian fields ζ and ψ.

(D) Use the results of (A) and (C) and Eq. (9.2.1) to calculate the post-Newtonian corrections to the trajectories $\mathbf{x}_n(t)$.

(E) And so on.

4 Multipole Fields

As a first example, let us calculate the gravitational field far away from an arbitrary finite distribution of energy and momentum. Let $T^{\mu\nu}(\mathbf{x}, t)$ vanish for $r > R$, where $r \equiv |\mathbf{x}|$. We may then expand the denominators $|\mathbf{x} - \mathbf{x}'|$ of Eqs. (9.1.59), (9.1.62), and (9.1.65) in inverse powers of r/R

$$|\mathbf{x} - \mathbf{x}'|^{-1} \to \frac{1}{r} + \frac{\mathbf{x} \cdot \mathbf{x}'}{r^3} + \cdots \tag{9.4.1}$$

and we find

$$\phi \to -\frac{G\overset{0}{M}}{r} - \frac{G\mathbf{x}\cdot\overset{0}{\mathbf{D}}}{r^3} + \bigcirc\left(\frac{1}{r^3}\right) \tag{9.4.2}$$

$$\zeta_i \to -\frac{4G\overset{1}{P_i}}{r} - \frac{2Gx^j\overset{1}{J}_{ji}}{r^3} + \bigcirc\left(\frac{1}{r^3}\right) \tag{9.4.3}$$

$$\psi \to -\frac{G\overset{2}{M}}{r} - \frac{G\mathbf{x}\cdot\overset{2}{\mathbf{D}}}{r^3} + \bigcirc\left(\frac{1}{r^3}\right) \tag{9.4.4}$$

where

$$\overset{0}{M} \equiv \int \overset{0}{T}{}^{00}\, d^3x \tag{9.4.5}$$

$$\overset{0}{\mathbf{D}} \equiv \int \mathbf{x}\overset{0}{T}{}^{00}\, d^3x \tag{9.4.6}$$

$$\overset{1}{P^i} \equiv \int \overset{1}{T}{}^{i0}\, d^3x \tag{9.4.7}$$

$$\overset{1}{J}_{ij} \equiv 2 \int x^i\overset{1}{T}{}^{j0}\, d^3x \tag{9.4.8}$$

$$\overset{2}{M} \equiv \int (\overset{2}{T}{}^{00} + \overset{2}{T}{}^{ii})\, d^3x \tag{9.4.9}$$

$$\overset{2}{\mathbf{D}} \equiv \int \mathbf{x}\left(\overset{2}{T}{}^{00} + \overset{2}{T}{}^{ii} + \frac{1}{4\pi G}\frac{\partial^2\phi}{\partial t^2}\right) d^3x \tag{9.4.10}$$

(The term $\partial^2\phi/\partial t^2$ does not contribute to $\overset{2}{M}$, because it equals $-\frac{1}{4}\mathbf{V}\cdot(\partial\zeta/\partial t)$, and hence vanishes upon integration.)

The field ψ has physical effects only through its presence in the expansion of g_{00}:

$$g_{00} = -1 - 2\phi - 2\psi - 2\phi^2 + \bigcirc(\bar{v}^6)$$

Evidently, we can take account of ψ by simply replacing ϕ everywhere with $\psi + \phi$. That is, within the accuracy of the post-Newtonian approximation we may write

$$g_{00} = -1 - 2(\phi + \psi) - 2(\phi + \psi)^2 + \bigcirc(\bar{v}^6) \tag{9.4.11}$$

Equations (9.4.2) and (9.4.4) give the physically significant field $\phi + \psi$ as

$$\phi + \psi \to -\frac{GM}{r} - \frac{G\mathbf{x}\cdot\mathbf{D}}{r^3} + \bigcirc\left(\frac{1}{r^3}\right) \tag{9.4.12}$$

where

$$M \equiv \overset{0}{M} + \overset{2}{M} \qquad \mathbf{D} \equiv \overset{0}{\mathbf{D}} + \overset{2}{\mathbf{D}} \tag{9.4.13}$$

The quantity **D** does not represent an effect of physical importance, but rather just a displacement of the whole field, for (9.4.12) may be written

$$\phi + \psi \rightarrow -\frac{GM}{|\mathbf{x} - \mathbf{D}/M|} + O\left(\frac{1}{r^3}\right) \tag{9.4.14}$$

We could have avoided the **D** term altogether by defining our coordinate system with its origin at the center of energy. On the other hand, the $1/r$ and $1/r^2$ terms in the expansion (9.4.3) for ζ are true physical effects of great interest.

We can derive a number of useful properties of the moments of $T^{\mu\nu}$ by making use of energy and momentum conservation. From the mass-conservation equation (9.3.2) it follows that in general

$$\frac{d\overset{0}{M}}{dt} = 0 \tag{9.4.15}$$

$$\frac{d\overset{0}{\mathbf{D}}}{dt} = \overset{1}{\mathbf{P}} \tag{9.4.16}$$

If the energy-momentum tensor is time independent then (9.3.2) reads

$$\frac{\partial}{\partial x^i} \overset{1}{T^{i0}} = 0$$

and therefore, integrating by parts,

$$0 = \int x^i \frac{\partial}{\partial x^j} \overset{1}{T^{j0}}\, d^3x = -\overset{1}{P^i} \tag{9.4.17}$$

$$0 = 2\int x^i x^j \frac{\partial}{\partial x^k} \overset{1}{T^{k0}}\, d^3x = -\overset{1}{J_{ij}} - \overset{1}{J_{ji}} \tag{9.4.18}$$

The result that $\overset{1}{\mathbf{P}}$ vanishes for a static system is hardly surprising. The result that $\overset{1}{J_{ij}}$ is antisymmetric is not so obvious, and allows us to write it as

$$\overset{1}{J_{ij}} = \varepsilon_{ijk}\overset{1}{J_k} \tag{9.4.19}$$

where $\overset{1}{J_k}$ is the angular momentum vector

$$\overset{1}{J_k} \equiv \tfrac{1}{2}\varepsilon_{ijk}\overset{1}{J_{ij}} = \int d^3x\, \varepsilon_{ijk} x^i \overset{1}{T^{j0}} \tag{9.4.20}$$

Using (9.4.17) and (9.4.19) in (9.4.3) gives

$$\zeta \rightarrow \frac{2G}{r^3}\,(\mathbf{x} \times \mathbf{J}) + O\left(\frac{1}{r^3}\right) \tag{9.4.21}$$

Our results (9.4.14) and (9.4.21) for $\phi + \psi$ and ζ hold generally only far from the gravitating mass. However, they also hold right down to the surface of a spherical distribution of energy and momentum. First suppose that $T^{\mu\nu}(\mathbf{x}, t)$ depends on the position $\mathbf{x}$ only through the radius $r \equiv |\mathbf{x}|$. Then the factor $|\mathbf{x} - \mathbf{x}'|^{-1}$ can be replaced in (9.1.59), (9.1.62), and (9.1.65) with its angular average, which for $r > r'$ is

$$\frac{1}{4\pi} \int \frac{d\Omega}{|\mathbf{x} - \mathbf{x}'|} = \frac{1}{2} \int_0^\pi \frac{\sin \theta \, d\theta}{[r^2 - 2rr' \cos \theta + r'^2]^{1/2}} = \frac{1}{r}$$

Hence *everywhere* outside the sphere the fields are

$$\phi = -\frac{G\overset{0}{M}}{r} \tag{9.4.22}$$

$$\zeta = -4G\frac{\overset{1}{\mathbf{P}}}{r} \tag{9.4.23}$$

$$\psi = -\frac{G\overset{2}{M}}{r} \tag{9.4.24}$$

If the sphere is at rest, then $\overset{1}{\mathbf{P}}$ vanishes; in this case (9.1.57), (9.1.60), (9.1.61), (9.1.63), and (9.4.13) give the metric as

$$g_{00} \simeq -1 + \frac{2MG}{r} - \frac{2M^2G^2}{r^2} \tag{9.4.25}$$

$$g_{i0} \simeq 0 \tag{9.4.26}$$

$$g_{ij} \simeq \delta_{ij} + 2\delta_{ij}\frac{MG}{r} \tag{9.4.27}$$

This is in agreement with the exact Schwarzschild solution, given in harmonic coordinates by Eq. (8.2.15):

$$g_{00} = -\frac{1 - MG/r}{1 + MG/r}$$

$$g_{i0} = 0$$

$$g_{ij} = \left(1 + \frac{MG}{r}\right)^2 \delta_{ij} + \left(\frac{MG}{r}\right)^2 \frac{1 + MG/r}{1 - MG/r}\left(\frac{x^i x^j}{r^2}\right)$$

However, there is an important difference in the two derivations, in that the exact Schwarzschild solution was derived in Section 7.2 for a static spherically symmetric system, while the post-Newtonian solution is valid for a system that can vary over times of order $\bar{r}/\bar{v}$. It will be shown in Section 11.7 that the Schwarzschild solution is actually valid outside any spherically symmetric system, static or not.

Now consider a system that is at rest and spherically symmetric, but rotates with angular frequency $\omega(r)$. The momentum density is now given by

$$\overset{1}{T}{}^{i0}(\mathbf{x}', t) = \overset{0}{T}{}^{00}(r')[\omega(r') \times \mathbf{x}']_i \tag{9.4.28}$$

Equation (9.1.62) then gives the field ζ as

$$\zeta(\mathbf{x}) = -4G \int \frac{[\omega(r') \times \mathbf{x}']}{|\mathbf{x} - \mathbf{x}'|} \overset{0}{T}{}^{00}(r') \, d^4x' \tag{9.4.29}$$

The solid-angle integral is now

$$\int \frac{d\Omega' \mathbf{x}'}{|\mathbf{x} - \mathbf{x}'|} = \begin{cases} \left(\dfrac{4\pi r'^2}{3r^3}\right) \mathbf{x} & \text{for } r' < r \tag{9.4.30} \\[2ex] \left(\dfrac{4\pi}{3r'}\right) \mathbf{x} & \text{for } r' > r \tag{9.4.31} \end{cases}$$

Thus, the field outside the sphere is

$$\zeta(\mathbf{x}) = \frac{16\pi G}{3r^3} \left[\mathbf{x} \times \int \omega(r')\overset{0}{T}{}^{00}(r')r'^4 \, dr' \right] \tag{9.4.32}$$

The integral may be expressed in terms of the angular momentum, given by Eqs. (9.4.20) and (9.4.28) as

$$\mathbf{J} = \int (\mathbf{x}' \times [\omega(r') \times \mathbf{x}'])\overset{0}{T}{}^{00}(r') \, d^4x'$$

$$= \int [r'^2\omega(r') - \mathbf{x}'(\mathbf{x}' \cdot \omega(r'))]\overset{0}{T}{}^{00}(r') \, d^3x'$$

$$= \frac{8\pi}{3} \int \omega(r')\overset{0}{T}{}^{00}(r')r'^4 \, dr' \tag{9.4.33}$$

Thus (9.4.32) gives, everywhere outside the sphere,

$$\zeta(\mathbf{x}) = \frac{2G}{r^3} (\mathbf{x} \times \mathbf{J}) \tag{9.4.34}$$

in agreement with the general asymptotic formula (9.4.21). The field *inside* a hollow spinning sphere is given by (9.4.29) and (9.4.31) as

$$\zeta(\mathbf{x}) = \mathbf{x} \times \mathbf{\Omega} \tag{9.4.35}$$

where

$$\mathbf{\Omega} \equiv \frac{16\pi G}{3} \int \omega(r')\overset{0}{T}{}^{00}(r')r' \, dr' \tag{9.4.36}$$

The implications of this result for Mach's principle are discussed in Section 9.7.

5   Precession of Perihelia

We shall now see how the post-Newtonian formalism developed in the last four sections can be used to calculate the precession of planetary orbits in the actual solar system, taking into account other planets, solar rotation, solar oblateness, and so on. The potential $\phi + \psi$ that determines g_{00} [see Eq. (9.4.11)] is overwhelmingly dominated by the spherically symmetric part $-GM_\odot/r$ of the sun's contribution, so it is convenient to write

$$\phi + \psi \equiv -\frac{GM_\odot}{r} + \varepsilon(\mathbf{x}, t) \tag{9.5.1}$$

with ε including not only the Newtonian potentials of the other planets but also any quadrupole or higher terms in the sun's contribution to $\phi + \psi$. The equation of motion (9.2.1) of a point particle now reads

$$\frac{d\mathbf{v}}{dt} = -\frac{GM_\odot \mathbf{x}}{r^3} + \boldsymbol{\eta} + O(\bar{v}^6) \tag{9.5.2}$$

with $\boldsymbol{\eta}$ a small perturbation:

$$\boldsymbol{\eta} = -\boldsymbol{\nabla}(\varepsilon + 2\phi^2) - \frac{\partial \boldsymbol{\zeta}}{\partial t} + \mathbf{v} \times (\boldsymbol{\nabla} \times \boldsymbol{\zeta})$$

$$+ 3\mathbf{v}\frac{\partial \phi}{\partial t} + 4\mathbf{v}(\mathbf{v} \cdot \boldsymbol{\nabla})\phi - v^2 \boldsymbol{\nabla}\phi \tag{9.5.3}$$

By far the most convenient technique for calculating the precession of perihelia is to compute the rate of change of the *Runge-Lenz vector*

$$\mathbf{A} = -M_\odot G \frac{\mathbf{x}}{r} + (\mathbf{v} \times \mathbf{h}) \tag{9.5.4}$$

Here $r \equiv |\mathbf{x}|$, $\mathbf{v} \equiv d\mathbf{x}/dt$, and $\mathbf{h}$ is the orbital angular momentum per unit mass:

$$\mathbf{h} \equiv \mathbf{x} \times \mathbf{v} \tag{9.5.5}$$

If the perturbation $\boldsymbol{\eta}$ in Eq. (9.5.2) were absent, then the orbit would be an ellipse, described by the familiar formulas

$$r = \frac{L}{1 + e\cos(\varphi - \varphi_0)} \tag{9.5.6}$$

$$\frac{d\varphi}{dt} = \frac{\sqrt{LM_\odot G}}{r^2} \tag{9.5.7}$$

$$\frac{dr}{dt} = e\sqrt{\frac{M_\odot G}{L}}\sin(\varphi - \varphi_0) \tag{9.5.8}$$

with e the eccentricity and L the semilatus rectum. (See Section 8.6. We are taking the orbit to lie in the plane $\theta = \pi/2$, with perihelia at an azimuthal angle φ_0.) Then $\mathbf{h}$ would be a constant vector normal to the orbit, with magnitude

$$|\mathbf{h}| = \sqrt{LM_\odot G} \tag{9.5.9}$$

and $\mathbf{A}$ would be a constant vector pointing toward *perihelion*, with magnitude

$$|\mathbf{A}| = eM_\odot G \tag{9.5.10}$$

Thus, the rate of precession of perihelia $d\varphi_0/dt$ caused by any perturbation is just the component of the change $d\hat{\mathbf{A}}/dt$ in the unit vector $\hat{\mathbf{A}} \equiv \mathbf{A}/|\mathbf{A}|$ along a direction perpendicular to both $\mathbf{A}$ and $\mathbf{h}$, that is

$$\frac{d\varphi_0}{dt} = (\hat{\mathbf{h}} \times \hat{\mathbf{A}}) \cdot \frac{d\hat{\mathbf{A}}}{dt} = (\mathbf{h} \times \mathbf{A}) \cdot \frac{\dfrac{d\mathbf{A}}{dt}}{|\mathbf{h}|\mathbf{A}^2} \tag{9.5.11}$$

(If $d\varphi_0/dt$ is positive, then the precession is in the same sense as the direction of the planet's motion.) A straightforward calculation gives for the rate of change of $\mathbf{A}$ produced by the perturbation $\boldsymbol{\eta}$ in (9.5.2) the value

$$\frac{d\mathbf{A}}{dt} = \boldsymbol{\eta} \times \mathbf{h} + \mathbf{v} \times (\mathbf{x} \times \boldsymbol{\eta}) \tag{9.5.12}$$

Note that $d\mathbf{A}/dt$ and hence $d\varphi_0/dt$ is linear in $\boldsymbol{\eta}$, so $d\varphi_0/dt$ is correctly calculated by adding up the precessions produced by each small term in $\boldsymbol{\eta}$.

The largest term in $\boldsymbol{\eta}$ is the part of $-\nabla\varepsilon$ arising from the Newtonian potentials of the other planets. We shall make no attempt to calculate this term; the experts tell us that it produces a precession $d\varphi_0/dt$, which for Mercury is about $532''$ per century. (See Section 8.6.) The next largest term is obtained from the relativistic corrections in Eq. (9.5.3), setting ϕ and ζ equal to the values they would have for a spherical nonrotating sun:

$$\phi_\odot = -\frac{GM_\odot}{r} \qquad \zeta_\odot = 0 \tag{9.5.13}$$

Then (9.5.3) gives

$$\boldsymbol{\eta} = -2\nabla\phi_\odot^2 + 4\mathbf{v}(\mathbf{v} \cdot \nabla)\phi_\odot - \mathbf{v}^2\nabla\phi_\odot \tag{9.5.14}$$

Using (9.5.12)–(9.5.14) and (9.5.6)–(9.5.10) in (9.5.11) gives the precession as

$$\frac{d\varphi_0}{dt} = 8M_\odot GhL^{-3}[1 + e\cos(\varphi - \varphi_0)]^3 \sin^2(\varphi - \varphi_0) - M_\odot Ge^{-1}hL^{-3}$$

$$\times \{7[1 + e\cos(\varphi - \varphi_0)]^2 + 4[1 + e\cos(\varphi - \varphi_0)]^3$$
$$+ [1 + e\cos(\varphi - \varphi_0)]^4\}\cos(\varphi - \varphi_0) \tag{9.5.15}$$

Since φ_0 changes slowly, the change in φ_0 in one revolution can be determined by integrating $d\varphi_0/dt$ over one period, keeping φ_0 fixed in the integrand, and using for $d\varphi/dt$ the Keplerian formulas (9.5.6)–(9.5.10). This gives for the precession per revolution

$$\Delta\varphi_0 = \int_0^{2\pi} \frac{d\varphi_0}{dt} \frac{dt}{d\varphi} \, d\varphi = \frac{L^2}{h} \int_0^{2\pi} \frac{d\varphi_0}{dt} [1 + e \cos (\varphi - \varphi_0)]^{-2} \, d\varphi \qquad (9.5.16)$$

Most terms drop out on performing the angular integration, and we are left with

$$\Delta\varphi_0 = 6\pi \frac{M_\odot G}{L} \qquad (9.5.17)$$

in agreement with our earlier result, Eq. (8.6.11).

As an example of another small term in the precession, let us calculate the effect of the field ζ produced by the sun's rotation. According to (9.4.34), this field is

$$\zeta = \frac{2G}{r^3} (\mathbf{x} \times \mathbf{J}_\odot) \qquad (9.5.18)$$

This contributes to the acceleration $d\mathbf{v}/dt$ an amount given by (9.5.3) as

$$\boldsymbol{\eta} = \mathbf{v} \times (\nabla \times \zeta) = 6Gh(\mathbf{x} \cdot \mathbf{J}_\odot)r^{-5} + 2G(\mathbf{v} \times \mathbf{J}_\odot)r^{-3} \qquad (9.5.19)$$

and (9.5.12) tells us that this causes $\mathbf{A}$ to change at a rate

$$\frac{d\mathbf{A}}{dt} = -6Gh(\mathbf{v} \cdot \mathbf{x})(\mathbf{x} \cdot \mathbf{J}_\odot)r^{-5} - 2G(\mathbf{v} \times \mathbf{J}_\odot)(\mathbf{x} \cdot \mathbf{v})r^{-3} - 2G\mathbf{v}(h \cdot \mathbf{J}_\odot)r^{-3} \qquad (9.5.20)$$

For simplicity we will take the sun's axis of rotation to be normal to the plane of the planet's orbit, so that $\mathbf{J}_\odot$ is parallel to $\mathbf{h}$. Using (9.5.20) and (9.5.6)–(9.5.10) in (9.5.11) gives for the rate of precession

$$\frac{d\varphi_0}{dt} = \frac{2J_\odot h^2}{M_\odot L^4 e} \{-[1 + e \cos (\varphi - \varphi_0)]^2 e \sin^2 (\varphi - \varphi_0)$$
$$- [1 + e \cos (\varphi - \varphi_0)]^3 [e + \cos (\varphi - \varphi_0)]\} \qquad (9.5.21)$$

and (9.5.16) then gives for the precession per revolution

$$\Delta\varphi_0 = \frac{-8\pi J_\odot h}{M_\odot L^2} \qquad (9.5.22)$$

The sun is generally supposed to have an angular momentum $J_\odot \simeq 1.7 \times 10^{48}$ g cm^2 sec^{-1}, and its mass is $M_\odot = 1.99 \times 10^{33}$ g, so in our natural units with 1 sec $= 3 \times 10^{10}$ cm we have

$$\frac{J_\odot}{M_\odot} \simeq 0.28 \text{ km}$$

Also, the orbit of Mercury has $L = 55.5 \times 10^6$ km and $h = 9.03 \times 10^3$ km, so the field ζ would contribute to the precession of Mercury's perihelia an amount

$$\Delta\varphi_0 \simeq -2.06 \times 10^{-11} \text{ radians/revolution}$$

or in more conventional terms

$$\Delta\varphi_0 \simeq -17.6 \times 10^{-4} \text{ arc-sec/century}$$

Even if Dicke and Goldenberg are right, and the sun has an angular momentum 25 times larger than generally believed, the precession caused by ζ is still only of order $0.04''$ per century, too small to be measured.

Perhaps it should be stressed again that the total precession is to be calculated by adding the Newtonian term $532''$ per century, plus the Einstein term (9.5.17), plus the ζ term (9.5.22), plus a Newtonian term arising from any oblateness of the sun, plus a term arising from the contribution of the sun's rotation to the anisotropic part of ψ, plus terms arising from post-Newtonian corrections to the perturbations caused by other planets. Only the Newtonian terms and the Einstein term (9.5.17) are large enough to be measured.

6   Precession of Orbiting Gyroscopes

We saw in Section 5.1 that the spin S_μ of a particle in free fall precesses according to the equation of parallel transport:

$$\frac{dS_\mu}{d\tau} = \Gamma^\lambda{}_{\mu\nu} S_\lambda \frac{dx^\nu}{d\tau} \tag{9.6.1}$$

A few years ago Pugh[2] and Schiff[3] suggested that a gyroscope might be placed in orbit around the earth, and the precession of its spin vector be used to measure the fine details of the earth's gravitational field. Schiff made use of a calculational method developed by Papapetrou[4] and Fock[5] in which the motion is first calculated for an extended spinning body and then evaluated in the limit as the body size goes to zero. We shall instead treat the gyroscope as a point particle from the beginning, because for such particles the Principle of Equivalence tells us that there is a locally inertial frame of reference in which the spin does *not* precess, and we can use Eq. (9.6.1) as the translation of this statement into a general coordinate system.

The spin four-vector S_μ is defined to remain orthogonal to the velocity $dx^\mu/d\tau$:

$$\frac{dx^\mu}{d\tau} S_\mu = 0$$

[See Eq. (5.1.9).] In other words

$$S_0 = -v^i S_i \tag{9.6.2}$$

We set $\mu = i$ in (9.6.1), multiply with $d\tau/dt$, and use (9.6.2) to eliminate S_0; this gives

$$\frac{dS_i}{dt} = \Gamma^j{}_{i0} S_j - \Gamma^0{}_{i0} v^j S_j + \Gamma^j{}_{ik} v^k S_j - \Gamma^0{}_{ik} v^k v^j S_j \tag{9.6.3}$$

The post-Newtonian approximation will allow us to evaluate the coefficient of S_j on the right-hand side to order $\bar{v}^3/\bar{r}$:

$$\frac{dS_i}{dt} \simeq [\overset{3}{\Gamma}{}^j{}_{i0} - \overset{2}{\Gamma}{}^0{}_{i0} v^j + \overset{2}{\Gamma}{}^j{}_{ik} v^k] S_j \tag{9.6.4}$$

(The last term drops out because there is no $\overset{1}{\Gamma}{}^0{}_{ik}$.) The components of the affine connection are provided by Eqs. (9.1.69), (9.1.70), and (9.1.72). We find

$$\frac{d\mathbf{S}}{dt} \simeq \tfrac{1}{2}\mathbf{S} \times (\mathbf{\nabla} \times \boldsymbol{\zeta}) - \mathbf{S}\frac{\partial\phi}{\partial t} - 2(\mathbf{v} \cdot \mathbf{S})\mathbf{\nabla}\phi - \mathbf{S}(\mathbf{v} \cdot \mathbf{\nabla}\phi) + \mathbf{v}(\mathbf{S} \cdot \mathbf{\nabla}\phi) \tag{9.6.5}$$

To solve (9.6.5), we make use of the fact that parallel transport preserves the value of $S_\mu S^\mu$, so that [see Eq. (5.1.10)]

$$\frac{d}{dt}(g^{\mu\nu} S_\mu S_\nu) = 0 \tag{9.6.6}$$

The rate of change of $\mathbf{S}$ is seen from (9.6.4) to be of order $\mathbf{S}$ times $\bar{v}^3/\bar{r}$, so we need only keep those terms in $g^{\mu\nu} - \eta^{\mu\nu}$ whose rate of change is comparable as seen by a particle moving with velocity $\bar{v}$, that is, those terms whose gradient is of order $(\bar{v}^2/\bar{r})$. Hence $g^{\mu\nu}$ may be replaced in Eq. (9.6.6) with $\eta^{\mu\nu} + \overset{2}{g}{}^{\mu\nu}$. Furthermore $(S_0)^2$ is already of order $\bar{v}^2$ with respect to $\mathbf{S}^2$, so we need not keep $\overset{2}{g}{}^{00}$. Thus, to the order needed here, we expect that (9.6.5) will have the integral

$$\mathbf{S}^2 + 2\phi\mathbf{S}^2 - (\mathbf{v} \cdot \mathbf{S})^2 = \text{constant} \tag{9.6.7}$$

This suggests that we should introduce a new spin vector $\mathscr{S}$ by

$$\mathbf{S} = (1 - \phi)\mathscr{S} + \tfrac{1}{2}\mathbf{v}(\mathbf{v} \cdot \mathscr{S}) \tag{9.6.8}$$

so that to order $\bar{v}^2 S^2$, (9.6.5) reads

$$\mathscr{S}^2 = \text{constant} \tag{9.6.9}$$

To the required order, we can invert (9.6.8) to read

$$\mathscr{S} = (1 + \phi)\mathbf{S} - \tfrac{1}{2}\mathbf{v}(\mathbf{v} \cdot \mathbf{S}) \tag{9.6.10}$$

The rate of change of $\mathscr{S}$ is given to order $(\bar{v}^3/\bar{r})S$ by treating $\mathbf{S}$ as constant everywhere it appears with coefficients of order $\bar{v}^2$, and setting $d\mathbf{v}/dt \simeq -\nabla\phi$. We then find

$$\frac{d\mathscr{S}}{dt} = \frac{d\mathbf{S}}{dt} + \mathbf{S}\left(\frac{\partial\phi}{\partial t} + \mathbf{v}\cdot\nabla\phi\right) + \tfrac{1}{2}\nabla\phi(\mathbf{v}\cdot\mathbf{S}) + \tfrac{1}{2}\mathbf{v}(\mathbf{S}\cdot\nabla\phi)$$

and inserting (9.6.5), we find to order $(\bar{v}^3/\bar{r})\mathscr{S}$:

$$\frac{d\mathscr{S}}{dt} = \boldsymbol{\Omega} \times \mathscr{S} \tag{9.6.11}$$

where

$$\boldsymbol{\Omega} = -\tfrac{1}{2}\nabla \times \boldsymbol{\zeta} - \tfrac{3}{2}\mathbf{v} \times \nabla\phi \tag{9.6.12}$$

Eq. (9.6.11) shows that $\mathscr{S}$ just precesses at a rate $|\boldsymbol{\Omega}|$ around the direction of $\boldsymbol{\Omega}$, with no change in magnitude, thus verifying (9.6.9).

What does this have to do with the measurement of the precession of a gyroscope in free fall? The answer as always is to be sought by reference to the method actually used to measure the effect. In the present case the spin direction of the gyroscope is monitored by measuring, *in the inertial frame moving with the gyroscope*, the angles θ between the spin $\mathbf{S}_g$ of the gyroscope in this frame and the velocity vectors $\mathbf{u}_g$ of light rays from one or more distant stars:

$$\cos\theta = \mathbf{S}_g \cdot \frac{\mathbf{u}_g}{|\mathbf{S}_g|\,|\mathbf{u}_g|} \tag{9.6.13}$$

(This angle can be measured by focusing the star's image on an array of photoelectric cells fixed to the gyroscope, in such a way that a change in θ causes the image to move over the cells, producing a change in the photoelectric current.) In the inertial frame of the gyroscope, light moves with unit velocity

$$|\mathbf{u}_g| = 1$$

the time component of $S_{g\mu}$ vanishes [see Eq. (9.6.2)]

$$S_{g0} = 0$$

and the vector $\mathbf{S}_g$ has constant magnitude

$$|\mathbf{S}_g| = (S_{g\mu}S_g{}^\mu)^{1/2}$$

Thus, the measured angles θ can be expressed in the form

$$\cos\theta = \frac{S_\mu u^\mu}{(S_\mu S^\mu)^{1/2}} \tag{9.6.14}$$

This is now an invariant, so we are no longer restricted to the rather inconvenient inertial coordinate system fixed to the gyroscope, but can use for S_μ and u^μ the spin and light-velocity four-vectors in any convenient coordinate system, such as the reference frame fixed to the earth. In this frame we have for the velocity four-vector of the starlight ray:

$$u^i = u^i_\infty + \delta u^i$$
$$u^0 = 1 + \delta u^0$$

where $\mathbf{u}_\infty$ is a fixed unit vector giving the light velocity far from the earth, and δu^μ is a correction of order $M_\oplus G/r \sim \bar{v}^2$, arising from the effect of the earth's gravitational field on the speed and direction of light rays. Also, Eqs. (9.6.10) and (9.6.2) give, to order $\bar{v}^2$,

$$S_i = \mathscr{S}_i - \phi\mathscr{S}_i + \tfrac{1}{2}v_i(\mathbf{v} \cdot \mathscr{S}) + \mathrm{O}(\bar{v}^4)$$
$$S_0 = -\mathbf{v} \cdot \mathscr{S} + \mathrm{O}(\bar{v}^3)$$

Thus (9.6.14) gives the measured angle θ as

$$\cos\theta \simeq \hat{\mathscr{S}} \cdot [\mathbf{u}_\infty - \mathbf{v} + \delta\mathbf{u} - \phi\mathbf{u}_\infty + \tfrac{1}{2}\mathbf{v}(\mathbf{v} \cdot \mathbf{u}_\infty)] \tag{9.6.15}$$

where $\hat{\mathscr{S}} = \mathscr{S}/|\mathscr{S}|$. The term $-\mathbf{v}$ represents the *aberration of starlight*, an important effect known since the eighteenth century, which certainly must be taken into account. Apart from this term, $\cos\theta$ evidently changes with time because $\delta\mathbf{u}$, ϕ, and $\mathbf{v}$ change as the gyroscope revolves about the earth, and also because $\hat{\mathscr{S}}$ precesses with angular frequency $\mathbf{\Omega}$. Indeed, these fractional changes in $\cos\theta$ produced by the variations in each of $\delta\mathbf{u}$, ϕ, $\mathbf{v}$, and $\mathscr{S}$ in the course of one revolution are, aside from aberration, all of order $\bar{v}^2$ [see Eqs. (8.5.8) and (9.6.12)], so in order to measure the precession of $\mathscr{S}$ in one revolution of the gyroscope it would be necessary to measure θ to an accuracy of order 10^{-10} radians, and even then we would have to disentangle the effect of the bending $\delta\mathbf{u}$ of starlight and the other terms in (9.6.15) in order to interpret the result as a spin precession. Fortunately, there is one property of the spin precession that distinguishes it from all other effects: it is *cumulative*. After a large number N of revolutions, the spin direction $\hat{\mathscr{S}}$ will have changed by an amount of order $N\bar{v}^2$, while $\delta\mathbf{u}$, ϕ, and $\mathbf{v}(\mathbf{v} \cdot \mathbf{u}_\infty)$ will still be of order $\bar{v}^2$, so to a good approximation the change in θ will, after aberration is taken into account, be just given by the change in $\hat{\mathscr{S}}$:

$$\Delta(\cos\theta) \simeq \mathbf{u}_\infty \cdot \Delta\hat{\mathscr{S}} \tag{9.6.16}$$

Our conclusion then is that the precession $\mathbf{\Omega}$ of $\mathscr{S}$ is a directly measurable effect, provided we have the patience to wait for the gyroscope to complete many revolutions about the earth.

Returning now to the problem of calculating $\mathbf{\Omega}$: if we regard the earth as a rotating sphere at rest, the fields ζ and ϕ can be taken from (9.4.34) and (9.4.22):

$$\phi = -\frac{GM_\oplus}{r} \qquad \zeta = \frac{2G}{r^3}(\mathbf{x} \times \mathbf{J}_\oplus)$$

The precession frequency (9.6.12) is therefore

$$\boldsymbol{\Omega} = 3G\mathbf{x}(\mathbf{x} \cdot \mathbf{J}_{\oplus})r^{-5} - G\mathbf{J}_{\oplus}r^{-3} + \frac{3GM_{\oplus}(\mathbf{x} \times \mathbf{v})}{2r^3} \qquad (9.6.17)$$

The last term, which depends only upon the mass of the earth but not its spin, is called the *geodetic precession*;[6] it is essentially just the Thomas precession caused by gravitation. (See Section 5.1.) The first two terms represent an interaction between the spin orbital angular momenta of the earth and the gyroscope, analogous to the hyperfine interaction of atomic physics. If for simplicity we take the gyroscope's orbit to be a circle of radius r with unit normal $\hat{\mathbf{h}}$, then the gyroscope's velocity is

$$\mathbf{v} = -\left(\frac{M_{\oplus}G}{r^3}\right)^{1/2} (\mathbf{x} \times \hat{\mathbf{h}}) \qquad (9.6.18)$$

and the precession rate, averaged over a revolution, is

$$\langle \boldsymbol{\Omega} \rangle = \frac{(\mathbf{J}_{\oplus} - \hat{\mathbf{h}}(\hat{\mathbf{h}} \cdot \mathbf{J}_{\oplus}))G}{2r^3} + 3(M_{\oplus}G)^{3/2}\frac{\hat{\mathbf{h}}}{2r^{5/2}} \qquad (9.6.19)$$

Both terms are maximized by taking r as small as possible, that is, about equal to the radius $R_{\oplus}$ of the earth. At these low altitudes the ratio of the first "hyperfine" term to the second "geodetic" term is of order

$$\frac{\text{hyperfine}}{\text{geodetic}} \approx \frac{J_{\oplus}G}{3(M_{\oplus}G)^{3/2}R_{\oplus}^{1/2}} = 6.5 \times 10^{-3} \qquad (9.6.20)$$

so the main effect is a precession of the spin around the orbital angular momentum $\mathbf{h}$, with average rate

$$\langle |\boldsymbol{\Omega}| \rangle \simeq \frac{3(M_{\oplus}G)^{3/2}}{2r^{5/2}} \simeq 8.4\left(\frac{R_{\oplus}}{r}\right)^{5/2} \text{sec/year} \qquad (9.6.21)$$

This should be measurable.[7] In order to detect the small "hyperfine" precession, one might direct the spin axis of the gyroscope along the direction $\hat{\mathbf{h}}$ normal to the plane of the orbit; in this case the terms in $\boldsymbol{\Omega}$ parallel to $\hat{\mathbf{h}}$ have no effect [see Eq. (9.6.11)] so the effective precession is just around $\mathbf{J}_{\oplus}$:

$$\langle \boldsymbol{\Omega} \rangle_{\text{eff}} = \frac{G\mathbf{J}_{\oplus}}{2r^3} \qquad (9.6.22)$$

with magnitude

$$|\langle \boldsymbol{\Omega} \rangle_{\text{eff}}| = 0.055\left(\frac{R_{\oplus}}{r}\right)^3 \text{sec/year} \qquad (9.6.23)$$

In order to maximize the effect of this tiny precession, one would like to have the spin axis of the gyroscope perpendicular to $\mathbf{J}_{\oplus}$; since it must also be perpendicular to the plane of the orbit, the best arrangement would be to place the gyroscope in a polar orbit, with its spin axis parallel to the equatorial plane of the earth.

As usual, the effect of putting the satellite in an eccentric orbit would simply be to replace the radius r everywhere with the semilatus rectum L. It is also easy to take into account the effect of a possible departure from Einstein's field equations. The Robertson expansion (8.3.1) for a general static spherically symmetric metric in isotropic coordinates gives (with $\alpha \equiv 1$)

$$\overset{2}{g_{00}} = -2\phi \qquad \overset{2}{g_{ij}} = -2\gamma\phi\delta_{ij} \qquad \overset{3}{g_{i0}} = 0$$

where ϕ as usual is $-GM/r$ and γ is a dimensionless constant that in Einstein's theory would be unity. Referring back to (9.1.18), (9.1.19), and (9.1.21), we see that now

$$\overset{2}{\Gamma}{}^{j}_{ik} = \gamma\left[-\frac{\partial\phi}{\partial x^k}\delta_{ij} - \frac{\partial\phi}{\partial x^i}\delta_{jk} + \frac{\partial\phi}{\partial x^j}\delta_{ik}\right]$$

$$\overset{2}{\Gamma}{}^{0}_{i0} = \frac{\partial\phi}{\partial x^i}$$

$$\overset{3}{\Gamma}{}^{j}_{i0} = 0$$

Using these in (9.6.4) gives for the rate of change of spin

$$\frac{d\mathbf{S}}{dt} = -(1 + \gamma)(\mathbf{v}\cdot\mathbf{S})\nabla\phi - \gamma(\mathbf{v}\cdot\nabla\phi)\mathbf{S} + \gamma(\mathbf{S}\cdot\nabla\phi)\mathbf{v}$$

As before, it is convenient to introduce a spin vector of constant magnitude, which now is

$$\mathscr{S} \equiv (1 + \gamma\phi)\mathbf{S} - \tfrac{1}{2}\mathbf{v}(\mathbf{v}\cdot\mathbf{S})$$

Again, $\mathscr{S}$ just precesses about a vector $\mathbf{\Omega}$

$$\frac{d\mathscr{S}}{dt} = \mathbf{\Omega}\times\mathscr{S}$$

but now $\mathbf{\Omega}$ is given by

$$\mathbf{\Omega} = -(\tfrac{1}{2} + \gamma)(\mathbf{v}\times\nabla\phi) \tag{9.6.24}$$

Thus, the effect of a modification of Einstein's field equations on the geodetic precession is simply to multiply it with a factor

$$\frac{(1 + 2\gamma)}{3}$$

In order to calculate the effects on $\mathbf{\Omega}$ of a modification of Einstein's field equations in a system that is not static and spherically symmetric, it would be necessary to know the details of the new theory; we shall return to this problem in Section 9.9.

7   Spin Precession and Mach's Principle*

The spin precession effects calculated in the last section have a remarkable interpretation in terms of the ideas of Ernst Mach discussed in Section 1.3. Recall that the spin of a freely falling gyroscope does *not* precess in an inertial coordinate system that moves along with the gyroscope; this after all is just the meaning of Eq. (9.6.1) of parallel transport. Thus, the precession $\mathbf{\Omega}$ of a gyroscope in another frame, say one fixed to the earth, arises entirely from a rotation of the inertial frame carried by the gyroscope with respect to earth and the distant stars, with angular frequency $\mathbf{\Omega}$. This is why $\mathbf{\Omega}$ does not depend on the rate of the gyroscope's spin; *any* vector that keeps a fixed direction in an inertial system will appear to precess in the "lab" or earth system with angular frequency $\mathbf{\Omega}$ given by Eq. (9.6.12).

Why should the inertial frame that falls with the gyroscope rotate with respect to the distant stars? Mach tells us that inertial forces arise from accelerations, including rotations, with respect to the total matter of the universe, so a reference frame will be inertial if it is not accelerating with respect to some average cosmic distribution of matter. Normally this means that the inertial frames do not rotate with respect to the distant stars. However, an observer on a gyroscope orbiting the earth sees a mass distribution consisting, not only of the distant stars, but also of a large sphere called the earth, which appears to revolve around the gyroscope once every 90 minutes or so, and which also rotates on its own axis. Thus the inertial frames fixed to the gyroscope have to reach some sort of compromise between following the distant stars and following the earth; it tries to rotate in the same direction as the rotation and apparent revolution of the earth, but lags far behind, the distant stars always winning the struggle.

The rather vague ideas of this sort that are suggested by Mach's principle find their concrete expression in detailed calculations based on the Principle of Equivalence. We saw in Eq. (9.6.19) that the precession of an orbiting gyroscope, and hence the rotation of the inertial frame it carries, consists of a small "geodetic" term parallel to the orbital angular momentum $\mathbf{h}$, and an even smaller "hyperfine" term parallel to the component of the earth's spin $\mathbf{J}_{\oplus}$ perpendicular to $\mathbf{h}$. Thus the rotation and apparent revolution of the earth about the gyroscope do seem slightly to drag along the inertial frame that falls with the gyroscope.

This effect may be seen more clearly in a thought-experiment discussed by Lense and Thirring[8] shortly after the advent of general relativity. They considered a hollow spherical shell that rotates rigidly with angular velocity ω. According to Eq. (9.4.35), the metric field ζ inside the sphere is

$$\zeta = \mathbf{x} \times \mathbf{\Omega}$$

where

$$\mathbf{\Omega} = -4\phi \frac{\omega}{3}$$

---

* This section lies somewhat out of the book's main line of development, and may be omitted in a first reading.

with ϕ the constant gravitational potential inside the sphere

$$\phi = -4\pi G \int_{\text{shell}}^{0} T^{00}(r')r' \, dr'$$

Equation (9.6.12) then tells us that any inertial frame inside the sphere rotates with angular velocity $\boldsymbol{\Omega}$.

We note that $\boldsymbol{\Omega}$ is parallel to $\boldsymbol{\omega}$, but smaller by the dimensionless factor $-4\phi/3$. It is therefore interesting to ask what would happen if the shell were made so massive that ϕ approached a value of order $-\frac{3}{4}$. Would the inertial frames inside the shell decouple entirely from the distant stars and follow the shell, rotating with frequency $\boldsymbol{\omega}$? (We catch an echo of Mach's remark about Newton's water bucket experiment quoted in Section 1.3, "No one is competent to say how the experiment would turn out if the sides of the vessel increased in thickness and mass until they were several leagues thick.") Unfortunately, the post-Newtonian method breaks down just when this problem becomes interesting, when $|\phi|$ is of order unity. An exact solution of Einstein's equations that *looks* like the metric outside a spinning sphere has been found by Kerr;[9] it is of the form

$$-d\tau^2 = -dt^2 + d\mathbf{x}^2 + \frac{2MG\rho}{(\rho^4 + (\mathbf{x} \cdot \mathbf{a})^2)(\rho^2 + \mathbf{a}^2)^2}$$
$$\times [\rho^2 \mathbf{x} \cdot d\mathbf{x} + \rho \, d\mathbf{x} \cdot (\mathbf{a} \times \mathbf{x}) + (\mathbf{a} \cdot \mathbf{x})(\mathbf{a} \cdot d\mathbf{x}) + (\rho^2 + \mathbf{a}^2)\rho \, dt]^2$$

where $\mathbf{x}$ is a quasi-Euclidean three-vector; $\mathbf{a}$ is a constant vector; scalar products $\mathbf{x} \cdot \mathbf{a}$, $\mathbf{x}^2$, and so on, are defined as in Euclidean geometry; and ρ is defined by

$$\rho^4 - (r^2 - \mathbf{a}^2)\rho^2 - (\mathbf{a} \cdot \mathbf{x})^2 = 0$$

where as usual, $r^2 \equiv \mathbf{x}^2$. For $r \to \infty$, we have $\rho \to r$, and the metric coefficients become

$$g_{00} \to -1 + \frac{2MG}{r} + O\left(\frac{1}{r^2}\right)$$

$$g_{0i} \to \frac{2MG}{r^2}\left\{x_i + \frac{1}{r}(\mathbf{a} \times \mathbf{x})_i\right\} + O\left(\frac{1}{r^3}\right)$$

$$g_{ij} \to \delta_{ij} + \frac{2MG}{r^3}x_i x_j + O\left(\frac{1}{r^2}\right)$$

A straightforward calculation, using Eqs. (7.6.22)–(7.6.24), shows that the total momentum, energy, and angular momentum of the system *and its gravitational field* are

$$\mathbf{P} = 0$$
$$P^0 = M$$
$$\mathbf{J} = M\mathbf{a}$$

Unfortunately, it has not yet been possible to show that this exact exterior solution fits smoothly to an exact solution inside a spinning sphere. Recently Brill and Cohen[10] have found a solution for a very thin rotating spherical shell, which is valid both inside and outside the shell to lowest order in the rotation frequency ω but to all orders in the shell mass M, and which satisfies the correct continuity conditions across the shell radius R. The solution is

$$-d\tau^2 = -H(r)\,dt^2 + J(r)\,[dr^2 + r^2\,d\theta^2 + r^2\sin^2\theta\,(d\varphi - \Omega(r)\,dt)^2]$$

where

$$H(r) = \begin{cases} \left(\dfrac{1 - 2MG/r}{1 + 2MG/r}\right)^2 & (r > R) \\[2ex] \left(\dfrac{1 - 2MG/R}{1 + 2MG/R}\right)^2 & (r < R) \end{cases}$$

$$J(r) = \begin{cases} (1 + 2MG/r)^4 & (r > R) \\ (1 + 2MG/R)^4 & (r < R) \end{cases}$$

Inside the sphere the angular velocity $\Omega(r)$ is a constant

$$\Omega = \omega\left[1 + \frac{3(R - 2MG)}{4MG(1 + \beta)}\right]^{-1} \qquad (r < R)$$

with β a dimensionless constant that depends on the relative contributions of T^{ij} and T^{00} to the shell's gravitational mass. We get an inertial coordinate system inside the sphere if we define new coordinates

$$t' = \sqrt{H}\,t \qquad r' = \sqrt{J}\,r \qquad \varphi' = \varphi - \Omega t$$

so Ω is the rotation frequency (in t units) of the inertial frames within the shell with respect to the Minkowski metric at infinity. When MG is small and β is small Ω/ω approaches the post-Newtonian value $4MG/3R$, but when MG is so large that the Schwarzschild radius $2MG$ of the shell approaches the shell radius R the ratio Ω/ω approaches unity, as Mach might have expected.

8 Post-Newtonian Hydrodynamics*

The post-Newtonian program outlined in Sections 9.1–9.3 would form an adequate basis for relativistic celestial mechanics if the sun and planets could be regarded as point particles. However, this is not the case; for instance, the tidal forces on the moon due to its finite size are very much larger than the effects of the post-Newtonian corrections to the earth's gravitational field. Often such finite-size effects may be calculated to sufficient accuracy if we treat astronomical bodies

---

* This section lies somewhat out of the book's main line of development, and may be omitted in a first reading.

as composed of perfect fluids.[11] The energy-momentum tensor is then given by Eq. (5.4.2):

$$T^{\mu\nu} = pg^{\mu\nu} + (p + \rho)U^{\mu}U^{\nu} \tag{9.8.1}$$

where p and ρ are the proper pressure and energy density, that is, those measured by a locally comoving and freely falling observer, and U^{μ} is the four-vector velocity $dx^{\mu}/d\tau$. (Of course, p and ρ vanish except within the sun and planets.) To calculate U^{μ} we set

$$\frac{U^{i}}{U^{0}} = \frac{dx^{i}}{dt} \equiv v^{i} \tag{9.8.2}$$

and we calculate U^{0} from (9.2.2):

$$U^{0} = \frac{dt}{d\tau} = 1 - \phi + \tfrac{1}{2}v^{2} + O(\bar{v}^{4}) \tag{9.8.3}$$

The program for calculating the motion of the fluid depends crucially on whether there is an equation of state giving p as a function of ρ, as is the case for the cold degenerate fluids studied in Chapter 11, or whether p depends on temperature as well. If the pressure is a function of ρ alone, then our program is essentially the same as in Section 9.3:

(A) First solve the Newtonian problem. The pressure is to be regarded as of order $\bar{v}^{2}\bar{M}/\bar{r}^{3}$, so the necessary components of the energy-momentum tensor are given by (9.8.1)–(9.8.3) as

$$\overset{0}{T}{}^{00} = \rho \tag{9.8.4}$$

$$\overset{1}{T}{}^{i0} = \rho v_{i} \tag{9.8.5}$$

$$\overset{2}{T}{}^{ij} = p\delta_{ij} + \rho v_{i}v_{j} \tag{9.8.6}$$

The Newtonian equations of motion are provided by using (9.8.4)–(9.8.6) in the mass- and momentum-conservation equations (9.3.2) and (9.3.3):

$$\frac{\partial\rho}{\partial t} + \nabla \cdot (\rho\mathbf{v}) = 0 \tag{9.8.7}$$

$$\frac{\partial}{\partial t}(\rho\mathbf{v}) + \nabla \cdot (\rho\mathbf{v}\mathbf{v}) = -\rho\nabla\phi - \nabla p \tag{9.8.8}$$

with p given as a function of ρ by the equation of state, and with ϕ determined by Poisson's equation (9.3.12):

$$\nabla^{2}\phi = 4\pi G\rho \tag{9.8.9}$$

(B) Use the values of ρ, p, $\mathbf{v}$, and ϕ determined in (A) to calculate the terms (9.8.4)–(9.8.6) of $T^{\mu\nu}$, and also to calculate

$$\overset{2}{T}{}^{00} = \rho(\mathbf{v}^{2} - 2\phi) \tag{9.8.10}$$

(C) Use the results of (A) and (B) and Eqs. (9.1.62) and (9.1.65), to compute the post-Newtonian fields ζ and ψ.

(D) Solve for ρ, p, and $\mathbf{v}$ in the post-Newtonian approximation. The energy-momentum tensor is given to the necessary order by (9.8.1)–(9.8.3) as

$$\overset{0}{T}{}^{00} + \overset{2}{T}{}^{00} = \rho(1 + \mathbf{v}^2 - 2\phi) \tag{9.8.11}$$

$$\overset{1}{T}{}^{i0} + \overset{3}{T}{}^{i0} = (\rho + p + \mathbf{v}^2\rho - 2\phi\rho)\mathbf{v} \tag{9.8.12}$$

$$\overset{2}{T}{}^{ij} + \overset{4}{T}{}^{ij} = p\delta_{ij}(1 + 2\phi) + v^i v^j(p + \rho - 2\phi\rho + \phi\mathbf{v}^2) \tag{9.8.13}$$

and the post-Newtonian equations of motion are obtained by using (9.8.11)–(9.8.13) in the energy and momentum conservation equations, (9.3.2) plus (9.3.4) and (9.3.3) plus (9.3.5):

$$\frac{\partial}{\partial t}[\rho(1 - \mathbf{v}^2 - 2\phi)] + \nabla \cdot [\mathbf{v}(\rho + p + \mathbf{v}^2\rho - 2\phi\rho)] = \rho\frac{\partial \phi}{\partial t} \tag{9.8.14}$$

$$\frac{\partial}{\partial t}[\mathbf{v}(\rho + p + \mathbf{v}^2\rho - 2\phi\rho)] + \nabla \cdot [\mathbf{v}\mathbf{v}(p + \rho - 2\phi\rho + \phi\mathbf{v}^2)]$$

$$= -\nabla[p(1 + 2\phi)] - \rho\nabla(\phi + 2\phi^2 + \psi) - \rho\frac{\partial \zeta}{\partial t}$$

$$- \rho(\mathbf{v}^2 - 2\phi)\nabla\phi + \rho\mathbf{v} \times (\nabla \times \zeta) + 4\rho\mathbf{v}\frac{\partial \phi}{\partial t}$$

$$- (3p + \rho\mathbf{v}^2)\nabla\phi + 4p\nabla\phi + 4\rho\mathbf{v}(\mathbf{v} \cdot \nabla\phi) \tag{9.8.15}$$

(E) And so on.

Matters are more complicated when the temperature is an independent variable. We now need one additional equation at every stage in the calculation, which is provided for us by an equation of continuity

$$\frac{\partial}{\partial x^\mu}(\sqrt{g}\,\mu U^\mu) = 0 \tag{9.8.16}$$

where μ is a rest-mass density proportional to the number density of particles in the fluid. [Compare Eq. (5.2.14).] It may be assumed that the pressure is given by the equation of state as a function of both μ and an energy density $\varepsilon = \bigcirc(\bar{v}^2)$ defined by

$$T^{00} \equiv \mu U^0 + \varepsilon \tag{9.8.17}$$

Our equations are then the continuity equation (9.8.16), the momentum-conservation equation $(T^{\mu i})_{;\mu} = 0$, and an energy-conservation equation, which after subtracting (9.8.16) may be written

$$\frac{\partial}{\partial t}\sqrt{g}\,\varepsilon + \frac{\partial}{\partial x^i}\sqrt{g}[T^{i0} - \mu U^i] = -\sqrt{g}\,\Gamma^0_{\mu\nu}T^{\mu\nu} \tag{9.8.18}$$

However, now we use the energy-conservation equation to one higher order in $\bar{v}^2$ at every step: In the Newtonian approximation we use the continuity equation to order $\bar{v}$, the momentum-conservation equation to order $\bar{v}^2$, and the energy-conservation equation to order $\bar{v}^3$, while in the post-Newtonian approximation we use continuity to order $\bar{v}^3$, momentum conservation to order $\bar{v}^4$, and energy conservation to order $\bar{v}^5$. Without writing out these equations in detail we may note that this program is possible, because in the Newtonian calculation we need $\overset{3}{\Gamma}{}^0_{00}$ and $\overset{2}{\Gamma}{}^0_{i0}$, which are given by (9.1.71) and (9.1.72) in terms of ϕ alone, while in the post-Newtonian calculation we also need $\overset{5}{\Gamma}{}^0_{00}$, $\overset{4}{\Gamma}{}^0_{i0}$, and $\overset{3}{\Gamma}{}^0_{ij}$, which are given by (9.1.73)–(9.1.75) in terms of the post-Newtonian fields.

9 Approximate Solutions to the Brans-Dicke Theory

In order to test general relativity, it is useful to have in mind some other theory with which to compare it. The Brans-Dicke theory described in Section 7.3 is identical with general relativity in the physical interpretation of the metric $g_{\mu\nu}$, and differs only in that a new scalar field ϕ enters the gravitational field equations. In order to avoid confusion with the Newtonian potential, we shall write the Brans-Dicke scalar field ϕ as $\mathscr{G}^{-1}(1 + \xi)$, with $\mathscr{G}$ a constant of order G, and ξ a scalar field defined by

$$\xi^{\,\mu}_{;\,;\mu} = \frac{8\pi\mathscr{G}}{3 + 2\omega}\, T^{\mu}_{\,\mu} \tag{9.9.1}$$

$$\xi \to 0 \qquad \text{for } r \to \infty \tag{9.9.2}$$

(See Eq. (7.3.13). We have dropped the subscript M, but $T^{\mu\nu}$ should be understood as the energy-momentum tensor of matter, excluding ξ. Also, ω is a dimensionless constant, perhaps of order 6.) The gravitational field equations are given by Eq. (7.3.14) as

$$\begin{aligned}
R_{\mu\nu} - \tfrac{1}{2}g_{\mu\nu}R = &-8\pi\mathscr{G}(1 + \xi)^{-1}T_{\mu\nu} \\
&- \omega(1 + \xi)^{-2}(\xi_{;\mu}\xi_{;\nu} - \tfrac{1}{2}g_{\mu\nu}\xi_{;\rho}\xi_{;}^{\;\rho}) \\
&- (1 + \xi)^{-1}(\xi_{;\mu;\nu} - g_{\mu\nu}\xi_{;}^{\;\rho}{}_{;\rho})
\end{aligned} \tag{9.9.3}$$

By using (9.9.1) to determine $\xi_{;}^{\;\rho}{}_{;\rho}$, and contracting (9.9.3) to find R, we can rewrite this in the form

$$\begin{aligned}
R_{\mu\nu} = &-8\pi\mathscr{G}(1 + \xi)^{-1}\left[T_{\mu\nu} - g_{\mu\nu}T^{\lambda}_{\;\lambda}\left(\frac{\omega + 1}{2\omega + 3}\right)\right] \\
&- \omega(1 + \xi)^{-2}\xi_{;\mu}\xi_{;\nu} - (1 + \xi)^{-1}\xi_{;\mu;\nu}
\end{aligned} \tag{9.9.4}$$

It follows from (9.9.1) and (9.9.2) that ξ has the expansion

$$\xi = \overset{2}{\xi} + \overset{4}{\xi} + \cdots \tag{9.9.5}$$

where $\overset{N}{\xi}$ is or order $\bar{v}^N$, and in particular

$$\nabla^2 \overset{2}{\xi} = -\frac{8\pi\mathscr{G}}{3 + 2\omega} \overset{0}{T}{}^{00} \tag{9.9.6}$$

The field equations are now given by (9.9.4)–(9.9.6) and (9.1.37)–(9.1.40) as

$$\nabla^2 \overset{2}{g}_{00} = -8\pi\mathscr{G}\left(\frac{2\omega + 4}{2\omega + 3}\right)\overset{0}{T}_{00} \tag{9.9.7}$$

$$
\begin{aligned}
\nabla^2 \overset{4}{g}_{00} = {} & \frac{\partial^2 \overset{2}{g}_{00}}{\partial t^2} + \overset{2}{g}_{ij}\frac{\partial^2 \overset{2}{g}_{00}}{\partial x^i\,\partial x^j} - (\nabla \overset{2}{g}_{00})^2 \\
& + 8\pi\mathscr{G}\left(\frac{2\omega + 4}{2\omega + 3}\right)\overset{2}{\xi}\overset{0}{T}{}^{00} - 8\pi\mathscr{G}\overset{2}{T}{}^{ii}\left(\frac{2\omega + 2}{2\omega + 3}\right) \\
& + 16\pi\mathscr{G}\overset{2}{g}_{00}\overset{0}{T}{}^{00}\left(\frac{2\omega + 4}{2\omega + 3}\right) - 8\pi\mathscr{G}\left(\frac{2\omega + 4}{2\omega + 3}\right)\overset{2}{T}{}^{00} \\
& - 2\omega\left(\frac{\partial \overset{2}{\xi}}{\partial t}\right)^2 - 2\frac{\partial^2 \overset{2}{\xi}}{\partial t^2} + 2\overset{2}{\Gamma}{}^i{}_{00}\frac{\partial \overset{2}{\xi}}{\partial x^i}
\end{aligned}
\tag{9.9.8}
$$

$$\nabla^2 \overset{3}{g}_{i0} = 16\pi\mathscr{G}\overset{1}{T}{}^{i0} - 2\frac{\partial^2 \overset{2}{\xi}}{\partial x^i\,\partial t} \tag{9.9.9}$$

$$\nabla^2 \overset{2}{g}_{ij} = -8\pi\mathscr{G}\overset{0}{T}{}^{00}\delta_{ij}\left(\frac{2\omega + 2}{2\omega + 3}\right) - 2\frac{\partial^2 \overset{2}{\xi}}{\partial x^i\,\partial x^j} \tag{9.9.10}$$

From (9.9.7) it follows that the gravitational constant measured by observation of slowly moving particles or in time dilation experiments is not $\mathscr{G}$, but rather is

$$G = \left(\frac{2\omega + 4}{2\omega + 3}\right)\mathscr{G} \tag{9.9.11}$$

That is, we have the usual relation between $\overset{2}{g}_{00}$ and the Newtonian potential ϕ

$$\overset{2}{g}_{00} = -2\phi \tag{9.9.12}$$

provided we define ϕ by

$$\nabla^2\phi = 4\pi G\overset{0}{T}{}^{00} \tag{9.9.13}$$

Also, it follows from (9.9.6) and (9.9.13) that

$$\overset{2}{\xi} = -(\omega + 2)^{-1}\phi$$

and the field equations for $\overset{4}{g}_{00}$, $\overset{3}{g}_{i0}$, and $\overset{2}{g}_{ij}$ are

$$\nabla^2 \overset{4}{g}_{00} = -2\left(\frac{\omega + 1}{\omega + 2}\right)\frac{\partial^2 \phi}{\partial t^2} - 2\overset{2}{g}_{ij}\frac{\partial^2 \phi}{\partial x^i \partial x^j} - 2\left(\frac{2\omega + 5}{\omega + 2}\right)(\nabla\phi)^2$$

$$- 8\pi G\left[4 + \frac{1}{\omega + 2}\right]\phi \overset{0}{T}{}^{00} - 8\pi G\left(\frac{2\omega + 2}{2\omega + 4}\right)\overset{2}{T}{}^{ii}$$

$$- 8\pi G \overset{2}{T}{}^{00} - \frac{2\omega}{(\omega + 2)^2}\left(\frac{\partial\phi}{\partial t}\right)^2 \tag{9.9.14}$$

$$\nabla^2 \overset{3}{g}_{i0} = 16\pi G\left(\frac{2\omega + 3}{2\omega + 4}\right)\overset{1}{T}{}^{i0} + \frac{2}{\omega + 2}\frac{\partial^2 \phi}{\partial x^i \partial t} \tag{9.9.15}$$

$$\nabla^2 \overset{2}{g}_{ij} = -8\pi G \overset{0}{T}{}^{00}\delta_{ij}\left(\frac{\omega + 1}{\omega + 2}\right) + \frac{2}{(\omega + 2)}\frac{\partial^2 \phi}{\partial x^i \partial x^j} \tag{9.9.16}$$

As an example, let us consider the field of a static spherically symmetric mass. The Newtonian potential is then a function of r alone, and (9.9.16) gives

$$\overset{2}{g}_{ij} = -2\delta_{ij}\left(\frac{\omega + 1}{\omega + 2}\right)\phi + \frac{2}{\omega + 2}\left\{\left(\delta_{ij} - \frac{3x_i x_j}{r^2}\right)\frac{1}{r^3}\int_0^r r^2 \phi(r)\,dr + \frac{x_i x_j \phi}{r^2}\right\} \tag{9.9.17}$$

Outside the mass, we have

$$\phi = -\frac{MG}{r} \tag{9.9.18}$$

and so (9.9.17) gives

$$\overset{2}{g}_{ij} = \left(\frac{2\omega + 1}{\omega + 2}\right)\frac{MG}{r}\delta_{ij} + \frac{MG}{\omega + 2}\frac{x_i x_j}{r^3}$$

$$+ \frac{2MGR^2}{\omega + 2}\left(\delta_{ij} - \frac{3x_i x_j}{r^2}\right)\frac{1}{r^3} \tag{9.9.19}$$

where R is an effective radius, defined by

$$MGR^2 \equiv \int_0^\infty \left[\phi(r) + \frac{MG}{r}\right]r^2\,dr \tag{9.9.20}$$

(The integrand vanishes outside the mass, so we are free to change its upper limit from r to ∞.) Using (9.9.18) and (9.9.19) in (9.9.14) gives

$$\nabla^2 \overset{4}{g}_{00} = -\frac{2(2\omega + 3)M^2G^2}{(\omega + 2)r^4} - \frac{24M^2G^2R^2}{(\omega + 2)r^6}$$

The solution is

$$\overset{4}{g}_{00} = -\frac{(2\omega + 3)M^2G^2}{(\omega + 2)r^2} - \frac{2M^2G^2R^2}{(\omega + 2)r^4} + \frac{\kappa M^2G^2}{rR} \tag{9.9.21}$$

where κ is a dimensionless constant, which must be determined by the condition that the exterior solution (9.9.21) fit smoothly onto a nonsingular internal solution.

The results (9.9.19)–(9.9.21) make it appear that the gravitational field outside a spherical static mass depends upon the size and distribution of the mass. However, this size-dependent effect can be eliminated by a suitable redefinition of M and $\mathbf{x}$

$$M' = M\left[1 - \frac{\kappa MG}{R}\right] \tag{9.9.22}$$

$$\mathbf{x'} = \mathbf{x}\left[1 + \frac{MGR^2}{(\omega + 2)r^3}\right] \tag{9.9.23}$$

The last two terms in (9.9.21) and the last term in (9.9.19) are then cancelled by the changes in $\overset{2}{g}_{00}$ and $\overset{0}{g}_{ij}$, and so, dropping primes, we now have

$$\overset{2}{g}_{00} = \frac{2MG}{r} \tag{9.9.24}$$

$$\overset{4}{g}_{00} = -\frac{(2\omega + 3)M^2G^2}{(\omega + 2)r^2} \tag{9.9.25}$$

$$\overset{2}{g}_{ij} = \left(\frac{2\omega + 1}{\omega + 2}\right)\frac{MG}{r}\delta_{ij} + \frac{MG}{\omega + 2}\frac{x_i x_j}{r^3} \tag{9.9.26}$$

Thus the Brans-Dicke theory shares the property of the Einstein theory, that the gravitational field outside a static spherically symmetric mass depends on M, but not on any other property of the mass.

This solution may be compared with the general Robertson expansion (8.3.7) in harmonic coordinates, which (with $\alpha \equiv 1$) gives

$$\overset{2}{g}_{00} = \frac{2MG}{r}$$

$$\overset{4}{g}_{00} = -(\gamma - 1 + 2\beta)\frac{M^2G^2}{r^2}$$

$$\overset{2}{g}_{ij} = (3\gamma - 1)\delta_{ij}\frac{MG}{r} + (1 - \gamma)\frac{MGx_i x_j}{r^3}$$

Thus the Brans-Dicke results (9.9.24)–(9.9.26) can be summarized by giving formulas for the Robertson parameters

$$\gamma = \frac{\omega + 1}{\omega + 2} \qquad \beta = 1 \tag{9.9.27}$$

These formulas were already used in the last chapter to compare the Brans-Dicke theory with experiment.

We also note that the element $\overset{3}{g_{i0}} \equiv \zeta_i$ of the metric tensor is given for a static system by Eq. (9.9.15) as

$$\zeta_i = -4G\left(\frac{2\omega + 3}{2\omega + 4}\right) \int \frac{\overset{1}{T^{i0}}(\mathbf{x}', t)}{|\mathbf{x} - \mathbf{x}'|}\, d^3x' \qquad (9.9.28)$$

Thus the effects of the rotation of a spherical mass on the precession of spins and perihelia are smaller in the Brans-Dicke theory (for $0 < \omega < \infty$) than in general relativity, by a factor $(2\omega + 3)/(2\omega + 4)$.

By far the most dramatic tests of the Brans-Dicke theory are those that also test the "very strong" Principle of Equivalence. At any point P in a gravitational field we can choose a locally inertial coordinate system, for which $g_{\mu\nu} = \eta_{\mu\nu}$ and $\Gamma^\lambda_{\mu\nu} = 0$ at that point. However, the Brans-Dicke field ξ is a scalar, and hence will not vanish at P, being given by Eq. (9.9.6) and (9.9.13) as

$$\xi \simeq \overset{2}{\xi} = -(\omega + 2)^{-1}\phi$$

where ϕ is the Newtonian gravitational potential. Equation (9.9.4) shows that in this coordinate system, the gravitational field of a small mass introduced at P can be calculated as usual, but with the gravitational constant G replaced with

$$G_{\text{eff}} = G(1 + \xi)^{-1} \simeq G[1 + (\omega + 2)^{-1}\phi] \qquad (9.9.29)$$

For instance, with $\omega = 6$ and ϕ at the surface of the earth equal to -6.9×10^{-10}, the effective gravitational constant measured by Cavendish experiments on the surface of the earth is *smaller* than the "true" gravitational coupling constant that would be measured on a satellite in a high orbit by a factor $[1 - 8 \times 10^{-11}]$.

9 BIBLIOGRAPHY

☐ S. Chandrasekhar, "The Post-Newtonian Equations of Hydrodynamics in General Relativity," "The Post-Newtonian Effects on the Equilibrium of the Maclaurin Spheroids," "The Stability of Gaseous Masses in the Post-Newtonian Approximation," in *Relativity Theory and Astrophysics*. **3**. *Stellar Structure*, ed. by J. Ehlers (American Mathematical Society, Providence, R. I., 1967).

☐ J. N. Goldberg, "The Equations of Motion," in *Gravitation: An Introduction to Current Research*, ed. by L. Witten (Wiley, New York, 1962), p. 102.

L. Infeld and J. Plebanski, *Motion and Relativity* (Pergamon Press, New York, 1960).

☐ V. Fock, *The Theory of Space, Time, and Gravitation*, trans. by N. Kemmer (2nd rev. ed., Macmillan, New York, 1964), Chapter VI.

☐ L. D. Landau and E. M. Lifshitz, *The Classical Theory of Fields*, trans. by M. Hamermesh (Pergamon Press, Oxford, 1962), Section 105.

9 REFERENCES

1. A. Einstein, L. Infeld, and B. Hoffmann, Ann. Math., **39**, 65 (1938); A. Einstein and L. Infeld, Ann. Math., **41**, 455 (1940); A. Einstein and L. Infeld, Canad. J. Math., **1**, 209 (1949).

2. G. E. Pugh, WSEG Research Memo 11, U. S. Dept. of Defense (1959).

3. L. I. Schiff, Proc. Nat. Acad. Sci., **46**, 871 (1960); Phys. Rev. Letters, **4**, 215 (1960).

4. A. Papapetrou, Proc. Roy. Soc., **A209**, 248 (1951); E. Corinaldesi and A. Papapetrou, Proc. Roy. Soc., **A209**, 259 (1951).

5. V. A. Fock, J. Phys. U.S.S.R., **1**, 81 (1939).

6. W. de Sitter, Mon. Nat. Roy. Astron. Soc., **77**, 155, 481 (1920); A. D. Fokker, Kon. Akad. Weten. Amsterdam, Proc., **23**, 729 (1920); F. A. E. Pirani, Acta Physica Polonica, **15**, 389 (1956).

7. C. W. F. Everitt and W. M. Fairbank, *Proceedings of the Tenth International Conference on Low Temperature Physics*, Moscow, August 1966; D. H. Frisch and J. F. Kasper, Jr., J. Applied Phys., Vol. 40, No. 8, 3376; D. I. Shalloway and D. H. Frisch, Astrophys. and Space Sci., **10**, 106 (1971). Also, reports at the Third "Cambridge" Conference on Relativity, June 8, 1970, by D. H. Frisch and W. M. Fairbank.

8. H. Thirring, Phys. Zeits., **19**, 33 (1918). J. Lense and H. Thirring, Phys. Zeitschr., **19**, 156 (1918).

9. R. Kerr, Phys. Rev. Letters, **11**, 237 (1963).

10. J. M. Cohen, in *Relativity Theory and Astrophysics. 1. Relativity and Cosmology*, ed. by J. Ehlers (American Mathematical Society, Providence, R. I., 1967), p. 200.

11. The post-Newtonian equations for a perfect fluid were worked out by S. Chandrasekhar, Ap. J., **142**, 1488 (1965); *ibid.*, **158**, 45 (1969). For post-post-Newtonian equations, see S. Chandrasekhar and Y. Nutku, Ap. J., **158**, 55 (1969). Radiation reaction effects are included by S. Chandrasekhar and F. P. Esposito, Ap. J., **160**, 153 (1970).

IO GRAVITATIONAL RADIATION

We have seen a great many similarities between gravitation and electro-magnetism. It should therefore come as no surprise that Einstein's equations, like Maxwell's equations, have radiative solutions.

No one has yet certainly detected gravitational radiation, but the reason for this is not hard to find; Einstein's theory predicts that gravitational radiation is produced in extremely small quantities in ordinary atomic processes. For instance, the probability that a transition between two atomic states will proceed by emission of gravitational, rather than electromagnetic, radiation is typically of order GE^2/e^2, where E is the energy released and e is the electronic charge. For $E \simeq 1$ eV, this probability is about 3×10^{-54}.

Why then study gravitational radiation? One reason is of course that some day we may find a strong source of gravitational radiation. Such a source may indeed already have been detected. (See Section 10.7.) However, gravitational radiation would be interesting even if there were no chance of ever detecting any, for the theory of gravitational radiation provides a crucial link between general relativity and the microscopic frontier of physics.

We have learned in recent years to describe the fundamental observables of microscopic phenomena in terms of elementary particles and their collisions. In classical electrodynamics it is the plane-wave solutions of Maxwell's equations that lead most naturally to an interpretation in terms of a particle, the photon. Similarly, it is the radiative solutions of Einstein's equations that will lead here to the concept of a particle of gravitational radiation, the *graviton*.

Unfortunately, the theory of gravitational radiation is complicated by the nonlinearity of Einstein's equations. In the spirit of Section 7.6, we may say that any gravitational wave is itself a distribution of energy and momentum that contributes to the gravitational field of the wave. This complication prevents our being able to find general radiative solutions of the exact Einstein equations.

There are two approaches to this difficulty. One is to study only the weak-field radiative solutions of Einstein equations, which describe waves carrying not enough energy and momentum to affect their own propagation. The other approach is to look long and hard for special solutions of the exact Einstein field equations. A great deal of mathematical ingenuity has gone into the second approach, with results of some elegance. However, this chapter deals with only the first, weak-field, approach to gravitational radiation. One reason is that any observable gravitational radiation is likely to be of very low intensity. A second, deeper, reason is that it is only possible to attach a precise meaning to the concept of an elementary particle when it is far away from all other particles, and for gravitons this corresponds to a weak-field solution of the field equations.

The reader should not conclude that there is any fundamental gap in our understanding of gravitation because we cannot find general exact solutions of the nonlinear field equations. Indeed, similar problems arise in electrodynamics: The problem of computing the exact electromagnetic field produced by a decaying current in an electrical oscillator is highly nonlinear, because the field acts back on the current that produces it. Even though this problem was not solved for many years after Maxwell's theory, still there was no doubt that electrical oscillators would produce the electromagnetic waves studied by Maxwell. Gravitational waves are more complicated than electromagnetic waves because they contribute to their own source outside the material gravitational antenna. However, the simple properties of both electromagnetic and gravitational waves emerge when we look far out into the wave zone, where the fields are weak.

1   The Weak-Field Approximation

We suppose the metric to be close to the Minkowski metric $\eta_{\mu\nu}$:

$$g_{\mu\nu} = \eta_{\mu\nu} + h_{\mu\nu} \tag{10.1.1}$$

where $|h_{\mu\nu}| \ll 1$. To first order in h, the Ricci tensor is then

$$R_{\mu\nu} \simeq \frac{\partial}{\partial x^\nu} \Gamma^\lambda_{\lambda\mu} - \frac{\partial}{\partial x^\lambda} \Gamma^\lambda_{\mu\nu} + \mathcal{O}(h^2) \tag{10.1.2}$$

and the affine connection is

$$\Gamma^\lambda_{\mu\nu} = \tfrac{1}{2}\eta^{\lambda\rho}\left[\frac{\partial}{\partial x^\mu} h_{\rho\nu} + \frac{\partial}{\partial x^\nu} h_{\rho\mu} - \frac{\partial}{\partial x^\rho} h_{\mu\nu}\right] + \mathcal{O}(h^2) \tag{10.1.3}$$

As long as we restrict ourselves to first order in h, we must raise and lower *all* indices using $\eta^{\mu\nu}$, not $g^{\mu\nu}$; that is,

$$\eta^{\lambda\rho}h_{\rho\nu} \equiv h^{\lambda}{}_{\nu} \qquad \eta^{\lambda\rho}\frac{\partial}{\partial x^{\rho}} \equiv \frac{\partial}{\partial x_{\lambda}}\,, \qquad \text{etc.}$$

With this understanding, Eqs. (10.1.2) and (10.1.3) yield the first-order Ricci tensor

$$R_{\mu\nu} \simeq R^{(1)}_{\mu\nu} \equiv \frac{1}{2}\left(\Box^{2}h_{\mu\nu} - \frac{\partial}{\partial x^{\lambda}\,\partial x^{\mu}}\,h^{\lambda}{}_{\nu} - \frac{\partial^{2}}{\partial x^{\lambda}\,\partial x^{\nu}}\,h^{\lambda}{}_{\mu} + \frac{\partial^{2}}{\partial x^{\mu}\,\partial x^{\nu}}\,h^{\lambda}{}_{\lambda}\right)$$

The Einstein field equations therefore read

$$\Box^{2}h_{\mu\nu} - \frac{\partial}{\partial x^{\lambda}\,\partial x^{\mu}}\,h^{\lambda}{}_{\nu} - \frac{\partial^{2}}{\partial x^{\lambda}\,\partial x^{\nu}}\,h^{\lambda}{}_{\mu} + \frac{\partial^{2}}{\partial x^{\mu}\,\partial x^{\nu}}\,h^{\lambda}{}_{\lambda} = -16\pi G S_{\mu\nu} \qquad (10.1.4)$$

$$S_{\mu\nu} \equiv T_{\mu\nu} - \tfrac{1}{2}\eta_{\mu\nu}T^{\lambda}{}_{\lambda} \qquad (10.1.5)$$

Here $T_{\mu\nu}$ is taken to lowest order in $h_{\mu\nu}$, so it is independent of $h_{\mu\nu}$, and satisfies the ordinary conservation conditions

$$\frac{\partial}{\partial x^{\mu}}\,T^{\mu}{}_{\nu} = 0 \qquad (10.1.6)$$

(If gravitational forces play an important role in the structure of the radiating system, then $\tau^{\mu\nu}$ should be used in place of $T^{\mu\nu}$; see Section 7.6.) Note that it is this form of the conservation law that is needed for the consistency of (10.1.4), because (10.1.6) implies

$$\frac{\partial}{\partial x^{\mu}}\,S^{\mu}{}_{\nu} = \frac{1}{2}\frac{\partial}{\partial x^{\nu}}\,S^{\lambda}{}_{\lambda}$$

whereas the linearized Ricci tensor satisfies Bianchi identities of the form

$$\frac{\partial}{\partial x^{\mu}}\,R^{(1)\mu}{}_{\nu} = \frac{1}{2}\frac{\partial}{\partial x^{\nu}}\left[\Box^{2}h^{\lambda}{}_{\lambda} - \frac{\partial^{2}h^{\lambda\nu}}{\partial x^{\lambda}\,\partial x^{\nu}}\right] = \frac{1}{2}\frac{\partial R^{(1)\lambda}{}_{\lambda}}{\partial x^{\nu}}$$

As already discussed in Section 7.4, we cannot expect a field equation such as (10.1.4) to yield unique solutions, because given any solution, we can always generate other solutions by performing coordinate transformations. The most general coordinate transformation that leaves the field weak is of the form

$$x^{\mu} \to x'^{\mu} = x^{\mu} + \varepsilon^{\mu}(x) \qquad (10.1.7)$$

where $\partial\varepsilon^{\mu}/\partial x^{\nu}$ is at most of the same order of magnitude as $h_{\mu\nu}$. The metric in the new coordinate system is given by

$$g'^{\mu\nu} = \frac{\partial x'^{\mu}}{\partial x^{\lambda}}\frac{\partial x'^{\nu}}{\partial x^{\rho}}\,g^{\lambda\rho}$$

or, since $g^{\mu\nu} \simeq \eta^{\mu\nu} - h^{\mu\nu}$,

$$h'^{\mu\nu} = h^{\mu\nu} - \frac{\partial\varepsilon^{\mu}}{\partial x^{\lambda}}\eta^{\lambda\nu} - \frac{\partial\varepsilon^{\nu}}{\partial x^{\rho}}\eta^{\rho\mu}$$

Thus, if $h_{\mu\nu}$ is a solution of (10.1.4), then so will be

$$h'_{\mu\nu} = h_{\mu\nu} - \frac{\partial\varepsilon_{\mu}}{\partial x^{\nu}} - \frac{\partial\varepsilon_{\nu}}{\partial x^{\mu}} \tag{10.1.8}$$

where $\varepsilon_{\mu} \equiv \varepsilon^{\nu}\eta_{\mu\nu}$ are four small but otherwise arbitrary functions of x^{μ}. That this is the case can be verified by direct inspection of Eq. (10.1.4); this property is called the *gauge invariance* of the field equation.

The gauge invariance of Eq. (10.1.4) is a nuisance when it comes to actually solving the field equations. However, the difficulty can be removed by choosing some particular gauge, that is, coordinate system. The most convenient choice is to work in a *harmonic coordinate system*, for which

$$g^{\mu\nu}\Gamma^{\lambda}_{\mu\nu} = 0$$

Using (10.1.3), this gives to first order

$$\frac{\partial}{\partial x^{\mu}}h^{\mu}{}_{\nu} = \frac{1}{2}\frac{\partial}{\partial x^{\nu}}h^{\mu}{}_{\mu} \tag{10.1.9}$$

That this choice is always possible follows from the general argument of Section 7.4; it can also be seen from (10.1.8) that if $h_{\mu\nu}$ does not satisfy (10.1.9), then we can find an $h'_{\mu\nu}$ that does, by performing the coordinate transformation (10.1.7) with

$$\Box^{2}\varepsilon_{\nu} \equiv \frac{\partial}{\partial x^{\mu}}h^{\mu}{}_{\nu} - \frac{1}{2}\frac{\partial}{\partial x^{\nu}}h^{\mu}{}_{\mu}$$

It will be assumed from now on that $h_{\mu\nu}$ does satisfy Eq. (10.1.9).

Using (10.1.9) in (10.1.4), the field equation now read

$$\Box^{2}h_{\mu\nu} = -16\pi G S_{\mu\nu} \tag{10.1.10}$$

One solution is the *retarded potential*

$$h_{\mu\nu}(\mathbf{x}, t) = 4G \int d^{3}\mathbf{x}' \frac{S_{\mu\nu}(\mathbf{x}', t - |\mathbf{x} - \mathbf{x}'|)}{|\mathbf{x} - \mathbf{x}'|} \tag{10.1.11}$$

We have already remarked that the conservation law (10.1.6) for $T^{\mu\nu}$ is equivalent to

$$\frac{\partial}{\partial x^{\mu}}S^{\mu}{}_{\nu} = \frac{1}{2}\frac{\partial}{\partial x^{\nu}}S^{\mu}{}_{\mu} \tag{10.1.12}$$

and in consequence the solution (10.1.11) for a source $S_{\mu\nu}$ confined to a finite volume automatically satisfies the harmonic coordinate conditions (10.1.9). (The proof is identical to that used in electrodynamics in the calculation of the vector potential in Lorentz gauge.) To (10.1.11) we can add any solution of the homogeneous equations

$$\Box^2 h_{\mu\nu} = 0 \tag{10.1.13}$$

$$\frac{\partial}{\partial x^\mu} h^\mu{}_\nu = \frac{1}{2} \frac{\partial}{\partial x^\nu} h^\mu{}_\mu \tag{10.1.14}$$

We interpret (10.1.11) as the gravitational radiation produced by the source $S_{\mu\nu}$, whereas any additional term satisfying (10.1.13), (10.1.14) represents the gravitational radiation coming in from infinity. The occurrence in (10.1.11) of the time argument $t - |\mathbf{x} - \mathbf{x}'|$ shows that gravitational effects propagate with unit velocity, that is, with the speed of light.

2 Plane Waves

We now consider the plane-wave solutions of the homogeneous equations (10.1.13) and (10.1.14), both because they are important in their own right and, as we shall see, because the retarded wave (10.1.11) approaches a plane wave as $r \to \infty$. The general solution of (10.1.13) and (10.1.14) is a linear superposition of solutions of the form

$$h_{\mu\nu}(x) = e_{\mu\nu} \exp(ik_\lambda x^\lambda) + e^*_{\mu\nu} \exp(-ik_\lambda x^\lambda) \tag{10.2.1}$$

This satisfies (10.1.13) if

$$k_\mu k^\mu = 0 \tag{10.2.2}$$

and satisfies (10.1.14) if

$$k_\mu e^\mu{}_\nu = \tfrac{1}{2} k_\nu e^\mu{}_\mu \tag{10.2.3}$$

(Of course we are still raising and lowering indices with $\eta_{\mu\nu}$, so that $k^\mu \equiv \eta^{\mu\nu} k_\nu$.) The matrix $e_{\mu\nu}$ is obviously symmetric:

$$e_{\mu\nu} = e_{\nu\mu} \tag{10.2.4}$$

It will be called the *polarization tensor*.

A symmetric 4×4 matrix would in general have ten independent components, and the four relations (10.2.3) would lower this number to six, but of these six only two represent physically significant degrees of freedom. By a change of coordinates $x^\mu \to x^\mu + \varepsilon^\mu(x)$ we can transform the metric $\eta_{\mu\nu} + h_{\mu\nu}$ into a new metric $\eta_{\mu\nu} + h'_{\mu\nu}$ with $h'_{\mu\nu}$ given by (10.1.8). Suppose that we choose

$$\varepsilon^\mu(x) = i\varepsilon^\mu \exp(ik_\lambda x^\lambda) - i\varepsilon^{\mu*} \exp(-ik_\lambda x^\lambda) \tag{10.2.5}$$

Then (10.1.8) gives

$$h'_{\mu\nu}(x) = e'_{\mu\nu} \exp{(ik_\lambda x^\lambda)} + e'^*_{\mu\nu} \exp{(-ik_\lambda x^\lambda)} \qquad (10.2.6)$$

where

$$e'_{\mu\nu} = e_{\mu\nu} + k_\mu \varepsilon_\nu + k_\nu \varepsilon_\mu \qquad (10.2.7)$$

[Note that the wave still satisfies the harmonic coordinate condition (10.2.3).] We conclude that $e'_{\mu\nu}$ and $e_{\mu\nu}$ represent the same physical situation for arbitrary values of the four parameters ε_μ, so of the six independent $e_{\mu\nu}$'s satisfying (10.2.3) and (10.2.4), only $6 - 4 = 2$ are physically significant. For instance, consider a wave traveling in the $+z$-direction, with wave vector

$$k^1 = k^2 = 0 \qquad k^3 = k^\circ \equiv k > 0 \qquad (10.2.8)$$

In this case (10.2.3) gives

$$e_{31} + e_{01} = e_{32} + e_{02} = 0$$
$$e_{33} + e_{03} = -e_{03} - e_{00} = \tfrac{1}{2}(e_{11} + e_{22} + e_{33} - e_{00})$$

These four equations allow us to express e_{i0} and e_{22} in terms of the other six $e_{\mu\nu}$:

$$e_{01} = -e_{31}; \qquad e_{02} = -e_{32}; \qquad e_{03} = -\tfrac{1}{2}(e_{33} + e_{00}); \qquad e_{22} = -e_{11}$$
$$(10.2.9)$$

When the coordinate system is subjected to the transformation defined by (10.1.7) and (10.2.5), these six independent components of $e_{\mu\nu}$ change according to Eq. (10.2.7):

$$e'_{11} = e_{11}, \qquad\qquad e'_{12} = e_{12}$$
$$e'_{13} = e_{13} + k\varepsilon_1, \qquad e'_{23} = e_{23} + k\varepsilon_2$$
$$e'_{33} = e_{33} + 2k\varepsilon_1, \qquad e'_{00} = e_{00} - 2k\varepsilon_0$$

Thus it is only e_{11} and e_{12} that have an absolute physical significance. Indeed, we can arrange that all components of $e'_{\mu\nu}$ vanish except for e'_{11}, e'_{12}, and $e'_{22} = -e'_{11}$, by performing a coordinate transformation with

$$\varepsilon_1 = -\frac{e_{13}}{k}, \qquad \varepsilon_2 = -\frac{e_{23}}{k}, \qquad \varepsilon_3 = -\frac{e_{33}}{2k}, \qquad \varepsilon_0 = \frac{e_{00}}{2k}$$

The distinction between the different components of the polarization tensor is clarified if we ask how $e_{\mu\nu}$ changes when we subject the coordinate system to a rotation about the z-axis. This is just a Lorentz transformation of the form

$$R_1{}^1 = \cos\theta \qquad\qquad R_1{}^2 = \sin\theta$$
$$R_2{}^1 = -\sin\theta \qquad\qquad R_2{}^2 = \cos\theta \qquad (10.2.10)$$
$$R_3{}^3 = R_0{}^0 = 1 \qquad \text{other } R_\mu{}^\nu = 0$$

and since it leaves k_μ invariant (i.e., $R_\mu{}^\nu k_\nu = k_\mu$), the only effect is to transform $e_{\mu\nu}$ into

$$e'_{\mu\nu} = R_\mu{}^\rho R_\nu{}^\sigma e_{\rho\sigma} \tag{10.2.11}$$

Using the relations (10.2.9), we find that

$$e'_\pm = \exp(\pm 2i\theta)e_\pm \tag{10.2.12}$$

$$f'_\pm = \exp(\pm i\theta)f_\pm \tag{10.2.13}$$

$$e'_{33} = e_{33}, \qquad e'_{00} = e_{00} \tag{10.2.14}$$

where

$$e_\pm \equiv e_{11} \mp ie_{12} = -e_{22} \mp ie_{12} \tag{10.2.15}$$

$$f_\pm \equiv e_{31} \mp ie_{32} = -e_{01} \pm ie_{02} \tag{10.2.16}$$

In general, any plane wave ψ, which is transformed by a rotation of any angle θ about the direction of propagation into

$$\psi' = e^{ih\theta}\psi \tag{10.2.17}$$

is said to have *helicity h*. We thus have shown that a gravitational plane wave can be decomposed into parts $e_\pm$ with helicity ± 2, parts $f_\pm$ with helicity ± 1, and parts e_{00} and e_{33} with helicity zero. However, we have also seen that the parts with helicity 0 and ± 1 can be made to vanish by a suitable choice of coordinates, so *the physically significant components are just those with helicity ± 2*.

Once again we find a fruitful analogy with electromagnetism. The Maxwell equations in Lorentz gauge are (2.7.12) and (2.7.13); in empty space they become

$$\Box^2 A_\alpha = 0 \qquad \frac{\partial A^\alpha}{\partial x^\alpha} = 0$$

in analogy with Eqs. (10.1.13) and (10.1.14) for the metric in harmonic coordinates. (We are now in an inertial coordinate system, and so $\Box^2 \equiv \eta^{\alpha\beta}\,\partial^2/\partial x^\alpha\,\partial x^\beta$.) We can find a plane-wave solution of the form

$$A_\alpha = e_\alpha \exp(ik_\beta x^\beta) + e_\alpha^* \exp(-ik_\beta x^\beta)$$

where

$$k_\alpha k^\alpha = 0$$
$$k_\alpha e^\alpha = 0$$

in analogy with Eqs. (10.2.1)–(10.2.3).

In general e^α would have four independent components, but the condition that $k_\alpha e^\alpha$ vanish reduces the number of independent components to three, just as (10.2.3) reduces the number of independent components of $e_{\mu\nu}$ from ten to six.

Furthermore, without changing the physical fields **E** and **B** and without leaving the Lorentz gauge, we can change A_α by a gauge transformation

$$A_\alpha \rightarrow A'_\alpha = A_\alpha + \frac{\partial \Phi}{\partial x_\alpha}$$

$$\Phi(x) = i\varepsilon \exp(ik_\beta x^\beta) - i\varepsilon^* \exp(-ik_\beta x^\beta)$$

in analogy with (10.1.8) and (10.2.5). The new potential can be written

$$A'_\alpha = e'_\alpha \exp(ik_\beta x^\beta) + e'^*_\alpha \exp(-ik_\beta x^\beta)$$

$$e'_\alpha = e_\alpha - \varepsilon k_\alpha$$

in analogy with (10.2.6) and (10.2.7). The parameter ε is arbitrary, so of the three algebraically independent components of e_α only $3 - 1 = 2$ are physically significant, just as general covariance renders only two of the six independent components of $e_{\mu\nu}$ physically significant. To identify the two significant components of e_α, we may consider a wave traveling in the z-direction, with k^α given by Eq. (10.2.8). Then the condition that $k_\alpha e^\alpha$ vanish allows us to determine e^0,

$$e_0 = -e_3$$

just as (10.2.3) allows us to determine e_{22} and e_{0i} in terms of the other six $e_{\mu\nu}$. Also, the preceding gauge transformation leaves e_1 and e_2 invariant but changes e_3 into

$$e'_3 = e_3 - \varepsilon k$$

Hence e'_3 can be made equal to zero by choosing $\varepsilon = e_3/k$, so it is only e_1 and e_2 that carry physical significance, just as it was only e_{11} and e_{12} that could not be made equal to zero by a suitable coordinate transformation. Finally, we can work out the meaning of these two components by subjecting the plane electromagnetic wave to the rotation (10.2.10). The polarization vector is then changed into

$$e'_\alpha = R_\alpha{}^\beta e_\beta$$

and therefore

$$e'_\pm = \exp(\pm i\theta)e_\pm$$
$$e'_3 = e_3$$

where

$$e_\pm \equiv e_1 \mp i e_2$$

Thus the electromagnetic wave can be decomposed into parts with helicity ± 1 and 0. However, the physically significant helicities are ± 1, not 0, just as for gravitational waves they are ± 2, not ± 1 or 0. This is what we mean when we say, speaking classically, that electromagnetism and gravitation are carried by waves of spin 1 and spin 2, respectively.

3 Energy and Momentum of Plane Waves

The physical significance of the plane-wave solution (10.2.1) is brought forward by calculating the energy and momentum it carries. According to Eq. (7.6.4), the energy-momentum tensor of gravitation is given to order h^2 by

$$t_{\mu\nu} \simeq \frac{1}{8\pi G} [-\tfrac{1}{2}h_{\mu\nu}\eta^{\lambda\rho}R^{(1)}_{\lambda\rho} + \tfrac{1}{2}\eta_{\mu\nu}h^{\lambda\rho}R^{(1)}_{\lambda\rho} + R^{(2)}_{\mu\nu} - \tfrac{1}{2}\eta_{\mu\nu}\eta^{\lambda\rho}R^{(2)}_{\lambda\rho}]$$

where $R^{(N)}_{\mu\nu}$ is the term in the Ricci tensor of order N in $h_{\mu\nu}$. The metric $g_{\mu\nu} = \eta_{\mu\nu} + h_{\mu\nu}$ satisfies the first-order Einstein equations $R^{(1)}_{\mu\nu} = 0$, so we can drop these terms in $t_{\mu\nu}$ and use

$$t_{\mu\nu} \simeq \frac{1}{8\pi G} [R^{(2)}_{\mu\nu} - \tfrac{1}{2}\eta_{\mu\nu}\eta^{\lambda\rho}R^{(2)}_{\lambda\rho}] \tag{10.3.1}$$

(For the actual metric it is $R_{\mu\nu}$ rather than $R^{(1)}_{\mu\nu}$ that vanishes, and $t_{\mu\nu}$ arises only from the first-order terms in Eq. (7.6.4). Here it is $R^{(1)}_{\mu\nu}$ rather than $R_{\mu\nu}$ that vanishes, because $g_{\mu\nu} = \eta_{\mu\nu} + h_{\mu\nu}$ satisfies the first-order Einstein equations rather than the exact equations. The difference is only of order h^3.) To calculate $R^{(2)}_{\mu\nu}$ we must use Eq. (10.2.1) in Eq. (7.6.15); the result is extremely complicated, but can be simplified if we average $t_{\mu\kappa}$ over a region of space and time much larger than $|\mathbf{k}|^{-1}$. (This is the way the energy and momentum of any wave are usually evaluated.) The averaging kills all terms proportional to $\exp(\pm 2ik_\lambda x^\lambda)$, and we are left with only the x^μ-independent cross-terms:

$$\langle R^{(2)}_{\mu\nu} \rangle = \text{Re } \{e^{\lambda\rho*}[k_\mu k_\nu e_{\lambda\rho} - k_\mu k_\lambda e_{\nu\rho} - k_\nu k_\rho e_{\mu\lambda} + k_\lambda k_\rho e_{\mu\nu}]$$
$$+ [e^\lambda{}_\rho k_\lambda - \tfrac{1}{2}e_\lambda{}^\lambda k_\rho]^*[k_\mu e^\rho{}_\nu + k_\nu e^\rho{}_\mu - k^\rho e_{\mu\nu}]$$
$$- \tfrac{1}{2}[k_\lambda e_{\rho\nu} + k_\nu e_{\rho\lambda} - k_\rho e_{\lambda\nu}]^*[k^\lambda e^\rho{}_\mu + k_\mu e^{\rho\lambda} - k^\rho e^\lambda{}_\mu]\} \tag{10.3.2}$$

(We have not yet made use of the conditions (10.2.2) and (10.2.3) appropriate to harmonic coordinates, so suppose for a moment that we leave the harmonic coordinate system by adding to $h_{\mu\nu}(x)$ a term

$$i(q_\mu \varepsilon_\nu + q_\nu \varepsilon_\mu) \exp(iq_\lambda x^\lambda) - i(q_\mu \varepsilon^*_\nu + q_\nu \varepsilon^*_\mu) \exp(-iq_\lambda x^\lambda) \tag{10.3.3}$$

where $q_\mu q^\mu \neq 0$. After averaging over space-time distances large compared with $|q - k|^{-1}$, the interference between (10.2.1) and (10.3.3) drops out, and we find for $\langle R^{(2)}_{\mu\kappa} \rangle$ the term (10.3.2), plus another term obtained by replacing k with q and $e_{\mu\nu}$ with $q_\mu \varepsilon_\nu + q_\nu \varepsilon_\mu$. Inspection of (10.3.2) shows immediately that this second term *vanishes*, so $\langle R^{(2)}_{\mu\kappa} \rangle$ and hence $\langle t_{\mu\kappa} \rangle$ may be calculated in harmonic coordinates with no loss of generality.)

If we now use in (10.3.2) the harmonic coordinate conditions (10.2.2) and (10.2.3), we find

$$\langle R^{(2)}_{\mu\nu} \rangle = \frac{k_\mu k_\nu}{2} (e^{\lambda\rho*}e_{\lambda\rho} - \tfrac{1}{2}|e^\lambda{}_\lambda|^2) \tag{10.3.4}$$

The quantity $\eta^{\lambda\rho}\langle R^{(2)}_{\lambda\rho}\rangle$ vanishes because $k^\rho k_\rho = 0$, so (10.3.1) now gives the average energy-momentum tensor of a plane wave as

$$\langle t_{\mu\nu}\rangle = \frac{k_\mu k_\nu}{16\pi G}\,(e^{\lambda\rho*}e_{\lambda\rho} - \tfrac{1}{2}|e^\lambda{}_\lambda|^2) \qquad (10.3.5)$$

Note that a "gauge transformation" (10.2.7) will change the terms in $\langle t_{\mu\kappa}\rangle$ into

$$e'^{\lambda\rho*}e'_{\lambda\rho} = e^{\lambda\rho*}e_{\lambda\rho} + 2\,\mathrm{Re}\,\varepsilon^*_\rho k^\rho e^\lambda{}_\lambda + 2|\varepsilon_\rho k^\rho|^2$$

$$e'^\lambda{}_\lambda = e^\lambda{}_\lambda + 2k^\lambda\varepsilon_\lambda$$

but $\langle t_{\mu\kappa}\rangle$ is gauge-invariant! Thus, as far as energy and momentum are concerned, the polarizations $e_{\mu\nu}$ and $e_{\mu\nu} + k_\mu\varepsilon_\nu + k_\nu\varepsilon_\mu$ represent the same physical wave, and we see again that there are not six but only two physically significant polarization parameters. In particular, a wave traveling in the z-direction, with wave vector and polarization tensor given by (10.2.8) and (10.2.9), has the energy-momentum tensor

$$\langle t_{\mu\nu}\rangle = \frac{k_\mu k_\nu}{8\pi G}\,(|e_{11}|^2 + |e_{12}|^2) \qquad (10.3.6)$$

or, in terms of the helicity amplitudes (10.2.15),

$$\langle t_{\mu\nu}\rangle = \frac{k_\mu k_\nu}{16\pi G}\,(|e_+|^2 + |e_-|^2) \qquad (10.3.7)$$

4 Generation of Gravitational Waves

We wish to calculate the energy emitted in the form of gravitational radiation by a system whose energy-momentum tensor can be expressed as a Fourier integral,

$$T_{\mu\nu}(\mathbf{x}, t) = \int_0^\infty d\omega\, T_{\mu\nu}(\mathbf{x}, \omega)e^{-i\omega t} + \text{c.c.} \qquad (10.4.1)$$

or as a sum of Fourier components,

$$T_{\mu\nu}(\mathbf{x}, t) = \sum_\omega e^{-i\omega t}T_{\mu\nu}(\mathbf{x}, \omega) + \text{c.c.} \qquad (10.4.2)$$

(Here "+ c.c." means "plus the complex conjugate of the preceding term.") We first do the calculation for a single Fourier component,

$$T_{\mu\nu}(\mathbf{x}, t) = T_{\mu\nu}(\mathbf{x}, \omega)e^{-i\omega t} + \text{c.c.} \qquad (10.4.3)$$

and will then return to the more general systems described by (10.4.1) and (10.4.2).

From (10.1.11) we find that the field emitted by the source (10.4.3) is

$$h_{\mu\nu}(\mathbf{x}, t) = 4G \int \frac{d^3x'}{|\mathbf{x} - \mathbf{x}'|} S_{\mu\nu}(\mathbf{x}', \omega) \exp\{-i\omega t + i\omega|\mathbf{x} - \mathbf{x}'|\} + \text{c.c.} \quad (10.4.4)$$

where

$$S_{\mu\nu}(\mathbf{x}, \omega) \equiv T_{\mu\nu}(\mathbf{x}, \omega) - \tfrac{1}{2}\eta_{\mu\nu}T^\lambda{}_\lambda(\mathbf{x}, \omega) \quad (10.4.5)$$

Suppose that we observe this radiation in the *wave zone*, that is, at distances $r \equiv |\mathbf{x}|$ much larger than the dimension $R = |\mathbf{x}'|_{\max}$ of the source, and also much larger than ωR^2 and $1/\omega$. Then the denominator $|\mathbf{x} - \mathbf{x}'|$ can be replaced with r, while in the exponent we may approximate

$$|\mathbf{x} - \mathbf{x}'| \simeq r - \mathbf{x}' \cdot \hat{\mathbf{x}} \qquad \hat{\mathbf{x}} \equiv \frac{\mathbf{x}}{r}$$

and the field becomes

$$h_{\mu\nu}(\mathbf{x}, t) = \frac{4G}{r} \exp(i\omega r - i\omega t) \int d^3x' S_{\mu\nu}(\mathbf{x}', \omega)e^{-i\omega\hat{\mathbf{x}}\cdot\mathbf{x}'} + \text{c.c.} \quad (10.4.6)$$

Since $r\omega$ is assumed large, this looks just like a plane wave,

$$h_{\mu\nu}(\mathbf{x}, t) = e_{\mu\nu}(\mathbf{x}, \omega) \exp(ik_\mu x^\mu) + \text{c.c.} \quad (10.4.7)$$

with "wave vector" and "polarization tensor" given by

$$\mathbf{k} \equiv \omega\hat{\mathbf{x}} \qquad k^0 \equiv \omega \quad (10.4.8)$$

$$e_{\mu\nu}(\mathbf{x}, \omega) \equiv \frac{4G}{r} \int d^3x' S_{\mu\nu}(\mathbf{x}', \omega)e^{-i\mathbf{k}\cdot\mathbf{x}'} \quad (10.4.9)$$

It will be convenient to write $e_{\mu\nu}$ explicitly in terms of the Fourier transform of $T_{\mu\nu}$:

$$e_{\mu\nu}(\mathbf{x}, \omega) = \frac{4G}{r} [T_{\mu\nu}(\mathbf{k}, \omega) - \tfrac{1}{2}\eta_{\mu\nu}T^\lambda{}_\lambda(\mathbf{k}, \omega)] \quad (10.4.10)$$

$$T_{\mu\nu}(\mathbf{k}, \omega) \equiv \int d^3x' T_{\mu\nu}(\mathbf{x}', \omega)e^{-i\mathbf{k}\cdot\mathbf{x}'} \quad (10.4.11)$$

The conservation equation for $T_{\mu\nu}(\mathbf{x}, t)$ is

$$\frac{\partial}{\partial x^\mu} T^\mu{}_\nu(\mathbf{x}, t) = 0$$

Applying this to (10.4.3) gives

$$\frac{\partial}{\partial x^i} T^i{}_\nu(\mathbf{x}, \omega) - i\omega T^0{}_\nu(\mathbf{x}, \omega) = 0$$

Multiplying with $e^{-i\mathbf{k}\cdot\mathbf{x}}$ and integrating over $\mathbf{x}$, we find that $T_{\mu\nu}(\mathbf{k}, \omega)$ is subject to the algebraic relations

$$k_\mu T^\mu{}_\nu(\mathbf{k}, \omega) = 0 \tag{10.4.12}$$

where k^μ is the vector given by (10.4.8). This, incidentally, verifies that (10.4.10) obeys the harmonic coordinate condition (10.2.3).

Now let us calculate the power per unit solid angle emitted in a direction $\hat{x}$. Since $r \gg 1/\omega$, we can use for the energy flux vector the value $\langle t^{i0} \rangle$ obtained by averaging over space-time dimensions large compared with $1/\omega$, so that the power per solid angle is

$$\frac{dP}{d\Omega} = r^2 \hat{x}^i \langle t^{i0} \rangle$$

We use for $\langle t^{\mu\nu} \rangle$ the value (10.3.5), so this gives

$$\frac{dP}{d\Omega} = \frac{r^2(\mathbf{k}\cdot\hat{\mathbf{x}})k^0}{16\pi G} \left[e^{\lambda\nu*}(\mathbf{x}, \omega)e_{\lambda\nu}(\mathbf{x}, \omega) - \tfrac{1}{2}|e^\lambda{}_\lambda(\mathbf{x}, \omega)|^2\right]$$

and inserting the values (10.4.8) and (10.4.10) for k^μ and $e_{\lambda\nu}$, we find that the r^2 factors cancel, and

$$\frac{dP}{d\Omega} = \frac{G\omega^2}{\pi} \left[T^{\lambda\nu*}(\mathbf{k}, \omega)T_{\lambda\nu}(\mathbf{k}, \omega) - \tfrac{1}{2}|T^\lambda{}_\lambda(\mathbf{k}, \omega)|^2\right] \tag{10.4.13}$$

The problem is thus solved once we have calculated the Fourier transform (10.4.11).

It will be convenient to express this result in terms of the purely spacelike components of $T^{\lambda\nu}(\mathbf{k}, \omega)$. From (10.4.12) we have

$$T_{0i}(\mathbf{k}, \omega) = -\hat{k}^j T_{ji}(\mathbf{k}, \omega)$$

$$T_{00}(\mathbf{k}, \omega) = \hat{k}^i \hat{k}^j T_{ji}(\mathbf{k}, \omega)$$

where $\hat{\mathbf{k}} \equiv \mathbf{k}/\omega \equiv \hat{\mathbf{x}}$. Using this in (10.4.13) gives

$$\frac{dP}{d\Omega} = \frac{G\omega^2}{\pi} \Lambda_{ij,lm}(\hat{k})T^{ij*}(\mathbf{k}, \omega)T^{lm}(\mathbf{k}, \omega) \tag{10.4.14}$$

where

$$\Lambda_{ij,lm}(\hat{k}) \equiv \delta_{il}\delta_{jm} - 2\hat{k}_j\hat{k}_m\delta_{il} + \tfrac{1}{2}\hat{k}_i\hat{k}_j\hat{k}_l\hat{k}_m$$
$$- \tfrac{1}{2}\delta_{ij}\delta_{lm} + \tfrac{1}{2}\delta_{ij}\hat{k}_l\hat{k}_m + \tfrac{1}{2}\delta_{lm}\hat{k}_i\hat{k}_j \tag{10.4.15}$$

If the energy-momentum tensor is a *sum* of individual Fourier components as in (10.4.2), then the field $h_{\mu\nu}$ in the wave zone will look like a sum of the plane waves (10.4.7). The gravitational energy-momentum tensor will then be given by a double sum over these Fourier components, but all cross-terms drop out when we average over a time interval long compared with the longest "beat period," that is, the reciprocal of the shortest frequency difference. The power is thus given by a sum of terms like (10.4.14), one for each frequency in the source.

Suppose, on the other hand, that the energy-momentum tensor is a Fourier *integral*, as in (10.4.1). Then $h_{\mu\nu}$ in the wave zone will look like an integral over ω of the individual plane waves (10.4.7), and the gravitational energy-momentum tensor will be given by a double integral $\iint d\omega\, d\omega'$ of products of these terms. The integrand again has time dependence $\exp(-i(\omega - \omega')t)$, but now there is no "longest beat period," so instead of computing the average power we calculate the total energy emitted. This is given by integrating the power over all time, and the effect is to replace the factors $e^{-i\omega t}e^{i\omega't}$ in the double integral for the power with

$$\int_{-\infty}^{\infty} \exp(-i(\omega - \omega')t)\, dt = 2\pi\, \delta(\omega - \omega')$$

The energy per solid angle emitted in a direction $\hat{\mathbf{k}}$ is thus a single integral:

$$\frac{dE}{d\Omega} = 2G \int_{0}^{\infty} \omega^2 [T^{\lambda\nu*}(\mathbf{k}, \omega)T_{\lambda\nu}(\mathbf{k}, \omega) - \tfrac{1}{2}|T^{\lambda}{}_{\lambda}(\mathbf{k}, \omega)|^2]\, d\omega \qquad (10.4.16)$$

or, in terms of the space-space components,

$$\frac{dE}{d\Omega} = 2G\Lambda_{ij,lm}(\hat{k}) \int_{0}^{\infty} \omega^2 T^{ij*}(\mathbf{k}, \omega)T^{lm}(\mathbf{k}, \omega)\, d\omega$$

As an example, consider a system of free particles n that initially move at constant velocity $\mathbf{v}_n$, collide at the origin at $t = 0$, and then move off again at velocities $\tilde{\mathbf{v}}_n$. The energy-momentum tensor is then

$$T^{\mu\nu}(\mathbf{x}, t) = \sum_{n} \frac{P_n{}^{\mu}P_n{}^{\nu}}{E_n} \delta^3(\mathbf{x} - \mathbf{v}_n t)\theta(-t)$$

$$+ \sum_{n} \frac{\tilde{P}_n{}^{\mu}\tilde{P}_n{}^{\nu}}{\tilde{E}_n} \delta^3(\mathbf{x} - \tilde{\mathbf{v}}_n t)\theta(t) \qquad (10.4.17)$$

where $P_n{}^0 = E_n = m_n(1 - \mathbf{v}_n{}^2)^{-1/2}$ and $\mathbf{P}_n = E_n\mathbf{v}_n$ are the energy and momentum of the nth incoming particle, $\tilde{P}_n{}^0 = \tilde{E}_n$ and $\tilde{\mathbf{P}}_n$ are the corresponding quantities for the outgoing particles, and θ is the step function

$$\theta(s) = \begin{cases} +1 & s > 0 \\ 0 & s < 0 \end{cases} \qquad (10.4.18)$$

The functions θ and δ^3 have the well-known integral representations

$$\theta(s) = \frac{1}{2\pi i} \int_{-\infty}^{\infty} \frac{e^{+i\omega s}}{\omega - i\varepsilon}\, d\omega \qquad \varepsilon \to 0+ \qquad (10.4.19)$$

$$\delta^3(\mathbf{x}) = \frac{1}{(2\pi)^3} \int d^3k\, e^{i\mathbf{k}\cdot\mathbf{x}} \qquad (10.4.20)$$

(To prove (10.4.19), note that the contour can be closed with a large semicircle in the lower or upper half-plane, according to whether $s < 0$ or $s > 0$. To prove (10.4.20), take the Fourier transform of both sides.) We see then that $T^{\mu\nu}(\mathbf{x}, t)$ is of the form (10.4.1), with

$$
T^{\mu\nu}(\mathbf{x}, \omega) = \frac{1}{(2\pi)^4 i} \left[\sum_n \frac{P_n{}^\mu P_n{}^\nu}{E_n} \int d^3\mathbf{k} \, \frac{e^{i\mathbf{k}\cdot\mathbf{x}}}{\omega - \mathbf{v}_n \cdot \mathbf{k} - i\varepsilon} \right.
$$
$$
\left. - \sum_n \frac{\tilde{P}_n{}^\mu \tilde{P}_n{}^\nu}{\tilde{E}_n} \int d^3\mathbf{k} \, \frac{e^{i\mathbf{k}\cdot\mathbf{x}}}{\omega - \tilde{\mathbf{v}}_n \cdot \mathbf{k} + i\varepsilon} \right]
$$

and the Fourier transform (10.4.11) is

$$
T^{\mu\nu}(\mathbf{k}, \omega) = \frac{1}{2\pi i} \left[\sum_n \frac{P_n{}^\mu P_n{}^\nu}{E_n(\omega - \mathbf{v}_n \cdot \mathbf{k} - i\varepsilon)} - \sum_n \frac{\tilde{P}_n{}^\mu \tilde{P}_n{}^\nu}{\tilde{E}_n(\omega - \tilde{\mathbf{v}}_n \cdot \mathbf{k} + i\varepsilon)} \right]
$$

We can now drop the $\pm i\varepsilon$ in the denominators, for $\omega - \mathbf{v}_n \cdot \mathbf{k}$ cannot vanish if $\omega = |\mathbf{k}|$ and $|\mathbf{v}_n| < 1$. (For the case of particles traveling at the speed of light, see below.) Also, $E_n(\mathbf{v}_n \cdot \mathbf{k} - \omega) = P_n{}^\lambda k_\lambda \equiv (P_n \cdot k)$, so we can write

$$
T^{\mu\nu}(\mathbf{k}, \omega) = \frac{1}{2\pi i} \sum_N \frac{P_N{}^\mu P_N{}^\nu \eta_N}{(P_N \cdot k)} \tag{10.4.21}
$$

where N runs over particles in both the initial and the final states, the sign factor η_N being

$$
\eta_N = \begin{cases} +1 & N \text{ in final state} \\ -1 & N \text{ in initial state} \end{cases}
$$

We note that (10.4.12) is satisfied, because

$$
k_\mu T^{\mu\nu}(\mathbf{k}, \omega) = \frac{1}{2\pi i} \sum_N P_N{}^\mu \eta_N
$$

and this must vanish because $\sum_N P_N{}^\mu \eta_N$ is simply the change in the total P^μ, which is conserved.

The gravitational energy per solid angle and per unit frequency interval emitted at frequency ω and direction $\hat{\mathbf{k}}$ is now given by (10.4.16) as

$$
\left(\frac{dE}{d\Omega \, d\omega} \right) = \frac{G\omega^2}{2\pi^2} \sum_{N,M} \frac{\eta_N \eta_M}{(P_N \cdot k)(P_M \cdot k)} \left[(P_N \cdot P_M)^2 - \tfrac{1}{2} m_N{}^2 m_M{}^2 \right] \tag{10.4.22}
$$

If we tried to compute the total emitted energy by integrating ω from 0 to ∞, we would get a result that diverges like $\int^\infty d\omega$. This is just due to our approximation that the collision occurs instantaneously; actually it must take up some finite time Δt, and the ω-integral will be cut off at ω of order $1/\Delta t$.

Note that if none of the momenta $P_N{}^\mu$ are changed by the collision, then the contributions of the incoming and outgoing particles in (10.4.21) will cancel, and so $T_{\mu\nu}(\mathbf{k}, \omega)$ will vanish. Gravitational radiation is only emitted when the particles actually undergo accelerations.

Note also that (10.4.22) seems to become infinite if one of the particles participating in the reaction (say, $N = 1$) has zero mass, and has momentum approaching a direction parallel to $\mathbf{k}$, since then $P_1 \cdot k = E_1 \omega(\hat{\mathbf{P}}_1 \cdot \hat{\mathbf{k}} - 1) \to 0$. However, this singularity is spurious, for when $\mathbf{P}_1$ becomes parallel to $\mathbf{k}$ we can treat $(P_1 \cdot P_M)$ in (10.4.22) as proportional for all $M \neq 1$ to $(k \cdot P_M)$, so the singular part of (10.4.22) is

$$\frac{G\omega^2}{\pi^2} \frac{\eta_1}{(P_1 \cdot k)} \sum_{M \neq 1} \frac{\eta_M}{(P_M \cdot k)} (P_1 \cdot P_M)^2 \propto \sum_{M \neq 1} \eta_M(P_1 \cdot P_M)$$

We have already remarked that $\sum_M \eta_M P_M{}^\mu$ must vanish when the sum is extended over all particles, so the right-hand side is simply $-\eta_1 P_1{}^2$, and this vanishes because particle 1 is assumed to have zero mass. Thus no difficulty is encountered in applying (10.4.22) to collisions involving photons, neutrinos, or even (to run ahead of ourselves a bit) gravitons.

The total energy per unit frequency interval emitted as gravitational radiation in a collision is obtained by integrating Eq. (10.4.22) over the directions of $\hat{\mathbf{k}}$. We then find

$$\frac{dE}{d\omega} = \frac{G}{2\pi} \sum_{NM} \eta_N \eta_M m_N m_M \frac{1 + \beta_{NM}^2}{\beta_{NM}(1 - \beta_{NM}^2)^{1/2}} \ln\left(\frac{1 + \beta_{NM}}{1 - \beta_{NM}}\right) \qquad (10.4.23)$$

with β_{NM} the relative speed of particles N and M:

$$\beta_{NM} \equiv \left[1 - \frac{m_N^2 m_M^2}{(P_N \cdot P_M)^2}\right]^{1/2}$$

For nonrelativistic two-body elastic scattering this reduces to

$$\frac{dE}{d\omega} = \frac{8G}{5\pi} \mu^2 v^4 \sin^2 \theta \qquad (10.4.24)$$

where μ is the reduced mass, v is the relative velocity, and θ is the scattering angle in the center-of-mass reference frame.

The gravitational radiation produced by the collisions occurring in a gas can be determined by summing up the radiated energies per collision given by Eq. (10.4.23) or (10.4.24), provided that there is enough time between collisions so that they do not interfere. This conditions can be expressed as

$$\omega \gg \omega_c \qquad (10.4.25)$$

where ω_c is the collision frequency of a typical gas particle. (If $\omega \ll \omega_c$, then the gas behaves as a fluid rather than as a collection of independent particles.) When (10.4.25) is satisfied, the power per unit volume and per unit frequency interval is

$$\frac{dP}{d\omega} = \frac{8G}{5\pi} \sum_{(a,b)} \mu_{ab}^2 n_a n_b \left\langle v_{ab}^5 \int \frac{d\sigma_{ab}}{d\Omega} \sin^2 \theta \, d\Omega \right\rangle \qquad (10.4.26)$$

where n_a is the number density of gas particles of type a, $d\sigma_{ab}/d\Omega$ is the center-of-mass-system differential scattering cross-section, the sum runs over all different pairs of particle types, and the average $\langle\cdots\rangle$ is taken over all collisions.

As an example, let us calculate the gravitational radiation emitted by Coulomb collisions in a plasma. The Rutherford scattering cross-section is

$$\frac{d\sigma_{ab}}{d\Omega} = \frac{e_a^2 e_b^2}{4v_{ab}^4 \mu_{ab}^2 \sin^4(\theta/2)} \tag{10.4.27}$$

The integral over θ must be cut off at a minimum angle $1/\Lambda$, with $\Lambda \gg 1$ determined by the Debye screening of the Coulomb force at large impact parameters; we have then

$$\int \frac{d\sigma_{ab}}{d\Omega} \sin^2\theta \, d\Omega \simeq \frac{4\pi e_a^2 e_b^2 \ln\Lambda}{\mu_{ab}^2 v_{ab}^4} \tag{10.4.28}$$

We are left with an average of v_{ab}, which for a Maxwell-Boltzmann distribution is

$$\langle v_{ab}\rangle = 2\left(\frac{2kT}{\pi\mu_{ab}}\right)^{1/2} \tag{10.4.29}$$

Putting (10.4.28) and (10.4.29) into (10.4.26) gives the power per unit volume and per unit frequency interval (in c.g.s. units) as

$$\frac{dP}{d\omega} = \frac{64G}{5c^5}\left(\frac{2kT}{\pi}\right)^{1/2} \ln\Lambda \sum_{(a,b)} \frac{n_a n_b e_a^2 e_b^2}{\sqrt{\mu_{ab}}} \tag{10.4.30}$$

Typically $\ln\Lambda$ is of order 10. For a plasma of completely ionized hydrogen we must take into account electron-electron and electron-proton collisions, and (10.4.30) gives

$$\frac{dP}{d\omega} = \frac{64Gn_e^2 e^4}{5c^5}\left(\frac{2kT}{\pi m_e}\right)^{1/2}(1+\sqrt{2})\ln\Lambda \tag{10.4.31}$$

The electron collision frequency may in this case be estimated as

$$\omega_c \approx \frac{e^4 n_e \langle v\rangle}{(kT)^2} \approx \frac{e^4 n_e}{(kT)^{3/2}\sqrt{m_e}} \tag{10.4.32}$$

Equation (10.4.30) or (10.4.31) holds for $\omega \gg \omega_c$ and $\hbar\omega \ll kT$.

These results can be applied to the hydrogen plasma in the solar core. Within a volume V of order 2×10^{31} cm^3 this plasma has $T \simeq 10^7\,^\circ$K, $n_e \simeq 3 \times 10^{25}$ cm^{-3}, and $\ln\Lambda \simeq 4$. The collision frequency (10.4.32) is 10^{15} sec^{-1}, three orders of magnitude less than the thermal frequency $kT/\hbar \approx 10^{18}$ sec^{-1}, so the total power produced in gravitational radiation can be estimated by multiplying (10.4.31) by $VkT/\hbar$. In this way we find that the thermal collisions in the solar core produce about 10^8 watts of gravitational radiation.

5   Quadrupole Radiation

Up to this point we have made no approximations beyond the basic assumption that the fields are weak. (Our use of the wave zone limits $r \gg R, r \gg 1/\omega, r \gg \omega R^2$ was not really an approximation, since we can always choose r large enough to make these assumptions true; and by the conservation of energy, the power passing through a sphere at large r must equal that passing through any surface enclosing the radiating system.) We now make a further approximation, and assume that the source radius R is much smaller than the wavelength $1/\omega$:

$$\omega R \ll 1 \tag{10.5.1}$$

Most of the radiation is emitted at frequencies of order $\bar{v}/R$, where $\bar{v}$ is some typical velocity within the system, so we are really making the same sort of approximation as that made in the previous chapter, that is, $\bar{v} \ll 1$.

When (10.5.1) holds we may approximate the Fourier transforms needed in (10.4.14) and (10.4.16) by the **k**-independent integral

$$T_{ij}(\mathbf{k}, \omega) \simeq \int d^3x\, T_{ij}(\mathbf{x}, \omega) \tag{10.5.2}$$

This can be rewritten in a useful way by using the conservation laws in the form

$$\frac{\partial^2}{\partial x^i\, \partial x^j}\, T^{ij}(\mathbf{x}, \omega) = -\omega^2 T^{00}(\mathbf{x}, \omega)$$

Multiplying with $x^i x^j$ and integrating over **x**, we find

$$T_{ij}(\mathbf{k}, \omega) \simeq -\frac{\omega^2}{2}\, D_{ij}(\omega) \tag{10.5.3}$$

$$D_{ij}(\omega) \equiv \int d^3x\, x^i x^j T^{00}(\mathbf{x}, \omega) \tag{10.5.4}$$

The power per solid angle is therefore

$$\frac{dP}{d\Omega} = \frac{G\omega^6}{4\pi}\, \Lambda_{ij,lm}(\hat{k}) D_{ij}^*(\omega) D_{lm}(\omega) \tag{10.5.5}$$

If the source is a sum of Fourier components, then the power radiated is a sum of terms such as (10.5.5). If the source is a Fourier integral like (10.4.1), then the energy emitted per solid angle is

$$\frac{dE}{d\Omega} = \tfrac{1}{2}G\Lambda_{ij,lm}(\hat{k}) \int_0^\infty \omega^6 D_{ij}^*(\omega) D_{lm}(\omega)\, d\omega \tag{10.5.6}$$

The coefficients $D_{ij}(\omega)$ in (10.5.5) and (10.5.6) do not depend on the direction $\hat{k}$ of the emitted radiation, so we can do the integral over solid angle once and for all. We use the formulas

$$\int d\Omega \hat{k}_i \hat{k}_j = \frac{4\pi}{3} \delta_{ij}$$

$$\int d\Omega \hat{k}_i \hat{k}_j \hat{k}_l \hat{k}_m = \frac{4\pi}{15} (\delta_{ij}\delta_{lm} + \delta_{il}\delta_{jm} + \delta_{im}\delta_{jl})$$

(The form of the right-hand sides is dictated by symmetry and rotational invariance; the numerical coefficients can be calculated by contracting i with j and l with m.) We then find

$$\int d\Omega \Lambda_{ij,lm}(\hat{k}) = \frac{2\pi}{15} [11\delta_{il}\delta_{jm} - 4\delta_{ij}\delta_{lm} + \delta_{im}\delta_{jl}]$$

so the power emitted at a single discrete frequency ω is

$$P = \frac{2G\omega^6}{5} [D_{ij}^*(\omega)D_{ij}(\omega) - \tfrac{1}{3}|D_{ii}(\omega)|^2] \tag{10.5.7}$$

whereas for a smooth distribution of frequencies the total emitted energy is

$$E = \frac{4\pi G}{5} \int_0^\infty \omega^6 [D_{ij}^*(\omega)D_{ij}(\omega) - \tfrac{1}{3}|D_{ii}(\omega)|^2] \, d\omega \tag{10.5.8}$$

Before going on to calculate the quadrupole radiation emitted in a few special cases, it will be necessary to pause for a few comments on the method of calculation:

(A) The quadrupole approximation is usually applicable to nonrelativistic systems, and for these systems the energy density $T^{00}(\mathbf{x}, \omega)$ is dominated by the rest-mass density of the system. It may be surprising that we do not need to take explicit account of the potential and kinetic energy terms in the full tensor $T^{\mu\nu}$, because such terms must be included if $T^{\mu\nu}$ is to be conserved! Indeed, for a system of particles bound by gravitational forces we should in principle take $T^{\mu\nu}$ as the total "tensor" $\tau^{\mu\nu}$ constructed in Section 7.6, including terms nonlinear in the gravitational fields. However, we have already exploited energy and momentum conservation in our derivation of Eqs. (10.5.3)–(10.5.6), and since this has given a result involving only T^{00}, we are now free to approximate T^{00} with the rest-mass density.

(B) For general systems of vibrating and/or rotating solids, it is often quite difficult to evaluate the Fourier transform $T^{00}(\mathbf{x}, \omega)$, defined by Eq. (10.4.1) or (10.4.2). It is much easier first to evaluate the moments

$$D_{ij}(t) \equiv \int d^3x \, x^i x^j T^{00}(\mathbf{x}, t) \tag{10.5.9}$$

and then evaluate $D_{ij}(\omega)$ by expressing $D_{ij}(t)$ as a Fourier integral,

$$D_{ij}(t) = \int_0^\infty d\omega D_{ij}(\omega)e^{-i\omega t} + \text{c.c.} \tag{10.5.10}$$

or as a sum of Fourier components,

$$D_{ij}(t) = \sum_\omega e^{-i\omega t}D_{ij}(\omega) + \text{c.c.} \tag{10.5.11}$$

(C) The question may arise: What origin should be taken for the coordinates x^i in the integral (10.5.4) for D_{ij}? In principle, it doesn't matter. When we shift the origin of coordinates by an amount a_i, we change D_{ij} into

$$\int (x^i - a^i)(x_j - a_j)T^{00}(\mathbf{x}, t)\, d^3x$$

$$= \int x^i x^j T^{00}(\mathbf{x}, t)\, d^3x - a^i \int x^j T^{00}(\mathbf{x}, t)\, d^3x - a^j \int x^i T^{00}(\mathbf{x}, t)\, d^3x$$

$$+ a^i a^j \int T^{00}(\mathbf{x}, t)\, d^3x$$

But conservation of energy and momentum tells us that the last three terms are at most linear functions of time, because

$$\frac{\partial}{\partial t} \int T^{00}(x, t)\, d^3x = -\int \frac{\partial}{\partial x^i} T^{i0}(x, t)\, d^3x = 0$$

$$\frac{\partial^2}{\partial t^2} \int x^i T^{00}(x, t)\, d^3x = \int x^i \frac{\partial^2}{\partial x^j\, \partial x^k} T^{jk}(x, t)\, d^3x$$

$$= -\int \frac{\partial}{\partial x^j} T^{ij}(x, t)\, d^3x = 0$$

Thus the shift in origin does not affect the Fourier components with $\omega \neq 0$, that is,

$$D_{ij}(\omega) \equiv \int x^i x^j T^{00}(\mathbf{x}, \omega)\, d^3x = \int (x^i - a^i)(x^j - a^j)T^{00}(\mathbf{x}, \omega)\, d^3x \tag{10.5.12}$$

However, it is only when we take T^{00} as the energy density of the entire system that we can shift origins freely in computing $D_{ij}(\omega)$.

As a first example, let us calculate the gravitational radiation produced by a sound wave in a tube lying in the z-direction. The density of the vibrating material can be written

$$\rho = \rho_0 + \rho_1$$

where ρ_0 is the constant unperturbed value and ρ_1 is a small perturbation. We also treat the material velocity v (in the z-direction) as a small perturbation, and neglect dissipative effects. The equations of motion are then

$$\frac{\partial \rho_1}{\partial t} + \rho_0 \frac{\partial v}{\partial z} = 0$$

$$\rho_0 \frac{\partial v}{\partial t} + v_s^2 \frac{\partial \rho_1}{\partial z} = 0$$

where v_s is the speed of sound. The tube is not supported at its ends (otherwise we would have to take into account the gravitational radiation emitted by the support!), so the pressure $v_s^2 \rho_1$ must vanish at the tube ends. With this boundary condition, the general solution for a tube extending from $z = 0$ to $z = L$ is a superposition of the normal modes

$$v = -\varepsilon v_s \cos kz \sin (\omega t + \phi) \tag{10.5.13}$$

$$\rho_1 = \varepsilon \rho_0 \sin kz \cos (\omega t + \phi) \tag{10.5.14}$$

where ε is a small dimensionless number, ϕ is an arbitrary phase, and

$$k = N \frac{\pi}{L} \qquad \omega = N\pi \frac{v_s}{L} \tag{10.5.15}$$

with N any positive integer. Since v is not constrained to vanish at the tube ends, these ends will in general be displaced by amounts $\delta(0, t)$ and $\delta(L, t)$, respectively, where

$$\delta(z, t) \equiv \int v(z, t) \, dt = \varepsilon v_s \omega^{-1} \cos kz \cos (\omega t + \phi)$$

The time-dependent part of the second-moment of the mass density is given here by

$$D_{ij}(t) = n_i n_j A \left(\int_0^L \rho_1(z, t) z^2 \, dz + L^2 \rho_0 \, \delta(L, t) \right)$$

where A is the cross-sectional area of the tube, and $\mathbf{n} = (0, 0, 1)$ is a unit vector in the z-direction. This vanishes for N even, whereas for N odd we find

$$D_{ij}(t) = - \left(\frac{4 n_i n_j M L^2 \varepsilon}{N^3 \pi^3} \right) \cos (\omega t + \phi)$$

where $M \equiv \rho_0 A L$ is the mass of the tube. (The reader can easily check that $D_{ij}(t)$ would be the same if the second moment of the mass distribution were evaluated using as origin some point other than $z = 0$.) Comparing with Eq. (10.5.11), we see that $D_{ij}(t)$ has a Fourier component

$$D_{ij} \left(N\pi \frac{v_s}{L} \right) = - \frac{2 n_i n_j M L^2 \varepsilon}{N^3 \pi^3} \tag{10.5.16}$$

The radiated power is thus given by Eq. (10.5.7) (in c.g.s. units) as

$$P = \frac{16GM^2 v_s{}^6 \varepsilon^2}{15L^2 c^5} \tag{10.5.17}$$

for each odd N. This may be compared with the total energy of the oscillation, which is simply the kinetic energy at the times when ρ_1 vanishes, that is, when v is greatest:

$$E = \tfrac{1}{2}\rho_0 A \int_0^L v^2{}_{max}(z)\, dz = \tfrac{1}{4}M v_s{}^2 \varepsilon^2$$

Evidently the emission of gravitational radiation will cause the oscillator to lose energy at a rate

$$\Gamma_{grav} \equiv \frac{P}{E} = \frac{64GM v_s{}^4}{15L^2 c^5} \tag{10.5.18}$$

For instance, let us calculate the rate of gravitational radiation by acoustic oscillations in the large aluminum cylinders used as antennas in Weber's experiments on gravitational radiation.[1] (As we shall see, the effective cross-section of such antennas is determined by Γ_{grav}.) Weber's cylinders have the parameters

$$L = 153 \text{ cm} \qquad v_s = 5.1 \times 10^5 \text{ cm/sec} \qquad M = 1.4 \times 10^6 \text{ gm}$$

Hence, if gravitational radiation were the only loss mechanism, the oscillations (10.5.13), (10.5.14) with N odd would lose energy at a rate

$$\Gamma_{grav} = 4.7 \times 10^{-35} \text{ sec}^{-1} \tag{10.5.19}$$

In contrast, the actual decay rate Γ of the $N = 1$ mode in this cylinder is about 0.15 sec^{-1}, owing primarily to viscous dissipation within the aluminum. Hence the "branching ratio" of gravitational radiation here is of order

$$\eta \equiv \frac{\Gamma_{grav}}{\Gamma} \simeq 3 \times 10^{-34} \qquad (N = 1) \tag{10.5.20}$$

Any ordinary mechanical oscillation will always give up vastly more of its energy to heat than to gravitational radiation.

As a second example, let us calculate the power radiated by a rotating body. If the body rotates rigidly about the 3-axis with angular frequency T, then the mass density T^{00} will take the form

$$T^{00}(\mathbf{x}, t) = \rho(\mathbf{x}')$$

where $\rho(\mathbf{x}')$ is the mass density expressed in coordinates $\mathbf{x}'$ fixed in the body, defined by

$$x_1 \equiv x_1' \cos \Omega t - x_2' \sin \Omega t$$
$$x_2 \equiv x_1' \sin \Omega t + x_2' \cos \Omega t$$
$$x_3 \equiv x_3'$$

Hence, by changing coordinates in (10.5.9), we may express $D_{ij}(t)$ in terms of the moment-of-inertia tensor in body-fixed coordinates

$$I_{ij} \equiv \int d^3x' x'_i x'_j \rho(\mathbf{x}')$$
(10.5.21)

For simplicity, let us consider rotation around one of the principal axes of the ellipsoid of inertia, so that $I_{13} = I_{23} = 0$. We can also choose the x'_1 and x'_2 axes along the two other principal axes, so that $I_{12} = 0$. With I_{ij} diagonal, we now find

$$D_{11}(t) = \tfrac{1}{2}(I_{11} + I_{22}) + \tfrac{1}{2}(I_{11} - I_{22}) \cos 2\Omega t$$
$$D_{12}(t) = \tfrac{1}{2}(I_{11} - I_{22}) \sin 2\Omega t$$
$$D_{22}(t) = \tfrac{1}{2}(I_{11} + I_{22}) - \tfrac{1}{2}(I_{11} - I_{22}) \cos 2\Omega t$$
$$D_{13}(t) = D_{23}(t) = 0$$
$$D_{33}(t) = I_{33}$$

The nonvanishing Fourier coefficients for $\omega = 2\Omega$ in Eq. (10.5.11) are then

$$D_{11}(2\Omega) = -D_{22}(2\Omega) = iD_{12}(2\Omega) = \tfrac{1}{4}(I_{11} - I_{22})$$

According to Eq. (10.5.7), the total power emitted at twice the rotation frequency is then (in c.g.s. units)

$$P(2\Omega) = \frac{32G\Omega^6 I^2 e^2}{5c^5}$$
(10.5.22)

where I and e are the moment of inertia and equatorial ellipticity,

$$I \equiv I_{11} + I_{22}$$
(10.5.23)

$$e \equiv \frac{I_{11} - I_{22}}{I}$$
(10.5.24)

A body with circular symmetry around the axis of rotation will have $e = 0$, and therefore will not emit gravitational radiation. (Indeed, this conclusion does not even depend on the quadrupole approximation, since such a body, though rotating, has a time-independent energy-momentum tensor.) On the other hand, for a point mass m fixed in the *rotating* coordinate system at a point $x'_1 = r$, $x'_2 = x'_3 = 0$, the only nonvanishing element of I_{ij} will be $I_{11} = mr^2$, so that $I = mr^2$ and $e = 1$, and Eq. (10.5.23) gives a radiated power

$$P(2\Omega) = \frac{32G\Omega^6 m^2 r^4}{5c^5}$$
(10.5.25)

For instance, for the orbital motion of the planet Jupiter, we have

$$\Omega = 1.68 \times 10^{-8} \text{ sec}^{-1}, \qquad m = 1.9 \times 10^{30} \text{ g}, \qquad r = 7.78 \times 10^{13} \text{ cm}$$

and Eq. (10.5.25) gives a gravitational radiation power of only 5.3 kW, less even than the solar thermal gravitation power calculated in the last section. At this rate, it would take very much longer than the age of the solar system to observe any effects of this energy loss on Jupiter's orbit.

The negligibility of gravitational radiation in celestial mechanics can be stated in more general terms. For a system consisting of particles with typical mass $\bar{M}$, typical separations $\bar{r}$, and typical velocities $\bar{v}$, the power radiated at a frequency ω of order $\bar{v}/\bar{r}$ will be of order [compare Eq. (10.5.7)]

$$P \sim G \left(\frac{\bar{v}}{\bar{r}} \right)^6 \bar{M}^2 \bar{r}^4$$

or, since $G\bar{M}/\bar{r}$ is of order $\bar{v}^2$,

$$P \sim \bar{M} \frac{\bar{v}^8}{\bar{r}}$$

The typical deacceleration $\bar{a}_{\mathrm{rad}}$ of particles in the system owing to this energy loss is given by the power P divided by the momentum $\bar{M}\bar{v}$, or

$$\bar{a}_{\mathrm{rad}} \sim \frac{\bar{v}^7}{\bar{r}}$$

This may be compared with the accelerations computed in Newtonian mechanics, which are of order $\bar{v}^2/\bar{r}$, and with the post-Newtonian corrections discussed in the last chapter, which are of order $\bar{v}^4/\bar{r}$. [Radiation effects enter with an *odd* power of $\bar{v}$ because they represent an irreversible process, as shown by our use of an outgoing wave solution in Eq. (10.4.4).] Since radiation reaction is smaller than the post-Newtonian effects by a factor $\bar{v}^3 < 10^{-12}$, the neglect of radiation reaction in the last section was perfectly justified. Indeed, if we had strength we could even compute the post-post-Newtonian accelerations,[2] which are of order $\bar{v}^6/\bar{r}$, without encountering the effects of gravitational radiation!

The discovery of the pulsars has provided us with a more promising source of gravitational radiation. As discussed in Section 11.4, pulsars are probably neutron stars,[3] with masses of the order of one solar mass, radii of the order of 10 km, and hence moments of inertia I of the order of 10^{45} g cm^2. A newborn pulsar, formed in a supernova, may be rotating with Ω of order 10^4 sec^{-1}, so according to Eq. (10.5.22), it would be emitting gravitational radiation at a rate of order $10^{55} e^2$ ergs/sec. For comparison, the total rotational energy of the pulsar would be about 10^{53} ergs, so most of the pulsar's kinetic energy would be radiated away as gravitational waves[4] within a few years, provided that the equatorial ellipticity e is greater than about 10^{-4}. This is too large a static ellipticity to be maintained in the huge gravitational field of a neutron star, but it might be possible for dynamical effects to produce a mean ellipticity this large, particularly in the early period before the pulsar has settled down to its equilibrium configuration. Eventually the pulsar will slow down sufficiently so that other loss mechanisms, such as magnetic dipole radiation (for which $P \propto \Omega^4$) become more important than gravitational radiation.

6   Scattering and Absorption of Gravitational Radiation

Consider a plane gravitational wave with polarization $e_{\mu\nu}$ and wave vector k^μ, impinging on a target at the origin. At great distances from the target, the gravitational wave will in general consist of the plane wave and an outgoing scattered wave,[5]

$$h_{\mu\nu}(\mathbf{x}, t) \xrightarrow[r \to \infty]{} \left[e_{\mu\nu} e^{i\mathbf{k} \cdot \mathbf{x}} + f_{\mu\nu}(\hat{\mathbf{x}}) \frac{e^{i\omega r}}{r} \right] e^{-i\omega t} \tag{10.6.1}$$

where $r \equiv |\mathbf{x}|$, $\hat{\mathbf{x}} \equiv \mathbf{x}/r$, $\omega \equiv |\mathbf{k}|$, and $f_{\mu\nu}$ is a *scattering amplitude*, which may depend on $\hat{\mathbf{x}}$ and ω, but not on r or t.

In order to analyze the energy balance between the gravitational wave and the target, it is necessary to decompose the wave (10.6.1) into incoming and outgoing parts. The plane-wave part has the Legendre expansion[6]

$$e^{i\mathbf{k} \cdot \mathbf{x}} = \sum_{l=0}^{\infty} (2l + 1)P_l(\hat{\mathbf{k}} \cdot \hat{\mathbf{x}})i^l j_l(\omega r)$$

where P_l is the usual Legendre polynomial and j_l is the spherical Bessel function[7] of order l. Asymptotically, we have[8]

$$i^l j_l(\omega r) \to \frac{1}{2i\omega r} [e^{i\omega r} - (-)^l e^{-i\omega r}]$$

so the sums over l become simply the Legendre expansions of delta functions[9]:

$$\sum_l (2l + 1)P_l(\mu) = 2\,\delta(1 - \mu)$$

$$\sum_l (2l + 1)(-)^l P_l(\mu) = 2\,\delta(1 + \mu)$$

The plane wave may therefore be asymptotically decomposed into outgoing and ingoing waves,

$$e^{i\mathbf{k} \cdot \mathbf{x}} \xrightarrow[r \to \infty]{} \frac{e^{i\omega r}}{i\omega r}\,\delta(1 - \hat{\mathbf{k}} \cdot \hat{\mathbf{x}}) - \frac{e^{-i\omega r}}{i\omega r}\,\delta(1 + \hat{\mathbf{k}} \cdot \hat{\mathbf{x}})$$

and the gravitational wave (10.6.1) has the corresponding decomposition

$$h_{\mu\nu} \xrightarrow[r \to \infty]{} [e_{\mu\nu}^{\text{out}} e^{i\omega r} + e_{\mu\nu}^{\text{in}} e^{-i\omega r}]e^{-i\omega t} + \text{c.c.} \tag{10.6.2}$$

where

$$e_{\mu\nu}^{\text{out}}(\mathbf{x}) = \frac{1}{i\omega r}[e_{\mu\nu}\,\delta(1 - \hat{\mathbf{k}} \cdot \hat{\mathbf{x}}) + i\omega f_{\mu\nu}(\hat{\mathbf{x}})] \tag{10.6.3}$$

$$e_{\mu\nu}^{\text{in}}(\mathbf{x}) = -\frac{1}{i\omega r} e_{\mu\nu}\,\delta(1 + \hat{\mathbf{k}} \cdot \hat{\mathbf{x}}) \tag{10.6.4}$$

The total power carried *out* of a large sphere of radius r by the outgoing wave part of (10.6.2) may be calculated, following the same reasoning as in Section 10.4, as

$$P_{\text{out}} = \int d\Omega \langle t_{\text{out}}^{0i} \rangle \hat{x}_i r^2 \tag{10.6.5}$$

where $\langle t_{\text{out}}^{0i} \rangle$ is the mean energy flux, obtained by using $e_{\mu\nu}^{\text{out}}$ in place of $e_{\mu\nu}$ in Eq. (10.3.5), and averaging over space-time dimensions large compared with $1/\omega$ and small compared with r. Inspection of Eq. (10.6.3) shows that P_{out} will consist of three terms,

$$P_{\text{out}} = P_{\text{scat}} + P_{\text{int}} + P_{\text{plane}} \tag{10.6.6}$$

which arise, respectively, from $f_{\mu\nu}$ alone, from the interference between $f_{\mu\nu}$ and $e_{\mu\nu}$, and from $e_{\mu\nu}$ alone. The first term, which represents the total power scattered away from the incident direction, is calculated by using (10.3.5) in (10.6.5), with $f_{\mu\nu}/r$ in place of $e_{\mu\nu}$:

$$P_{\text{scat}} = \frac{\omega^2}{16\pi G} \int d\Omega [f^{\lambda\nu*}(\hat{x})f_{\lambda\nu}(\hat{x}) - \tfrac{1}{2}|f^{\lambda}{}_{\lambda}(\hat{x})|^2] \tag{10.6.7}$$

The interference term is similarly calculated as

$$P_{\text{int}} = \frac{\omega^2}{8\pi G} \operatorname{Re} \left\{ -\frac{1}{i\omega} \int d\Omega \, \delta(1 - \hat{k} \cdot \hat{x})[e^{\lambda\nu*}f_{\lambda\nu}(\hat{x}) - \tfrac{1}{2}e^{\lambda}{}_{\lambda}*f^{\nu}{}_{\nu}(\hat{x})] \right\}$$

or, integrating over the delta function,

$$P_{\text{int}} = -\frac{\omega}{4G} \operatorname{Im} \{e^{\lambda\nu*}f_{\lambda\nu}(\hat{k}) - \tfrac{1}{2}e^{\lambda}{}_{\lambda}*f^{\nu}{}_{\nu}(\hat{k})\} \tag{10.6.8}$$

The last term in (10.6.6), which represents the power carried out of the sphere by the plane wave, is formally infinite for $r \to \infty$. However, a plane wave carries as much power into any volume as it carries out of it, so the power brought into a sphere of radius r by the ingoing wave (10.6.4) is also equal to this term

$$P_{\text{in}} = P_{\text{plane}} \tag{10.6.9}$$

Thus P_{plane} cancels out of the equation of energy conservation, which gives the power absorbed by the target as

$$P_{\text{abs}} = P_{\text{in}} - P_{\text{out}} = -P_{\text{scat}} - P_{\text{int}} \tag{10.6.10}$$

The energy flux in the incident wave is given by Eq. (10.3.5) as

$$\Phi \equiv \langle t^{0i} \rangle \hat{k}_i = \frac{\omega^2}{16\pi G} (e^{\lambda\nu*}e_{\lambda\nu} - \tfrac{1}{2}|e^{\nu}{}_{\nu}|^2) \tag{10.6.11}$$

Thus the effective cross-section for elastic scattering of the gravitational wave is

$$\sigma_{\text{scat}} \equiv \frac{P_{\text{scat}}}{\Phi}$$

$$= \frac{\int d\Omega[f^{\lambda\nu*}(\hat{\mathbf{x}})f_{\lambda\nu}(\hat{\mathbf{x}}) - \frac{1}{2}|f^{\lambda}{}_{\lambda}(\hat{\mathbf{x}})|^2]}{[e^{\lambda\nu*}e_{\lambda\nu} - \frac{1}{2}|e^{\nu}{}_{\nu}|^2]} \tag{10.6.12}$$

This must be distinguished from the *total cross-section* for scattering or absorption of the wave:

$$\sigma_{\text{tot}} \equiv \frac{P_{\text{scat}} + P_{\text{abs}}}{\Phi} \tag{10.6.13}$$

According to (10.6.10), the total cross-section can be expressed in terms of the interference between the incident and scattered waves,

$$\sigma_{\text{tot}} = -\frac{P_{\text{int}}}{\Phi} \tag{10.6.14}$$

or, using (10.6.8) and (10.6.11),

$$\sigma_{\text{tot}} = \frac{4\pi \ \text{Im} \ \{e^{\lambda\nu*}f_{\lambda\nu}(\hat{\mathbf{k}}) - \frac{1}{2}e^{\lambda}{}_{\lambda}*f^{\nu}{}_{\nu}(\hat{k})\}}{\omega(e^{\lambda\nu*}e_{\lambda\nu} - \frac{1}{2}|e^{\lambda}{}_{\lambda}|^2)} \tag{10.6.15}$$

This result, that the total cross-section is $4\pi/\omega$ times the imaginary part of a *forward* scattering amplitude, was first derived in classical electrodynamics,[10] and is therefore known as the *optical theorem*. Here and in electrodynamics it is a consequence of the conservation of energy, whereas in quantum mechanics there is a similar theorem based on the conservation of probability.[11]

Since the incident wave is weak, the scattering amplitude $f_{\lambda\nu}$ is a linear combination of the components of the incident polarization tensor $e_{\rho\sigma}$. It follows that the cross-sections (10.6.12) and (10.6.15) are independent of the *normalization* of $e_{\mu\nu}$, though they may depend on **k** and on the *form* of the polarization tensor. The aim of gravitational scattering theory is to calculate $f_{\lambda\nu}$; following this, the various cross-sections can be determined from (10.6.12) and (10.6.15).

7   Detection of Gravitational Radiation

Experiments that aim at the detection of gravitational radiation have been carried out by Weber over the last decade,[1] and are presently being planned in laboratories throughout the world. Most of these experiments make use of *resonant quadrupole antennas*, which can be any "small" mechanical or hydrodynamical system with a natural mode of free oscillation. It happens that the effective cross-sections of these antennas can be evaluated by use of the optical theorem derived in the last section, without any need for a detailed analysis of the interaction between the gravitational wave and the antenna.

Our first assumption is that the antenna is much smaller than the wavelength $2\pi/\omega$, so that the scattered gravitational wave is pure quadrupole radiation. By the same reasoning that earlier led to Eqs. (10.4.10), (10.4.12), (10.5.3), and (10.5.4), we may now conclude that the scattering amplitude in Eq. (10.6.1) takes the form

$$f_{\mu\nu}(\hat{\mathbf{x}}) = t_{\mu\nu}(\hat{\mathbf{x}}) - \tfrac{1}{2}\eta_{\mu\nu}t^{\lambda}{}_{\lambda}(\hat{\mathbf{x}}) \tag{10.7.1}$$

where $t_{\mu\nu}$ is proportional to the Fourier transform of the perturbation in $T_{\mu\nu}$ caused by the wave; the conservation of energy and momentum give, as before,

$$t_{0i}(\hat{\mathbf{x}}) = -\hat{x}_j t_{jk} \qquad t_{00}(\hat{\mathbf{x}}) = \hat{x}_i \hat{x}_j t_{ij} \tag{10.7.2}$$

where t_{ij} is independent of $\mathbf{x}$, though depending of course on ω, $e_{\mu\nu}$, and on the detailed interaction between the incident wave and the antenna. Adopting a coordinate system in which the incident propagation vector $\mathbf{k}$ is in the 3-direction, and a gauge in which the only nonvanishing elements of the polarization tensor are $e_{11} = -e_{22}$ and $e_{12} = e_{21}$, the total cross-section (10.6.15) is now given by

$$\sigma_{\text{tot}} = \frac{2\pi \,\text{Im}\, \{e_{11}^*(t_{11} - t_{22}) + 2e_{12}^* t_{12}\}}{\omega[|e_{11}|^2 + |e_{12}|^2]} \tag{10.7.3}$$

Also, the angular integral in (10.6.12) can now be calculated by the same method as in Section 10.5, and we find for the elastic scattering cross-section the value

$$\sigma_{\text{scat}} = \frac{4\pi[t_{ij}^* t_{ij} - \tfrac{1}{3}|t_{ii}|^2]}{5[|e_{11}|^2 + |e_{12}|^2]} \tag{10.7.4}$$

Our other assumption is that the scattering is resonant, that is, that the frequency ω of the incident wave is close to a natural frequency ω_0 of free oscillation of the antenna system. We can think of the incident wave as merely serving to excite this free oscillation, which then loses energy through reradiation of gravitational waves or into other channels, corresponding to elastic scattering or to absorption of the incident wave, respectively.

One consequence of this assumption is that the ratio of the elastic scattering cross-section to the total cross-section is simply equal to the fraction η of the energy of the free oscillation that is dissipated as gravitational radiation rather than heat, light, and so on,

$$\sigma_{\text{scat}} = \eta\sigma_{\text{tot}} \tag{10.7.5}$$

where

$$\eta \equiv \frac{\Gamma_{\text{grav}}}{\Gamma}$$

with Γ the total decay rate of the free oscillation and Γ_{grav} the decay rate owing to the emission of gravitational radiation. Since η is a parameter characterizing the free oscillation of the antenna, and has nothing to do with how the oscillation is excited, it is independent of $e_{\mu\nu}$.

Another consequence of the assumption of resonant scattering is that the *form* of the matrix t_{ij} is given by some fixed matrix n_{ij}, depending only on the geometric properties of the oscillation being excited. That is, t_{ij} must equal n_{ij} times some function of the polarization components e_{11} and e_{12}. The incident field is assumed here to be weak, so this latter function must be linear, and therefore

$$t_{ij} = n_{ij}(\alpha e_{11} + \beta e_{12}) \tag{10.7.6}$$

with n_{ij}, α, and β all independent of e_{11} and e_{12}. For instance, if the antenna has an axis of symmetry along some direction $\mathbf{n}$, then n_{ij} is a linear combination of δ_{ij} and $n_i n_j$; the term proportional to δ_{ij} does not contribute to (10.7.3) or (10.7.4), so in this case we could take

$$n_{ij} = n_i n_j \tag{10.7.7}$$

The two requirements, (10.7.5) and (10.7.6), impose stringent conditions on the scattering amplitude. Using (10.7.3), (10.7.4), and (10.7.6) in (10.7.5), we find

$$\frac{\eta}{\omega} \operatorname{Im} \{[e_{11}^*(n_{11} - n_{22}) + 2e_{12}^* n_{12}][\alpha e_{11} + \beta e_{12}]\}$$
$$= \tfrac{2}{5}|\alpha e_{11} + \beta e_{12}|^2[n_{ij}^* n_{ij} - \tfrac{1}{3}|n_{ii}|^2]$$

This must hold for all $e_{\mu\nu}$, so by equating the coefficients of $|e_{11}|^2$, $e_{11}^* e_{12}$, and $|e_{12}|^2$, we obtain the conditions

$$\frac{1}{|\alpha|^2} \operatorname{Im} \{(n_{11} - n_{22})\alpha\} = \frac{1}{2i\alpha^*\beta} \{(n_{11} - n_{22})\beta - 2n_{12}^*\alpha^*\}$$

$$= \frac{2}{|\beta|^2} \operatorname{Im} \{n_{12}\beta\}$$

$$= \frac{2\omega}{5\eta} [n_{ij}^* n_{ij} - \tfrac{1}{3}|n_{ii}|^2]$$

The solution of these equations takes the form

$$\alpha = \frac{5\eta g(n_{11}^* - n_{22}^*)}{2\omega[n_{ij}^* n_{ij} - \tfrac{1}{3}|n_{ii}|^2]}$$

$$\beta = \frac{5\eta g n_{12}^*}{\omega[n_{ij}^* n_{ij} - \tfrac{1}{3}|n_{ii}|^2]}$$

where g is a complex number with

$$\operatorname{Im} g = |g|^2 \tag{10.7.8}$$

The scattering amplitude (10.7.6) is now

$$t_{ij} = \frac{5g\eta n_{ij}[(n_{11}^* - n_{22}^*)e_{11} + 2n_{12}^* e_{12}]}{2\omega[n_{lm}^* n_{lm} - \frac{1}{3}|n_{ll}|^2]}$$

(10.7.9)

Note that this depends only on the form of the matrix n_{ij}, not on its normalization.

The final consequence of the assumption of resonant scattering is that the frequency dependence of the scattering amplitude t_{ij} is given by the Fourier transform of a function with the time-dependence

$$e^{-i\omega_0 t} e^{-\Gamma t/2}$$

which oscillates at frequency ω_0, and decays in amplitude and energy at the rates $\Gamma/2$ and Γ, respectively. That is, t_{ij} must have the frequency dependence

$$t_{ij} \propto \left[\omega - \omega_0 + i\frac{\Gamma}{2}\right]^{-1}$$

(10.7.10)

Since η and n_{ij} depend on the properties of the free oscillation, and not on how it is produced, this frequency dependence can only arise from the factor g. In order to satisfy the "unitarity" condition (10.7.8) for all ω, we must then have

$$g = \frac{-\Gamma/2}{\omega - \omega_0 + i\Gamma/2}$$

(10.7.11)

The total cross-section (10.7.3) for absorption or scattering of gravitational waves by the antenna is then, in c.g.s. units,

$$\sigma_{\text{tot}} = \left(\frac{5\pi\eta c^2}{\omega^2}\right)\left(\frac{\Gamma^2/4}{(\omega - \omega_0)^2 + \Gamma^2/4}\right)$$
$$\times \frac{|(n_{11}^* - n_{22}^*)e_{11} + 2n_{12}^* e_{12}|^2}{[n_{ij}^* n_{ij} - \frac{1}{3}|n_{ii}|^2][|e_{11}|^2 + |e_{12}|^2]}$$

(10.7.12)

It is truly remarkable that this cross-section is entirely determined as a function of $e_{\mu\nu}$ and ω by the parameters ω_0, Γ, and η, and by the form of the matrix n_{ij}, whether the resonant oscillation is mechanical, acoustical, electrical, or anything else.

In the special case of an antenna with an axis of circular symmetry **n**, the matrix n_{ij} has the simple form (10.7.7). Taking the symmetry axis to lie in the $1 - 3$ plane at an angle θ to the incident 3-direction, the nonvanishing elements of this matrix are

$$n_{11} = \sin^2\theta \qquad n_{13} = \cos\theta\sin\theta \qquad n_{33} = \cos^2\theta$$

The total cross-section (10.7.12) is then

$$\sigma_{\text{tot}} = \left(\frac{15\pi\eta c^2}{2\omega^2}\right)\left(\frac{\Gamma^2/4}{(\omega - \omega_0)^2 + \Gamma^2/4}\right)\sin^4\theta$$

$$\times \left(\frac{|e_{11}|^2}{|e_{11}|^2 + |e_{12}|^2}\right) \tag{10.7.13}$$

The factor $\sin^4\theta$ makes the cross-section greatest when the antenna axis is oriented at right angles to the direction of wave propagation, that is, for $\theta = \pi/2$. This is just a reflection of the fact that gravitational waves, like electromagnetic waves, are *transverse*.

When the polarization of the gravitational wave is not measured, the quantity of interest is the average of (10.7.12) over the helicities ± 2, that is, over polarization tensors with $e_{11} = \mp i e_{12}$:

$$\bar{\sigma}_{\text{tot}} = \left(\frac{5\pi\eta c^2}{2\omega^2}\right)\left(\frac{\Gamma^2/4}{(\omega - \omega_0)^2 + \Gamma^2/4}\right)\left(\frac{|n_{11} - n_{22}|^2 + 4|n_{12}|^2}{n_{ij}^*n_{ij} - \frac{1}{3}|n_{ii}|^2}\right) \tag{10.7.14}$$

For an antenna with an axis of circular symmetry, the effect of averaging over helicities is just to replace the last factor in Eq. (10.7.13) with $\frac{1}{2}$.

The above analysis is strictly applicable only where there is a single nondegenerate resonant oscillation. When there are several degenerate modes, the particular linear combination excited by a gravitational wave can depend on the polarization of the wave, so that t_{ij} need not be proportional to a fixed matrix n_{ij}. For instance, if the antenna is an elastic sphere, then any quadrupole oscillation will consist of five independent modes. In this case, t_{ij} must be a linear combination of δ_{ij} and e_{ij}, but again the term proportional to δ_{ij} does not contribute to (10.7.3) or (10.7.4), so we can take

$$t_{ij} = \gamma e_{ij}$$

Equations (10.7.3)–(10.7.5) now give

$$\text{Im } \gamma = \frac{2\omega}{5\eta}|\gamma|^2$$

so, since γ must have the frequency dependence (10.7.10), it is given by

$$\gamma = \left(\frac{5\eta}{2\omega}\right)\left(\frac{-\Gamma/2}{\omega - \omega_0 + i\Gamma/2}\right)$$

The total cross-section (10.7.3) is now

$$\sigma_{\text{tot}} = \left(\frac{10\pi\eta c^2}{\omega^2}\right)\left(\frac{\Gamma^2/4}{(\omega - \omega_0)^2 + \Gamma^2/4}\right) \tag{10.7.15}$$

for any incident polarization.

In all cases the effective cross-section has a maximum when the antenna is *tuned* so that the resonant frequency ω_0 is equal to the frequency ω of the incident wave. Inspection of (10.7.12)–(10.7.15) shows that this maximum cross-section is of the order

$$\sigma_{max} \approx \eta \lambda^2 \qquad (10.7.16)$$

where λ is the wavelength $2\pi c/\omega$. In the ideal case of an oscillation that decays purely through the emission of gravitational radiation, we would have $\eta = 1$, and σ_{max} would have the very large value λ^2. Of course, this ideal case is never even approached in practice; for instance, we found in Section 10.5 that Weber's large aluminum cylinders have $\eta \simeq 3 \times 10^{-34}$! Generally the rate Γ_{grav} at which a resonant oscillator emits gravitational radiation depends on the gross dimensions of the antenna, and is difficult to increase; thus, in order to make σ_{max} as large as possible, it is necessary to reduce the total loss rate Γ in the ratio $\eta \equiv \Gamma_{grav}/\Gamma$ as much as possible, possibly by employing some sort of oscillation in a superfluid.

However, tuning our antenna does no good unless we have some strong source of gravitational radiation with a known frequency to which we can tune. Perhaps the most promising source[12] is the pulsar NP 0532 in the Crab nebula. This object is observed to emit pulses of electromagnetic radiation at optical, X-ray, and radio frequencies with a period $2\pi/\Omega = 0.03309$ sec. As discussed in Section 10.5, the pulsars are believed to be rotating neutron stars[3] with moments of inertia of order 10^{45} gcm^2 and unknown equatorial ellipticities e. The Crab pulsar is therefore presumably emitting gravitational radiation, with $\omega = 2\Omega = 379.8$ Hz, at a rate of about $10^{45} e^2$ ergs/sec. Since the Crab nebula is at a distance of 6500 light years, or 6.2×10^{21} cm, the flux of gravitational radiation passing the earth should be about $\Phi \simeq e^2$ ergs/sec-cm^2. A resonant linear quadrupole antenna that is "aimed" at and tuned to the Crab pulsar will have a cross-section $\bar{\sigma}_{tot}$ given by (10.7.13) as $7.4 \times 10^{16}\eta$ cm^2. The power absorbed by the antenna will thus be of order $10^{16}e^2\eta$ ergs/sec. For instance, if η is 10^{-32} and e is of order 10^{-4}, this power is of order 10^{-24} ergs/sec, which might perhaps be detectable. Unfortunately, in order to use an aluminum cylinder of the sort discussed in Section 10.5 as an antenna tuned to the Crab, the cylinder would have to have the rather ungainly length $\pi v_s/\omega$ of 42 meters. To get around this difficulty, one can use as antennas hoops, forks, and so on, which have lower natural frequencies for a given size than bars or cylinders. A group at Rochester[13] is planning a hoop antenna that could be tuned to the Crab.

All of the experiments carried out so far by Weber have made use of resonant quadrupole antennas that are not tuned to any particular source. Since it is too much to ask that a monochromatic source like a pulsar should just happen to fall within the bandwidth of the antenna, these experiments really aim at the detection of broad-band gravitational radiation, with an energy flux $\Phi(\omega)\,d\omega$ between frequencies ω and $\omega + d\omega$. If exposed to such radiation, the power absorbed by a resonant antenna will be

$$P = \sigma_{max} \int \left[\frac{\Gamma^2/4}{(\omega - \omega_0)^2 + \Gamma^2/4} \right] \Phi(\omega)\,d\omega$$

where σ_{max} is the effective cross-section of the antenna at resonance, given by setting $\omega = \omega_0$ in (10.7.12), (10.7.13), (10.7.14), or (10.7.15). If $\Phi(\omega)$ is roughly constant over the frequency range $\omega_0 - \Gamma$ to $\omega_0 + \Gamma$, then it can be taken out of the integral, and we have

$$P = \pi \sigma_{max} \Phi(\omega_0) \frac{\Gamma}{2} \qquad (10.7.17)$$

For a source that radiates for a time much longer than the antenna relaxation time $1/\Gamma$, a quasi-steady state will be reached in which the mean energy E in the resonant mode is such that the loss rate $E\Gamma$ just balances the absorbed power P:

$$E = \frac{P}{\Gamma} = \pi \sigma_{max} \frac{\Phi(\omega_0)}{2} \qquad (10.7.18)$$

In this case, a measurement of the mean excitation energy of the resonant mode serves to measure, or at least to set an upper limit on, the power flux at the resonant frequency. For instance, the earth has a fundamental spheroidal oscillation mode[14] $_0S_2$, with a period $2\pi/\omega$ of 54 min and a decay rate Γ of order 5×10^{-6} sec^{-1}, in which the mass density perturbation is of the form $\rho_1(r) Y_2{}^m(\theta, \varphi)$. The gravitational decay rate Γ_{grav} of this mode will be roughly of order $GM_\oplus R_\oplus{}^2 \omega^4/c^5$ [compare Eq. (10.5.18)], or about 10^{-25} sec^{-1}, so the branching ratio η is of order 10^{-20}. The cross-section (10.7.15) at resonance is here $7.5 \times 10^{27}\eta$ cm^2, or roughly 10^7 to 10^8 cm^2. From seismic measurements of the mean strain in the earth's crust during quiet periods, Forward et al.[15] in 1961 set an upper limit on $\Phi(\omega_0)$ of roughly 20 watts/cm^2-Hz. It is hoped that a much better upper limit on Φ can be set by placing a gravimeter on the moon,[16] which is very much quieter seismically than the earth.

For a "burst" source that radiates for a time τ less than the antenna relaxation time $1/\Gamma$, the total energy picked up by the antenna will be

$$\Delta E = P\tau = \pi \sigma_{max} \Phi(\omega_0)\tau \frac{\Gamma}{2}$$

Thus the energy per unit area in the burst reaching the antenna *within the beam width* Γ may be determined as

$$\mathscr{E} \equiv \Phi(\omega_0)\Gamma\tau = \frac{2\Delta E}{\pi \sigma_{max}} \qquad (10.7.19)$$

However, if the source radiates for a time $\tau < 1/\Gamma$, its bandwidth must be greater than $1/\tau$, so the *total* energy per unit area in the burst must be larger than $\mathscr{E}$ by a factor greater than $(\tau\Gamma)^{-1}$.

The only positive indication so far of the presence of gravitational radiation in the universe comes from the experiments of Weber,[1] which use as antennas the aluminum cylinders described in Section 10.5. These antennas have the frequency and "branching ratio"

$$\omega_0/2\pi = 1660 \text{ Hz} \qquad \eta = 3 \times 10^{-34}$$

[see Eq. (10.5.20)] so by setting $\omega = \omega_0$ and averaging over helicities in Eq. (10.7.13), we find a cross-section at resonance

$$\bar{\sigma}_{max} = 2.9 \times 10^{-20} \sin^4 \theta \text{ cm}^2$$

If the smallest energy increment ΔE that can be distinguished from thermal fluctuations is kT, or 4×10^{-14} ergs at room temperature, then according to (10.7.19), a burst of gravitational radiation will be detectable if the energy $\mathscr{E}$ per unit area within the beam width satisfies the condition

$$\mathscr{E} \gtrsim 9 \times 10^5 \text{ ergs/cm}^2 \qquad \text{for } \theta = \pi/2$$

(It is actually possible to do a little better than this by careful data processing.) The mere observation of a number of pulses in a single cylinder would leave open the possibility that these pulses were due to nonthermal noise, such as seismic disturbances, electric storms, or cosmic rays, so Weber looked for coincident pulses in aluminum cylinders 1000 km apart, at College Park in Maryland and the Argonne National Laboratory in Illinois. In 1969 Weber reported over 100 coincident pulses, occurring at a rate that indicates a mean gravitational radiation flux (within the bandwidth $\Gamma \sim 0.1$ Hz) of about 0.1 erg cm^{-2} sec^{-1}.[17]

Shortly thereafter,[18] Weber found that the rate of coincident pulses was correlated with sidereal time, in a manner consistent with the expected $\sin^4 \theta$ antenna pattern if the gravitational radiation is coming from the center of our galaxy. (See Figure 10.1.) The galactic center is about 2.5×10^{22} cm from the earth, so an observed flux of 0.1 erg cm^{-2} sec^{-1} would indicate an energy production of about 8×10^{44} ergs/sec, or 0.013 $M_\odot c^2$/year. This would not in itself be so remarkable, but since Weber's antennas are not tuned to any particular frequency, an energy production of 0.01 $M_\odot c^2$ in a bandwidth of 0.1 Hz at 1660 Hz presumably represents a total energy production 10^3 to 10^5 times larger, or 10 to 10^3 $M_\odot c^2$/year. At this rate, the whole mass of the galaxy would be used up in 10^8 to 10^{10} years! If Weber is really observing gravitational radiation from the galactic center, then either he accidentally picked the precise frequency at which most of this radiation is emitted, or else he has discovered an incredibly powerful new source of energy.

Weber has also looked for scalar radiation, using a disk with a monopole mode of oscillation having the same frequency, 1660 Hz, as the cylinders. The coincidence rate is observed to be much less than for the pair of cylinders; the apparent correlation of coincidences with sidereal time agrees with a pure tensor theory.[19]

Plans are now in train to repeat Weber's experiments with much greater sensitivity. One important improvement that is being planned at Stanford[20] is

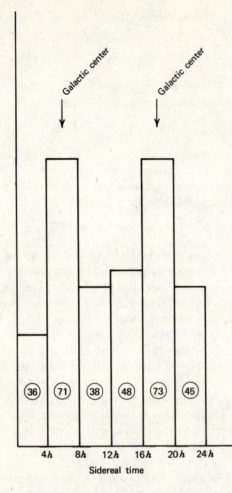

Figure 10.1 Evidence for gravitational radiation emanating from the center of the galaxy.[18] The detector intensity observed by Weber is plotted here (in arbitrary units) against sidereal time. Arrows mark the sidereal times at which the antenna is most nearly perpendicular to the line of sight to the center of the galaxy. Numbers in circles give the observed numbers of coincidences in each time interval.

to operate a cylindrical antenna at a very low temperature, in the range of milli-degrees Kelvin. If the antenna is limited by thermal noise, then lowering the temperature by a factor 10^{-5} would increase the sensitivity by a factor 10^{5}. A group at Moscow[21] is carrying out gravitational radiation experiments with improved instrumentation, and is designing novel kinds of gravitational wave antennas.[22] Weber is continuing his observations, using new antennas and instrumentation. At present, the best a theorist can do is to wait for the experimentalists to reach some sort of consensus on whether gravitational radiation has indeed been observed.

8 Quantum Theory of Gravitation*

At present there does not exist any complete and self-consistent quantum theory of gravitation, and it would be out of place in this book to describe in detail the attempts that have been made to construct such a theory. However, it will be possible and it may be useful to give the reader some taste of what a quantum theory of gravitation would be like.

To start at the simplest level, we would interpret a gravitational plane wave, with wave vector k_μ and helicity ± 2, as consisting of *gravitons*: quanta with energy-momentum vector $p^\mu = \hbar k^\mu$ and spin component in the direction of motion $\pm 2\hbar$. (Here $\hbar = 1.054 \times 10^{-27}$ erg sec). Since $k_\mu k^\mu = 0$, the graviton is a particle of zero mass, like the photon and neutrino. According to Eq. (2.8.4), the energy-momentum tensor of an assembly of gravitons, all of which have four-momenta $p^\mu = \hbar k^\mu$, is

$$T_{\mu\nu} = \frac{\hbar k_\mu k_\nu}{\omega} \mathcal{N} \qquad (10.8.1)$$

where $\mathcal{N}$ is the number of gravitons per unit volume. Comparing this with our result for a gravitational plane wave,

$$\langle t_{\mu\nu} \rangle = \frac{k_\mu k_\nu}{16\pi G} \left(|e_+|^2 + |e_-|^2 \right) \qquad (10.8.2)$$

we conclude that the number density of gravitons with helicity ± 2 in a plane wave is

$$\mathcal{N}_\pm = \frac{\omega}{16\pi\hbar G} |e_\pm|^2 \qquad (10.8.3)$$

The total number density is

$$\mathcal{N} = \mathcal{N}_+ + \mathcal{N}_- = \frac{\omega}{16\pi\hbar G} \left(e^{\lambda\nu *} e_{\lambda\nu} - \tfrac{1}{2} |e^\lambda{}_\lambda|^2 \right) \qquad (10.8.4)$$

In the same way, we can interpret our formula (10.4.13) for the power emitted as gravitational radiation by an arbitrary system as giving the rate $d\Gamma$ of emitting gravitons of energy $\hbar\omega$ into the solid angle $d\Omega$:

$$d\Gamma = \frac{dP}{\hbar\omega} = \frac{G\omega\, d\Omega}{\hbar\pi} \left[T^{\lambda\nu *}(k, \omega) T_{\lambda\nu}(k, \omega) - \tfrac{1}{2} |T^\lambda{}_\lambda(k, \omega)|^2 \right] \qquad (10.8.5)$$

However, the energy-momentum tensor $T^{\lambda\nu}(k, \omega)$ must now be interpreted as a matrix element of an energy-momentum tensor operator between final and initial

---

* This section lies somewhat out of the book's main line of development, and may be omitted in a first reading.

states. In particular, in the quadrupole approximation the total rate for an atom
to make a transition $a \rightarrow b$ by emitting gravitational radiation is

$$\Gamma(a \rightarrow b) = \frac{2G\omega^5}{5\hbar} \left[D_{ij}^*(a \rightarrow b) D_{ij}(a \rightarrow b) - \tfrac{1}{3} |D_{ij}(a \rightarrow b)|^2 \right]$$

$$(10.8.6)$$

where

$$D_{ij}(a \rightarrow b) \equiv m_e \int \psi_b^*(\mathbf{x}) x_i x_j \psi_a(\mathbf{x}) \, d^3\mathbf{x} \qquad (10.8.7)$$

with ψ_a, ψ_b the initial and final state wave functions. For instance, the rate for
decay of the $3d(m = 2)$ state of the hydrogen atom into the $1s$ state with emission
of one graviton is

$$\Gamma(3d \rightarrow 1s) = \frac{2^{23} G m_e^3 c}{3^7 5^{15} (137)^6 \hbar^2} = 2.5 \times 10^{-44} \ \text{sec}^{-1}$$

Needless to say, there is no chance of observing such a transition.

The above estimates apply to a process in which a transition occurs *because*
a graviton is emitted, so that the graviton has a definite frequency $\omega = (E_a - E_b)/\hbar$. We can also consider a process that is going on anyway, such as a collision
between particles, and ask what is the probability of a graviton being emitted
during the process. Here the possible graviton frequencies form a continuum, so
we use our formula (10.4.22) for the emitted energy, and divide by $\hbar\omega$. The probability of emitting a graviton in the solid angle $d\Omega$ and in a frequency range $d\omega$
is then

$$dP = \frac{G\omega^2 \, d\omega \, d\Omega \, P_c}{2\pi^2 \hbar\omega} \sum_{N,M} \frac{\eta_N \eta_M}{(P_N \cdot k)(P_M \cdot k)} \left[(P_N \cdot P_M)^2 - \tfrac{1}{2} m_N^2 m_M^2 \right]$$

$$(10.8.8)$$

where P_c is the probability of the collision occurring without graviton emission,
and where once again the sums over N and M run over all particles in the initial
($\eta = -1$) or final ($\eta = +1$) states. This formula has also been derived by purely
quantum mechanical methods.[23]

It should be noted that the emission probability dP is proportional to $d\omega/\omega$
(the factors $P \cdot k$ in the denominator being proportional to ω), so the total probability for emission of gravitational radiation in a collision diverges logarithmically,
both at $\omega \rightarrow \infty$ and $\omega \rightarrow 0$. The first, or "ultraviolet," divergence was encountered
classically, and arises just because of our approximation that the collision occurs
instantaneously; it is to be eliminated by cutting off the ω-integral at $\omega \sim 1/\Delta t \sim$
$E/\hbar$, where Δt is the duration of the collision and, via the uncertainty principle,
$\bar{E}$ is some typical energy characteristic of the collision. The second, or "infrared,"
divergence at $\omega = 0$ is a purely quantum mechanical problem; it enters here
only because we divided the emitted energy dE by $\hbar\omega$ to get the emission probability. It is removed by recognizing that P_c, the probability for the collision to

occur without gravitational radiation, is itself logarithmically divergent because of emission and reabsorption of virtual gravitons, and that the divergences *cancel.*[24] We see that once we have accepted the most elementary ideas about the quantum nature of gravitational radiation, we are inevitably led to the full infrastructure of real and virtual gravitons.

The quantum interpretation of gravitational radiation allows a simple derivation of the relations between absorption and emission of gravitons. Imagine a black-body cavity in a body of temperature T that is so large and dense that it is opaque to gravitational radiation. The cavity will be filled with both electromagnetic and gravitational radiation in equilibrium with the container. By using the same statistical arguments that give the Planck distribution law for electromagnetic radiation,[25] we may conclude that the number of gravitons per unit volume with frequencies between ω and $\omega + d\omega$ is

$$n(\omega)\, d\omega = \frac{\omega^2\, d\omega}{\pi^2}\left[\exp\!\left(\frac{\hbar\omega}{kT}\right) - 1\right]^{-1} \tag{10.8.9}$$

where $k = 1.38 \times 10^{-16}$ erg/$^\circ$K is Boltzmann's constant. (It is crucial in the derivation of this result that gravitons, like photons, have two independent polarization states.) In order for equilibrium to be maintained, it is necessary that the absorption rate $A(\omega)$ of a single graviton in the container wall be related to the rate per unit volume $E(\omega)\, d\omega$ of graviton emission between frequencies ω and $\omega + d\omega$ by

$$A(\omega)n(\omega)\, d\omega = E(\omega)\, d\omega \tag{10.8.10}$$

This can also be written[26]

$$E(\omega) = I(\omega) + S(\omega) \tag{10.8.11}$$

where

$$S(\omega) = \left(\frac{\omega^2}{\pi^2}\right)\exp\!\left(-\frac{\hbar\omega}{kT}\right)A(\omega) \tag{10.8.12}$$

$$I(\omega) = n(\omega)\exp\!\left(-\frac{\hbar\omega}{kT}\right)A(\omega) \tag{10.8.13}$$

We interpret $S(\omega)$ as the rate per unit volume and per unit frequency interval of *spontaneous* emission of gravitational radiation. [Equation (10.8.12) can also be derived from the "crossing symmetry" between emission and absorption; ω^2/π^2 is a "phase-space" factor, and $\exp(-\hbar\omega/kT)$ is a Boltzmann factor representing the relative probability that an atom is in an upper level waiting to emit a graviton or a lower level waiting to absorb a graviton.] The remaining term $I(\omega)$, which is proportional to $n(\omega)$, is interpreted as the rate per unit volume and per unit frequency interval of *induced* emission of gravitational radiation, an effect due to the Bose statistics of the gas of gravitons.[27]

The useful thing about Eqs. (10.8.12) and (10.8.13) is that they remain valid even if the gravitational radiation is not in equilibrium with matter, so that $n(\omega)$ is not given by Eq. (10.8.9). It is only necessary that the matter be in thermal equilibrium at temperature T. For instance, we can calculate the rate $S(\omega)$ of spontaneous emission of gravitons per unit volume and per unit frequency interval in a nonrelativistic gas by dividing Eq. (10.4.26) by $\hbar\omega$, provided that the graviton frequency ω is in the range $\omega_c \ll \omega \ll kT/\hbar$. Applying Eq. (10.8.12) then gives the absorption rate of such gravitons as

$$A(\omega) = \frac{8\pi G}{5\hbar\omega^3} \sum_{(a,b)} \mu_{ab}^2 n_a n_b \left\langle v_{ab}^5 \int \frac{d\sigma_{ab}}{d\Omega} \sin^2\theta \, d\Omega \right\rangle$$

This ω^{-3} behavior can make $A(\omega)$ surprisingly large for low-frequency gravitons in gases at high temperature. However, the effect of induced emission is to reduce the effective absorption rate by a factor $\hbar\omega/kT$. There does not appear to be any situation in the present universe where the absorption of gravitational radiation plays any important role.

The preceding remarks describe what may be called a *semiclassical* theory of gravitation. The development of a true quantum theory of gravitation is unfortunately much more difficult. One approach is to construct an interaction Hamiltonian that can create and destroy gravitons, and then calculate transition probabilities as a power series in this interaction. Usually the Hamiltonian would be built up out of quantum *fields*, of the form

$$h_{\rho\nu}(x) = \sum_\mu \int d^3k \{a(\mathbf{k}, \mu)e_{\rho\nu}(\mathbf{k}, \mu)\exp(ik_\lambda x^\lambda)$$

$$+ \, a^\dagger(\mathbf{k}, \mu)e_{\rho\nu}^*(\mathbf{k}, \mu)\exp(-ik_\lambda x^\lambda)\} \tag{10.8.14}$$

where $e_{\rho\nu}(\mathbf{k}, \mu)$ is a polarization tensor for a graviton of momentum $\hbar\mathbf{k}$ and helicity μ, and $a(\mathbf{k}, \mu)$ and $a^\dagger(\mathbf{k}, \mu)$ are the corresponding *annihilation and creation operators*, characterized by the commutation relations

$$[a(\mathbf{k}, \mu), a^\dagger(\mathbf{k}', \mu')] = \delta^3(\mathbf{k} - \mathbf{k}')\delta_{\mu\mu'} \tag{10.8.15}$$

$$[a(\mathbf{k}, \mu), a(\mathbf{k}', \mu')] = [a^\dagger(\mathbf{k}, \mu), a^\dagger(\mathbf{k}', \mu')] = 0 \tag{10.8.16}$$

The difficulty in this approach comes from the fact that the operator (10.8.15) cannot be a Lorentz tensor as long as the helicity sum is limited to the physical values $\mu = \pm 2$; as we saw in Section 10.2, a true tensor would have helicities 0 and ± 1 and well as ± 2. It is true that we can start with a true tensor and then subject $e_{\mu\nu}$ to a gauge transformation that will eliminate the unphysical helicities 0 and ± 1, but once we choose a gauge in this way, $h_{\mu\nu}$ is no longer a tensor. To put this another way, a gauge condition, such as the statement that e_{13}, e_{23}, e_{10}, e_{20},

e_{00}, e_{03}, and e_{33} vanish for **k** in the 3-direction, is not Lorentz invariant, so if we define these components to vanish, then under a Lorentz transformation $\Lambda^\mu{}_\nu$, $h_{\mu\nu}$ will not simply transform into $\Lambda_\mu{}^\rho \Lambda_\nu{}^\sigma h_{\rho\sigma}$, but will be subjected to an additional gauge transformation[28]:

$$h_{\mu\nu} \to \Lambda_\mu{}^\rho \Lambda_\nu{}^\sigma h_{\rho\sigma} + \frac{\partial \varepsilon_\mu}{\partial x^\nu} + \frac{\partial \varepsilon_\nu}{\partial x^\mu}$$

It is no easy task to construct a Hamiltonian out of such an object in such a way as to obtain Lorentz-invariant transition probabilities.

There are two possible ways out of this difficulty. One possibility is to accept the nontensor character of $h_{\mu\nu}$, and use the noncovariant Hamiltonian formalism to derive Lorentz-invariant rules for the calculation of transition amplitudes.[29] This works fairly easily in electrodynamics, but the self-interaction of the gravitational field has so far prevented the completion of this program in general relativity. A different method, pioneered by Feynman,[30] is to start out with manifestly Lorentz-invariant calculational rules, and then tinker with them to prevent the appearance of unphysical particles with helicities 0 or ± 1 in physical states. This program has been successfully carried through to completion in the work of Fadeev,[31] Mandelstam,[32] and DeWitt.[33]

Unfortunately, the formulation of general rules for the calculation of transition probabilities in the quantum theory of gravitation has only confirmed the presence of another difficulty: The theory contains infinities, arising from integrals over large virtual momenta. Quantum electrodynamics contains similar infinities, but only in three or four special places, where they can be dealt with by a renormalization of mass, charge, and wave functions.[34] In contrast, the quantum theory of gravitation contains an infinite variety of infinities, as can be seen by an elementary dimensional argument: The gravitational constant has dimensions $\hbar/m^2$, so a term in a dimensionless probability amplitude of order G^n will diverge like a momentum-space integral $\int p^{2n-1} \, dp$. In this respect, the theory of gravitation is more like other nonrenormalizable theories, such as the Fermi theory of beta decay, than it is like quantum electrodynamics.

Despite these difficulties, there is one very important conclusion that can already be drawn from the quantum theory of gravitation: It is quite impossible to construct a Lorentz invariant quantum theory of particles of mass zero and helicity ± 2 without building some sort of gauge invariance into the theory,[23, 28] because only in this way can the interaction of the nontensor field $h_{\mu\nu}$ generate Lorentz-invariant transition amplitudes. However, we saw in Section 10.2 that the theory of gravitational radiation is gauge-invariant because general relativity is generally covariant, and, as argued in Section 4.1, general covariance is but the mathematical expression of the Principle of Equivalence. It therefore appears that the Principle of Equivalence, on which the whole of classical general relativity is based, is itself a consequence of the requirement that the quantum theory of gravitation should be Lorentz invariant.

9 Gravitational Disturbances in Gravitational Fields*

The foregoing sections have described a Lorentz-invariant theory for the behavior of weak gravitational waves in a Minkowskian space-time. It will be useful later, in Chapter 15, on cosmology, to have available a generally covariant theory for the propagation of weak gravitational disturbances in a preexisting gravitational field $g_{\mu\nu}$.

According to Eq. (6.1.5), if $g_{\mu\nu}$ is changed by some disturbance to $g_{\mu\nu} + \delta g_{\mu\nu}$, with $\delta g_{\mu\nu}$ small, then to first order in $\delta g_{\mu\nu}$,

$$\delta R_{\mu\kappa} = \frac{\partial \delta \Gamma^\lambda_{\mu\lambda}}{\partial x^\kappa} - \frac{\partial \delta \Gamma^\lambda_{\mu\kappa}}{\partial x^\lambda} + \delta \Gamma^\eta_{\mu\lambda} \Gamma^\lambda_{\kappa\eta}$$

$$+ \delta \Gamma^\lambda_{\kappa\nu} \Gamma^\eta_{\mu\lambda} - \delta \Gamma^\eta_{\mu\kappa} \Gamma^\lambda_{\lambda\eta} - \delta \Gamma^\lambda_{\lambda\eta} \Gamma^\eta_{\mu\kappa}$$

where $\delta \Gamma^\lambda_{\kappa\eta}$ is the change in the affine connection:

$$\delta \Gamma^\lambda_{\mu\nu} = -g^{\lambda\rho} \delta g_{\rho\sigma} \Gamma^\sigma_{\mu\nu} + \tfrac{1}{2} g^{\lambda\rho} \left[\frac{\partial \delta g_{\rho\mu}}{\partial x^\nu} + \frac{\partial \delta g_{\rho\nu}}{\partial x^\mu} - \frac{\partial \delta g_{\mu\nu}}{\partial x^\rho} \right]$$

We note that $\delta \Gamma^\lambda_{\mu\nu}$ can be expressed as a tensor:

$$\delta \Gamma^\lambda_{\mu\nu} = \tfrac{1}{2} g^{\lambda\rho} [(\delta g_{\rho\mu})_{;\nu} + (\delta g_{\rho\nu})_{;\mu} - (\delta g_{\mu\nu})_{;\rho}] \tag{10.9.1}$$

the covariant derivatives being of course constructed using the unperturbed affine connection $\Gamma^\lambda_{\mu\nu}$. Since $\delta \Gamma^\lambda_{\mu\nu}$ is a tensor, the change in the Ricci tensor can also be written in terms of covariant derivatives:

$$\delta R_{\mu\kappa} = (\delta \Gamma^\lambda_{\mu\lambda})_{;\kappa} - (\delta \Gamma^\lambda_{\mu\kappa})_{;\lambda} \tag{10.9.2}$$

This is known as the *Palatini identity*. In terms of $\delta g_{\mu\nu}$, it reads:

$$\delta R_{\mu\kappa} = \tfrac{1}{2} g^{\lambda\rho} [(\delta g_{\lambda\rho})_{;\mu;\kappa} - (\delta g_{\rho\mu})_{;\kappa;\lambda} \tag{10.9.3}$$

$$- (\delta g_{\rho\kappa})_{;\mu;\lambda} + (\delta g_{\mu\kappa})_{;\rho;\lambda}]$$

The Einstein field equations are here presumed to be satisfied for the undisturbed gravitational field $g_{\mu\nu}$ and energy-momentum tensor $T_{\mu\nu}$. The condition that they should also be satisfied for $g_{\mu\nu} + \delta g_{\mu\nu}$ and $T_{\mu\nu} + \delta T_{\mu\nu}$ is then

$$\tfrac{1}{2} g^{\lambda\rho} [(\delta g_{\lambda\rho})_{;\mu;\kappa} - (\delta g_{\rho\mu})_{;\kappa;\lambda} - (\delta g_{\rho\kappa})_{;\mu;\lambda} + (\delta g_{\mu\kappa})_{;\rho;\lambda}]$$

$$= -8\pi G [\delta T_{\mu\nu} - \tfrac{1}{2} g_{\mu\nu} g^{\rho\sigma} \delta T_{\rho\sigma} + \tfrac{1}{2} g_{\mu\nu} \delta g_{\lambda\eta} T^{\lambda\eta} - \tfrac{1}{2} \delta g_{\mu\nu} T^\lambda_{\ \lambda}] \tag{10.9.4}$$

Also, the source term $\delta T_{\mu\nu}$ obeys the conservation law:

$$0 = (\delta T^{\nu\mu})_{;\mu} + T^{\nu\lambda} \delta \Gamma^\mu_{\mu\lambda} + T^{\lambda\mu} \delta \Gamma^\nu_{\mu\lambda} \tag{10.9.5}$$

The general covariance of these equations is manifest.

* This section lies somewhat out of the book's main line of development, and may be omitted in a first reading.

Just as for gravitational waves in a Minkowskian space-time, it is important here to distinguish physical disturbances from mere changes in the coordinate system. To this end, let us consider a general infinitesimal coordinate transformation

$$x^\mu \rightarrow x'^\mu = x^\mu - \varepsilon^\mu(x) \tag{10.9.6}$$

where $\varepsilon^\mu(x)$ is an arbitrary infinitesimal vector field. The partial derivatives occurring in the tensor transformation rules are here

$$\frac{\partial x'^\mu}{\partial x^\nu} = \delta^\mu_\nu - \frac{\partial \varepsilon^\mu(x)}{\partial x^\nu}$$

$$\frac{\partial x^\nu}{\partial x'^\mu} = \delta^\nu_\mu + \frac{\partial \varepsilon^\nu(x)}{\partial x^\mu} + 0(\varepsilon^2)$$

Since Einstein's equations are generally covariant, and $g_{\mu\nu}(x)$ is a solution for an energy-momentum tensor $T_{\mu\nu}(x)$, it follows that $g'_{\mu\nu}(x)$ is a solution for $T'_{\mu\nu}(x)$, where

$$g'_{\mu\nu}(x) = g'_{\mu\nu}(x') + \frac{\partial g_{\mu\nu}(x)}{\partial x^\lambda} \varepsilon^\lambda(x) + 0(\varepsilon^2)$$

$$= g_{\mu\nu}(x) + g_{\lambda\nu}(x) \frac{\partial \varepsilon^\lambda(x)}{\partial x^\mu} + g_{\lambda\mu}(x) \frac{\partial \varepsilon^\lambda(x)}{\partial x^\nu} + \frac{\partial g_{\mu\nu}(x)}{\partial x^\lambda} \varepsilon^\lambda(x)$$

and likewise for $T'_{\mu\nu}(x)$. In covariant terms, we conclude that

$$g'_{\mu\nu}(x) = g_{\mu\nu}(x) + \Delta_\varepsilon g_{\mu\nu}(x) \tag{10.9.7}$$

is a solution of Einstein's equations for an energy-momentum tensor

$$T'_{\mu\nu}(x) = T_{\mu\nu}(x) + \Delta_\varepsilon T_{\mu\nu}(x) \tag{10.9.8}$$

where

$$\Delta_\varepsilon g_{\mu\nu} \equiv \varepsilon_{\mu;\nu} + \varepsilon_{\nu;\mu} \tag{10.9.9}$$

$$\Delta_\varepsilon T_{\mu\nu} \equiv T^\lambda{}_\mu \varepsilon_{\lambda;\nu} + T^\lambda{}_\nu \varepsilon_{\lambda;\mu} + T_{\mu\nu;\lambda} \varepsilon^\lambda \tag{10.9.10}$$

(Note that $\Delta_\varepsilon g_{\mu\nu}$ has the same form as $\Delta_\varepsilon T_{\mu\nu}$, except that $g_{\mu\nu}$ has vanishing covariant derivatives, whereas $T_{\mu\nu}$ does not.) It follows, and it is straightforward to verify directly, that $\delta g_{\mu\nu} = \Delta_\varepsilon g_{\mu\nu}$ is a solution of the field equation (10.9.4) for a source perturbation $\delta T_{\mu\nu} = \Delta_\varepsilon T_{\mu\nu}$. But Eq. (10.9.4) is a *linear* differential equation, and so, given any solution $\delta g_{\mu\nu}$, we can always find other solutions of the form $\delta g_{\mu\nu} + \Delta_\varepsilon g_{\mu\nu}$ with precisely the same physical content. The freedom to add terms $\Delta_\varepsilon g_{\mu\nu}$ for arbitrary functions $\varepsilon^\mu(x)$ corresponds to the "gauge invariance" discussed in Section 10.1.

The operator Δ_ε introduced in Eqs. (10.9.9) and (10.9.10) can be generalized to arbitrary tensors by specifying that a term involving the contraction of the tensor with the covariant derivative of ε should be included with a $+$ sign for each covariant index and a $-$ sign for each contravariant index. That is, for scalars we define

$$\Delta_\varepsilon S \equiv S_{;\lambda}\varepsilon^\lambda$$

for vectors we define

$$\Delta_\varepsilon V_\mu \equiv V^\lambda \varepsilon_{\lambda;\mu} + V_{\mu;\lambda}\varepsilon^\lambda$$

$$\Delta_\varepsilon U^\mu \equiv -U^\lambda \varepsilon^\mu_{;\lambda} + U^\mu_{;\lambda}\varepsilon^\lambda$$

for contravariant and mixed tensors of second rank we define

$$\Delta_\varepsilon T^{\mu\nu} \equiv -T^{\lambda\nu}\varepsilon^\mu_{;\lambda} - T^{\mu\lambda}\varepsilon^\nu_{;\lambda} + T^{\mu\nu}_{;\lambda}\varepsilon^\lambda$$

$$\Delta_\varepsilon T^\mu_{\ \nu} \equiv -T^\lambda_{\ \nu}\varepsilon^\mu_{;\lambda} + T^\mu_{\ \lambda}\varepsilon^\lambda_{;\nu} + T^\mu_{\ \nu;\lambda}\varepsilon^\lambda$$

and so on. The operator Δ_ε defined in this way is known as the *Lie derivative*. In general, the effect of an infinitesimal coordinate transformation on any tensor T is that the new tensor equals the old tensor *at the same coordinate point*, plus the Lie derivative $\Delta_\varepsilon T$. It is easy to show that the operator Δ_ε has the same abstract properties as ordinary derivatives or covariant derivatives: It is linear,

$$\Delta_\varepsilon[aA^\mu_{\ \nu} + bB^\mu_{\ \nu}] = a\Delta_\varepsilon A^\mu_{\ \nu} + b\Delta_\varepsilon B^\mu_{\ \nu}$$

(for a, b constant scalars)

it obeys the Leibniz rule,

$$\Delta_\varepsilon(A^\mu_{\ \nu}B^\lambda) = B^\lambda\Delta_\varepsilon A^\mu_{\ \nu} + A^\mu_{\ \nu}\Delta_\varepsilon B^\lambda$$

and it commutes with the operation of contraction,

$$\delta^\lambda_\nu\Delta_\varepsilon T^{\mu\nu}_{\ \ \lambda} = \Delta_\varepsilon T^{\mu\lambda}_{\ \ \lambda} \equiv -T^{\nu\lambda}_{\ \ \lambda}\varepsilon^\mu_{;\nu} + T^{\mu\lambda}_{\ \ \lambda;\nu}\varepsilon^\nu$$

In particular, the Lie derivative of the energy-momentum tensor for a perfect fluid is

$$\Delta_\varepsilon T_{\mu\nu} = p\Delta_\varepsilon g_{\mu\nu} + g_{\mu\nu}\Delta_\varepsilon p + (p + \rho)[U_\mu\Delta_\varepsilon U_\nu + U_\nu\Delta_\varepsilon U_\mu] + U_\mu U_\nu[\Delta_\varepsilon p + \Delta_\varepsilon\rho]$$

so $\Delta_\varepsilon g_{\mu\nu}$ is a solution of Einstein's equations for a fluid whose velocity, pressure, and density are perturbed by $\Delta_\varepsilon U_\mu$, $\Delta_\varepsilon p$, and $\Delta_\varepsilon\rho$, respectively.

The solution of the field equations (10.9.4) is quite complicated, except for the simple case of a homogeneous and isotropic unperturbed metric $g_{\mu\nu}$. This case will be considered in Section 15.10.

IO BIBLIOGRAPHY

On Gravitational Radiation in General

☐ L. D. Landau and E. M. Lifshitz, *The Classical Theory of Fields* (Addison-Wesley Publishing Co., Reading, Mass., 1962), Section 101.

☐ J. Weber, *General Relativity and Gravitational Waves* (Interscience Publishers, New York, 1961), Chapters 7 and 8.

Exact Solutions of the Einstein Equations

☐ H. Bondi, "Some Special Solutions of the Einstein Equations," in *Lectures on General Relativity* (Prentice-Hall, Englewood Cliffs, N. J., 1965), p. 375.

☐ D. R. Brill, "General Relativity: Selected Topics of Current Interest," Nuovo Cimento Suppl., **2**, No. 1 (1964).

☐ J. Ehlers and W. Kundt, "Exact Solutions of the Gravitational Field Equations," in *Gravitation: An Introduction to Current Research*, ed. by L. Witten (Wiley, New York, 1962), p. 49.

☐ A. Z. Petrov, *Einstein Spaces*, trans. by R. F. Kelleher (Pergamon Press, Oxford, 1969).

☐ F. A. E. Pirani, "Gravitational Radiation," in *Gravitation: An Introduction to Current Research*, op. cit., p. 199.

☐ F. A. E. Pirani, "Introduction to Gravitational Radiation Theory," in *Lectures on General Relativity*, op. cit., p. 249.

☐ F. A. E. Pirani, "Survey of Gravitational Radiation Theory," in *Recent Developments in General Relativity* (Pergamon Press, Oxford, 1962), p. 89.

☐ R. K. Sachs, "Gravitational Radiation," in *Relativity, Groups, and Topology*, ed. by C. DeWitt and B. DeWitt (Gordon and Breach Science Publishers, New York, 1964), p. 523.

☐ R. K. Sachs, "Gravitational Waves," in *Relativity Theory and Astrophysics. 1. Relativity and Cosmology*, ed. by J. Ehlers (American Mathematical Society, Providence, R. I., 1967), p. 129.

Quantum Theory of Gravitation

☐ B. S. DeWitt, "Dynamical Theory of Groups and Fields," in *Relativity, Groups, and Topology*, op. cit., p. 587.

☐ B. S. De Witt, "The Quantization of Geometry," in *Gravitation: An Introduction to Current Research*, op. cit., p. 266.

☐ P. A. M. Dirac, "The Quantization of the Gravitational Field," in *Contemporary Physics—Trieste Symposium 1968*, ed. by A. Salam, Vol. 1 (International Atomic Energy Agency, Vienna, 1969), p. 539.

☐ A. Komar, "The Quantization Program for General Relativity," in *Relativity—Proceedings of the Relativity Conference in the Midwest*, ed. by M. Carmeli, S. I. Fickler, and L. Witten (Plenum Press, New York, 1970).

☐ S. Weinberg, "The Quantum Theory of Massless Particles," in *Lectures on Particles and Field Theory* (Prentice-Hall, Englewood Cliffs, N. J., 1965).

10 REFERENCES

1. J. Weber, refs. 17, 18. Also see J. Weber, Phys. Rev., **117**, 306 (1960); Phys. Rev. Letters, **17**, 1228 (1966); Phys. Rev. Letters, **20**, 1307 (1968); Physics Today, **21**, 34 (1968); in *Relativity—Proceedings of the Relativity Conference in the Midwest*, ed. by M. Carmeli, S. I. Fickler, and L. Witten (Plenum Press, New York, 1970), p. 133; Nuovo Cimento Letters, Ser. I, **4**, 653 (1970).

2. S. Chandrasekhar and Y. Nutku, Ap. J., **158**, 55 (1969). Gravitational radiation reaction is included in S. Chandrasekhar and F. P. Esposito, Ap. J., **160**, 153 (1970).

3. T. Gold, Nature, **218**, 731 (1968); *ibid.*, **221**, 25 (1968).

4. The slowing-down of pulsars by gravitational radiation reaction has been considered by J. E. Gunn and J. P. Ostriker, Nature, **221**, 454 (1969); Phys. Rev. Letters, **22**, 728 (1969); J. P. Ostriker and J. E. Gunn, Ap. J., **157**, 1395 (1969).

5. See, for example, L. I. Schiff, *Quantum Mechanics* (3rd ed., McGraw-Hill, New York, 1968), Section 18.

6. G. N. Watson, *Theory of Bessel Functions* (rev. ed., Macmillan, New York, 1944), p. 128.

7. *Ibid.*, p. 52.

8. *Ibid.*, p. 44.

9. These formulas can be verified by multiplying with $P_l(\mu)$ and integrating over μ. The necessary integrals are given, for example, by L. Schiff, *op. cit.*, Section 14.

10. See, for example, H. A. Kramers, Atti. Congr. Intern. Fisici, Como (1927); reprinted in H. A. Kramers, *Collected Scientific Papers* (North-Holland, Amsterdam, 1956).

11. E. Feenberg, Phys. Rev., **40**, 40 (1932); N. Bohr, R. E. Peierls, and G. Placzek, Nature, **144**, 200 (1939).

12. J. Weber, Phys. Rev. Letters, **21**, 395 (1968).

13. D. H. Douglass and J. A. Tyson, report at the Third "Cambridge" Conference on General Relativity, June 8, 1970 (unpublished).

14. For a description of the normal modes of the earth and moon, see B. A. Bolt, in *Physics and Chemistry of the Earth*, ed. by L. A. Ahrens, F. Press, and S. K. Runcorn (Pergamon Press, New York, 1964), p. 55.

15. R. L. Forward, D. Zipoy, J. Weber, S. Smith, and H. Benioff, Nature, **189**, 473 (1961); also see J. Weber and J. V. Larson, Jour. Geophys. Res., **71**, 6005 (1966); R. A. Wiggins and F. Press, Jour. Geophys. Res., **74**, 5351 (1969); F. J. Dyson, Ap. J., **156**, 529 (1969).

16. J. Weber, in *Physics of the Moon*, ed. by S. F. Singer (American Astronautical Society, Hawthorne, Cal., 1967), p. 199.

17. J. Weber, Phys. Rev. Lett., **22**, 1320 (1969); *ibid.*, **24**, 276 (1970).

18. J. Weber, Phys. Rev. Lett., **25**, 180 (1970); also see *Proceedings of the Midwest Conference on Theoretical Physics*, Notre Dame, Indiana, April 1970 (unpublished), p. 118. For a comment on the astronomical significance of mass loss from galaxies by gravitational radiation, see G. B. Field, M. J. Rees, and D. W. Sciama, Comments on Astrophys. and Space Phys., **1**, 187 (1969).

19. J. Weber, to be published; also, report at the Third "Cambridge" Conference on General Relativity, June 8, 1970 (unpublished).

20. W. M. Fairbank, report at the Third "Cambridge" Conference on General Relativity, June 8, 1970 (unpublished).

21. V. B. Braginski, report at the Third "Cambridge" Conference on General Relativity, June 8, 1970 (unpublished).

22. V. B. Braginski, Ya. B. Zeldovich, and V. N. Rudenko, JETP Lett., **10**, 280 (1969).

23. S. Weinberg, Phys. Letters, **9**, 357 (1964); Phys. Rev., **135**, B1049 (1964).

24. S. Weinberg, Phys. Rev., **140**, B516 (1965).

25. See, for example, K. Huang, *Statistical Mechanics* (Wiley, 1963), Section 12.1.

26. A. Einstein, Phys. Z., **18**, 121 (1917).

27. See, for example, L. I. Schiff, *op. cit.*, p. 531.

28. S. Weinberg, Phys. Rev., **138**, B988 (1965).

29. R. L. Arnowitt and S. Deser, Phys. Rev., **113**, 745 (1959); R. L. Arnowitt, S. Deser, and C. W. Misner, Phys. Rev., **116**, 1322 (1959); *ibid.*, **117**, 1595 (1960); J. Math. Phys., **1**, 434 (1960); Phys. Rev., **118**, 1100 (1960); Nuovo Cimento, **19**, 668 (1961); Phys. Rev., **121**, 1556 (1961); *ibid.*, **122**, 997 (1961), *ibid.*, **120**, 313 (1960); *ibid.*, **120**, 321 (1960); Ann. Phys. (N.Y.), **11**, 116 (1960); P. A. M. Dirac, Phys. Rev., **114**, 924 (1959).

30. R. P. Feynman, Acta Phys. Polon., **24**, 697 (1963).

31. L. D. Fadeev and V. N. Popov, Phys. Letters, **25B**, 29 (1967).

32. S. Mandelstam, Phys. Rev., **175**, 1604 (1968).

33. B. S. DeWitt, Phys. Rev., **162**, 1195, 1239 (1967); erratum, Phys. Rev., **171**, 1834 (1968).

34. See, for example, J. D. Bjorken and S. D. Drell, *Relativistic Quantum Fields* (McGraw-Hill, New York, 1965), Chapter 19.

II STELLAR EQUILIBRIUM AND COLLAPSE

Gravitational fields are so weak that the practicing astrophysicist can usually ignore general relativity. This chapter deals with various sorts of objects in which relativistic effects play an important, or in some cases a dominant, role. One of these is the neutron star, a "cold" star composed primarily of neutrons and supported against collapse by neutron degeneracy pressure. Another is the super-massive star, a giant object supported by radiation pressure, in which general-relativistic effects can tip the balance between stability and instability. Most impressive of all is the black hole, a body caught in an inexorable gravitational collapse.

The existence of neutron stars and black holes was suggested in the 1930's on purely theoretical grounds, chiefly through the work of J. Robert Oppenheimer and his collaborators. However, these exotic objects remained a textbook curiosity until the 1960's, when the cooperative efforts of radio and optical astronomers began to reveal a great many strange new things in the sky.

First came the quasi-stellar objects (QSO's), objects with starlike optical images, often containing powerful compact radio sources, and with red shifts $\Delta\lambda/\lambda$ ranging from 0.131 to nearly 3. (See Figure 11.1.) One can suppose three different sorts of explanations for these red shifts: They can arise from a Döppler effect, caused either by a local explosion or by the general cosmological recession of very distant objects (see Chapter 14), or they can arise from powerful gravitational fields within the objects themselves. In any case, it is likely that general-relativistic effects will play an important role in the explanation of the QSO's. If these objects are relatively near but moving at relativistic velocities, then some source of energy must be found that could convert mass into kinetic energy, with nearly 100% efficiency. If the QSO's are at cosmological distances, then their

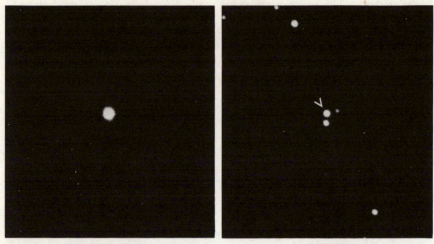

3C 48   3C 147

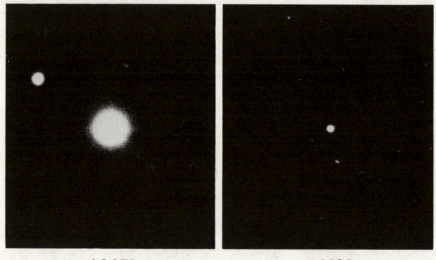

3C 273   3C 196

Figure 11.1 Four quasi-stellar objects. These photographs were taken with the 200-in. telescope at Mt. Palomar, following position determinations by radio astronomers. (Courtesy Mt. Wilson and Palomar Observatories.)

apparent optical luminosity indicates an absolute luminosity much greater than that of the largest galaxies, so again a powerful new source of energy is required. Only gravitational attraction seems to offer an adequate energy source, and for this reason the discovery of the QSO's reawakened general interest in the phenomenon of gravitational collapse. Finally, if the QSO red shifts are gravitational, then these objects must be so highly compressed that their structure would have to be understood in terms of general relativity rather than Newtonian mechanics.

The quasi-stellar objects are only the most spectacular end of a continuum of ill-understood objects discovered in recent years, including Seyfert galaxies, giant elliptical galaxies with powerful compact radio sources, X-ray sources, galactic nuclei that seem in some cases to be exploding, and so on. It is not clear what, if anything, general relativity has to do with these objects.

In the last few years a new species of astronomical exotica was discovered— the pulsars, radio sources that pulse at regular frequencies ranging from a few tenths Hz to 30 Hz. The pulsars are often associated with optical and even X-ray sources that pulse at the same rate. There appears now to be a general consensus that pulsars are the neutron stars discovered theoretically in the 1930's, but with a rapid rate of rotation that somehow or other produces the observed pulses.

A realistic discussion of quasi-stellar objects, galactic nuclei, pulsars, and so on, would require that we consider the effects of radiative energy transport, neutrino energy transport, turbulence, nuclear forces, magnetic fields and, above all, rotation. It would also require the discussion of massive calculations using automatic computers. In preparing this chapter, I have tried to restrict myself to the simplest calculations, which can be carried out analytically without too much trouble. These simple calculations are not very useful for a detailed understanding of astronomical observations, but they provide a valuable insight into the possible roles that general relativity can play in astrophysical phenomena.

1 Differential Equations for Stellar Structure

We first set up the general-relativistic machinery for computing the pressure, density, and gravitational fields within a spherically symmetric static star.

The metric will be taken in the "standard" form discussed in Section 8.1:

$$g_{rr} = A(r), \qquad g_{\theta\theta} = r^2, \qquad g_{\varphi\varphi} = r^2 \sin^2\theta, \qquad g_{tt} = -B(r)$$

(11.1.1)

$$g_{\mu\nu} = 0 \qquad \text{for } \mu \neq \nu$$

The energy-momentum tensor is assumed to be that for a perfect fluid (see Section 5.4):

$$T_{\mu\nu} = pg_{\mu\nu} + (p + \rho)U_\mu U_\nu$$

(11.1.2)

with p the proper pressure, ρ the proper total energy density, and U^μ the velocity four-vector, defined so that

$$g^{\mu\nu}U_\mu U_\nu = -1$$

(11.1.3)

Since the fluid is at rest, we take

$$U_r = U_\theta = U_\varphi = 0; \qquad U_t = -(-g^{tt})^{-1/2} = -\sqrt{B(r)}$$

(11.1.4)

Our assumptions of time independence and spherical symmetry imply that p and ρ are functions only of the radial coordinate r.

By making use of Eqs. (11.1.1)–(11.1.4) and the Ricci tensor components given by Eq. (8.1.13), we find that the Einstein equations (7.1.15) read

$$R_{rr} = \frac{B''}{2B} - \frac{B'}{4B}\left(\frac{A'}{A} + \frac{B'}{B}\right) - \frac{A'}{rA} = -4\pi G(\rho - p)\mathrm{A} \qquad (11.1.5)$$

$$R_{\theta\theta} = -1 + \frac{r}{2A}\left(-\frac{A'}{A} + \frac{B'}{B}\right) + \frac{1}{A} = -4\pi G(\rho - p)r^2 \qquad (11.1.6)$$

$$R_{tt} = -\frac{B''}{2A} + \frac{B'}{4A}\left(\frac{A'}{A} + \frac{B'}{B}\right) - \frac{B'}{rA} = -4\pi G(\rho + 3p)B \qquad (11.1.7)$$

A prime denotes d/dr. (We do not need to write down the equation for $R_{\varphi\varphi}$, which is identical to that for $R_{\theta\theta}$, or the equations for off-diagonal elements of $R_{\mu\nu}$, which simply say that zero equals zero.) In addition, we may recall the equation (5.4.5) for hydrostatic equilibrium,

$$\frac{B'}{B} = -\frac{2p'}{p + \rho} \qquad (11.1.8)$$

Our first step in solving these equations is to derive an equation for $A(r)$ alone, by forming the quantity

$$\frac{R_{rr}}{2A} + \frac{R_{\theta\theta}}{r^2} + \frac{R_{tt}}{2B} = -\frac{A'}{rA^2} - \frac{1}{r^2} + \frac{1}{Ar^2} = -8\pi G\rho \qquad (11.1.9)$$

This equation can be written

$$\left(\frac{r}{A}\right)' = 1 - 8\pi G\rho r^2 \qquad (11.1.10)$$

The solution with $A(0)$ finite is

$$A(r) = \left[1 - \frac{2G\mathcal{M}(r)}{r}\right]^{-1} \qquad (11.1.11)$$

where

$$\mathcal{M}(r) \equiv \int_0^r 4\pi r'^2 \rho(r')\, dr' \qquad (11.1.12)$$

We can now use (11.1.11) and (11.1.8) to eliminate the gravitational fields $A(r)$, $B(r)$ from Eq. (11.1.6), which becomes

$$-1 + \left[1 - \frac{2G\mathcal{M}}{r}\right]\left[1 - \frac{rp'}{p + \rho}\right] + \frac{G\mathcal{M}}{r} - 4\pi G\rho r^2 = -4\pi G(\rho - p)r^2$$

We rewrite this as

$$-r^2 p'(r) = G\mathscr{M}(r)\rho(r)\left[1 + \frac{p(r)}{\rho(r)}\right]\left[1 + \frac{4\pi r^3 p(r)}{\mathscr{M}(r)}\right]\left[1 - \frac{2G\mathscr{M}(r)}{r}\right]^{-1}$$

$$(11.1.13)$$

The reader may recognize this differential equation as the fundamental equation of Newtonian astrophysics (see Section 11.3), with general-relativistic corrections supplied by the last three factors.

We are primarily concerned in this chapter with stars that are isentropic, that is, in which the entropy per nucleon s does not vary throughout the star. This is the case for two very different kinds of star:

(A) **Stars at Absolute Zero.** When a star exhausts its thermonuclear fuel it can become a white dwarf (Section 11.3), or a neutron star (Section 11.4), in which the temperature is essentially at absolute zero. According to Nernst's theorem, the entropy per nucleon will then be zero throughout the star.

(B) **Stars in Convective Equilibrium.** If the most efficient mechanism for energy transfer within the star is convection, then in equilibrium the entropy per nucleon must be nearly constant throughout the star, because otherwise a small element of fluid containing A nucleons could gain or lose an energy $A\Delta s/T$ when transported from one part of the star to another, and convection would therefore disturb the energy distribution. The supermassive "stars" discussed in Section 11.5 are generally presumed to be in convective equilibrium.

We also assume that the stars we consider have a chemical composition that is constant throughout.

The importance of the preceding assumptions lies in the fact that the pressure p may in general be expressed as a function of the density ρ, the entropy per nucleon s, and the chemical composition. Hence, with s and the chemical composition constant throughout the star, $p(r)$ may be regarded as a function of $\rho(r)$ alone, with no *explicit* dependence on r.

Given $p(r)$ as a function $p(\rho(r))$, we now formulate our problem as a pair of first-order differential equations for $\rho(r)$ and $\mathscr{M}(r)$. One of these is Eq. (11.1.13); the other is the derivative of Eq. (11.1.12):

$$\mathscr{M}'(r) = 4\pi r^2 \rho(r) \qquad (11.1.14)$$

In addition, Eq. (11.1.12) provides an initial condition:

$$\mathscr{M}(0) = 0 \qquad (11.1.15)$$

Equations (11.1.13)–(11.1.15), together with an equation of state giving $p(\rho)$, serve to determine $\rho(r)$, $\mathscr{M}(r)$, $p(r)$, and so on, throughout the star, once we specify the *other* initial condition, that is, the value of $\rho(0)$. The differential equations (11.1.13) and (11.1.14) must be integrated out from the center of the star, until

$p(\rho(r))$ drops to zero at some point $r = R$, which we then interpret as the radius of the particular star with central density $\rho(0)$.

Let us return to the problem of calculating the metric. Once we compute $\rho(r)$, $\mathcal{M}(r)$, and $p(r)$, we can immediately obtain $A(r)$ from Eq. (11.1.11); to find $B(r)$ we use Eq. (11.1.13) to rewrite (11.1.8) as

$$\frac{B'}{B} = \frac{2G}{r^2} [\mathcal{M} + 4\pi r^3 p] \left[1 - \frac{2G\mathcal{M}}{r} \right]^{-1}$$

The solution with $B(\infty) = 1$ is

$$B(r) = \exp\left\{ -\int_r^\infty \frac{2G}{r^2} [\mathcal{M}(r') + 4\pi r'^3 p(r')] \left[1 - \frac{2G\mathcal{M}(r')}{r'} \right]^{-1} dr' \right\}$$

(11.1.16)

Our solution is now complete. (Incidentally, we did not need to use Eqs. (11.1.5) and (11.1.7) for R_{rr} and R_{tt} separately, because these equations follow from (11.1.6), (11.1.8), and (11.1.9), which *were* used in our calculation. This should not be surprising, because Eq. (11.1.8), which is really just the equation for momentum conservation, follows from the Einstein equations (11.1.5)–(11.1.7) *via* the Bianchi identities.)

Outside the star, $p(r)$ and $\rho(r)$ vanish, and $\mathcal{M}(r)$ is the constant $\mathcal{M}(R)$, so Eqs. (11.1.11) and (11.1.16) give

$$B(r) = A^{-1}(r) = 1 - \frac{2G\mathcal{M}(R)}{r} \qquad \text{for } r \geq R \qquad (11.1.17)$$

The discussion of Section 8.2 shows that the constant $\mathcal{M}(R)$ that appears in the asymptotic gravitational field (11.1.17) must equal the mass M of the star, defined as the total energy of the star and its gravitational field, that is,

$$M = \mathcal{M}(R) \equiv \int_0^R 4\pi r^2 \rho(r) \, dr \qquad (11.1.18)$$

Thus (11.1.17) is just the familiar exterior Schwarzschild solution.

It may appear paradoxical that M, which must include the energy of the gravitational field, is given in (11.1.18) as the integral of the energy density $\rho(r)$ of matter (including radiation) alone. The resolution is that (11.1.18) does *not* say that M is the total energy of the matter. The total material energy is not really well defined, but it might be computed by splitting up the star into small volume elements and adding up the energies of each element as measured in a locally inertial reference frame; this would give the material energy as

$$M_{\text{matter}} \equiv \int \sqrt{g}\, \rho \, dr \, d\theta \, d\phi = \int_0^R 4\pi r^2 \sqrt{A(r)B(r)}\, \rho(r) \, dr \qquad (11.1.19)$$

The difference between (11.1.18) and (11.1.19) can be regarded as the energy of the gravitational field. However, this decomposition is not particularly useful, and will not be employed here.

It is more informative to compare (11.1.18) with the energy M_0 that the matter of the star would have if dispersed to infinity. This is simply

$$M_0 = m_N N \tag{11.1.20}$$

where $m_N = 1.66 \times 10^{-24}$ g is the rest-mass of a nucleon and N is the number of nucleons in the star. The nucleon number is given by

$$N = \int \sqrt{g} \, J_N{}^0 \, dr \, d\theta \, d\varphi = \int_0^R 4\pi r^2 \sqrt{A(r)B(r)} \, J_N{}^0(r) \, dr \tag{11.1.21}$$

where $J_N{}^\mu$ is the conserved nucleon number current. It is convenient to express $J_N{}^0$ in terms of the proper nucleon number density n, that is, the nucleon number density measured in a locally inertial reference frame at rest in the star, which is

$$n = -U_\mu J_N{}^\mu = \sqrt{B} J_N{}^0 \tag{11.1.22}$$

(See Eq. (11.1.4), and recall that in a locally inertial coordinate frame $U_0 = -1$.) Equation (11.1.21) then becomes

$$N = \int_0^R 4\pi r^2 \sqrt{A(r)} \, n(r) \, dr = \int_0^R 4\pi r^2 \left[1 - \frac{2G\mathcal{M}(r)}{r} \right]^{-1/2} n(r) \, dr \tag{11.1.23}$$

The proper number density $n(r)$ is in general a function of the proper density $\rho(r)$, the chemical composition, and the entropy per nucleon s, so $n(r)$ and N are fixed for a star with a given constant s and chemical composition, once we choose $\rho(0)$.

The internal energy of the star is now given by

$$E \equiv M - m_N N \tag{11.1.24}$$

We can also define a proper internal material energy density as

$$e(r) \equiv \rho(r) - m_N n(r) \tag{11.1.25}$$

and write (11.1.24) as

$$E = T + V \tag{11.1.26}$$

where T and V are the thermal and gravitational energies, respectively, of the star:

$$T \equiv \int_0^R 4\pi r^2 \left[1 - \frac{2G\mathcal{M}(r)}{r} \right]^{-1/2} e(r) \, dr \tag{11.1.27}$$

$$V \equiv \int_0^R 4\pi r^2 \left\{ 1 - \left[1 - \frac{2G\mathcal{M}(r)}{r} \right]^{-1/2} \right\} \rho(r) \, dr \tag{11.1.28}$$

Expanding the square roots gives

$$T = \int_0^R 4\pi r^2 \left\{ 1 + \frac{G\mathcal{M}(r)}{r} + \cdots \right\} e(r)\, dr \tag{11.1.29}$$

$$V = - \int_0^R 4\pi r^2 \left\{ \frac{G\mathcal{M}(r)}{r} + \frac{3G^2\mathcal{M}^2(r)}{2r^2} + \cdots \right\} \rho(r)\, dr \tag{11.1.30}$$

The first terms in T and V are recognizable as the Newtonian values for the thermal and gravitational energies of the star; in particular, note that the first term in V may be written

$$-G \int_0^R 4\pi r \mathcal{M}(r)\rho(r)\, dr = -\frac{G}{2} \int_0^R \frac{1}{r}\, d(\mathcal{M}^2(r))$$

$$= -\frac{GM^2}{2R} - \frac{G}{2} \int_0^R \frac{\mathcal{M}^2(r)}{r^2}\, dr = \frac{\phi(R)\mathcal{M}(R)}{2} - \frac{1}{2} \int_0^R \mathcal{M}(r)\, d\phi(r)$$

$$= \frac{1}{2} \int_0^R \phi(r)\, d\mathcal{M}(r) \tag{11.1.31}$$

where ϕ is the Newtonian potential, given inside the star by

$$\phi(r) = -\frac{GM}{R} - G \int_r^R \frac{\mathcal{M}(r')}{r'^2}\, dr'$$

The higher terms in T and V are discussed in Section 11.5.

To repeat our main conclusion: Once we specify that a star has a definite uniform entropy per nucleon and chemical composition, all properties of the star, including $\rho(r)$, $p(r)$, $n(r)$, $e(r)$, M, N, and E, are determined as function of the central density $\rho(0)$. This is *not* the case for ordinary stars like the sun, in which the entropy distribution is not uniform and has to be determined from the equations of radiative equilibrium. However, the considerations of this section do provide an adequate basis for the study of the exotic structures discussed in this chapter.

2   Stability

Our work is not done when we obtain a solution of the fundamental equations (11.1.13), (11.1.14). Such a solution represents an equilibrium state of the star, but it may be a state of stable or of unstable equilibrium. For most purposes it is only the stable solutions that interest the astrophysicist.

In order to tell whether a particular configuration is unstable it would in general be necessary to compute the frequencies ω_n of all normal modes of the configuration and check whether any ω_n has a positive imaginary part; in this case the factor $\exp(-i\omega_n t)$, which gives the time variation of this mode, would grow exponentially and the system would be unstable. However, it is often possible to tell from the equilibrium solution alone whether the corresponding configuration is stable, by making use of the following theorem[1]:

Theorem 1. A star, consisting of a perfect fluid with constant chemical composition and entropy per nucleon, can only pass from stability to instability with respect to some particular radial normal mode, at a value of the central density $\rho(0)$ for which the equilibrium energy E and nucleon number N are stationary, that is,

$$\frac{\partial E(\rho(0); s, \ldots)}{\partial \rho(0)} = 0$$

$$\frac{\partial N(\rho(0); s, \ldots)}{\partial \rho(0)} = 0$$

By a "radial" normal mode is meant a mode of oscillation in which the density perturbation $\delta\rho$ is a function of r and t alone, and in which nuclear reactions, viscosity, heat conduction, and radiative energy transfer play no role.

The first step in the proof is to note that dissipative forces are absent here, so the dynamical equations are time-reversal-invariant, and give the squared frequencies $\omega_n{}^2$ of the various normal modes as *real* continuous functions of $\rho(0)$, just as in an electrical circuit without resistors. For each $\omega_n{}^2 > 0$ there are two modes that undergo stable oscillation. For each $\omega_n{}^2 < 0$ there are two modes, one of which is exponentially damped and one of which grows exponentially, as $\exp(-|\omega_n|t)$ and $\exp(+|\omega_n|t)$, respectively. Thus the transition from stability to instability can only occur at a value of $\rho(0)$ for which $\omega_n{}^2$ *vanishes*.

Consider some value of $\rho(0)$ for which a particular frequency ω_n is nearly zero. Then it takes a long time for the oscillation or growth of this mode to change the equilibrium configuration into some neighboring configuration $\rho(r) + \delta\rho(r)$. Since this is going on so slowly, $\rho(r) + \delta\rho(r)$ must also be essentially an equilibrium configuration. In the absence of nuclear reactions, the new configuration will have the same uniform chemical composition as the old one. In the absence of viscosity, heat conduction, or radiative energy transfer, the new configuration will also have the same entropy per nucleon as the old one. Moreover, the conservation of energy and of nucleon number tells us that the new configuration will have the same energy E and baryon number N as the old one. However, $\delta\rho(0)$ cannot vanish, because an equilibrium configuration is entirely specified (for a given uniform s and chemical composition) by the value of $\rho(0)$; if $\delta\rho(0)$ were zero, then $\delta\rho(r)$ would be zero for all r, and the normal mode would be absent. Thus at a point of transition from stability to instability there are neighboring equilibrium configurations with different values of $\rho(0)$, but with the same uniform entropy per nucleon and chemical composition, and with the same E and N, as was to be proven.

This theorem is particularly valuable because we can often use qualitative arguments to show that an equilibrium configuration is stable for $\rho(0)$ sufficiently small (or large) and unstable for $\rho(0)$ sufficiently large (or small); the theorem tells us precisely where the transition from stability to instability occurs. As a guide in such qualitative considerations, it is helpful to reformulate the fundamental equations of stellar structure in a variational principle[2]:

Theorem 2.   A particular stellar configuration, with uniform entropy per nucleon and chemical composition, will satisfy the equations (11.1.12), (11.1.13) for equilibrium, if and only if the quantity M, defined by

$$M \equiv \int 4\pi r^2 \rho(r)\, dr$$

is stationary with respect to all variations of $\rho(r)$ that leave unchanged the quantity

$$N \equiv \int 4\pi r^2 n(r) \left[1 - \frac{2G\mathcal{M}(r)}{r} \right]^{-1/2} dr$$

and that leave the entropy per nucleon and the chemical composition uniform and unchanged. [It is understood here that with the entropy per nucleon and chemical composition fixed, the equation of state gives both $p(r)$ and $n(r)$ as functions of $\rho(r)$.] The equilibrium is stable with respect to radial oscillations if and only if M, or equivalently E, is a *minimum* with respect to all such variations.

To prove this theorem we use the Lagrange multiplier method[3]: M will be stationary with respect to all variations that leave N fixed if and only if there exists a constant λ for which $M - \lambda N$ is stationary with respect to all variations. In general, the change in $M - \lambda N$ for a given variation $\delta\rho(r)$ is

$$\delta M - \lambda \delta N = \int_0^\infty 4\pi r^2 \delta\rho(r)\, dr$$

$$- \lambda \int_0^\infty 4\pi r^2 \left[1 - \frac{2G\mathcal{M}(r)}{r} \right]^{-1/2} \delta n(r)\, dr$$

$$- \lambda G \int_0^\infty 4\pi r \left[1 - \frac{2G\mathcal{M}(r)}{r} \right]^{-3/2} n(r) \delta\mathcal{M}(r)\, dr$$

(The integrals are carried to infinity for notational convenience; actually the integrands vanish outside a radius $R + \delta R$.) These variations are supposed not to change the entropy per nucleon, so

$$0 = \delta\left(\frac{\rho}{n}\right) + p\delta\left(\frac{1}{n}\right)$$

and therefore

$$\delta n(r) = \frac{n(r)}{p(r) + \rho(r)} \delta\rho(r)$$

Also,

$$\delta\mathscr{M}(r) = \int_0^r 4\pi r'^2 \delta\rho(r')\, dr'$$

Interchanging the r and r' integrals in the last term, we now have

$$\delta M - \lambda\delta N = \int_0^\infty 4\pi r^2 \left\{ 1 - \frac{\lambda n(r)}{p(r) + \rho(r)} \left[1 - \frac{2G\mathscr{M}(r)}{r} \right]^{-1/2} \right.$$
$$\left. - \lambda G \int_r^\infty 4\pi r' n(r') \left[1 - \frac{2G\mathscr{M}(r')}{r'} \right]^{-3/2} dr' \right\} \delta\rho(r)\, dr$$

Thus $\delta M - \lambda\delta N$ will vanish for all $\delta\rho(r)$ if and only if

$$\frac{1}{\lambda} = \frac{n(r)}{p(r) + \rho(r)} \left[1 - \frac{2G\mathscr{M}(r)}{r} \right]^{-1/2}$$
$$+ G \int_r^\infty 4\pi r' n(r') \left[1 - \frac{2G\mathscr{M}(r')}{r'} \right]^{-3/2} dr'$$

This will be the case for some Lagrange multiplier λ if and only if the right-hand side is independent of r, that is, if and only if

$$0 = \left\{ \frac{n'}{p + \rho} - \frac{n(p' + \rho')}{(p + \rho)^2} \right\} \left[1 - \frac{2G\mathscr{M}}{r} \right]^{-1/2}$$
$$+ \frac{Gn}{p + \rho} \left\{ 4\pi r\rho - \frac{\mathscr{M}}{r^2} \right\} \left[1 - \frac{2G\mathscr{M}}{r} \right]^{-3/2}$$
$$- 4\pi Grn \left[1 - \frac{2G\mathscr{M}}{r} \right]^{-3/2}$$

The condition of uniform entropy per nucleon gives

$$0 = \frac{d}{dr}\left(\frac{\rho}{n}\right) + p\frac{d}{dr}\left(\frac{1}{n}\right)$$

and therefore

$$n'(r) = \frac{n(r)\rho'(r)}{p(r) + \rho(r)}$$

Therefore δM vanishes for all $\delta\rho(r)$ that give $\delta N = 0$, if and only if

$$-r^2 p' = G\left[1 - \frac{2G\mathscr{M}}{r} \right]^{-1} [p + \rho][\mathscr{M} + 4\pi r^3 p]$$

as was to be proved. If the term in δM of *second* order in $\delta\rho(r)$ is positive-definite for all perturbations, then energy must be supplied in order to produce any perturbation, and the star is stable. On the other hand, if δM can in second order be negative for some perturbation $\delta\rho(r)$, then this perturbation can grow with an increase in kinetic energy, and the star is unstable.

3   Newtonian Stars: Polytropes and White Dwarfs

Most of the stars in the sky are adequately described by Newtonian physics, without taking account of general relativity. Such Newtonian stars deserve some attention here, both because they serve us limiting cases for the more exotic objects that interest the general relativist, and because they can guide us in understanding the qualitative properties of these objects.

In Newtonian astrophysics the internal energy and pressure are very much less than the rest-mass density,

$$e \ll m_N n \qquad p \ll m_N n \qquad\qquad (11.3.1)$$

so that total density is dominated by the density of rest-mass,

$$\rho \simeq m_N n \qquad\qquad (11.3.2)$$

and also

$$p \ll \rho \qquad 4\pi r^3 p \ll \mathcal{M}$$

In addition, the gravitational potential is everywhere small, so

$$\frac{2G\mathcal{M}}{r} \ll 1 \qquad\qquad (11.3.3)$$

The fundamental equation (11.1.13) thus simplifies to

$$-r^2 p'(r) = G\mathcal{M}(r)\rho(r) \qquad\qquad (11.3.4)$$

with $\mathcal{M}(r)$ still defined by

$$\mathcal{M}(r) \equiv \int_0^r 4\pi r'^2 \rho(r')\, dr' \qquad\qquad (11.3.5)$$

Dividing (11.3.4) by $\rho(r)$ and differentiating allows us to combine both (11.3.4) and (11.3.5) in a single second-order differential equation:

$$\frac{d}{dr}\frac{r^2}{\rho(r)}\frac{dp(r)}{dr} = -4\pi G r^2 \rho(r) \qquad\qquad (11.3.6)$$

In order that $\rho(0)$ be finite, it is necessary that $p'(0)$ vanish. Thus, given an equation of state $p = p(\rho)$ (with $dp/d\rho \neq 0$), we can obtain $\rho(r)$ by solving Eq. (11.3.6) with the initial conditions that $\rho(0)$ have some given value and that

$$\rho'(0) = 0 \tag{11.3.7}$$

(Eq. (11.3.7) also follows from the requirement that $\rho(r)$ be an analytic function of x, y, and z at $x = y = z = 0$.)

We still need to prescribe an equation of state. It is often the case that the internal energy density is proportional to the pressure, that is,

$$e \equiv \rho - m_N n = (\gamma - 1)^{-1} p \tag{11.3.8}$$

(Here $(\gamma - 1)^{-1}$ is just a constant proportionality coefficient; γ will not be the ratio of specific heats unless e and p are proportional to the temperature.) The condition of uniform entropy per nucleon then reads

$$0 = \frac{d}{dr}\left(\frac{\rho}{n}\right) + p \frac{d}{dr}\left(\frac{1}{n}\right) = \frac{d}{dr}\left(\frac{e}{n}\right) + p \frac{d}{dr}\left(\frac{1}{n}\right)$$

$$= \frac{1}{\gamma - 1}\left\{\gamma p \frac{d}{dr}\left(\frac{1}{n}\right) + \left(\frac{1}{n}\right)\frac{dp}{dr}\right\}$$

and therefore

$$p \propto n^\gamma$$

or, since $\rho \simeq m_N n$,

$$p = K\rho^\gamma \tag{11.3.9}$$

The proportionality constant K depends on the entropy per nucleon and chemical composition, but it does not depend on r or on $\rho(0)$. Any star for which the equation of state takes the form (11.3.9) is called a *polytrope*.

The fundamental equation (11.3.6) can, in the case of a polytrope, be transformed into a convenient dimensionless form. Define a new independent variable ξ, by

$$r = \left(\frac{K\gamma}{4\pi G(\gamma - 1)}\right)^{1/2} \rho(0)^{(\gamma - 2)/2}\xi \tag{11.3.10}$$

and a new dependent variable θ, by

$$\rho = \rho(0)\theta^{1/(\gamma - 1)} \qquad p = K\rho(0)^\gamma \theta^{\gamma/(\gamma - 1)} \tag{11.3.11}$$

Equation (11.3.6) then takes the form

$$\frac{1}{\xi^2}\frac{d}{d\xi}\xi^2\frac{d\theta}{d\xi} + \theta^{1/(\gamma - 1)} = 0 \tag{11.3.12}$$

The boundary conditions are

$$\theta(0) = 1 \qquad \theta'(0) = 0 \tag{11.3.13}$$

[See Eq. (11.3.7).] The function $\theta(\xi)$ defined by (11.3.12), (11.3.13) is known as the *Lane-Emden function*[4] of index $(\gamma - 1)^{-1}$. For ξ near zero, Eq. (11.3.12) gives

$$\theta(\xi) = 1 - \frac{\xi^2}{6} + \frac{\xi^4}{120(\gamma - 1)} - \cdots \tag{11.3.14}$$

Also, it can be shown that for $\gamma > 6/5$, $\theta(\xi)$ vanishes at some finite ξ_1:

$$\theta(\xi_1) = 0 \tag{11.3.15}$$

The radius of the star is thus given by Eq. (11.3.10) as

$$R = \left(\frac{K\gamma}{4\pi G(\gamma - 1)}\right)^{1/2} \rho(0)^{(\gamma-2)/2}\xi_1 \tag{11.3.16}$$

We can also use the Lane-Emden solutions to calculate the stellar mass:

$$M \equiv \int_0^R 4\pi r^2 \rho(r)\, dr$$

$$= 4\pi\rho(0)^{(3\gamma-4)/2}\left(\frac{K\gamma}{4\pi G(\gamma - 1)}\right)^{3/2}\int_0^{\xi_1}\xi^2\theta^{1/(\gamma-1)}(\xi)\, d\xi$$

$$= 4\pi\rho(0)^{(3\gamma-4)/2}\left(\frac{K\gamma}{4\pi G(\gamma - 1)}\right)^{3/2}\xi_1^2|\theta'(\xi_1)| \tag{11.3.17}$$

By eliminating $\rho(0)$ in (11.3.16) and (11.3.17), we obtain a relation between M and R:

$$M = 4\pi R^{(3\gamma-4)/(\gamma-2)}\left(\frac{K\gamma}{4\pi G(\gamma - 1)}\right)^{-1/(\gamma-2)}\xi_1^{-(3\gamma-4)/(\gamma-2)}\xi_1^2|\theta'(\xi_1)| \tag{11.3.18}$$

Values[5] of the numerical constants ξ_1 and $\xi_1^2|\theta'(\xi_1)|$ are tabulated in Table 11.1.

Table 11.1. Values[5] of the Numerical Parameters ξ_1 and $-\xi_1^2\theta'(\xi_1)$ for Various Newtonian Polytropes

γ	ξ_1	$-\xi_1^2\theta'(\xi_1)$	Examples
6/5	∞	1.73205	
11/9	31.83646	1.73780	
5/4	14.97155	1.79723	
9/7	9.53581	1.89056	
4/3	6.89685	2.01824	Largest mass white dwarfs
7/5	5.35528	2.18720	
3/2	4.35287	2.41105	
5/3	3.65375	2.71406	Small mass white dwarfs
2	π	π	
3	2.7528	3.7871	
∞	$\sqrt{6}$	$2\sqrt{6}$	Incompressible stars

For Newtonian stars, M is dominated by the total rest-mass Nm_N, so the nucleon number of the star is given to a good approximation by

$$N \simeq \frac{M}{m_N} \tag{11.3.19}$$

We also want to know the internal energy $E \equiv M - Nm_N$. For general Newtonian stars this is given by Eqs. (11.1.26), (11.1.29), and (11.1.30) as

$$E = T + V \tag{11.3.20}$$

with the thermal energy T and the gravitational energy V given by

$$T = \int_0^R 4\pi r^2 e(r) \, dr \tag{11.3.21}$$

$$V = -\int_0^R 4\pi r G \mathcal{M}(r) \rho(r) \, dr \tag{11.3.22}$$

We now show that for polytropes, T and V are given by the remarkably simple formulas[6]

$$T = \frac{1}{(5\gamma - 6)} \frac{GM^2}{R} \tag{11.3.23}$$

$$V = -\frac{3(\gamma - 1)}{(5\gamma - 6)} \frac{GM^2}{R} \tag{11.3.24}$$

so the total internal energy is

$$E = -\frac{(3\gamma - 4)}{(5\gamma - 6)} \frac{GM^2}{R} \tag{11.3.25}$$

To prove the formula for V, we use Eq. (11.3.4) to rewrite (11.3.22) as

$$V = 4\pi \int_0^R r^3 \frac{dp(r)}{dr} \, dr = -12\pi \int_0^R r^2 p(r) \, dr \tag{11.3.26}$$

Multiplying and dividing in the integrand by $\rho(r)$, we have

$$V = -3 \int_0^R \frac{p(r)}{\rho(r)} \, d\mathcal{M}(r) = 3 \int_0^R \mathcal{M}(r) \, d\left(\frac{p(r)}{\rho(r)}\right)$$

(We assume here that $\gamma > 1$, so that p/ρ vanishes at R.) This can be evaluated by using the equation of state to calculate

$$\frac{d}{dr}\left(\frac{p(r)}{\rho(r)}\right) = \left(\frac{\gamma - 1}{\gamma}\right) \frac{p'(r)}{\rho(r)} = -\left(\frac{\gamma - 1}{\gamma}\right) \frac{G\mathcal{M}(r)}{r^2}$$

so

$$V = -3\left(\frac{\gamma - 1}{\gamma}\right)\int_0^R \frac{G\mathcal{M}^2(r)}{r^2}\,dr \tag{11.3.27}$$

Since $dr/r^2 = -d(1/r)$, we can integrate by parts once again, and find

$$V = 3\left(\frac{\gamma - 1}{\gamma}\right)\left\{\frac{GM^2}{R} - 2\int_0^R 4\pi r G\mathcal{M}(r)\rho(r)\,dr\right\}$$

$$= 3\left(\frac{\gamma - 1}{\gamma}\right)\left\{\frac{GM^2}{R} + 2V\right\}$$

Solving for V then gives the desired result (11.3.24). To calculate T we use (11.3.8) in (11.3.26), which gives

$$V = -3(\gamma - 1)T \tag{11.3.28}$$

Equations (11.3.24) and (11.3.28) then give the desired result (11.3.23).

Inspection of (11.3.17) and (11.3.19) shows that the nucleon number N behaves like $\rho(0)^{(3\gamma-4)/2}$ whereas (11.3.25), (11.3.16), and (11.3.17) shown that the internal energy E behaves like $\rho(0)^{(5\gamma-6)/2}$. Thus $\partial N/\partial \rho(0)$ and $\partial E/\partial \rho(0)$ can never vanish together. Theorem 1 of the last section tells us then that each polytrope is either stable or unstable for all $\rho(0)$, depending on the value of γ. But which?

In order to answer this question, we turn to Theorem 2 of the last section, which tells us that the star will be stable if and only if E is a minimum with respect to all variations in $\rho(r)$ that leave N (and the equation of state) unchanged. It is intuitively likely that the first instability to occur will correspond to a uniform implosion of the whole star, and since we are only trying to answer the yes-or-no question about stability with respect to this mode, it will hopefully be sufficient for us to consider only trial configurations with $\rho(r)$ *constant*.[7] In any such configuration, (11.3.19), (11.3.21), (11.3.22), and (11.3.8) give

$$N = \frac{4\pi}{3m_N}\rho R^3 \tag{11.3.29}$$

$$T = \frac{4\pi}{3}(\gamma - 1)^{-1}K\rho^\gamma R^3 \tag{11.3.30}$$

$$V = -\frac{16\pi^2}{15}G\rho^2 R^5 \tag{11.3.31}$$

so, eliminating R,

$$E = T + V = a\rho^{\gamma-1} - b\rho^{1/3} \tag{11.3.32}$$

where

$$a = \frac{KM}{(\gamma - 1)} \tag{11.3.33}$$

$$b = \frac{3}{5}\left(\frac{4\pi}{3}\right)^{1/3} GM^{5/3} \tag{11.3.34}$$

For $\gamma > 4/3$, E has a *minimum* at

$$\rho = \left(\frac{b}{3a(\gamma - 1)}\right)^{1/(\gamma - 4/3)} = \left(\frac{M^{2/3}G(4\pi/3)^{1/3}}{5K}\right)^{1/(\gamma - 4/3)} \tag{11.3.35}$$

corresponding to a configuration of *stable* equilibrium. For $\gamma = 4/3$, E is stationary with respect to ρ only if it vanishes everywhere, which requires that $a = b$, or

$$M = \left(\frac{5K}{G}\right)^{3/2} \left(\frac{4\pi}{3}\right)^{-1/2} \tag{11.3.36}$$

For $\gamma < 4/3$, E has a *maximum* at the point (11.3.35), corresponding to a state of *unstable* equilibrium.

Incidentally, Eq. (11.3.35) gives an estimate for the mass

$$M \simeq \frac{4\pi}{3} \rho^{(3\gamma - 4)/2} \left(\frac{15K}{4\pi G}\right)^{3/2}$$

which may be compared with the exact result (11.3.17). The ratio of these two expressions is

$$\frac{M(\text{variational})}{M(\text{exact})} = \frac{(15(\gamma - 1)/\gamma)^{3/2}}{3\xi_1^2|\theta'(\xi_1)|}$$

For $\gamma = 5/3$ this ratio is 1.8; for $\gamma = 4/3$ it is 1.2. Not only does the variational method give the correct dependence of M on ρ (including the fact that for $\gamma = 4/3$, M is independent of ρ, and E vanishes), but it even provides a fair approximation to the exact numerical results. We can accept with confidence its prediction that a polytrope is stable or unstable according to whether $\gamma > 4/3$ or $\gamma < 4/3$.[7]

The variational approach also provides a simple method for estimating the oscillation frequency for dilation and contraction of the star. Equations (11.3.29)–(11.3.31) show that for fixed N,

$$T \propto R^{3(1-\gamma)} \qquad V \propto R^{-1}$$

We can use Eqs. (11.3.23) and (11.3.24) to fix the correct values of T and V at the equilibrium radius (which we shall now write as R_{eq}, to distinguish it from the instantaneous radius R of an oscillating configuration). This gives then

$$E = \frac{1}{(5\gamma - 6)} \frac{GM^2}{R_{\text{eq}}^{(4-3\gamma)}} R^{3(1-\gamma)} - \frac{3(\gamma - 1)}{5\gamma - 6} GM^2 R^{-1}$$

For $\gamma > 4/3$, this has a minimum at $R = R_{eq}$, as it should. For R near R_{eq}, E behaves like

$$E \to E_{eq} + \frac{3(\gamma - 1)(3\gamma - 4)}{2(5\gamma - 6)} \frac{GM^2}{R_{eq}{}^3} (R - R_{eq})^2$$

The uniform dilation of a sphere with uniform density will give it a kinetic energy

$$U = \frac{3}{10} M\dot{R}^2$$

so the condition of energy conservation, that $U + E$ be constant, leads to modes with

$$R - R_{eq} \propto \sin \omega_0 t$$

$$\omega_0 \approx \left[\frac{5(\gamma - 1)(3\gamma - 4)}{5\gamma - 6} \frac{GM^2}{R_{eq}{}^3} \right]^{1/2} \tag{11.3.37}$$

Finally, we note that a uniform sphere rotating with angular velocity Ω will have kinetic energy

$$U = \tfrac{1}{5} M R_{eq}{}^2 \Omega^2$$

This must be less than the binding energy $-E$, so the maximum angular velocity with which a star can rotate is of order

$$\Omega_{max} \approx \left[\frac{5(3\gamma - 4)}{(5\gamma - 6)} \frac{GM}{R_{eq}{}^3} \right]^{1/2} \approx \frac{\omega_0}{\sqrt{\gamma - 1}} \tag{11.3.38}$$

Of course a star rotating this fast will no longer be a sphere, and (11.3.38) only gives an order-of-magnitude estimate of the actual maximum rotation frequency.

Now let us apply what we have learned to the stars known as *white dwarfs*. Imagine an aged star that exhausts its nuclear fuel and begins to cool and contract. When the temperature is sufficiently low (see below for just how low), the electrons will be frozen into the lowest available energy levels. The Pauli principle tells us that there will be two electrons in each level (because of the two spin states available) and there are $4\pi k^2 (2\pi\hbar)^{-3} dk$ levels per unit volume with momenta between k and $k + dk$, so the number of electrons per unit volume will be related to the maximum momentum k_F by

$$n = \frac{8\pi}{(2\pi\hbar)^3} \int_0^{k_F} k^2 \, dk = \frac{k_F{}^3}{3\pi^2 \hbar^3} \tag{11.3.39}$$

The mass density is

$$\rho = n m_N \mu \tag{11.3.40}$$

where μ is the number of nucleons per electron; $\mu \simeq 2$ for stars that have used up their hydrogen. This gives

$$k_F = \hbar \left(\frac{3\pi^2 \rho}{m_N \mu}\right)^{1/3} \qquad (11.3.41)$$

The condition that the temperature is negligible is

$$kT \ll [k_F{}^2 + m_e{}^2]^{1/2} - m_e$$

The kinetic energy density and pressure of these electrons are

$$e = \frac{8\pi}{(2\pi\hbar)^3} \int_0^{k_F} [(k^2 + m_e{}^2)^{1/2} - m_e] k^2 \, dk \qquad (11.3.42)$$

$$p = \frac{8\pi}{3(2\pi\hbar)^3} \int_0^{k_F} \frac{k^2}{(k^2 + m_e{}^2)^{1/2}} k^2 \, dk \qquad (11.3.43)$$

[See Eq. (2.8.4).] The equation of state can be made explicit by using (11.3.41) in (11.3.43).

The equation of state here is not simple, but it reduces to a polytrope in two extreme cases, distinguished by the criteria $\rho \ll \rho_c$ or $\rho \gg \rho_c$, where ρ_c is the critical density at which k_F becomes equal to m_e (in c.g.s. units):

$$\rho_c = \frac{m_N \mu m_e{}^3 c^3}{3\pi^2 \hbar^3} = 0.97 \times 10^6 \mu \text{ gm/cm}^3 \qquad (11.3.44)$$

(A) $\rho \ll \rho_c$. In this case $k_F \ll m_e$, so Eqs. (11.3.42) and (11.3.43) give

$$e = \tfrac{3}{2} p$$

$$p = \frac{8\pi k_F{}^5}{15 m_e (2\pi\hbar)^3} = \frac{\hbar^2}{15 m_e \pi^2} \left(\frac{3\pi^2 \rho}{m_N \mu}\right)^{5/3}$$

This is a polytrope, with

$$\gamma = \tfrac{5}{3}, \qquad K = \frac{\hbar^2}{15 m_e \pi^2} \left(\frac{3\pi^2}{m_N \mu}\right)^{5/3} \qquad (11.3.45)$$

Equation (11.3.17) then gives a mass (in c.g.s. units)

$$M = \frac{1}{2} \left(\frac{3\pi}{8}\right)^{1/2} (2.71406) \left(\frac{\hbar^{3/2} c^{3/2}}{m_N{}^2 \mu^2 G^{3/2}}\right) \left(\frac{\rho(0)}{\rho_c}\right)^{1/2} \qquad (11.3.46)$$

$$= 2.79 \mu^{-2} \left(\frac{\rho(0)}{\rho_c}\right)^{1/2} M_\odot$$

whereas Eq. (11.3.16) gives a radius (in c.g.s. units)

$$R = \left(\frac{3\pi}{8}\right)^{1/2} (3.65375) \left(\frac{\hbar^{3/2}}{c^{1/2}G^{1/2}m_e m_N \mu}\right) \left(\frac{\rho(0)}{\rho_c}\right)^{-1/6} \tag{11.3.47}$$

$$= 2.0 \times 10^4 \mu^{-1} \left(\frac{\rho(0)}{\rho_c}\right)^{-1/6} \text{ km}$$

(B) $\rho \gg \rho_c$. In this case $k_F \gg m_e$, so Eqs. (11.3.42) and (11.3.43) give

$$e = 3p$$

$$p = \frac{8\pi k_F^4}{12(2\pi\hbar)^3} = \frac{\hbar}{12\pi^2} \left(\frac{3\pi^2 \rho}{m_N \mu}\right)^{4/3}$$

This is a polytrope, with

$$\gamma = \tfrac{4}{3} \qquad K = \frac{\hbar}{12\pi^2} \left(\frac{3\pi^2}{m_N \mu}\right)^{4/3} \tag{11.3.48}$$

Equation (11.3.17) then gives a *unique* mass (in c.g.s. units)

$$M = \tfrac{1}{2}(3\pi)^{1/2}(2.01824) \left(\frac{\hbar^{3/2}c^{3/2}}{G^{3/2}m_N^2 \mu^2}\right) \tag{11.3.49}$$

$$= 5.87\mu^{-2}M_\odot$$

whereas Eq. (11.3.16) gives the radius (in c.g.s. units)

$$R = \tfrac{1}{2}(3\pi)^{1/2}(6.89685) \left(\frac{\hbar^{3/2}}{c^{1/2}G^{1/2}m_e m_N \mu}\right) \left(\frac{\rho_c}{\rho(0)}\right)^{1/3} \tag{11.3.50}$$

$$= 5.3 \times 10^4 \mu^{-1} \left(\frac{\rho_c}{\rho(0)}\right)^{1/3} \text{ km}$$

We note that $\gamma > 4/3$ for $\rho(0) \ll \rho_c$, so the least massive white dwarfs are definitely stable. We also see that M appears to grow monotonically with increasing central density, reaching a maximum (11.3.49) when $\rho(0) \to \infty$, so there is no point where the star can become unstable. Our tentative conclusion is that stable white dwarfs can exist for any mass less than (11.3.49). This maximum mass is known as the *Chandrasekhar limit*.[8]

Actually, matters are not so simple. When $k_F \simeq 5m_e$, it becomes energetically favorable for electrons to be captured by nucleii, turning protons into neutrons, and producing neutrinos that escape forthwith. The effect is to increase the number μ of nucleons per electron, and according to Eq. (11.3.46) this will reduce the mass M for a given central density. We therefore expect M to increase toward the Chandrasekhar limit until $\rho(0) \simeq 5^3 \rho_c$ [see Eqs. (11.3.41) and (11.3.44)], where M reaches a maximum and then begins to decrease. Detailed calculations[9] show that the maximum mass is $1.2M_\odot$, nearly equal to the Chandrasekhar limit, which is $1.26M_\odot$ for $\mu = 56/26$. The radius of a star with this maximum mass is about 4×10^3 km. Theorem 2 of Section 11.2 suggests that this maximum is a point of transition from stability to instability, so stable iron white dwarfs can exist only for $M < 1.2M_\odot$.

To the student of general relativity, the most interesting parameter characterizing a white dwarf is the absolute value GM/R of the gravitational potential at its surface. For $\rho(0) \ll \rho_c$, this is given by (11.3.46) and (11.3.47) as

$$\frac{GM}{R} = \frac{1}{2}\left(\frac{2.71406}{3.65375}\right)\mu^{-1}\left(\frac{m_e}{m_N}\right)\left(\frac{\rho(0)}{\rho_c}\right)^{2/3} \qquad (11.3.51)$$

whereas for $\rho(0) \gg \rho_c$, it is given by Eqs. (11.3.49) and (11.3.50) as

$$\frac{GM}{R} = \left(\frac{2.01824}{6.89685}\right)\mu^{-1}\left(\frac{m_e}{m_N}\right)\left(\frac{\rho(0)}{\rho_c}\right)^{1/3} \qquad (11.3.52)$$

We see that GM/R is always going to be quite small, because $m_e/m_N = 5.4 \times 10^{-4}$. Thus general relativity plays no important role in the structure of white dwarfs. The quantity GM/R increases with increasing central density, so it is largest at the maximum mass $1.2M_\odot$, where it takes the value 4×10^{-4}. Our old friend 40 Eridani *B* had $GM/R \simeq 6 \times 10^{-5}$ (see Section 3.5), so it is not going to be possible to improve astronomical red-shift experiments very dramatically by finding white dwarfs with much larger red shifts.

4 Neutron Stars

We saw in the last section that a white dwarf star supported by the pressure of cold degenerate electrons cannot be in equilibrium if its mass is greater than the Chandrasekhar limit, about $\hbar^{3/2}/m_N^2 G^{3/2}$. Also, the gravitational potential at the surface of such a star cannot be greater than about m_e/m_N, so general relativity plays no role in its structure.

Continuing our search for astrophysical applications of general relativity, let us ask what happens when a star whose mass is above the Chandrasekhar limit reaches the end of its thermonuclear evolution and grows cold. Its internal pressure then fails to support it, and it collapses. One possibility is that the star will simply go on collapsing forever, in which case general relativity will certainly come into play. Another possibility is that the star will become so heated during its collapse that it will explode, becoming a supernova. It might then blow off enough matter so that its mass drops below the Chandrasekhar limit. It is believed that in this case the highly compressed remnant does not find its quietus as a white dwarf, but rather becomes a superdense *neutron star*.[10] (See Figure 11.2.)

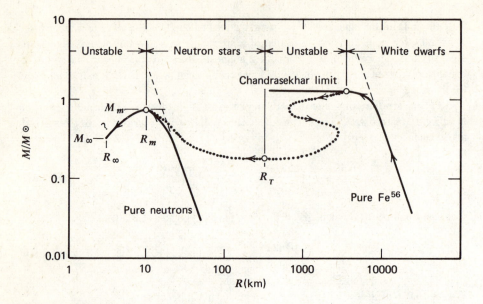

Figure 11.2 Configurations of stellar equilibrium..The solid curves on the left and right represent the Oppenheimer-Volkoff[10] solution for a pure neutron star and the Chandrasekhar[8] solution for a pure Fe^{56} white dwarf star, respectively. The dashed lines give the extrapolated nonrelativistic solutions in these two cases. The dotted line represents the interpolating solution of Harrison, Thorne, Wakano, and Wheeler,[12] which takes into account the shift in chemical composition from Fe^{56} to neutrons. Arrows indicate the direction of increasing central density. As shown in Theorem 1, the various transitions between stability and instability occur at the maxima and minima of M, marked here with small circles.

A neutron star is like a white dwarf, except that it consists almost entirely of "cold" degenerate neutrons, all electrons and protons having been converted into neutrons through the reaction

$$p + e^- \rightarrow n + \nu$$

the neutrinos escaping the star. Enough electrons and protons must remain so that the Pauli principle prevents neutron beta decay, $n \rightarrow p + e^- + \bar{\nu}$; this sets a lower limit on the mass of stable neutron stars, to be evaluated below.

Neutron stars of low mass are much like white dwarfs of the same mass, except that neutron degeneracy pressure replaces electron degeneracy pressure, and thus m_e should be replaced in all formulas with m_n (and μ should be set equal to unity). Thus, by noting how m_e enters in the formulas (11.3.44)–(11.3.47) for small white dwarfs, we can immediately conclude that a neutron star of small mass will have a central density higher than that of a white dwarf with the same mass (and $\mu = 2$) by a factor $\frac{1}{2}(m_n/m_e)^3 = 3.1 \times 10^9$, and will have a radius smaller by a factor $m_n/2m_e = 920$.

The electrons in a white dwarf will begin to be relativistic when its mass becomes comparable to the theoretical upper limit, given by Eq. (11.3.49). Since m_e does not enter in (11.3.49), we expect that the neutrons in a neutron star will begin to be relativistic at just such masses, that is, when M is of order $M_\odot$. However, at this point the analogy between white dwarfs and neutron stars breaks down. For one thing, the total energy density ρ of a white dwarf is always dominated by the rest-mass density of its nonrelativistic *nucleons*, whereas a neutron star with mass of order $M_\odot$ will consist of nucleons whose kinetic energies are comparable with their rest-masses. Another difference that is even more interesting is that, whereas a white dwarf whose electrons are moderately relativistic will have a surface gravitational potential GM/R of order m_e/m_n, a neutron star of equal mass will have a surface potential roughly of order unity. Thus general relativity will necessarily play a role in the theory of the more massive neutron stars.

In order to formulate the quantitative theory of neutron stars, we begin by writing down expressions for the total energy density and pressure of an ideal Fermi gas of neutrons with maximum momentum k_F:

$$\rho = \frac{8\pi}{(2\pi\hbar)^3} \int_0^{k_F} (k^2 + m_n^2)^{1/2} k^2 \, dk = 3\rho_c \int_0^{k_F/m_n} (u^2 + 1)^{1/2} u^2 \, du$$

$$(11.4.1)$$

$$p = \frac{8\pi}{3(2\pi\hbar)^3} \int_0^{k_F} \frac{k^2}{(k^2 + m_n^2)^{1/2}} k^2 \, dk = \rho_c \int_0^{k_F/m_n} (u^2 + 1)^{-1/2} u^4 \, du$$

$$(11.4.2)$$

where now (in c.g.s. units)

$$\rho_c \equiv \frac{8\pi m_n^4 c^3}{3(2\pi\hbar)^3} = 6.11 \times 10^{15} \text{ gm/cm}^3 \qquad (11.4.3)$$

By eliminating k_F/m_n in Eqs. (11.4.1) and (11.4.2), we obtain the equation of state in the form

$$\frac{p}{\rho_c} = F\left(\frac{\rho}{\rho_c}\right) \qquad (11.4.4)$$

with F a definite transcendental function. The structure of a neutron star with given central density $\rho(0)$ is to be calculated by solving (11.1.13) with p given as a function of ρ by (11.4.4). Since the only dimensional quantities in these equations are $\rho(0)$, ρ_c, and G, the solution must give the mass and radius as functions of $\rho(0)$ of the form

$$M = M_0 f\left(\frac{\rho(0)}{\rho_c}\right) \qquad (11.4.5)$$

$$R = R_0 g\left(\frac{\rho(0)}{\rho_c}\right) \qquad (11.4.6)$$

where (in c.g.s. units)

$$R_0 \equiv c(8\pi G \rho_c)^{-1/2} = 3.0 \text{ km} \tag{11.4.7}$$

$$M_0 \equiv \frac{c^2 R_0}{G} = 2.0 M_\odot \tag{11.4.8}$$

and f and g are unknown dimensionless functions. This problem, like that of the white dwarfs, is analytically tractable only for very large and very small central densities.

For $\rho(0) \ll \rho_c$ we can use the analogy with white dwarfs discussed above, and conclude from Eqs. (11.3.46) and (11.3.47) that

$$\begin{aligned} M &= \frac{1}{2} \left(\frac{3\pi}{8} \right)^{1/2} (2.71406) \left(\frac{\hbar^{3/2}}{m_n{}^2 G^{3/2}} \right) \left(\frac{\rho(0)}{\rho_c} \right)^{1/2} \\ &= \frac{1}{2} (2.71406) M_0 \left(\frac{\rho(0)}{\rho_c} \right)^{1/2} \end{aligned} \tag{11.4.9}$$

$$\begin{aligned} R &= \left(\frac{3\pi}{8} \right)^{1/2} (3.65375) \left(\frac{\hbar^{3/2}}{m_n{}^2 G^{1/2}} \right) \left(\frac{\rho_c}{\rho(0)} \right)^{1/6} \\ &= (3.65375) R_0 \left(\frac{\rho_c}{\rho(0)} \right)^{1/6} \end{aligned} \tag{11.4.10}$$

with ρ_c now given by Eq. (11.4.3).

For $\rho(0) \gg \rho_c$, the neutrons near the center of the star have $k_F \gg m_n$, so (11.4.1) and (11.4.2) give

$$\rho = \frac{3\rho_c}{4} \left(\frac{k_F}{m_n} \right)^5 \qquad p = \frac{\rho_c}{4} \left(\frac{k_F}{m_n} \right)^5$$

and therefore

$$p = \frac{\rho}{3} \tag{11.4.11}$$

as would be expected for a gas of highly relativistic particles. Using this equation of state in the fundamental differential equation (11.1.13) gives

$$-r^2 \rho'(r) = 4G \mathcal{M}(r) \rho(r) \left[1 + \frac{4\pi r^3 \rho(r)}{3\mathcal{M}(r)} \right] \left[1 - \frac{2G\mathcal{M}(r)}{r} \right]^{-1} \tag{11.4.12}$$

Amazingly, we can find an *exact* solution of this equation[11]:

$$\rho(r) = \frac{3}{56\pi G r^2} \tag{11.4.13}$$

corresponding to the limit $\rho(0) \to \infty$. However, even in the limit of infinite central density, this $\rho(r)$ will drop below ρ_0 at a radius r of order R_0, so that the equation of state (11.4.11) is not valid for the outer layers of *any* neutron star. To deal with the crust of nonrelativistic neutrons, it is necessary to solve the full equation (11.1.13) using the equation of state (11.4.4); the condition of infinite central density is imposed by (11.4.13) for $r \ll R_0$. We shall not do this here; the important points are that the solution has a finite radius R where ρ vanishes, and that the mass M within this radius is finite, because the singularity in Eq. (11.4.13) is integrable at $r = 0$. Thus the mass and radius of a neutron star approach finite limits as $\rho(0) \to \infty$. Numerical solution of the fundamental equation (11.1.13) gives these limits as[10]

$$M_\infty = 0.171 M_0 \qquad R_\infty = 1.06 R_0 \qquad (11.4.14)$$

There remains the question of stability. For $\rho(0) \ll \rho_c$, a pure neutron star is simply a Newtonian polytrope with $\gamma = 5/3$ (like a small white dwarf) and is therefore stable. (See Section 11.3.) Equation (11.4.9) shows that M is a monotonically increasing function of $\rho(0)$ for these small central densities. If M continues to increase monotonically to the value M_∞, then no transition to instability can occur, according to Theorem 1 of Section 11.2. But (11.4.9) shows that when $\rho(0) = 0.016 \rho_c$ (which is small enough for (11.4.9) to be a good approximation), the mass M is already greater than M_∞. Thus we expect that M rises to a maximum value $M > M_\infty$ at some central density ρ_m of order ρ_c, and then drops to the value M_∞ at infinite central density. This expectation is confirmed by detailed calculation[10] using Eqs. (11.1.13) and (11.4.1)–(11.4.3). The mass M of a pure ideal-gas neutron star reaches a maximum

$$M_m = 0.36 M_0 = 0.7 M_\odot \qquad (11.4.15)$$

at a radius

$$R_m = 3.2 R_0 = 9.6 \text{ km} \qquad (11.4.16)$$

Since this is a point where $\partial M / \partial \rho(0)$ vanishes, we expect a transition here from stability to instability with respect to radial oscillations. Thus (11.4.15) and (11.4.16) characterize a neutron star with the greatest mass and central density allowed by the requirement that the star be stable. The mass (11.4.15) is known as the *Oppenheimer-Volkoff limit*. Note that the fractional red shift of a spectral line emitted from the surface of such a neutron star is

$$z \equiv \frac{\Delta\lambda}{\lambda} = B^{-1/2}(R_m) - 1 = \left(1 - \frac{2M_m G}{R_m}\right)^{-1/2} - 1 = 0.13$$

$$(11.4.17)$$

[See Eqs. (3.5.3), (11.1.1), and (11.1.17).] Evidently general relativity is just beginning to be important for the most massive stable neutron stars.

Of course, a neutron star cannot consist purely of neutrons, if only because we need a Fermi sea of electrons so that the Pauli exclusion principle can block the neutrons' beta decay. In order to get a first taste of the chemical composition in a neutron star, let us consider the equilibrium among neutrons, protons, and electrons. The energy density and number density of each one of these three Fermi gases are given (for $i = n, p, e$) by

$$\rho_i = \frac{8\pi}{(2\pi\hbar)^3} \int_0^{k_{F,i}} \sqrt{k^2 + m_i^2} \, k^2 \, dk \tag{11.4.18}$$

$$n_i = \frac{8\pi}{(2\pi\hbar)^3} \int_0^{k_{F,i}} k^2 \, dk = \frac{k_{F,i}^3}{3\pi^2\hbar^3} \tag{11.4.19}$$

At any given point in the star, the reactions $n \rightarrow p + e + \bar{v}$ and $p + e \rightarrow n + v$ can convert neutrons into protons and vice versa. (The neutrinos escape.) These reactions preserve the total number density of baryons,

$$n_n + n_p = n_B \quad \text{(fixed)} \tag{11.4.20}$$

and preserve charge neutrality,

$$n_p - n_e = 0 \tag{11.4.21}$$

But with n_B fixed, the total energy density may be expressed in terms of n_n alone:

$$\rho \equiv \rho_n + \rho_e + \rho_p$$
$$= 3C^{-3} \left\{ \int_0^{Cn_n^{1/3}} \sqrt{k^2 + m_n^2} \, k^2 \, dk + \int_0^{C[n_B - n_n]^{1/3}} \sqrt{k^2 + m_p^2} \, k^2 \, dk \right.$$
$$\left. + \int_0^{C[n_B - n_n]^{1/3}} \sqrt{k^2 + m_e^2} \, k^2 \, dk \right\} \tag{11.4.22}$$

where

$$C \equiv (3\pi^2\hbar^3)^{1/3}$$

Chemical equilibrium is reached when this function is a minimum, that is, at

$$0 = \frac{d\rho}{dn_n} = (C^2 n_n^{2/3} + m_n^2)^{1/2} - (C^2[n_B - n_n]^{2/3} + m_p^2)^{1/2}$$
$$- (C^2[n_B - n_n]^{2/3} + m_e^2)^{1/2}$$

We can solve for $n_p = n_B - n_n$ as a function of n_n, and find

$$\frac{n_p}{n_n} = \frac{1}{8} \left\{ \frac{1 + \dfrac{2(m_n^2 - m_p^2 - m_e^2)}{C^2 n_n^{2/3}} + \dfrac{(m_n^2 - m_p^2)^2 - 2m_e^2(m_n^2 + m_p^2) + m_e^4}{C^4 n_n^{4/3}}}{1 + \dfrac{m_n^2}{C^2 n_n^{2/3}}} \right\}^{3/2}$$

The nucleon mass difference $Q \equiv m_n - m_p$ and the electron mass m_e are of comparable magnitude and very much less than m_n, so this result can be written more simply as

$$\frac{n_p}{n_n} = \frac{1}{8} \left\{ \frac{1 + \dfrac{4Q}{m_n}\left(\dfrac{\rho_c}{m_n n_n}\right)^{2/3} + \dfrac{4(Q^2 - m_e^2)}{m_n^2}\left(\dfrac{\rho_c}{m_n n_n}\right)^{4/3}}{1 + \left(\dfrac{\rho_c}{m_n n_n}\right)^{2/3}} \right\}^{3/2} \tag{11.4.23}$$

where $\rho_c = m_n^4/C^3$ is the critical density previously defined in Eq. (11.4.3).

The condition for the neutrons to be stable against beta decay is that the electron Fermi sea should be filled up to a momentum greater than the maximum momentum $k_{\max}$ of the electron emitted in neutron beta decay:

$$k_{F,e} > k_{\max} \tag{11.4.24}$$

where

$$k_{\max} = \frac{[(m_n^2 - m_p^2)^2 - 2m_e^2(m_n^2 + m_p^2) + m_e^4]^{1/2}}{2m_n}$$

$$\simeq [Q^2 - m_e^2]^{1/2} = 1.19 \text{ MeV} \tag{11.4.25}$$

The electron Fermi momentum is given by (11.4.19) and (11.4.21) as

$$k_{F,e}^2 = C^2 n_e^{1/3} = C^2 n_p^{2/3} = m_n^2 \left(\frac{m_n n_n}{\rho_c}\right)^{2/3} \left(\frac{n_p}{n_n}\right)^{2/3}$$

$$= \left\{ \frac{\dfrac{m_n^2}{4}\left(\dfrac{m_n n_n}{\rho_c}\right)^{4/3} + Q m_n \left(\dfrac{m_n n_n}{\rho_c}\right)^{2/3} + Q^2 - m_e^2}{\left(\dfrac{m_n n_n}{\rho_c}\right)^{2/3} + 1} \right\} \tag{11.4.26}$$

This is smallest at $n_n = 0$, where $k_{F,e}$ barely equals the value $k_{\max}$. Hence the condition (11.4.24) for neutron beta stability is indeed satisfied for any positive neutron density.

The proton-neutron ratio (11.4.23) is large and decreasing for very small neutron densities, reaches a minimum for $m_n n_n$ equal to the transition density

$$\rho_T \simeq \rho_c \left[\frac{4(Q^2 - m_e^2)}{m_n^2}\right]^{3/4} = 1.28 \times 10^{-4} \rho_c \tag{11.4.27}$$

where

$$\left(\frac{n_p}{n_n}\right)_{\min} \simeq \left(\frac{Q + \frac{1}{2}(Q^2 - m_e^2)^{1/2}}{m_n}\right)^{3/2} = 0.002 \tag{11.4.28}$$

and then rises monotonically, reaching the value $1/8$ for $n_n m_n \gg \rho_c$. Stars with a central density somewhat less than the transition value (11.4.27) are not really neutron stars at all, but belong to the extreme high-density branch of the white dwarf equilibrium solutions, and are therefore unstable. (See Section 11.3.) Thus we expect there to be some minimum central density of order ρ_T, and some minimum mass of order $3M_\odot (\rho_T/\rho_c)^{1/2} \simeq 0.03 M_\odot$ [see Eq. (11.4.9)], below which stable neutron stars could not exist. Detailed calculations[2] show that the minimum mass of a neutron star is actually about $0.2 M_\odot$.

The small hydrogen contamination in a neutron star is more interesting than might be thought, because the filling of proton and electron as well as neutron energy levels will block the decay of various particles besides the neutron that are normally unstable. For instance, the μ^--meson becomes stable when $k_{F,e} > 53$ MeV, because then the Pauli principle will block the emission of electrons in the decay process $\mu^- \to e^- + \nu + \bar{\nu}$. According to Eq. (11.4.26), this happens when the density $\rho \simeq m_n n_n$ reaches the value $0.038\rho_c$, when $n_p = 0.005n_n$. When the density reaches $0.107\rho_c$, with $n_p = 0.013n_n$, the electron Fermi momentum reaches the μ-meson mass 105 MeV, and it becomes energetically favorable for electrons at the top of the Fermi sea to be converted (say by collisions) into μ^--mesons, with a neutrino-antineutrino pair escaping from the star. Thus neutron stars of even moderate mass will be contaminated with μ^--mesons as well as hydrogen. The same reasoning leads us to expect that hyperons and various excited states of the nucleons and hyperons also will be stable and present in small amounts.

This raises an interesting question of principle. For instance, the famous 3–3 resonance in pion-nucleon scattering may be thought of either as a manifestation of the pion-nucleon force, or as a particle, the Δ baryon, with a mass 1236 MeV and the very short lifetime, 5.5×10^{-20} sec. Should we include the Δ in ideal-gas models of neutron stars? Normally one would think not, but for high enough nucleon density the Pauli principle will block the decays $\Delta \to N + \pi$, $\Delta \to N + \gamma$, and so on, and energetic considerations will favor the conversion of some neutrons and protons into Δ's. Of course, it is possible that the strong interactions among nucleons simply rule out any ideal-gas model of a dense neutron star, but it is also possible that the effects of these forces can be taken into account by treating a neutron star as an ideal gas of neutrons, protons, electrons, μ^--mesons, hyperons, and nucleon and hyperon resonances. (See Section 15.11.)

In any case, it should be clear that the Oppenheimer-Volkoff calculation, which treats a neutron star as a pure ideal gas of neutrons, must be used with a good deal of caution when $\rho(0)$ is comparable with or greater than ρ_c. Merely including protons and electrons as well as neutrons in an ideal-gas model does not by itself have a serious effect on the structure of a neutron star,[12] but nuclear forces can be quite important: For instance, various detailed calculations yield values of the maximum stable mass equal to $0.37M_\odot$,[13] $1.95M_\odot$,[14] and $2.4M_\odot$.[15] Even these models are still highly idealized; a real neutron star is expected to have a crystalline crust,[16] a superfluid interior,[17] powerful magnetic fields,[18] and often a very rapid rate of rotation.[19]

The discovery[20] in 1967 of "pulsars," stars that emit radiation at various wavelengths in regular pulses separated by intervals from a few seconds down to 0.033 sec, suggests that we should look into the possible rotation and vibration periods of neutron stars and white dwarfs. Equations (11.3.37) and (11.3.38) show that for all γ except 6/5, 4/3, or 1, the maximum rotation frequency and the fundamental vibration frequency of any Newtonian polytrope are both of order $\sqrt{GM/R^3}$. Presumably this result holds to within an order of magnitude for any stable neutron star; in this case the characteristic frequency is greatest when M and R take the values (11.4.15) and (11.4.16), where it has the value

$$\left(\frac{GM_m}{R_m^3}\right)^{1/2} = 10^4 \text{ sec}^{-1} \tag{11.4.29}$$

This is considerably faster than any observed pulsar frequency. It is currently believed that pulsars are rotating neutron stars,[21] which begin life with a rotation frequency near the maximum, of order 10^4 sec^{-1}, but which subsequently slow down, losing their energy through gravitational and electromagnetic radiation and through electromagnetic acceleration of charged particles. (In order to account for the existence of this radiation as well as for the observed pulses, it is necessary to suppose that the star does not have circular symmetry about the axis of rotation, as would be the case if its magnetic poles are offset from its rotation poles.) This interpretation is supported by the observation that several pulsars are slowing down.[22]

A white dwarf with the same mass as a neutron star will have a radius larger by a factor $m_n/\mu m_e \simeq 900$, so the fundamental vibration frequency and the maximum rotation frequency will be smaller than for a neutron star by a factor 3×10^{-5}. For M near M_m this gives a characteristic frequency smaller than (11.4.29) by this factor, or about 0.3 sec^{-1}. This is slower than the observed pulse rate of most pulsars. Pulsars are probably neutron stars, not white dwarfs.

5 Supermassive Stars

We now turn to a different kind of "star,"[23] in which general relativity enters in quite a different way. Let us consider a Newtonian star that is supported by the pressure of radiation rather than of matter; the conditions under which this occurs will be discovered as we go along. Let us also assume that the star is in convective equilibrium (see Section 11.1) and has uniform chemical composition. Radiation has an energy density $e = 3p$, so this star will be a polytrope with $\gamma = 4/3$, that is,

$$p = K\rho^{4/3} \tag{11.5.1}$$

The radiation pressure is given by the Stefan-Boltzmann law,

$$p_r = \frac{\pi^2 (kT)^4}{45\hbar^3} \tag{11.5.2}$$

so with $p \simeq p_r$, the temperature is given by

$$kT = \left(\frac{45\hbar^3 K}{\pi^2}\right)^{1/4} \rho^{1/3} \tag{11.5.3}$$

The pressure of matter here is given by the ideal-gas law:

$$p_m = \rho \frac{kT}{\overline{m}} \tag{11.5.4}$$

where $\overline{m}$ is the mean mass of the gas particles. Thus the ratio of matter to radiation pressure is

$$\beta \equiv \frac{p_m}{p_r} = \frac{45\hbar^3}{\pi^2 \overline{m}} \frac{\rho}{(kT)^3} = \frac{1}{\overline{m}} \left(\frac{45\hbar^3}{\pi^2 K^3} \right)^{1/4} \tag{11.5.5}$$

This is a constant throughout the star, so we can use β instead of K (or the entropy per nucleon, on which they both depend) to define the equation of state, writing

$$K = \left(\frac{45\hbar^3}{\overline{m}^4 \pi^2 \beta^4} \right)^{1/3} \tag{11.5.6}$$

The mass of a polytrope with $\gamma = 4/3$ is given by Eq. (11.3.17) and Table 11.1 as

$$M = 4\pi (2.01824) \left(\frac{K}{\pi G} \right)^{3/2} \tag{11.5.7}$$

and, using (11.5.6), this becomes

$$M = \frac{12\sqrt{5}}{\pi^{3/2}} (2.01824) \frac{\hbar^{3/2}}{\overline{m}^2 G^{3/2} \beta^2}$$

$$= 18 M_\odot \left(\frac{m_N}{\overline{m}} \right)^2 \beta^{-2} \tag{11.5.8}$$

For ionized hydrogen at temperatures between $10^{5\circ}$ K and $10^{10\circ}$ K, $\overline{m}$ is the average of the proton and the electron mass, so $\overline{m} \simeq m_N/2$. Thus in this case the condition for radiation pressure to dominate material pressure by, say, a factor 10, is that $M \gtrsim 7200 M_\odot$. No such supermassive star has been definitely observed, but they have been considered as possible sites for the production of radiant energy through gravitational collapse.[23] The structure of a supermassive star is entirely determined by the equations for a Newtonian polytrope with $\gamma \simeq 4/3$. In particular, Eq. (11.3.16) gives the radius of the star as

$$R = 6.89685 \left(\frac{K}{\pi G} \right)^{1/2} \rho(0)^{-1/3}$$

and, using (11.5.6), this is

$$R = \left(\frac{45}{\pi^5} \right)^{1/6} (6.89685) \frac{\hbar^{1/2}}{\overline{m}^{2/3} G^{1/2} \beta^{2/3}} \rho(0)^{-1/3} \tag{11.5.9}$$

The radius is restricted by our assumption that the rest-mass energy of the star be much greater than its radiant energy, and, *a fortiori*, than the thermal energy of its matter. This condition reads

$$\frac{\pi^2 (kT)^4}{15\hbar^3} \ll \rho$$

or, using (11.5.3) and (11.5.6),

$$\rho \ll \frac{\pi^2}{1215} \frac{\beta^4 \overline{m}^4}{\hbar^3} \tag{11.5.10}$$

The density ρ is greatest at the center, so this can be regarded as a condition on $\rho(0)$; using (11.5.8) and (11.5.9) to express β and $\rho(0)$ in terms of M and R, the condition (11.5.10) becomes

$$\frac{MG}{R} \ll \frac{4}{3} \left(\frac{2.01824}{6.89685} \right) = 0.39 \tag{11.5.11}$$

This is essentially equivalent to the statement that the gravitational potential is small, which also was assumed. For $M = 10^4 M_\odot$, Eq. (11.5.11) requires that $R \gg 4 \times 10^4$ km.

Although we do not need general relativity to understand the structure of these supermassive stars, we shall need it to settle the question of stability. A polytrope with $\gamma = 4/3$ is trembling between stability and instability, so it is necessary to take into account the small effects of the matter pressure and of general relativity, which play no appreciable role in structure calculations.

We use Theorem 1 of Section 11.2, which tells us that the transition from stability to instability will occur at a value of $\rho(0)$ for which the internal energy E is stationary. To calculate E, we use Eqs. (11.1.29)–(11.1.31), which to first order in GM/R give

$$E \simeq \int_0^R 4\pi r^2 e(r)\, dr + \int_0^R 4\pi G r \mathcal{M}(r) e(r)\, dr$$

$$- \int_0^R 4\pi G r \mathcal{M}(r)\, dr - \int_0^R 6\pi G^2 \mathcal{M}^2(r) \rho(r)\, dr \tag{11.5.12}$$

The internal energy density e is

$$e = \frac{\pi^2}{15} \frac{(kT)^4}{\hbar^3} + \frac{1}{\Gamma - 1} \frac{\rho kT}{\overline{m}} = 3p_r \left[1 + \frac{\beta}{3(\Gamma - 1)} \right]$$

where Γ is the specific heat ratio of the matter. (For ionized hydrogen, $\Gamma = 5/3$.) The total pressure is

$$p = p_r + p_m = p_r(1 + \beta)$$

Therefore, to first order in the small parameter β, the ratio of energy density to pressure is given by

$$e \simeq 3p\left[1 - \frac{(3\Gamma - 4)}{3(\Gamma - 1)}\beta + O(\beta^2)\right] \tag{11.5.13}$$

The small correction of order β can be ignored in the second term in (11.5.12), which is already smaller than the first term by a factor of order GM/R, but it must be kept in the large first term, and therefore

$$E \simeq \left[1 - \frac{(3\Gamma - 4)}{3(\Gamma - 1)}\beta\right]\int_0^R 12\pi r^2 p(r)\,dr + \int_0^R 12\pi Gr\mathcal{M}(r)p(r)\,dr$$

$$-\int_0^R 4\pi Gr\mathcal{M}(r)\,dr - \int_0^R 6\pi G^2\mathcal{M}^2(r)\rho(r)\,dr - \cdots \tag{11.5.14}$$

The first integral can be rewritten by integrating by parts:

$$\int_0^R 12\pi r^2 p(r)\,dr = \int_0^R p(r)\,d(4\pi r^3) = -\int_0^R 4\pi r^3 p'(r)\,dr$$

To calculate $p'(r)$, we expand the fundamental equation (11.1.13) to first order in GM/R:

$$-r^2 p'(r) \simeq G\mathcal{M}(r)\rho(r)\left[1 + \frac{p(r)}{\rho(r)} + \frac{4\pi r^3 p(r)}{\mathcal{M}(r)} + \frac{2G\mathcal{M}(r)}{r}\right]$$

so

$$\int_0^R 12\pi r^2 p(r)\,dr \simeq \int_0^R 4\pi Gr\mathcal{M}(r)\rho(r)\,dr + \int_0^R 4\pi Gr\mathcal{M}(r)p(r)\,dr$$

$$+ \int_0^R 16\pi^2 Gr^4\rho(r)p(r)\,dr + \int_0^R 8\pi G^2\rho(r)\mathcal{M}^2(r)\,dr$$

The β-correction needs to be kept in only the first term, which is larger than the others by a factor of order R/MG, so to first order in β and GM/R, Eq. (11.5.14) reads

$$E \simeq -\frac{(3\Gamma - 4)}{3(\Gamma - 1)}\beta\int_0^R 4\pi Gr\mathcal{M}(r)\rho(r)\,dr + \int_0^R 16\pi Gr\mathcal{M}(r)p(r)\,dr$$

$$+ \int_0^R 16\pi^2 Gr^4\rho(r)p(r)\,dr + \int_0^R 2\pi G^2\mathcal{M}^2(r)\rho(r)\,dr \tag{11.5.15}$$

Now every term is small, so they can all be evaluated using for ρ, p, and $\mathcal{M}$ the values obtained by solving the Newtonian equation

$$-r^2 p'(r) \simeq G\mathcal{M}(r)\rho(r)$$

for a Newtonian polytrope with $\gamma = 4/3$. In particular, the first integral in Eq. (11.5.15) is given by setting $\gamma = 4/3$ in Eq. (11.3.24),

$$\int_0^R 4\pi Gr\mathscr{M}(r)\rho(r)\,dr = -V = \frac{3GM^2}{2R}$$

whereas an integration by parts lets us write the third term as

$$\int_0^R 16\pi^2 Gr^4\rho(r)p(r)\,dr = \int_0^R 4\pi r^2 p(r)\,d\mathscr{M}(r)$$

$$= -\int_0^R 4\pi Gr^2 p'(r)\mathscr{M}(r)\,dr - \int_0^R 8\pi Grp(r)\mathscr{M}(r)\,dr$$

$$= \int_0^R 4\pi G^2\mathscr{M}^2(r)\rho(r)\,dr - \int_0^R 8\pi Grp(r)\mathscr{M}(r)\,dr$$

Equation (11.5.15) now reads

$$E \simeq -\frac{(3\Gamma - 4)}{2(\Gamma - 1)}\beta\frac{GM^2}{R} + \int_0^R 8\pi Gr\mathscr{M}(r)p(r)\,dr + \int_0^R 6\pi G^2\mathscr{M}^2(r)\rho(r)\,dr$$

The last two integrals may be calculated in terms of the Lane-Emden function $\theta(\xi)$ for $\gamma = 4/3$:

$$\int_0^R 6\pi G^2\mathscr{M}(r)\rho(r)\,dr = \frac{6K^{7/2}\rho(0)^{2/3}}{\pi^{5/2}G^{3/2}}\int_0^{\xi_1}\xi^4\theta'^2(\xi)\theta^3(\xi)\,d\xi$$

$$\int_0^R 8\pi G\mathscr{M}(r)p(r)r\,dr = \frac{8K^{7/2}\rho(0)^{2/3}}{\pi^{3/2}G^{3/2}}\int_0^{\xi_1}\xi^3|\theta'(\xi)|\theta^4(\xi)\,d\xi$$

whereas K and $\rho(0)$ can be expressed in terms of M and R by using (11.3.16) and (11.3.17), yielding

$$\frac{K^{7/2}\rho(0)^{2/3}}{G^{3/2}} = \frac{\sqrt{\pi}}{64\xi_1{}^4|\theta'(\xi_1)|^3}\frac{GM^2}{R}$$

A numerical integration gives[24]

$$\frac{1}{8\pi\xi_1{}^4|\theta'(\xi_1)|^3}\left\{\int_0^{\xi_1}\xi^3|\theta'(\xi)|\theta^4(\xi)\,d\xi + \frac{3}{4\pi}\int_0^{\xi_1}\xi^4\theta'^2(\xi)\theta^3(\xi)\,d\xi\right\} = 5.1$$

so, putting this all together, we have at last

$$E \simeq -\frac{(3\Gamma - 4)}{2(\Gamma - 1)}\beta\frac{GM^2}{R} + 5.1\frac{G^2M^3}{R^2} \tag{11.5.16}$$

The star is certainly stable when R is so large that general relativity can be neglected, for then the star behaves like a Newtonian polytrope with

$$\gamma \equiv 1 + \frac{p}{e} \simeq \frac{4}{3} + \frac{(3\Gamma - 4)}{9(\Gamma - 1)}\beta > \frac{4}{3}$$

[See Eq. (11.5.13).] The transition from stability to instability will occur when R decreases to a value where

$$\frac{\partial E}{\partial R} = \frac{\partial E}{\partial \rho(0)}\frac{\partial \rho(0)}{\partial R} = 0$$

The derivative must be taken with constant entropy per nucleon, and hence in this case with β fixed and M fixed. [See Eqs. (11.5.6) and (11.5.7).] Thus the *minimum* radius for stability is

$$R_{\min} = \frac{20.4\,(\Gamma - 1)}{(3\Gamma - 4)}\frac{GM}{\beta} \tag{11.5.17}$$

The maximum energy that can be released by letting the star shrink slowly (through radiation at its surface) to this minimum stable radius is

$$-E(R_{\min}) = \frac{(3\Gamma - 4)^2\beta^2 M}{81.6\,(\Gamma - 1)^2} \tag{11.5.18}$$

For instance, a star with $\beta = 0.1$ will have $M \simeq 7200 M_\odot$; if $\Gamma = 5/3$ then the minimum radius is 1.45×10^6 km, and the fraction of its rest-mass that can be released by assembling the star is 0.03%. The maximum value of the surface potential MG/R for $\Gamma = 5/3$ is 0.0735β, well under the limit (11.5.11).

6 Stars of Uniform Density

General relativity finds an interesting application to one other class of stable stars, those consisting of incompressible fluids, with equation of state

$$\rho = \text{constant} \tag{11.6.1}$$

These stars are of interest, not because they actually exist (they don't), but because they are simple enough to allow an exact solution of Einstein's equations,[25] and because they set an upper limit to the gravitational red shift of spectral lines from the surface of *any* star.[26]

With ρ constant, the fundamental equation (11.1.13) may be written

$$\frac{-p'(r)}{[\rho + p(r)][(\rho/3) + p(r)]} = 4\pi Gr\left[1 - \frac{8\pi G\rho r^2}{3}\right]^{-1} \tag{11.6.2}$$

The pressure must now be determined by integrating *inward* from the surface where $p = 0$, rather than outward, as for more realistic models. This gives

$$\frac{p(r) + \rho}{3p(r) + \rho} = \left[\frac{1 - 8\pi G\rho R^2/3}{1 - 8\pi G\rho r^2/3}\right]^{1/2}$$

Solving for $p(r)$, and expressing ρ in terms of the stellar mass,

$$\rho = \frac{3M}{4\pi R^3} \qquad \text{for } r < R \tag{11.6.3}$$

we find

$$p(r) = \frac{3M}{4\pi R^3} \left\{\frac{[1 - (2MG/R)]^{1/2} - [1 - (2MGr^2/R^3)]^{1/2}}{[1 - (2MGr^2/R^3)]^{1/2} - 3[1 - (2MG/R)]^{1/2}}\right\} \tag{11.6.4}$$

The metric component $A(r)$ is immediately given by Eq. (11.1.11):

$$A(r) = \left[\frac{1 - 2MGr^2}{R^3}\right]^{-1} \tag{11.6.5}$$

whereas $B(r)$ can be calculated by using (11.6.4) in the integral (11.1.16):

$$B(r) = \frac{1}{4}\left[3\left(1 - \frac{2MG}{R}\right)^{1/2} - \left(1 - \frac{2MGr^2}{R^3}\right)^{1/2}\right]^2 \tag{11.6.6}$$

The most interesting feature of this solution is that it does not make sense for all values of M and R. The pressure given by Eq. (11.6.4) will become infinite at a point r_∞ where

$$r_\infty^2 = 9R^2 - \frac{4R^3}{MG} \tag{11.6.7}$$

(Also, the metric becomes singular at r_∞ because $B(r_\infty)$ vanishes.) But the pressure is a scalar, and so an infinity in $p(r)$ cannot be blamed on an injudicious choice of coordinate system. We must see to it that $p(r)$ is not singular for any real r, and the only way to accomplish this is to have r_∞^2 negative, or

$$\frac{MG}{R} < \frac{4}{9} \tag{11.6.8}$$

Note that the Schwarzschild radius $2MG$ is then less than 8/9 the actual radius R, so there is no singularity in either the exterior solution (11.1.17) or the interior solution (11.6.5), (11.6.6).

This is not the first time that we have discovered an upper bound on the absolute value MG/R of the gravitational potential of a star. We learned in Section 11.4 that for a stable ideal-gas neutron star, MG/R is never greater than 0.36/3.2, or 0.11. [See Eqs. (11.4.15) and (11.4.16).] Is there then an absolute upper limit to MG/R imposed by the structure of the Einstein equations, irrespective of the equation of state?

To frame this question as a mathematical problem, we consider ρ as an arbitrary finite positive function, subject only to these general requirements:

(A) The radius R is fixed, with

$$\rho(r) = 0 \qquad \text{for } r > R \tag{11.6.9}$$

(B) The mass M is fixed, with

$$\int_0^R 4\pi r^2 \rho(r) \, dr = M \tag{11.6.10}$$

(C) The metric coefficient $A(r)$ given by (11.1.11) must not be singular, so

$$\mathscr{M}(r) < \frac{r}{2G} \tag{11.6.11}$$

where

$$\mathscr{M}(r) \equiv \int_0^r 4\pi r'^2 \rho(r') \, dr'$$

(D) The density $\rho(r)$ must not increase outward:

$$\rho'(r) \le 0 \tag{11.6.12}$$

(It is difficult to imagine that a fluid sphere with a larger density near the surface than near the center could be stable.) Given any function $\rho(r)$, satisfying these conditions, we can calculate $A(r)$ from Eq. (11.1.11); we can determine $p(r)$ by integrating Eq. (11.1.13) inward from the surface (with the boundary condition that $p(R) = 0$); and we can then calculate $B(r)$ from Eq. (11.1.16). Equation (11.6.11) guarantees that $A(r)$ is well behaved, and as long as $p(r)$ is finite, Eq. (11.1.13) will give $p(r) \ge 0$, and Eq. (11.1.16) will give a finite positive-definite $B(r)$. Thus any absolute limitations on the input function $\rho(r)$ (such as an upper bound on MG/R) can only come from the condition that Eq. (11.1.13) must yield a finite solution for the pressure $p(r)$.

We shall exploit this condition rather indirectly, by concentrating on the metric coefficient $B(r)$ rather than on $p(r)$ itself. We first derive an equation that allows $B(r)$ to be calculated for a given density function $\rho(r)$, without having to solve for $p(r)$; from (11.1.5) and (11.1.7), we have

$$3R_{rr}B + R_{tt}A = B'' - \frac{B'}{2}\left(\frac{A'}{A} + \frac{B'}{B}\right) - \frac{3BA'}{rA} - \frac{B'}{r} = -16\pi G\rho AB$$

or

$$B'' - \frac{B'}{2}\left(\frac{A'}{A} + \frac{B'}{B} + \frac{2}{r}\right) = \frac{B}{rA}[3A' - 16\pi G\rho r A^2]$$

This equation can be linearized by defining

$$B \equiv \zeta^2 \qquad (11.6.13)$$

Introducing Eq. (11.1.11) for $A(r)$, and rearranging a bit, we find

$$\frac{d}{dr}\left[\frac{1}{r}\left(1 - \frac{2G\mathscr{M}(r)}{r}\right)^{1/2}\frac{d\zeta(r)}{dr}\right] = G\left(1 - \frac{2G\mathscr{M}(r)}{r}\right)^{-1/2}\left(\frac{\mathscr{M}(r)}{r^3}\right)'\zeta(r)$$

$$(11.6.14)$$

The initial conditions at $r = R$ can be determined directly from Eq. (11.1.16), or from the condition that $B(r)$ fit smoothly to the exterior solution (11.1.17); either way, we find that

$$\zeta(R) = \left[1 - \frac{2MG}{R}\right]^{1/2} \qquad (11.6.15)$$

$$\zeta'(R) = \frac{MG}{R^2}\left[1 - \frac{2MG}{R}\right]^{-1/2} \qquad (11.6.16)$$

The solution for $\zeta(r)$ must be *positive*, because $\zeta(r)$ can become negative only if it passes through the value zero, at which point B would vanish, and, according to Eq. (11.1.16), B can vanish only if the pressure $p(r)$ has a singularity.

We next proceed to derive an upper bound for $\zeta(0)$. If ζ is positive, then the right-hand side of (11.6.14) is negative, because $3\mathscr{M}(r)/4\pi r^3$ is the mean density within the radius r, and the mean density cannot increase with r if the density does not. Thus (11.6.14) gives

$$\frac{d}{dr}\left[\frac{1}{r}\left(1 - \frac{2G\mathscr{M}(r)}{r}\right)^{1/2}\frac{d\zeta(r)}{dr}\right] \leq 0$$

the equality being attained only for uniform density. Integrating this inequality from r to R and using (11.6.16), we have

$$\zeta'(r) \geq \frac{MGr}{R^3}\left(1 - \frac{2G\mathscr{M}(r)}{r}\right)^{-1/2}$$

Integrating again from 0 to R and using (11.6.15) gives

$$\zeta(0) \leq \left[1 - \frac{2MG}{R}\right]^{1/2} - \frac{MG}{R^3}\int_0^R \frac{r\,dr}{[1 - (2G\mathscr{M}(r)/r)]^{1/2}}$$

The right-hand side is largest when $\mathscr{M}(r)$ is as small as possible. For a given mass M and radius R, the density distribution with $\rho'(r) \leq 0$ that gives an $\mathscr{M}(r)$ that is everywhere as small as possible has $\rho(r)$ constant, in which case

$$\mathscr{M}(r) = \frac{Mr^3}{R^3}$$

Using this in the integral, our inequality is

$$\zeta(0) \le \frac{3}{2}\left[1 - \frac{2MG}{R}\right]^{1/2} - \frac{1}{2} \tag{11.6.17}$$

We have already noted that $\zeta(r)$ must be positive-definite; hence (11.6.17) implies that

$$\frac{MG}{R} < \frac{4}{9} \tag{11.6.18}$$

This is just the upper limit found earlier for stars of uniform density, but now we know that (11.6.18) holds for all stars, uniform or not.

It can also be proved that for a given mass and radius, the stable stars with smallest central pressure are those with uniform density. Hence the central pressure of any star is not less than the value obtained by setting $r = 0$ in Eq. (11.6.4), that is,

$$p(0) \ge \frac{3M}{4\pi R^3}\left\{\frac{[1 - (2MG/R)]^{1/2} - 1}{1 - 3[1 - (2MG/R)]^{1/2}}\right\} \tag{11.6.19}$$

This again shows that MG/R can never equal the forbidden value 4/9.

Our result can be immediately translated into a statement about the red shift of spectral lines from the surface of any star. According to Eqs. (3.5.3), (11.1.1), and (11.1.17), this is

$$z \equiv \frac{\Delta\lambda}{\lambda} = B^{-1/2}(R) - 1 = \left(1 - \frac{2MG}{R}\right)^{-1/2} - 1$$

Equation (11.6.18) imposes on z the upper bound

$$z < 2 \tag{11.6.20}$$

In fact, there seems to be a large concentration of quasi-stellar radio sources (see Chapter 14) whose spectral lines show red shifts close to 1.95! However, we should not jump to the conclusion that these red shifts are necessarily due to strong gravitational fields, for red shifts near $z = 2$ require the star to be composed of a nearly incompressible fluid, with $\partial\rho/\partial p$ very small. This would seem unphysical, since we do not want the speed of sound $(\partial p/\partial\rho)^{1/2}$ to become larger than the speed of light![26] Bondi[27] has shown that for a stable star with $(\partial p/\partial\rho) < 1$ and with $p/\rho \le 1/3$ (as is the case for particles that interact only electromagnetically and/or in localized collisions; see Section 2.10) the red shift of spectral lines emitted from the surface is bounded by $z \le 0.615$. In any case, there *are* quasi-stellar objects with red shifts $z > 2$, such as 4C25.5, with $z = 2.358$.

However, there is no theorem that limits the red shifts of light signals from the *interior* of static spherically symmetric bodies.[28] For instance, a light signal from the center of a transparent uniform star would have a red shift given by Eqs. (3.5.3), (11.1.1), and (11.6.6):

$$1 + z = B^{-1/2}(0) = \frac{2}{3(1 - (2MG/R))^{1/2} - 1}$$

As MG/R approaches the maximum value 4/9, this red shift becomes infinite. Hoyle and Fowler[29] have suggested that a quasi-stellar object can consist of a cluster of small dense stars, with the red shifts arising from emission and absorption in a hot cloud of gas trapped near the cluster center. It is not yet clear whether the red shifts of the QSO's arise internally, or from some other cause, such as the general cosmological recession of distant objects discussed in Chapter 14.

7 Time-Dependent Spherically Symmetric Fields

We now turn to the problems of stellar dynamics, and begin by writing down the metric and Ricci tensor for a spherically symmetric but time-dependent system. Spherical symmetry requires the proper time interval $d\tau^2$ to depend only on the rotational invariants

$$t, dt, r, \mathbf{x} \cdot d\mathbf{x} = r \, dr, \quad d\mathbf{x}^2 = dr^2 + r^2(d\theta^2 + \sin^2 \theta \, d\varphi^2)$$

so it can be written

$$d\tau^2 = C(r, t) \, dt^2 - D(r, t) \, dr^2 - 2E(r, t) \, dr \, dt - F(r, t)r^2(d\theta^2 + \sin^2 \theta \, d\varphi^2)$$

The function F can be removed by defining a new radial variable

$$r' \equiv rF^{1/2}(r, t)$$

The metric will then be of the same form, but with new functions C', D', E' in place of C, D, E, and of course with r' in place of r and no factor F. Dropping primes, we have then

$$d\tau^2 = C(r, t) \, dt^2 - D(r, t) \, dr^2 - 2E(r, t) \, dr \, dt - r^2(d\theta^2 + \sin^2 \theta \, d\varphi^2)$$

We next remove E, by defining a new time

$$dt' = \eta(r, t)[C(r, t) \, dt - E(r, t) \, dr]$$

where η is an integrating factor defined to make the right-hand side a perfect differential, that is, so that

$$\frac{\partial}{\partial r} [\eta(r, t)C(r, t)] = -\frac{\partial}{\partial t} [\eta(r, t)E(r, t)]$$

(This equation can be solved by treating it as an initial value problem; given $\eta(r, t_0)$ for all r, we can solve for $\partial \eta(r, t)/\partial t$ at $t = t_0$ and thus determine $\eta(r, t_0 + dt)$ for all r.) The proper time is then

$$d\tau^2 = \eta^{-2} C^{-1} dt'^2 - (D + C^{-1} E^2) dr^2 - r^2(d\theta^2 + \sin^2 \theta \, d\varphi^2)$$

or, introducing new functions A and B in place of $D + C^{-1} E^2$ and $\eta^{-2} C^{-1}$ and dropping the prime on t,

$$d\tau^2 = B(r, t) \, dt^2 - A(r, t) \, dr^2 - r^2(d\theta^2 + \sin^2 \theta \, d\varphi^2) \tag{11.7.1}$$

Thus we can use the metric in its familiar "standard" form, the only new feature being that A and B now depend on t as well as r.

The nonvanishing elements of the metric tensor and its inverse are

$$g_{rr} = A \qquad g_{\theta\theta} = r^2 \qquad g_{\varphi\varphi} = r^2 \sin^2 \theta \qquad g_{tt} = -B$$

$$g^{rr} = A^{-1} \qquad g^{\theta\theta} = r^{-2} \qquad g^{\varphi\varphi} = r^{-2}(\sin \theta)^{-2} \qquad g^{tt} = -B^{-1} \tag{11.7.2}$$

It follows that the nonvanishing elements of the affine connection are

$$\Gamma^r_{rr} = \frac{A'}{2A} \qquad \Gamma^r_{\theta\theta} = -\frac{r}{A} \qquad \Gamma^r_{\varphi\varphi} = -\frac{r \sin^2 \theta}{A}$$

$$\Gamma^r_{tt} = \frac{B'}{2A} \qquad \Gamma^r_{rt} = \Gamma^r_{tr} = \frac{\dot{A}}{2A} \qquad \Gamma^\theta_{\theta r} = \Gamma^\theta_{r\theta} = \frac{1}{r}$$

$$\Gamma^\theta_{\varphi\varphi} = -\sin \theta \cos \theta \qquad \Gamma^\varphi_{\varphi r} = \Gamma^\varphi_{r\varphi} = \frac{1}{r} \qquad \Gamma^\varphi_{\varphi\theta} = \Gamma^\varphi_{\theta\varphi} = \cot \theta$$

$$\Gamma^t_{rr} = +\frac{\dot{A}}{2B} \qquad \Gamma^t_{tt} = \frac{\dot{B}}{2B} \qquad \Gamma^t_{tr} = \Gamma^t_{rt} = \frac{B'}{2B} \tag{11.7.3}$$

(A prime or a dot now denotes $\partial/\partial r$ or $\partial/\partial t$, respectively.) From (6.1.5) we obtain the independent nonzero components of the Ricci tensor:

$$R_{rr} = \frac{B''}{2B} - \frac{B'^2}{4B^2} - \frac{A'B'}{4AB} - \frac{A'}{Ar} - \frac{\ddot{A}}{2B} + \frac{\dot{A}\dot{B}}{4B^2} + \frac{\dot{A}^2}{4AB} \tag{11.7.4}$$

$$R_{\theta\theta} = -1 + \frac{1}{A} - \frac{rA'}{2A^2} + \frac{rB'}{2AB} \tag{11.7.5}$$

$$R_{tt} = -\frac{B''}{2A} + \frac{B'A'}{4A^2} - \frac{B'}{Ar} + \frac{B'^2}{4AB} + \frac{\ddot{A}}{2A} - \frac{\dot{A}^2}{4A^2} - \frac{\dot{B}\dot{A}}{4AB} \tag{11.7.6}$$

$$R_{tr} = -\frac{\dot{A}}{Ar} \tag{11.7.7}$$

Also, it follows from the spherical symmetry of the metric that

$$R_{\varphi\varphi} = \sin^2\theta R_{\theta\theta} \tag{11.7.8}$$

$$R_{r\theta} = R_{r\varphi} = R_{\theta\varphi} = R_{\theta t} = R_{\varphi t} = 0 \tag{11.7.9}$$

As a simple but important application of these results, let us consider a spherically symmetric but not necessarily static field in *empty space*, where the field equations read $R_{\mu\nu} = 0$. According to (11.7.7), the field equation $R_{tr} = 0$ just tells us that A is time-independent:

$$\dot{A} = 0$$

Inspection of (11.7.4)–(11.7.6) then shows that *all* time derivatives drop out of the field equations, and they become identical with the equations for a static isotropic gravitational field in empty space. [See Eq. (8.1.13).] We can then repeat the arguments of Section 8.2; the vanishing of R_{rr} and R_{tt} gives

$$(AB)' = 0$$

and the vanishing of $R_{\theta\theta}$ gives

$$\left(\frac{r}{A}\right)' = 1$$

Since A is time-independent, the general solution is

$$A = \left(1 - \frac{2MG}{r}\right)^{-1} \qquad B = f(t)\left(1 - \frac{2MG}{r}\right)$$

with GM a time-independent integration constant, and $f(t)$ an unknown function of t. The function $f(t)$ can be made to equal unity by defining a new time coordinate:

$$t' = \int^t f^{1/2}(t)\, dt$$

The metric is now entirely time-independent, and agrees with the Schwarzschild solution (8.2.12). We have thus proved the *Birkhoff theorem*,[30] that a spherically symmetric gravitational field in empty space must be static, with a metric given by the Schwarzschild solution.

The Birkhoff theorem is analogous to the result proved by Newton in his theory of the lunar motion, that the gravitational field outside a spherically symmetric body behaves as if the whole mass of the body were concentrated at the center. It is a little surprising that this result should apply in general relativity as well as in Newton's theory, for in general relativity a nonstatic body will usually radiate gravitational waves. The Birkhoff theorem tells us that, although a pulsating spherically symmetric body can of course produce nonstatic gravitational fields within its mass, no gravitational radiation can escape into empty space. In this sense, the Birkhoff theorem is analogous to the well-known result of atomic theory, that a photon cannot be emitted in a quantum transition between two states of zero spin.

The Birkhoff theorem may be applied, not only to the gravitational field outside a body, but also to the field *inside* an empty spherical cavity at the center of a spherically symmetric (but not necessarily static) body. In this case the metric is again given by the Schwarzschild solution, but since the point $r = 0$ is here in empty space, there can be no singularity, so the integration constant MG must vanish. The Birkhoff theorem thus has the corollary that *the metric inside an empty spherical cavity at the center of a spherically symmetric system must be equivalent to the flat-space Minkowski metric $\eta_{\mu\nu}$*. This corollary is analogous to another famous result of Newtonian theory, that the gravitational field of a spherical shell vanishes inside the shell. Stars do not usually have holes at their centers, so this corollary will not be of much use to us in this chapter. Its importance arises from the fact that the Birkhoff theorem is a local theorem, not depending on any conditions on the metric for $r \to \infty$ (aside from spherical symmetry), so that space must be flat in a spherical cavity at the center of a spherically symmetric system, even if the system is infinite—even, in fact, if the system is the whole universe. We shall see in Section 15.1 that the corollary to Birkhoff's theorem can be used to justify a limited use of Newtonian mechanics in cosmological problems.

8   Comoving Coordinates

As a further preparation for our treatment of gravitational collapse, and also to lay a groundwork for our discussion of cosmology in Chapter 14, we now construct a very useful set of coordinates, the *comoving coordinate system*,[31] which incorporates a more natural separation between space and time than that provided by the standard coordinates used in the last section.

Imagine a finite region of space filled with a dense cloud of freely falling particles. Each particle is assumed to carry along a little clock, and is given a fixed set of spatial coordinates, which can be defined as the coordinates x^i of the particle, in some arbitrary system, when its own clock reads $t = 0$. (The rules for setting these different clocks are discussed below.) The space-time coordinates $\mathbf{x}$, t of any event are defined by taking $\mathbf{x}$ as the spatial coordinate label of the particle that is just going by when and where the event occurs, and by taking t as the time then shown on that particle's clock. We may think of the coordinate mesh as being dragged along by the cloud of particles, with time defined by clocks stuck on the mesh. This coordinate system will be useful throughout the region occupied by the particle cloud, for whatever interval of time in which particle trajectories do not cross.

The metric $g_{\mu\nu}$ in comoving coordinates is characterized by certain specially simple features. First, we note that the clocks are in free fall and therefore tell proper time, so the proper time interval between two points $\mathbf{x}$, t and $\mathbf{x}$, $t + dt$ on a given particle's trajectory is just dt, that is,

$$d\tau^2 = -g_{\mu\nu} \, dx^\mu \, dx^\nu = -g_{tt} \, dt^2$$

and therefore

$$g_{tt} = -1 \tag{11.8.1}$$

Also, we note that the particle trajectory $\mathbf{x} = $ constant, $t = \tau$ satisfies the equation of free fall, so

$$0 = \frac{d^2 x^i}{d\tau^2} + \Gamma^i_{\mu\nu} \frac{dx^\mu}{d\tau} \frac{dx^\nu}{d\tau} = \Gamma^i_{tt}$$

Using (11.8.1), this gives

$$0 = g^{ij} \frac{\partial g_{jt}}{\partial t}$$

or, since g^{ij} is generally a nonsingular matrix,

$$0 = \frac{\partial g_{jt}}{\partial t} \tag{11.8.2}$$

We have kept open the option of setting the clocks attached to the different particles in an arbitrary fashion. Suppose that we reset these clocks by a transformation

$$t' = t + f(\mathbf{x}) \qquad \mathbf{x}' = \mathbf{x} \tag{11.8.3}$$

The new metric will have the elements

$$g'_{tt} = -1 \tag{11.8.4}$$

$$g'_{ti} = g_{ti} + \frac{\partial f}{\partial x^i} \tag{11.8.5}$$

$$g'_{ij} = g_{ij} - g_{ti} \frac{\partial f}{\partial x^j} - g_{tj} \frac{\partial f}{\partial x^i} - \frac{\partial f}{\partial x^i} \frac{\partial f}{\partial x^j} \tag{11.8.6}$$

It would be a great simplification if the function f could be chosen so that the two terms in Eq. (11.8.5) cancel, giving $g'_{it} = 0$. There are two important cases where this *is* possible:

(A) Suppose that we can reset all clocks so that all particles are at rest at a time $t = 0$. This assumption can be given an absolute physical significance by intepreting it to mean that for each particle P at $t = 0$, it is possible to find a locally inertial coordinate system $\tilde{x}^\mu$ in which the separation between P and neighboring particles is purely spatial,

$$\left(\frac{\partial \tilde{x}^0}{\partial x^i} \right)_{t=0, \, \mathbf{x}=\mathbf{x}_P} = 0$$

and in which the movement of P in a time interval dt is purely temporal,

$$\left(\frac{\partial \tilde{x}^i}{\partial t}\right)_{t=0, \mathbf{x}=\mathbf{x}_p} = 0$$

The metric in this locally inertial system is the Minkowski metric $\eta_{\mu\nu}$, so the space-time components of the metric in the comoving system at $t = 0$ are

$$g_{ti}(\mathbf{x}_p, 0) = \left[\eta_{\mu\nu} \frac{\partial \tilde{x}^\mu}{\partial x^i} \frac{\partial \tilde{x}^\nu}{\partial t}\right]_{t=0, \mathbf{x}=\mathbf{x}_p} = 0$$

With (11.8.2), it follows that g_{ti} vanishes everywhere, so the metric is given by

$$d\tau^2 = dt^2 - g_{ij}(\mathbf{x}, t)\, dx^i\, dx^j \tag{11.8.7}$$

(B) If the metric is manifestly spherically symmetric, then the line element must have the general form with which we started in the last section, that is,

$$d\tau^2 = C(r, t)\, dt^2 - D(r, t)\, dr^2 - 2E(r, t)\, dr\, dt - F(r, t)r^2(d\theta^2 + \sin^2\theta\, d\varphi^2)$$

The only nonvanishing time-space component g_{tj} is $g_{tr} = 2E$, and (11.8.2) then tells us that E is time-independent, so

$$g_{tr} = 2E(r)$$

$$g_{t\theta} = g_{t\varphi} = 0$$

We can therefore eliminate the components g_{tj} by resetting the clocks as in (11.8.3), with

$$f = -2 \int^r E(r)\, dr$$

Using (11.8.4) and dropping primes, the metric is now of the form

$$d\tau^2 = dt^2 - U(r, t)\, dr^2 - V(r, t)(d\theta^2 + \sin^2\theta\, d\varphi^2) \tag{11.8.8}$$

with U and V new unknown functions that replace D and F.

It is of course possible to construct coordinate systems of this sort even if the cloud of freely falling particles is purely imaginary. In differential geometry, coordinate systems satisfying (11.8.1) and (11.8.2) are called *Gaussian*, and if g_{ti} vanishes, so that the line element takes the form (11.8.7), then we call the coordinates *Gaussian normal*. However, these coordinate systems find their most important applications to systems that actually do consist of a freely falling fluid. In this case the fluid velocity four-vector by definition has zero space component,

$$U^i = 0 \tag{11.8.9}$$

and since U^μ is normalized so that

$$g_{\mu\nu}U^\mu U^\nu = -1 \tag{11.8.10}$$

[see Eq. (5.4.4)] the time component of U^μ must be

$$U^t = (-g_{tt})^{-1/2} = 1 \tag{11.8.11}$$

We shall be working only with spherically symmetric comoving coordinate systems, with line element (11.8.8). The nonvanishing elements of the metric tensor are

$$g_{rr} = U \qquad g_{\theta\theta} = V \qquad g_{\varphi\varphi} = V \sin^2 \theta \qquad g_{tt} = -1$$
$$g^{rr} = U^{-1} \qquad g^{\theta\theta} = V^{-1} \qquad g^{\varphi\varphi} = (V \sin^2 \theta)^{-1} \qquad g^{tt} = -1$$
$$\tag{11.8.12}$$

The nonvanishing elements of the affine connection are readily calculated as

$$\Gamma^r_{rr} = \frac{U'}{2U} \qquad \Gamma^r_{\theta\theta} = -\frac{V'}{2U} \qquad \Gamma^r_{\varphi\varphi} = -\frac{V'}{2U} \sin^2 \theta \qquad \Gamma^r_{rt} = \Gamma^r_{tr} = \frac{\dot{U}}{2U}$$

$$\Gamma^\theta_{r\theta} = \Gamma^\theta_{\theta r} = \frac{V'}{2V} \qquad \Gamma^\theta_{\theta t} = \Gamma^\theta_{\theta t} = \frac{\dot{V}}{2V} \qquad \Gamma^\theta_{\varphi\varphi} = -\sin\theta\cos\theta$$

$$\Gamma^\varphi_{r\varphi} = \Gamma^\varphi_{\varphi r} = \frac{V'}{2V} \qquad \Gamma^\varphi_{t\varphi} = \Gamma^\varphi_{\varphi t} = \frac{\dot{V}}{2V} \qquad \Gamma^\varphi_{\theta\varphi} = \cot\theta$$

$$\Gamma^t_{rr} = \frac{\dot{U}}{2} \qquad\qquad \Gamma^t_{\theta\theta} = \frac{\dot{V}}{2} \qquad\qquad \Gamma^t_{\varphi\varphi} = \frac{\dot{V}}{2} \sin^2 \theta \tag{11.8.13}$$

(A prime or dot denotes $\partial/\partial r$ or $\partial/\partial t$, respectively.) From (6.1.5) we obtain the independent nonzero components of the Ricci tensor:

$$R_{rr} = \frac{V''}{V} - \frac{V'^2}{2V^2} - \frac{U'V'}{2UV} - \frac{\ddot{U}}{2} + \frac{\dot{U}^2}{4U} - \frac{\dot{U}\dot{V}}{2V} \tag{11.8.14}$$

$$R_{\theta\theta} = -1 + \frac{V''}{2U} - \frac{V'U'}{4U^2} - \frac{\ddot{V}}{2} - \frac{\dot{V}\dot{U}}{4U} \tag{11.8.15}$$

$$R_{tt} = \frac{\ddot{U}}{2U} + \frac{\ddot{V}}{V} - \frac{\dot{U}^2}{4U^2} - \frac{\dot{V}^2}{2V^2} \tag{11.8.16}$$

$$R_{tr} = \frac{\dot{V}'}{V} - \frac{V'\dot{V}}{2V^2} - \frac{\dot{U}V'}{2UV} \tag{11.8.17}$$

Also, it again follows from the spherical symmetry of the metric that

$$R_{\varphi\varphi} = R_{\theta\theta} \sin^2 \theta \tag{11.8.18}$$

$$R_{r\theta} = R_{r\varphi} = R_{\theta\varphi} = R_{\theta t} = R_{\varphi t} = 0 \tag{11.8.19}$$

9   Gravitational Collapse

We saw in Sections 11.3 and 11.4 that a cooling star of mass greater than a few solar masses cannot reach equilibrium as either a white dwarf or a neutron star. It may be that a massive star will always eject enough matter by the time it reaches the end of its thermonuclear evolution so that its mass drops below the Chandrasekhar or the Oppenheimer-Volkoff limits. If not, then it will collapse.

A proper treatment of gravitational collapse would be prohibitively complicated for this book. In order to get some feeling for what can happen during collapse, we consider only the simplest case,[32] the spherically symmetric collapse of "dust" with negligible pressure. Since the dust particles are acted on by purely gravitational forces, they fall freely, and we can use them as the physical basis of a comoving coordinate system of the sort discussed in the last section. The metric is then given by Eq. (11.8.8):

$$d\tau^2 = dt^2 - U(r, t)\, dr^2 - V(r, t)(d\theta^2 + \sin^2\theta\, d\varphi^2) \qquad (11.9.1)$$

The energy-momentum tensor for a fluid of negligible pressure is given by Eq. (5.4.2) as

$$T^{\mu\nu} = \rho U^\mu U^\nu \qquad (11.9.2)$$

where $\rho(r, t)$ is the proper energy density and U^μ is the velocity four-vector, given for a comoving coordinate system by Eqs. (11.8.9) and (11.8.11):

$$U^r = U^\theta = U^\varphi = 0, \qquad U^t = 1 \qquad (11.9.3)$$

The equations of momentum conservation $(T^\mu{}_i)_{;\mu} = 0$ are automatically satisfied, and the equation for energy conservation reads

$$0 = (T^\mu{}_t)_{;\mu} = -\frac{\partial\rho}{\partial t} - \rho\Gamma^\lambda_{\lambda t} = -\frac{\partial\rho}{\partial t} - \rho\left(\frac{\dot U}{2U} + \frac{\dot V}{V}\right)$$

or in other words

$$\frac{\partial}{\partial t}(\rho V \sqrt{U}) = 0 \qquad (11.9.4)$$

The Einstein field equations can be written

$$R_{\mu\nu} = -8\pi G S_{\mu\nu} \qquad (11.9.5)$$

where

$$S_{\mu\nu} = T_{\mu\nu} - \tfrac{1}{2}g_{\mu\nu}T^\lambda{}_\lambda = \rho[\tfrac{1}{2}g_{\mu\nu} + U_\mu U_\nu] \qquad (11.9.6)$$

This may be evaluated with the aid of Eqs. (11.9.1) and (11.9.3); we find that the only nonvanishing components of $S_{\mu\nu}$ are

$$S_{rr} = \rho\,\frac{U}{2} \qquad S_{\theta\theta} = \rho\,\frac{V}{2} \qquad S_{\varphi\varphi} = S_{\theta\theta}\sin^2\theta \qquad S_{tt} = \frac{\rho}{2}$$

$$(11.9.7)$$

In particular,

$$S_{tr} = 0 \qquad\qquad (11.9.8)$$

Using (11.9.7)–(11.9.8) and (11.8.14)–(11.8.17) in (11.9.5) yields four field equations:

$$\frac{1}{U}\left[\frac{V''}{V} - \frac{V'^2}{2V^2} - \frac{U'V'}{2UV}\right] - \frac{\ddot{U}}{2U} + \frac{\dot{U}^2}{4U^2} - \frac{\dot{U}\dot{V}}{2UV} = -4\pi G\rho \qquad (11.9.9)$$

$$-\frac{1}{V} + \frac{1}{U}\left[\frac{V''}{2V} - \frac{U'V'}{4UV}\right] - \frac{\ddot{V}}{2V} - \frac{\dot{V}\dot{U}}{4VU} = -4\pi G\rho \qquad (11.9.10)$$

$$\frac{\ddot{U}}{2U} + \frac{\ddot{V}}{V} - \frac{\dot{U}^2}{4U^2} - \frac{\dot{V}^2}{2V^2} = -4\pi G\rho \qquad (11.9.11)$$

$$\frac{\dot{V}'}{V} - \frac{V'\dot{V}}{2V^2} - \frac{\dot{U}V'}{2UV} = 0 \qquad (11.9.12)$$

Let us simplify our model even further, and assume that ρ is independent of position.[32] We can now seek a separable solution, with

$$U = R^2(t)f(r) \qquad V = S^2(t)g(r)$$

Then (11.9.12) requires that $\dot{S}/S$ equal $\dot{R}/R$, so we can normalize f and g so that

$$S(t) = R(t)$$

Also, we are still free to redefine the radial coordinate as an arbitrary function $\tilde{r}$ of r, and in particular we can choose $\tilde{r} = \sqrt{g(r)}$, so f and g are replaced with $\tilde{f} = fg'^2/4g$ and $\tilde{g} = \tilde{r}^2$. Dropping the tildas, we have then

$$U = R^2(t)f(r) \qquad V = R^2(t)r^2 \qquad\qquad (11.9.13)$$

Equations (11.9.9) and (11.9.10) then read

$$-\frac{f'(r)}{rf^2(r)} - \ddot{R}(t)R(t) - 2\dot{R}^2(t) = -4\pi GR^2(t)\rho(t) \qquad (11.9.14)$$

$$\left[-\frac{1}{r^2} + \frac{1}{rf^2(r)} - \frac{f'(r)}{2rf^2(r)}\right] - \ddot{R}(t)R(t) - 2\dot{R}^2(t) = -4\pi GR^2(t)\rho(t)$$

$$(11.9.15)$$

The first terms in (11.9.14) and (11.9.15) must evidently be equal constants, which we shall call $-2k$:

$$-\frac{f'(r)}{rf^2(r)} = -\frac{1}{r^2} + \frac{1}{r^2f(r)} - \frac{f'(r)}{2rf^2(r)} = -2k$$

The unique solution is

$$f(r) = [1 - kr^2]^{-1}$$

so the metric takes the form

$$d\tau^2 = dt^2 - R^2(t)\left[\frac{dr^2}{1 - kr^2} + r^2\,d\theta^2 + r^2\sin^2\theta\,d\varphi^2\right] \quad (11.9.16)$$

(Incidentally, this metric is spatially homogeneous as well as isotropic, and for this reason it will provide the kinematic framework for our treatment of relativistic cosmology in Chapter 14.)

Our remaining problem is to calculate the functions $\rho(t)$ and $R(t)$. Using (11.9.13) and (11.9.14) in the energy-conservation equation (11.9.4), we find that $\rho(t)R^3(t)$ is constant. We normalize the radial coordinate r so that

$$R(0) = 1 \quad (11.9.17)$$

and therefore

$$\rho(t) = \rho(0)R^{-3}(t) \quad (11.9.18)$$

The field equations (11.9.14) or (11.9.15) and (11.9.11) are now ordinary differential equations:

$$-2k - \ddot{R}(t)R(t) - 2\dot{R}^2(t) = -4\pi G\rho(0)R^{-1}(t) \quad (11.9.19)$$

$$\ddot{R}(t)R(t) = -\frac{4\pi G}{3}\,\rho(0)R^{-1}(t) \quad (11.9.20)$$

We can eliminate $\ddot{R}(t)$ by adding these two equations, and find

$$\dot{R}^2(t) = -k + \frac{8\pi G}{3}\,\rho(0)R^{-1}(t) \quad (11.9.21)$$

Equations (11.9.19) and (11.9.20) can be recovered from (11.9.21) and its time derivative, so we can forget about them and simply use (11.9.21) to calculate $R(t)$.

We shall now assume that the fluid is at rest (in standard coordinates) at $t = 0$, so

$$\dot{R}(0) = 0 \quad (11.9.22)$$

and therefore (11.9.21) and (11.9.17) give

$$k = \frac{8\pi G}{3}\,\rho(0) \quad (11.9.23)$$

Thus Eq. (11.9.21) can be written

$$\dot{R}^2(t) = k[R^{-1}(t) - 1] \quad (11.9.24)$$

The solution is given by the parametric equations of a *cycloid*:

$$t = \left(\frac{\psi + \sin \psi}{2\sqrt{k}}\right)$$

$$R = \tfrac{1}{2}(1 + \cos \psi) \tag{11.9.25}$$

Note that $R(t)$ vanishes when $\psi = \pi$, and hence when $t = T$, where

$$T = \frac{\pi}{2\sqrt{k}} = \frac{\pi}{2}\left(\frac{3}{8\pi G\rho(0)}\right)^{1/2} \tag{11.9.26}$$

Thus a fluid sphere of initial density $\rho(0)$ and zero pressure will collapse from rest to a state of infinite proper energy density in the finite time T.

Although the collapse is complete at a finite coordinate time $t = T$, any light signal coming to us from the sphere's surface will be delayed by its gravitational field (see Section 8.7), so we on earth will *not* see the star suddenly vanish. To make this more specific, we have to complete our calculation by finding the metric outside the star.

The Birkhoff theorem proved in Section 11.7 shows that it is always possible to find a "standard" coordinate system $\bar{r}, \bar{\theta}, \bar{\varphi}, \bar{t}$ in which the metric outside the sphere takes the form

$$d\tau^2 = \left(1 - \frac{2MG}{\bar{r}}\right) d\bar{t}^2 - \left(1 - \frac{2MG}{\bar{r}}\right)^{-1} d\bar{r}^2 - \bar{r}^2\, d\bar{\theta}^2 - \bar{r}^2 \sin^2 \bar{\theta}\, d\bar{\varphi}^2 \tag{11.9.27}$$

But this metric is not in the Gaussian normal form (11.9.1), so in order to match solutions at the surface we either have to convert the interior solution (11.9.16) into standard coordinates, or the exterior solution (11.9.27) into Gaussian normal coordinates. We choose the former course.[32]

Inspection of Eq. (11.9.16) shows immediately that the standard spatial coordinate $\bar{r}, \bar{\theta}, \bar{\varphi}$ must be chosen as

$$\bar{r} = rR(t), \qquad \bar{\theta} = \theta, \qquad \bar{\varphi} = \varphi \tag{11.9.28}$$

In order to define a standard time coordinate such that $d\tau^2$ does not contain a cross-term $d\bar{r}\, d\bar{t}$, we employ the "integrating factor" technique described in Section 11.7, which gives

$$\bar{t} = \left(\frac{1 - ka^2}{k}\right)^{1/2} \int_{S(r,t)}^{1} \frac{dR}{(1 - ka^2/R)} \left(\frac{R}{1 - R}\right)^{1/2} \tag{11.9.29}$$

where

$$S(r, t) = 1 - \left(\frac{1 - kr^2}{1 - ka^2}\right)^{1/2} (1 - R(t)) \tag{11.9.30}$$

The constant a is arbitrary, but may conveniently be chosen as the radius of the sphere in comoving coordinates. It is straightforward to check that the metric in the coordinate system $\bar{r}, \bar{\theta}, \bar{\varphi}, \bar{t}$ takes the standard form

$$d\tau^2 = B(\bar{r}, \bar{t}) \, d\bar{t}^2 - A(\bar{r}, \bar{t}) \, d\bar{r}^2 - \bar{r}^2(d\bar{\theta}^2 + \sin^2\bar{\theta} \, d\bar{\varphi}^2)$$

with

$$B = \frac{R}{S} \left(\frac{1 - kr^2}{1 - ka^2}\right)^{1/2} \frac{(1 - ka^2/S)^2}{(1 - kr^2/R)} \tag{11.9.31}$$

$$A = \left(1 - \frac{kr^2}{R}\right)^{-1} \tag{11.9.32}$$

it now being understood that S is a function of $\bar{t}$ defined by Eq. (11.9.29) and that r and $R(t)$ are functions of $\bar{r}$ and S, or $\bar{r}$ and $\bar{t}$, defined by solving Eqs. (11.9.28) and (11.9.30). This is a mess, but at the radius $r = a$ of the star (a constant, since r is a comoving coordinate) we have

$$\bar{r} = \bar{a}(t) \equiv aR(t) \tag{11.9.33}$$

$$\bar{t} = \left(\frac{1 - ka^2}{k}\right)^{1/2} \int_{R(t)}^{1} \frac{dR}{(1 - ka^2/R)} \left(\frac{R}{1 - R}\right)^{1/2} \tag{11.9.34}$$

$$B(\bar{a}, \bar{t}) = \left(1 - \frac{ka^2}{R(t)}\right) \tag{11.9.35}$$

$$A(\bar{a}, \bar{t}) = \left(1 - \frac{ka^2}{R(t)}\right)^{-1} \tag{11.9.36}$$

(Eq. (11.9.34) could have been obtained by integrating the equations for free fall given in Section 8.4.) Comparing with (11.9.27), we see that the interior and exterior solutions fit continuously at $\bar{r} = aR(t)$ if

$$k = \frac{2MG}{a^3} \tag{11.9.37}$$

With (11.9.23), this just says that

$$M = \frac{4\pi}{3} \rho(0)a^3 \tag{11.9.38}$$

not a surprising result!

Now we return to the problem of calculating the behavior of light signals emitted from the surface of the collapsing sphere. A light signal emitted in a radial direction at a standard time $\bar{t}$ will have $d\bar{r}/d\bar{t}$ given by Eq. (11.9.27) and the condition $d\tau = 0$, so it will arrive at a distant point $\bar{r}$ at a time

$$\bar{t}' = \bar{t} + \int_{aR(t)}^{\bar{r}'} \left(1 - \frac{2MG}{r}\right)^{-1} dr \tag{11.9.39}$$

The most striking consequence of Eqs. (11.9.39) and (11.9.34) is that both $\bar{t}$ and $\bar{t}'$ approach infinity when the radius (11.9.33) of the sphere approaches the Schwarzschild radius $2GM$, that is, when

$$R(t) \to \frac{2GM}{a} = ka^2 \qquad (11.9.40)$$

The collapse to the Schwarzschild radius therefore appears to an outside observer to take an infinite time, and the collapse to $R = 0$ is utterly unobservable from outside.

Although the collapsing sphere does not suddenly disappear, it does fade out of sight, because light from its surface is subject to an increasing red shift. The proper time for a light source on the sphere's surface is just the comoving time t, so the comoving time interval between emission of wave crests at the surface equals the natural wavelength λ_0 that would be emitted by the source in the absence of gravitation. The standard time interval $d\bar{t}'$ between arrivals of wave crests at $\bar{r}'$ is the observed wavelength λ'; thus the fractional change of wavelength is

$$z \equiv \frac{\lambda' - \lambda_0}{\lambda_0} = \frac{d\bar{t}'}{dt} - 1 = \frac{d\bar{t}}{dt} - a\dot{R}(t)\left(1 - \frac{2MG}{aR(t)}\right)^{-1} - 1$$

$$= -\dot{R}(t)\left(1 - \frac{ka^2}{R(t)}\right)^{-1}\left[\left(\frac{1 - ka^2}{k}\right)^{1/2}\left(\frac{R(t)}{1 - R(t)}\right)^{1/2} + a\right] - 1$$

Using (11.9.24) to determine $\dot{R}(t)$, this is

$$z = \left(1 - \frac{ka^2}{R(t)}\right)^{-1}\left[(1 - ka^2)^{1/2} + a\sqrt{k}\left(\frac{1 - R(t)}{R(t)}\right)^{1/2}\right] - 1$$

$$(11.9.41)$$

In order to see how the red shift z varies with $\bar{t}'$, let us assume that the sphere is initially very much larger than its Schwarzschild radius

$$ka^2 = \frac{2GM}{a} \ll 1 \qquad (11.9.42)$$

and distinguish two periods in the history of the collapse:

(A) Until t gets close to T, we have

$$\frac{ka^2}{R(t)} \ll 1 \qquad (11.9.43)$$

Using (11.9.42) and (11.9.43) in (11.9.34), (11.9.39), and (11.9.41) gives (with $\bar{r}' \gg a$)

$$\bar{t} \simeq t$$
$$\bar{t}' \simeq \bar{t} + \bar{r}' - aR(t) \simeq t + \bar{r}' - aR(t) \simeq t + \bar{r}'$$
$$z \simeq a\sqrt{k}\left(\frac{1 - R(t)}{R(t)}\right)^{1/2} \simeq a\sqrt{k}\left(\frac{1 - R(\bar{t}' - \bar{r}')}{R(\bar{t}' - \bar{r}')}\right)^{1/2} \qquad (11.9.44)$$

(B) Eventually we have

$$\frac{ka^2}{R(t)} \to 1$$

at a time t_1 given by (11.9.25) as

$$t_1 \simeq \frac{1}{2\sqrt{k}} [\pi - \tfrac{4}{3}(ka^2)^{3/2}] \tag{11.9.45}$$

Now (11.9.34), (11.9.39), and (11.9.41) give

$$\bar{t} \simeq -ka^3 \ln\left[1 - \frac{ka^2}{R(t)}\right] + \text{constant}$$

$$\bar{t}' \simeq \bar{t} - ka^3 \ln\left[1 - \frac{ka^2}{R(t)}\right] + \text{constant}$$

$$\simeq -2ka^3 \ln\left[1 - \frac{ka^2}{R(t)}\right] + \text{constant}$$

$$z \simeq 2\left(1 - \frac{ka^2}{R(t)}\right)^{-1} \propto \exp\left(\frac{\bar{t}'}{2ka^3}\right) \tag{11.9.46}$$

Putting (A) and (B) together, we conclude that the red shift z seen by an observer at $\bar{r}'$ is zero when the collapse is observed to begin, increases gradually but remains of order $a\sqrt{k} \ll 1$ until a time very close to $T = \pi/2\sqrt{k}$ has passed, and then grows exponentially with a rate $1/2ka^3$. For example, a collapsing sphere with a mass $M = 10^8 M_\odot$ and radius $a = 100$ light years will have a red shift z of order 10^{-3} for a period of order 10^5 years, after which the red shift suddenly begins growing exponentially with an e-folding time of order 1 min. For practical purposes, the collapsing sphere is suddenly and completely cut off from communication with the rest of the universe.

Completely cut off? Even if a collapsing body does fade out of sight, it still has a gravitational field, and, as shown in Section 7.6, the measurement of this field at great distances can be used to determine the energy, momentum, and angular momentum of the body. If the body has a net electric charge, then measurement of the electric field at great distances will, via Gauss's theorem, also tell us the charge. It is interesting to ask whether measurements of the gravitational and/or electromagnetic fields outside a collapsing body can yield any information about the body *beyond* the energy, momentum, angular momentum, and charge. In the case of a spherically symmetric electrically neutral body, which we have been considering in this chapter, the answer is provided by Birkhoff's theorem: The gravitational field outside a spherically symmetric body must be of the Schwarzschild form, so all we can ever learn about the body is its mass. (Spherical symmetry, of course, implies zero momentum and zero angular momentum.)

Carter[33] has shown that when the gravitational field of an *axially symmetric* collapsing body settles down to a stationary state, its exterior metric belongs to a *two-parameter* family of solutions, such as the Kerr metrics (see Section 11.7) that are completely specified by the total mass and angular momentum. It is widely believed that the gravitational field of any electrically neutral collapsing body will eventually approach the Kerr form.

As mentioned in the introduction to this chapter, interest in the phenomenon of gravitational collapse was rekindled in the last decade by the discovery of quasi-stellar sources, which appear to require some powerful new source of energy. The maximum energy available from fusion of hydrogen into the most stable nuclei, say iron, is only 8 MeV per nucleon, or less than 1% of the rest-mass. Matter-antimatter annihilation could have 100% efficiency (apart from neutrino energy losses), but this process can be important only if there is some abundant natural source of antinucleons. Otherwise the only likely mechanism for conversion of mass into energy with high efficiency is gravitational collapse.[34]

A cloud of dust that is collapsing as in the Oppenheimer-Snyder model will obviously release no energy to the outside world. To extract the growing kinetic energy of the falling particles, something must slow them on the way down— either a macroscopic "bounce" of the whole system, or particle-particle collisions that heat the collapsing gas. Detailed calculations reveal a discouragingly low efficiency for conversion of mass into available energy in the gravitational collapse of an *isolated* body.[35] However, particles falling into a Kerr metric can reemerge with a higher energy, acquired at the expense of the rotational energy of the collapsing body.[36]

Whether or not gravitational collapse has anything to do with quasi-stellar sources, the question remains: What happens to a real cooling star whose mass is above the Chandrasekhar and Oppenheimer-Volkoff limits? In recent years topological methods have been used by Penrose and Hawking to prove a number of powerful theorems,[37] to the effect that under reasonable conditions (validity of general relativity, positivity of energy, ubiquity of matter, causality) collapse becomes inevitable once a *trapped surface* forms. A trapped surface is a closed spacelike two-dimensional surface for which both the outgoing and the ingoing families of future-directed null geodesics orthogonal to the surface are converging. (For the Schwarzschild metric, the spheres with r and t constant are trapped surfaces for r within the Schwarzschild radius $2MG$.) However, it is not known whether a real massive star will actually develop a trapped surface, or merely explode into fragments with small enough mass to form stable neutron stars or white dwarfs.

If gravitational collapse is indeed the inevitable fate of massive bodies, then we must expect that the universe is full of *black holes*, collapsing bodies whose presence is betrayed only through their gravitational fields or through the energy released when matter is drawn in.[38] Our best hope of observing gravitational collapse would be to find a binary star, one member an ordinary visible star, and the other member a black hole.[39]

11 BIBLIOGRAPHY

Relativistic Astrophysics in General

☐ *Quasars and High Energy Astronomy*, ed. by K. N. Douglas, I. Robinson, A. Schild, E. L. Schucking, J. A. Wheeler, and N. J. Woolf (Second "Texas" Symposium on Relativistic Astrophysics, Gordon and Breach, New York, 1969).

☐ *High Energy Astrophysics*, ed. by L. Gratton (Proceedings of the International School of Physics "Enrico Fermi," Course XXXV, Academic Press, New York, 1966).

☐ *Quasi-Stellar Sources and Gravitational Collapse*, ed. by I. Robinson, A. Schild, and E. L. Schucking (First "Texas" Symposium on Relativistic Astrophysics, University of Chicago Press, Chicago, 1965).

☐ Ya. B. Zeldovich and I. D. Novikov, "Relativistic Astrophysics I," Usp. Fiz. Nauk, **84**, 377 (1964) (trans. Soviet Physics Uspekhi, May–June 1965).

☐ Ya. B. Zeldovich and I. D. Novikov, "Relativistic Astrophysics II," Usp. Fiz. Nauk., **86**, 447 (1965) [trans. Soviet Physics Uspekhi, **8**, 522 (1965)].

☐ Ya. B. Zeldovich and I. D. Novikov, *Relativistic Astrophysics: Volume 1, Stars and Relativity*, translated by K. S. Thorne and W. D. Arnett (University of Chicago Press, Chicago, 1971). I regret that this comprehensive work was not yet available during the preparation of this chapter.

Nonrelativistic Theory of Stellar Structure

☐ S. Chandrasekhar, *An Introduction to the Study of Stellar Structure* (Dover Publications, New York, 1939).

☐ E. E. Salpeter, "Stellar Structure Leading up to White Dwarfs and Neutron Stars," in *Relativity Theory and Astrophysics. 3. Stellar Structure*, ed. by J. Ehlers (American Mathematical Society, Providence, R. I., 1967), p. 1.

☐ M. Schwarzschild, *Structure and Evolution of the Stars* (Princeton University Press, Princeton, N. J., 1958).

Pulsars and Neutron Stars

☐ A. G. W. Cameron, "Neutron Stars," in *Annual Review of Astronomy and Astrophysics*, Vol. 8, ed. by L. Goldberg (Annual Reviews, Inc., Palo Alto, 1970), p. 179.

☐ A. G. W. Cameron, "How Are Neutron Stars Formed?", Comments Astrophys. and Space Phys., **1**, 172 (1969).

☐ S. Frautschi, J. N. Bahcall, G. Steigman, and J. C. Wheeler, "Ultradense Matter," Comments Astrophys. and Space Phys., to be published.

☐ V. L. Ginzburg, "Superfluidity and Superconductivity in Astrophysics," Comments Astrophys. and Space Phys., **1**, 81 (1969).

☐ T. Gold, "The Nature of Pulsars," in *Contemporary Physics—Trieste Sym-*

posium 1968, ed. by A. Salam, Vol. 1 (International Atomic Energy Agency, Vienna, 1969), p. 477.

☐ A. Hewish, "Pulsars," in *Annual Review of Astronomy and Astrophysics*, Vol. 8, ed. by L. Goldberg (Annual Reviews, Inc., Palo Alto, Cal., 1970), p. 265.

☐ L. D. Landau and E. M. Lifshitz, *Statistical Physics*, trans. by E. Peierls and R. F. Peierls (Pergamon Press, London, 1958), Chapter XI.

☐ J. P. Ostriker, "The Nature of Pulsars," Scientific American, January, 1971, p. 48.

☐ M. A. Ruderman, "Solid Stars," Scientific American, March 1971, p. 24.

☐ Symposium on the Crab Pulsar, Pub. Astron. Soc. Pac., **82**, No. 486 (1970).

☐ J. A. Wheeler, "Superdense Stars," *Annual Review of Astronomy and Astrophysics*, Vol. 4, ed. by L. Goldberg (Annual Reviews, Inc., Palo Alto, Cal., 1966), p. 393.

Supermassive Objects

☐ R. V. Wagoner, "Physics of Massive Objects," in *Annual Review of Astronomy and Astrophysics*, Vol. 7, ed. by L. Goldberg (Annual Reviews, Inc., Palo Alto, Cal., 1969), p. 553.

Gravitational Collapse

☐ B. K. Harrison, K. S. Thorne, M. Wakano, and J. A. Wheeler, *Gravitational Theory and Gravitational Collapse* (University of Chicago Press, Chicago, 1965).

☐ S. W. Hawking and D. W. Sciama, "Singularities in Collapsing Stars and Expanding Universes," Comments Astrophys. and Space Phys., **1**, 1 (1969).

☐ R. Geroch, "Singularities," in *Relativity—Proceedings of the Relativity Conference in the Midwest*, ed. by M. Carmeli, S. I. Fickler, and L. Witten (Plenum Press, New York, 1970), p. 259.

☐ M. M. May and R. H. White, "Hydrodynamic Calculations of General Relativistic Collapse," in *Relativity Theory and Astrophysics. 3. Stellar Structure*, *op. cit.*, p. 96.

☐ C. W. Misner, "Gravitational Collapse," in *Astrophysics and General Relativity* (1968 Brandeis University Summer Institute in Theorerical Physics), Vol. 1, ed. by M. Chretien, S. Deser, and J. Goldstein (Gordon and Breach Science Publishers, New York, 1969).

☐ R. Penrose, "On Gravitational Collapse," in *Contemporary Physics—Trieste Symposium* 1968, ed. by A. Salam, Vol. 1 (International Atomic Energy Agency, Vienna, 1969), p. 545.

☐ R. Penrose, "Structure of Space-Time," in *Batelle Rencontres*, ed. by C. M. DeWitt and J. A. Wheeler (W. A. Benjamin, New York, 1968), p. 121.

☐ K. S. Thorne, "Nonspherical Gravitational Collapse: Does it Produce Black Holes?," Comments Astrophys. and Spaec Phys., **2**, 191 (1970).

☐ R. Ruffini and J. A. Wheeler, "Introducing the Black Hole," Physics Today, January 1971, p. 30.

☐ J. A. Wheeler, "Geometrodynamics and the Issue of the Final State," in *Relativity, Groups, and Topology*, ed. by C. DeWitt and B. DeWitt (Gordon and Breach Science Publishers, New York, 1964), p. 317.

For material on the quasi-stellar objects, see the bibliography to Chapter 14.

11 REFERENCES

1. B. K. Harrison, K. S. Thorne, M. Wakano, and J. A. Wheeler, *Gravitation Theory and Gravitational Collapse* (University of Chicago Press, Chicago, 1965), Appendix B; J. M. Bardeen, unpublished Ph.D. thesis, California Institute of Technology, 1965. For the extension of this theorem to slowly rotating stars, see J. B. Hartle and K. S. Thorne, Ap. J., **158**, 179 (1969).

2. Harrison, Thorne, Wakano, and Wheeler, *op. cit.*, Chapter 3.

3. See, for example, P. M. Morse and H. Feshbach, *Methods of Mathematical Physics* (McGraw-Hill, New York, 1953), p. 278.

4. The Lane-Emden functions are extensively discussed by S. Chandrasekhar, *Stellar Structure* (Dover Publications, New York, 1939), Chapter IV.

5. Chandrasekhar, *op. cit.*, Table 4.

6. A. Ritter, Wiedemann Ann., **11**, 332 (1880); E. Betti, Nuovo Cimento, **7**, 26 (1880).

7. For a detailed discussion of stellar stability, see P. Ledoux, in *Stars and Stellar Structure VIII: Stellar Structure*, ed. by L. H. Aller and D. B. McLaughlin (University of Chicago Press, Chicago, 1965), Chapter 10. Relativistic effects are considered by S. Chandrasekhar, Ap. J., **140**, 417 (1964).

8. S. Chandrasekhar, Mon. Not. Roy. Astron. Soc., **95**, 207 (1935). Also see L. D. Landau, Phys. Z. Sowjetunion, **1**, 285 (1932).

9. Harrison, Thorne, Wakano, and Wheeler, ref. 1, Figure 5, and Chapter 10.

10. J. R. Oppenheimer and G. M. Volkoff, Phys. Rev., **55**, 374 (1939). Some earlier references include L. Landau, ref. 8; W. Baade and F. Zwicky, Proc. Nat. Acad. Sci. U.S., **20**, 254 (1934); J. R. Oppenheimer and R. Serber, Phys. Rev., **54**, 540 (1938); R. C. Tolman, Phys. Rev., **55**, 364 (1939).

11. C. W. Misner and H. S. Zapolsky, Phys. Rev. Letters, **12**, 635 (1964).

12. Harrison, Thorne, Wakano, and Wheeler, ref. 1, Appendix A.

13. Y. C. Leung and C. G. Wang, to be published. Also see C. G. Wang, W. K. Rose, and S. L. Schlenker, Ap. J., **160**, L17 (1970). H. Lee, Y. C. Leung, and C. G. Wang, Ap. J., **166**, 387 (1971).

14. S. Tsuruta and A. G. W. Cameron, Canadian J. Phys., **44**, 1895 (1966).

15. A. G. W. Cameron, Ann. Rev. Astron. and Astrophys., **8**, 179 (1970).

16. M. Ruderman, Nature, **223**, 597 (1969).

17. A. B. Midgal, Nucl. Phys., **13**, 655 (1959). For other references, see Cameron, ref. 15.

18. The effects of high magnetic fields are considered by V. Canuto and H. Y. Chiu, Phys. Rev., **173**, 1210, 1220, 1229 (1968). For other references, see Cameron, ref. 15.

19. The effects of rotation are considered by J. B. Hartle, Ap. J., **150**, 1005 (1967); J. B. Hartle and K. S. Thorne, Ap. J., **153**, 807 (1968); *ibid.*, **158**, 719 (1969).

20. A. Hewish, S. J. Bell, J. D. H. Pilkington, P. F. Scott, R. A. Collins, Nature, **217**, 709 (1968).

21. T. Gold, Nature, **218**, 731 (1968).

22. For a summary, see A. Hewish, Ann. Rev. Astron. and Astrophys., **8**, 265 (1970).

23. F. Hoyle and W. A. Fowler, Mon. Not. Roy. Astron. Soc., **125**, 169 (1963); Nature, **197**, 533 (1963); F. Hoyle, W. A. Fowler, G. R. Burbidge, and E. M. Burbidge, Ap. J., **139**, 909 (1964).

24. This is obtained by comparison with W. A. Fowler, in *Quasi-Stellar Sources and Gravitational Collapse* (University of Chicago Press, Chicago, 1965), p. 56, Eq. (24).

25. K. Schwarzschild, Sitzungsberichte Preuss. Akad. Wiss., 424, 1916.

26. H. Bondi, Proc. Roy. Soc. (London), **A281**, 39 (1964); also, in *Lectures on General Relativity*, ed. by S. Deser and K. W. Ford (Prentice-Hall, Englewood Cliffs, New Jersey, 1964), p. 375.

27. However, see S. A. Bludman and M. Ruderman, Phys. Rev., **170**, 1176 (1968); M. A. Ruderman, Phys. Rev., **172**, 1286 (1968); S. A. Bludman and M. A. Ruderman, Phys. Rev., **D1**, 3243 (1970).

28. G. S. Bisnovatyi-Kogan and Ya. B. Zeldovich, Astrofizika, **5**, 223 (1969). For a discussion of stability of relativistic gas spheres and clusters of point masses with arbitrarily large central redshifts, see G. S. Bisnovatyi-Hogan and K. S. Thorne, Ap. J., **160**, 875 (1970); E. D. Fackerell, J. R. Ipser, and K. S. Thorne, Comments Astrophys. and Space Phys., **1**, 140 (1969).

29. F. Hoyle and W. A. Fowler, Nature, **213**, 373 (1967): also see H. S. Zapolsky, Ap. J., **153**, L163 (1968).

30. G. Birkhoff, *Relativity and Modern Physics* (Harvard University Press, Cambridge, Mass., 1923), p. 253. Also see S. Deser and B. E. Laurent, Am. J. Phys., **36**, 789 (1968).

31. R. C. Tolman, Proc. Nat. Acad. Sci. U.S.A., **20**, 3 (1934).

32. J. R. Oppenheimer and H. Snyder, Phys. Rev., **56**, 455 (1939). For further details on the Oppenheimer-Snyder solution and other spherically symmetric solutions, see O. Klein, in *Werner Heisenberg und die Physik unserer Zeit* (Vieweg, Braunschweig, Germany, 1961); F. Hoyle, W. A. Fowler, G. R. Burbidge, and E. M. Burbidge, Ap. J., **139**, 909 (1964);

F. Hoyle and W. A. Fowler, in *Quasi-Stellar Sources and Gravitational Collapse*, ed. by I. Robinson, A. Schild, and E. L. Schucking (University of Chicago Press, Chicago, 1965); C. W. Misner and D. H. Sharp, Phys. Rev., **136**, B571 (1964); G. C. McVittie, Ap. J., **140**, 401 (1964); G. C. McVittie, Ann. Inst. Henri Poincaré, **6**, No. 1 (1967); M. M. May and R. H. White, Phys. Rev., **141**, 1232 (1966); S. A. Colgate and R. H. White, Ap. J., **143**, 626 (1966); For asymmetric collapse, see J. M. Cohen, Phys. Rev., **173**, 1258 (1966); M. Fujimoto, Ap. J., **152**, 523 (1968); V. de la Cruz, J. E. Chase, and W. Israel, Phys. Rev. Letters, **24**, 423 (1970); and so on.

33. B. Carter, Phys. Rev. Letters, **26**, 331 (1971). For other rigorous theorems on the possible forms of exterior metrics under various conditions, see A. Lichnerowicz, *Théories relativistes de la gravitation* (Masson, Paris, 1955); S. Deser, C. R. Acad. Sci. Paris, **264**, 805 (1967); W. Israel, Phys. Rev., **164**, 1776 (1967); A. G. Doroshkevich, Ya. B. Zeldovich, and I. D. Novikov, Zh. Eksp. Teor. Fiz., **49**, 170 (1965) [trans. Sov. Phys. JETP, **22**, 122 (1966)]; R. M. Wald, Phys. Rev. Letters **26**, 1653 (1971); and so on.

34. F. Hoyle and W. Fowler, Nature, **197**, 533 (1963).

35. F. J. Dyson, Comments Astrophys. and Space Phys., **1**, 75 (1969); C. Leibovitz and W. Israel, Phys. Rev., **1**, 3226 (1970).

36. R. Penrose, Riv. Nuovo Cimento, **1**, Numero Speciale, 252 (1969).

37. R. Penrose, Phys. Rev. Letters, **14**, 57 (1965); S. W. Hawking, Proc. Roy. Soc., **294A**, 511 (1966); *ibid.*, **295A**, 490 (1966); *ibid.*, **300A**, 187 (1967); *ibid.*, **308A**, 433 (1967); S. W. Hawking and R. Penrose, Proc. Roy. Soc., **314A**, 529 (1970); and ref. 36.

38. Gravitational radiation from vibrating black holes has been considered by W. H. Press, Ap. J. **170**, L105 (1971). Gravitational radiation from matter falling into black holes has been considered by M. Davis, R. Ruffini, W. H. Press, and R. H. Price, Phys. Rev. Letters **27**, 1466 (1971).

39. See A. G. W. Cameron, Nature **229**, 178 (1971); R. E. Wilson, Ap. J. **170**, 529 (1971).

PART FOUR
FORMAL DEVELOPMENTS

12 THE ACTION PRINCIPLE

There are a great many physical systems whose dynamic equations can be derived from a "principle of least action," that is, from a statement that some functional of the dynamical variables, the "action," is stationary with respect to small variations of these variables. This formulation of the dynamic equations has one great advantage: It allows us to establish an immediate connection between symmetry principles and conservation laws.

The symmetry of the action that concerns us most in this book is general covariance. In this section we shall develop a general definition of the energy-momentum tensor for any material system, as a functional derivative of the action for that system. The use of the action principle and general covariance will then allow us to show that this tensor is indeed conserved.

To achieve a truly general formulation of general relativity in terms of an action principle, it is necessary to uncover a question that has been carefully buried until now: How can we incorporate the effects of gravitation into the field theories of particles with half-integer spin? The answer requires the development of an approach to general relativity, the "tetrad formalism," based directly on the families of locally inertial frames that were our starting point in Chapter 3.

Although the proof is more complicated, the energy-momentum tensor in this formalism is still symmetric and conserved.

1 The Matter Action: An Example

We shall begin by displaying one example of a physical system whose equations of motion can be derived from a principle of stationary action. The system is a collisionless plasma, consisting of particles n of mass m_n and charge e_n, together with the electromagnetic field $F_{\mu\nu}(x)$ they produce. The equations of motion in an arbitrary external gravitational field $g_{\mu\nu}$ are

$$\frac{d^2 x_n{}^\mu}{d\tau_n} + \Gamma^\mu_{\nu\lambda}(x_n) \frac{dx_n{}^\nu}{d\tau_n} \frac{dx_n{}^\lambda}{d\tau_n} = \left(\frac{e_n}{m_n}\right) F^\mu{}_\nu(x_n) \frac{dx_n{}^\nu}{d\tau_n} \tag{12.1.1}$$

$$d\tau_n \equiv (-g_{\mu\nu}\, dx_n{}^\mu\, dx_n{}^\nu)^{1/2} \tag{12.1.2}$$

$$\frac{\partial}{\partial x^\mu} [\sqrt{g(x)}\, F^{\mu\nu}(x)] = -\sum_n e_n \int \delta^4(x - x_n) \frac{dx_n{}^\nu}{d\tau_n}\, d\tau_n \tag{12.1.3}$$

$$\frac{\partial}{\partial x^\lambda} F_{\mu\nu}(x) + \frac{\partial}{\partial x^\nu} F_{\lambda\mu}(x) + \frac{\partial}{\partial x^\mu} F_{\nu\lambda}(x) = 0 \tag{12.1.4}$$

[See Eqs. (5.2.9), (5.1.11), (5.2.6), (5.2.13), (5.2.7).] In order to satisfy (12.1.4), we introduce a vector potential $A_\mu(x)$:

$$F_{\mu\nu}(x) = \frac{\partial A_\nu(x)}{\partial x^\mu} - \frac{\partial A_\mu(x)}{\partial x^\nu} \tag{12.1.5}$$

The independent dynamical variables may then be taken as $A_\mu(x)$ and $x_n{}^\nu(p)$, where p is some quantity that simultaneously parameterizes all the space-time trajectories of the various particles.

We tentatively take the action for this system as

$$I_M = -\sum_n m_n \int_{-\infty}^\infty dp \left[-g_{\mu\nu}(x_n(p)) \frac{dx_n{}^\mu(p)}{dp} \frac{dx_n{}^\nu(p)}{dp} \right]^{1/2}$$

$$-\tfrac{1}{4} \int d^4x\, g^{1/2}(x) F_{\mu\nu}(x) F^{\mu\nu}(x)$$

$$+ \sum_n e_n \int_{-\infty}^\infty dp\, \frac{dx_n{}^\mu(p)}{dp} A_\mu(x_n(p)) \tag{12.1.6}$$

(The subscript M is to remind us that this is the action only for matter and radiation, with $g_{\mu\nu}(x)$ taken as a prescribed external gravitational field.) It is understood here that $F_{\mu\nu}$ is given by Eq. (12.1.5), and the indices on $F^{\mu\nu}$ are raised with the contravariant metric tensor as usual.

The principle of stationary action says that the action I_M will not be changed by an infinitesimal variation in the dynamical variables

$$x^\mu(p) \rightarrow x^\mu(p) + \delta x^\mu(p)$$
$$A_\mu(x) \rightarrow A_\mu(x) + \delta A_\mu(x)$$

where

$$\delta x^\mu(p) \rightarrow 0 \qquad \text{for } |p| \rightarrow \infty$$
$$\delta A_\mu(x) \rightarrow 0 \qquad \text{for } |x^\lambda| \rightarrow \infty$$

if and only if $x^\mu(p)$ and $A_\mu(x)$ obey the dynamical equations (12.1.1)–(12.1.3). To check that this is correct, let us compute the change in I_M produced by this variation, without yet assuming that (12.1.1)–(12.1.3) are satisfied. We find that

$$\delta I_M = \tfrac{1}{2} \sum_n m_n \int_{-\infty}^{\infty} dp \left[-g_{\mu\nu}(x_n(p)) \frac{dx_n{}^\mu(p)}{dp} \frac{dx_n{}^\nu(p)}{dp} \right]^{-1/2}$$

$$\times \left\{ 2g_{\mu\nu}(x_n(p)) \frac{dx_n{}^\mu(p)}{dp} \frac{d\delta x_n{}^\nu(p)}{dp} + \left(\frac{\partial g_{\mu\nu}(x)}{\partial x^\lambda} \right)_{x=x_n(p)} \frac{dx_n{}^\mu(p)}{dp} \frac{dx_n{}^\nu(p)}{dp} \delta x_n{}^\lambda(p) \right\}$$

$$- \int d^4x \, g^{1/2}(x) F^{\mu\nu}(x) \frac{\partial}{\partial x^\mu} \delta A_\nu(x)$$

$$+ \sum_n e_n \int_{-\infty}^{\infty} dp \left\{ \frac{d\delta x_n{}^\mu(p)}{dp} A_\mu(x_n(p)) + \frac{dx_n{}^\mu(p)}{dp} \left(\frac{\partial A_\mu(x)}{\partial x^\lambda} \right)_{x=x_n(p)} \delta x_n{}^\lambda(p) \right.$$

$$\left. + \frac{dx_n{}^\mu(p)}{dp} \delta A_\mu(x_n(p)) \right\}$$

It is convenient at this point to change variables of integration from p to the τ_n defined by Eq. (12.1.2). This gives

$$\delta I_M = \tfrac{1}{2} \sum_n m_n \int_{-\infty}^{\infty} d\tau_n \left\{ 2g_{\mu\lambda}(x_n) \frac{dx_n{}^\mu}{d\tau_n} \frac{d\delta x_n{}^\lambda}{d\tau_n} + \frac{\partial g_{\mu\nu}(x_n)}{\partial x_n{}^\lambda} \frac{dx_n{}^\mu}{d\tau_n} \frac{dx_n{}^\nu}{d\tau_n} \delta x_n{}^\lambda \right\}$$

$$- \int d^4x \, g^{1/2}(x) F^{\mu\nu}(x) \frac{\partial}{\partial x^\mu} \delta A_\nu(x)$$

$$+ \sum_n e_n \int_{-\infty}^{\infty} d\tau_n \left\{ \frac{d\delta x_n{}^\mu}{d\tau_n} A_\mu(x_n) + \frac{dx_n{}^\mu}{d\tau_n} \frac{\partial A_\mu(x_n)}{\partial x_n{}^\lambda} \delta x_n{}^\lambda + \frac{dx_n{}^\mu}{d\tau_n} \delta A_\mu(x_n) \right\}$$

The condition that $\delta x^\mu(\tau_n)$ and $\delta A_\mu(x)$ vanish on the boundaries of the region of integration allows us to integrate by parts, and we obtain

$$\delta I_M = \sum_n \int_{-\infty}^{\infty} d\tau_n g_{\mu\lambda}(x_n) \left\{ - m_n \left[\frac{d^2 x_n{}^\mu}{d\tau_n{}^2} + \Gamma^\mu_{\rho\sigma}(x_n) \frac{dx_n{}^\rho}{d\tau_n} \frac{dx_n{}^\sigma}{d\tau_n} \right] \right.$$

$$\left. + e_n \frac{dx_n{}^\rho}{d\tau_n} F^\mu{}_\rho(x_n) \right\} \delta x_n{}^\lambda$$

$$+ \int d^4x \left\{ \frac{\partial}{\partial x^\mu} [g^{1/2}(x) F^{\mu\nu}(x)] + \sum_n e_n \int_{-\infty}^{\infty} d\tau_n \delta^4(x - x_n) \frac{dx_n{}^\nu}{d\tau_n} \right\} \delta A_\nu(x)$$

Evidently δI_M vanishes for general variations $\delta x_n{}^\lambda$ and δA_ν if and only if $x_n{}^\lambda$ and A_ν obey the dynamical equations (12.1.1) and (12.1.3), and we therefore conclude that (12.1.6) does qualify as a suitable action for this system.

2   General Definition of $T^{\mu\nu}$

We are going to define the energy-momentum tensor for a material system described by an action I_M as the "functional derivative" of I_M with respect to $g_{\mu\nu}$. That is, we imagine $g_{\mu\nu}(x)$ to be subject to an infinitesimal variation

$$g_{\mu\nu} \to g_{\mu\nu} + \delta g_{\mu\nu} \tag{12.2.1}$$

where $\delta g_{\mu\nu}$ is arbitrary, except that it is required to vanish as $|x^\lambda| \to \infty$. The action I_M will not be stationary with respect to this variation, because for the moment we are regarding $g_{\mu\nu}(x)$ not as a dynamical variable like $x_n{}^\mu$ or A_μ but as an external field. Rather, δI_M will be some linear functional of the infinitesimal $\delta g_{\mu\nu}(x)$, and therefore takes the form

$$\delta I_M = \tfrac{1}{2} \int d^4x \sqrt{g(x)} \; T^{\mu\nu}(x)\delta g_{\mu\nu}(x) \tag{12.2.2}$$

The coefficient $T^{\mu\nu}(x)$ is *defined* to be the energy-momentum tensor of this system.

A general proof that $T^{\mu\nu}$ is a conserved symmetric tensor will be given in the next section. However, let us first check that (12.2.2) gives the correct energy-momentum tensor for the collisionless plasma described by the action (12.1.6). We are varying $g_{\mu\nu}$ with A_μ held fixed, so

$$\delta F^{\mu\nu} = F_{\rho\sigma}\delta(g^{\mu\rho}g^{\nu\sigma}) = F_{\rho\sigma}g^{\mu\rho}\delta g^{\nu\sigma} + F_{\rho\sigma}g^{\nu\sigma}\delta g^{\mu\rho}$$

To calculate $\delta g^{\nu\sigma}$, we note that

$$0 = \delta(g_{\lambda\kappa}g^{\kappa\sigma}) = g^{\kappa\sigma}\delta g_{\lambda\kappa} + g_{\lambda\kappa}\delta g^{\kappa\sigma}$$

so

$$\delta g^{\nu\sigma} = -g^{\nu\lambda}g^{\kappa\sigma}\delta g_{\lambda\kappa}$$

and therefore

$$\delta F^{\mu\nu} = -F^{\mu\kappa}g^{\nu\lambda}\delta g_{\lambda\kappa} + F^{\nu\lambda}g^{\mu\kappa}\delta g_{\lambda\kappa}$$

Also, we showed in Section 4.7 that

$$\delta g = gg^{\lambda\kappa}\delta g_{\lambda\kappa}$$

A straightforward calculation gives then

$$\delta I_M = \tfrac{1}{2}\sum_n m_n \int_{-\infty}^{\infty} dp \left[-g_{\mu\nu}(x_n(p))\frac{dx_n{}^\mu(p)}{dp}\frac{dx_n{}^\nu(p)}{dp} \right]^{-1/2} \frac{dx_n{}^\lambda(p)}{dp}\frac{dx_n{}^\kappa(p)}{dp}\delta g_{\lambda\kappa}(x_n(p))$$

$$+ \tfrac{1}{2}\int d^4x\, g^{1/2}(x)\{F_\mu{}^\lambda(x)F^{\mu\kappa}(x) - \tfrac{1}{4}g^{\lambda\kappa}(x)F_{\mu\nu}(x)F^{\mu\nu}(x)\}\delta g_{\lambda\kappa}(x)$$

This is of the form (12.2.2), with

$$T^{\lambda\kappa}(x) = g^{-1/2}(x) \sum_n m_n \int_{-\infty}^{\infty} d\tau_n \frac{dx_n{}^{\lambda}}{d\tau_n} \frac{dx_n{}^{\kappa}}{d\tau_n} \delta^4(x - x_n)$$

$$+ F_{\mu}{}^{\lambda}(x)F^{\mu\kappa}(x) - \tfrac{1}{4}g^{\lambda\kappa}(x)F_{\mu\nu}(x)F^{\mu\nu}(x)$$

in agreement with the previously obtained energy-momentum tensor, given by Eqs. (5.3.5) and (5.3.7).

The definition (12.2.2) is closely analogous to a similar definition of the electric current J^{μ}. We can break up the total matter action into a purely electromagnetic term I_E and another term I_M' that describes the charged particles and their electromagnetic interactions

$$I_M \equiv I_E + I_M' \qquad (12.2.3)$$

$$I_E \equiv -\tfrac{1}{4} \int d^4x\, g^{1/2}(x) F_{\mu\nu}(x) F^{\mu\nu}(x) \qquad (12.2.4)$$

Consider the effect on I_M' of an infinitesimal variation in the vector potential

$$A_{\mu} \to A_{\mu} + \delta A_{\mu} \qquad (12.2.5)$$

Since I_M' is not the whole action, the change in I_M due to this variation in A_{μ} does not vanish, but it is necessarily a linear functional of δA_{μ}:

$$\delta I_M' = \int d^4x\, \sqrt{g(x)}\, J^{\mu}(x)\delta A_{\mu}(x) \qquad (12.2.6)$$

and the coefficient $J^{\mu}(x)$ is *defined* to be the electromagnetic current of the system. For instance, for the collisionless plasma described by Eq. (12.1.6), the term I_M' is given as the sum of the first and third terms in Eq. (12.1.6), and we immediately find that

$$\delta I_M' = \sum_n e_n \int_{-\infty}^{\infty} dx_n{}^{\mu}\delta A_{\mu}(x_n)$$

This is of the form (12.2.6), with

$$J^{\mu}(x) = g^{-1/2}(x) \sum_n e_n \int \delta^4(x - x_n)\, dx_n{}^{\mu}$$

in agreement with Eq. (5.2.13). The proof that (12.2.6) always yields a *conserved* current $J^{\mu}(x)$ is given in the next section.

3   General Covariance and Energy-Momentum Conservation

If the action I_M for a material system is a scalar, then the statement that δI_M vanishes is generally covariant, and so also are the dynamical equations derived from this statement. For instance, a glance at the action (12.1.6) for a collisionless plasma shows that this I_M is a scalar, and this ensures the general covariance of the dynamical equations (12.1.1)–(12.1.3) that follow from (12.1.6) upon application of the principle of stationary action.

We shall therefore assume that I_M is a scalar. This means that I_M will be unchanged if we simultaneously make the replacements

$$d^4x \rightarrow d^4x'$$

$$\frac{\partial}{\partial x^\mu} \rightarrow \frac{\partial}{\partial x'^\mu}$$

$$x_n{}^\mu(p) \rightarrow x_n'{}^\mu(p)$$

$$A^\mu(x) \rightarrow A_\mu'(x') \equiv A_\nu(x) \frac{\partial x^\nu}{\partial x'^\mu}$$

$$g_{\mu\nu}(x) \rightarrow g_{\mu\nu}'(x') \equiv g_{\rho\sigma}(x) \frac{\partial x^\rho}{\partial x'^\mu} \frac{\partial x^\sigma}{\partial x'^\nu}$$

However, x'^μ is a mere variable of integration (as opposed to $x_n{}^\mu$, which is a dynamical variable) so we can change x'^μ back to x^μ without changing I_M. We conclude then that I_M is unchanged by the replacements

$$x_n{}^\mu(p) \rightarrow x_n'{}^\mu(p)$$

$$A_\mu(x) \rightarrow A_\mu'(x) = A_\nu(x) \frac{\partial x^\nu}{\partial x'^\mu} - [A_\mu'(x') - A_\mu'(x)]$$

$$g_{\mu\nu}(x) \rightarrow g_{\mu\nu}'(x) = g_{\rho\sigma}(x) \frac{\partial x^\rho}{\partial x'^\mu} \frac{\partial x^\sigma}{\partial x'^\nu} - [g_{\mu\nu}'(x') - g_{\mu\nu}'(x)]$$

with d^4x and $\partial/\partial x^\mu$ now left alone. If the original transformation $x^\mu \rightarrow x'^\mu$ was infinitesimal

$$x'^\mu = x^\mu + \varepsilon^\mu(x)$$

then the change in the dynamical variables is

$$\delta x_n{}^\mu(p) = \varepsilon^\mu(x_n(p))$$

$$\delta A_\mu(x) = -A_\nu(x) \frac{\partial \varepsilon^\nu(x)}{\partial x^\mu} - \frac{\partial A_\mu(x)}{\partial x^\nu} \varepsilon^\nu(x)$$

$$\delta g_{\mu\nu}(x) = -g_{\mu\lambda}(x) \frac{\partial \varepsilon^\lambda(x)}{\partial x^\nu} - g_{\lambda\nu}(x) \frac{\partial \varepsilon^\lambda(x)}{\partial x^\mu} - \frac{\partial g_{\mu\nu}(x)}{\partial x^\lambda} \varepsilon^\lambda(x) \qquad (12.3.1)$$

(This change in A or g is just the Lie derivative; see Section 10.9.) The important point is that this is now an infinitesimal transformation of the dynamical variables alone, not of the coordinates over which we integrate, so the principle of stationary action tells us that when the dynamical equations for $x_n{}^\mu$, A_μ, and so on, are satisfied the change in these quantities produces no change in the matter action I_M. The only change in I_M comes from the variation in the external field $g_{\mu\nu}$, and (12.2.2) gives for this change

$$\delta I_M = -\tfrac{1}{2} \int d^4x \sqrt{g}\; T^{\mu\nu} \left[g_{\mu\lambda} \frac{\partial \varepsilon^\lambda}{\partial x^\nu} + g_{\lambda\nu} \frac{\partial \varepsilon^\lambda}{\partial x^\mu} + \frac{\partial g_{\mu\nu}}{\partial x^\lambda} \varepsilon^\lambda \right]$$

If I_M is a scalar, then this must vanish; integrating by parts gives then

$$0 = \delta I_M = \int d^4x\, \varepsilon^\lambda \left[\frac{\partial}{\partial x^\nu} (\sqrt{g}\; T^\nu{}_\lambda) - \frac{1}{2} \left(\frac{\partial g_{\mu\nu}}{\partial x^\lambda} \right) \sqrt{g}\; T^{\mu\nu} \right]$$

and since $\varepsilon^\lambda(x)$ is arbitrary

$$0 = \frac{\partial}{\partial x^\nu} (\sqrt{g}\; T^\nu{}_\lambda) - \frac{1}{2} \left(\frac{\partial g_{\mu\nu}}{\partial x^\lambda} \right) \sqrt{g}\; T^{\mu\nu}$$

or, recalling (4.7.6):

$$0 = (T^\nu{}_\lambda)_{;\nu} \qquad\qquad (12.3.2)$$

Thus the energy-momentum tensor defined by Eq. (12.2.2) is conserved (in the covariant sense) if and only if the matter action is a scalar. Also, with I_M a scalar, (12.2.2) shows immediately that $T^{\mu\nu}$ is a symmetric *tensor*, so this definition of the energy-momentum tensor has all the properties for which one could wish.

This proof, that general covariance implies energy-momentum conservation, has an exact analog in the proof that gauge invariance implies charge conservation. The change in the action I'_M defined by (12.2.3) caused by an arbitrary gauge transformation can arise only from the change in A_μ, since I'_M is stationary with respect to all other dynamical variables. A general infinitesimal gauge transformation ε will produce in A_μ the change (see Sections 4.11 and 10.2)

$$\delta A_\mu = \frac{\partial \varepsilon}{\partial x^\mu}$$

Using this in (12.2.6), we see that I'_M is gauge invariant if and only if

$$0 = \delta I'_M = \int d^4x \sqrt{g}\; J^\mu \frac{\partial \varepsilon}{\partial x^\mu}$$

Integrating by parts gives

$$0 = \int d^4x\, \varepsilon\, \frac{\partial}{\partial x^\mu} (\sqrt{g}\; J^\mu)$$

or, since ε is arbitrary

$$0 = \frac{1}{\sqrt{g}} \frac{\partial}{\partial x^\mu} \sqrt{g}\; J^\mu = J^\mu{}_{;\mu}$$

We see again how closely analogous are gauge invariance and general covariance.

4   The Gravitational Action

So far, in this chapter the gravitational field $g_{\mu\nu}$ has been an external field that could be prescribed at will. (Indeed, (12.2.2) usually provides the most convenient definition of the energy-momentum tensor even in the absence of gravitation.) We will now give $g_{\mu\nu}$ field equations of its own, by adding to the total action I a purely gravitational term I_G

$$I = I_M + I_G \tag{12.4.1}$$

$$I_G \equiv -\frac{1}{16\pi G} \int \sqrt{g(x)}\; R(x)\; d^4x \tag{12.4.2}$$

Clearly I_G is a scalar, so this would be a good candidate for a theory of gravitation even if we had no experience with gravitational phenomena. We shall now show that the application to I of the principle of stationary action does in fact yield Einstein's theory.

The curvature scalar R is defined as $g^{\mu\nu}R_{\mu\nu}$, so a variation $\delta g_{\mu\nu}$ in the metric produces in the integrand of (12.4.2) a change

$$\delta(\sqrt{g}\,R) = \sqrt{g}\,R_{\mu\nu}\delta g^{\mu\nu} + R\delta\sqrt{g} + \sqrt{g}\,g^{\mu\nu}\delta R_{\mu\nu}$$

According to Eq. (10.9.2), the change in the Ricci tensor is

$$\delta R_{\mu\nu} = (\delta\Gamma^\lambda_{\mu\lambda})_{;\nu} - (\delta\Gamma^\lambda_{\mu\nu})_{;\lambda}$$

the covariant derivatives being defined as if $\delta\Gamma^\lambda_{\mu\nu}$ were a tensor (as, in fact, it is). Thus the last term in $\delta(\sqrt{g}\,R)$ is

$$\sqrt{g}\,g^{\mu\nu}\delta R_{\mu\nu} = \sqrt{g}[(g^{\mu\nu}\delta\Gamma^\lambda_{\mu\lambda})_{;\nu} - (g^{\mu\nu}\delta\Gamma^\lambda_{\mu\nu})_{;\lambda}]$$

or, using (4.7.7),

$$\sqrt{g}\,g^{\mu\nu}\delta R_{\mu\nu} = \frac{\partial}{\partial x^\nu}(\sqrt{g}\,g^{\mu\nu}\delta\Gamma^\lambda_{\mu\lambda}) - \frac{\partial}{\partial x^\lambda}(\sqrt{g}\,g^{\mu\nu}\delta\Gamma^\lambda_{\mu\nu})$$

This term therefore drops out when we integrate over all space. Also,

$$\delta\sqrt{g} = \tfrac{1}{2}\sqrt{g}\,g^{\mu\nu}\delta g_{\mu\nu} \qquad \delta g^{\mu\nu} = -g^{\mu\rho}g^{\nu\sigma}\delta g_{\rho\sigma}$$

so the change in the action (12.4.2) is

$$\delta I_G = \frac{1}{16\pi G}\int \sqrt{g}\,[R^{\mu\nu} - \tfrac{1}{2}g^{\mu\nu}R]\delta g_{\mu\nu}\,d^4x \tag{12.4.3}$$

Combining (12.4.3) with (12.2.2), we see that the total action I is stationary with respect to arbitrary variations in $g_{\mu\nu}$ if and only if

$$R^{\mu\nu} - \tfrac{1}{2}g^{\mu\nu}R + 8\pi G T^{\mu\nu} = 0$$

which, of course, is the Einstein field equation.

[As another application of (12.4.3), we may use it to derive the contracted Bianchi identity. Since I_G is a scalar it must be stationary with respect to the variation (12.3.1) in $g_{\mu\nu}$. Repeating the reasoning that led before to Eq. (12.3.2), we now find that

$$[R^{\nu}{}_{\lambda} - \tfrac{1}{2}\delta^{\nu}{}_{\lambda}R]_{;\nu} = 0$$

which we recognize as the contracted Bianchi identity (6.8.3).]

This formalism suggests that Einstein's theory might be modified by adding to R in Eq. (12.4.2) terms proportional to R^2, R^3, etc. As discussed in Section 7.1, such terms would only show up on a sufficiently small space-time scale.

5 The Tetrad Formalism*

Until now, we have followed only one approach in determining the effects of gravitation on general physical systems. Given the special-relativistic equations that govern the system in the absence of gravitation, we replace all Lorentz tensors $T^{\alpha\cdots}_{\beta\cdots}$ with objects $T^{\mu\cdots}_{\nu\cdots}$ that behave like tensors (or tensor densities) under general coordinate transformations. Also, we replace all derivatives $\partial/\partial x^{\alpha}$ with covariant derivatives, and replace $\eta_{\alpha\beta}$ everywhere with $g_{\mu\nu}$. The equations of motion are then generally covariant. (See Chapter 5.)

This method actually works *only* for objects that behave like tensors under Lorentz transformation, and not for the spinor fields discussed in Section 2.12. (Mathematically, this is because the tensor representations of the group GL(4) of general linear 4×4 matrices behave like tensors under the subgroup of Lorentz transformations, but there are no representations of GL(4), or even "representations up to a sign," which behave like spinors under the Lorentz subgroup.) How then can we incorporate spinors into general relativity?

The answer lies in a different approach to the problem of determining the effects of gravitation on physical systems, an approach that is rather interesting in its own right, quite apart from the problem of dealing with spinors.

To start, let us take advantage of the Principle of Equivalence, and at every point X erect a set of coordinates $\xi_X{}^{\alpha}$ that are locally inertial at X. (Of course, it will not be possible to erect any *single* coordinate system that is locally inertial everywhere, unless the space-time continuum is "flat.") As shown in Sections 3.2 and 3.3, the metric in any general noninertial coordinate system is then

$$g_{\mu\nu}(x) = V^{\alpha}{}_{\mu}(x)V^{\beta}{}_{\nu}(x)\eta_{\alpha\beta} \tag{12.5.1}$$

where

$$V^{\alpha}{}_{\mu}(X) \equiv \left(\frac{\partial\xi_X{}^{\alpha}(x)}{\partial x^{\mu}}\right)_{x=X} \tag{12.5.2}$$

---

* This section lies somewhat out of the book's main line of development, and may be omitted in a first reading.

Note that we fix the locally inertial coordinates ξ_X^α once and for all at each physical point X, so when we change our general noninertial coordinates from x^μ to x'^μ, the partial derivatives $V^\alpha{}_\mu$ change according to the rule

$$V^\alpha{}_\mu \rightarrow V'^\alpha{}_\mu = \frac{\partial x^\nu}{\partial x'^\mu} V^\alpha{}_\nu \tag{12.5.3}$$

Thus, we are to think of $V^\alpha{}_\mu$ as forming *four* covariant vector fields, *not* as a single tensor: This set of four vectors is known as a *tetrad*, or *vierbein*.

Now, given any contravariant vector field $A^\mu(x)$, we can use the tetrad to refer its components at x to the coordinate system $\xi_x{}^\alpha$ locally inertial at x:

$$*A^\alpha \equiv V^\alpha{}_\mu A^\mu \tag{12.5.4}$$

We are contracting a contravariant vector A^μ with four covariant vectors $V^\alpha{}_\mu$, so this has the effect of replacing the single four-vector A^μ with the *four* scalars $*A^\alpha$. We can do the same with covariant vector fields, and indeed with general tensor fields:

$$\begin{aligned} *A_\alpha &\equiv V_\alpha{}^\mu A_\mu \\ *B^\alpha{}_\beta &\equiv V^\alpha{}_\mu V_\beta{}^\nu B^\mu{}_\nu, \text{ etc.} \end{aligned} \tag{12.5.5}$$

Here $V_\beta{}^\nu$ is just the tetrad (12.5.2), but with α-index lowered with the Minkowski tensor and μ-index raised with the metric tensor:

$$V_\beta{}^\nu \equiv \eta_{\alpha\beta} g^{\mu\nu} V^\alpha{}_\mu \tag{12.5.6}$$

Note that according to Eq. (12.5.1) this is just the inverse of the tetrad

$$\delta^\mu{}_\nu = V_\beta{}^\mu V^\beta{}_\nu \tag{12.5.7}$$

and hence also

$$\delta^\alpha{}_\beta = V^\alpha{}_\mu V_\beta{}^\mu \tag{12.5.8}$$

The scalar components of the metric tensor are then simply

$$*g_{\alpha\beta} \equiv V_\alpha{}^\mu V_\beta{}^\nu g_{\mu\nu} = \eta_{\alpha\beta} \tag{12.5.9}$$

Now that we have shown how to make any tensor field into a set of scalars, we can forget the original tensors V^μ, $T_{\mu\nu}$, and so on, with which we started, and consider how we would construct an action if we worked from the beginning with the scalars $*V^\alpha$, $*T_{\alpha\beta}$, and so on. In this way, a spinor field, like Dirac's electron field, can be brought into our formalism in precisely the same way as any other field, and its peculiar Lorentz transformation properties cause no particular trouble. There are now *two* invariance principles which must be met in constructing a suitable matter action I_M:

(A) The action must be generally covariant, with all fields treated as scalars, except for the tetrad itself.

(B) The Principle of Equivalence requires that special relativity should apply in locally inertial frames, and in particular, that it should make no difference which locally inertial frame we choose at each point. Thus since our scalar field components $*V^\alpha$, $*T_{\alpha\beta}$, and so on, are defined with respect to an arbitrarily chosen locally inertial coordinate system, the field equations and the action must be invariant with respect to a redefinition of these locally inertial coordinate systems at each point, or in other words, with respect to Lorentz transformations $\Lambda^\alpha{}_\beta(x)$ that can depend on position in space-time:

$$*A^\alpha(x) \to \Lambda^\alpha{}_\beta(x) *A^\beta(x)$$

$$*T_{\alpha\beta}(x) \to \Lambda_\alpha{}^\gamma(x) \Lambda_\beta{}^\delta(x) *T_{\gamma\delta}(x), \text{ etc.}$$

where

$$\eta_{\alpha\beta} \Lambda^\alpha{}_\gamma(x) \Lambda^\beta{}_\delta(x) = \eta_{\gamma\delta} \tag{12.5.10}$$

The tetrad (12.5.2) changes according to the same rule as $*A^\alpha$:

$$V^\alpha{}_\mu(x) \to \Lambda^\alpha{}_\beta(x) V^\beta{}_\mu(x) \tag{12.5.11}$$

and in general an arbitrary field $*\psi_n(x)$ will change according to the rule:

$$*\psi_n(x) \to \sum_m [D(\Lambda(x))]_{nm} *\psi_m(x) \tag{12.5.12}$$

where $D(\Lambda)$ is a matrix representation of the Lorentz group (or at least of the infinitesimal Lorentz group), of the sort discussed in Section 2.12.

These two invariance principles lead to a dual classification of physical quantities. A *coordinate scalar* or *coordinate tensor* transforms as a scalar or a tensor under changes in the coordinate system. A *Lorentz scalar* or *Lorentz tensor* or *Lorentz spinor* transforms according to a rule like (12.5.12), with $D(\Lambda)$ the identity or a tensor representation or a spinor representation of the infinitesimal Lorentz group, under changes in the choice of the locally inertial coordinate frame. For instance, a field such as (12.5.4) is a coordinate scalar and a Lorentz vector, the Dirac field of the electron is a coordinate scalar and a Lorentz spinor, and the tetrad $V^\alpha{}_\mu$ is a coordinate vector and a Lorentz vector. To be physically acceptable, the matter action I_M must be both a coordinate scalar and a Lorentz scalar.

At this point, the reader may be becoming uneasy. How is the gravitational field going to get into this sort of theory, when the coordinate-scalar components (12.5.9) of the metric tensor are just the constants $\eta_{\alpha\beta}$? The answer is that gravitational tensor fields appear in the action because, and only because, of the necessity of introducing *derivatives* into the theory. If it made sense to construct an action I_M solely from fields, and not their derivatives, then it would only be necessary to choose some arbitrary Lorentz-invariant function $\mathscr{L}(*\psi(x))$ of various fields $*\psi_n(x)$ (but not the tetrad), call them all coordinate scalars, and take the action as

$$I_M = \int d^4x \sqrt{g(x)} \, \mathscr{L}(*\psi(x))$$

This would then automatically be a coordinate scalar and a Lorentz scalar. However, the examples discussed in the previous sections of this chapter show that any physically sensible action must involve derivatives of physical quantities as well as the quantities themselves. The tetrad field must enter into the action in such a way as to keep it a coordinate scalar and a Lorentz scalar despite the presence of derivatives.

An ordinary derivative is of course a coordinate vector, in the sense that under a coordinate transformation $x \rightarrow x'$, it transforms according to the rule

$$\frac{\partial}{\partial x^\mu} \rightarrow \frac{\partial}{\partial x'^\mu} = \frac{\partial x^\nu}{\partial x'^\mu} \frac{\partial}{\partial x^\nu}$$

If all the fields appearing in the action were coordinate scalars, there would be no contravariant indices with which to contract the covariant index μ; hence, in order to make the action a coordinate scalar, it is necessary to introduce the tetrad field, and incorporate derivatives into the action in the form

$$V_\alpha{}^\mu \frac{\partial}{\partial x^\mu} \tag{12.5.13}$$

However, although this is a coordinate scalar, it does not have simple transformation properties under position-dependent Lorentz transformations. Acting on a general field $*\psi$ with the Lorentz transformation rule (12.5.12), the coordinate-scalar derivative has the transformation rule

$$V_\alpha{}^\mu(x) \frac{\partial}{\partial x^\mu} *\psi(x) \rightarrow \Lambda_\alpha{}^\beta(x) V_\beta{}^\mu(x) \frac{\partial}{\partial x^\mu} \{D(\Lambda(x))*\psi(x)\}$$

$$= \Lambda_\alpha{}^\beta(x) V_\beta{}^\mu(x) \left\{ D(\Lambda(x)) \frac{\partial}{\partial x^\mu} *\psi(x) + \left[\frac{\partial}{\partial x^\mu} D(\Lambda(x)) \right] *\psi(x) \right\}$$

$$\tag{12.5.14}$$

However, what we need is to incorporate derivatives into the action in the form of an operator $\mathscr{D}_\alpha$ that is not only a coordinate scalar, but also, unlike (12.5.13), is a Lorentz vector, in the sense that for a position-dependent Lorentz transformation $\Lambda^\alpha{}_\beta(x)$,

$$\mathscr{D}_\alpha *\psi(x) \rightarrow \Lambda_\alpha{}^\beta(x) D(\Lambda(x)) \mathscr{D}_\beta *\psi(x) \tag{12.5.15}$$

Any action, which depends only on various fields $*\psi$ and on their "derivatives" $\mathscr{D}_\alpha *\psi$, will then automatically be independent of the choice of locally inertial frames if it is invariant under ordinary constant Lorentz transformations. Inspection of Eq. (12.5.14) shows that we can construct a coordinate-scalar Lorentz vector derivative[1] $\mathscr{D}_\alpha$ of the form

$$\mathscr{D}_\alpha \equiv V_\alpha{}^\mu \left[\frac{\partial}{\partial x^\mu} + \Gamma_\mu \right] \tag{12.5.16}$$

where Γ_μ is a matrix with the Lorentz transformation rule

$$\Gamma_\mu(x) \to D(\Lambda(x))\Gamma_\mu(x)D^{-1}(\Lambda(x))$$

$$- \left[\frac{\partial}{\partial x^\mu} D(\Lambda(x))\right] D^{-1}(\Lambda(x)) \tag{12.5.17}$$

The inhomogeneous term in (12.5.17) will then cancel the second term in (12.5.14), giving $\mathscr{D}_\alpha$ the desired transformation property (12.5.15).

In order to determine the structure of the matrix $\Gamma_\mu(x)$, it will be sufficient to consider Lorentz transformations that are infinitesimally close to the identity. Such transformations must be of the form (2.12.5), (2.12.6):

$$\Lambda^\alpha{}_\beta(x) = \delta^\alpha{}_\beta + \omega^\alpha{}_\beta(x) \tag{12.5.18}$$

with

$$\omega_{\alpha\beta}(x) = -\omega_{\beta\alpha}(x) \tag{12.5.19}$$

In this case, the matrix D has the form (2.12.7):

$$D(1 + \omega(x)) = 1 + \tfrac{1}{2}\omega^{\alpha\beta}(x)\sigma_{\alpha\beta} \tag{12.5.20}$$

where $\sigma_{\alpha\beta}$ are a set of constant matrices that are antisymmetric in α and β

$$\sigma_{\alpha\beta} = -\sigma_{\beta\alpha} \tag{12.5.21}$$

and that satisfy the commutation relations (2.12.12):

$$[\sigma_{\alpha\beta}, \sigma_{\gamma\delta}] = \eta_{\gamma\beta}\sigma_{\alpha\delta} - \eta_{\gamma\alpha}\sigma_{\beta\delta} + \eta_{\delta\beta}\sigma_{\gamma\alpha} - \eta_{\delta\alpha}\sigma_{\gamma\beta} \tag{12.5.22}$$

The condition (12.5.17) tells us that under the infinitesimal Lorentz transformation (12.5.18), the matrix $\Gamma_\mu(x)$ transforms according to

$$\Gamma_\mu(x) \to \Gamma_\mu(x) + \tfrac{1}{2}\omega^{\alpha\beta}(x)[\sigma_{\alpha\beta}, \Gamma_\mu(x)]$$

$$- \tfrac{1}{2}\sigma_{\alpha\beta}\frac{\partial}{\partial x^\mu}\omega^{\alpha\beta}(x) \tag{12.5.23}$$

Note that $V^\alpha{}_\mu(x)$ transforms into

$$V^\alpha{}_\nu(x) \to V^\alpha{}_\nu(x) + \omega^\alpha{}_\beta(x)V^\beta{}_\nu(x)$$

and therefore, using (12.5.8)

$$V_\beta{}^\nu(x)\frac{\partial}{\partial x^\mu}V_{\alpha\nu}(x) \to V_\beta{}^\nu(x)\frac{\partial}{\partial x^\mu}V_{\alpha\nu}(x)$$

$$+ \omega_\beta{}^\gamma(x)V_\gamma{}^\nu(x)\frac{\partial}{\partial x^\mu}V_{\alpha\nu}(x) + \omega_\alpha{}^\gamma(x)V_\beta{}^\nu(x)\frac{\partial}{\partial x^\mu}V_{\gamma\nu}(x)$$

$$+ \frac{\partial}{\partial x^\mu}\omega_{\alpha\beta}(x)$$

Multiplying this with $\sigma^{\alpha\beta}$ and using the commutation rules (12.5.22), we find that the transformation condition (12.5.23) is satisfied by the matrix

$$\Gamma_\mu(x) = \tfrac{1}{2}\sigma^{\alpha\beta} V_\alpha{}^\nu(x)\, V_{\beta\nu;\mu} \tag{12.5.24}$$

To summarize: The effects of gravitation on any physical system can be taken into account by writing down the matter action or the field equations that hold in special relativity, and then replacing all derivatives $\partial/\partial x^\alpha$ with the "covariant" derivatives

$$\mathscr{D}_\alpha \equiv V_\alpha{}^\mu \frac{\partial}{\partial x^\mu} + \tfrac{1}{2}\sigma^{\beta\gamma} V_\beta{}^\nu V_\alpha{}^\mu\, V_{\gamma\nu;\mu} \tag{12.5.25}$$

This prescription yields a matter action or field equations that are invariant under general coordinate transformations, with $V_\alpha{}^\mu$ regarded as a four-vector and with all other fields regarded as scalars, and that also do not depend on how we choose locally inertial frames in defining the tetrad.

How are we to define the energy-momentum tensor in this formalism? The variation $\delta V_\alpha{}^\mu$ in the tetrad field will produce in the matter action a change

$$\delta I_M = \int d^4x \sqrt{g}\; U^\alpha{}_\mu \delta V_\alpha{}^\mu \tag{12.5.26}$$

where $U^\alpha{}_\mu$ is a coordinate vector and a Lorentz vector. Let us tentatively define the energy-momentum tensor by

$$T_{\mu\nu} \equiv V_{\alpha\mu} U^\alpha{}_\nu \tag{12.5.27}$$

As required, this is manifestly a coordinate tensor and a Lorentz scalar. To verify that (12.5.27) is a suitable energy-momentum tensor, we must also check that it is symmetric

$$T_{\mu\nu} = T_{\nu\mu} \tag{12.5.28}$$

and that it is conserved

$$(T^\nu{}_\lambda)_{;\nu} = 0 \tag{12.5.29}$$

The symmetry property of the energy-momentum tensor is not at all obvious in the tetrad formalism, but must be derived from the invariance of the matter action under the infinitesimal Lorentz transformations

$$\Lambda^\alpha{}_\beta(x) = \delta^\alpha{}_\beta + \omega^\alpha{}_\beta(x)$$

with

$$|\omega^\alpha{}_\beta(x)| \ll 1$$

These transformations will produce small changes in all the dynamic variables, but the matter action I_M is supposed to be stationary with respect to variations in each of these variables except the tetrad, which enters in I_M as an external field.

Hence we only need to take account of the change (12.5.11) in the tetrad field

$$\delta V_\alpha{}^\mu(x) = \omega_\alpha{}^\beta(x) V_\beta{}^\mu(x) \tag{12.5.30}$$

Using (12.5.30) in (12.5.26), we find that the invariance of the matter action under position-dependent Lorentz transformations requires that

$$0 = \int d^4x \sqrt{g(x)}\, U^\alpha{}_\mu(x) V^{\beta\mu}(x) \omega_{\alpha\beta}(x)$$

But $\omega_{\alpha\beta}(x)$ is arbitrary, except for the antisymmetry condition (12.5.19), so the coefficient of $\omega_{\alpha\beta}(x)$ must be symmetric:

$$U^\alpha{}_\mu V^{\beta\mu} = U^\beta{}_\mu V^{\alpha\mu}$$

Multiplying this equation with $V_{\beta\nu} V_{\alpha\lambda}$, and using (12.5.7), we find that

$$U^\alpha{}_\nu V_{\alpha\lambda} = U^\beta{}_\lambda V_{\beta\nu} \tag{12.5.31}$$

which is the same as the expected symmetry condition (12.5.28).

To show that the energy-momentum tensor defined by Eq. (12.5.27) is conserved, we must use the invariance of the matter action under infinitesimal coordinate transformations:

$$x'^\mu = x^\mu + \varepsilon^\mu(x)$$

with $|\varepsilon^\mu|$ very small. Such transformations alter the tetrad field by an infinitesimal amount

$$\delta V_\alpha{}^\mu(x) \equiv V'_\alpha{}^\mu(x) - V_\alpha{}^\mu(x)$$

$$= V_\alpha{}^\nu(x) \frac{\partial \varepsilon^\mu(x)}{\partial x^\nu} - \frac{\partial V_\alpha{}^\mu(x)}{\partial x^\lambda} \varepsilon^\lambda(x) \tag{12.5.32}$$

[Compare Eq. (12.3.1).] Also, all other coordinate-scalar fields $\psi(x)$ change by the amounts

$$\delta\psi(x) \equiv \psi'(x) - \psi(x) = -\frac{\partial \psi(x)}{\partial x^\lambda} \varepsilon^\lambda(x)$$

but again, the matter action I_M is stationary with respect to variations in these fields, so it is only the variation in the tetrad field that matters here. Using (12.5.32) in (12.5.26), we find that the invariance of the matter action I_M under general coordinate transformations requires that

$$0 = \int d^4x \sqrt{g}\, U^\alpha{}_\mu \left\{ V_\alpha{}^\nu \frac{\partial \varepsilon^\mu}{\partial x^\nu} - \frac{\partial V_\alpha{}^\mu}{\partial x^\lambda} \varepsilon^\lambda \right\}$$

But $\varepsilon^\lambda(x)$ is arbitrary, so after integrating by parts we can set the coefficient of $\varepsilon^\lambda(x)$ equal to zero, and find

$$0 = \frac{\partial}{\partial x^\nu} \left(\sqrt{g}\, U^\alpha{}_\lambda V_\alpha{}^\nu \right) + \sqrt{g}\, U^\alpha{}_\mu \frac{\partial V_\alpha{}^\mu}{\partial x^\lambda}$$

Using Eqs. (12.5.27) and (12.5.8) let us write this as

$$0 = \frac{\partial}{\partial x^\nu} (\sqrt{g}\, T^\nu{}_\lambda) + \sqrt{g}\, T_{\nu\mu} V^{\alpha\nu} \frac{\partial V_\alpha{}^\mu}{\partial x^\lambda} \qquad (12.5.33)$$

According to (12.5.1), the metric tensor is related to the tetrad by

$$g^{\mu\nu} = V_\alpha{}^\mu V^{\alpha\nu}$$

and therefore

$$\frac{\partial g^{\mu\nu}}{\partial x^\lambda} = V^{\alpha\nu} \frac{\partial V_\alpha{}^\mu}{\partial x^\lambda} + V^{\alpha\mu} \frac{\partial V_\alpha{}^\nu}{\partial x^\lambda}$$

Since $T_{\mu\nu}$ is symmetric, Eq. (12.5.33) may now be written

$$0 = \frac{\partial}{\partial x^\nu} (\sqrt{g}\, T^\nu{}_\lambda) + \tfrac{1}{2} \sqrt{g}\, T_{\mu\nu} \frac{\partial g^{\mu\nu}}{\partial x^\lambda} \qquad (12.5.34)$$

But Eqs. (3.3.1) and (4.7.6) give

$$\frac{\partial g^{\mu\nu}}{\partial x^\lambda} = -g^{\rho\mu} g^{\sigma\nu} \frac{\partial g_{\rho\sigma}}{\partial x^\lambda} = -g^{\sigma\nu} \Gamma^\mu_{\sigma\lambda} - g^{\rho\mu} \Gamma^\nu_{\rho\lambda}$$

$$\frac{\partial}{\partial x^\nu} \ln \sqrt{g} = \Gamma^\lambda_{\nu\lambda}$$

so (12.5.34) is the same as the usual conservation law (12.5.29).

Our definition (12.5.27) of the energy-momentum tensor is thus completely satisfactory. Note, however, that if the matter action were not invariant under position-dependent Lorentz transformations, then not only would $T_{\mu\nu}$ not be symmetric, but in consequence, it also would not be conserved.

The total action for matter *and* gravitation is again of the form

$$I = I_M + I_G$$

with I_G of the form (12.4.2). A variation in the tetrad will produce in the metric a change given by (12.5.1) as

$$\delta g_{\mu\nu} = V^\alpha{}_\mu \delta V_{\alpha\nu} + V^\alpha{}_\nu \delta V_{\alpha\mu}$$
$$= -[g_{\mu\lambda} V^\alpha{}_\nu + g_{\nu\lambda} V^\alpha{}_\mu] \delta V_\alpha{}^\lambda$$

so (12.4.3) gives the change in the gravitational action as

$$\delta I_G = -\frac{1}{8\pi G} \int \sqrt{g}\, [R^\mu{}_\lambda - \tfrac{1}{2} \delta^\mu{}_\lambda R] V^\alpha{}_\mu \delta V_\alpha{}^\lambda \, d^4x \qquad (12.5.35)$$

The total action must be stationary with respect to variations in the tetrad, so (12.5.26) and (12.5.35) yield the field equation

$$(R^\mu{}_\lambda - \tfrac{1}{2} \delta^\mu{}_\lambda R) V^\alpha{}_\mu = -8\pi G U^\alpha{}_\lambda$$

Contracting this with $V_{\alpha \nu}$ and using (12.5.1) and (12.5.27) then yields the familiar Einstein field equation

$$R_{\nu \lambda} - \tfrac{1}{2}g_{\nu \lambda}R = -8\pi G T_{\nu \lambda} \tag{12.5.36}$$

These equations serve to determine only $g_{\mu \nu}$, leaving the tetrad determined only up to a Lorentz transformation (12.5.11). However, the invariance of the matter action under such position-dependent Lorentz transformations ensures that all tetrads associated with a given metric have the same physical effects.

I2 REFERENCES

1. R. Utiyama, Phys. Rev., **101**, 1597 (1956); T. W. B. Kibble, J. Math. Phys., **2**, 212 (1961).

> "Symmetry, as wide or as narrow as you may define it, is one idea by which man through the ages has tried to comprehend and create order, beauty, and perfection." *Hermann Weyl, Symmetry*

13 SYMMETRIC SPACES

Euclid implicitly assumed that metric relations are unaffected by translations or rotations. Real gravitational fields do not usually have such a high degree of symmetry, but they often admit some group of approximate symmetry transformations, and when they do, we can use this information to help solve the Einstein equations, or even to do without a solution. I shall give only a very brief introduction to the elaborate mathematical theory of symmetric spaces, with special attention to the maximally symmetric spaces that are of special interest in cosmology.

The initial difficulty here is: How can we use some supposed symmetry of a metric space to gain information about the metric, when we need to know the metric before we can establish a coordinate system in which to define the symmetry? In order to avoid this impasse, we shall have to learn ways of describing symmetries in a covariant language, which does not depend on any particular choice of coordinate system. Once this language is established, it becomes a matter of mathematical manipulation to determine those properties of a metric that follow from its symmetries.

1 Killing Vectors

A metric $g_{\mu\nu}(x)$ is said to be *form-invariant* under a given coordinate transformation $x \to x'$, when the transformed metric $g'_{\mu\nu}(x')$ is the same function of its argument x'^{μ} as the original metric $g_{\mu\nu}(x)$ was of *its* argument x^{μ}, that is,

$$g'_{\mu\nu}(y) = g_{\mu\nu}(y) \qquad \text{for all } y \tag{13.1.1}$$

[This is different from the condition for a scalar, which is that $S'(x') = S(x)$.]
At any given *point* the transformed metric is given by the relation

$$g'_{\mu\nu}(x') = \frac{\partial x^\rho}{\partial x'^\mu} \frac{\partial x^\sigma}{\partial x'^\nu} g_{\rho\sigma}(x)$$

or equivalently

$$g_{\mu\nu}(x) = \frac{\partial x'^\rho}{\partial x^\mu} \frac{\partial x'^\sigma}{\partial x^\nu} g'_{\rho\sigma}(x')$$

When (13.1.1) is valid, we can replace $g'_{\rho\sigma}(x')$ with $g_{\rho\sigma}(x')$ and so obtain the fundamental requirement for a form invariance of the metric:

$$g_{\mu\nu}(x) = \frac{\partial x'^\rho}{\partial x^\mu} \frac{\partial x'^\sigma}{\partial x^\nu} g_{\rho\sigma}(x') \tag{13.1.2}$$

Any transformation $x \to x'$ that satisfies (13.1.2) is called an *isometry*.

In general, Eq. (13.1.2) is a very complicated restriction on the function $x'^\mu(x)$. It can be greatly simplified by descending to the special case of an infinitesimal coordinate transformation:

$$x'^\mu = x^\mu + \varepsilon \xi^\mu(x) \qquad \text{with } |\varepsilon| \ll 1 \tag{13.1.3}$$

To first order in ε, Eq. (13.1.2) now reads

$$0 = \frac{\partial \xi^\mu(x)}{\partial x^\rho} g_{\mu\sigma}(x) + \frac{\partial \xi^\nu(x)}{\partial x^\sigma} g_{\rho\nu}(x) + \xi^\mu(x) \frac{\partial g_{\rho\sigma}(x)}{\partial x^\mu} \tag{13.1.4}$$

This can be rewritten in terms of derivatives of the covariant components $\xi_\sigma \equiv g_{\mu\sigma}\xi^\mu$:

$$0 = \frac{\partial \xi_\sigma}{\partial x^\rho} + \frac{\partial \xi_\rho}{\partial x^\sigma} + \xi^\mu \left[\frac{\partial g_{\rho\sigma}}{\partial x^\mu} - \frac{\partial g_{\mu\sigma}}{\partial x^\rho} - \frac{\partial g_{\rho\mu}}{\partial x^\sigma} \right]$$

$$= \frac{\partial \xi_\sigma}{\partial x^\rho} + \frac{\partial \xi_\rho}{\partial x^\sigma} - 2\xi_\mu \Gamma^\mu_{\rho\sigma}$$

or, more compactly,

$$0 = \xi_{\sigma;\rho} + \xi_{\rho;\sigma} \tag{13.1.5}$$

Any four-vector field $\xi_\sigma(x)$ that satisfies Eq. (13.1.5) will be said to form a *Killing vector*[1] of the metric $g_{\mu\nu}(x)$. The problem of determining all infinitesimal isometries of a given metric is now reduced to the problem of determining all Killing vectors of the metric. Any linear combination of Killing vectors (with constant coefficients) is a Killing vector, so it is the space of vector fields spanned by the Killing vectors that really determines the infinitesimal isometries of a metric.

The Killing condition (13.1.5) is much more restrictive than it looks, for it allows us to determine the whole function $\xi_\mu(x)$ from given values of ξ_σ and $\xi_{\sigma;\rho}$

at some point X. To see this, we need only recall our formula (6.5.1) for the commutator of two covariant derivatives,

$$\xi_{\sigma;\rho;\mu} - \xi_{\sigma;\mu;\rho} = -R^{\lambda}_{\sigma\rho\mu}\xi_{\lambda} \tag{13.1.6}$$

and the cyclic sum rule (6.6.5) for the curvature tensor,

$$R^{\lambda}_{\sigma\rho\mu} + R^{\lambda}_{\mu\sigma\rho} + R^{\lambda}_{\rho\mu\sigma} = 0 \tag{13.1.7}$$

By adding (13.1.6) and its two cyclic permutations, we find that any vector ξ_{μ} must satisfy the relation

$$0 = \xi_{\sigma;\rho;\mu} - \xi_{\sigma;\mu;\rho} + \xi_{\mu;\sigma;\rho} - \xi_{\mu;\rho;\sigma} + \xi_{\rho;\mu;\sigma} - \xi_{\rho;\sigma;\mu} \tag{13.1.8}$$

For a Killing vector, (13.1.5) and (13.1.8) give

$$0 = \xi_{\sigma;\rho;\mu} - \xi_{\sigma;\mu;\rho} - \xi_{\mu;\rho;\sigma}$$

and thus (13.1.6) becomes

$$\xi_{\mu;\rho;\sigma} = -R^{\lambda}_{\sigma\rho\mu}\xi_{\lambda} \tag{13.1.9}$$

Hence, given ξ_{λ} and $\xi_{\lambda;\nu}$ at some point X, we can determine the second derivatives of $\xi_{\lambda}(x)$ at X from Eq. (13.1.9), and we can find successively higher derivatives of ξ_{λ} at X by taking derivatives of Eq. (13.1.9). All the derivatives of ξ_{λ} at X will thus be determined as linear combinations of $\xi_{\lambda}(X)$ and $\xi_{\lambda;\nu}(X)$. The function $\xi_{\lambda}(x)$ can then (when it exists) be constructed as a Taylor series in $x^{\lambda} - X^{\lambda}$ within some finite neighborhood of X, and will again be linear in the initial values $\xi_{\lambda}(X)$, $\xi_{\lambda;\nu}(X)$. Thus any particular Killing vector $\xi_{\rho}{}^{n}(x)$ of the metric $g_{\mu\nu}(x)$ can be expressed as

$$\xi_{\rho}{}^{n}(x) = A_{\rho}{}^{\lambda}(x;X)\xi_{\lambda}{}^{n}(X) + B_{\rho}{}^{\lambda\nu}(x;X)\xi_{\lambda;\nu}{}^{n}(X) \tag{13.1.10}$$

where $A_{\rho}{}^{\lambda}$ and $B_{\rho}{}^{\lambda\nu}$ are functions that of course depend on the metric and on X, but do not depend on the initial values $\xi_{\lambda}(X)$ and $\xi_{\lambda;\nu}(X)$, and hence are the same for all Killing vectors. Each Killing vector $\xi_{\rho}(x)$ of a given metric is uniquely specified by the values of $\xi_{\rho}(X)$ and $\xi_{\rho;\sigma}(X)$ at any particular point X.

A set of Killing vectors $\xi_{\rho}{}^{n}(x)$ is said to be *independent* if they do not satisfy any linear relations of the form

$$\sum_{n} c_{n}\xi_{\rho}{}^{n}(x) = 0 \tag{13.1.11}$$

with *constant* coefficients c_{n}. Equation (13.1.10) tells us that *there can be at most* $N(N+1)/2$ *independent Killing vectors in N dimensions.* For consider any M Killing vectors $\xi_{\rho}{}^{n}(x)$. For each n, there are N quantities $\xi_{\rho}{}^{n}(X)$ and $N(N-1)/2$ independent quantities $\xi^{n}_{\rho;\nu}(X)$ [recall Eq. (13.1.5)], so we can think of the quantities $\xi_{\rho}{}^{n}(X)$ and $\xi^{n}_{\rho;\nu}(X)$ as the components of M vectors in an $N(N+1)/2$

dimensional space. If $M > N(N + 1)/2$, then these M vectors cannot be linearly independent, so they must satisfy relations of the form

$$\sum_n c_n \xi_\rho{}^n(X) = \sum_n c_n \xi_{\rho;\nu}^n(X) = 0$$

Equation (13.1.10) then tells us that the Killing vectors $\xi_\rho{}^n(x)$ satisfy the relations (13.1.11) everywhere, and are therefore not independent Killing vectors.

This result is significant only because we defined independent Killing vectors as vectors that are not subject to any linear relations with *constant* coefficients. At some given point X in an N-dimensional space, any set of more than N Killing vectors will of course be subject to one or more linear relations such as (13.1.11). However, the coefficients c_n in these linear relations need not be constant in x^μ. The above theorem says that any set of more than $N(N + 1)/2$ Killing vectors will be subject to linear relations with constant coefficients.

A metric space is said to be *homogeneous* if there exist infinitesimal isometries (13.1.3) that carry any given point X into any other point in its immediate neighborhood. That is, the metric must admit Killing vectors that at any given point take all possible values. In particular, in an N-dimensional space we can choose a set of N Killing vectors $\xi_\lambda^{(\mu)}(x; X)$ with

$$\xi_\lambda^{(\mu)}(X; X) = \delta_\lambda^\mu$$

These are evidently independent, because any relation of the form $c_\mu \xi_\nu^{(\mu)}(x; X) = 0$ would at $x = X$ imply that all c_λ vanish.

A metric space is said to be *isotropic* about a given point X if there exist infinitesimal isometries (13.1.3) that leave the point X fixed, so that $\xi^\lambda(X) = 0$, and for which the first derivatives $\xi_{\lambda;\nu}(X)$ take all possible values, subject only to the antisymmetry condition (13.1.5). In particular, in N dimensions we can choose a set of $N(N - 1)/2$ Killing vectors $\xi_\lambda^{(\mu\nu)}(x; X)$ with

$$\xi_\lambda^{(\mu\nu)}(x; X) \equiv -\xi_\lambda^{(\nu\mu)}(x; X)$$

$$\xi_\lambda^{(\mu\nu)}(X; X) \equiv 0$$

$$\xi_{\lambda;\rho}^{(\mu\nu)}(X; X) \equiv \left[\frac{\partial}{\partial x^\rho} \xi_\lambda^{(\mu\nu)}(x; X)\right]_{x=X} \equiv \delta_\lambda^\mu \delta_\rho{}^\nu - \delta_\rho{}^\mu \delta_\lambda{}^\nu$$

These are independent, because any relation of the form $c_{\mu\nu} \xi_\lambda^{(\mu\nu)}(x; X) = 0$ with $c_{\mu\nu} = -c_{\nu\mu}$ would at X imply that $c_{\lambda\rho} - c_{\rho\lambda} = 2c_{\lambda\rho} = 0$.

We shall also have to deal with spaces that are isotropic about *every* point. In this case there are Killing vectors $\xi_\lambda^{(\mu\nu)}(x; X)$ and $\xi_\lambda^{(\mu\nu)}(x; X + dX)$ that satisfy the above initial conditions at X and at $X + dX$, respectively. Any linear combination of these will be a Killing vector, and so $\partial \xi_\lambda^{(\mu\nu)}(x; X)/\partial X^\rho$ will also be a Killing vector of the metric. In order to evaluate this Killing vector at $x = X$ we need only recall that $\xi_\lambda^{(\mu\nu)}(X; X)$ vanishes, and therefore

$$0 = \frac{\partial}{\partial X^\rho} \xi_\lambda^{(\mu\nu)}(X; X) = \left[\frac{\partial \xi_\lambda^{(\mu\nu)}(x; X)}{\partial x^\rho}\right]_{x=X} + \left[\frac{\partial \xi_\lambda^{(\mu\nu)}(x; X)}{\partial X^\rho}\right]_{x=X}$$

This gives

$$\left[\frac{\partial}{\partial X^\rho} \xi_\lambda^{(\mu\nu)}(x; X)\right]_{x=X} = -\delta_\lambda{}^\mu \delta_\rho{}^\nu + \delta_\rho{}^\mu \delta_\lambda{}^\nu$$

It is now obvious that we can construct a Killing vector $\xi_\lambda(x)$ that takes any arbitrary value a_λ at $x = X$; we need only take

$$\xi_\lambda(x) = \frac{a_\nu}{N-1} \frac{\partial}{\partial X^\rho} \xi_\lambda^{(\rho\nu)}(x; X)$$

Hence *any space that is isotropic about every point is also homogeneous.*

A metric that admits the maximum number $N(N+1)/2$ of Killing vectors is said to be *maximally symmetric*. In particular, a space that is both homogeneous and isotropic about some given point X will admit the $N(N+1)/2$ Killing vectors $\xi_\lambda^{(\mu)}(x; X)$ and $\xi_\lambda^{(\mu\nu)}(x; X)$. These Killing vectors are obviously independent, for if they satisfy a linear relation

$$0 = c_\mu \xi_\lambda^{(\mu)}(x; X) + c_{\mu\nu} \xi_\lambda^{(\mu\nu)}(x; X)$$

$$c_{\mu\nu} = -c_{\nu\mu}$$

then differentiating with respect to x^ρ and setting $x = X$ gives $c_{\lambda\rho} = 0$, and setting $x = X$ then gives $c_\lambda = 0$. Thus *a homogeneous space that is isotropic about some point is maximally symmetric.* It then also follows that *any space that is isotropic about every point is maximally symmetric.*

We can also prove the converse, that *a maximally symmetric space is necessarily homogeneous and isotropic about all points.* If there are $N(N+1)/2$ independent Killing vectors $\xi_\lambda^n(x)$, then we can think of the quantities $\xi_\rho^n(X), \xi_{\lambda;\nu}^n(X)$ as forming a square matrix, with $N(N+1)/2$ rows labeled by n, and $N(N+1)/2$ columns labeled by the N values of ρ and the $N(N-1)/2$ values of λ and ν with $\lambda > \nu$. Furthermore, this matrix must have a nonvanishing determinant, because any relation of the form

$$\sum_n c_n \xi_\rho^n(X) = \sum_n c_n \xi_{\lambda;\nu}^n(X) = 0$$

would with (13.1.10) imply that $\sum_n c_n \xi_\rho^n(x)$ vanishes, contrary to our assumption that these Killing vectors are independent. It must therefore be possible, for any "row vector" with "components" a_μ and $b_{\mu\nu} = -b_{\nu\mu}$, to find a solution of the equations

$$\sum_n d_n \xi_\mu^n(X) = a_\mu$$

$$\sum_n d_n \xi_{\mu;\nu}^n(X) = b_{\mu\nu}$$

Hence we can find a Killing vector $\xi_\mu(x)$ for which $\xi_\mu(X)$ takes the value a_μ and $\xi_{\mu;\nu}(X)$ takes the value $b_{\mu\nu}$, by choosing

$$\xi_\mu(x) = \sum_n d_n \xi_\mu^n(x)$$

But a_μ is arbitrary, so the space is homogeneous, and $b_{\mu\nu}$ is arbitrary (except that $b_{\mu\nu} = -b_{\nu\mu}$), so the space is isotropic about X.

As an example of a maximally symmetric space, consider an N-dimensional flat space, with vanishing curvature tensor. We can then choose Cartesian coordinates with a constant metric and vanishing affine connection. In this coordinate system, Eq. (13.1.9) reads

$$\frac{\partial^2 \xi_\mu}{\partial x^\rho \partial x^\sigma} = 0$$

The solution is

$$\xi_\mu(x) = a_\mu + b_{\mu\nu} x^\nu$$

with a_μ and $b_{\mu\nu}$ constant. This satisfies the Killing vector condition (13.1.5) if and only if

$$b_{\mu\nu} = -b_{\nu\mu}$$

We can thus choose a set of $N(N + 1)/2$ Killing vectors as follows:

$$\xi_\mu^{(\nu)}(x) \equiv \delta_\mu{}^\nu$$

$$\xi_\mu^{(\nu\lambda)}(x) \equiv \delta_\mu{}^\nu x^\lambda - \delta_\mu{}^\lambda x^\nu$$

and the general Killing vector is

$$\xi_\mu(x) = a_\nu \xi_\mu^{(\nu)}(x) + b_{\nu\lambda} \xi_\mu^{(\nu\lambda)}(x)$$

The N vectors $\xi_\mu^{(\nu)}(x)$ represent translations, whereas the $N(N - 1)/2$ vectors $\xi_\mu^{(\nu\lambda)}$ represent infinitesimal rotations (or, for a Minkowski space, Lorentz transformations). Thus any flat metric admits $N(N + 1)/2$ independent Killing vectors, and is therefore maximally symmetric.

Of course, not all metrics admit the maximum number of Killing vectors. Whether (13.1.9) is soluble for a given set of initial data $\xi_\lambda(X)$, $\xi_{\lambda;\rho}(X)$ depends on the integrability of this equation, which in turn depends on the metric. One integrability condition we shall use below follows from the general formula for commutators of covariant derivatives of tensors:

$$\xi_{\rho;\mu;\sigma;\nu} - \xi_{\rho;\mu;\nu;\sigma} = -R^\lambda{}_{\rho\sigma\nu} \xi_{\lambda;\mu} - R^\lambda{}_{\mu\sigma\nu} \xi_{\rho;\lambda}$$

Equation (13.1.9) will satisfy this condition if and only if

$$R^\lambda{}_{\sigma\rho\mu} \xi_{\lambda;\nu} - R^\lambda{}_{\nu\rho\mu} \xi_{\lambda;\sigma} + (R^\lambda{}_{\sigma\rho\mu;\nu} - R^\lambda{}_{\nu\rho\mu;\sigma}) \xi_\lambda$$
$$= -R^\lambda{}_{\rho\sigma\nu} \xi_{\lambda;\mu} - R^\lambda{}_{\mu\sigma\nu} \xi_{\rho;\lambda}$$

or, using (13.1.5),

$$[-R^\lambda{}_{\rho\sigma\nu} \delta_\mu^\varkappa + R^\lambda{}_{\mu\sigma\nu} \delta_\rho^\varkappa - R^\lambda{}_{\sigma\rho\mu} \delta_\nu^\varkappa + R^\lambda{}_{\nu\rho\mu} \delta_\sigma^\varkappa] \xi_{\lambda;\varkappa} = [R^\lambda{}_{\sigma\rho\mu;\nu} - R^\lambda{}_{\nu\rho\mu;\sigma}] \xi_\lambda \quad (13.1.12)$$

These conditions are of course empty for a flat space, but in general they will impose linear relations among the ξ_λ and $\xi_{\lambda;\varkappa}$ at any given point. Alternatively, if we know something about the Killing vectors admitted by an unknown metric, then we can use (13.1.12) to learn something about its curvature tensor. In this way, we shall be able in the following sections to deduce the form of a maximally symmetric metric from its isometries.

It should be emphasized that the existence of a definite number of independent Killing vectors does not depend on a particular choice of coordinate system. If $\xi^\mu(x)$ is a Killing vector of a metric $g_{\mu\nu}(x)$, then by performing a coordinate transformation $x^\mu \rightarrow x'^\mu$ we obtain a metric

$$g'_{\mu\nu}(x') = \frac{\partial x^\rho}{\partial x'^\mu} \frac{\partial x^\sigma}{\partial x'^\nu} g_{\rho\sigma}(x)$$

and, since (13.1.5) is generally covariant, this obviously has a Killing vector

$$\xi'^\mu(x') = \frac{\partial x'^\mu}{\partial x^\nu} \xi^\nu(x)$$

If M Killing vectors $\xi_\mu{}^n(x)$ are independent, then so are the M Killing vectors $\xi_\mu{}^{n\prime}(x')$, for any linear relation among the $\xi^{n\prime}$ would imply a linear relation among the ξ^n. Thus the maximal symmetry of a given space is an inner property, not depending on how we choose the coordinate system. In particular, it follows that any space with vanishing curvature tensor is maximally symmetric; the converse, however, is not true. It is also easy to see that the homogeneity or isotropy of a given space is independent of the choice of coordinates. As far as these simple symmetries are concerned, we have accomplished the task laid out in the introduction to this chapter, that of describing symmetries of the metric in a generally covariant language.

2 Maximally Symmetric Spaces: Uniqueness

We now show that the maximally symmetric spaces are uniquely specified by a "curvature constant" K, and by the numbers of eigenvalues of the metric that are positive or negative. That is, given *two maximally symmetric metrics with the same K and the same numbers of eigenvalues of each sign, it will always be possible to find a coordinate transformation that carries one metric into the other*. Armed with this theorem, we shall be able in the next section to carry out an exhaustive study of maximally symmetric spaces by simply constructing such metrics in one convenient coordinate system.

We showed in the last section that at any given point x in a maximally symmetric space, we can find Killing vectors for which $\xi_\lambda(x)$ vanishes and for which $\xi_{\lambda;\kappa}(x)$ is an arbitrary antisymmetric matrix. It follows then that the coefficient of $\xi_{\lambda;\kappa}(x)$ in Eq. (13.1.12) must have a vanishing antisymmetric part, that is,

$$-R^\lambda_{\rho\sigma v}\delta^\kappa_\mu + R^\lambda_{\mu\sigma v}\delta^\kappa_\rho - R^\lambda_{\sigma\rho\mu}\delta^\kappa_v + R^\lambda_{v\rho\mu}\delta^\kappa_\sigma$$
$$= -R^\kappa_{\rho\sigma v}\delta^\lambda_\mu + R^\kappa_{\mu\sigma v}\delta^\lambda_\rho - R^\kappa_{\sigma\rho\mu}\delta^\lambda_v + R^\kappa_{v\rho\mu}\delta^\lambda_\sigma \qquad (13.2.1)$$

We also showed that at any given point x in a maximally symmetric space, there exist Killing vectors for which $\xi_\lambda(x)$ takes any values we like, so (13.1.12) and (13.2.1) require that

$$R^\lambda_{\sigma\rho\mu;v} = R^\lambda_{v\rho\mu;\sigma} \qquad (13.2.2)$$

We actually only need to use (13.2.1), because we have shown in the last section that a space that is isotropic about every point, and hence satisfies (13.2.1), must also be homogeneous, and hence must also satisfy (13.2.2).

Our first step in the proof is to use Eq. (13.2.1) to derive a formula for the curvature tensor. Contracting κ with μ yields

$$-NR^\lambda_{\rho\sigma v} + R^\lambda_{\rho\sigma v} - R^\lambda_{\sigma\rho v} + R^\lambda_{v\rho\sigma} = -R^\lambda_{\rho\sigma v} + R_{\sigma\rho}\delta^\lambda_v - R_{v\rho}\delta^\lambda_\sigma$$

(Recall that $R^\kappa_{\kappa\sigma v}$ vanishes, $-R^\kappa_{\sigma\rho\kappa}$ is the Ricci tensor $R_{\sigma\rho}$, and in N dimensions, $\delta^\kappa_\kappa = N$.) Using the cyclic sum rule (6.6.5) and the antisymmetry of $R^\lambda_{\sigma\rho v}$, we find

$$(N - 1)R_{\lambda\rho\sigma v} = R_{v\rho}g_{\lambda\sigma} - R_{\sigma\rho}g_{\lambda v} \qquad (13.2.3)$$

But this must be antisymmetric in λ and ρ, so

$$R_{v\rho}g_{\lambda\sigma} - R_{\sigma\rho}g_{\lambda v} = -R_{v\lambda}g_{\rho\sigma} + R_{\sigma\lambda}g_{\rho v}$$

Contracting λ with v, we find

$$R_{\sigma\rho} - NR_{\sigma\rho} = -R^\lambda_\lambda g_{\sigma\rho} + R_{\rho\sigma}$$

The Ricci tensor thus takes the form

$$R_{\sigma\rho} = \frac{1}{N} g_{\sigma\rho}R^\lambda_\lambda \qquad (13.2.4)$$

Inserting this in (13.2.3) gives our formula for the curvature tensor

$$R_{\lambda\rho\sigma v} = \frac{R^\lambda_\lambda}{N(N - 1)} \{g_{v\rho}g_{\lambda\sigma} - g_{\sigma\rho}g_{\lambda v}\} \qquad (13.2.5)$$

This formula satisfies (13.2.1), so there is nothing further to be learned from that condition.

In a space that is isotropic about *every* point, Eqs. (13.2.4) and (13.2.5) will hold everywhere, and we can use the Bianchi identities to say something about the dependence of the curvature scalar R^λ_λ on position. Using (13.2.4) in (6.8.4), we have

$$0 = [R^\sigma_\rho - \tfrac{1}{2}\delta^\sigma_\rho R^\lambda_\lambda]_{;\sigma} = \left(\frac{1}{N} - \frac{1}{2}\right)(R^\lambda_\lambda)_{;\sigma}$$

or

$$0 = \left(\frac{1}{N} - \frac{1}{2}\right)\frac{\partial}{\partial x^\sigma} R^\lambda_{\ \lambda} \tag{13.2.6}$$

Hence any space of three or more dimensions, in which (13.2.4) holds everywhere, will have $R^\lambda_{\ \lambda}$ *constant*. It is convenient to introduce a curvature constant K in place of $R^\lambda_{\ \lambda}$, with

$$R^\lambda_{\ \lambda} \equiv -N(N-1)K \tag{13.2.7}$$

Using this in (13.2.4) gives the Ricci tensor and the Riemann-Christoffel tensor here as

$$R_{\sigma\rho} = -(N-1)Kg_{\sigma\rho} \tag{13.2.8}$$

$$R_{\lambda\rho\sigma\nu} = K\{g_{\sigma\rho}g_{\lambda\nu} - g_{\nu\rho}g_{\lambda\sigma}\} \tag{13.2.9}$$

In differential geometry a space with these properties is called a *space of constant curvature*.

Incidentally, we showed in Section 6.7 that the curvature tensor in two dimensions is always of the form (13.2.5), so it is not surprising that in this case (13.2.6) does not allow us to draw any conclusions about the constancy of $R^\lambda_{\ \lambda}$. However, by using (13.2.2) one can show that the quantity K in (13.2.9) is also constant for maximally symmetric spaces of dimensionality $N = 2$.

Now suppose that we are given two metrics $g_{\mu\nu}(x)$ and $g'_{\mu\nu}(x')$, both having the same numbers of positive and negative eigenvalues, and both satisfying the condition (13.2.9) for a maximally symmetric space, that is,

$$R_{\lambda\rho\sigma\nu} = K(g_{\sigma\rho}g_{\lambda\nu} - g_{\nu\rho}g_{\lambda\sigma}) \tag{13.2.10}$$

$$R'_{\lambda\rho\sigma\nu} = K(g'_{\sigma\rho}g'_{\lambda\nu} - g'_{\nu\rho}g'_{\lambda\sigma}) \tag{13.2.11}$$

with the same curvature constant K. We shall show that $g_{\mu\nu}(x)$ and $g'_{\mu\nu}(x')$ must be equivalent, in the sense that there is a transformation $x \rightarrow x'$ that converts $g_{\mu\nu}(x)$ into $g'_{\mu\nu}(x')$, that is, for which

$$g'_{\mu\nu}(x')\frac{\partial x'^\mu}{\partial x^\rho}\frac{\partial x'^\nu}{\partial x^\sigma} = g_{\rho\sigma}(x) \tag{13.2.12}$$

We shall prove this by actually constructing $x'^\mu(x)$ as a power series in x^μ. First, note that the equality in the numbers of positive and negative eigenvalues of $g_{\mu\nu}$ and $g'_{\mu\nu}$ means that we can find a nonsingular matrix $d^\mu_{\ \rho}$ for which

$$g'_{\mu\nu}(0)\, d^\mu_{\ \rho}\, d^\nu_{\ \sigma} = g_{\rho\sigma}(0) \tag{13.2.13}$$

(The argument here is the same as in Section 6.4.) Thus we can satisfy (13.2.12) to *zero* order in x with

$$x'^\mu = d^\mu_{\ \rho}x^\rho$$

Now we proceed by mathematical induction. Suppose that we succeed in satisfying (13.2.12) to order $n - 1$ in x^μ with a polynomial

$$x'^\mu(x) = d^\mu_{\ \rho} x^\rho + \sum_{m=2}^{n} \frac{1}{m!} d^\mu_{\ \rho_1 \cdots \rho_m} x^{\rho_1} \cdots x^{\rho_m} \qquad (13.2.14)$$

We want to add a term of order $n + 1$ in x^μ so that (13.2.12) holds to order n. This condition will be satisfied if the derivative of (13.2.12) holds in order $n - 1$, that is, if

$$\frac{\partial^2 x'^\mu}{\partial x^\rho \partial x^\lambda} \frac{\partial x'^\nu}{\partial x^\sigma} g'_{\mu\nu}(x') + \frac{\partial^2 x'^\nu}{\partial x^\sigma \partial x^\lambda} \frac{\partial x'^\mu}{\partial x^\rho} g'_{\mu\nu}(x')$$

$$+ \frac{\partial x'^\mu}{\partial x^\rho} \frac{\partial x'^\nu}{\partial x^\sigma} \frac{\partial x'^\kappa}{\partial x^\lambda} \frac{\partial g'_{\mu\nu}(x')}{\partial x'^\kappa} = \frac{\partial g_{\rho\sigma}(x)}{\partial x^\lambda} \qquad \text{in order } x^{n-1}$$

This will be satisfied if (and, in fact, only if)

$$\frac{\partial^2 x'^\mu}{\partial x^\rho \partial x^\lambda} \frac{\partial x'^\nu}{\partial x^\sigma} g'_{\mu\nu}(x') = g_{\sigma\tau}(x)\Gamma^\tau_{\lambda\rho}(x)$$

$$- \frac{\partial x'^\mu}{\partial x^\rho} \frac{\partial x'^\nu}{\partial x^\sigma} \frac{\partial x'^\kappa}{\partial x^\lambda} g'_{\nu\eta}(x')\Gamma'^\eta_{\mu\kappa}(x') \qquad \text{in order } x^{n-1}$$

This only needs to hold in order $n - 1$ in x^μ, so we can use (13.2.12), which was assumed to hold to this order, to convert it into an equivalent requirement

$$\frac{\partial^2 x'^\mu}{\partial x^\rho \partial x^\lambda} = \frac{\partial x'^\mu}{\partial x^\kappa} \Gamma^\kappa_{\lambda\rho}(x) - \frac{\partial x'^\nu}{\partial x^\rho} \frac{\partial x'^\kappa}{\partial x^\lambda} \Gamma'^\mu_{\nu\kappa}(x') \qquad \text{in order } x^{n-1}$$

$$(13.2.15)$$

We can use (13.2.14), which is correct to order x^n, to calculate the term on the right-hand side of order x^{n-1}. Let us write the result as

$$\left[\frac{\partial x'^\mu}{\partial x^\kappa} \Gamma^\kappa_{\lambda\rho}(x) - \frac{\partial x'^\nu}{\partial x^\rho} \frac{\partial x'^\kappa}{\partial x^\lambda} \Gamma'^\mu_{\nu\kappa}(x') \right]_{\text{order } n-1}$$

$$= \frac{1}{(n-1)!} c^\mu_{\lambda\rho\sigma_1 \cdots \sigma_{n-1}} x^{\sigma_1} \cdots x^{\sigma_{n-1}} \qquad (13.2.16)$$

the coefficients $c^\mu_{\lambda\rho} \cdots$ depending in a complicated way on the functions $g_{\mu\nu}(x)$ and $g'_{\mu\nu}(x')$ and on the previously determined coefficients $d^\mu_{\ \rho_1 \cdots \rho_m}$. Then (13.2.15) will be satisfied in order $n - 1$ if we add to (13.2.14) a term

$$[x'^\mu(x)]_{\text{order } n+1} = \frac{1}{(n+1)!} c^\mu_{\lambda\rho\sigma_1 \cdots \sigma_{n-1}} x^\lambda x^\rho x^{\sigma_1} \cdots x^{\sigma_{n-1}} \qquad (13.2.17)$$

provided that the coefficient $c^{\mu}_{\lambda\rho\sigma_1\cdots\sigma_{n-1}}$ is totally symmetric in all its lower indices. These coefficients are obviously symmetric under interchange of λ and ρ or among the σ_m indices, so the only condition that needs to be satisfied is that they are symmetric between λ and any σ_m, or equivalently, that the derivative of (13.2.16) with respect to x^{σ} should be symmetric between λ and σ:

$$\frac{\partial}{\partial x^{\sigma}}\left(\frac{\partial x'^{\mu}}{\partial x^{\kappa}}\,\Gamma^{\kappa}_{\lambda\rho}(x) - \frac{\partial x'^{\nu}}{\partial x^{\rho}}\frac{\partial x'^{\kappa}}{\partial x^{\lambda}}\,\Gamma'^{\mu}_{\nu\kappa}(x')\right)$$

$$= \frac{\partial}{\partial x^{\lambda}}\left(\frac{\partial x'^{\mu}}{\partial x^{\kappa}}\,\Gamma^{\kappa}_{\sigma\rho}(x) - \frac{\partial x'^{\nu}}{\partial x^{\rho}}\frac{\partial x'^{\kappa}}{\partial x^{\sigma}}\,\Gamma'^{\mu}_{\nu\kappa}(x')\right) \quad \text{in order } x^{n-2}$$

$$(13.2.18)$$

Since (13.2.12) is assumed to hold to order x^{n-1}, its derivative, Eq. (13.2.15), will hold to order x^{n-2}, so we can use (13.2.12) and (13.2.15) to rewrite (13.2.18) as the equivalent requirement

$$\frac{\partial x'^{\mu}}{\partial x^{\kappa}}\,R^{\kappa}_{\rho\lambda\eta}(x) = \frac{\partial x'^{\nu}}{\partial x^{\rho}}\frac{\partial x'^{\kappa}}{\partial x^{\lambda}}\frac{\partial x'^{\sigma}}{\partial x^{\eta}}\,R'^{\mu}_{\nu\kappa\sigma}(x') \quad \text{in order } x^{n-2} \qquad (13.2.19)$$

Now for the first time we use Eqs. (13.2.10) and (13.2.11), which allow (13.2.19) to be replaced with the equivalent requirement

$$\frac{\partial x'^{\mu}}{\partial x^{\eta}}\,g_{\lambda\rho}(x) - \frac{\partial x'^{\mu}}{\partial x^{\lambda}}\,g_{\rho\eta}(x)$$

$$= \frac{\partial x'^{\nu}}{\partial x^{\rho}}\left(\frac{\partial x'^{\kappa}}{\partial x^{\lambda}}\frac{\partial x'^{\mu}}{\partial x^{\eta}}\,g'_{\nu\kappa}(x') - \frac{\partial x'^{\mu}}{\partial x^{\lambda}}\frac{\partial x'^{\sigma}}{\partial x^{\eta}}\,g'_{\nu\sigma}(x')\right) \quad \text{in order } x^{n-2}$$

$$(13.2.20)$$

This condition *is* satisfied, because (13.2.12) was assumed to hold to order x^{n-1}. To recapitulate, this implies that (13.2.19) holds in order x^{n-2}, which implies that (13.2.18) holds in order x^{n-2}, which implies that the coefficients $c^{\mu}_{\lambda\rho\sigma_1\cdots\sigma_m}$ are totally symmetric in their lower indices, which implies that (13.2.17) satisfies (13.2.15), which implies that by adding (13.2.17) to (13.2.14) we can satisfy (13.2.12) to order x^n. Thus, if (13.2.12) can be satisfied to order x^{n-1} by a polynomial $x'(x)$ of order n, it can be satisfied to order x^n by a polynomial $x'(x)$ of order $n + 1$, and therefore a function $x'(x)$ satisfying (13.2.12) exactly can be built up as a power series, as was to be proven.

3   Maximally Symmetric Spaces: Construction

Maximally symmetric spaces are essentially unique, so we can learn all about them by constructing examples with arbitrary curvature K in any way we like.

This is one rather obvious way to carry out this construction. (See Figure 13.1.) Consider a flat $(N + 1)$-dimensional space, with metric given by

$$- d\tau^2 \equiv g_{AB} \, dx^A \, dx^B = C_{\mu\nu} \, dx^\mu \, dx^\nu + K^{-1} \, dz^2 \tag{13.3.1}$$

where $C_{\mu\nu}$ is a constant $N \times N$ matrix and K is some constant. We can embed a non-Euclidean N-dimensional space in this larger space by restricting the variables x^μ and z to the surface of a sphere (or pseudosphere):

$$KC_{\mu\nu} x^\mu x^\nu + z^2 = 1 \tag{13.3.2}$$

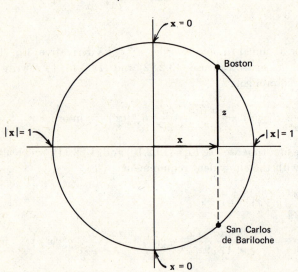

Figure 13.1   Representation of points on a sphere by projection onto the equatorial plane. Note that two points on the sphere correspond to each projected point with given coordinates x^i.

On this surface, dz^2 is given by

$$dz^2 = \frac{K^2 (C_{\mu\nu} x^\mu \, dx^\nu)^2}{z^2}$$

$$= \frac{K^2 (C_{\mu\nu} x^\mu \, dx^\nu)^2}{(1 - KC_{\mu\nu} x^\mu x^\nu)}$$

and therefore (13.3.1) gives

$$- d\tau^2 = C_{\mu\nu} \, dx^\mu \, dx^\nu + \frac{K (C_{\mu\nu} x^\mu \, dx^\nu)^2}{(1 - KC_{\rho\sigma} x^\rho x^\sigma)} \tag{13.3.3}$$

The metric is then

$$g_{\mu\nu}(x) = C_{\mu\nu} + \frac{K}{(1 - KC_{\rho\sigma}x^\rho x^\sigma)} C_{\mu\lambda}x^\lambda C_{\nu\kappa}x^\kappa \qquad (13.3.4)$$

A flat space appears here as the special case $K = 0$.

This construction makes it obvious that (13.3.4) admits an $(N(N+1)/2)$-parameter group of isometries, for both the $(N+1)$-dimensional line element (13.3.1) and the "embedding" condition (13.3.2) are manifestly invariant under rigid "rotations" of the $(N+1)$-dimensional space, that is, under the transformations

$$x^\mu \to x'^\mu = R^\mu{}_\nu x^\nu + R^\mu{}_z z \qquad (13.3.5)$$

$$z \to z' = R^z{}_\mu x^\mu + R^z{}_z z \qquad (13.3.6)$$

where the $R^A{}_B$ are constants, with

$$C_{\mu\nu}R^\mu{}_\rho R^\nu{}_\sigma + K^{-1}R^z{}_\rho R^z{}_\sigma = C_{\rho\sigma} \qquad (13.3.7)$$

$$C_{\mu\nu}R^\mu{}_\rho R^\nu{}_z + K^{-1}R^z{}_\rho R^z{}_z = 0 \qquad (13.3.8)$$

$$C_{\mu\nu}R^\mu{}_z R^\nu{}_z + K^{-1}(R^z{}_z)^2 = K^{-1} \qquad (13.3.9)$$

It is convenient to distinguish two classes of simple transformations satisfying (13.3.7)–(13.3.9):

(A) $\qquad R^\mu{}_\nu = \mathscr{R}^\mu{}_\nu \qquad R^\mu{}_z = R^z{}_\mu = 0 \qquad R^z{}_z = 1 \qquad (13.3.10)$

where $\mathscr{R}^\mu{}_\nu$ is any $N \times N$ matrix with

$$C_{\mu\nu}\mathscr{R}^\mu{}_\rho \mathscr{R}^\nu{}_\sigma = C_{\rho\sigma} \qquad (13.3.11)$$

These are just rigid "rotations" about the origin:

$$x'^\mu = \mathscr{R}^\mu{}_\nu x^\nu \qquad (13.3.12)$$

(B) $\qquad R^\mu{}_z = a^\mu \qquad R^z{}_\mu = -KC_{\mu\nu}a^\nu \qquad R^z{}_z = (1 - KC_{\rho\sigma}a^\rho a^\sigma)^{1/2} \quad (13.3.13)$

$$R^\mu{}_\nu = \delta^\mu{}_\nu - bKC_{\nu\rho}a^\rho a^\mu \qquad (13.3.14)$$

where a^μ is arbitrary except that $R^z{}_z$ must be real, that is,

$$KC_{\rho\sigma}a^\rho a^\sigma \le 1 \qquad (13.3.15)$$

and

$$b \equiv \frac{1 - (1 - KC_{\rho\sigma}a^\rho a^\sigma)^{1/2}}{KC_{\rho\sigma}a^\rho a^\sigma} \qquad (13.3.16)$$

These are "quasitranslations," with

$$x'^\mu = x^\mu + a^\mu[(1 - KC_{\rho\sigma}x^\rho x^\sigma)^{1/2} - bKC_{\rho\sigma}x^\rho a^\sigma] \qquad (13.3.17)$$

In particular, these transformations take the origin $x^\mu = 0$ into a^μ.

The existence of isometries (13.3.17) that take the origin into any point (at least within a finite region) means that this space is *homogeneous*; any point is geometrically like any other point. (Our coordinate system hides this property, just as a polar projection map of the earth hides the fact that the curvature of the earth is about the same in Massachusetts as at the North Pole.) Also, the existence of isometries (13.3.10) that include all rigid "rotations" about the origin means that this space is *isotropic* about the origin. Since the metric is homogeneous, and isotropic about the origin, it is isotropic about every point, and maximally symmetric.

We can construct the Killing vectors for this metric by letting the finite transformations (13.2.5), (13.2.6) approach the unit transformation. First, consider the transformations (A), and let

$$\mathscr{R}^\mu{}_\nu = \delta^\mu{}_\nu + \varepsilon\Omega^\mu{}_\nu, \qquad |\varepsilon| \ll 1$$

$$C_{\mu\sigma}\Omega^\mu{}_\rho + C_{\rho\mu}\Omega^\mu{}_\sigma = 0 \tag{13.3.18}$$

Comparing with (13.1.3), the corresponding Killing vectors are

$$\zeta^\mu{}_\Omega(x) = \Omega^\mu{}_\nu x^\nu \tag{13.3.19}$$

Next, consider the transformations (B), and let

$$a^\mu = \varepsilon\alpha^\mu, \qquad |\varepsilon| \ll 1$$

Comparing with (13.1.3), the corresponding Killing vectors are

$$\zeta^\mu{}_\alpha(x) = \alpha^\mu[1 - KC_{\mu\nu}x^\mu x^\nu]^{1/2} \tag{13.3.20}$$

The reader may check that (13.3.19) and (13.3.20) do satisfy the Killing conditions (13.1.5). There are $N(N-1)/2$ independent parameters $\Omega^\mu{}_\nu$ [that is, N^2 elements $\Omega^\mu{}_\nu$, subject to the $N(N+1)/2$ conditions (13.3.18)] and N parameters α^μ, so this metric admits $N(N+1)/2$ independent Killing vectors, verifying maximal symmetry.

The geodesics of this metric take a remarkably simple form. From (13.3.4) we can readily calculate that the affine connection is

$$\Gamma^\mu_{\nu\lambda} = Kx^\mu g_{\nu\lambda} \tag{13.3.21}$$

so the differential equation for a geodesic is

$$\frac{d^2x^\mu}{d\tau^2} + Kx^\mu = 0 \tag{13.3.22}$$

The solutions are thus linear combinations of $\sin(\tau\sqrt{K})$ and $\cos(\tau\sqrt{K})$ for $K > 0$, or of $\sinh(\tau\sqrt{-K})$ and $\cosh(\tau\sqrt{-K})$ for $K < 0$.

We can uncover the inner properties of this space by calculating the curvature tensor; a straightforward computation using Eqs. (6.6.2) and (13.2.21) gives the Riemann-Christoffel tensor for the metric (13.3.4) as

$$R_{\kappa\nu\rho\sigma} = K[C_{\kappa\sigma}C_{\nu\rho} - C_{\kappa\rho}C_{\nu\sigma}]$$
$$+ K^2[1 - KC_{\mu\lambda}x^\mu x^\lambda]^{-1} [C_{\kappa\sigma}x_\nu x_\rho - C_{\kappa\rho}x_\nu x_\sigma + C_{\nu\rho}x_\kappa x_\sigma - C_{\nu\sigma}x_\rho x_\kappa]$$

(where $X_\nu \equiv C_{\nu\mu}X^\mu$), or

$$R_{\kappa\nu\rho\sigma} = K[g_{\rho\nu}g_{\kappa\sigma} - g_{\sigma\nu}g_{\kappa\rho}]$$

in agreement with Eq. (13.2.9). Hence the constant K introduced in Eqs. (13.3.1) and (13.3.2) is the same as the curvature constant introduced in the last section.

Since K is an invariant parameter, we cannot by a coordinate transformation convert the metric (13.3.4) into a similar metric with a different K. In contrast, Eq. (13.3.3) makes it obvious that by a linear transformation

$$x^\mu = A^\mu{}_\nu x'^\nu$$

we can convert the metric (13.3.4) into a similar metric with the same K and with $C_{\mu\nu}$ changed into

$$C'_{\mu\nu} = A^\rho{}_\mu A^\sigma{}_\nu C_{\rho\sigma}$$

Our discussion in Section 3.6 shows that in this way $C_{\mu\nu}$ can be changed into any real symmetric matrix we like, as long as we do not change the numbers of its positive and negative eigenvalues. Also, the numbers of eigenvalues of each sign of the matrix $C_{\mu\nu}$ are the same as for the matrix $g_{\mu\nu}$ at the point $x = 0$, and hence the same everywhere, since all points are equivalent.

An N-dimensional metric that allows the introduction of locally Euclidean (as opposed, say, to Minkowskian) coordinate systems will have all its eigenvalues positive, so for $K \neq 0$ we can take $C_{\mu\nu}$ as $|K|^{-1}$ times the unit matrix, in which case (13.3.3) becomes

$$ds^2 = K^{-1}\left[d\mathbf{x}^2 + \frac{(\mathbf{x}\cdot d\mathbf{x})^2}{1 - \mathbf{x}^2}\right] \qquad \text{for } K > 0 \qquad (13.3.23)$$

or

$$ds^2 = |K|^{-1}\left[d\mathbf{x}^2 - \frac{(\mathbf{x}\cdot d\mathbf{x})^2}{1 + \mathbf{x}^2}\right] \qquad \text{for } K < 0 \qquad (13.3.24)$$

For $K = 0$, we take $C_{\mu\nu}$ as just the unit matrix, and (13.3.3) gives

$$ds^2 = d\mathbf{x}^2 \qquad \text{for } K = 0 \qquad (13.3.25)$$

(We are using an obvious N-dimensional vector notation. Also, we have replaced $-d\tau^2$ with a proper length ds^2, because for the moment we are doing geometry rather than physics.) Let us explore the *global* properties of these spaces.

For $K > 0$, our most convenient approach is to go back to the interpretation of (13.3.23) as the metric of the curved space embedded by Eq. (13.3.2) in the flat space (13.3.1); that is, (13.3.23) describes the surface

$$\mathbf{x}^2 + z^2 = 1 \qquad (13.3.26)$$

in the flat space with

$$ds^2 = K^{-1}[d\mathbf{x}^2 + dz^2] \qquad (13.3.27)$$

Obviously this metric simply describes the surface of a sphere of radius $K^{-1/2}$ in an $(N + 1)$-dimensional Euclidean space. (To make the coordinates $\mathbf{x}$ and z truly Euclidean, we should define $\mathbf{x}' = K^{-1/2}\mathbf{x}$ and $z' = K^{-1/2}z$, in which case (13.3.26) reads $\mathbf{x}'^2 + z'^2 = K^{-1}$.) Indeed, in two dimensions we can introduce angular coordinates θ, φ by:

$$x_1 = \sin\theta\cos\varphi \qquad x_2 = \sin\theta\sin\varphi$$

and (13.3.27) then becomes the familiar line element on a sphere of radius $K^{-1/2}$:

$$ds^2 = K^{-1}[d\theta^2 + \sin^2\theta\, d\varphi^2] \qquad (13.3.28)$$

In general, the range of the variables $\mathbf{x}$ is

$$\mathbf{x}^2 \leq 1$$

However, each $\mathbf{x}$ actually corresponds to *two* points, corresponding to the two roots of Eq. (13.3.26) for z. (For instance, in two dimensions the components of $\mathbf{x}$ are the coordinates of points on a sphere projected on a tangent plane; in a polar projection map of the earth, Boston will appear at the same point as San Carlos de Bariloche, Argentina.) The volume of the N-dimensional space described by (13.3.23) is therefore

$$V_N = 2\int_{\mathbf{x}^2 \leq 1} \sqrt{g}\, dx_1 \cdots dx_N = 2K^{-N/2}\int_{\mathbf{x}^2 \leq 1} \frac{dx_1 \cdots dx_N}{[1 - \mathbf{x}^2]^{1/2}}$$

A straightforward calculation gives

$$V_N = \frac{2\pi^{(N+1)/2}}{\Gamma((N+1)/2)} K^{-N/2} \qquad (13.3.29)$$

For instance, $V_1 = 2\pi K^{-1/2}$, which is just the perimeter of a circle of radius $K^{-1/2}$, and $V_2 = 4\pi K^{-1}$, which is just the area of a sphere with radius $K^{-1/2}$. A three-dimensional space of constant positive curvature has the volume

$$V_3 = 2\pi^2 K^{-3/2}$$

We can also calculate the circumference of such spaces, using for the geodesics the solutions of Eq. (13.3.22), which now reads

$$\frac{d^2\mathbf{x}}{ds^2} + K\mathbf{x} = 0 \qquad (13.3.30)$$

The solutions that pass through the point $\mathbf{x} = 0$ are

$$\mathbf{x} = \mathbf{e} \sin(sK^{1/2}) \tag{13.3.31}$$

where, in order to satisfy (13.3.23),

$$\mathbf{e}^2 = 1 \tag{13.3.32}$$

As we go out along a geodesic from the "North pole" $\mathbf{x} = 0$, we reach the "equator" $\mathbf{x} = \mathbf{e}$ at $s = \pi K^{-1/2}/2$, we reach the "South pole" $\mathbf{x} = 0$ at $s = \pi K^{-1/2}$, we reach the opposite point $\mathbf{x} = -\mathbf{e}$ of the "equator" at $s = 3\pi K^{-1/2}/2$, and we return to our starting point at $s = 2\pi K^{-1/2}$. Thus the distance from any point around the whole space and back to itself along a geodesic is

$$L = 2\pi K^{-1/2} \tag{13.3.33}$$

for spaces of constant positive curvature and arbitrary dimensionality. This calculation shows very clearly that the space described by (13.3.23) is *finite*, but it is not *bounded*; when we come to the apparent singularity at $\mathbf{x}^2 = 1$, we continue right through, but with z given by the root of Eq. (13.3.26) of opposite sign.

For $K < 0$ the metric (13.3.24) does not even have an apparent singularity, and there is nothing to restrict the coordinates $\mathbf{x}$ to any finite range. This can be seen even more definitely by calculating the geodesics, which are now given by Eqs. (13.3.30) and (13.3.24) as

$$\mathbf{x} = \mathbf{e} \sinh\left(s(-K)^{1/2}\right) \tag{13.3.34}$$

$$\mathbf{e}^2 = 1 \tag{13.3.35}$$

We can obviously go out along this geodesic an unlimited distance from the origin. For $N = 2$, this space is just that discovered by Gauss, Bólyai, and Lobachevski. [See Section 1.1. In order to put the metric in the form (1.1.9) of Klein's model, it is necessary to introduce a new set of coordinates x'^i, defined by $\mathbf{x}' = \mathbf{x}(1 + \mathbf{x}^2)^{-1/2}$.] We see from (13.3.1) and (13.3.2) that this geometry describes the surface

$$-\mathbf{x}^2 + z^2 = 1 \tag{13.3.36}$$

in a flat space, with

$$ds^2 = |K|^{-1}[d\mathbf{x}^2 - dz^2] \tag{13.3.37}$$

The minus sign in (13.3.37) means that this flat space is *not* Euclidean. It is therefore understandable that the Gauss-Bólyai-Lobachevski geometry could not be discovered until geometers had learned to think of curved surfaces, not as subspaces of an ordinary Euclidean space, but as spaces characterized by their own inner metric relations.

Finally, let us return to space-time, and consider the structure of a four-dimensional maximally symmetric metric with three positive and one negative eigenvalue. In this case, we can set

$$C_{\mu\nu} = \eta_{\mu\nu} \tag{13.3.38}$$

and the metric is

$$-d\tau^2 = d\mathbf{x}^2 - dt^2 + \frac{K(\mathbf{x} \cdot d\mathbf{x} - t\,dt)^2}{1 - K(\mathbf{x}^2 - t^2)} \tag{13.3.39}$$

For $K > 0$, we can introduce coordinates in which the metric appears *spatially* flat, by setting

$$t = \frac{1}{\sqrt{K}} \left[\frac{Kr'^2}{2} \cosh(K^{1/2}t') + \left(1 + \frac{Kr'^2}{2}\right) \sinh(K^{1/2}t') \right]$$

$$\mathbf{x} = \mathbf{x}' \exp(K^{1/2}t') \tag{13.3.40}$$

Then (13.3.39) becomes

$$d\tau^2 = dt'^2 - \exp(2K^{1/2}t')\,d\mathbf{x}'^2 \tag{13.3.41}$$

We can also introduce coordinates in which the metric appears time-independent, by setting

$$t'' = t' - \frac{1}{2K^{1/2}} \ln\left[1 - K\mathbf{x}'^2 \exp(2K^{1/2}t')\right]$$

$$\mathbf{x}'' = \mathbf{x}' \exp(K^{1/2}t') \tag{13.3.42}$$

Then (13.3.41) becomes

$$d\tau^2 = (1 - K\mathbf{x}''^2)\,dt''^2 - d\mathbf{x}''^2 - \frac{K(\mathbf{x}'' \cdot d\mathbf{x}'')^2}{1 - K\mathbf{x}''^2} \tag{13.3.43}$$

This metric was first discussed in this form by deSitter;[2] it will provide the basis for our treatment of the steady state cosmology in Chapter 14.

Once again, it should be stressed that the maximally symmetric metric (13.3.4), although derived by an apparently arbitrary procedure, actually represents the most general possible maximally symmetric metric, because the uniqueness theorem of the last section tells us that any other maximally symmetric metric can be converted into the form (13.3.4) by a suitable coordinate transformation.

4   Tensors in a Maximally Symmetric Space

The assumption of maximal symmetry can be applied, not only to the metric of a space, but to any tensor fields that inhabit the space. A tensor field $T_{\mu\nu}$...

is said to be *form-invariant* under a transformation $x \rightarrow x'$ if $T'_{\mu\nu}\ldots(x')$ is the same function of its argument x'^{μ} as $T_{\mu\nu}\ldots(x)$ was of its argument x^{μ}, that is,

$$T'_{\mu\nu}\ldots(y) = T_{\mu\nu}\ldots(y) \qquad \text{for all } y \tag{13.4.1}$$

At any given point, the transformed tensor is given by the usual formula

$$T_{\mu\nu}\ldots(x) = \frac{\partial x'^{\rho}}{\partial x^{\mu}} \frac{\partial x'^{\sigma}}{\partial x^{\nu}} \cdots T'_{\rho\sigma}\ldots(x')$$

so the form-invariance condition (13.4.1) reads

$$T_{\mu\nu}\ldots(x) = \frac{\partial x'^{\rho}}{\partial x^{\mu}} \frac{\partial x'^{\sigma}}{\partial x^{\nu}} \cdots T_{\rho\sigma}\ldots(x') \tag{13.4.2}$$

For an infinitesimal transformation

$$x'^{\mu} = x^{\mu} + \varepsilon \xi^{\mu}(x) \qquad |\varepsilon| \ll 1$$

the condition (13.4.2) becomes, to first order in ε,

$$0 = \frac{\partial \xi^{\rho}(x)}{\partial x^{\mu}} T_{\rho\nu}\ldots(x) + \frac{\partial \xi^{\sigma}(x)}{\partial x^{\nu}} T_{\mu\sigma}\ldots(x) + \cdots + \xi^{\lambda}(x) \frac{\partial}{\partial x^{\lambda}} T_{\mu\nu}\ldots(x) \tag{13.4.3}$$

(That is, the *Lie derivative* of $T_{\mu\nu}\ldots$ with respect to ξ^{λ} vanishes; see Section 10.9.) A tensor in a maximally symmetric space, which satisfies (13.4.3) for all $N(N+1)/2$ independent Killing vectors $\xi^{\lambda}(x)$, will be called *maximally form-invariant*.

For a scalar $S(x)$, Eq. (13.4.3) reads simply

$$\xi^{\lambda}(x) \frac{\partial}{\partial x^{\lambda}} S(x) = 0 \tag{13.4.4}$$

If the scalar is maximally form-invariant, then $\xi^{\lambda}(x)$ can at any given point be chosen to have any value we like, and (13.4.4) therefore requires that S be constant:

$$\frac{\partial S}{\partial x^{\lambda}} = 0 \tag{13.4.5}$$

For any other maximally form-invariant tensor, it is convenient first to choose a Killing vector $\xi^{\lambda}(x)$ that at a given point X satisfies

$$\xi^{\lambda}(X) = 0$$

and for which the quantities

$$\xi_{\sigma;\mu}(X) = g_{\sigma\rho}(X) \left(\frac{\partial \xi^{\rho}(x)}{\partial x^{\mu}} \right)_{x=X}$$

form an arbitrary antisymmetric matrix. Equation (13.4.3) then reads, at $x = X$:

$$0 = \xi_{\sigma;\tau}\{\delta_\mu^\tau T^\sigma{}_\nu \ldots + \delta_\nu^\tau T_\mu{}^\sigma \ldots + \cdots\}$$

Since $\xi_{\sigma;\tau}$ is an arbitrary antisymmetric matrix, its coefficient must be symmetric in σ and τ:

$$\delta_\mu^\tau T^\sigma{}_\nu \ldots + \delta_\nu^\tau T_\mu{}^\sigma \ldots + \cdots = \delta_\mu^\sigma T^\tau{}_\nu \ldots + \delta_\nu^\sigma T_\mu{}^\tau \ldots + \cdots$$

$$(13.4.6)$$

Since X was arbitrary, this must hold everywhere.

For a maximally form-invariant vector $A_\mu(x)$, Eq. (13.4.6) reads

$$\delta_\mu^\tau A^\sigma = \delta_\mu^\sigma A^\tau$$

Contracting τ with μ, we find that in N dimensions

$$N A^\sigma = A^\sigma$$

so, except for the trivial case $N = 1$, we must have

$$A^\sigma = 0 \tag{13.4.7}$$

For a maximally form-invariant tensor $B_{\mu\nu}$ of second rank, Eq. (13.4.6) reads

$$\delta_\mu{}^\tau B^\sigma{}_\nu + \delta_\nu{}^\tau B_\mu{}^\sigma = \delta_\mu{}^\sigma B^\tau{}_\nu + \delta^\sigma{}_\nu B_\mu{}^\tau$$

Contracting τ with μ gives

$$N B^\sigma{}_\nu + B_\nu{}^\sigma = B^\sigma{}_\nu + \delta^\sigma{}_\nu B_\mu{}^\mu$$

or, lowering the σ index,

$$(N - 1)B_{\sigma\nu} + B_{\nu\sigma} = g_{\sigma\nu} B_\mu{}^\mu \tag{13.4.8}$$

Subtracting the same equation with ν and σ interchanged yields

$$(N - 2)(B_{\sigma\nu} - B_{\nu\sigma}) = 0$$

so as long as $N \neq 2$, the tensor $B_{\sigma\nu}$ must be symmetric:

$$B_{\sigma\nu} = B_{\nu\sigma} \tag{13.4.9}$$

(In two dimensions, $B_{\sigma\nu}$ can have an antisymmetric part proportional to $g^{-1/2}\varepsilon_{\sigma\nu}$; see Section 4.4.) Using (13.4.9) in (13.4.8) gives now for $N \geq 3$ (and for the symmetric part of $B_{\sigma\nu}$ for $N = 2$)

$$B_{\sigma\nu} = f g_{\sigma\nu} \tag{13.4.10}$$

where

$$f \equiv \frac{1}{N} B_\mu{}^\mu$$

To determine the dependence of f on the coordinates, we can use (13.4.10) back in the form-invariance condition (13.4.3):

$$0 = \frac{\partial \xi^\rho}{\partial x^\mu} f g_{\rho\nu} + \frac{\partial \xi^\sigma}{\partial x^\nu} f g_{\mu\sigma} + \xi^\lambda \frac{\partial}{\partial x^\lambda} (f g_{\mu\nu})$$

But $g_{\mu\varphi}$ satisfies the Killing condition (13.1.4), so this becomes

$$0 = g_{\mu\nu} \xi^\lambda \frac{\partial f}{\partial x^\lambda}$$

In a maximally symmetric space we can at any given point choose ξ^λ to have any value we like, and therefore

$$\frac{\partial f}{\partial x^\lambda} = 0 \tag{13.4.11}$$

Thus *the only maximally form-invariant tensor of second rank is the metric tensor, times a possible constant.*

5 Spaces with Maximally Symmetric Subspaces

In many cases of physical importance, the whole space (or space-time) is not maximally symmetric, but it can be decomposed into maximally symmetric subspaces. For instance, a spherically symmetric three-dimensional space can be decomposed into a family of spherical surfaces centered on the origin, each of which is described by a metric of the form (13.3.28). Also, in Chapter 14 we shall deal with space-times in which the metric is spherically symmetric and homogeneous on each "plane" of constant time.

We shall see here that the maximal symmetry of a family of subspaces imposes very strong constraints on the metric of the whole space. In order to state and prove these results, let us first adopt a suitable coordinate system. If the whole space has N dimensions and its maximally symmetric subspaces have M dimensions, then we can distinguish these subspaces from each other with $N - M$ coordinate labels v^a, and locate points within each subspace with M coordinates u^i. Some illustrations are given in Table 13.1.

Table 13.1   Examples of Spaces with Maximally Symmetric Subspaces

Example	v-Coordinates	u-Coordinates
Spherically symmetric space	r	θ, φ
Spherically symmetric space-time	r, t	θ, φ
Spherically symmetric and homogeneous space-time	t	r, θ, φ

We say that the subspaces with constant v^a are maximally symmetric if the metric of the whole space is invariant under a group of infinitesimal transformations

$$u^i \rightarrow u'^i = u^i + \varepsilon \xi^i(u, v) \tag{13.5.1}$$

$$v^a \rightarrow v'^a = v^a \tag{13.5.2}$$

with $M(M + 1)/2$ independent Killing vectors ξ^i. These are transformations of the general form (13.1.3), but with the special feature that the v^a are invariant, so that

$$\xi^a(u, v) = 0 \tag{13.5.3}$$

Note that, although these transformations affect only the u-variables, there is no reason why the transformation rules cannot depend parametrically on the labels v^a of the particular subspace being transformed. Also, our statement that there are $M(M + 1)/2$ "independent" Killing vectors should be construed to mean that there are $M(M + 1)/2$ Killing vectors that are not subject to any linear relations with u-independent coefficients.

The general result that governs the structure of such spaces is contained in the following theorem: It is always possible to choose the u-coordinates so that the metric of the whole space is given by

$$- d\tau^2 \equiv g_{\mu\nu} \, dx^\mu \, dx^\nu = g_{ab}(v) \, dv^a \, dv^b + f(v)\tilde{g}_{ij}(u) \, du^i \, du^j \tag{13.5.4}$$

where $g_{ab}(v)$ and $f(v)$ are functions of the v-coordinates alone, and $\tilde{g}_{ij}(u)$ is a function of the u-coordinates alone that is by itself the metric of an M-dimensional maximally symmetric space. (The summation convention is in force, with $a, b, \ldots$ running over the $N - M$ labels of the v-coordinates, and $i, j, k, l, \ldots$ running over the M labels of the u-coordinates.)

To begin our proof, we set down the condition that (13.5.1) is an isometry of the whole metric $g_{\mu\nu}(x)$. It is convenient to use this condition here in its original form (13.1.4) rather than in the more elegant covariant form (13.1.5). Each index $\mu, \nu, \rho \ldots$ in (13.1.4) now runs over the $N - M$ coordinate labels of the v^a and the M coordinate labels of the u^i, so (13.1.4) now yields three separate equations: For $\rho = i, \sigma = j$ we have

$$0 = \frac{\partial \xi^k(u, v)}{\partial u^i} \, g_{kj}(u, v) + \frac{\partial \xi^k(u, v)}{\partial u^j} \, g_{ki}(u, v) + \xi^k(u, v) \frac{\partial g_{ij}(u, v)}{\partial u^k} \tag{13.5.5}$$

for $\rho = i, \sigma = a$ we have

$$0 = \frac{\partial \xi^k(u, v)}{\partial u^i} \, g_{ka}(u, v) + \frac{\partial \xi^k(u, v)}{\partial v^a} \, g_{ik}(u, v) + \xi^k(u, v) \frac{\partial g_{ia}(u, v)}{\partial u^k} \tag{13.5.6}$$

and for $\rho = a$, $\sigma = b$ we have

$$0 = \frac{\partial \xi^k(u, v)}{\partial v^a} g_{kb}(u, v) + \frac{\partial \xi^k(u, v)}{\partial v^b} g_{ka}(u, v) + \xi^k(u, v) \frac{\partial g_{ab}(u, v)}{\partial u^k}$$

$$(13.5.7)$$

Of these three equations, the first simply tells us that $g_{ij}(u, v)$ must, for each fixed set of v^a, be the metric of an M-dimensional space, with coordinates u^i, which admits the Killing vector ξ^i. We assume here that there are $M(M + 1)/2$ such independent Killing vectors, so this means that the submatrix $g_{ij}(u, v)$ is by itself a maximally symmetric metric for each set of fixed v^a. According to the arguments of Section 13.1, it also follows that at any given point u_0 we can find Killing vectors $\xi^k(u, v)$ for which $\xi^k(u_0, v)$ and $\xi_{k;l}(u_0, v)$ take arbitrary values, subject only to the requirement that $\xi_{k;l} = -\xi_{l;k}$. Thus the metric $g_{ij}(u, v)$ is for each v both homogeneous in u and isotropic about any point.

The other two equations contain information about the other elements g_{ai} and g_{ab}, and also about the v-dependence of the Killing vectors. (This v-dependence is not entirely arbitrary. For instance, it is true that by redefining the u-coordinates we can always arrange that the metric $g_{ij}(u, v)$ has v-independent Killing vectors $\xi^i(u)$, but the Killing vectors $\xi^i(u, v)$ of the whole space will then in general be linear combinations of the $\xi^i(u)$, with coefficients that can depend on the v-coordinates.) In order to disentangle the different information contained in (13.5.6) and (13.5.7), it is extremely useful to choose a new set of coordinates $u'^i(u, v)$ of the maximally symmetric subspaces, so that g'_{ja} vanishes. Suppose for a moment that we can find a function $U^k(v; u_0)$ that satisfies the differential equation

$$g_{ik}(U, v) \frac{\partial U^k}{\partial v^a} = -g_{ia}(U, v)$$

$$(13.5.8)$$

with the initial condition that

$$U^k(v_0; u_0) \equiv u_0{}^k$$

$$(13.5.9)$$

at some point $v_0{}^a$. The coordinates u'^i, v'^a are then defined by

$$u^i = U^i(v'; u')$$

$$(13.5.10)$$

$$v^a = v'^a$$

$$(13.5.11)$$

In this coordinate system, the metric has

$$g'_{ja}(u', v') = \frac{\partial u^l}{\partial u'^j} \frac{\partial u^k}{\partial v'^a} g_{lk}(u, v) + \frac{\partial u^l}{\partial u'^j} g_{la}(u, v)$$

$$= \frac{\partial U^l(v'; u')}{\partial u'^j} \left\{ \frac{\partial U^k(v'; u')}{\partial v'^a} g_{lk}(U, v') + g_{la}(U, v') \right\}$$

and thus (13.5.8) gives

$$g'_{ia} = 0$$

$$(13.5.12)$$

Thus we can construct u'-coordinates in which g'_{ia} vanishes, if we can find solutions of the differential equation (13.5.8) with arbitrary initial conditions (13.5.9).

We can rewrite (13.5.8) in the equivalent form

$$\frac{\partial U^k}{\partial v^a} = -F^k{}_a(U, v) \tag{13.5.13}$$

where

$$F^k{}_a(U, v) \equiv \bar{g}^{ki}(U, v)g_{ia}(U, v) \tag{13.5.14}$$

and $\bar{g}^{ij}$ is the matrix reciprocal to g_{ij}, that is,

$$\bar{g}^{ij}g_{jk} = \delta^i_k \tag{13.5.15}$$

(The bar is to remind us that the ij-element $\bar{g}^{ij}$ of the matrix reciprocal to g_{ij} is not the same as the ij-element g^{ij} of the matrix $g^{\mu\nu}$ reciprocal to $g_{\mu\nu}$.) When there is only one v-coordinate, as will be the case in our chapters on cosmology, it is obvious that (13.5.13) can be solved with arbitrary initial conditions. In the general case, we shall have to do some work to prove that (13.5.13) is integrable. Our method is the same as in Section 13.2; we try to solve (13.5.13) in a neighborhood of v_0 with a power series in $v - v_0$:

$$U^k = \sum_{n=0}^{\infty} \frac{1}{n!} c^k{}_{a_1 \cdots a_n}(v - v_0)^{a_1} \cdots (v - v_0)^{a_n} \tag{13.5.16}$$

Clearly the initial conditions (13.5.9) are satisfied if we choose the $n = 0$ coefficient as

$$c^k = u_0{}^k$$

and Eq. (13.5.13) is satisfied to zero order in $v - v_0$ if we choose

$$c^k{}_a = -F^k{}_a(u_0, v_0)$$

Now, proceeding by mathematical induction, suppose that we are able to choose the terms in (13.5.16) up to order $(v - v_0)^n$ so that (13.5.13) is satisfied to order $(v - v_0)^{n-1}$. Then we can use these terms to calculate the term in $F^k{}_a(U, v)$ of order $(v - v_0)^n$. Let us write this term as

$$[F^k{}_a(U(v; u_0), v)]_{\text{order } n} = \frac{1}{n!} f^k{}_{ab_1 \cdots b_n}(v - v_0)^{b_1} \cdots (v - v_0)^{b_n}$$

Then (13.5.16) will satisfy (13.5.13) to order $(v - v_0)^n$ if we choose the term in U of order $n + 1$ as

$$[U^k(v; u_0)]_{n+1} = \frac{1}{(n + 1)!} f^k{}_{ab_1 \cdots b_n}(v - v_0)^a(v - v_0)^{b_1} \cdots (v - v_0)^{b_n}$$

providing that f is symmetric in all its subscripts. Since $f^k{}_{ab_1 \cdots b_n}$ can obviously be chosen symmetric in the b's, it is sufficient to require that it should also be symmetric between a and any b, or equivalently that

$$\left[\frac{\partial}{\partial v^b} F^k{}_a(U(v; u_0), v)\right]_{\text{order } n-1}$$

should be symmetric in a and b. But U is assumed to satisfy (13.5.13) to order $(v - v_0)^{n-1}$, so this condition is satisfied if

$$\left[-\frac{\partial F^k{}_a(u, v)}{\partial u^l} F^l{}_b(u, v) + \frac{\partial F^k{}_a(u, v)}{\partial v^b} \right]_{u \,=\, U(v;\,u_0)}$$

is symmetric in a and b. We thus conclude that (13.5.13) is integrable if

$$\frac{\partial F^k{}_a(u, v)}{\partial u^l} F^l{}_b(u, v) - \frac{\partial F^k{}_a(u, v)}{\partial v^b} = \frac{\partial F^k{}_b(u, v)}{\partial u^l} F^l{}_a(u, v) - \frac{\partial F^k{}_b(u, v)}{\partial v^a}$$

$$(13.5.17)$$

for all u, v.

In order to prove that (13.5.17) is indeed satisfied, we return to the Killing vector condition (13.5.6). Multiplying with $\bar{g}^{il}$, we have

$$\frac{\partial \xi^l}{\partial v^a} = -\bar{g}^{il} \frac{\partial \zeta^m}{\partial u^i} g_{ma} - \bar{g}^{il} \zeta^k \frac{\partial g_{ia}}{\partial u^k}$$

Also, multiplying (13.5.5) with $\bar{g}^{il} \cdot \bar{g}^{jm}$ gives

$$\bar{g}^{il} \frac{\partial \zeta^m}{\partial u^i} + \bar{g}^{jm} \frac{\partial \xi^l}{\partial u^j} = -\zeta^k \bar{g}^{il} \bar{g}^{jm} \frac{\partial g_{ij}}{\partial u^k}$$

$$= \zeta^k \frac{\partial \bar{g}^{lm}}{\partial u^k}$$

so

$$\frac{\partial \xi^l}{\partial v^a} = \bar{g}^{jm} \frac{\partial \xi^l}{\partial u^j} g_{ma} - \zeta^k \frac{\partial \bar{g}^{lm}}{\partial u^k} g_{ma} - \zeta^k \bar{g}^{lm} \frac{\partial g_{ma}}{\partial u^k}$$

Recalling (13.5.14), we can write this as

$$\frac{\partial \xi^l}{\partial v^a} = F^j{}_a \frac{\partial \xi^l}{\partial u^j} - \zeta^k \frac{\partial F^l{}_a}{\partial u^k} \tag{13.5.18}$$

Now differentiate with respect to v^b; this gives

$$\frac{\partial^2 \xi^l}{\partial v^b \, \partial v^a} = F^j{}_a \frac{\partial}{\partial u^j}\left(\frac{\partial \xi^l}{\partial v^b}\right) + \frac{\partial F^j{}_a}{\partial v^b} \frac{\partial \xi^l}{\partial u^j} - \frac{\partial \zeta^k}{\partial v^b} \frac{\partial F^l{}_a}{\partial u^k} - \zeta^k \frac{\partial^2 F^l{}_a}{\partial v^b \, \partial u^k}$$

or, using (13.5.18) on the right-hand side,

$$\frac{\partial^2 \xi^l}{\partial v^b \, \partial v^a} = F^j{}_a F^i{}_b \frac{\partial^2 \xi^l}{\partial u^j \, \partial u^i} + F^j{}_a \frac{\partial F^i{}_b}{\partial u^j} \frac{\partial \xi^l}{\partial u^i} - F^j{}_a \frac{\partial F^l{}_b}{\partial u^k} \frac{\partial \xi^k}{\partial u^j} - F^j{}_a \frac{\partial^2 F^l{}_b}{\partial u^k \, \partial u^j} \zeta^k$$

$$+ \frac{\partial F^j{}_a}{\partial v^b} \frac{\partial \xi^l}{\partial u^j} - F^i{}_b \frac{\partial F^l{}_a}{\partial u^k} \frac{\partial \xi^k}{\partial u^i} + \frac{\partial F^k{}_b}{\partial u^i} \frac{\partial F^l{}_a}{\partial u^k} \xi^i - \frac{\partial^2 F^l{}_a}{\partial v^b \, \partial u^k} \zeta^k$$

But this must be symmetric between a and b, so

$$
\begin{aligned}
0 = & \left\{ F^j{}_a \frac{\partial F^i{}_b}{\partial u^j} - F^j{}_b \frac{\partial F^i{}_a}{\partial u^j} + \frac{\partial F^i{}_a}{\partial v^b} - \frac{\partial F^i{}_b}{\partial v^a} \right\} \frac{\partial \xi^l}{\partial u^i} \\
& + \left\{ -F^j{}_a \frac{\partial^2 F^l{}_b}{\partial u^k \, \partial u^j} + F^j{}_b \frac{\partial^2 F^l{}_a}{\partial u^k \, \partial u^j} + \frac{\partial F^i{}_b}{\partial u^k} \frac{\partial F^l{}_a}{\partial u^i} - \frac{\partial F^i{}_a}{\partial u^k} \frac{\partial F^l{}_b}{\partial u^i} \right. \\
& \left. - \frac{\partial^2 F^l{}_a}{\partial v^b \, \partial u^k} + \frac{\partial^2 F^l{}_b}{\partial v^a \, \partial u^k} \right\} \xi^k
\end{aligned}
\tag{13.5.19}
$$

We have already remarked that our assumption that there are $M(M+1)/2$ independent Killing vectors allows us at any given point to find Killing vectors for which ξ^k vanishes and for which $\xi_{k;j} = g_{kl} \, \partial \xi^l / \partial u^i$ is an arbitrary antisymmetric matrix. In particular, we can at any given point choose ξ^i so that

$$
\xi^k = 0
$$

$$
\xi_{k;i} = g_{kl} \frac{\partial \xi^l}{\partial x^i} = \delta_{km} \delta_{in} - \delta_{kn} \delta_{im}
$$

Hence multiplying (13.4.19) with g_{kl} and setting $k = n \neq m$, we find that

$$
F^j{}_a \frac{\partial F^m{}_b}{\partial u^j} - F^j{}_b \frac{\partial F^m{}_a}{\partial u^j} = \frac{\partial F^m{}_b}{\partial v^a} - \frac{\partial F^m{}_a}{\partial v^b}
$$

which is just the desired relation (13.5.17). The coefficient of ξ^i in (13.5.19) must also vanish, but we do not need this information here.

To return now to the main line of our proof: Having proved (13.5.17), we know that (13.5.13) is integrable, so we can construct the coordinates u'^i, v'^a defined by (13.5.10) and (13.5.11), in which the metric components g'_{ia} vanish. Let us do so, and drop the primes, so that now

$$
g_{ia} = 0
\tag{13.5.20}
$$

The Killing vector conditions (13.5.6), (13.5.7) now read

$$
0 = \frac{\partial \xi^k}{\partial v^a} g_{ik}
\tag{13.5.21}
$$

$$
0 = \xi^k \frac{\partial g_{ab}}{\partial u^k}
\tag{13.5.22}
$$

Since g_{ik} is nonsingular, it follows from (13.5.21) that

$$
\frac{\partial \xi^k}{\partial v^a} = 0
\tag{13.5.23}
$$

Also, we have noted that at each point we can find Killing vectors for which ξ^k takes any arbitrary value, so the coefficient of ξ^k in (13.5.22) must vanish:

$$\frac{\partial g_{ab}}{\partial u^k} = 0 \qquad (13.5.24)$$

It only remains to show that $g_{ij}(u, v)$ is v-independent, except for a possible factor $f(v)$. We use the fact that for any fixed v_0 there are $M(M + 1)/2$ independent Killing vectors of $g_{ij}(u, v_0)$ that, according to (13.5.23), are also Killing vectors of $g_{ij}(u, v)$ for any v. Each one of these Killing vectors $\xi^i(u)$ will then satisfy (13.5.5) at $v = v_0$ and for general v:

$$0 = \frac{\partial \xi^k(u)}{\partial u^i} g_{kj}(u, v_0) + \frac{\partial \xi^k(u)}{\partial u^j} g_{ki}(u, v_0) + \xi^k(u) \frac{\partial g_{ij}(u, v_0)}{\partial u^k}$$

$$0 = \frac{\partial \xi^k(u)}{\partial u^i} g_{kj}(u, v) + \frac{\partial \xi^k(u)}{\partial u^j} g_{ki}(u, v) + \xi^k(u) \frac{\partial g_{ij}(u, v)}{\partial u^k}$$

We can interpret these two equations as saying that $g_{ij}(u, v)$ is a maximally form-invariant tensor [in the sense of Eq. (13.4.3)] in the maximally symmetric space with metric $g_{ij}(u, v_0)$. It follows then, according to Eqs. (13.4.10) and (13.4.11), that the tensor $g_{ij}(u, v)$ is proportional to the metric $g_{ij}(u, v_0)$, with a u-independent coefficient:

$$g_{ij}(u, v) = f(v, v_0)g_{ij}(u, v_0)$$

The valve of v_0 can be fixed in any way we like, so we can suppress the label v_0, and write this as

$$g_{ij}(u, v) = f(v)\tilde{g}_{ij}(u) \qquad (13.5.25)$$

with

$$f(v) \equiv f(v, v_0) \qquad \tilde{g}_{ij}(u) \equiv g_{ij}(u, v_0) \qquad (13.5.26)$$

Putting together (13.5.20), (13.5.24), and (13.5.25) now shows that the metric $g_{\mu\nu}(u, v)$ does have the form given by (13.5.4), and (13.5.26) and (13.5.5) with $v = v_0$ show that $\tilde{g}_{ij}(u)$ is a maximally symmetric metric, as was to be proved.

This theorem could also have been proved under the apparently weaker assumption, that the whole space can be decomposed into subspaces that are isotropic about every point. This assumption means that any point u_0, v we can find Killing vectors of the whole space with $\xi^a \equiv 0$, for which ξ^i vanishes at u_0, v, and for which $\xi_{i;k}$ at u_0, v is an arbitrary antisymmetric matrix. In particular, we can find $M(M - 1)/2$ Killing vectors $\xi^{(lm)}(u, v; u_0)$ with

$$\xi^{a(lm)}(u, v; u_0) = 0$$

$$\xi^{i(lm)}(u, v; u_0) = -\xi^{i(ml)}(u, v; u_0)$$

for which

$$\xi^{(lm)}_{i;j}(u_0, v; u_0) \equiv g_{ik}(u_0, v) \left(\frac{\partial \xi^{k(lm)}(u, v; u_0)}{\partial u^j}\right)_{u=u_0}$$

$$= \delta_i{}^l \delta_j{}^m - \delta_i{}^m \delta_j{}^l$$

We can then define

$$\xi^{\mu(l)}(u, v; u_0) \equiv \frac{\partial}{\partial u_0{}^m} \xi^{\mu(lm)}(u, v; u_0)$$

and the arguments of Section 13.1 show that these are Killing vectors *of the whole space*, with

$$\xi^{a(l)}(u, v; u_0) = 0$$

and with

$$\xi^{i(l)}(u_0, v; u_0) = - \frac{1}{(N-1)} \bar{g}^{il}(u_0, v)$$

The existence of the $M(M+1)/2$ independent Killing vectors $\xi^{\mu(lm)}$ and $\xi^{\mu(l)}$ shows that the space does have maximally symmetric subspaces after all.

In all cases of practical importance, the maximally symmetric subspaces are *spaces*, as opposed to space-times, so all eigenvalues of the submatrix g_{ij} are positive. In this case, we can use (13.3.23), (13.3.24), or (13.3.25) to evaluate $\tilde{g}_{ij} \, du^i \, du^j$, and (13.5.3) then gives

$$- d\tau^2 = g_{ab}(v) \, dv^a \, dv^b + f(v) \left\{ d\mathbf{u}^2 + \frac{k(\mathbf{u} \cdot d\mathbf{u})^2}{1 - k\mathbf{u}^2} \right\} \tag{13.5.27}$$

where $f(v)$ is positive and

$$k = \begin{cases} +1 & \text{if max. sym. subsp. has } K > 0 \\ -1 & \text{if max. sym. subsp. has } K < 0 \\ 0 & \text{if max. sym. subsp. has } K = 0 \end{cases} \tag{13.5.28}$$

[We have absorbed the curvature constant $|K|^{-1}$ appearing in (13.3.23) and (13.3.24) into the function $f(v)$.] Let us now use this formula to treat the special cases listed in Table 13.1:

(A) **Spherically Symmetric Space.** Suppose that the dimensionality of the whole space is $N = 3$, that all eigenvalues of its metric are positive, and that it has maximally symmetric two-dimensional subspaces with positive curvature. Then there is one v-coordinate, which we can call r, and two u-coordinates, which we can replace with angles θ, φ defined by

$$u^1 = \sin\theta \cos\varphi \qquad u^2 = \sin\theta \sin\varphi \tag{13.5.29}$$

Eq. (13.5.27), with $k = 1$, then gives

$$ds^2 = g(r)\,dr^2 + f(r)\,\{d\theta^2 + \sin^2\theta\,d\varphi^2\} \tag{13.5.30}$$

with $f(r)$ and $g(r)$ positive functions of r.

(B) **Spherically Symmetric Space-Time.** Suppose that the dimensionality of the whole space-time is $N = 4$, that three of the eigenvalues of its metric are positive and one is negative, and that it has maximally symmetric two-dimensional subspaces whose metric has positive eigenvalues and positive curvature. Then there are two v-coordinates, which we can call r and t, and two u-coordinates, which can be replaced with θ and φ as in (13.5.29). Eq. (13.5.27), with $k = 1$, then gives

$$- d\tau^2 = g_{tt}(r, t)\,dt^2 + 2g_{rt}(r, t)\,dr\,dt + g_{rr}(r, t)\,dr^2$$
$$+ f(r, t)\,\{d\theta^2 + \sin^2\theta\,d\varphi^2\} \tag{13.5.31}$$

where $f(r, t)$ is a positive function and $g_{ij}(r, t)$ is a 2×2 matrix with one positive and one negative eigenvalue.

(C) **Spherically Symmetric Homogeneous Space-Time.** Suppose that the dimensionality of the whole space-time is $N = 4$, that three of the eigenvalues of its metric are positive and one is negative, and that it has maximally symmetric *three*-dimensional subspaces whose metric has positive eigenvalues and arbitrary curvature. Then there is one v-coordinate and three u-coordinates, and (13.5.27) gives

$$- d\tau^2 = g(v)\,dv^2 + f(v)\left\{d\mathbf{u}^2 + \frac{k(\mathbf{u} \cdot d\mathbf{u})^2}{1 - k\mathbf{u}^2}\right\}$$

where $f(v)$ is a positive function and $g(v)$ is a negative function of v. It is very convenient to define new coordinates t, v, θ, φ by

$$\int (- g(v))^{1/2}\,dv \equiv t$$

$$u^1 \equiv r \sin\theta \cos\varphi$$
$$u^2 \equiv r \sin\theta \sin\varphi$$
$$u^3 \equiv r \cos\theta$$

We then have

$$d\tau^2 = dt^2 - R^2(t)\left\{\frac{dr^2}{1 - kr^2} + r^2\,d\theta^2 + r^2 \sin^2\theta\,d\varphi^2\right\} \tag{13.5.32}$$

where $R(t) \equiv \sqrt{f(v)}$.

The first two examples show how it is possible to capture the essence of spherical symmetry by giving a qualitative description of a space or space-time in terms of dimensionalities, signs of eigenvalues and curvatures, and the maximal symmetry of its subspaces. The metrics (13.5.30) and (13.5.31) are just what we would have expected on more elementary grounds; indeed, (13.5.31) was our starting point in Section 11.7.

On the other hand, our third example leads to a result that could not easily have been anticipated. It is true that Eq. (13.5.32) was already derived in Section 11.9 as the metric inside a spherically symmetric collapsing star of uniform density and zero pressure. The beautiful new thing we have learned in this chapter is that this metric can be derived solely from the assumption of homogeneity and isotropy, with no use of the Einstein field equations.

13 BIBLIOGRAPHY

☐ L. P. Eisenhart, *Continuous Groups of Transformations* (corrected ed. Dover Publications, New York, 1961).

☐ L. P. Eisenhart, *Riemannian Geometry* (Princeton University Press, Princeton, N. J., 1926).

☐ S. Helgason, *Differential Geometry and Symmetric Spaces* (Academic Press, New York, 1962).

13 REFERENCES

1. W. Killing, J. f. d. reine u. angew. Math. (Crelle), **109**, 121 (1892).
2. W. de Sitter, Proc. Roy. Acad. Sci. (Amsterdam), **19**, 1217 (1917); **20**, 229 (1917); **20**, 1309 (1917); Mon. Not. Roy. Astron. Soc., **78**, 3 (1917).

PART FIVE
COSMOLOGY

"We wonder, Oh, we wonder,
what on earth the world
may be?"
W. S. Gilbert, The Mikado

14 COSMOGRAPHY

Modern science began with the discovery that the earth is not at the center of the universe. Antianthropocentrism has become incorporated into the scientific mentality, and no one now would seriously suggest that the earth, or the solar system, or our galaxy, or our local group of galaxies, occupies any specially favored position in the cosmos. Rather, our intuition now runs in precisely the opposite direction. A large portion of modern cosmological theory is built on the *Cosmological Principle*, the hypothesis that all positions in the universe are essentially equivalent. Of course, the homogeneity of the universe has to be understood in the same sense as the homogeneity of a gas: It does not apply to the universe in detail, but only to a "smeared-out" universe averaged over cells of diameter 10^8 to 10^9 light years, which are large enough to include many clusters of galaxies. Also, it appears that the universe is spherically symmetric about us, so included in the Cosmological Principle is the assumption that the "smeared" universe is isotropic about every point.

The question still remains, whether the universe is spherically symmetric and homogeneous at all times, or merely over some temporary present phase of its history. There has been an interesting suggestion, to be discussed in Section 15.11, that the universe may have been highly anisotropic during some dense early phase, but that the anisotropies have since been largely smoothed out through the action of neutrino viscosity and other dissipative effects. However, even in such theories, the universe has been highly isotropic and homogeneous over all of that part of its history which is directly accessible to astronomical observation.

This chapter will outline and apply a mathematical framework for the description of the universe, based entirely on the Cosmological Principle, and on those parts of general relativity that follow directly from the Principle of Equivalence. (This includes Chapters 2 to 6 and Chapter 13.) I shall first show that the Cosmological Principle allows the specification of the cosmic metric entirely in terms of a "radius" $R(t)$ and a trichotomic constant k [as in Eq. (13.5.32)] and we shall then see how astronomical observations can be interpreted as measurements of $R(t)$ and k.

This inherently kinematic approach, pioneered in the 1930's by H. P. Robertson[1] and A. G. Walker,[2] is incomplete, in that it does not provide an *a priori* prediction of the function $R(t)$. To calculate $R(t)$ we need to make some assumption about the material content of the universe, and then derive the Robertson-Walker metric as a solution of the Einstein field equations, as first done by Alexandre Friedmann[3] in 1922. Our discussion of the contents of the universe, and the use of the Einstein field equations, will be postponed until the next chapter, on cosmology.

Why make this distinction between cosmography and cosmology? The reason is simply that we do not know the equation of state of the matter and radiation of the universe throughout its history, and even if we did, we could not be sure that the Einstein equations really hold over cosmic times and distances. A modification of the field equations or the equation of state, such as the introduction of a Brans-Dicke field, a cosmological constant, or a large population of neutrinos or gravitons, would affect the function $R(t)$ and invalidate the simplest Friedmann solution, but it would not require us to make any change in the descriptive framework assembled in this chapter.

There remains the possibility that the universe is not homogeneous and isotropic after all. It might be homogeneous but not isotropic, as in the model of K. Gödel.[4] However, the cosmic microwave radiation discussed in Chapter 15 appears to be highly isotropic. (The universe cannot be isotropic about every point without also being homogeneous, as shown in the last chapter.) A more radical notion is that there is no "smeared" universe at all, but only clusters of galaxies, and clusters of clusters, and clusters of clusters of clusters, and so on, as in the hierarchical model proposed in 1908 by C. V. I. Charlier.[5] Empirical arguments for such super-clustering have been offered by G. de Vaucouleurs,[6] but the work of F. Zwicky,[7] G. O. Abell,[8] and J. H. Oort[9] indicates that the hierarchy stops at clusters of galaxies or at most clusters of clusters of galaxies, and shows no evidence of inhomogeneities of larger scale.

The real reason, though, for our adherence here to the Cosmological Principle is not that it is surely correct, but rather, that it allows us to make use of the extremely limited data provided to cosmology by observational astronomy. If we make any weaker assumptions, as in the anisotropic or hierarchical models, then the metric would contain so many undetermined functions (whether or not we use the field equations) that the data would be hopelessly inadequate to determine the metric. On the other hand, by adopting the rather restrictive mathematical framework described in this chapter, we have a real chance of confronting theory with observation. If the data will not fit into this framework, we shall be able to conclude that either the Cosmological Principle or the Principle of Equivalence is wrong. Nothing could be more interesting.

1 The Cosmological Principle

The Cosmological Principle is the hypothesis that the universe is spatially homogeneous and isotropic. Before applying this principle, we shall have to formulate our intuitive ideas of homogeneity and isotropy in precise mathematical terms.

First, let us fix our attention on one particular space-time coordinate system, of the sort that might be used by terrestrial cosmographers. Spatial coordinates x^i can be constructed with origin $x^i = 0$ at the center of the Milky Way, with coordinate directions fixed by the line of sight from the Milky Way to some typical distant galaxies, and with a scale of distances defined by the apparent luminosities of distant galaxies, or other suitable objects, as seen from the Milky Way. To define a time coordinate, it is convenient to use the evolving universe itself as a clock. It is believed that several cosmic scalar fields, such as the proper energy density ρ, or the black-body radiation temperature T_γ (see Chapter 15) are everywhere decreasing monotonically; choose any one of these, say a scalar S, and let the time of any event be any definite decreasing function $t(S)$ of the chosen scalar, when *and where* the event occurs. (We shall have to reopen the question of how to define the time when we consider a steady state universe, in Section 14.8.) The coordinates **x**, t so defined will be called the *cosmic standard coordinate system*.

The Cosmological Principle can be formulated as a statement about the existence of equivalent coordinate systems. Suppose that we use the cosmic standard coordinate system to carry out astronomical observations, determining (never mind how!) the metric tensor $g_{\mu\nu}$, the energy-momentum tensor $T_{\mu\nu}$, and all other cosmic fields, as functions of the cosmic standard coordinates x^μ. A different set of space-time coordinates x'^μ may be considered *equivalent* to the cosmic standard coordinates, if the whole history of the universe appears the same in the x'^μ coordinate system as in the cosmic standard coordinate system. This requires that every cosmic field $g'_{\mu\nu}(x')$, $T'_{\mu\nu}(x')$, and so on, must be the same function of the x'^μ as the corresponding quantities $g_{\mu\nu}(x)$, $T_{\mu\nu}(x)$, and so on, are of the standard coordinates x^μ. That is, at any *coordinate* point y^μ, we must have

$$g_{\mu\nu}(y) = g'_{\mu\nu}(y) \tag{14.1.1}$$

$$T_{\mu\nu}(y) = T'_{\mu\nu}(y) \quad \text{etc.} \tag{14.1.2}$$

In the language of the last chapter, Eq. (14.1.1) says that the coordinate transformation $x \to x'$ must be an *isometry*, and (14.1.2) says that $T_{\mu\nu}$, and so on, must be *form-invariant* under this transformation.

In particular, Eq. (14.1.2) will have to hold for the scalar S used to define our cosmic standard time t. Since S is by definition a function only of t, and a scalar, Eq. (14.1.2) for S reads, at $y = x'$,

$$S(t') = S'(x') \equiv S(x) = S(t)$$

and so

$$t' = t \tag{14.1.3}$$

All coordinate systems that are equivalent to the cosmic standard system necessarily use cosmic standard time.

The assumption of spatial isotropy can now be formulated as the requirement that there exists a family of coordinate systems $x'^{\mu}(x; \theta)$, depending on three independent parameters θ^1, θ^2, θ^3, which are equivalent to the cosmic standard coordinates, and which have the same origin, that is,

$$x'^{i}(0, t; \theta) = 0 \tag{14.1.4}$$

We can intuitively think of the three parameters θ^n as Euler angles that specify the orientation of the x'^i coordinate axes relative to the x^i coordinate axes, but it is unnecessary to be so specific; the important thing is that there be *three* independent parameters. (In formulating this assumption, we have tacitly assumed that the privileged Lorentz frame in which the universe appears isotropic happens to coincide more or less with our own galaxy.)

It is a little trickier to formulate the assumption of homogeneity. Clearly, homogeneity does not mean that *any* object can be chosen as the origin of a coordinate system equivalent to our cosmic standard coordinates—after all, the universe looks different to an observer moving away from the Milky Way at half the speed of light than it does to us! The most we can expect is that every point x^{μ} in space-time is on some "fundamental trajectory" $x^i = X^i(t)$, which can serve as the origin of a coordinate system x'^{μ} equivalent to the cosmic standard system. (This is closely related to a postulate called Weyl's principle, used in some formulations of cosmology.) The Milky Way appears to be a rather ordinary galaxy, more or less at rest with respect to its nearest neighbors, so we can expect that the fundamental trajectories $\mathbf{X}(t)$ are pretty well defined by the motions of typical members of the cosmic gas of galaxies, but this is by no means an essential part of the assumption of homogeneity. The important point is that, since the $\mathbf{X}(t)$ at any time t fill up all space, they are determined by *three* independent parameters a^i, which can be taken, for instance, as the values $a^i \equiv X^i(T)$ of $X^i(t)$ at some particular time $t = T$. Thus homogeneity means that there is a three-parameter set of coordinates $\bar{x}'^{\mu}(x; a)$, which are equivalent to the cosmic standard coordinates x^{μ}, and which have origin on the trajectory $x^i = X^i(t; a)$, that is,

$$\bar{x}'^{i}(\mathbf{X}(t; a), t; a) = 0 \tag{14.1.5}$$

To be more precise the $\mathbf{X}(t; a)$ are the trajectories of the privileged observers to whom the universe appears isotropic.

Putting this together, we see that the Cosmological Principle entails the existence of two independent three-parameter families of coordinate transformations $x \rightarrow x'$, $x \rightarrow \bar{x}'$, which are isometries in the sense of Eq. (14.1.1), and which according to Eq. (14.1.3), leave the time coordinate invariant. The universe therefore satisfies the requirements assumed in Section 13.5 for a four-dimensional space with three-dimensional maximally symmetric subspaces $t = $ const.

(To see this in detail, we can descend to the case of infinitesimal transformations, letting θ^i and a^i approach zero. There are then six "Killing vectors" $\xi_j^\mu(x)$ and $\bar{\xi}_j^\mu(x)$, defined by

$$\xi_j^i(x) \equiv \left.\frac{\partial x'^i(x\,;\,\theta)}{\partial \theta^j}\right|_{\theta=0} \qquad \xi_j^t(x) \equiv 0 \tag{14.1.6}$$

$$\bar{\xi}_j^i(x) \equiv \left.\frac{\partial \bar{x}'^i(x\,;\,a)}{\partial a^j}\right|_{a=0} \qquad \bar{\xi}_j^t(x) \equiv 0 \tag{14.1.7}$$

It is only necessary to show that these six vectors are independent. Suppose that they satisfy a linear relation

$$\sum_j c^j(t)\xi_j^i(x) + \sum_j \bar{c}^j(t)\bar{\xi}_j^i(x) = 0 \tag{14.1.8}$$

At the origin, Eqs. (14.1.4) and (14.1.5) give

$$\xi_j^i(0,\,t) = 0 \tag{14.1.9}$$

$$\bar{\xi}_j^i(0,\,t) = -\left.\frac{\partial X^i(t,\,a)}{\partial a^j}\right|_{a=0} \tag{14.1.10}$$

so at $x^i = 0$, Eq. (14.1.8) gives

$$\sum_j \bar{c}^j(t)\left(\frac{\partial X^i(t,\,a)}{\partial a^j}\right)_{a=0} = 0$$

Since the a^i are independent parameters, this requires that

$$\bar{c}^j(t) = 0 \tag{14.1.11}$$

Going back to (14.1.8) and (14.1.6), we then have

$$\sum_j c^j(t)\left(\frac{\partial x'^i(x\,;\,\theta)}{\partial \theta^j}\right)_{\theta=0} = 0$$

and since the θ^i are independent parameters, this requires that

$$c^j(t) = 0 \tag{14.1.12}$$

Thus there are six independent Killing vectors with $\xi^t = 0$, the maximum number possible (see Section 13.1) in three dimensions.)

In summary, the Cosmological Principle can be formulated in the language of Chapter 13, as follows:

(i) The hypersurfaces with constant cosmic standard time are maximally symmetric subspaces of the whole of space-time.

(ii) Not only the metric $g_{\mu\nu}$, but all cosmic tensors such as $T_{\mu\nu}$, are form-invariant with respect to the isometries of these subspaces.

2   The Robertson-Walker Metric

The formulation of the Cosmological Principle given in the last section allows us to apply the results of Section 13.5 for spaces with maximally symmetric subspaces. We see immediately that it must be possible to choose coordinates r, θ, ϕ, t, for which the metric takes the form given in Eq. (13.5.32):

$$d\tau^2 = dt^2 - R^2(t)\left\{\frac{dr^2}{1 - kr^2} + r^2\,d\theta^2 + r^2\sin^2\theta\,d\phi^2\right\} \qquad (14.2.1)$$

where $R(t)$ is an unknown function of time, and k is a constant, which by a suitable choice of units for r can be chosen to have the value $+1$, 0, or -1. (These are not necessarily the same as the cosmic standard coordinates introduced in the last section, although t in Eq. (14.2.1) is the cosmic standard time, or a function of it.) The metric (14.2.1) is known in cosmology as the *Robertson-Walker metric*.

It is interesting to consider the geometrical properties of the three-dimensional spaces of constant t. These have metric

$$^3g_{rr} = \frac{R^2(t)}{1 - kr^2} \qquad ^3g_{\theta\theta} = r^2 R^2(t) \qquad ^3g_{\phi\phi} = r^2\sin^2\theta R^2(t) \qquad (14.2.2)$$

with $^3g_{\mu\nu}$ vanishing for $\mu \neq \nu$. Comparing with (13.3.23)–(13.3.25) shows that the *three-dimensional* curvature scalar is

$$^3K(t) = kR^{-2}(t) \qquad (14.2.3)$$

For $k = -1$ or $k = 0$ the space is infinite, while for $k = +1$ it is finite (though unbounded), with proper circumference given by Eq. (13.3.33) as

$$^3L = 2\pi R(t) \qquad (14.2.4)$$

and proper volume given by Eq. (13.3.29) as

$$^3V = 2\pi^2 R^3(t) \qquad (14.2.5)$$

For $k = +1$ the spatial universe can be regarded as the surface of a sphere of radius $R(t)$ in four-dimensional Euclidean space (see Section 13.3), and $R(t)$ can justly be called the "radius of the universe." For $k = -1$ and $k = 0$ no such interpretation is possible, but $R(t)$ still sets the scale of the geometry of space, so $R(t)$ will in all cases be called the *cosmic scale factor*.

The construction in Section 13.5 of the coordinates r, θ, ϕ, t was carried out in such a way that the coordinate transformations, which leave the four-dimensional metric (14.2.1) form-invariant, are just the purely spatial transformations that leave (14.2.2) form-invariant. These include the rigid rotations

$$x'^i = R^i{}_j x^j \qquad (i, j = 1, 2, 3) \tag{14.2.6}$$

where R is an arbitrary orthogonal matrix (and as usual $x^1 \equiv r \sin \theta \cos \phi$, $x^2 \equiv r \sin \theta \sin \phi$, $x^3 \equiv r \cos \theta$), together with "quasitranslations," given by setting KC equal to k times the unit matrix in Eq. (13.3.17):

$$\mathbf{x}' = \mathbf{x} + \mathbf{a} \left\{ (1 - k\mathbf{x}^2)^{1/2} - [1 - (1 - k\mathbf{a}^2)^{1/2}] \left(\frac{\mathbf{x} \cdot \mathbf{a}}{\mathbf{a}^2} \right) \right\} \tag{14.2.7}$$

where $\mathbf{a}$ is an arbitrary three-vector.

The transformation (14.2.7) carries the origin into the point $\mathbf{a}$, so we can conclude that any *fixed* point can serve as the origin of a system of coordinates equivalent to the coordinate system of Eq. (14.2.1). That is, the "fundamental trajectories" of observers, to whom the universe looks the same as it does to us, are just $\mathbf{X}(t; \mathbf{a}) = \mathbf{a}$. We have already noted in the last section that the fundamental trajectories ought to be close to the paths of "typical" galaxies, so we can tentatively conclude that the spatial coordinates r, θ, ϕ form a *comoving system*, in the sense that *typical galaxies have constant spatial coordinates r, θ, ϕ*. One can imagine the comoving coordinate mesh to be like lines painted on the surface of a balloon, on which dots represent typical galaxies. As the balloon is inflated or deflated the dots will move, but the lines will move with them, so each dot will keep the same coordinates.

It is important to note that the fundamental trajectories $\mathbf{x} = \mathrm{const}$ are geodesics, because Eq. (14.2.1) gives

$$\Gamma^\mu_{tt} = 0 \tag{14.2.8}$$

Thus the statement that a galaxy has constant r, θ, ϕ is perfectly consistent with the supposition that galaxies are in free fall. Note also that the time coordinate t in (14.2.1) is not only a possible "cosmic standard" time in the sense of the last section; it is also the proper time told by a clock at rest in any typical freely falling galaxy. The coordinates x, t are thus co-moving in precisely the same sense as the "Gaussian normal" coordinates introduced in Section 11.8.

We can obtain a deeper insight into the behavior of matter in a Robertson-Walker universe by applying the Cosmological Principle to the tensors that describe the average state of cosmic matter, such as the energy-momentum tensor $T^{\mu\nu}$, and the current $J_G{}^\mu$ of galaxies. ($J_G{}^\mu$ is defined exactly like the electric current (5.2.13), but the sum runs over galaxies instead of particles, and a factor 1 replaces e_n.) All such tensors are required to be form-invariant [in the sense of Section 13.4 or

Eq. (14.1.2)] with respect to those coordinate transformations, such as (14.2.6) and (14.2.7), which leave the metric (14.2.1) form-invariant. These "isometries" are purely spatial, so they transform $J_G{}^t$ and T^{tt} as three-scalars; $J_G{}^i$ and T^{it} as three-vectors; and T^{ij} as a three-tensor. According to the theorems proved in Section 13.4, this then requires that

$$J_G{}^t = n_G(t) \qquad J_G{}^i = 0 \tag{14.2.9}$$

and

$$T_{tt} = \rho(t) \qquad T_{it} = 0 \qquad T_{ij} = {}^3g_{ij}p(t) \tag{14.2.10}$$

where n_G, ρ, and p are unknown quantities that may depend on t, but *not* on r, θ, or ϕ. These results can be written more elegantly as

$$J_G{}^\mu = n_G U^\mu \tag{14.2.11}$$

$$T_{\mu\nu} = (\rho + p)U_\mu U_\nu + pg_{\mu\nu} \tag{14.2.12}$$

where U_μ is a "velocity four-vector"

$$U^t \equiv 1 \tag{14.2.13}$$

$$U^i \equiv 0 \tag{14.2.14}$$

Equation (14.2.14) shows that *the contents of the universe are, on the average, at rest in the coordinate system* r, θ, ϕ, as expected. In addition, comparison of Eq. (14.2.12) with Eq. (5.4.2) shows that *the energy-momentum tensor of the universe necessarily takes the same form as for a perfect fluid.*

It will be useful to have on hand the differential equations for $n_G(t)$, $\rho(t)$, and $p(t)$ that are provided by conservation principles. If galaxies are neither created nor destroyed, then $J_G{}^\mu$ obeys the conservation equation (5.2.14):

$$0 = (J_G{}^\mu)_{;\mu} = g^{-1/2} \frac{\partial}{\partial x^\mu} (g^{1/2} J_G{}^\mu) = g^{-1/2} \frac{\partial}{\partial t} (g^{1/2} n_G) \tag{14.2.15}$$

The metric (14.2.1) has a determinant $-g$ given by

$$g = R^6(t) r^4 (1 - kr^2)^{-1} \sin^2 \theta \tag{14.2.16}$$

and therefore the conservation of galaxies yields the relation

$$n_G(t) R^3(t) = \text{constant} \tag{14.2.17}$$

(Note that n_G is the number density per unit *proper* volume, and therefore increases or decreases according as the universe shrinks or expands, while $n_G R^3$ is the number density per unit *coordinate* volume, and therefore remains constant in a comoving coordinate system.) The energy-momentum tensor (14.2.12) obeys the conservation equation (5.4.3):

$$0 = T^{\mu\nu}{}_{;\nu}$$

$$= \frac{\partial p}{\partial x^\nu} g^{\mu\nu} + g^{-1/2} \frac{\partial}{\partial x^\nu} [g^{1/2}(\rho + p)U^\mu U^\nu] + \Gamma^\mu_{\nu\lambda}(p + \rho)U^\nu U^\lambda \tag{14.2.18}$$

Using Eqs. (14.2.8) and (14.2.14), we find that this equation is trivially satisfied for $\mu = r, \theta, \phi$, while for $\mu = t$ it reads

$$R^3(t) \frac{dp(t)}{dt} = \frac{d}{dt} \{R^3(t)[\rho(t) + p(t)]\} \qquad (14.2.19)$$

For instance, if the pressure of cosmic matter is negligible then (14.2.19) gives a result analogous to Eq. (14.2.17):

$$\rho(t)R^3(t) = \text{constant} \qquad (14.2.20)$$

The very great convenience of a comoving coordinate system should not blind us to the fact that the typical galaxies actually do move further apart or closer together when $R(t)$ increases or decreases. To see this clearly, we need to consider what we mean by the distance between galaxies. Imagine a chain of typical galaxies lying close together on the line of sight between us and a distant galaxy at r_1, θ_1, ϕ_1, and suppose that *at the same cosmic time t*, observers in each galaxy measured the distance to the next galaxy, say by measuring the travel time for light signals. (Note that this is not the same as measuring the time for a single light signal to go from $r = 0$ to $r = r_1$.) Adding up all these subdistances gives the *proper distance*

$$d_{\text{prop}}(t) = \int_0^{r_1} \sqrt{g_{rr}} \, dr = R(t) \int_0^{r_1} \frac{dr}{\sqrt{1 - kr^2}} \qquad (14.2.21)$$

Obviously, no one is going to organize this sort of cosmic conspiracy, so the proper distance is not very relevant to observational cosmology. However, we shall see in Section 14.4 that the more relevant measures of distance, based on apparent luminosities and angular diameters, all approach the proper distance (14.2.21) for $r_1 \ll 1$. Thus, in one sense or another, galaxies do move apart when $R(t)$ increases, or together when $R(t)$ decreases.

Cosmological theory presents observational astronomy with the challenge of measuring the function $R(t)$ and determining whether k is $+1$, or 0, or -1. This is not all there is to cosmology, but it is a central problem that must be solved if we are to understand the universe. The balance of this chapter will describe how well this challenge has been met.

3 The Red Shift

Our most important information about the cosmic scale factor $R(t)$ comes to us through the observation of shifts in frequency of light emitted by distant sources. To calculate such frequency shifts, we shall place ourselves at the origin $r = 0$ of coordinates (according to the Cosmological Principle, this is a mere convention) and consider an electromagnetic wave traveling to us along the $-r$ direction, with θ and ϕ fixed. The equation of motion of a given wave crest is then

$$0 = d\tau^2 = dt^2 - R^2(t)\frac{dr^2}{1 - kr^2}$$

Hence if the wave leaves a typical galaxy, located at r_1, θ_1, ϕ_1, at time t_1, then it will reach us at a time t_0 given by

$$\int_{t_1}^{t_0} \frac{dt}{R(t)} = f(r_1) \tag{14.3.1}$$

where

$$f(r_1) \equiv \int_0^{r_1} \frac{dr}{\sqrt{1 - kr^2}} = \begin{cases} \sin^{-1} r_1 & k = +1 \\ r_1 & k = 0 \\ \sinh^{-1} r_1 & k = -1 \end{cases} \tag{14.3.2}$$

We saw in the last section that typical galaxies will have constant coordinates r_1, θ_1, ϕ_1, so $f(r_1)$ is time-independent. Hence, if the next wave crest leaves r_1 at time $t_1 + \delta t_1$, it will arrive here at a time $t_0 + \delta t_0$, which again is given by a relation like (14.3.1)

$$\int_{t_1 + \delta t_1}^{t_0 + \delta t_0} \frac{dt}{R(t)} = f(r_1) \tag{14.3.3}$$

Subtracting Eq. (14.3.1) from (14.3.3), and noting that $R(t)$ changes very little during the period 10^{-14} sec of a typical light signal, we find that

$$\frac{\delta t_0}{R(t_0)} = \frac{\delta t_1}{R(t_1)}$$

The frequency ν_0 observed here is thus related to the frequency ν_1 when emitted by

$$\frac{\nu_0}{\nu_1} = \frac{\delta t_1}{\delta t_0} = \frac{R(t_1)}{R(t_0)} \tag{14.3.4}$$

This is conventionally expressed in terms of a *red-shift parameter* z, defined as the fractional increase in wavelength

$$z \equiv \frac{\lambda_0 - \lambda_1}{\lambda_1} \tag{14.3.5}$$

Since λ_0/λ_1 equals ν_1/ν_0, (14.3.4) gives

$$z = \frac{R(t_0)}{R(t_1)} - 1 \tag{14.3.6}$$

To avoid confusion, it should be kept in mind that ν_1 and λ_1 are the frequency and wavelength of the light if observed near the place and time of emission, and hence presumably take the values measured when the same atomic transition occurs in terrestrial laboratories, while ν_0 and λ_0 are the frequency and wavelength of the light observed after its long journey to us. If $z > 0$ then $\lambda_0 > \lambda_1$, and we speak of a *red* shift; if $z < 0$ then $\lambda_0 < \lambda_1$, and we speak of a *blue* shift.

If the universe is expanding, then $R(t_0) > R(t_1)$, and (14.3.6) gives a red shift, while if the universe is contracting, then $R(t_0) < R(t_1)$, and (14.3.6) gives a blue shift. Such frequency shifts find a natural explanation in terms of the Döppler effect discussed in Section 2.2. Equation (14.2.21) shows that a relatively close galaxy will move away from or toward the Milky Way, with a radial velocity

$$v_r \simeq \dot{R}(t_0)r_1 \qquad (14.3.7)$$

The frequency shift is given for $r_1 \to 0$ and $t_0 \to t_1$ by Eqs. (14.3.6) and (14.3.1) as

$$z \to \frac{\dot{R}(t_0)\,(t_0 - t_1)}{R(t_0)} \to r_1\dot{R}(t_0) \to v_r \qquad (14.3.8)$$

in agreement with Eq. (2.2.2). However, the frequency of light is also affected by the gravitational field of the universe, and it is neither useful nor strictly correct to interpret the frequency shifts of light from very distant sources in terms of a special-relativistic Döppler effect alone. [The reader should be warned though, that astronomers conventionally report even large frequency shifts in terms of a recessional velocity, a "red shift" of v km/sec meaning that $z = v/(3 \times 10^5)$.]

The first evidence for a systematic red shift of spectral lines from distant objects was provided by a program of observations carried out by Vesto Melvin Slipher with the Lowell Observatory 24-in. refractor from about 1910 to the mid-1920's. In a 1922 summary,[10] he gave data for 41 spiral nebulae, of which 36 had absorption lines shifted to the red by amounts up to $z \simeq 0.006$, and only five showed blue shifts, the largest being that of the Andromeda nebula, with $z \simeq -0.001$. From the beginning these frequency shifts were interpreted as due to the Döppler effect, but at first it was expected that they could be accounted for by the notion of the solar system, rather than the galaxies. The preponderance of *red* shifts in all parts of the sky made this interpretation increasingly untenable, and by 1918 Wirtz[11] suggested that in addition to the solar motion there was a general recession of spiral nebulae (then called the "K-term") away from us in all directions. Of course, other explanations were possible, such as a gravitational red shift caused by very strong *local* gravitational fields. (Perhaps the triumph of general relativity in the 1919 eclipse expedition made this explanation particularly attractive.) However, in a series of papers[12] written in the 1920's, Wirtz and K. Lundmark showed that Slipher's red shifts increased with the distance of the spiral nebulae, and therefore could most easily be understood in terms of a general recession of distant galaxies, the furthest naturally being those moving fastest. The announcement by Edwin Hubble[13] in 1929 of a "roughly linear relation between velocities and distances" established in most astronomer's minds the interpretation of the red shift as a cosmological Döppler effect, and this interpretation has survived through the decades until the present.

It is not possible to carry this discussion further, without first sharpening our understanding of how cosmological distances are defined, and how they relate to the coordinate distance r_1. The red shifts will be taken up again in Section 14.6.

4 Measures of Distance

There are at present only two practical methods (not counting the measurement of red shifts) for determining the distance of an object outside our galaxy. If we know its absolute luminosity, we can compare it to the observed apparent luminosity; or if we know its true diameter, we can compare it to the observed angular diameter. In addition, the distance to a near enough object can be determined by measuring its *parallax*, the shift in apparent position in the sky caused by the earth's revolution about the sun, or its *proper motion*, the shift in apparent position in the sky caused by the object's actual motion relative to the sun.

The "distances" measured by these four methods are identical for objects that are nearer than about 10^9 light years, but beyond this range they differ from each other, and also from the "proper distance" defined in Section 14.2. Thus, in order to use the correlation between red shifts and apparent luminosities or angular diameters to measure $R(t)$ and k, it will first be necessary to express the distances, determined from apparent luminosities or angular diameters, in terms of r_1 and t_0. It will be instructive, if largely academic, to do the same for distances determined from measurements of parallax or proper motion.

In order to calculate parallaxes and apparent luminosities, we must know the paths of light rays that leave a source at r_1, θ_1, ϕ_1 and pass near $r = 0$. (See Figure 14.1.) In a coordinate system x'^μ in which the light source is at the origin, the ray path is given by the very simple equation:

$$\mathbf{x}'(\rho) = \mathbf{n}\rho \qquad (14.4.1)$$

where $\mathbf{n}$ is a *fixed* vector, ρ is a variable positive parameter describing positions along the path (with $\rho = 0$ at the source), and $\mathbf{x}'$ is a three-vector formed from the comoving coordinates $r'\theta'\phi'$ in the usual way:

$$\mathbf{x}' \equiv (r' \sin \theta' \cos \phi', \; r' \sin \theta' \sin \phi', \; r' \cos \theta')$$

The transformation between the x'^μ coordinates, and another coordinate system in which the light source is at $\mathbf{x}_1$, is given by setting $\mathbf{a} = \mathbf{x}_1$ and interchanging $\mathbf{x}$ and $\mathbf{x}'$ in Eq. (14.2.7):

$$\mathbf{x} = \mathbf{x}' + \mathbf{x}_1 \left((1 - k\mathbf{x}'^2)^{1/2} - \{1 - (1 - k\mathbf{x}_1{}^2)^{1/2}\} \frac{(\mathbf{x}' \cdot \mathbf{x}_1)}{\mathbf{x}_1{}^2} \right) \qquad (14.4.2)$$

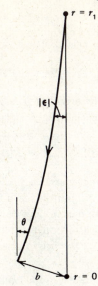

Figure 14.1 Quantities used in the calculation of parallaxes and apparent luminosities. The angles and the curvature of the light ray are greatly exaggerated.

Here again, we use a vector notation, with

$$\mathbf{x} \equiv (r \sin \theta \cos \phi, \, r \sin \theta \sin \phi, \, r \cos \phi)$$

and we define scalar products as in Euclidean geometry. There is no loss of generality in taking $\mathbf{n}$ to be a unit vector, with $\mathbf{n}^2 = 1$. The parametric equation of the light paths, given by substituting (14.4.1) in (14.4.2), is then

$$\mathbf{x}(\rho) = \mathbf{n}\rho + \mathbf{x}_1 \left[(1 - k\rho^2)^{1/2} - \{1 - (1 - kr_1{}^2)^{1/2}\}(n \cdot \mathbf{x}_1) \frac{\rho}{r_1{}^2} \right] \qquad (14.4.3)$$

where $r_1 \equiv (\mathbf{x}_1{}^2)^{1/2}$.

We shall now specify that the origin of the x^μ coordinate system is some definite point in the solar system, such as the center of the sun or the center of the 200 in. mirror at Palomar, and we shall restrict ourselves to light paths that pass close to this origin. In this case, the unit vector $\mathbf{n}$ must point nearly in the $-\mathbf{x}_1$ direction, so

$$\mathbf{n} \simeq -\hat{\mathbf{x}}_1 + \varepsilon \qquad (14.4.4)$$

where $\hat{\mathbf{x}}_1$ is the unit vector $\mathbf{x}_1/r_1$, and ε is a very small vector perpendicular to $\mathbf{x}_1$. (Here and below, $\simeq$ means that an equation is valid to first order in ε.) Recalling Eq. (14.4.1), we note for future reference that $|\varepsilon|$ is the angle between the light path and the $-\mathbf{x}_1$ direction, as measured in the coordinate system x'^μ, which is locally inertial *at the light source*. The light path, given by inserting (14.4.4) in (14.4.3) and discarding terms of order ε^2, is

$$\mathbf{x}(\rho) \simeq -\hat{\mathbf{x}}_1[\rho(1 - kr_1^2)^{1/2} - r_1(1 - k\rho^2)^{1/2}] + \varepsilon\rho \qquad (14.4.5)$$

The light path comes closest to the origin at $\rho \simeq r_1$. The *impact parameter b* is the *proper* distance of the path from the origin at this point, given by (14.2.1) and (14.4.5) as

$$b \simeq R(t_0)|\mathbf{x}(r_1)| \simeq R(t_0)r_1|\varepsilon| \qquad (14.4.6)$$

where t_0 is the time that the light ray arrives near the origin.

Measurements of astronomical parallaxes amount to measurements of the direction of light paths as a function of impact parameter, which in this case is the projection of the earth–sun separation on the plane normal to the line of sight. The light path has a direction near the origin given by

$$\left.\frac{d\mathbf{x}(\rho)}{d\rho}\right|_{\rho = r_1} \simeq \varepsilon - (1 - kr_1^2)^{-1/2}\hat{\mathbf{x}}_1$$

so the line of sight is given by a unit vector in the opposite direction

$$\hat{\mathbf{u}} \simeq -(1 - kr_1^2)^{1/2}\left.\frac{d\mathbf{x}(\rho)}{d\rho}\right|_{\rho = r_1} = \hat{\mathbf{x}}_1 - (1 - kr_1^2)^{1/2}\varepsilon \qquad (14.4.7)$$

Hence the angle between the actual line of sight, and the line of sight $\hat{\mathbf{x}}_1$ that would be observed at the origin, is

$$\theta \simeq |\hat{\mathbf{u}} - \hat{\mathbf{x}}_1| \simeq (1 - kr_1^2)^{1/2}|\varepsilon| \simeq (1 - kr_1^2)^{1/2}\frac{b}{R(t_0)r_1} \qquad (14.4.8)$$

In Euclidean geometry, a source at distance d would have a parallactic angle $\theta \simeq b/d$, so in general we may define the *parallax distance* d_P of a light source as

$$d_P \equiv \frac{b}{\theta} \qquad \text{for } \theta \to 0, \ b \to 0 \qquad (14.4.9)$$

and (14.4.8) may therefore be written

$$d_P = R(t_0)\frac{r_1}{(1 - kr_1^2)^{1/2}} \qquad (14.4.10)$$

In a universe with $k = +1$, objects at $r_1 = 1$ have infinite parallax distance and further objects (with $r_1 < 1$) have decreasing parallax distances, as noted first in 1900 by K. Schwarzschild.[14]

In order to calculate apparent luminosities, consider a circular telescope mirror of radius b, placed with its center at the origin and its normal along the line of sight $\hat{\mathbf{x}}_1$ to the light source. The light rays that just graze the mirror edge form a cone at the light source that, in the coordinate system x'^μ locally inertial at the source, has a half-angle $|\varepsilon|$ given by Eq. (14.4.6). The solid angle of this cone is

$$\pi|\varepsilon|^2 = \frac{\pi b^2}{R^2(t_0)r_1{}^2}$$

and the fraction of all isotropically emitted photons that reach the mirror is the ratio of this solid angle to 4π, or

$$\frac{|\varepsilon|^2}{4} = \frac{A}{4\pi R^2(t_0)r_1{}^2} \tag{14.4.11}$$

where A is the proper area of the mirror

$$A \equiv \pi b^2$$

However, each photon emitted with energy $h\nu_1$ will be red-shifted to energy $h\nu_1 R(t_1)/R(t_0)$, and photons emitted at time intervals δt_1 will arrive at time intervals $\delta t_1 R(t_0)/R(t_1)$, where as always t_1 is the time the light leaves the source, and t_0 is the time the light arrives at the mirror. Thus the total power P received by the mirror is the total power emitted by the source, its *absolute luminosity* L, times a factor $R^2(t_1)/R^2(t_0)$, times the fraction (14.4.11):

$$P = L\left(\frac{R^2(t_1)}{R^2(t_0)}\right)\left(\frac{A}{4\pi R^2(t_0)r_1{}^2}\right)$$

The apparent luminosity l is the power per unit mirror area, so

$$l \equiv \frac{P}{A} = \frac{LR^2(t_1)}{4\pi R^4(t_0)r_1{}^2} \tag{14.4.12}$$

In a Euclidean space the apparent luminosity of a source at rest at distance d would be $L/4\pi d^2$, so in general we may define the *luminosity distance* d_L of a light source as

$$d_L \equiv \left(\frac{L}{4\pi l}\right)^{1/2} \tag{14.4.13}$$

and (14.4.12) may therefore be written

$$d_L = R^2(t_0)\frac{r_1}{R(t_1)} \tag{14.4.14}$$

(This calculation could also have been carried out without using quantum theory, by applying the energy-conservation equation $(T^{\mu\nu})_{;\nu} = 0$ to the radiation emitted by the source.[15])

Next, let us calculate the angular diameter, observed at $r = 0$, $t = t_0$, of a light source of true proper diameter D at $r = r_1$, $t = t_1$. The light rays from the edge of the source travel to the origin along fixed directions $\mathbf{x}/r$. Without loss of generality, we can rotate the coordinate system so that the center of the light source is at $\theta = 0$, and suppose that light from its edges travels to the origin on a cone with half-angle $\theta = \delta/2$. (See Figure 14.2.) The proper distance across the source is then given by Eq. (14.2.1) as

$$D = R(t_1)r_1\delta \qquad \text{for } \delta \ll 1$$

so the angular diameter of the source is thus

$$\delta = \frac{D}{R(t_1)r_1} \tag{14.4.15}$$

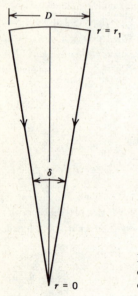

Figure 14.2 Quantities used in the calculation of angular diameters and proper motions. The angle δ is greatly exaggerated.

In Euclidean geometry, the angular diameter of a source of diameter D at a distance d is $\delta = D/d$, so in general we may define the *angular diameter distance* d_A of a light source as

$$d_A \equiv \frac{D}{\delta} \tag{14.4.16}$$

and (14.4.15) may therefore be written

$$d_A = R(t_1)r_1 \tag{14.4.17}$$

Note that $R(t_1)$ decreases as r_1 increases, so in some models d_A can have a maximum, with objects at very large distances having angular diameters that *increase* with increasing luminosity distance.

Finally, let us consider the determination of distances from proper motions. A source with a true velocity $V_\perp$ transverse to the line of sight will, in a time Δt_0, move a proper distance

$$\Delta D = V_\perp \Delta t_1 = V_\perp \Delta t_0 \frac{R(t_1)}{R(t_0)}$$

so, by the same reasoning that led to (14.4.15), the source will appear to move an angular distance

$$\Delta \delta = \frac{\Delta D}{R(t_1)r_1} = \frac{V_\perp \Delta t_0}{R(t_0)r_1} \tag{14.4.18}$$

In a Euclidean space the change in apparent position on the celestial sphere of a source at distance d would be $V_\perp \Delta t_0/d$, so we may define the *proper-motion distance* of a light source as

$$d_M \equiv \frac{V_\perp}{\mu} \tag{14.4.19}$$

where μ is the *proper motion*

$$\mu \equiv \frac{\Delta \delta}{\Delta t_0} \tag{14.4.20}$$

Equation (14.4.18) may therefore be written

$$d_M = R(t_0)r_1 \tag{14.4.21}$$

Of course, we can use (14.4.19) to measure the proper-motion distance only if we have some *a priori* knowledge of the transverse velocity, a point to which we return in the next section.

The luminosity distance d_L, angular diameter distance d_A, and proper-motion distance d_M of a light source with red shift z are related by the simple formulas

$$\frac{d_A}{d_L} = \frac{R^2(t_1)}{R^2(t_0)} = (1+z)^{-2} \tag{14.4.22}$$

$$\frac{d_M}{d_L} = \frac{R(t_1)}{R(t_0)} = (1+z)^{-1} \tag{14.4.23}$$

If one can measure z accurately, there is no point in attempting a separate determination of d_L, d_A, and d_M, except perhaps as a check of the Robertson-Walker metric or of the cosmological origin of red shifts. In contrast, the measurement of the parallax distance d_P could in principle give information beyond what could be learned from a measurement of d_L and z, but of course at present it is only possible to measure parallaxes for very close objects, with $z \ll 1$ and $r_1 \ll 1$. In this case all these observable distances become essentially equal to each other, and also to the proper distance (14.2.21):

$$d_A \simeq d_L \simeq d_M \simeq d_P \simeq d_{prop}(t_0) \simeq R(t_0)r_1 \qquad (14.4.24)$$

The distinction among different measures of distance only becomes important for objects that are billions of light years away.

Actually, measurements of luminosity distance, angular diameter distance, and red shift are inextricably mixed, for at least two reasons:

(A) Light sources such as galaxies have smooth luminosity distributions, without sharp edges. Let $L(D)$ be the absolute luminosity of that part of a light source within a circle (in the plane transverse to the line of sight) of diameter D. Then (14.4.12) and (14.4.15) give the apparent luminosity within an angular diameter δ as

$$l(\delta) = \frac{L(r_1 R(t_1)\delta)R^2(t_1)}{4\pi R^4(t_0){r_1}^2} \qquad (14.4.25)$$

It is more convenient to write this formula in terms of an absolute luminosity per unit transverse area

$$B(D) \equiv \frac{L'(D)}{2\pi D} \qquad (14.4.26)$$

and an apparent luminosity per unit solid angle, or *brightness*:

$$b(\delta) \equiv \frac{l'(\delta)}{2\pi\delta} \qquad (14.4.27)$$

Using (14.4.26), (14.4.27), and (14.3.6) in (14.4.25) then gives the brightness as

$$b(\delta) = \frac{B(r_1 R(t_1)\delta)}{4\pi(1 + z)^4} \qquad (14.4.28)$$

The *isophotal angular diameter* is the angle δ_b at which the brightness (14.4.28) falls below some fixed threshold value b:

$$\delta_b \equiv \frac{D_b}{r_1 R(t_1)} \qquad (14.4.29)$$

where D_b is defined by the implicit equation

$$B(D_b) \equiv 4\pi b(1 + z)^4 \qquad (14.4.30)$$

For instance, Hubble has suggested that $B(D)$ is well represented for most galaxies by a function that near the galactic edge has the approximate form[16]

$$B(D) \simeq \frac{\alpha L}{D^2} \qquad (14.4.31)$$

where α is a dimensionless constant, of order unity. Then (14.4.29)–(14.4.31) and (14.4.12) give

$$D_b \simeq \left(\frac{\alpha L}{4\pi b(1+z)^4}\right)^{1/2} \tag{14.4.32}$$

and

$$\delta_b \simeq \left(\frac{\alpha l}{b}\right)^{1/2} \tag{14.4.33}$$

In this particular case, the measurement of an isophotal angular diameter is just tantamount to a measurement of apparent luminosity.

(B) Most detectors of radiation respond only to photons in a narrow range of wavelengths. Thus it is necessary to distinguish between the *bolometric luminosities* L or l discussed above, which take account of radiation emitted or received at all wavelengths, and the *ultraviolet, blue, photographic, visual*, and *infrared* luminosities, which represent the average power or flux in various wavelength bands. If a source emits a radiant power $L(v_1)$ at all frequencies *less* than v_1, then Eqs. (14.4.12) and (14.3.4) give an apparent luminosity for all frequencies less than v_0 as

$$l(v_0) = \frac{L[v_0 R(t_0)/R(t_1)]R^2(t_1)}{4\pi R^4(t_0)r_1{}^2} \tag{14.4.34}$$

The frequency distributions of received and emitted power are therefore related by the formula:

$$l'(v_0) = \frac{L'[v_0 R(t_0)/R(t_1)]R(t_1)}{4\pi R^3(t_0)r_1{}^2} \tag{14.4.35}$$

For a black body, $L'(v)$ is given by the Planck formula

$$L'(v) = \frac{15L}{\pi^4 v}\left(\frac{hv}{kT_1}\right)^4\left(\exp\left(\frac{hv}{kT_1}\right)-1\right)^{-1} \tag{14.4.36}$$

where T_1 is the source temperature, k is the Boltzmann constant, and h is Planck's constant. The frequency distribution of received radiation is then

$$l'(v_0) = \frac{15l}{\pi^4 v_0}\left(\frac{hv_0}{kT_0}\right)^4\left(\exp\left(\frac{hv_0}{kT_0}\right)-1\right)^{-1} \tag{14.4.37}$$

where l is the bolometric apparent magnitude (14.4.12), and T_0 is the red-shifted temperature

$$T_0 = T_1\frac{R(t_1)}{R(t_0)} \tag{14.4.38}$$

If we know the temperatures T_1 or T_0, it is easy to convert the absolute luminosity $L'(v_1)\,\Delta v_1$ or the apparent luminosity $l(v_0)\,\Delta v_0$ in any narrow frequency band into a bolometric absolute or apparent luminosity.

A word must be said about the time-honored language used by astronomers to describe astronomical distances and luminosities. The astronomical unit (abbreviated a.u.) is the mean distance of the earth from the sun

$$1 \text{ a.u.} = 1.49598 \times 10^8 \text{ km} \tag{14.4.39}$$

Treating the earth's orbit as a circle, the projection of the earth–sun separation vector on the plane normal to the line of sight to any fixed star reaches a maximum value b_{max} equal to 1 a.u. during the course of a year, so the position of a star traces out an ellipse of maximum radius π given by Eq. (14.4.9) as

$$\pi \text{ (in radians)} = \frac{1}{d_P} \text{ (in a.u.)} \tag{14.4.40}$$

We shall call π the *trigonometric parallax*. One *parsec* (abbreviated pc) is defined as the distance d_P at which a star would have a trigonometric parallax of $1''$; there are 206,264.8 seconds in one radian, so

$$1 \text{ pc} = 206{,}264.8 \text{ a.u.} = 3.0856 \times 10^{13} \text{ km}$$
$$= 3.2615 \text{ light years} \tag{14.4.41}$$

Thus (14.4.40) may in general be expressed as

$$\pi \text{ (in seconds)} = \frac{1}{d_P} \text{ (in pc)} \tag{14.4.42}$$

Only the nearest stars have measurable trigonometric parallaxes, but such is the power of tradition, that all astronomical distances outside our solar system are conventionally given in parsecs, and sometimes these distances, however measured, are even described in terms of an equivalent parallax.

The apparent bolometric luminosity l is usually expressed in terms of an apparent bolometric magnitude m_{bol}, or simply m, which for historical reasons is defined so that

$$l = 10^{-2m/5} \times 2.52 \times 10^{-5} \text{ erg/cm}^2\text{-sec} \tag{14.4.43}$$

The absolute bolometric magnitude M is defined as the apparent bolometric magnitude the source would have at a distance 10 pc, so

$$L = 10^{-2M/5} \times 3.02 \times 10^{35} \text{ erg/sec} \tag{14.4.44}$$

Equation (14.4.13) may be expressed as a formula for the luminosity distance d_L in terms of the *distance modulus* $m - M$:

$$d_L = 10^{1+(m-M)/5} \text{ pc} \tag{14.4.45}$$

The apparent magnitudes m_U, m_B, and so on, in the ultraviolet, blue, photographic, visual, and infrared wavelength bands are related to the corresponding apparent luminosities by formulas like (14.4.43), but with different normalization constants, chosen so that all apparent magnitudes will be the same for stars of the Ao spectral type between fifth and sixth magnitude. The corresponding absolute magnitudes are defined so that the distance moduli $m_U - M_U$, $m_B - M_B$, and so on, are all equal to $m - M$. (Often the ultraviolet, blue, and visual apparent magnitudes m_U, m_B, m_V are denoted U, B, and V.) The *color index* is the quantity $m_B - m_V = M_B - M_V$; stars with negative color index are bluer than stars with positive color index. For purposes of comparison, the sun has absolute magnitudes

$$M \text{ (bolometric)} = +4.72 \qquad M_U = 5.51 \qquad M_B = 5.41 \qquad M_V = 4.79$$

and apparent magnitudes

$$m \text{ (bolometric)} = -26.85 \qquad m_U = -26.06 \qquad m_B = -26.16 \qquad m_V = -26.78$$

so that its distance modulus is -31.57 and its color index is 0.62.

5 The Cosmic Distance Ladder

If we know the absolute luminosity L of a light source, then we can determine its luminosity distance d_L by measuring its apparent luminosity l and using (14.4.13). The difficult problem is to determine L. At present, there is a ladder of distance determinations, with five distinct rungs, that must be climbed to get out to cosmologically interesting distances. (See Figure 14.3.)

Kinematic Methods

It is possible to measure the distance of some of the nearest stars by methods that do not require prior knowledge of the absolute luminosity L. One such star is the sun. Its distance, the astronomical unit, was first measured with tolerable accuracy in 1672 by Jean Richer and Giovanni Domenico Cassini. They determined the distance to Mars, and hence to the sun, by measuring the difference between the directions to Mars as seen from Paris and Cayenne, a known baseline of 6000 miles. Of course, our knowledge of the astronomical unit has in the ensuing three centuries been enormously improved, most recently by the use of radar astronomy.

A few thousand other stars are close enough so that their distances can be determined from the shift in their apparent positions caused by the earth's revolution about the sun. We have defined the trigonometric parallax π of a star as the maximum angular radius of the ellipse traced out annually by the star's apparent motion in the sky; the star's distance in parsecs is $1/\pi$, with π expressed in seconds of arc. (The adjective "trigonometric" is used here because astronomers have the habit of expressing stellar distances, however measured, in terms of a parallax, so that one encounters photometric parallaxes, moving-cluster parallaxes, and so on.)

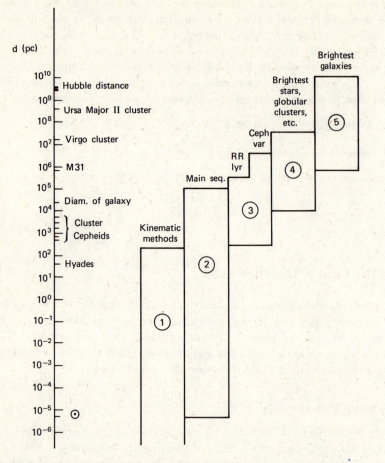

Figure 14.3 The cosmic distance ladder. The position and height of the vertical bars mark roughly the range of distance over which each class of distance indicators may be used.

The first star whose distance was measured in this way was 61 Cygni; in 1838 Friedrich Wilhelm Bessel determined its trigonometric parallax as about 0.3″, and hence its distance as about 3 pc. (Thomas Henderson had measured the trigonometric parallax of α Centaurus in 1832, but his calculations were not published until 1838.) Generally it is possible to determine stellar distances from trigonometric parallaxes only when π is greater than about 0.03″, that is, only for stars closer than about 30 pc.

In recent years it has become possible to measure the distance to some nearby clusters of stars by a method based ultimately on our knowledge of the speed of light rather than the astronomical unit. These *moving clusters* consist of stars moving through the galaxy with equal and parallel velocities, as shown by the fact that their proper motions across the sky seem to converge to a common point. The radial velocities v_r of the stars can be determined from the Döppler shifts $\Delta v/v$ of their spectra (and the known speed of light), while the velocity components transverse to the line of sight can be expressed as the product of the distance to the cluster times the proper motion (in radians per unit time) of the star across the sky. [See Eq. (14.4.19).] Thus observations of Döppler shifts and proper motions give us a complete kinematic model of the cluster, the single unknown being its distance. The distance can then be determined by imposing on this model the condition that all stars move with equal and parallel velocity. The best studied moving cluster is the Hyades, which contains about 100 stars within a radius of about 5 pc. Its distance has been measured by this "moving cluster method" as about 40.8 pc.

It is sometimes possible to estimate the distances of stars, which are neither in moving clusters nor close enough for measurement of trigonometric parallaxes, by a statistical analysis of proper motions and radial velocities. Suppose that we know the *relative* distances of a sample of stars, that is, that we know the ratios d/d_0, where d_0 is some unknown distance scale. (This would be the case, for example, if we knew that all stars in the sample had the *same* unknown absolute luminosity L, for then the apparent luminosities l would give us the relative distances through the formula $d = (L/4\pi l)^{1/2}$. Even if different stars in the sample have different absolute luminosities, measurement of their apparent luminosities will still give their relative distances, if we know the *ratios* of their absolute luminosities.) The transverse velocity is related to the radial velocity by

$$v_\perp = v_r \tan \phi$$

where ϕ is the unknown angle between the star's velocity and the line of sight. Equation (14.4.19) can thus be written

$$\frac{\mu}{v_r} \frac{d}{d_0} = \frac{\tan \phi}{d_0}$$

By measuring the quantities on the left-hand side for a large sample of stars, and making some reasonable guess as to the *distribution* in ϕ, it is then possible to deduce the unknown constant d_0. Although this method can be used at distances beyond 200 pc, it is intrinsically inaccurate, and can be thrown off badly if the sample of stars studied does not have the assumed distribution in ϕ.

It hardly needs to be mentioned that all of the above kinematic distance measurements can be used only for stars within our galaxy, where cosmological effects are surely negligible. Thus they can be regarded as determinations of the luminosity distance d_L, or the proper distance d_{prop}, or what you will. (It has occasionally been proposed that trigonometric parallaxes might be measured out to distances of order 10^8 pc by interferometric radio observations, using as baseline the distance from the earth to an artificial satellite in orbit around the sun. If this could be accomplished, then the problems of cosmography could be solved by determination of trigonometric parallax as a function of red shift.)

Main-Sequence Photometry ($\lesssim 10^5$ pc)

Once we know the distance of a star by one of the above kinematic methods, we can determine its absolute luminosity L by measuring its apparent luminosity l and using the formula $L = 4\pi d^2 l$. In this way it was independently discovered by Ejnar Hertzsprung and Henry Norris Russell during the decade 1905–1915 that a large proportion of nearby stars, the *main sequence*, obey a rather strict relation between absolute luminosity and spectral type. (The spectral type, which is actually a measure of surface temperature, is usually denoted by one of the letters O, B, A, F, G, K, M, R, N, S, with O very hot and S comparatively cold. (See Figure 14.4.) The canonical mnemonic is "Oh be a fine girl, kiss me right now sweetheart!") Astrophysical theory[17] explains the main sequence as a rather long initial phase in the thermonuclear evolution of almost all stars.

Figure 14.4 Spectra of stars belonging to various spectral classes. (Courtesy Mt. Wilson and Mt. Palomar observatories.)

Given the Hertzsprung-Russell relation between absolute luminosity and spectral type, it became possible for astronomers to determine the distance to any main-sequence star whose spectral type and apparent luminosity could be measured. The method works best when applied to a *cluster* of stars, all of which are about at the same distance from the earth, for then the main sequence can be picked out by plotting *apparent* luminosity versus spectral type for a large sample of cluster stars. It also works well only for the lower part of the main sequence, where the Hertzsprung-Russell relation is best known.

The catalogued clusters of our galaxy are divided between some 650 *open clusters*, like the Hyades and Pleiades, each containing 20 to 1000 stars, and some 130 *globular clusters*, such as the great cluster M13 in Hercules, each containing 10^5 to 10^7 stars. (See Figures 14.5 and 14.6.) In determining the distance of these

Figure 14.5   The open cluster NGC2682, in Cancer; photographed with the 200-in. telescope at Mt. Palomar. (Courtesy Mt. Wilson and Mt. Palomar observatories.)

clusters it is important to recognize a distinction between their stellar populations, first pointed out by Walter Baade[18] in 1944. (See Figure 14.7.) Stars in open clusters, as well as most nearby stars like the sun, generally belong to *Population I*, which is characterized by high metal content and relative youth, and is limited in our galaxy to the spiral arms. Stars in globular clusters belong to *Population II*, which is characterized by lower metal content and greater age, and pervades the whole galaxy. There are differences between the main sequences of Populations I and II, so the use of a main sequence, calibrated from nearby stars, to determine

Figure 14.6 The globular cluster NGC6205 (M13) in Hercules; photographed with the 200-in. telescope at Mt. Palomar. (Courtesy Mt. Wilson and Mt. Palomar observatories.)

I (M31 Spiral Arms, Blue Light)

II (NGC 205, Yellow Light)

Figure 14.7 Examples of stellar populations I and II; photographed with the 200-in. telescope at Mt. Palomar. On the left are Pop. I stars in the spiral arms of M31; on the right Pop. II stars in the M31 satellite NGC 205. (Courtesy Mt. Wilson and Mt. Palomar observatories.)

the distance of the globular clusters is beset with technical complications, which need not concern us here.

The method of determining distances by main-sequence photometry is limited because typical main-sequence stars are not particularly bright. For instance, the Hale reflector at Palomar has difficulty in resolving stars that are fainter than $m = 22.7$, so it can resolve a star with the absolute magnitude $M = 4.7$ of the sun only out to a distance modulus $m - M = 18$, which according to Eq. (14.4.45) corresponds to a distance of 40,000 pc.

At present, it is primarily the stars of the Hyades that are used to calibrate the Hertzsprung-Russell relation, so the whole scale of galactic and extragalactic distances rests on our knowledge of the distance to the Hyades, as determined by the "moving-cluster method" discussed above. Recently Hodge and Wallerstein[19] have noted that both the mean trigonometric parallax of the Hyades stars, and the comparison of their apparent magnitudes with the Hertzsprung-Russell relation obtained from other nearby stars, suggest that the distance to the Hyades may be about 50 pc rather than 40.8 pc. If this is so, then all galactic and extragalactic distances must be increased by about 20%.

Variable Stars ($< 4 \times 10^6$ pc)

There are about 10,000 catalogued stars whose apparent luminosity is observed to vary more or less regularly with time. In setting the extragalactic distance scale, an important role is presently played by two families of variable stars, the *cluster variables*, or *RR Lyrae* stars, and the *classical Cepheids*, or *δ Cephei stars*. The RR Lyrae stars have periods ranging from a few hours to a day, and belong to Population II, while the classical Cepheids have periods ranging from 2 to 40 days, and belong to Population I. (In addition there is another family of variable stars, the *W Virginis stars*, which belong to Population II, but have long periods like the classical Cepheids. As we shall see, W Virginis stars were confused with Cepheids before Baade distinguished the two stellar populations.)

The absolute magnitudes of the RR Lyrae stars are presently best known both through direct statistical studies of proper motion and parallax and through their presence in globular clusters, whose distance can be determined by main-sequence photometry. In this way it has been found[20] that the RR Lyrae stars all have roughly the same absolute magnitude, somewhere between $M_v \simeq 0.2$ and $M_v \simeq 1.0$. Thus, once we recognize an RR Lyrae star by its short-period pulsation, we can estimate its distance from its apparent magnitude. However, RR Lyrae stars are not bright enough to be used at distances beyond about 3×10^5 pc. For this reason, much more attention has been devoted to the brighter classical Cepheids.

Unfortunately, the classical Cepheids differ widely in absolute luminosity. However, in 1912 it was noted by Henrietta Swan Leavitt[21] that the 25 classical Cepheids then known in the smaller Magellenic cloud have apparent luminosities given by a smooth function $l_{SMC}(P)$ of the period P. (Roughly, $l \propto P$.) The stars of this cloud are all at about the same distance from the earth, so Leavitt could

Figure 14.8 The great galaxy M31 (NGC 224) in Andromeda, with satellite galaxies NGC205 and 221. This photograph was taken with the 48-in. Schmidt telescope at Mt. Palomar. (Courtesy Mt. Wilson and Mt. Palomar observatories.)

conclude that the absolute luminosity of a classical Cepheid of period P is a smooth function $L(P)$, proportional to $l_{SMC}(P)$. However, she did not know the distance to the smaller Magellenic cloud, and there are no Cepheids near enough to the earth to have measurable trigonometric parallax, so Leavitt could not determine the constant of proportionality.

The laborious task of calibrating the Cepheid *P-L* relation was carried out between the two world wars, first by Russell[22] and Hertzsprung,[23] then by Harlow Shapley,[24] and eventually by Ralph E. Wilson.[25] They did not then make use of main-sequence photometry; rather, the main tool was the statistical analysis of proper motions and radial velocities for the Cepheids nearest the sun, as described above under "Kinematic Methods," with ratios of Cepheid absolute luminosities provided by the *P-l* relation of the Magellenic clouds. Cepheids were discovered in the great nebula M31 in Andromeda by Edwin Hubble[26] in 1923, and their observed periods and apparent luminosities were used, together with the Cepheid *P-L* relation, to estimate the distance of M31 as 280,000 pc. (See Figures 14.8 and 14.9.) It was this measurement that definitely established the status of the "spiral nebulae" as islands of stars comparable with our own galaxy, as suggested by Immanuel Kant, rather than mere clouds or clusters within our galaxy. Taking

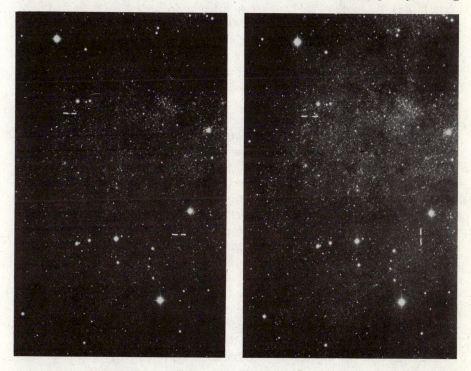

Figure 14.9 Variable stars in a portion of M31. Two of the variables are marked. This photograph was taken with the 200-in. telescope at Mt. Palomar. (Courtesy Mt. Wilson and Mt. Palomar observatories.)

interstellar absorption into account later lowered this figure to 230,000 pc, but otherwise the extragalactic distance scale remained essentially unchanged until Palomar started operations in 1950.

By 1952 it had become apparent that something was seriously wrong with the Cepheid *P-L* relation of Shapley *et al*. Photographs of M31, taken with a 30 minute exposure at Palomar, showed only the brightest stars of Population II, but no RR Lyrae variables. This indicated that the brightest Population II stars of M31 had apparent photographic magnitude $m \simeq 22.4$, and since the RR Lyrae stars were known to have an absolute luminosity about four times fainter, their apparent magnitudes in M31 would have to be about $m \simeq 23.9$, beyond the reach of Palomar. However, the absolute magnitude of the RR Lyrae stars was by then reasonably well known through the photometric determination by Allan Sandage of the distance of the globular cluster M3. If M31 were really at a distance of 230,000 pc, then the RR Lyrae stars ought to have shown up at $m \simeq 22.4$ at least in their maximum phase, and the brightest stars of Population II ought to have had apparent magnitude $m \simeq 20.9$, rather than $m \simeq 22.4$. This discrepancy was interpreted by Baade[27] to mean that M31 was not at 230,000 pc, but at a distance about *twice* as great (a difference of 1.5 apparent magnitudes corresponds to a factor of 2 in distance), so that the classical Cepheids in its spiral arms had to be about four times brighter than had been estimated.

The source of this error is somewhat obscure. The *P-L* relation, as calibrated by Shapley et al., actually works rather well for the Population II W Virginis variables, but fails for the Population I classical Cepheids, which are generally four times brighter than W Virginis stars of the same period. However, it should not be thought that Shapley, not knowing of the distinction between stellar populations, had based his calibration on W Virginis stars rather than classical Cepheids. Indeed, the 11 variables considered by Shapley[24] in 1918 were Population I stars, and even included the eponymous classical Cepheid, δ Cephei! (In any case, W Virginis stars are both less luminous than classical Cepheids, and rarer near the sun, so it would have been remarkable if they had played a large part in the statistical proper-motion studies used to calibrate the Cepheids.) A recent reanalysis[28] of the same 11 classical Cepheids used in Shapley's calibration reveals that Shapley's calibration contained errors of about 0.7 magnitudes due to the neglect of interstellar absorption, 0.6 magnitudes due to systematic errors in proper motions, and 0.1 or 0.2 magnitudes due to galactic rotation, which introduces anisotropies in the distribution of stellar velocities. All these errors tended in the same direction, and led to the famous 1.5 magnitude underestimate in Cepheid absolute luminosities discovered by Baade in 1952. Thus it was a pure coincidence that Shapley's original *P-L* curve, though not valid for the classical Cepheids of Population I, was actually valid for the W Virginis stars[29] of Population II.

One may also ask why Shapley's calibration was repeatedly confirmed during the third of a century from 1918 to 1952. One simple reason is that interstellar absorption was persistently underestimated. Thus, when Ralph Wilson[25] attempted to improve the statistical analysis of proper motions and radial velocities by using more Cepheids, 74 in 1923 and 157 in 1939, he had to include more and more distant stars, so that his improvement in statistical accuracy was cancelled by the increasing effects of absorption. Another reason does have to do with the confusion of populations. Shapley[24] had immediately applied his *P-L* relation to what he thought were classical Cepheids in the globular clusters ω Cen, M3, and M5, and in this way he was able to determine the distance to these globular clusters, and thereby to calculate the absolute magnitudes of their short-period variables, the RR Lyrae stars. This procedure actually gave the right answer, because the stars in the globular clusters that Shapley thought were classical Cepheids really were W Virginis stars, and Shapley's *P-L* calibration, though seriously in error for the classical Cepheids from which it had been derived, was accidentally more or less correct for the W Virginis stars! Hence, when statistical studies of the proper motions and radial velocities of nearly RR Lyrae stars were carried out a few years later, they tended to confirm Shapley's estimate of the absolute magnitudes of the RR Lyrae stars, and this naturally appeared as a confirmation of the Cepheid *P-L* relation. The argument was then turned around: Taking the ratio of RR Lyrae to "Cepheid" luminosities from the globular clusters, Wilson included 10 RR Lyrae stars along with the 74 Cepheids in his 1923 analysis of proper motions and radial velocities, and in 1939 he included 67 RR Lyrae stars along with 157 Cepheids. Oddly enough, the RR Lyrae stars did not, like the Cepheids, introduce large errors through the neglect of absorption, because RR Lyrae stars, being of Population II, are mostly found outside the plane of the galaxy. Rather, the trouble was that the classical Cepheids do not, like the W Virginis stars, fall on a *P-L* curve that extrapolates smoothly down to the RR Lyrae stars, but instead lie 1.4 magnitudes higher.

It should be noted that Baade's 1952 revision of the Cepheid *P-L* relation doubled the extragalactic distance scale, but did not affect the estimated size of our own galaxy, because the galactic distance scale was determined from the distances of the globular clusters, which, as we have seen, had partly by accident been determined correctly. Before 1952, it appeared that all neighboring galaxies were distinctly smaller than our own. After 1952, it was clear that many other galaxies are as large or larger than our own, a highly satisfactory if sobering result.

The calibration of the classical Cepheids has since been put on a much firmer footing by the discovery of five classical Cepheids in the galactic open clusters NGC6087, NGC129, M25, NGC7790, and NGC6664, together with four more classical Cepheids in the "association" $h + \chi$ Persei. The distance to these clusters is known through the photometric study of their main-sequence stars by Kraft,[30] and these nine Cepheids of known absolute magnitude have been used by Kraft, and more recently by Sandage and Tammann,[31] to fix the absolute scale of the Cepheid *P-L* relation. (The *form* of this relation which, to be precise, actually relates period, luminosity, and *color*, is of course determined by a much larger sample of Cepheids, taken from both Magellenic clouds, the Andromeda galaxy M31, and the small Fornax galaxy NGC6822.) At present the distance of M31, as determined by its classical Cepheids, is given as 700,000 pc, about three times greater than the distance accepted during the 1930's.

The RR Lyrae stars and classical Cepheids can be used to determine the distances for all members of the association of nearby galaxies and stellar systems known as *the local group*. Of these, it is only the closest objects, such as the Magellenic clouds and the Ursa Minor, Draco, and Sculptor systems, in which RR Lyrae stars can be employed. For all the major galaxies of the local group, such as M31 and M33, it is necessary to use the classical Cepheid P-L relation, as calibrated by the nine known Cepheids in the open clusters and associations of our galaxies. The classical Cepheids are bright enough at maximum light ($M_{v, \max} \simeq -5.3$) to be used at distances of order 4×10^6 pc, which is far enough to reach some galaxies outside our own local group, such as the beautiful spiral M81. However, the Cepheids are not bright enough to be used to determine the distance of the nearest cluster of galaxies, the *Virgo cluster*.

Novae, H II Regions, Brightest Stars, Globular Clusters, and so on ($\lesssim 3 \times 10^7$ pc)

The next rung may at present be the weakest.[32] In order to estimate the distances of objects far outside our local group of galaxies, it is necessary to find some distance indicators that are brighter than the Cepheids, but are present in sufficient numbers in our local group of galaxies (whose distances are known *via* the Cepheids) to permit an accurate calibration of their properties.

Novae are sudden increases by four to six orders of magnitude in the luminosity of a star, and occur in typical galaxies at a rate of 40 per year. They have been used as distance indicators since 1917, when a nova was found in the spiral nebula NGC6946. The brightest novae reach $M_v \simeq -7.5$, so they can in principle be used as distance indicators out to about 10^7 pc, but they tend to occur in the bright central regions of galaxies, and are therefore difficult to resolve.

Until recently, the primary distance indicators used to reach beyond our local group were the *brightest stars* of galaxies. A survey of the local group reveals that the stars of each galaxy generally have a well-defined maximum absolute luminosity, about $M_v \simeq -9.3$. They can therefore be used as distance indicators out to about 3×10^7 pc, but at distances beyond 10^7 pc, it is difficult to distinguish between brightest stars and nonstellar objects, such as associations or emission regions. (Indeed, it is believed that Hubble's 1936 calibration[33] of the distance scale was in error, partly because he confused such objects with brightest stars.)

It is also possible to use certain nonstellar objects as distance indicators. Among these are the *H II regions*, large clouds of interstellar hydrogen that are ionized and made luminous by the presence of O and B stars. They are hundreds of parsecs in diameter, so their angular diameters might be used to estimate their distances out to about 10^8 pc.

Recently Sandage[34] has developed the use of *globular clusters* as a distance indicator that may prove more reliable than any of the above. The hundreds of globular clusters in our galaxy have absolute magnitudes M_v that are typically about -8, but vary widely about this mean. However, the study[35] of 2000 globular clusters in the large E (elliptical) galaxy M87 of the Virgo cluster (see Figure 14.10)

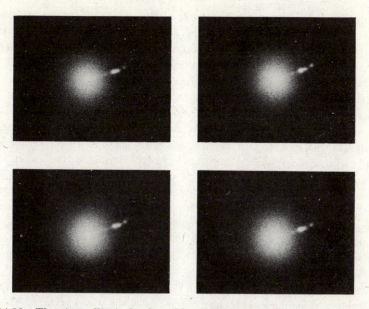

Figure 14.10 The giant elliptical galaxy M87 (NGC4486) in the Virgo cluster, for four different directions of polarization. These photographs, taken with the 200-in. telescope at Mt. Palomar, are underexposed in order to reveal the galactic nucleus and the remarkable "jet." (Courtesy Mt. Wilson and Mt. Palomar observatories.)

reveals a sharp cutoff in their luminosity distribution, with $m_B(\text{max}) \simeq 21.3$. Sandage suggests that the absolute magnitude of the brightest globular clusters in M87 be identified with the absolute magnitude of the brightest globular cluster B282 of the Andromeda galaxy M31, which is known to have absolute magnitude $M_B(\text{B282}) \simeq -9.83$. The distance modulus of M87 is thus 21.3 minus -9.8, or 31.1, giving a distance for M87, and hence for the Virgo cluster, of 1.7×10^7 pc. Of course it is not definitely known that the luminosity distribution of the globular clusters has a sharp cutoff rather than a smooth tail at high luminosities. De Vaucouleurs[36] has examined the latter possibility, and concludes that the Virgo cluster is at a distance of 2×10^7 pc, 20% further than calculated by Sandage.

Brightest Galaxies ($\lesssim 10^{10}$ pc)

The Virgo cluster has a small mean red shift, $z = 0.0038$, corresponding to a radial velocity of about 1100 km/sec. This is not much larger than the mean random velocity of typical galaxies, and it is only when we get out beyond the Virgo cluster that the cosmological expansion dominates the velocity field. In order to get out to these cosmologically interesting distances, it is generally necessary to use whole galaxies as distance indicators.

Clusters of galaxies contain hundreds to thousands of distinct galaxies (the Virgo cluster contains 2500), so if there is any natural upper limit to the absolute luminosity of individual galaxies, the absolute luminosity of the brightest galaxy in a rich cluster ought to be near that maximum. For this reason Edwin Hubble[33] in 1936 suggested the use of the brightest galaxies in clusters as distance indicators. (He actually used the *fifth* brightest, to minimize observational errors.) This procedure was validated when it was found that the use of brightest galaxies as distance indicators gave a good linear relation between luminosity distance and z for 10 clusters with $z \ll 1$. (See the next section.) These brightest cluster galaxies are usually the elliptical galaxies known as type E in Hubble's classification scheme. (See Appendix.)

According to Sandage,[34] the brightest E galaxy in the Virgo cluster is NGC4472, with absolute magnitude $M_B \simeq -21.68$, determined by using the globular clusters in M87 to give the distance to the Virgo cluster. If all brightest E galaxies have absolute magnitude $M_B \simeq -21.7$, then they can be used as distance indicators out to a distance modulus $m - M$ of about 44.5, or a luminosity distance of about 10^{10} pc.

However, it is possible that there is no sharp cutoff to the luminosity distribution function of galaxies in clusters. In this case, the use of brightest galaxies as distance indicators would be complicated by the *Scott effect*,[37] first discussed in this connection by Elizabeth L. Scott. As we look out to greater and greater distances, we tend to select increasingly rich clusters of galaxies for study, and if there is no absolute upper limit to galactic luminosity, the brightest galaxies in these clusters will have greater and greater absolute luminosity. If we mistakenly assume that these distant galaxies have the same absolute luminosity as NGC4472, we will underestimate their true luminosity distance. The existence or nonexistence of a Scott effect is still a matter of controversy.[38] Other problems affecting the use of brightest galaxies as distance indicators are discussed in the next section.

One need only summarize the rungs of the cosmic distance ladder to see how shaky it is. At the time of writing, the distance to the Hyades is determined by observation of its stars' proper motions and radial velocities; the distance to five open galactic clusters and the $h + \chi$ Persei association is determined by photometry of their main-sequence stars, whose absolute magnitude is known by study of the Hyades; the distance to the Andromeda nebula M31 is determined from observation of classical Cepheids, whose P-L relation is calibrated using the nine known Cepheids in open clusters and the $h + \chi$ Persei association; the distance to the Virgo cluster is determined by assuming that the brightest globular cluster in M87 has the same absolute luminosity as the brightest globular cluster B282 of M31; and the distances to more distant clusters of galaxies are determined by assuming that their brightest E galaxies have the same absolute luminosity as the brightest galaxy NGC4472 in the Virgo cluster. It is entirely possible that new errors may be found at any rung of the ladder, in which case adjustments would have to be made at all higher rungs.

6 The Red-Shift Versus Distance Relation

We shall now consider how the correlation of red shifts with distances can be used to gain information about the cosmic scale factor $R(t)$. For our present purposes, only luminosity distances will be considered; Eqs. (14.4.22) and (14.4.23) show that no new information can be gained by studying the correlation of red shifts with angular diameters or proper motions instead of apparent luminosities.

Suppose then that astronomers are able to define some family of objects, like the brightest E galaxies discussed at the end of the last section, whose absolute luminosities L are all known. By measuring their apparent luminosities, their luminosity distances can be calculated from Eq. (14.4.13):

$$d_L = \left(\frac{L}{4\pi l}\right)^{1/2}$$

Suppose also that the red shifts z of these objects are measured, so that an empirical curve for $d_L(z)$ is known. What does this tell us about $R(t)$?

The observables d_L and z are related to the unknown coordinates of the light source by the theoretical relations (14.3.1), (14.3.6), and (14.4.14):

$$\int_{t_1}^{t_0} \frac{dt}{R(t)} = \int_0^{r_1} [1 - kr^2]^{-1/2} \, dr$$

$$z = \frac{R(t_0)}{R(t_1)} - 1$$

$$d_L = r_1 \frac{R^2(t_0)}{R(t_1)} = r_1 R(t_0)(1 + z)$$

At present, the curve $d_L(z)$ is tolerably well known only for small z, so we are primarily concerned with the case where $t_0 - t_1$ and r_1 are small. The cosmic scale factor $R(t)$ may then be most usefully expressed as a power series

$$R(t) = R(t_0)[1 + H_0(t - t_0) - \tfrac{1}{2}q_0 H_0^2(t - t_0)^2 + \cdots] \qquad (14.6.1)$$

where t_0 is the present moment, and H_0 and q_0 are parameters known as *Hubble's constant* and the *deacceleration parameter*

$$H_0 \equiv \frac{\dot{R}(t_0)}{R(t_0)} \qquad (14.6.2)$$

$$q_0 \equiv -\ddot{R}(t_0) \frac{R(t_0)}{\dot{R}^2(t_0)} \qquad (14.6.3)$$

(Dots denote derivatives with respect to time.) We shall see in the next chapter that the whole function $R(t)$ can be calculated by using Einstein's field equations if we know the values of H_0 and q_0, with $k > 0$ if $q_0 > \tfrac{1}{2}$ and $k < 0$ if $q_0 < \tfrac{1}{2}$. Our present discussion will therefore be directed to the measurement of these two critical parameters.

The use of Eq. (14.6.1) in (14.3.6) yields a power series for the red shift as a function of the time of flight $t_0 - t_1$:

$$z = H_0(t_0 - t_1) + \left(1 + \frac{q_0}{2}\right) H_0^2(t_0 - t_1)^2 + \cdots \tag{14.6.4}$$

Inverting this power series, we obtain a formula for the time of flight in terms of the red shift:

$$(t_0 - t_1) = \frac{1}{H_0}\left[z - \left(1 + \frac{q_0}{2}\right)z^2 + \cdots\right] \tag{14.6.5}$$

To find r_1, we expand (14.3.1):

$$\frac{1}{R(t_0)}\int_{t_1}^{t_0}\left[1 + H_0(t_0 - t) + \left(1 + \frac{q_0}{2}\right) H_0^2(t_0 - t)^2 + \cdots\right] dt = r_1 + 0(r_1{}^3)$$

so

$$r_1 = \frac{1}{R(t_0)}\left[t_0 - t_1 + \tfrac{1}{2}H_0(t_0 - t_1)^2 + \cdots\right] \tag{14.6.6}$$

Using (14.6.5) in (14.6.6) gives r_1 in terms of the red shift:

$$r_1 = \frac{1}{R(t_0)H_0}\left[z - \tfrac{1}{2}(1 + q_0)z^2 + \cdots\right] \tag{14.6.7}$$

and (14.4.14) then gives the luminosity distance as a power series:

$$d_L = H_0^{-1}[z + \tfrac{1}{2}(1 - q_0)z^2 + \cdots] \tag{14.6.8}$$

This can also be written as a formula for apparent luminosity:

$$l = \frac{L}{4\pi d_L{}^2} = \frac{LH_0{}^2}{4\pi z^2}[1 + (q_0 - 1)z + \cdots] \tag{14.6.9}$$

or equivalently, for the distance modulus:

$$m - M = 25 - 5\ln_{10}H_0 \text{ (km/sec/Mpc)} + 5\ln_{10} cz \text{ (km/sec)}$$
$$+ 1.086(1 - q_0)z + \cdots \tag{14.6.10}$$

[One Mpc is 10^6 pc. Note that 100 km/sec/Mpc equals $(9.78 \times 10^9 \text{ years})^{-1}$.] The program is then to compare either (14.6.8), (14.6.9), or (14.6.10) with astronomical data, and thereby to determine the critical parameters q_0 and H_0.

In order to measure q_0, we need to go out to large values of z (say, $z \gtrsim 0.1$), where only brightest cluster galaxies (or possibly supernovae) can be used as distance indicators. However, it is only the *shape* of the curve of d_L or l or m versus z that is needed.

In order to measure H_0, a single object with $z \lesssim 0.1$ is all we need, but we have to know its absolute luminosity as well as its red shift and apparent luminosity. Also, its red shift must be large enough (say, $z \gtrsim 0.01$) so that its radial velocity reflects the general expansion of the universe, rather than a local velocity anomaly. Unfortunately, the Virgo cluster, whose luminosity is known from observation of its brightest stars and globular clusters (as in "rung 4" of the cosmic distance ladder described in the last section), has a radial velocity of only about 1000 km/sec, which is not large enough to ensure the dominance of the cosmological recession. There is a possibility that "rung 4" can be employed out to larger red shifts, for example, by using the angular diameters of H II regions. However, at present the only way to extend measurements of H_0 as well as q_0 out to large red shifts is to use all five rungs of the cosmic distance ladder, taking as distance indicators the brightest galaxies of rich clusters.

This program is subject to a great many complications. Some, which are now taken into account by applying well-understood corrections to the data, include:

(A) **Galactic Rotation.** The rotation of our galaxy gives the sun a velocity of about 215 km/sec. This produces systematic red or blue shifts in the spectra of distant galaxies, which are routinely subtracted from the observed red shifts in calculating the "cosmological" red shift z.

(B) **Aperture.** Since the edges of galaxies fade gradually into the background light of the sky, it is necessary to refer all measurements of galactic apparent luminosities to a standard telescope aperture.

(C) **k-Term.** As discussed in Section 14.4, the red shift will distort the frequency distribution of light from distant objects, so that their visual or blue magnitudes reflect their absolute luminosities at higher frequencies than for near objects. If we know the intrinsic frequency distribution, we can correct for this effect by using Eq. (14.4.35), with the result that the left-hand side of Eq. (14.6.10) is replaced with $m_B - M_B - k_B(z)$, where $k_B(z)$ is an explicitly known function of z, calculated by Oke and Sandage.[39] In an alternative procedure developed by Baum,[40] the luminosity distribution is measured directly for each galaxy studied, so that all apparent magnitudes can be referred to the same emission frequency, and no k-term is needed.

(D) **Absorption.** Our galaxy is known to absorb a certain fraction of the light coming to us from extragalactic objects. Treating our galaxy as an infinite flat slab, the distance through the galaxy that a light ray must travel on its way to us is proportional to cosec b, where b is the angle between the line of sight and the plane of the galaxy. The light will therefore be dimmed by a factor $\exp(-\lambda \csc b)$, with λ some constant. Taking λ from studies of the nearer extragalactic objects, the result is then that the left-hand side of Eq. (14.6.10) is replaced with the corrected distance modulus:

$$(m - M)_{\text{corr}} = m_B - M_B - k_B(z) - A_B(b)$$

where, roughly,

$$A_B(b) \simeq 0.25 \operatorname{cosec} b$$

(This is somewhat of an oversimplification. Sandage[41] first applies an absorption correction $A_v(b) = 0.18 (\operatorname{cosec} b - 1)$ to the visual magnitudes, then converts to blue magnitudes, and then applies an additional correction $A_B = 0.25$.) No correction is generally made for extragalactic absorption.[42] (See Section 15.4.)

In addition to the above complications, which are pretty well understood, there are a number of other possible sources of error, whose status is much more in doubt.

(E) **Uncertainty in L.**   As emphasized in Section 14.5, a new correction at any rung of the cosmic distance ladder, such as a change in the distance to the Hyades or the Virgo cluster, would require a corresponding correction to the estimated absolute luminosity of the brightest E galaxies. Inspection of Eq. (14.6.9) or (14.6.10) shows that this would affect the value of Hubble's constant, but would not change the deacceleration parameter q_0.

(F) **Scott Effect.**   It was also emphasized in the last section that, if there is no sharp upper limit to the absolute luminosity of cluster galaxies, then the tendency to select richer clusters at greater distances would mean that the absolute luminosities of their brightest galaxies would increase with z. According to Eq. (14.6.9), this selection effect would lead to an *overestimate* of the deacceleration parameter q_0. However, the Scott effect, if real, would only enter at very great distances, and therefore would have little effect on the value of H_0.

(G) **Shear Field.**   De Vaucouleurs[43] has suggested the existence of a local anisotropy in the galactic velocity field, encompassing our own local group and the Virgo cluster. If this is true, it could mean that red shifts with cz less than about 4000 km/sec do not accurately reflect the general expansion of the universe.

(H) **Galactic Evolution.**   As we look further and further out into space, we see galaxies that are presumably younger and younger. It may be that the luminosity of the brightest E galaxies is a function $L(t_1)$ of the time the light was emitted. Equation (14.6.5) tells us then that L in Eq. (14.6.9) should be replaced with

$$L(t_1) = L(t_0)[1 - E_0(t_0 - t_1) + \cdots]$$
$$= L(t_0)\left[1 - \frac{E_0 z}{H_0} + \cdots\right]$$

where

$$E_0 \equiv \frac{\dot{L}(t_0)}{L(t_0)} \qquad (14.6.11)$$

The effect would be that q_0 in Eq. (14.6.9) would be replaced with an *effective deacceleration parameter*

$$q_0^{\text{eff}} = q_0 - \frac{E_0}{H_0} \qquad (14.6.12)$$

so that astronomical observations would really measure q_0^{eff}, not q_0. Sandage has recently given two different estimates for the rate of change of L for brightest E galaxies—in our notation, they are[44]

$$E_0 = 0.04 \pm 0.02/10^9 \text{ years} \tag{14.6.13}$$

and[45]

$$E_0 = 0.00 \pm 0.05/10^9 \text{ years} \tag{14.6.14}$$

As we shall see, a rate of evolution E_0 of order $0.04/10^9$ years would have an important effect on the value of q_0^{eff}.

Returning now to the observations of red shifts and luminosity distances, we must pick up our story where we left it in Section 14.3, with Hubble's 1929 discovery[13] of a linear relation between d_L and z. Hubble estimated the distance to 18 nearby galaxies from the apparent magnitudes of their brightest stars, and plotted the results against Slipher's red shifts for these objects. (The absolute magnitude of the brightest stars was known from studies of our local group of galaxies, whose distance was known from observation of their Cepheid variables, whose *P-L* curve had been calibrated by Shapley from statistical studies of proper motions and radial velocities. For details, see the last section.) The most distant galaxies used by Hubble were members of the Virgo cluster, with a radial velocity of 1000 km/sec. This is not much greater than the r.m.s. random galactic velocity, and Hubble's data points were consequently spread all over the d_L versus z plot. Nevertheless, he was somehow able to deduce a "roughly linear" relation between cz and d_L, with slope

$$H_0 \simeq 500 \text{ km/sec/Mpc} \simeq [2 \times 10^9 \text{ years}]^{-1}$$

At this very time, Milton L. Humason was beginning a program of red-shift measurements at much greater distances, using the 100 in. reflector at Mount Wilson to study the brightest galaxies in clusters. His first definite result, a radial velocity $cz = 3779$ km/sec for the galaxy NGC7619, was used by Hubble in his 1929 paper[13] to check the linearity of the relation between cz and d_L. Assuming this relation to be linear, with slope 500 km/sec/Mpc, Hubble could deduce a distance 7.8 Mpc for NGC7619, so that its apparent magnitude $m = 11.8$ implied an absolute magnitude $M = -17.65$. Hubble also calculated the absolute magnitudes of the 18 galaxies used in his determination of H_0 (plus six additional members of our local group) from their distances and apparent magnitudes, and found M to range from -12.7 to -17.7. Since NGC7619, as the brightest galaxy in a cluster, is presumably brighter than average, this could be considered reasonably good agreement, indicating that cz is indeed roughly proportional to d_L out to $z \simeq 0.013$.

Continuing their collaboration, Hubble and Humason[46] by 1931 had verified the linearity of the relation between cz and d_L out to 20,000 km/sec ($z = 0.067$), and H_0 was revised to 550 km/sec/Mpc. The limit of the Mount Wilson telescope was reached in 1936, when Humason[47] recorded the radial velocity of the Ursa Major II cluster as 42,000 km/sec ($z = 0.14$). By plotting $\ln_{10} z$ against the apparent photographic magnitude (with absorption and k-term corrections) of the fifth brightest galaxy in 10 clusters, ranging from Virgo to UMaII, Hubble[48] could verify that the slope was close to $\frac{1}{5}$, as would be expected [see Eq. (14.6.10)] if the red shift is a linear function of luminosity distance out to $z \simeq 0.14$. A definite measurement of q_0 had to wait until completion of the 200-in. reflector at Mount Palomar.

Hubble also at this time[48] gave a new estimate of H_0, using 109 field galaxies, with cz ranging up to 19,070 km/sec, as the fifth rung of the cosmic distance ladder. The absolute magnitudes of these field galaxies were assumed to be the same as the average, $\bar{M} = -15.18$, for 145 resolved galaxies (only 29 of which belonged to the sample of 109 field galaxies), whose distances could be determined from the apparent magnitude of their brightest stars. Plotting $m - \bar{M}$ against $\ln_{10} cz$ gave $H_0 = 520$ km/sec/Mpc. A separate determination, based directly on the brightest stars in 29 resolved field galaxies, gave $H_0 = 526$ km/sec/Mpc.

In 1950, Palomar was ready, and Hubble's program was taken up again. As we saw in the last section, the first consequence of the observations at Palomar was the recalibration by Baade[27] of the Cepheid period-luminosity relation. This immediately doubled the extragalactic distance scale, and hence halved the value of Hubble's constant, to about 260 km/sec/Mpc. In 1956, Humason, Mayall, and Sandage[49] published an exhaustive survey of the information then available on red shifts and distances. Under the assumption that the brightest galaxies in clusters have the same absolute magnitude as M31, the intercept of the plot of $m_v - k_v - A_v$ against $\ln_{10} cz$ for the brightest galaxies in 18 clusters (out to $z = 0.18$) gave $H_0 = 180$ km/sec/Mpc. (A separate determination, based in the average red shift of the Virgo cluster and the apparent magnitudes of the brightest stars in the Virgo cluster galaxy NGC4321, gave $H_0 = 176$ km/sec/Mpc.) Also, with no evolutionary correction, the curvature of the graph of $m_v - k_v - A_v$ versus $\ln_{10} z$ gave

$$q_0 = 3.7 \pm 0.8$$

A year later, Baum[40] reported a study of eight clusters, using eight-color photometry to avoid the need for a k-term correction. His result was

$$q_0 = 1 \pm \tfrac{1}{2}$$

Next, Sandage[50] reexamined Hubble's use of brightest stars at "rung 4" of the distance ladder, and in 1958 concluded that some of these "brightest" stars are H II regions, which are 1.8 magnitudes brighter than the true brightest stars. The cosmic distance scale expanded again, and H_0 dropped to 75 km/sec/Mpc. Further analysis led Sandage[51] to give a value for H_0 of 98 km/sec/Mpc in 1961. Also, preliminary calculations of galactic evolution led Sandage[52] to estimate that galaxies are *decreasing* in luminosity, with $E_0 \simeq -0.8 H_0$, so that Baum's value of q_0^{eff} led to $q_0 = 0.2 \pm 0.5$.

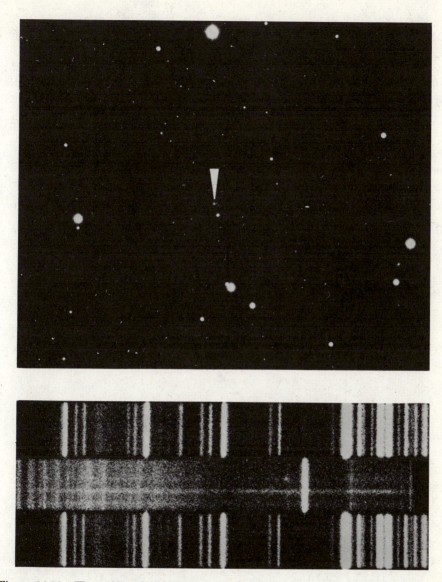

Figure 14.11   The radio galaxy 3C295 in Boötes. The spectrum of this galaxy, shown below, reveals a red shift $z = 0.46$, the largest yet observed for any galaxy. This photograph and spectrograph were taken with the 200-in. telescope at Mt. Palomar. (Courtesy Mt. Wilson and Mt. Palomar observatories.)

Meanwhile, red shifts were becoming available for radio galaxies. In 1960, R. Minkowski[53] discovered that one of these, 3C295, has red shift $z = 0.46$, the largest yet known for any galaxy. (See Figure 14.11.) Sandage[34] in 1968 included these new red shifts, along with the data considered earlier by Humason, Mayall, and Sandage[48] and Baum,[40] in a study of 41 first-ranked cluster members. His data, transformed to the blue magnitude system, could be well fit by the relation

$$m_{B,\,corr} \equiv m_B - k_B - A_B = 5 \ln_{10} cz - 6.06 \qquad (14.6.15)$$

where $c \equiv 3 \times 10^5$. (See Figure 14.12.) The dispersion of points about this curve was only about ± 0.3 magnitudes, indicating that these brightest galaxies really do have a uniform absolute magnitude M_B. The Hubble constant is thus reliably given by (14.6.10) and (14.6.15) as

$$5 \ln_{10} H_0 \, (\text{km/sec/Mpc}) = M_B + 31.06 \qquad (14.6.16)$$

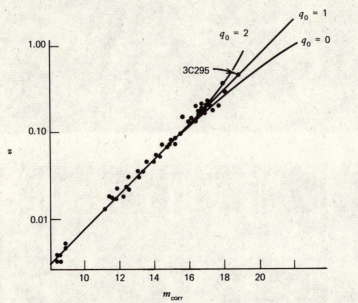

Figure 14.12   Red shifts and corrected apparent magnitudes for 42 first-ranked cluster galaxies. Data are taken from a 1970 review of Sandage.[44] Curves represent fits of Eq. (14.6.10) to the data.

Using globular clusters instead of brightest stars to fix the distance of the Virgo cluster, Sandage[34] estimated that brightest E galaxies have $M_B = -21.68$, so that

$$H_0 = 75.3_{-15}^{+19} \, \text{km/sec/Mpc} = [13.0_{-2.7}^{+3.7} \times 10^9 \, \text{years}]^{-1}$$

$$(14.6.17)$$

[The error quoted here includes uncertainties of ± 0.3 magnitudes in the apparent magnitude of the brightest globular cluster in M87; ± 0.2 magnitudes in the apparent magnitude of the first-ranked Virgo galaxy NGC4472; and ± 0.3 magnitudes in fitting the data with (14.6.15).] The persistence of a straight-line fit out to $z = 0.46$ indicates according to (14.6.10) that q_0^{eff} cannot be too different from unity. Peach[54] gives a value, with evolution neglected:

$$q_0 = 1.5 \pm 0.4 \qquad (14.6.18)$$

while Sandage gives[55]

$$q_0 = 1.2 \pm 0.4 \qquad (14.6.19)$$

What have we really learned from this forty-year program of astronomical observations? There is little doubt that (14.6.15) is a good fit to the data for small z, so that H_0 is given by Eq. (14.6.16). These results have hardly changed at all since Hubble's work in 1936. What has changed dramatically is the distance scale, which controls our estimates of M_B for first-ranked cluster galaxies, and hence plays a crucial role in the determination of H_0. A recent survey by Sandage gives[44]

$$50 \text{ km/sec/Mpc} \leq H_0 \leq 130 \text{ km/sec/Mpc}$$

or

$$20 \times 10^9 \text{ years} \geq H_0^{-1} \geq 7.5 \times 10^9 \text{ years}$$

and this probably represents a fair estimate of the range of possible error in H_0 due to uncertainties in the distance scale. Another change since 1936 is a tripling of the available range of red shifts. We can now be reasonably confident that q_0^{eff} is between $\frac{1}{2}$ and $\frac{3}{2}$. However, the role of evolutionary and selection effects is still very much in doubt. If $H_0 = 75$ km/sec/Mpc, and if galactic luminosities increase at the rate (14.6.13), then the true deacceleration parameter q_0 is related to the observed quantity q_0^{eff} by

$$q_0 = q_0^{\text{eff}} + 0.5$$

This correction is highly uncertain; remember that it was believed to have the opposite sign a few years ago! Thus we now know H_0 to within a factor of 2, and it seems likely that $q_0 > 0$, indicating gravitational *braking*, but about the precise value of q_0 we know almost as little as we did in 1931. (As this book goes to press, rumor has it that H_0 is going down again, perhaps even below 50 km/sec/Mpc.)

In 1963 a discovery was made by Maarten Schmidt,[56] which at first seemed to offer hope of a tremendous improvement in our knowledge of the cosmic scale factor. Since 1960, a number of radio sources had been identified with *quasi-stellar objects*, optical sources whose angular diameters are too small to be resolved at Palomar. Schmidt discovered that one of them, 3C273, had a red shift $z = 0.158$, corresponding to a luminosity distance (if $H_0 = 75$ km/sec/Mpc) of 630 Mpc. At this distance, its absolute luminosity would have to be greater than that of a whole galaxy, even though its small angular diameter ($<0.5''$) implied a size less than

1500 pc. From 1963 until the present, several hundred quasi-stellar objects have been discovered,[57] of which a good fraction have $z > 1$, and a few have $z > 2$. At the same time, the use of lunar occultation and long-base-line radio interferometry, and the observation of short-period time variations, made it clear that much of the enormous energy output of these objects comes from regions much less than 1 pc in diameter. The discovery of the quasi-stellar objects therefore revived interest in theories of gravitational collapse, already discussed in Chapter 11. It also opened up the possibility of extending the empirical relation between d_L and z out to really large distances and red shifts, provided that some method could be found to determine the absolute luminosities of the quasi-stellar objects.

Unfortunately, the plot of m_v versus $\ln z$ reveals no clear correlation of apparent magnitude with red shift.[57, 58] If quasi-stellar objects are indeed at cosmological distances (and about this there remains some doubt[59, 68]), then they must have an extremely wide spread in absolute luminosities. The comparison of red shifts with apparent magnitudes will become cosmologically interesting for quasi-stellar objects only when we learn how to distinguish quasi-stellar objects of different absolute luminosity.

It is nevertheless an interesting question of principle, to ask what could be learned about k and $R(t)$, if the luminosity distance could be exactly determined as a function $d_L(z)$ of red shift? It seems to be generally believed that knowledge of $d_L(z)$ would allow a unique determination of k and $R(t)$. However, this is not the case.[60] The governing theoretical equations here are (14.3.1), (14.3.6), and (14.4.14). Equation (14.3.1) can be replaced with an equivalent differential equation

$$-\frac{1}{R(t_1)}\frac{dt_1}{dz} = (1 - kr_1^2)^{-1/2}\frac{dr_1}{dz} \qquad (14.6.20)$$

with the initial condition that

$$t_1 = t_0 \qquad \text{for } r_1 = 0 \qquad (14.6.21)$$

Equations (14.3.6) and (14.4.14) simply serve to eliminate the unknowns $R(t_1)$ and r_1, so that (14.6.20) becomes

$$(1 + z)\frac{dt_1}{dz} = -[1 - kR^{-2}(t_0)(1 + z)^{-2}d_L^2(z)]^{-1/2}\frac{d}{dz}[(1 + z)^{-1}d_L(z)]$$

Thus t_1 can be calculated as a function of z by a single integration

$$t_1(z) = t_0 - \int_0^z (1 + z')^{-1}[1 - kR^{-2}(t_0)(1 + z')^{-2}d_L^2(z')]^{-1/2}$$

$$\times \frac{d}{dz'}[(1 + z')^{-1}d_L(z')]\,dz'$$

and the function $R(t)$ can then be determined by solving the functional equation:

$$t = t_0 - \int_0^{[R(t_0)/R(t)-1]} (1 + z)^{-1}[1 - kR^{-2}(t_0)(1 + z)^{-2}d_L^{\,2}(z)]^{-1/2}$$

$$\times \frac{d}{dz}[(1 + z)^{-1}d_L(z)]\, dz \qquad (14.6.22)$$

Note that this procedure will give a solution for any assumed values of the constants k, $R(t_0)$, or t_0. Hence *there is no way that measurements of luminosity distances and red shifts can determine k or $R(t_0)$*, unless we supplement the Robertson-Walker metric with dynamical equations for $R(t)$, as will be done in Chapter 15. This curious ambiguity can also be observed in calculating the expansion of $d_L(z)$ in powers of z; the first-order term depends on $\dot{R}(t_0)/R(t_0)$; the second-order term depends on $\dot{R}(t_0)/R(t_0)$ and $\ddot{R}(t_0)/R(r_0)$; the third-order term depends on $\dot{R}(t_0)/R(t_0)$, $\ddot{R}(t_0)/R(t_0)$, $\dddot{R}(t_0)/R(t_0)$, and $k/R^2(t_0)$; and terms of order z^N with $N > 3$ depend on $k/R^2(t_0)$ and the first N logarithmic derivatives of $R(t)$ at t_0. Thus no measurement of any number of derivatives of $d_L(z)$ can ever allow us to determine $k/R^2(t_0)$. However, once we *assume* values for k and $R(t_0)$, Eq. (14.6.22) will allow us to compute $R(t)$ as a function of $t - t_0$ from the empirical relation between d_L and z.

In principle, we could also determine the form of the function $R(t)$ by observing a single spectral line for a long enough time. According to Eqs. (14.3.6) and (14.3.1), the red shift of a comoving source changes at a rate

$$\frac{dz}{dt_0} = \frac{\dot{R}(t_0)}{R(t_1)} - \frac{R(t_0)\dot{R}(t_1)}{R^2(t_1)}\left(\frac{dt_1}{dt_0}\right)$$

$$= \frac{\dot{R}(t_0) - \dot{R}(t_1)}{R(t_1)} \qquad (14.6.23)$$

For $z \ll 1$, we can approximate $t_0 - t_1$ by the first term in the series (14.6.5), and (14.6.23) reads

$$\frac{1}{z}\frac{dz}{dt_0} \simeq \frac{\ddot{R}(t_0)}{H_0 R(t_0)} = -q_0 H_0 \qquad (14.6.24)$$

It does not seem possible to measure this very slow change in red shift with present techniques.[61]

7 Number Counts

Since the Hubble program has not yet succeeded in telling us very much about the cosmic scale factor $R(t)$, it is natural to widen our scope, and consider *numbers* of optical or radio sources, as functions of apparent luminosity and/or red shift.

The study of number counts offers two potential advantages over the Hubble program:

(A) The development of radio telescopes of very large aperture and sensitivity has led to the detection and resolution of thousands of faint radio sources, most of which are presumably at very great distances. The majority of these sources have not yet been identified with optical objects, so their red shifts are as yet unknown. (No radio *lines* have been observed from resolved radio sources, and their red shifts can therefore only be measured optically.) Not knowing z, the best use to which cosmologists can put these sources is to plot their number as a function of their strength.

(B) The quasi-stellar objects discussed in the last section have measured red shifts ranging up to $z \approx 2$, but have too wide a spread in absolute luminosity to allow determination of the luminosity distance d_L. By plotting the number of quasi-stellar objects as a function of z alone, or z and l, we can eliminate some of the problems caused by the spread in L.

To begin in a very general way, let us assume that at time t_1 there are $n(L, t_1) \, dL$ sources per unit volume with absolute luminosity between L and $L + dL$. The proper volume element is

$$dV = \sqrt{g} \, dr_1 \, d\theta_1 \, d\phi_1 = R^3(t_1)(1 - kr_1{}^2)^{-1/2} r_1{}^2 \, dr_1 \, \sin \theta_1 \, d\theta_1 \, d\phi_1$$

so the number of sources between r_1 and $r_1 + dr_1$ with absolute luminosity between L and $L + dL$ is

$$dN = 4\pi R^3(t_1)(1 - kr_1{}^2)^{-1/2} r_1{}^2 n(t_1, L) \, dr_1 \, dL \tag{14.7.1}$$

The coordinates r_1 and t_1 are related according to Eq. (14.3.1), which we can write as

$$r_1 = r(t_1) \tag{14.7.2}$$

where $r(t)$ is a function defined by the formula

$$\int_t^{t_0} \frac{dt'}{R(t')} \equiv \int_0^{r(t)} (1 - kr^2)^{-1/2} \, dr \tag{14.7.3}$$

Differentiation of Eq. (14.7.3) gives

$$dr_1 = -(1 - kr_1{}^2)^{1/2} \frac{dt_1}{R(t_1)}$$

and Eq. (14.7.1) can therefore be written

$$dN = 4\pi R^2(t_1) r^2(t_1) n(t_1, L) \, |dt_1| \, dL \tag{14.7.4}$$

The red shift and apparent luminosity of a source of absolute luminosity L at r_1, t_1 are given by Eq. (14.3.6) and (14.4.12):

$$z = \frac{R(t_0)}{R(t_1)} - 1 \qquad (14.7.5)$$

$$l = \frac{LR^2(t_1)}{4\pi r_1^2 R^4(t_0)} \qquad (14.7.6)$$

Hence the number of sources, with red shift *less* than z and apparent luminosity *greater* than l, is given by the integral of (14.7.4) over all L and a finite range of t_1:

$$N(<z, >l) = \int_0^\infty dL \int_{\max\{t_z,\, t_l(L)\}}^{t_0} 4\pi r^2(t_1) R^2(t_1) n(t_1, L)\, dt_1 \qquad (14.7.7)$$

where the lower limits, set by the conditions on red shift and apparent luminosity, are defined by

$$R(t_z) \equiv \frac{R(t_0)}{(1 + z)} \qquad (14.7.8)$$

$$\frac{r^2(t_l)}{R^2(t_l)} \equiv \frac{L}{4\pi l R^4(t_0)} \qquad (14.7.9)$$

If red shifts are not observed, then the quantity of interest is the number $N(>l)$ of sources with apparent luminosity greater than l, which can be calculated by taking the lower limit in (14.7.7) to be just $t_l(L)$. If apparent luminosities are not observed, then the quantity of interest is the number $N(<z)$ of sources with red shift less than z, which can be calculated by taking the lower limit in (14.7.7) to be just t_z. (However, the observed number counts can only be used to put a lower bound on $N(z)$, not to measure it, because any given optical or radio telescope will only detect sources above some minimum brightness.)

Radio telescopes do not measure total apparent luminosities, but instead measure the *flux density* S, the power per unit antenna area and per unit frequency interval, at a fixed frequency. The flux density of a source at r_1, t_1 is

$$S(v) = \frac{P(vR(t_0)/R(t_1))R(t_1)}{R^3(t_0)r_1^2} \qquad (14.7.10)$$

where P is the *intrinsic power*, the power emitted per unit solid angle and per unit frequency interval. (See Eq. (14.4.35), with $S \equiv l'$, $P \equiv L'/4\pi$.)

Following the same derivation that led to Eq. (14.7.7), we find that the number of sources, with red shift less than z and flux density at frequency v greater than S, is given by

$$N(<z, >S; v) = \int_0^\infty dP \int_{\max\{t_z,\, t_S(P)\}}^{t_0} 4\pi r^2(t_1) R^2(t_1) n\left(t_1, P, v\,\frac{R(t_0)}{R(t_1)}\right) dt_1 \qquad (14.7.11)$$

where $t_S(P)$ is defined by

$$\frac{r^2(t_S)}{R(t_S)} = \frac{P}{SR^3(t_0)} \tag{14.7.12}$$

The analysis of radio source counts is very much simplified by the observation[62] that such radio sources generally have "straight" spectra,

$$P \propto v^{-\alpha} \tag{14.7.13}$$

where α, the *spectral index*, is about 0.7 to 0.8. In this case, the number of sources with intrinsic power at frequency v between P and $P + dP$ is of the form

$$n(t, P, v)\, dP = n\left(t, P\left[\frac{v}{v_0}\right]^\alpha, v_0\right) d\left(P\left[\frac{v}{v_0}\right]^\alpha\right)$$

where v_0 is any arbitrary fixed frequency. The number density then obeys the scaling rule

$$n(t, P, v) = \left[\frac{v}{v_0}\right]^\alpha n\left(t, P\left[\frac{v}{v_0}\right]^\alpha, v_0\right) \tag{14.7.14}$$

By changing the variable of integration in Eq. (14.7.11) from P to $P[R(t_0)/R(t_1)]^\alpha$, we can refer the number density within the integral to the fixed frequency v, and find

$$N(<z, >S; v) = \int_0^\infty dP \int_{\max\{t_z,\, t_{S\alpha}(P)\}}^{t_0} 4\pi r^2(t_1) R^2(t_1) n(t_1, P, v)\, dt_1 \tag{14.7.15}$$

where $t_{S\alpha}(P)$ is defined by

$$r^2(t_{S\alpha})\left(\frac{R(t_{S\alpha})}{R(t_0)}\right)^{-1-\alpha} = \frac{P}{SR^2(t_0)} \tag{14.7.16}$$

The number counts will now obey the scaling rule

$$N(<z, >S; v) = N\left(<z, >S\left[\frac{v}{v_0}\right]^\alpha; v_0\right) \tag{14.7.17}$$

To the extent that (14.7.17) is verified by observation, we can conclude that all sources do have the "straight" spectrum (14.7.13), with the same spectral index α.

If there were no creation, destruction, or evolution of the sources during the time it takes for light to reach us from the furthest observed source, then $n(t, L)$ and $n(t, P, v)$ would have the simple time dependence (14.2.17):

$$n(t, L) = \left[\frac{R(t_0)}{R(t)}\right]^3 n(t_0, L) \tag{14.7.18}$$

$$n(t, P, v) = \left[\frac{R(t_0)}{R(t)}\right]^3 n(t_0, P, v) \tag{14.7.19}$$

In this case, the observed number counts could be used to gain information about k and $R(t)$. Alternatively, if we had a cosmological model for k and $R(t)$, we could use the observed number counts to deduce the functional dependence of the number density n on t and L or P.

A good deal of insight into the results to be expected from these two different modes of analysis can be gained by concentrating on the special cases where z is small or l or S is large. The lower limits on the t_1-integral in Eq. (14.7.7) and (14.7.11) are then close to t_0, so we can use the general expansions (14.6.1) and (14.6.6):

$$R(t_1) = R(t_0)\{1 - H_0(t_0 - t_1) + \cdots\}$$

$$r(t_1) = R^{-1}(t_0)(t_0 - t_1)\{1 + \tfrac{1}{2}H_0(t_0 - t_1) + \cdots\}$$

We will also express the number densities as expansions in $t_0 - t_1$:

$$n(t_1, L) = n(t_0, L)\{1 - \beta_0(L)H_0(t_0 - t_1) + \cdots\} \tag{14.7.20}$$

$$n\left(t_1, P, v\,\frac{R(t_0)}{R(t_1)}\right) = n(t_0, P, v)\{1 - [\beta_0(P, v) + 2\alpha_0(P, v)]H_0(t_0 - t_1) + \cdots\} \tag{14.7.21}$$

where β_0 measures the rate of change of source density

$$\beta_0(L) \equiv H_0^{-1}\left(\frac{\partial}{\partial t}\ln n(t, L)\right)_{t=t_0} \tag{14.7.22}$$

$$\beta_0(P, v) \equiv H_0^{-1}\left(\frac{\partial}{\partial t}\ln n(t, P, v)\right)_{t=t_0} \tag{14.7.23}$$

and α_0 is an *effective spectral index*

$$2\alpha_0(P, v) \equiv -v\,\frac{\partial}{\partial v}\ln n(t_0, P, v) \tag{14.7.24}$$

(The motivation for this last definition will be made clear below.) Then, for t close to t_0,

$$\int_t^{t_0} 4\pi r^2(t_1)R^2(t_1)n(t_1, L)\,dt_1$$

$$= \frac{4\pi}{3}\,n(t_0, L)(t_0 - t)^3\{1 - \tfrac{3}{4}[\beta_0(L) + 1]H_0(t_0 - t) + \cdots\} \tag{14.7.25}$$

$$\int_t^{t_0} 4\pi r^2(t_1)R^2(t_1)n\left(t_1, P, v\,\frac{R(t_0)}{R(t_1)}\right)dt_1$$

$$= \frac{4\pi}{3}\,n(t_0, P, v)(t_0 - t)^3\{1 - \tfrac{3}{4}[\beta_0(P, v) + 2\alpha_0(P, v) + 1]H_0(t_0 - t) + \cdots\} \tag{14.7.26}$$

If z is small, then the lower limit on the integral (14.7.25) is determined by (14.7.8), which gives

$$H_0(t_0 - t_z) = z - \left(1 + \frac{q_0}{2}\right)z^2 + \cdots$$

Hence (14.7.7) gives the number of sources with red shift less than z as

$$N(<z) = \frac{4\pi}{3} H_0^{-3} z^3 \int_0^\infty n(t_0, L)\{1 - \tfrac{3}{4}[\beta_0(L) + 2q_0 + 5]z + \cdots\} \, dL$$

$$(14.7.27)$$

If l is large, then the lower limit on the integral (14.7.25) is determined by (14.7.9), which gives

$$t_0 - t_l(L) = \left(\frac{L}{4\pi l}\right)^{1/2} \left\{1 - \frac{3}{2}\left(\frac{LH_0^2}{4\pi l}\right)^{1/2} + \cdots\right\}$$

Now (14.7.7) gives the number of sources with apparent luminosity greater than l as

$$N(>l) = \frac{4\pi}{3} (4\pi l)^{-3/2} \int_0^\infty n(t_0, L)$$

$$\times \left\{1 - \tfrac{3}{4}[\beta_0(L) + 7]\left(\frac{LH_0^2}{4\pi l}\right)^{1/2} + \cdots\right\} L^{3/2} \, dL \qquad (14.7.28)$$

Finally, if S is large, then the lower limit on the integral (14.7.26) is determined by (14.7.12), which gives

$$t_0 - t_S(P) = \left(\frac{P}{S}\right)^{1/2} \left\{1 - \left(\frac{PH_0^2}{S}\right)^{1/2} + \cdots\right\}$$

Thus (14.7.11) gives the numbers of sources, with flux density at frequency v greater than S, as

$$N(>S, v) = \frac{4\pi}{3} S^{-3/2} \int_0^\infty n(t_0, P, v) P^{3/2}$$

$$\times \left\{1 - \tfrac{3}{4}[\beta_0(P, v) + 2\alpha_0(P, v) + 5]\left(\frac{PH_0^2}{S}\right)^{1/2} + \cdots\right\} dP$$

$$(14.7.29)$$

If all sources have the "straight" spectrum (14.7.13), then (14.7.14) and (14.7.24) give

$$\alpha_0(P, v) = -\frac{\alpha}{2}\left[1 + P\frac{\partial}{\partial P} \ln n(t_0, P, v)\right] \qquad (14.7.30)$$

so integration by parts allows us to make the substitution

$$\alpha_0(P, v) \to \alpha \qquad (14.7.31)$$

That is, the "effective spectral index" α_0 may be replaced with α in Eq. (14.7.29), provided that sources have the spectrum $P \propto v^{-\alpha}$.

Inspection of these results shows that measured values of $N(<z)$ for $z \ll 1$ could be used to deduce the deacceleration parameter q_0 if we knew the evolution-

ary parameter β_0, while, in contrast, measurement of $N(>l)$ or $N(>S, v)$ for large l or S cannot tell us anything about q_0, whatever we assume for β_0.

If we assume no evolution, then n has a time dependence given by (14.7.18) or (14.7.19) and (14.6.1), so (14.7.22) and (14.7.23) give

$$\beta_0(L) = \beta_0(P, v) = -3 \qquad \text{(no evolution)} \qquad (14.7.32)$$

In this case, (14.7.27)–(14.7.29) give

$$N(<z) = \frac{4\pi}{3} H_0^{-3} z^3 \int_0^\infty n(t_0, L) \, dL\{1 - \tfrac{3}{2}(q_0 + 1)z + \cdots\} \qquad (14.7.33)$$

$$N(>l) = \frac{4\pi}{3} (4\pi l)^{-3/2} \int_0^\infty n(t_0, L) \left\{1 - 3\left(\frac{LH_0^2}{4\pi l}\right)^{1/2} + \cdots\right\} L^{3/2} \, dL \qquad (14.7.34)$$

and, for straight spectra,

$$N(>S, v) = \frac{4\pi}{3} S^{-3/2} \int_0^\infty n(t_0, P, v) \left\{1 - \tfrac{3}{2}(\alpha + 1)\left(\frac{PH_0^2}{S}\right)^{1/2} + \cdots\right\} P^{3/2} \, dP$$

$$(14.7.35)$$

Thus the neglect of evolution leads to the definite predictions that $N(>l)$ must decrease with l more *slowly* than $l^{-3/2}$, and, since α is positive, $N(>S, v)$ must decrease with S more slowly than $S^{-3/2}$.

However, this result appears to be contradicted by observation[69]. The relevant radio source surveys are listed in Table 14.1. Jointly and severally, they yield a number count function[63] $N(>S, v)$ that decreases with S (for $S > 5 \times 10^{-26}$ W m^{-2} × Hz^{-1}) roughly like $S^{-1.8}$, and definitely more *rapidly* than $S^{-3/2}$. We are forced to the conclusion that evolution *is* important. According to Eq. (14.7.29), the decrease of $S^{3/2}N(>S, v)$ with S requires that

$$\beta_0 < -2\,\alpha_0 - 5 \simeq -6.5 \qquad (14.7.36)$$

so that the number density of sources must be decreasing faster than $R(t)^{-6.5}$.

A similar conclusion is reached from the study of number counts of radio sources as a function of their angular diameters. From a study of the distribution of the sizes of the radio sources in the 3C catalogue, Longair and Pooley[64] calculated the distribution in angular diameters to be expected for the fainter sources in the 5C catalogue if no evolutionary effects are important. Their result does not agree with observation for any q_0, indicating an evolutionary decrease of the proper source density.

If evolution is as substantial as indicated by these source counts, we evidently cannot use the source counts to learn much about $R(t)$. We shall return to the program of using source counts to learn about source evolution in the next chapter, when we have a dynamical model for $R(t)$.

Table 14.1 The Major Radio Source Surveys[a]

Observatory	Survey	ν (MHz)	Sources	S_{min} (10^{-26} Wm^{-2} Hz^{-1})
Cambridge	3C	159	471	8
	3CR	178	—	9
	4C	178	4843	2
	5C	408	276	0.025
	WKB	38	1069	14
	RN	178	87	0.25
	NB	81.5	558	1
Mills Cross	MSH	86	2270	7
Parkes	PKS	408,1410,2650	297	4
	PKS	408,1410	247	0.5
	PKS	408,1410	564	0.3
	PKS	408,1410	628	0.4
	PKS	635,1410,2650	397	1.5
Owens	CTA	906	106	—
Valley	CTB, CTBR	960	110	—
	CTD	1421	—	1.15
National Radio	NRAO	750,1400	726	(3C and 3CR)
Observa- tory	NRAO	750,1400	458	0.5
Bologna	B1	408	629	1
	B2	408	3235	0.2
Ohio State	O	1415	128	2, 0.5
	O	1415	236	0.37
	O	1415	1199	0.3
	O	1415	2101	0.2
Vermillion	VRO	610.5	239	0.8
River	VRO	610.5	625	0.8
Dominion Radio Observa- tory	DA	1420	615	2
Dwingeloo- NRAO	DW	1417	188	2.3
Arecibo	AO	430	25	—

[a] The different surveys cover different, partially overlapping regions of the sky, and not all are complete within their region and flux range. For further details and references, see A. G. Pacholczyk *Radio Astrophysics* (W. H. Freeman and Co., San Francisco, 1970), pp. 241 ff.

8   The Steady State Cosmology

Our work so far has been based on the "Cosmological Principle" that the universe is spatially isotropic and homogeneous. Hermann Bondi and Thomas Gold[65] have gone one step further, and have suggested that the universe obeys a "perfect Cosmological Principle," that it looks the same not only at all points and in all directions, but *at all times*. This assumption leads to a *sieady state model* of the universe, which was suggested at about the same time by Fred Hoyle,[66] on the basis of an alteration in the structure of the energy-momentum tensor appearing in Einstein's field equations. We shall follow the Bondi-Gold approach here, as more suited to the spirit of the present chapter, and will come back to Hoyle's theory in the last chapter.

The work of Section 14.6 shows that the Hubble "constant" $\dot{R}(t_0)/R(t_0)$ is an observable parameter, so that it must be independent of the present time t_0 in a steady state model. Letting H denote the permanent value of the Hubble constant, we have then

$$\frac{\dot{R}(t)}{R(t)} = H \qquad \text{for all } t$$

and therefore

$$R(t) = R(t_0) \exp \{H(t - t_0)\} \tag{14.8.1}$$

In this model the deacceleration parameter takes the permanent value

$$q \equiv - \frac{\ddot{R}R}{\dot{R}^2} = -1 \tag{14.8.2}$$

To determine k, we return to the general relation (14.6.22) between $R(t)$ and the luminosity-distance versus red-shift function $d_L(z)$, which now reads

$$t_0 - t = \int_0^{[\exp\{H(t_0 - t)\} - 1]} (1 + z)^{-1}[1 - kR^{-2}(t_0)(1 + z)^{-2} d_L^2(z)]^{1/2}$$
$$\times \frac{d}{dz} [(1 + z)^{-1} d_L(z)] \, dz \tag{14.8.3}$$

Since $d_L(z)$ is observable, it must now be independent of t_0. Hence, in order that the integral depend only on $t - t_0$, not t or t_0 separately, it is necessary that

$$k = 0 \tag{14.8.4}$$

The metric is then

$$d\tau^2 = dt^2 - R^2(t_0)e^{2H(t - t_0)}\{dr^2 + r^2 \, d\theta^2 + r^2 \sin^2 \theta \, d\phi^2\} \tag{14.8.5}$$

This derivation might be challenged on the grounds that the metric (14.8.5) was obtained as a special case of the Robertson-Walker metric, which was derived in Sections 14.1 and 14.2 on the basis of a definition of cosmic time that makes no sense in an unevolving universe. We could avoid this difficulty by viewing (14.8.5) as the limiting metric for a universe with an extremely slow rate of evolution. A more satisfactory approach is to derive (14.8.5) directly from the assumption that the whole four-dimensional space-time is maximally symmetric. This assumption is shown in Section 13.3 to lead to the metric (13.3.41), which is identical to (14.8.5), except that a factor $R(t_0) \exp(-Ht_0)$ must be absorbed into the radial coordinate r. Comparing (13.3.41) and (14.8.5), we note that the curvature constant of the four-dimensional space-time of the steady state model is

$$K = H^2 \qquad (14.8.6)$$

The *space-time* is curved, even though the *space* is flat.

The most remarkable feature of the steady state cosmology is not its space-time metric, but rather, the necessity of *continuous creation* of matter. According to Eq. (14.2.21), the proper distance between any two comoving galaxies increases as $R(t)$, so if the average number of galaxies per unit proper volume is to remain constant, new galaxies must appear to fill up the holes in the widening comoving coordinate mesh. To put this formally, we recall that in the comoving coordinate system $r\theta\phi t$, the current vector of the galaxies and the total energy-momentum tensor are given by (14.2.11)–(14.2.14) as

$$J_G{}^\mu = n_G U^\mu$$

$$T^{\mu\nu} = (\rho + p)U^\mu U^\nu + pg^{\mu\nu}$$

with

$$U^t = 1 \qquad U^r = U^\theta = U^\phi = 0$$

In accordance with the spirit of the steady state model, we now take n_G, p, and ρ to be constant in time as well as space. Then $J_G{}^\mu$ and $T^{\mu\nu}$ are *not* conserved, but rather

$$J_G{}^\mu{}_{;\mu} = R^{-3}(t)\frac{\partial}{\partial t}(R^3(t)J_G{}^t) = 3n_G H \qquad (14.8.7)$$

$$T^{\mu t}{}_{;\mu} = R^{-3}(t)\frac{\partial}{\partial t}(R^3(t)[p + \rho]) = 3(p + \rho)H \qquad (14.8.8)$$

That is, a comoving observer using a locally inertial coordinate system will see galaxies created at a rate $3H$ per existing galaxy, and will see energy created at a rate $3H$ per existing mass plus enthalpy. The present density of the universe is roughly of the order 10^{-6} nucleons/cm^3, so with $H^{-1} = 10^{10}$ years this would require an average creation of order 10^{-16} nucleons/cm^3/year. The steady state model is silent as to whether this new matter is created as hydrogen, or protons plus electrons, or neutrons, and it does not tell us whether this new matter appears near the old matter or in the depths of intergalactic space. However, violent events do seem to be occurring in the nuclei of many galaxies, so galactic nuclei seem like natural candidates for the location of continuous creation.

The steady state cosmology makes very definite predictions as to the correlation of luminosity distance with red shift. According to (14.3.1), if light leaves a comoving source at time t_1 and arrives at the origin at time t_0, the source must be at a coordinate r_1 given for $k = 0$ by

$$r_1 = r(t_1) \equiv \int_{t_1}^{t_0} \frac{dt}{R(t)} = H^{-1}R^{-1}(t_0)\{\exp[H(t_0 - t_1)] - 1\} \quad (14.8.9)$$

Also, (14.3.6) gives the red shift of such a source as

$$z = \exp[H(t_0 - t_1)] - 1 \quad (14.8.10)$$

so the luminosity distance of the source is given by (14.4.14) as

$$d_L(z) = H^{-1}z(1 + z) \quad (14.8.11)$$

As a check, note that (14.8.9)–(14.8.11) agree with (14.6.6), (14.6.4), and (14.6.8) for the appropriate deacceleration parameter $q = -1$. This value does *not* seem to agree with the q_0 determined from the observed d_L versus z relation, as discussed in Section 14.6.

The "angular diameter" distance d_A is given in this model by (14.4.22) and (14.8.11) as

$$d_A(z) = \frac{H^{-1}z}{(1 + z)} \quad (14.8.12)$$

Note that $d_A(z)$ approaches the finite constant H^{-1} as $z \to \infty$. Hence objects with large red shifts look very faint, but their angular diameters do not shrink below a minimum value. If H^{-1} is 3×10^9 pc, then a galaxy of diameter 10^4 pc will never appear smaller than about 0.6″.

If we count the number of sources with red shift less than z, we must look back to a time t_z given by (14.7.8) as

$$t_z = t_0 - H^{-1}\ln(1 + z) \quad (14.8.13)$$

To count the number of sources with apparent luminosity greater than l, we must look back to a time given by (14.7.9), which together with Eq. (14.8.9) now reads

$$\exp[H(t_0 - t_l)]\{\exp[H(t_0 - t_l)] - 1\} = \left(\frac{LH^2}{4\pi l}\right)^{1/2}$$

The solution is

$$t_l(L) = t_0 - H^{-1}\ln\left[\frac{1}{2} + \left(\frac{1}{4} + \left(\frac{LH^2}{4\pi l}\right)^{1/2}\right)^{1/2}\right] \quad (14.8.14)$$

The t_1-integral in Eq. (14.7.7) can be done explicitly, and we find the number of sources with red shift less than z and apparent luminosity greater than l as

$$N(<z, >l) = \int_0^\infty n(L)\min\{V(t_z), V(t_l(L))\}\,dL \quad (14.8.15)$$

where V is the volume

$$V(t) = \int_t^{t_0} 4\pi r^2(t_1) R^2(t_1) \, dt_1$$

$$= 4\pi H^{-3}\{H(t_0 - t) - \tfrac{3}{2} + 2 \exp\left[-H(t_0 - t)\right] - \tfrac{1}{2}\exp\left[-2H(t_0 - t)\right]\}$$

$$(14.8.16)$$

and $n(L) \, dL$ is the time-independent proper number density of sources with absolute luminosity between L and $L + dL$.

As a special case of Eq. (14.8.15), we note that the number of sources with red shift less than z is

$$N(<z) = 4\pi H^{-3} n \left\{ \ln(1 + z) - \frac{z(1 + 3z/2)}{(1 + z)^2} \right\} \qquad (14.8.17)$$

where n is the total number density of sources

$$n \equiv \int_0^\infty n(L) \, dL$$

This result is independent of any assumptions concerning the luminosity distribution of sources, and of course no evolution of the source density or luminosity distribution is possible in a steady state universe. However, with the limited statistics now available, it appears that (14.8.17) does *not* agree with the observed . red-shift distribution of the quasi-stellar radio sources.[67] In particular, the observed red-shift distribution of the quasi-stellar sources shows a pronounced peak[57] near $z = 1.95$, which is absent in Eq. (14.8.17). It should be noted though that the observed value of $N(>z)$ should in general be smaller than the theoretical prediction (14.8.17), because some sources are not counted if their optical or radio strength is too small.

As another special case of Eq. (14.8.15), we note that the number of sources with apparent luminosity greater than l is

$$N(>l) = 4\pi H^{-3} \int_0^\infty n(L) \left(\ln\left(\frac{1}{2} + \left[\frac{1}{4} + \left(\frac{LH^2}{4\pi l} \right)^{1/2} \right]^{1/2} \right) - \frac{3}{2} \right.$$

$$+ 2 \left(\frac{1}{2} + \left[\frac{1}{4} + \left(\frac{LH^2}{4\pi l} \right)^{1/2} \right]^{1/2} \right)^{-1}$$

$$\left. - \frac{1}{2} \left(\frac{1}{2} + \left[\frac{1}{4} + \left(\frac{LH^2}{4\pi l} \right)^{1/2} \right]^{1/2} \right)^{-2} \right) dL \qquad (14.8.18)$$

In contrast with $N(<z)$, the result here depends on the details of the distribution function $n(L)$.

A quantity of greater observational interest is the number of radio sources with strength at frequency v greater than S. If all sources have the same "straight" spectrum $P \propto v^{-\alpha}$, then (14.7.15) gives the number of such sources as

$$N(>S; v) = \int_0^\infty n(P, v) V(t_{S\alpha}(P)) \, dP \tag{14.8.19}$$

where $n(P, v) \, dP$ is the number of sources with intrinsic power at frequency v between P and $P + dP$; $V(t)$ is given by (14.8.16); and $t_{S\alpha}(P)$ is determined by (14.7.16), which now reads

$$\exp\left[\tfrac{1}{2}(3 + \alpha)H(t_0 - t_{S\alpha})\right] - \exp\left[\tfrac{1}{2}(1 + \alpha)H(t_0 - t_{S\alpha})\right] = \left(\frac{PH^2}{S}\right)^{1/2}$$
$$\tag{14.8.20}$$

This equation cannot be solved analytically for the observed spectral index $\alpha \simeq 0.7$. However, it follows from (14.8.20), (14.8.19), and (14.8.16) that $N(>S, v)$ decreases more slowly than $S^{-3/2}$ for all source strengths, in contradiction with observed number counts, which decrease more rapidly[63] than $S^{-3/2}$ for S greater than about $4 \times 10^{-26} \, \text{Wm}^{-2} \, \text{Hz}^{-1}$, and only then begin to decrease more slowly than $S^{-3/2}$. In the last section we saw that these observations are also inconsistent with the results of nonsteady state cosmologies, but in that case the discrepancy could be removed by assuming an evolution of source densities, while in the steady state cosmology no evolution of the source density is allowed.

As a check, we note that the quantities $\beta_0(L)$ and $\beta_0(P, v)$ defined by Eqs. (14.7.22) and (14.7.23) must vanish in the steady state model

$$\beta_0(L) = \beta_0(P, v) = 0$$

Hence (14.7.27), (14.7.28), and (14.7.29), with $q_0 = -1$ and straight spectra, now give the number counts for "nearby" sources as

$$N(<z) = \frac{4\pi}{3} H^{-3} z^3 n\{1 - \tfrac{9}{4}z + \cdots\} \tag{14.8.21}$$

$$N(>l) = \frac{4\pi}{3} (4\pi l)^{-3/2} \int_0^\infty n(L) \left\{1 - \frac{21}{4}\left(\frac{LH^2}{4\pi l}\right)^{1/2} + \cdots\right\} L^{3/2} \, dL \tag{14.8.22}$$

$$N(>S, v) = \frac{4\pi}{3} S^{-3/2} \int_0^\infty n(P, v) \left\{1 - \tfrac{3}{4}(2\alpha + 5)\left(\frac{PH^2}{S}\right)^{1/2} + \cdots\right\} P^{3/2} \, dP \tag{14.8.23}$$

in agreement with the power-series expansions of the general formulas (14.8.17), (14.8.18), and (14.8.19).

The steady state model does not appear to agree with the observed d_L versus z relation or with the source counts for $N(<z)$ and $N(>S, v)$. In a sense, this disagreement is a credit to the model; alone among all cosmologies, the steady state model makes such definite predictions that it can be disproved even with the limited observational evidence at our disposal. The steady state model is so attractive that many of its adherents still retain hope that the evidence against it will eventually disappear as observations improve. However, if the cosmic microwave radiation discussed in the next chapter is really black-body radiation, it will be difficult to doubt that the universe has evolved from a hotter denser early stage.

14 BIBLIOGRAPHY

Where no other references are indicated, astronomical data are usually taken from C. W. Allen, *Astrophysical Quantities* (2nd ed., The Athlone Press, London, 1955).

Cosmology in General

☐ H. Bondi, *Cosmology* (Cambridge University Press, Cambridge, 1960).

☐ W. Davidson and J. V. Narlikar, "Cosmological Models and Their Observational Validation," reprinted in *Astrophysics* (W. A. Benjamin, New York, 1969).

☐ O. Heckmann and E. Schucking, "Relativistic Cosmology," in *Gravitation: An Introduction to Current Research*, ed. by L. Witten (Wiley, New York, 1962), p. 438.

☐ P. W. Hodge, *Galaxies and Cosmology* (McGraw-Hill, New York, 1966).

☐ *La Structure et l'Evolution de l'Univers* (Eleventh Solvay Conference, R. Stoops, ed., Brussels, 1958).

☐ G. C. McVittie, *General Relativity and Cosmology* (University of Illinois Press, Urbana, Ill., 1965).

☐ W. Rindler, "Relativistic Cosmology," Physics Today, November, 1967, p. 23.

☐ H. P. Robertson and T. W. Noonan, *Relativity and Cosmology* (W. B. Saunders Co., Philadelphia, 1968).

☐ E. L. Schucking, "Cosmology," in *Relativistic Theory and Astrophysics. 1. Relativity and Cosmology*, ed. by J. Ehlers (American Mathematical Society, Providence, R. I., 1967), p. 218.

☐ D. W. Sciama, *Modern Cosmology* (Cambridge University Press, Cambridge, 1971).

☐ R. C. Tolman, *Relativity, Thermodynamics, and Cosmology* (Clarendon Press, Oxford, 1934).

☐ Ya. B. Zeld'ovich, "Survey of Modern Cosmology," in *Advances in Astronomy and Astrophysics*, Vol. 3, ed. by Z. Kopal (Academic Press, New York, 1965).

History of Astronomy and Cosmology in the Twentieth Century

☐ W. Baade, *Evolution of Stars and Galaxies*, ed. by C. Payne-Gaposchkin (Harvard University Press, Cambridge, Mass., 1963).

☐ F. P. Dickson, *The Bowl of Night* (M.I.T. Press, Cambridge, Mass., 1968).

☐ J. D. Fernie, "The Period-Luminosity Relation: A Historical Review," Pub. Astron. Soc. Pac., **81,** 707 (1969).

☐ J. D. North, *The Measure of the Universe* (Clarendon Press, Oxford, 1965). *Source Book in Astronomy 1900–1950*, ed. by H. Shapley (Harvard University Press, Cambridge, Mass., 1960).

The Cosmic Distance Scale and the Hubble Program

☐ *Basic Astronomical Data*, ed. by K. Aa. Strand (University of Chicago Press, Chicago, 1963).

☐ A. Sandage, "Cosmology—A Search for Two Numbers," Physics Today, February, 1970, p. 34.

☐ A. Sandage, in *Problems of Extragalactic Research* (Macmillan, New York, 1962), p. 359.

☐ A. Sandage, "The Ability of the 200-Inch Telescope to Discriminate between Selected World Models," Ap. J., **133,** 355 (1961).

Quasi-Stellar Objects

☐ G. Burbidge and M. Burbidge, *Quasi-Stellar Objects* (W. H. Freeman and Co., San Francisco, 1967).

☐ L. C. Green, "Quasars Six Years Later," Sky and Telescope, May 1969.

☐ M. Schmidt, "Lectures on Quasi-Stellar Objects," in *Relativity Theory and Astrophysics. 1. Relativity and Cosmology, op. cit.*, p. 203.

☐ M. Schmidt, "Quasi-Stellar Objects," in *Annual Review of Astronomy and Astrophysics*, Vol. 7, ed. by L. Goldberg (Annual Reviews, Inc., Palo Alto, 1969), p. 527.

Radio Source Counts

☐ K. Brecher, G. Burbidge, and P. A. Strittmayer, "Counts of Sources and Theories," Comments on Astrophysics and Space Physics **3,** 99 (1971).

☐ M. Ryle, "The Counts of Radio Sources," in *Annual Review of Astronomy and Astrophysics*, Vol. 6, ed. by L. Goldberg (Annual Reviews, Inc., Palo Alto, 1968), p. 249.

☐ P. A. G. Scheuer, "Radio Astronomy and Cosmology," in *Stars and Stellar Systems, Vol. IX: Galaxies and the Universe*, ed. by A. and M. Sandage, to be published.

☐ F. G. Smith, "Radio Galaxies and Quasars," in *Contemporary Physics—Trieste Symposium 1968*, ed. by A. Salam, Vol. 1 (International Atomic Energy Agency, Vienna, 1969), p. 459.

14 REFERENCES

1. H. P. Robertson, Ap. J., **82**, 284 (1935); *ibid.*, **83**, 187, 257 (1936).
2. A. G. Walker, Proc. Lond. Math. Soc. (2), **42**, 90 (1936).
3. A. Friedmann, Z. Phys., **10**, 377 (1922); *ibid.*, **21**, 326 (1924).
4. K. Gödel, Rev. Mod. Phys., **21**, 447 (1949). For a general classification of homogeneous anisotropic spaces, see A. Taub, Ann. Math., **53**, 474 (1951).
5. C. V. I. Charlier, Arkiv. Mat. Astr. Fys., **4**, No. 24 (1908); *ibid.*, **16**, No. 22 (1922).
6. G. de Vaucouleurs, Science, **167**, 1203 (1970).
7. F. Zwicky, Pub. Ast. Soc. Pacific, **50, 218** (1938); F. Zwicky and K. Rudnicki, Ap. J., **137, 707** (1963); Z. Astrophys., **64**, 246 (1966).
8. G. O. Abell, Ap. J., Suppl. **3**, 211 (1958).
9. J. H. Oort, XI Solvay Conference, *La Structure et l'Evolution de l'Univers* (Brussels, 1958), p. 163.
10. Table prepared for A. S. Eddington, *The Mathematical Theory of Relativity* (2nd ed., Cambridge University Press, London, 1924), p. 162.
11. C. Wirtz, Astr. Nachr., **206**, 109 (1918).
12. C. Wirtz, Astr. Nachr., **215**, 349 (1921); *ibid.*, **216**, 451 (1922); *ibid.*, **222**, 21 (1924); Scientia, **38**, 303 (1925). K. Lundmark, Stock. Acad. Hand., **50**, No. 8 (1920); Mon. Not. Roy. Astron. Soc., **84**, 747 (1924); *ibid.*, **85**, 865 (1925).
13. E. P. Hubble, Proc. Nat. Acad. Sci., **15**, 168 (1927).
14. K. Schwarzschild, Vjschr. Astr. Geo. Lpz., **35, 337** (1900).
15. H. P. Robertson, Z. Astrophys., **15**, 69 (1937); Z. f. Astrophys., **15**, 69 (1938).
16. E. Hubble, Ap. J., **71**, 231 (1930).
17. See, for example, M. Schwarzschild, *Structure and Evolution of the Stars* (Princeton University Press, Princeton, N. J., 1958), Chapter IV.
18. W. Baade, Astron. J., **100**, 137 (1944).
19. P. W. Hodge and G. Wallerstein, Pub. Astron. Soc. Pacific, **78**, 411 (1966).
20. H. Arp, Ap. J., **135**, 311, 971 (1962); A. Sandage, Ap. J., **135**, 349 (1962); R. Christy, Ap. J., **144**, 108 (1966); L. Plaut, in *Galactic Structure*, ed. by A. Blaauw and M. Schmidt (University of Chicago Press, Chicago, Ill., 1965), p. 267.
21. H. S. Leavitt, Harvard Circular No. 173 (1912); reprinted in *Source Book of Astronomy*, ed. by H. Shapley (Harvard University Press, Cambridge, 1966).
22. H. N. Russell, Science, **37**, 651 (1913).
23. E. Hertzsprung, Astron. Nachr., **196**, 201 (1913).
24. H. Shapley, Ap. J., **48**, 89 (1918).
25. R. E. Wilson, Ap. J., **35**, 35 (1923); *ibid.*, **89**, 218 (1939).
26. E. P. Hubble, *Annual Reports of the Mount Wilson Observatory*, 1923–1924; Observatory, **48**, 139 (1925).

27. W. Baade, Trans. Int. Astron. Un., **8,** 397 (1952).

28. J. D. Fernie, Pub. Ast. Soc. Pac., **81,** 707 (1969).

29. G. Wallerstein, Ap. J., **127,** 583 (1958).

30. R. P. Kraft, Ap. J., **134,** 616 (1961). Also see R. P. Kraft and M. Schmidt, Ap. J., **137,** 249 (1962); J. D. Fernie, Astron. J., **72,** 1327 (1967).

31. A. Sandage and G. A. Tammann, Ap. J., **151,** 531 (1968). A revision of this calibration has been suggested by J. Jung, Astron. and Astrophys., **6,** 130 (1970). Also see A. Sandage and G. A. Tammann, Ap. J. **167,** 293 (1971).

32. For a summary, see P. W. Hodge, *Galaxies and Cosmology* (McGraw-Hill, New York, 1966), Chapter 12.

33. E. Hubble, Ap. J., **84,** 270 (1936).

34. A. Sandage, Ap. J., **152,** L149 (1968); Observatory, **88,** 91 (1968).

35. R. Racine, J.R.A.S. (Canada), in press.

36. G. de Vaucouleurs, Ap. J., **159,** 435 (1970).

37. E. L. Scott, Ap. J., **62,** 248 (1957).

38. P. J. E. Peebles, Ap. J., **153,** 13 (1968); J. V. Peach, Nature, **223,** 1140 (1969); P. J. E. Peebles, Nature, **224,** 1093 (1969); B. A. Peterson, Ap. J., **159,** 333 (1970); B. A. Peterson, Nature, **227,** 54 (1970): and so on.

39. J. B. Oke and A. Sandage, Ap. J., **154,** 21 (1968).

40. W. A. Baum, Ap. J., **62,** 6 (1957).

41. A. Sandage, Ap. J., **152,** L149 (1968).

42. J. N. Bahcall and R. M. May, Ap. J., **152,** 89 (1968).

43. G. de Vaucouleurs, Ap. J., **63,** 253 (1968); J. Kristian and R. K. Sachs, Ap. J., **143,** 379 (1966).

44. A. Sandage, Physics Today, February 1970, p. 34.

45. A. Sandage, Observatory, **88,** 91 (1968).

46. E. P. Hubble and M. L. Humason, Ap. J., **74,** 43 (1931).

47. M. L. Humason, Ap. J., **83,** 10 (1936).

48. E. P. Hubble, Ap. J., **84,** 158, 270, 516 (1936).

49. M. L. Humason, N. U. Mayall, and A. R. Sandage, Astron. J., **61,** 97 (1956).

50. A. Sandage, Ap. J., **127,** 513 (1958).

51. A. Sandage, I. A. U. Symposium No. 15, 359 (1961).

52. A. Sandage, Ap. J., **134,** 916 (1961).

53. R. Minkowski, Ap. J., **132,** 908 (1960).

54. J. V. Peach, Ap. J., **159,** 753 (1970).

55. A. Sandage, Yearbook of the Carnegie Institute of Washington, **65,** 163 (1966).

56. M. Schmidt, Nature, **197,** 1040 (1963).

57. For a review, see G. Burbidge and M. Burbidge, *Quasi-Stellar Objects* (W. H. Freeman and Co., San Francisco, 1967).

58. F. Hoyle and G. R. Burbidge, Nature, **210,** 1346 (1966).

59. G. R. Burbidge and M. Burbidge, Ap. J., **148,** L107 (1967).

60. S. Weinberg, Ap. J., **161,** L233 (1970).

61. A. Sandage, Ap. J., **136,** 319 (1962).

62. K. I. Kellerman, Ap. J., **140**, 969 (1964); I. I. Pauliny-Toth, C. M. Wade, D. S. Heeschen, Ap. J., Suppl., **13**, 65 (1966).

63. J. F. R. Gower, Mon. Not. Roy. Astron. Soc., **133**, 151 (1966); M. Ryle, Ann. Rev. Astron. and Astrophys., **6**, 249 (1968). There are source counts at higher frequency that seem to agree with an $S^{-3/2}$ law; see K. I. Kellermann, M. M. Davis, and I. I. Pauliny-Toth, Ap. J. Lett. **170**, L1 (1971), and A. T. Shimmins, J. Bolton, and J. V. Wall, Nature **217**, 818 (1968).

64. M. S. Longair and G. G. Pooley, Mon. Not. Roy. Astron. Soc., **145**, 121 (1969).

65. H. Bondi and T. Gold, Mon. Not. Roy. Astron. Soc., **108**, 252 (1948).

66. F. Hoyle, Mon. Not. Roy. Astron. Soc., **108**, 372 (1948); *ibid.*, **109**, 365 (1949).

67. M. Schmidt, Ann. Rev. Astron. and Astrophys., **7**, 527 (1969); Ap. J., **151**, 393 (1968); *ibid.*, **162**, 371 (1970).

68. Evidence for the association of the quasi-stellar object PKS 2251 + 11 with a galaxy of essentially equal red shift ($z = 0.323$) is presented by J. E. Gunn, Ap. J. **164**, L113 (1971).

69. For a discussion of number counts of galaxies in optical astronomy, see P. J. E. Peebles, Comments Ap. and Sp. Phys. **3**, 173 (1971).

I5 COSMOLOGY: THE STANDARD MODEL

In the last chapter we laid out the coordinates for a map of the universe in space and time. Now we must begin to fill in this map with the islands of matter and the seas of radiation that make up the physical contents of the universe.

For the most part, we shall continue to base our discussion on the assumption of isotropy and homogeneity, now supplemented with Einstein's field equations. The future of the universe then depends critically on its curvature: If the universe is open, it will go on expanding forever, whereas if it is closed, its present expansion will eventually cease and be succeeded by a general contraction. The curvature in turn depends critically on the present energy density ρ_0; the universe is open or closed according to whether ρ_0 is less or greater than a critical value ρ_c, of order 10^{-29} g/cm^3. It appears that ρ_0 mostly arises from the rest-mass of ordinary matter—neutrons and protons. In this case, the universe is open and ρ_0 is less than ρ_c if the deacceleration parameter q_0 is less than $\frac{1}{2}$, whereas the universe is closed and ρ_0 is greater than ρ_c if q_0 is greater than $\frac{1}{2}$; this justifies the emphasis on the measurement of q_0 throughout the last chapter. However, the observation that $q_0 \approx 1$ conflicts with the mass density observed in galaxies, which is considerably less than ρ_c. This discrepancy has led to an intensive search for signs of an inter-galactic gas, a search that has so far been quite unsuccessful.

Looking back in time, we find that any isotropic homogeneous universe governed by Einstein's equations must have started with a singularity of infinite density. Dating from this singularity, the age of the universe must be less than H_0^{-1}, and less than $\frac{2}{3}H_0^{-1}$ if $q_0 > \frac{1}{2}$. Radioactive dating and the theory of stellar evolution give uncertain ages, ranging from 7×10^9 to 16×10^9 years, but it would be difficult to admit an age much less than $\frac{2}{3}H_0^{-1}$.

The most prominent relic of the hot early universe is the 2.7°K microwave radiation background, predicted in 1950 and observed in 1965. The weight of the data is so far consistent with the expectation that this radiation has a Planck black-body spectrum and is perfectly isotropic. Knowing the present radiation temperature, we can trace the thermal history of the universe back to the first few minutes, and calculate the production of complex nuclei in the primordial fireball. A fairly definite prediction emerges, that about 27% of the nucleons in the early universe should have been fused into He^4. This is in agreement with some measurements of the present cosmic helium abundance, but in disagreement with others. Another relic of the early universe is our present cosmic morphology: Stars form galaxies, galaxies form clusters, and clusters form a more or less homogeneous gas. Our present theoretical understanding of how this structure evolved is not in good shape, but it is clear that the radiation background played an important role. We can also speculate on the first few seconds of cosmic history, when the temperature was high enough to produce mesons, baryons, and antibaryons in large numbers; so far there does not seem to be any way to check the results of these speculations.

The preceding summary describes what may be called the "standard model" of the universe, based on the Cosmological Principle and Einstein's field equations. One other "standard" assumption, which plays an important role in Sections 15.7–15.11, is that the distant galaxies are, like our own, composed of baryons rather than antibaryons. It has often been suggested that since baryon number is, like charge, exactly conserved, the universe ought to contain equal numbers of baryons and antibaryons as well as positive and negative charges. However, it should be kept in mind that baryon number is really unlike charge; There are long-range forces associated with charge but, as far as we know, not with baryon number. Indeed, in a finite universe the total charge *must* be zero, as can be immediately seen by integrating the Maxwell equation $\mathbf{V} \cdot \mathbf{E} = \varepsilon$ over the volume of the universe; no such conclusion can be derived for baryon number. In any case, even if the net baryon number of the universe is zero, baryons and antibaryons must somehow have become separated at some time in the past, and most of the considerations of this chapter are applicable to the evolution of the universe after that time.

Of course, the standard model may be partly or wholly wrong. However, its importance lies not in its certain truth, but in the common meeting ground that it provides for an enormous variety of observational data. By discussing these data in the context of a standard cosmological model, we can begin to appreciate their cosmological relevance, whatever model ultimately proves correct. Some other possible models are discussed in the next chapter.

1   Einstein's Equations

Let us begin our discussion of dynamical cosmology by considering the constraints imposed by Einstein's field equations on the metric for a general isotropic and homogeneous universe. According to the results of Section 14.2, this metric can be chosen to have the Robertson-Walker form:

$$g_{tt} = -1 \qquad g_{it} = 0 \qquad g_{kj} = R^2(t)\tilde{g}_{ij}(x) \tag{15.1.1}$$

Here t is a cosmic time coordinate; i and j run over three comoving spatial coordinates r, θ, and φ; and $\tilde{g}_{ij}$ is the metric for a three-dimensional maximally symmetric space:

$$\tilde{g}_{rr} = (1 - kr^2)^{-1} \qquad \tilde{g}_{\theta\theta} = r^2 \qquad \tilde{g}_{\varphi\varphi} = r^2 \sin^2\theta$$
$$\tilde{g}_{ij} = 0 \qquad \text{for } i \neq j \tag{15.1.2}$$

with k equal to $+1$, -1, or 0.

The only nonvanishing elements of the affine connection for this metric are

$$\Gamma^t_{ij} = R\dot{R}\tilde{g}_{ij} \tag{15.1.3}$$

$$\Gamma^i_{tj} = \frac{\dot{R}}{R}\delta^i_j \tag{15.1.4}$$

$$\Gamma^i_{jk} = \tfrac{1}{2}(\tilde{g}^{-1})^{il}\left(\frac{\partial \tilde{g}_{lj}}{\partial x^k} + \frac{\partial \tilde{g}_{lk}}{\partial x^j} - \frac{\partial \tilde{g}_{jk}}{\partial x^l}\right) \equiv \tilde{\Gamma}^i_{jk} \tag{15.1.5}$$

Its Ricci tensor then has the elements

$$R_{tt} = \frac{3\ddot{R}}{R} \tag{15.1.6}$$

$$R_{ti} = 0 \tag{15.1.7}$$

$$R_{ij} = \cdot\tilde{R}_{ij} - (R\ddot{R} + 2\dot{R}^2)\tilde{g}_{ij} \tag{15.1.8}$$

where $\tilde{R}_{ij}$ is the *spatial* Ricci tensor calculated from the metric $\tilde{g}_{ij}$:

$$\tilde{R}_{ij} = \frac{\partial \tilde{\Gamma}^k_{ki}}{\partial x^j} - \frac{\partial \tilde{\Gamma}^k_{ij}}{\partial x^k} + \tilde{\Gamma}^k_{li}\tilde{\Gamma}^l_{kj} - \tilde{\Gamma}^k_{ij}\tilde{\Gamma}^l_{kl} \tag{15.1.9}$$

Instead of calculating $\tilde{R}_{ij}$ directly, we can save ourselves a good deal of work by recalling that $\tilde{g}_{ij}$, as the metric of a maximally symmetric space, must necessarily have a Ricci tensor of the form (13.2.4):

$$\tilde{R}_{ij} = -2k\tilde{g}_{ij} \tag{15.1.10}$$

Together with (15.1.8), this gives the space-space components of the *space-time* Ricci tensor as

$$R_{ij} = -(R\ddot{R} + 2\dot{R}^2 + 2k)\tilde{g}_{ij} \tag{15.1.11}$$

As shown in Section 14.2, the energy-momentum tensor here must have the perfect-fluid form

$$T_{\mu\nu} = pg_{\mu\nu} + (p + \rho)U_\mu U_\nu \tag{15.1.12}$$

where p and ρ are functions of t alone, and U^μ is given by Eqs. (14.2.13) and (14.2.14):

$$U^t = 1 \qquad U^i = 0 \qquad (15.1.13)$$

The source term in the Einstein equations is then

$$
\begin{aligned}
S_{\mu\nu} &\equiv T_{\mu\nu} - \tfrac{1}{2}g_{\mu\nu}T^\lambda{}_\lambda \\
&= \tfrac{1}{2}(\rho - p)g_{\mu\nu} + (p + \rho)U_\mu U_\nu \qquad (15.1.14)
\end{aligned}
$$

so (15.1.1), (15.1.13), and (15.1.14) give

$$S_{tt} = \tfrac{1}{2}(\rho + 3p) \qquad (15.1.15)$$

$$S_{it} = 0 \qquad (15.1.16)$$

$$S_{ij} = \tfrac{1}{2}(\rho - p)R^2 \tilde{g}_{ij} \qquad (15.1.17)$$

The Einstein equations read

$$R_{\mu\nu} = -8\pi G S_{\mu\nu}$$

With (15.1.6), (15.1.7), (15.1.11), and (15.1.15)–(15.1.17), the time-time component gives

$$3\ddot{R} = -4\pi G(\rho + 3p)R \qquad (15.1.18)$$

the space-space components give the single equation

$$R\ddot{R} + 2\dot{R}^2 + 2k = 4\pi G(\rho - p)R^2 \qquad (15.1.19)$$

and the space-time components give $0 = 0$.

By eliminating $\ddot{R}$ from (15.1.18) and (15.1.19), we find a first-order differential equation for $R(t)$:

$$\dot{R}^2 + k = \frac{8\pi G}{3}\rho R^2 \qquad (15.1.20)$$

In addition we have the equation of energy conservation (14.2.19),

$$\dot{p}R^3 = \frac{d}{dt}\{R^3[\rho + p]\}$$

or, equivalently,

$$\frac{d}{dR}(\rho R^3) = -3pR^2 \qquad (15.1.21)$$

Given an equation of state $p = p(\rho)$, we can use this equation to determine ρ as a function of R. For instance, if the energy density of the universe is dominated by nonrelativistic matter with negligible pressure, then (15.1.21) gives

$$\rho \propto R^{-3} \qquad \text{for } p \ll \rho \qquad (15.1.22)$$

whereas if the energy density is dominated by relativistic particles, such as photons, then $p = \rho/3$, and (15.1.21) gives

$$\rho \propto R^{-4} \qquad \text{for } p = \frac{\rho}{3} \tag{15.1.23}$$

Knowing ρ as a function of R, we can determine $R(t)$ for all time by solving Eq. (15.1.20). *The fundamental equations of dynamical cosmology are thus the Einstein equations* (15.1.20), *the energy-conservation equation* (15.1.21), *and the equation of state.* The cosmological models, based on a Robertson-Walker metric, in which $R(t)$ is derived in this way, are known as Friedmann models.[1]

[Incidentally, the solution $R(t)$ determined in this way will automatically satisfy (15.1.18) and (15.1.19), for by differentiating (15.1.20) with respect to time and using (15.1.21), we find

$$2\dot{R}\ddot{R} = \frac{8\pi G \dot{R}}{3R} \left[-\rho R^2 + \frac{d}{dR}(\rho R^3) \right]$$

$$= \frac{8\pi G \dot{R}}{3R} (-\rho R^2 - 3p R^2)$$

which is just the same as (15.1.18). Equation (15.1.19) then follows trivially from (15.1.18) and (15.1.20). The reason we can make do with the single field equation (15.1.20), instead of the two equations (15.1.18) and (15.1.19), is of course that these two equations are not functionally independent, being related by the Bianchi identities to the energy-conservation equation (15.1.21).]

It is possible to learn a good deal about the past and future expansion of the universe by simply inspecting Eqs. (15.1.18)–(15.1.21), even without specifying a definite equation of state. Equation (15.1.18) shows that as long as the quantity $\rho + 3p$ remains positive, the "acceleration" $\ddot{R}/R$ is negative. Since at present $R > 0$ (by definition), and $\dot{R}/R > 0$ (because we see red shifts, not blue shifts), it follows that the curve of $R(t)$ versus t must be concave downward, and *must have reached $R(t) = 0$ at some finite time in the past.* Let us call this time $t = 0$, so that

$$R(0) = 0 \tag{15.1.24}$$

The present time t_0 is then the time elapsed since this singularity, and may justly be called the age of the universe. If $\ddot{R}(t)$ vanished for $0 < t < t_0$, then $R(t)$ would be just $R(t_0)t/t_0$, and so the age t_0 would just equal the Hubble time $H_0^{-1} = R(t_0)/\dot{R}(t_0)$. With $\ddot{R}$ negative for $0 < t < t_0$, *the age of the universe must be less than the Hubble time*:

$$t_0 < H_0^{-1} \tag{15.1.25}$$

Looking into the future, we see from Eq. (15.1.21) that as long as the pressure p does not become negative, the density ρ must decrease with increasing R at least as fast as R^{-3}, so that for $R \to \infty$, the right-hand side of Eq. (15.1.20) vanishes at least as fast as R^{-1}. For $k = -1$, $\dot{R}^2(t)$ remains positive-definite, so $R(t)$ goes on increasing, with

$$R(t) \to t \quad \text{as} \quad t \to \infty \qquad \text{for } k = -1$$

For $k = 0$, $\dot{R}^2(t)$ remains positive-definite so $R(t)$ goes on increasing, but more slowly than t. For $k = +1$, $\dot{R}^2(t)$ will reach zero when ρR^2 drops to the value $3/8\pi G$. Since $\ddot{R}$ is negative-definite, $R(t)$ will then begin to decrease again, and eventually must again reach $R = 0$ at some finite time in the future. Hence the qualitative course of cosmic history is determined by the sign of the spatial curvature: *If $k = -1$ or $k = 0$, then the universe will go on expanding forever, whereas if $k = +1$, then the expansion will eventually cease and be followed by a contraction back to a singular state with $R(t) = 0$.*

The combination of the Cosmological Principle with the Einstein field equations illuminates some of the profound questions raised by Newton and Mach. (See Section 1.3). Suppose that we want to study some physical system S, such as the solar system or the rotating bucket of Newton, whose size is much less than the cosmic scale factor R. We can imagine S to be placed in a spherical cavity, cut out of the expanding universe, and so long as the size of this cavity is much less than R, we can safely consider this cavity to be empty apart from the system S. If S were absent, the gravitational field inside the cavity would be a spherically symmetric field with $R_{\mu\nu} = 0$, and hence, according to the Birkhoff theorem (see Section 11.7), it would have a flat-space metric equivalent to the Minkowski metric $\eta_{\mu\nu}$. As long as the system S is not too big, we can then calculate its gravitational field as a perturbation on $\eta_{\mu\nu}$, ignoring all matter outside our cavity, and we can determine the behavior of the system by using Newtonian or special-relativistic mechanics. The question of what determines the inertial frames is now answered, for the only reference frames in which the whole universe appears spherically symmetric, so that Birkhoff's theorem applies, are the frames at the center of our cavity, which do not rotate with respect to the expanding cloud of "typical" galaxies. The inertial frames are any reference frames that move at constant velocity, and without rotation, relative to the frames in which the universe appears spherically symmetric.

These remarks lead to an alternative derivation[2] of the dynamical equations for an expanding universe. If we mentally draw a comoving spherical surface anywhere in the universe, then as long as its proper radius is much less than $R(t)$, the galaxies within this sphere will move under the influence of their own gravitational field, and the gravitational field of the rest of the universe may be neglected. We can then think of the universe as consisting of a Newtonian gas in a state of everywhere-uniform expansion. Any given gas particle will have a trajectory

$$\mathbf{x}(t) = \mathbf{x}(t_0)\, \frac{R(t)}{R(t_0)}$$

with $R(t)$ a scale factor common to the whole gas. [Note that the gas appears the same to an observer mounted on any gas particle as it does to an observer at the origin. Also, the "comoving" coordinates of a given gas particle are not $x^i(t)$, but rather $r^i \equiv x^i(t_0)$.] The gravitational potential energy V of such a particle just arises from the matter within a sphere of radius $|\mathbf{x}(t)|$ and center at the origin, so

$$V(t) = -\frac{4\pi}{3} |\mathbf{x}(t)|^3 \rho(t) \frac{mG}{|\mathbf{x}(t)|}$$

$$= -\frac{4\pi}{3} mG|\mathbf{x}(t_0)|^2 \rho(t) \frac{R^2(t)}{R^2(t_0)}$$

where m is the particle mass and $\rho(t)$ is the uniform mass density of the gas. The kinetic energy of this particle is

$$T(t) = \tfrac{1}{2}m|\dot{\mathbf{x}}(t)|^2 = \tfrac{1}{2}m|\mathbf{x}(t_0)|^2 \frac{\dot{R}^2(t)}{R^2(t_0)}$$

and its total energy is

$$E \equiv T(t) + V(t) = \tfrac{1}{2}m \frac{|\mathbf{x}(t_0)|^2}{R^2(t_0)} \left[\dot{R}^2(t) - \frac{8\pi G}{3} \rho(t)R^2(t) \right]$$

With E constant, this is just the same as Eq. (15.1.20), provided that we identify the energy of a particle as

$$E = -\tfrac{1}{2}m \frac{|\mathbf{x}(t_0)|^2}{R^2(t_0)} k \tag{15.1.26}$$

For $k = -1$, E is positive-definite, so gravitation cannot prevent the gas from dispersing to infinity, with a finite asymptotic velocity. For $k = 0$, E vanishes, and the gas is just barely able to expand indefinitely. For $k = +1$, E is negative, and the explosion must ultimately cease and be followed by an implosion.

Although Newtonian cosmology can reproduce the chief results derived from Einstein's equations, it is essentially incomplete, for several reasons. We need general relativity to justify the neglect of all the matter outside a sphere of radius $|\mathbf{x}(t)|$ in calculating the gravitational potential at $\mathbf{x}(t)$. We cannot use Newtonian mechanics when the medium itself consists of particles with relativistic *local* velocities. Finally, it is only through the use of general relativity that we are able to interpret the observation of light signals correctly in terms of the cosmic scale factor $R(t)$.

2 Density and Pressure of the Present Universe

At the present instant, the pressure and energy density of the universe are given by Eqs. (15.1.18) and (15.1.19) as

$$\rho_0 = \frac{3}{8\pi G} \left(\frac{k}{R_0^2} + H_0^2 \right) \tag{15.2.1}$$

$$p_0 = -\frac{1}{8\pi G}\left[\frac{k}{R_0{}^2} + H_0{}^2(1 - 2q_0)\right] \tag{15.2.2}$$

Here R_0 is the present value of the cosmic scale factor $R(t)$, and H_0 and q_0 are the Hubble constant and the deacceleration parameter, defined in Section 14.3 as the present values of $\dot R/R$ and $-R\ddot R/\dot R^2$. From (15.2.1) it follows that the spatial curvature k/R^2 is positive or negative, according to whether ρ_0 is greater or less than a *critical density*

$$\rho_c \equiv \frac{3H_0{}^2}{8\pi G} = 1.1 \times 10^{-29}\left(\frac{H_0}{75\ \text{km/sec/Mpc}}\right)^2\ \text{g/cm}^3 \tag{15.2.3}$$

As we shall see, there are good grounds to believe that the energy density of the present universe is dominated by nonrelativistic matter, with

$$p_0 \ll \rho_0 \tag{15.2.4}$$

If this is the case, then (15.2.2) yields a formula for the spatial curvature in terms of the observable parameters H_0 and q_0:

$$\frac{k}{R_0{}^2} = (2q_0 - 1)H_0{}^2 \tag{15.2.5}$$

and (15.2.1) gives the ratio of the present density to the critical density (15.2.3) as

$$\frac{\rho_0}{\rho_c} = 2q_0 \tag{15.2.6}$$

For $q_0 > \frac{1}{2}$ the universe is positively curved, with $\rho_0 > \rho_c$, whereas for $q_0 < \frac{1}{2}$ the universe is negatively curved, with $\rho_0 < \rho_c$. If we give credence to the values $q_0 \simeq 1$ and $H_0 \simeq 75$ km/sec/Mpc deduced from the red shift versus luminosity relation (see Section 14.6), then we must conclude that the density of the universe is about $2\rho_c$, or about 2×10^{-29} g/cm^3.

Unfortunately, this result does not agree with the observed density of galactic mass.[3] The masses of spiral galaxies within about 15 Mpc can be determined by a dynamical analysis of their rotational velocities as functions of distance from the galactic centers. The masses of a half-dozen or so elliptical galaxies can be calculated from the virial theorem,[4] which gives a mass

$$M = \frac{2\langle v^2\rangle}{G\langle d^{-1}\rangle} \tag{15.2.7}$$

where $\langle v^2\rangle$ is the mean-square velocity relative to the center of mass, and $\langle d^{-1}\rangle$ is the mean reciprocal separation between stars. The total masses of pairs of galaxies can be determined statistically from their relative velocities and separations, under the assumption that the pairs are oriented at random with respect to the line of sight.

In all three of the above methods, the galactic mass is given by a formula of form

$$M = \frac{\mu V^2 D}{G} \qquad (15.2.8)$$

where V is some characteristic internal velocity, D is some characteristic dimension of the object under study, and μ is a dimensionless number of order unity, which depends on the details of the method used and the object studied. The characteristic distance D is measured from the corresponding angular dimension δ and the cosmological red shift z, using Eqs. (14.4.15) and (14.6.7), which for $z \ll 1$ give

$$D = \frac{z\delta}{H_0} \qquad (15.2.9)$$

(For nearby galaxies, the "angular diameter distance" D/δ might be determined from the apparent magnitude of brightest stars, brightest globular clusters, and so on, rather than from the red shift. However, if such distance determinations form part of the cosmic distance ladder used to measure the Hubble constant, then any error in these distances would also show up in the Hubble constant, so that D would still scale like $1/H_0$.) The internal velocities V are measured directly from the distribution in red shift around the average value z for the galaxy. It is convenient to describe the masses determined in this way in terms of a mass-to-luminosity ratio M/L, the absolute luminosity L being given in terms of the apparent luminosity l by Eqs. (14.4.12) and (14.6.7), which for a small red shift z yield

$$L = 4\pi l z^2 H_0^{-2} \qquad (15.2.10)$$

From (15.2.8)–(15.2.10), it follows that the ratio M/L determined by the three methods described above is *proportional* to the value assumed for the Hubble constant H_0.

With H_0 taken as 75 km/sec/Mpc, it appears[3] that the galactic mass-to-light ratio M/L for elliptical galaxies is about 50 times the solar ratio $M_\odot/L_\odot$, whereas for spiral galaxies estimates of M/L range from 1 to 20 times $M_\odot/L_\odot$. According to a survey of these M/L values by Oort,[5] the overall mass-to-light ratio for all galaxies is about 21 $M_\odot/L_\odot$. Since the Hubble constant may well be different from 75 km/sec/Mpc, this result should be written

$$\frac{M}{L} \approx 21 \, \frac{M_\odot}{L_\odot} \left(\frac{H_0}{75 \text{ km/sec/Mpc}} \right) \qquad (15.2.11)$$

(For instance, van den Bergh[6] carried out an analysis of galactic masses similar to Oort's, but assumed that $H_0 = 120$ km/sec/Mpc, and therefore obtained for M/L the result 30 $M_\odot/L_\odot$.) Oort also used number counts of galaxies to estimate the luminosity density of the universe to be $2.2 \times 10^{-10} \, L_\odot/\text{pc}^3$; this value would scale with H_0 like L/D^3, which according to (15.2.9) and (15.2.10) scales like H_0, so for a general Hubble's constant Oort's estimate of the luminosity density would be

$$\mathscr{L} \simeq 2.2 \times 10^{-10}\, L_{\odot}/\text{pc}^3 \left(\frac{H_0}{75\ \text{km/sec/Mpc}}\right) \tag{15.2.12}$$

The *galactic* mass density of the universe can now be obtained as

$$\rho_G = \left(\frac{\mathscr{L}}{L_{\odot}}\right)\left(\frac{M/L}{M_{\odot}/L_{\odot}}\right) M_{\odot}$$

$$= 4.6 \times 10^{-9}\, M_{\odot}/\text{pc}^3 \left(\frac{H_0}{75\ \text{km/sec/Mpc}}\right)^2$$

$$= 3.1 \times 10^{-31}\, \text{g/cm}^3 \left(\frac{H_0}{75\ \text{km/sec/Mpc}}\right)^2 \tag{15.2.13}$$

This is *smaller* than the critical density (15.2.3) by a factor

$$\frac{\rho_G}{\rho_c} \simeq 0.028 \tag{15.2.14}$$

(More recently, Noonan[7] and S. L. Shapiro[7a] have given estimates of 0.016 and 0.010 for this ratio.) Note that such results are independent of the true value of Hubble's constant. Note also that although ρ_G and ρ_c do not turn out to be equal, they are close enough to reassure us that gravitation does have something to do with the expansion of the universe.

If the mass of the universe were primarily concentrated in galaxies, then Eqs. (15.2.14) and (15.2.6) would yield a deacceleration parameter

$$q_0 \simeq 0.014 \quad \text{if } \rho_0 \approx \rho_G \tag{15.2.15}$$

which would imply that the universe is negatively curved and open, with $R_0 \simeq H_0^{-1}$. This value of q_0 is not in agreement with the result found from red shifts and luminosities, which give $q_0 \simeq 1$, apart from possible corrections for evolution or selection effects. Of course, evolution or selection effects may have an appreciable effect on the measurements of q_0. However, if one tentatively accepts the result that q_0 is of order unity, then one is forced to the conclusion that the mass density of about 2×10^{-29} g/cm^3 must be found somewhere outside the normal galaxies. But where?

One place to look for the missing mass is in the intergalactic space within clusters of galaxies. In Coma there is a rich cluster of elliptical galaxies that appears from its smooth shape to be gravitationally bound. If bound, its mass is given by the virial formula (15.2.7). The values of M/L obtained in this way range[8] from 4 to 20 times the M/L ratio for individual elliptical galaxies. (These values are for $H_0 = 75$ km/sec/Mpc.) If there actually is 20 times more matter within clusters than within their individual galaxies, then the density of the universe is raised to near the critical density (15.2.3). In fact, an X-ray source has been discovered[8a] filling the Coma cluster, which suggests the presence of an intergalactic gas of ionized hydrogen, at a temperature of order $7 \times 10^{7\circ}$K. However, the strength of this source indicates that this gas has a mass only a percent or so of the mass required by the virial theorem. It must be kept in mind that the Coma cluster may not be bound at all,[9] in which case the virial theorem overestimates its mass. Many rich clusters, like those in Virgo or Hercules, are highly irregular, and do not appear at all stable.

If the missing mass is not within clusters of galaxies, then we must look for it in the space between the clusters. One reasonable requirement is that the total density of intercluster space must be less than the density within clusters, so that the clusters represent appreciable condensations. The total volume outside clusters is roughly 500 times greater than the volume within clusters, so the density within clusters is roughly 500 times (15.2.13), or about 10^{-28} g/cm^3. Hence, even if the density outside clusters is an order of magnitude less than the density within clusters, there is still plenty of room in intercluster space for all the missing mass we need.

It is possible that the missing mass might be contained in normal stars that happen to lie in intergalactic space (inside or outside clusters) or in dwarf galaxies which are too faint to have been observed. From limits on the extragalactic contribution to the night-sky brightness, Peebles and Partridge[9a] estimate that the total mass density in normal stars, wherever located, must be less than 0.13 ρ_c. This estimate does not rule out the possibility that the missing mass is contained in dark stars, with very high values of M/L, either in dwarf galaxies or in intergalactic space. One immediately thinks of the "black holes" discussed in Section 11.9. However, the estimates of galactic mass discussed above show that typical galaxies do not contain overwhelming numbers of dark stars, so why should dark stars predominate anywhere else? Another possibility is that the missing mass is contained in whole galaxies that have undergone gravitational collapse. It is difficult to see how this hypothesis could ever be verified, except through observations of galaxies in the throes of collapse, or through observation of the deflection of light rays that happen to pass close to a collapsed galaxy.

The missing mass might be found in the form of highly relativistic particles, such as cosmic rays, photons, neutrinos, or gravitons. It is easy to see that photons and neutrinos produced in ordinary thermonuclear processes cannot have an energy density comparable with that of ordinary nonrelativistic rest-mass, for even if the universe started out as pure hydrogen and has "cooked" all the way to iron, the energy released would be at most about 9 MeV per nucleon, which is 1% of the nucleon rest-mass. If highly relativistic particles dominate the mass density of the universe, then they must be produced in exotic processes like matter-anti-matter annihilation or gravitational collapse, or be left over from the early universe. The observed total flux density of faint discrete *radio* sources at frequency ν is of the order[10]

$$\mathscr{S}(\nu) \simeq 10^{-21} \text{ Wm}^{-2} \text{ Hz}^{-1} \left(\frac{\nu}{408 \text{ MHz}} \right)^{-0.7}$$

so the total energy density of the radio emission at wavelengths longer than 75 cm from these sources is roughly

$$\rho_{\text{radio}} = \int_0^{400\,\text{MHz}} \mathcal{S}(\nu)\,d\nu \simeq 10^{-12}\,\text{Wm}^{-2} \simeq 10^{-40}\,\text{g/cm}^3$$

The isotropic background at these wavelengths is not more than an order of magnitude greater.[11] For *microwave and far-infrared* wavelengths between 75 cm and 0.05 cm, the radiation flux is dominated by the 2.7°K background (see Section 15.5), with an energy density given by the Stefan-Boltzmann law as 4.4×10^{-34} g/cm³. The total energy density of starlight at *optical* frequencies is estimated[12] to be no more than about 10^{-35} g/cm³. The observed *X-ray* background has a flux density at energy E of order[13]

$$\Phi(E) \simeq 20 \text{ photons cm}^{-2} \text{ sec}^{-1} \text{ sr}^{-1} \text{ keV}^{-1} \, (E(\text{keV}))^{-2}$$

If this background is extragalactic, then it contains an energy density between 0.1 keV and 1 MeV given by

$$\rho_{\text{x-ray}} = \int_{0.1\,\text{keV}}^{1\,\text{MeV}} 4\pi\Phi(E)E\,dE \simeq 3 \times 10^3 \text{ keV cm}^{-2} \text{ sec}^{-1}$$

$$\simeq 10^{-37}\,\text{g/cm}^3$$

The energy density in *γ-rays* above 100 MeV is estimated[11] to be less than 3×10^{-38} g/cm³. The observed[14] energy density of *cosmic ray particles* is not more than about 10^{-35} g/cm³.

These estimates indicate that the largest contribution of relativistic particles to the total cosmic energy density is provided by the 2.7°K microwave background, to be discussed in Section 15.5. Its density is less than one-hundredth the density (15.2.13) of galactic rest-mass, which justifies our tentative neglect of pressure in the Einstein and conservation equations.

However, it is possible that the missing mass is made up by neutrinos or gravitons,[14a] which interact too weakly with matter to have been detected. In particular, the neutrino energy density is expected to be at least comparable with that of microwave electromagnetic radiation, but may well be many orders of magnitude greater (see Section 15.6). If the energy density of the universe *is* dominated by highly relativistic particles, then the pressure is

$$p_0 = \frac{\rho_0}{3} \tag{15.2.16}$$

and in place of (15.2.5) and (15.2.6), the Einstein equations now give

$$\frac{k}{R_0^2} = H_0^2(q_0 - 1) \tag{15.2.17}$$

$$\frac{\rho_0}{\rho_c} = q_0 \tag{15.2.18}$$

where ρ_c is the same critical density (15.2.3) as before. The critical deacceleration parameter, for which $k = 0$ and $\rho_0 = \rho_c$, is now $q_0 = 1$ rather than $q_0 = \frac{1}{2}$, and the density required for a given q_0 and H_0 is half of that needed for a dust-filled universe.

Although a photon-, neutrino-, or graviton-dominated universe cannot be ruled out on observational grounds at present, it is more conservative to suppose that the missing mass takes the form of a tenuous hydrogen gas, ionized or neutral, filling all space. The various methods that have been proposed to detect this gas depend on electromagnetic signals that reach us from cosmological distances, so we must defer our discussion of this gas until Section 15.4, and turn now to the solution of the equations of dynamical cosmology.

3   The Matter-Dominated Era

We have noted that the energy density of the *known* forms of radiation in the present universe is less than one-hundredth the density of rest-mass. According to Eqs. (15.1.22) and (15.1.23), the energy density of rest-mass scales as R^{-3}, and the energy density of radiation scales as R^{-4}, so we may conclude with some confidence that the expansion of the universe has been governed by its nonrelativistic matter content at least since the time when $R(t)$ was one-hundredth its present value. This period certainly goes back long before the emission of any of the light collected at Mt. Palomar, for the most distant galaxies and quasi-stellar objects observed have red shifts z that are much less than 100, and in fact less than 3! The study of the empirical relations between red shifts, luminosities, numbers, angular diameters, and so on, can therefore reveal only the matter-dominated era of the history of the universe.

The dynamical equation governing the universe during this era is Einstein's equation (15.1.20):

$$\dot{R}^2 + k = \frac{8\pi G}{3} \rho R^2 \qquad (15.3.1)$$

with ρ taking the form (15.1.22) appropriate to a matter-dominated universe:

$$\frac{\rho}{\rho_0} = \left(\frac{R}{R_0}\right)^{-3} \qquad (15.3.2)$$

It is convenient to make use of equations (15.2.5) and (15.2.6) to write ρ_0 and $k/R_0{}^2$ in terms of q_0 and H_0:

$$\frac{k}{R_0{}^2} = (2q_0 - 1)H_0{}^2$$

$$\frac{8\pi G \rho_0}{3} = 2q_0 H_0{}^2$$

Equations (15.3.1) and (15.3.2) then give

$$\left(\frac{\dot{R}}{R_0}\right)^2 = H_0{}^2\left[1 - 2q_0 + 2q_0\left(\frac{R_0}{R}\right)\right] \tag{15.3.3}$$

The solution may in general be expressed as a formula for t in terms of R:

$$t = \frac{1}{H_0}\int_0^{R/R_0}\left[1 - 2q_0 + \frac{2q_0}{x}\right]^{-1/2}dx \tag{15.3.4}$$

with $t = 0$ defined as the time when $R \ll R_0$. In particular, the present age of the universe is

$$t_0 = \frac{1}{H_0}\int_0^1\left[1 - 2q_0 + \frac{2q_0}{x}\right]^{-1/2}dx \tag{15.3.5}$$

For any positive q_0, the age of the universe must be less than the Hubble time,

$$t_0 < \frac{1}{H_0} \tag{15.3.6}$$

as already remarked in Section 15.1.

The behavior of the result (15.3.4) may conveniently be discussed under three special cases (see Figure 15.1):

(A) $q_0 > \frac{1}{2}$ ($k = +1, \rho_0 > \rho_c$). It is convenient here to define a *development angle* θ by

$$1 - \cos\theta = \left(\frac{2q_0 - 1}{q_0}\right)\frac{R(t)}{R_0} \tag{15.3.7}$$

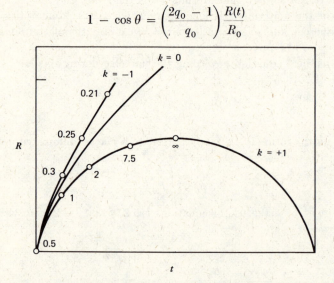

Figure 15.1   Solutions of Einstein's equations for a Robertson-Walker universe with curvature $k = +1$, $k = 0$, and $k = -1$. The numbers along the curves $k = \pm 1$ give the values of the deacceleration parameter q_0 at various epochs.

Then (15.3.4) gives

$$H_0 t = q_0 (2q_0 - 1)^{-3/2} [\theta - \sin \theta] \qquad (15.3.8)$$

This is the equation of a cycloid; $R(t)$ increases from zero at $\theta = 0$, $t = 0$; reaches a maximum at

$$\theta_m = \pi \qquad t_m = \frac{\pi q_0}{H_0 (2q_0 - 1)^{3/2}} \qquad R(t_m) = \frac{2q_0 R_0}{2q_0 - 1} \qquad (15.3.9)$$

and then returns to zero at $\theta = 2\pi$, $t = 2t_m$. The present instant is defined by setting $R(t)$ equal to R_0 in Eq. (15.3.7); the present value of the development angle θ is then given by

$$\cos \theta_0 = \frac{1}{q_0} - 1 \qquad (15.3.10)$$

and so the age of the universe is

$$t_0 = H_0^{-1} q_0 (2q_0 - 1)^{-3/2} \left[\cos^{-1}\left(\frac{1}{q_0} - 1 \right) - \frac{1}{q_0} (2q_0 - 1)^{1/2} \right] \qquad (15.3.11)$$

For example, if we believe that $q_0 \approx 1$ and $H_0^{-1} = 13 \times 10^9$ years, then Eq. (15.3.10) gives $\theta_0 \approx \pi/2$, Eq. (15.3.11) gives the age of the universe now as

$$t_0 \approx \left(\frac{\pi}{2} - 1 \right) H_0^{-1} \approx 7.5 \times 10^9 \text{ years} \qquad (15.3.12)$$

and (15.3.9) shows that the universe will reach its maximum radius $R(t_m) \approx 2R_0$ at a time

$$t_m \approx \pi H_0^{-1} \approx 40 \times 10^9 \text{ years} \qquad (15.3.13)$$

The whole life cycle of the universe takes a time $2t_m$, or about 80×10^9 years.

(B) $q_0 = \frac{1}{2}$ ($k = 0$, $\rho_0 = \rho_c$). Here (15.3.4) gives

$$\frac{R(t)}{R_0} = \left(\frac{3H_0 t}{2} \right)^{2/3} \qquad (15.3.14)$$

so $R(t)$ increases without limit. The age of the universe is given for $H_0^{-1} \approx 13 \times 10^9$ years as

$$t_0 = \tfrac{2}{3} H_0^{-1} \approx 9 \times 10^9 \text{ years} \qquad (15.3.15)$$

This is known as the *Einstein-deSitter model.*

(C) $0 < q_0 < \frac{1}{2}$ ($k = -1$, $\rho_0 < \rho_c$). The results (15.3.7) and (15.3.8) can be applied here, except that now the development angle θ is imaginary,

$$\theta = i\Psi$$

and so

$$H_0 t = q_0 (1 - 2q_0)^{-3/2} [\sinh \Psi - \Psi] \qquad (15.3.16)$$

with Ψ given by

$$\cosh \Psi - 1 = \frac{1 - 2q_0}{q_0} \frac{R(t)}{R_0} \tag{15.3.17}$$

Just as in case (B), the scale factor $R(t)$ here increases without limit; for $t \to \infty$, we have

$$\frac{R(t)}{R_0} \to \tfrac{1}{2}q_0(1 - 2q_0)^{-1} e^{\Psi} \to (1 - 2q_0)^{1/2} H_0 t \tag{15.3.18}$$

The present moment is defined by setting $R(t)$ equal to R_0 in Eq. (15.3.17):

$$\cosh \Psi_0 = \left(\frac{1}{q_0} - 1 \right) \tag{15.3.19}$$

and the age of the universe is

$$t_0 = H_0^{-1} \left[(1 - 2q_0)^{-1} - q_0(1 - 2q_0)^{-3/2} \cosh^{-1} \left(\frac{1}{q_0} - 1 \right) \right] \tag{15.3.20}$$

For instance, if we take the mass density of the universe to be that contained within the galaxies, then according to Eq. (15.2.15), q_0 is about 0.014, so $\Psi_0 \approx 5$, and the age of the universe is nearly equal to the Hubble time

$$t_0 \approx 0.96 \, H_0^{-1} \approx 13 \times 10^9 \text{ years} \tag{15.3.21}$$

It is worth mentioning here that the deacceleration parameter $q \equiv -\ddot{R}R/\dot{R}^2$ will in general change with time. For $k = +1$, $q(t)$ is given by the analogue of Eq. (15.3.10),

$$q = (1 + \cos \theta)^{-1}$$

so as θ goes from 0 to 2π during one cosmic cycle, q rises from $\tfrac{1}{2}$ to ∞ and then drops again to $\tfrac{1}{2}$. For $k = -1$, $q(t)$ is given by the analogue of Eq. (15.3.19):

$$q = (1 + \cosh \Psi)^{-1}$$

so as Ψ goes from 0 to ∞, q drops steadily from $\tfrac{1}{2}$ to 0. Only in the case $k = 0$ does q remain constant, at the value $q = \tfrac{1}{2}$. Thus no special significance can be attached to any particular value of q_0 other than $q_0 = \tfrac{1}{2}$. For instance, the straight-line fit of luminosity distance versus red shift indicates that $q_0 \simeq 1$ (unless evolutionary or selection effects are important; see Section 14.6), but in a matter-dominated universe this must be an accident, for if at present $q_0 = 1$, then previously $q_0 < 1$, and in the future $q_0 > 1$. It is only in a radiation-dominated universe that $k = 0$ entails $q_0 = 1$ [see Eq. (15.2.17)], so that a deacceleration parameter of unity is stable.

The formulas for $R(t)$ derived above may be used to extend the phenomenological analysis of the last chapter out to arbitrarily large red shifts. According to

Eq. (14.3.6), light, which arrives at time t_0 with red shift z, was emitted when the scale factor had the value

$$R_1 = \frac{R_0}{1 + z} \tag{15.3.22}$$

The comoving radial coordinate of the light source is given by Eqs. (14.3.1), (14.3.2), and (15.3.3):

$$\int_0^{r_1} [1 - kr^2]^{-1/2} \, dr = \int_{t_1}^{t_0} \frac{dt}{R(t)} = \int_{R_1}^{R_0} \frac{dR}{R\dot{R}}$$

$$= \frac{1}{R_0 H_0} \int_{(1+z)^{-1}}^{1} \left[1 - 2q_0 + \frac{2q_0}{x} \right]^{-1/2} x^{-1} \, dx$$

With the aid of Eq. (15.2.5), it is straightforward to show that for all three possible values of k, the formula for r_1 is the same:

$$r_1 = \frac{zq_0 + (q_0 - 1)(-1 + \sqrt{2q_0 z + 1})}{H_0 R_0 q_0^2 (1 + z)} \tag{15.3.23}$$

The "luminosity distance," measured by comparison of apparent with absolute luminosity, is then given by (14.4.14) as

$$d_L = R_0 r_1 (1 + z) = \frac{1}{H_0 q_0^2} [zq_0 + (q_0 - 1)(-1 + \sqrt{2q_0 z + 1})] \tag{15.3.24}$$

In some determinations of q_0, the observed d_L versus z curve is compared with this exact formula, rather than the model-independent approximation (14.6.8),

$$d_L \simeq H_0^{-1}[z + \tfrac{1}{2}(1 - q_0)z^2] \tag{15.3.25}$$

which is strictly valid only in the limit $z \to 0$. The difference between (15.3.24) and (15.3.25) vanishes for $q_0 = 0$ and $q_0 = 1$, and is less than 10% for $0 < z < 0.5$ (which includes all galaxies with known red shifts) and $0 < q_0 < 1.5$. Hence, as long as q_0 is not very large, there is no substantial difference between using the exact formula (15.3.24) and the approximation (15.3.25) to determine q_0.

The "angular diameter distance" d_A and the "proper motion distance" d_M are immediately given in terms of d_L by Eqs. (14.4.22) and (14.4.23):

$$d_A = (1 + z)^{-2} d_L \qquad d_M = (1 + z)^{-1} d_L$$

and the "parallax distance" d_P is given by (14.4.10) as

$$d_P = \frac{[zq_0 + (q_0 - 1)(-1 + \sqrt{2q_0 z + 1})]}{H_0 [q_0^4 (1 + z)^2 - (2q_0 - 1)\{zq_0 + (q_0 - 1)(-1 + \sqrt{2q_0 z + 1})\}^2]^{1/2}} \tag{15.3.26}$$

The number counts discussed in Section 14.7 can now be expressed more explicitly as functionals of the source density n. Shifting variables from t_1 to z, and using (15.3.4), (15.3.22), (15.3.23), and (14.7.7)–(14.7.9) gives the number of sources with red shift less than z and apparent luminosity greater than l as

$$N(<z, >l) = \int_0^\infty dL \int_0^{\min(z, z_l(L))} 4\pi H_0^{-3} q_0^{-4} (1 + z')^{-6} (1 + 2q_0 z')^{-1/2}$$

$$\times [z'q_0 + (q_0 - 1)(-1 + \sqrt{2q_0 z' + 1})]^2 n(z', L)\, dz' \qquad (15.3.27)$$

where

$$z_l(L) \equiv q_0 \left(\frac{LH_0^2}{4\pi l}\right)^{1/2} + (1 - q_0)\left(-1 + \left(1 + 2\left(\frac{LH_0^2}{4\pi l}\right)^{1/2}\right)^{1/2}\right) \qquad (15.3.28)$$

and $n(z, L)\, dL$ is the proper number density of sources at red shift z with absolute luminosity between L and $L + dL$. For radio sources with the spectrum (14.7.13), Eqs. (15.3.4), (15.3.22), (15.3.23), (14.7.15), (14.7.16), and (14.7.8) give the number of sources with red shift less than z and intrinsic power of frequency ν greater than S as

$$N(<z, >S; \nu) = \int_0^\infty dP \int_0^{\min(z, z_{S\alpha}(P))} 4\pi H_0^{-3} q_0^{-4} (1 + z')^{-6} (1 + 2q_0 z')^{-1/2}$$

$$\times (z'q_0 + (q_0 - 1)(-1 + \sqrt{2q_0 z' + 1}))^2 n(z', P; \nu)\, dz'$$

$$(15.3.29)$$

where $z_{S\alpha}(P)$ is the solution of the equation

$$(1 + z)^{(\alpha - 1)/2}(zq_0 + (q_0 - 1)(-1 + \sqrt{2q_0 z + 1})) = q_0^2 H_0 \left(\frac{P}{S}\right)^{1/2}$$

$$(15.3.30)$$

and $n(z, P; \nu)\, dP$ is the proper number density of sources at red shift z with intrinsic power at frequency ν between P and $P + dP$. If there were no evolution of sources, then (14.7.18) and (14.7.19) would give n the z-dependence,

$$n(z, L) = n(0, L)(1 + z)^3$$

$$n(z, P, \nu) = n(0, P, \nu)(1 + z)^3$$

and the z' integrals in (15.3.27) and (15.3.29) could be done explicitly. However, we have already seen in Section 14.7 that this hypothesis does not agree with measurements of $N(>S; \nu)$ for radio sources or $N(<z)$ for quasi-stellar sources. Thus (15.3.27) and (15.3.29) can best be used to gain information about the z and L or P dependence of $n(z, L)$ or $n(z, P; \nu)$. In this way, Longair[15] has concluded that either the number density of radio sources has been decreasing (apart from the expansion of the universe) like $t^{-2.5}$, or the mean source power has been decreasing like $t^{-3.5}$. In addition, a sharp cutoff seems to be needed at early times, although

this conclusion is not definitely established.[16] A study of the quasi-stellar sources in the 3C catalogue by Schmidt[17] shows the same general features—a proper number density that increases with z much faster than $(1 + z)^3$ for $0 < z < 1$, and falls off sharply for $z > 2$. It may be that this cutoff marks the epoch of galaxy or quasi-stellar source formation.

The age of the universe is one more important datum that can help us to decide among different cosmological models. A firm lower limit on the age of the universe is provided by the age of the earth, determined from the relative abundances of radioactive elements and their decay products in the earth's crust. In 1929 Lord Rutherford[18] calculated this age to be about 3.4×10^9 years. Modern studies[19] give for the age of the earth a reliable value of 4.5×10^9 years. If the Hubble time H_0^{-1} is 13×10^9 years, then according to Eq. (15.3.11), the lower limit $t_0 > 4.5 \times 10^9$ years requires that $q_0 < 5$.

Radioactive dating can also be applied to our galaxy. The basic work on the stellar synthesis of the heavy elements is a 1957 paper by the Burbidges, Fowler, and Hoyle.[20] (The purpose of this paper was, at least in part, to defend the steady state model, by showing that the elements could be formed in stars, without needing a "big bang." As discussed in Section 15.7, it is generally accepted today that the elements were mostly formed in stars, with the very important exception that helium may have been formed in the hot early universe.) According to this work, the isotopes of uranium were formed by a rapid process of neutron addition, the *r-process*, in an earlier generation of stars. The initial abundance ratio is calculated as[21]

$$\left[\frac{U^{235}}{U^{238}} \right]_1 = 1.65 \pm 0.15 \qquad \text{(at formation)}$$

The decay rates of these isotopes are accurately known to be

$$\lambda(U^{235}) = 0.971 \times 10^{-9}/\text{year}$$

$$\lambda(U^{238}) = 0.154 \times 10^{-9}/\text{year}$$

and at present their abundance ratio is

$$\left[\frac{U^{235}}{U^{238}} \right]_0 = 0.00723 \qquad \text{(at present)}$$

If all the uranium was formed promptly after the birth of the galaxy at a time t_G, then the age of the galaxy must be[20]

$$t_0 - t_G = \frac{\ln [U^{235}/U^{238}]_1 - \ln [U^{235}/U^{238}]_0}{\lambda(U^{235}) - \lambda(U^{238})}$$

$$\simeq 6.6 \times 10^9 \text{ years}$$

Any society that developed during an earlier epoch in the history of the galaxy would have found a larger proportion of the fissionable isotope U^{235} than now available on earth, and could therefore have moved toward nuclear destruction even faster than our own civilization.

An error of 20% in the estimated initial abundance ratio of U^{235} and U^{238} would produce only a 4% error in the age of the galaxy. A much greater source of uncertainty arises from the possibility that appreciable amounts of uranium were formed well after the galaxy. In this case, the galaxy must be appreciably *older* than 6.6×10^9 years. In order to settle this question, other abundance ratios have been used in conjunction with the ratio of U^{235} to U^{238}, the duration of the period of element synthesis being taken as a free parameter along with the time when this period began. Using the Th^{232}/U^{238} and U^{235}/U^{238} ratios, Fowler and Hoyle[21] estimated that the age of the oldest r-process elements is between 9.6×10^9 and 15.6×10^9 years. Clayton[22] has included the Re^{187}/Os^{187} ratio in his analysis, with results similar to those of Fowler and Hoyle. However, chemical separation effects may be important here, so these results are subject to possible large systematic errors. Dicke[23] has persistently argued that the bulk of the r-process elements were produced within a few hundred million years of the formation of the galaxy, in which case the age of the galaxy would be close to 7×10^9 years. It is safe to conclude that the galaxy and hence the universe is *at least* 7×10^9 years old, so that $q_0 < 2.3$ for $H_0^{-1} \simeq 13 \times 10^9$ years, but radioactive dating cannot yet be considered to give a precise age for the galaxy.

It is also possible to estimate the age of the galaxy by study of its globular clusters. These are large compact clusters containing thousands of individual stars, so their Hertzsprung-Russell diagrams (the relation between luminosity and spectral type) can be determined with some precision. Also, the low metal content of the globular cluster stars indicates that they belong to the first generation of stars to condense out of the protogalaxy (called Population II; see Section 14.5) and are therefore among the oldest objects in our galaxy. If all stars in a globular cluster have the same initial chemical composition and age, and differ only in mass, then these stars must fall on a locus in the Hertzsprung-Russell diagram, whose shape depends only on the age and initial chemical composition. By comparing computer solutions of the equations of stellar evolution with the densities of stars in the observed Hertzsprung-Russell diagrams for a large number of globular clusters, Iben[24] has deduced cluster ages ranging from 8×10^9 to 18×10^9 years, corresponding to initial helium abundances (by mass) ranging from 33 to 24%. It is not ruled out that all clusters have the same age, which would most probably lie in the range from 9.5×10^9 to 15.5×10^9 years. If the age of the universe is really greater than 9×10^9 years, and if the Hubble time H_0^{-1} is 13×10^9 years, then q_0 must be less than $\frac{1}{2}$, and the universe must be negatively curved and infinite, as also indicated by the mass-density estimates discussed in the last section.

It would certainly be premature to reach any definite conclusions about the curvature of space from these estimates of the age of the universe. However, the fact that the uranium and globular-cluster ages are roughly comparable with the Hubble time H_0^{-1} provides a strong argument that the observed correlation of red-shift with luminosity-distance really does have something to do with the evolution of the universe.

The explicit solution for $R(t)$ in the matter-dominated era can be used to illustrate the horizons that limit our vision of the universe. The speed of light sets an upper limit to the local propagation velocity of any signal, so at a given time t an observer at $r = 0$ can receive signals emitted at time t_1 only from radial co-ordinates $r < r_1$, where r_1 is the radial coordinate from which light signals emitted at time t_1 would just reach $r = 0$ at time t. According to Eq. (14.3.1), r_1 is determined by the formula

$$\int_0^{r_1} \frac{dr}{\sqrt{1 - kr^2}} = \int_{t_1}^t \frac{dt'}{R(t')} \tag{15.3.31}$$

If the t'-integral diverges as $t_1 \to 0$, then it is in principle possible to receive signals emitted at sufficiently early times from any comoving particle (such as a "typical galaxy") in the universe. On the other hand, if the t'-integral converges for $t_1 \to 0$ (or, in singularity-free models, for $t_1 \to -\infty$), then our vision is limited by what Rindler[25] has called a *particle horizon*: It is possible to receive signals at time t only from comoving particles that lie within the radial coordinate $r_H(t)$, where

$$\int_0^{r_H(t)} \frac{dr}{\sqrt{1 - kr^2}} = \int_0^t \frac{dt'}{R(t')}$$

The proper distance (14.2.21) of this horizon is

$$d_H(t) = R(t) \int_0^{r_H(t)} \frac{dr}{\sqrt{1 - kr^2}} = R(t) \int_0^t \frac{dt'}{R(t')} \tag{15.3.32}$$

It is easy to see from Eq. (15.1.20) that a particle horizon will be present if ρ grows faster than $R^{-2-\varepsilon}$ as $R \to 0$, as would generally be expected. In particular, if the greatest part of the t'-integral comes from the matter-dominated era, then (15.3.4) can be used to express dt' in terms of $R(t')$ and $dR(t')$, and we find

$$d_H(t) = \begin{cases} \dfrac{R(t)}{R_0 H_0 \sqrt{2q_0 - 1}} \cos^{-1}\left\{1 - \dfrac{(2q_0 - 1)R(t)}{q_0 R_0}\right\} & q_0 > \tfrac{1}{2} \quad (k = +1) \\[4mm] \dfrac{2}{H_0}\left(\dfrac{R(t)}{R_0}\right)^{3/2} & q_0 = \tfrac{1}{2} \quad (k = 0) \\[4mm] \dfrac{R(t)}{R_0 H_0 \sqrt{1 - 2q_0}} \cosh^{-1}\left\{1 + \dfrac{(1 - 2q_0)R(t)}{q_0 R_0}\right\} & q_0 < \tfrac{1}{2} \quad (k = -1) \end{cases}$$

$$\tag{15.3.33}$$

In the early part of the matter-dominated era, R was much less than R_0, so the particle horizon was at a small proper distance:

$$d_H(t) \rightarrow H_0^{-1} \left(\frac{q_0}{2}\right)^{-1/2} \left(\frac{R}{R_0}\right)^{3/2} \simeq \frac{t}{3} \tag{15.3.34}$$

For $q_0 \leq \frac{1}{2}$, $R(t)$ increases without limit as $t \rightarrow \infty$, so $d_H(t)$ increases faster than $R(t)$, and the particle horizon will thus eventually expand to include any given comoving particle. For $q_0 > \frac{1}{2}$, the universe is spatially finite, with a circumference given by Eq. (14.2.4):

$$L(t) = 2\pi R(t) \tag{15.3.35}$$

Looking out in any given direction, we can see comoving particles out to a fraction of this circumference, given by Eqs. (15.3.33) and (15.2.5) as

$$\frac{d_H(t)}{L(t)} = \frac{1}{2\pi} \cos^{-1} \left\{ 1 - \frac{(2q_0 - 1)R(t)}{q_0 R_0} \right\} \tag{15.3.36}$$

When $R(t)$ expands to its maximum value (15.3.9), this fraction will be $\frac{1}{2}$, and we shall be able to see all the way to the "antipodes." However, this fraction remains less than unity until $R(t)$ shrinks once again to zero, so we shall not be able to see all the way around the universe until then. If $q_0 = 1$ and $H_0^{-1} = 13 \times 10^9$ years, then the present circumference (15.3.35) is given by Eq. (15.2.5) as 82×10^9 light years and the particle horizon is at one-quarter this distance, or 20×10^9 light years.

Just as there are some comoving particles that we cannot now see, there may in some cosmological models be events that we never shall see. An event that occurs at r_1 at time t_1 will become visible at $r = 0$, at a time t given by Eq. (15.3.31). If the t'-integral diverges as $t \rightarrow \infty$ (or at the time of the next contraction to $R = 0$), then it will in principle be possible to receive signals from any event if we wait long enough. On the other hand, if the t'-integral converges for large t, then it will only be possible to receive signals from events for which

$$\int_0^{r_1} \frac{dr}{\sqrt{1 - kr^2}} \leq \int_{t_1}^{t_{\max}} \frac{dt'}{R(t')}$$

where $t_{\max}$ is either infinity or the time of the next contraction to $R = 0$. Rindler[25] calls this an *event horizon*. For $q_0 < \frac{1}{2}$ or $q_0 = \frac{1}{2}$, $R(t)$ grows as $t \rightarrow \infty$ like t or $t^{2/3}$, so the t'-integral diverges at $t = \infty$, and there is no event horizon. For $q_0 > \frac{1}{2}$, the t'-integral converges at $t_{\max}$, so there is an event horizon: The only events occurring at time t_1 that will be visible before the collapse of the universe are those within a proper distance

$$d_E(t_1) = R(t_1) \int_{t_1}^{t_{\max}} \frac{dt'}{R(t')}$$

$$= \frac{R(t_1)}{R_0 H_0 \sqrt{2q_0 - 1}} \left[2\pi - \cos^{-1} \left\{ 1 - \frac{(2q_0 - 1)R(t_1)}{q_0 R_0} \right\} \right] \tag{15.3.37}$$

If $q_0 = 1$ and $H_0^{-1} = 13 \times 10^9$ years, then the only events occurring now that will ever become visible to us are those that occur within a proper distance of 61×10^9 light years.

4   Intergalactic Emission and Absorption Processes

Up to now we have dealt only with light signals, which are emitted by distant discrete sources, and propagate to us through essentially empty space. However, we saw in Section 15.2 that Einstein's equations require a cosmic energy density (if $H_0 \simeq 75$ km/sec/Mpc and $q_0 \approx 1$) about equal to 2×10^{-29} gm/cm^3, which is some 70 times larger than the observed density of galactic mass. If the missing mass takes the form of an ionized or neutral gas filling intergalactic space, then we can hope to measure the mass density, and distinguish between cosmological models, by observing the absorption or time delay of light as it passes through the intergalactic gas, or by observing the background radiation emitted by this gas. The absorption of light signals, and the emission and absorption of background radiation, become even more important when we turn our attention back to the early universe, when the density and opacity of matter was enormously greater than it is today.

To lay a foundation for our treatment of these problems, let us first consider the effects of absorption and emission on a ray of light, which leaves a source at time t_1 with frequency ν_1, and arrives at the earth at time t_0. If no emission occurs in the intervening medium, then the loss of flux of the light ray is given by an equation of form

$$\dot{N}(t) = -\Lambda\left(\nu_1 \frac{R(t_1)}{R(t)}, t\right) N(t) \tag{15.4.1}$$

where N is the photon number density in the light ray, and $\Lambda(\nu, t)$ is the absorption rate (per unit proper time) of light with frequency ν. [It is implcitily understood here that at time t the photons in the ray have red-shifted frequency $\nu_1 R(t_1)/R(t)$.] The solution is usually written in the form

$$N(t_0) = e^{-\tau} N(t_1) \tag{15.4.2}$$

where τ is the *optical depth*:

$$\tau = \int_{t_1}^{t_0} \Lambda\left(\nu_1 \frac{R(t_1)}{R(t)}, t\right) dt \tag{15.4.3}$$

Now suppose that the medium itself, apart from effects of the light ray, isotropically emits $\Gamma(\nu, t)$ photons per unit proper volume, per unit proper time, and per unit frequency interval at frequency ν. These photons do not become part of the light ray, but join the isotropic background radiation, to be discussed below. However, Bose statistics require that photons will be added to the light ray through the process of *stimulated emission*,[26] at a rate, per photon in the light ray, given rigorously by

$$\Omega(v, t) = \frac{\Gamma(v, t)}{8\pi v^2} \tag{15.4.4}$$

In place of Eq. (15.4.2), the rate of change of photon number density in the light ray is now given by the formula

$$\dot{N}(t) = -\Lambda\left(v_1 \frac{R(t_1)}{R(t)}, t\right) N(t) + \Omega\left(v_1 \frac{R(t_1)}{R(t)}, t\right) N(t) \tag{15.4.5}$$

and so the optical depth in Eq. (15.4.2) must now be written as

$$\tau = \int_{t_1}^{t_0} \left(\Lambda\left(v_1 \frac{R(t_1)}{R(t)}, t\right) - \Omega\left(v_1 \frac{R(t_1)}{R(t)}, t\right)\right) dt \tag{15.4.6}$$

If the medium is in thermal equilibrium, *not* necessarily in equilibrium with radiation, then Ω and Λ are related by the Einstein formula[27]:

$$\Omega(v, t) = \exp\left[-\frac{hv}{kT(t)}\right] \Lambda(v, t) \tag{15.4.7}$$

where h is Planck's constant, k is Boltzmann's constant, and $T(t)$ is the temperature of the *medium* at time t. [This result simply follows from the principle of detailed balance. The rate of spontaneous emission of photons per unit volume of phase space in any given transition within the medium is equal to the rate of absorption of photons in the inverse transition, times the ratio of the populations of the upper and lower states, which is simply given by the Boltzmann factor $\exp(-hv/kT)$. This factor depends only on v and T, so the total rate of spontaneous emission per unit volume of phase space, which according to (15.4.4) is just Ω, is equal to the total absorption rate Λ, times $\exp(-hv/kT)$.] The optical depth is then

$$\tau = \int_{t_1}^{t_0} \left(1 - \exp\left(-\frac{hv_1 R(t_1)}{kT(t)R(t)}\right)\right) \Lambda\left(v_1 \frac{R(t_1)}{R(t)}, t\right) dt \tag{15.4.8}$$

Even if the medium is not strictly in thermal equilibrium, it is often a good approximation to use (15.4.7) and (15.4.8), with $T(t)$ taken as an *effective temperature*. Normally, T is positive, so $e^{-\tau} < 1$, and a light ray is weakened as it passes through the medium. However, it is sometimes possible to have a population inversion in the medium, with a negative effective temperature. In this case τ is negative, so $e^{-\tau} > 1$, and a light ray is amplified by the medium. Such *maser* phenomena have been detected within our galaxy, but not yet in intergalactic space.

Apart from the images of discrete sources, there is also an *isotropic background* of radiation produced by the universe as a whole. Let $\mathcal{N}(v_0, t)\, dv_0$ be the number density of photons at time t, which *at time* t_0 would have frequency between v_0

and $v_0 + dv_0$. If no absorption or emission occurs, then by the same reasoning that led to Eq. (14.2.17), the time dependence of $\mathscr{N}(v_0, t)$ would simply be given by a factor $R^{-3}(t)$, arising from the general expansion of the universe. In order to calculate the rate of change of $\mathscr{N}(v_0, t)R^3(t)$ owing to spontaneous emission processes, we note that photons, which at time t_0 are in the frequency range from v_0 to $v_0 + dv_0$, are in the frequency range from $v_0 R(t_0)/R(t)$ to $(v_0 + dv_0)R(t_0)/R(t)$ at time t; hence the rate of change of the number of photons $\mathscr{N}(v_0, t)R^3(t)\, dv_0$ in a proper volume $R^3(t)$ and a frequency interval dv_0 at t_0 is

$$\Gamma\left(v_0\,\frac{R(t_0)}{R(t)}\,,\;t\right)R^3(t)\left(\frac{R(t_0)\,dv_0}{R(t)}\right)$$

when Γ is again the rate of spontaneous emission per unit proper volume and per unit frequency interval, now including all discrete sources as well as the medium itself. The rate of change of $\mathscr{N}(v_0, t)R^3(t)\, dv_0$ due to induced emission and absorption is

$$\left(\Omega\left(v_0\,\frac{R(t_0)}{R(t)}\,,\;t\right) - \Lambda\left(v_0\,\frac{R(t_0)}{R(t)}\,,\;t\right)\right)\mathscr{N}(v_0, t)R^3(t)\, dv_0$$

just as in Eq. (15.4.5). Hence the effect of spontaneous and induced emission and absorption is to give $\mathscr{N}(v_0, t)R^3(t)$ the rate of change

$$\frac{d}{dt}\{\mathscr{N}(v_0, t)R^3(t)\} = \Gamma\left(v_0\,\frac{R(t_0)}{R(t)}\,,\;t\right)R^2(t)R(t_0)$$

$$+ \left(\Omega\left(v_0\,\frac{R(t_0)}{R(t)}\,,\;t\right) - \Lambda\left(v_0\,\frac{R(t_0)}{R(t)}\,,\;t\right)\right)\mathscr{N}(v_0, t)R^3(t)$$

Using (15.4.4) for Γ, the solution is

$$\mathscr{N}(v_0, t)R^3(t)$$

$$= \exp\left\{-\int_{t_1}^{t}\left[\Lambda\left(v_0\,\frac{R(t_0)}{R(t')}\,,\;t'\right) - \Omega\left(v_0\,\frac{R(t_0)}{R(t')}\,,\;t'\right)\right]dt'\right\}\mathscr{N}(v_0, t_1)R^3(t_1)$$

$$+ 8\pi v_0^2 R^3(t_0)\int_{t_1}^{t}\exp\left\{-\int_{t'}^{t}\left[\Lambda\left(v_0\,\frac{R(t_0)}{R(t'')}\,,\;t''\right) - \Omega\left(v_0\,\frac{R(t_0)}{R(t'')}\,,\;t''\right)\right]dt''\right\}$$

$$\times\,\Omega\left(v_0\,\frac{R(t_0)}{R(t')}\,,\;t'\right)dt'$$

with t_1 arbitrary. The first term just gives the number of photons left over from before t_1, and the second gives the number of photons emitted since t_1; in both cases the exponential factors represent the subsequent effects of absorption and induced emission. This result simplifies if we take t as the present instant t_0, and take t_1 sufficiently far in the past so that essentially all the background radiation was emitted since then. The present number density of photons per unit frequency interval is then

$$
n_{\gamma 0}(\nu_0) \equiv \mathscr{N}(\nu_0, t_0)
$$

$$
= 8\pi\nu_0{}^2 \int_{t_1}^{t_0} \exp\left\{ -\int_t^{t_0} \left[\Lambda\left(\nu_0 \frac{R(t_0)}{R(t')}, t' \right) - \Omega\left(\nu_0 \frac{R(t_0)}{R(t')}, t' \right) \right] dt' \right\}
$$

$$
\times \, \Omega\left(\nu_0 \frac{R(t_0)}{R(t)}, t \right) dt \tag{15.4.9}
$$

If the *medium* is in thermal equilibrium, then (15.4.7) can be used to express Ω in terms of Λ, and the present photon number density becomes

$$
n_{\gamma 0}(\nu_0) = 8\pi\nu_0{}^2 \int_{t_1}^{t_0} \exp\left(-\frac{h\nu_0 R(t_0)}{kT(t)R(t)} \right) \Lambda\left(\nu_0 \frac{R(t_0)}{R(t)}, t \right)
$$

$$
\times \exp\left\{ -\int_t^{t_0} \left[1 - \exp\left(-\frac{h\nu_0 R(t_0)}{kT(t')R(t')} \right) \right] \Lambda\left(\nu_0 \frac{R(t_0)}{R(t')}, t' \right) dt' \right\} dt \tag{15.4.10}
$$

We have not yet considered the effects of photon *scattering*. In calculating the optical depth of a discrete source, any sort of scattering will remove photons from the light rays, with no stimulated emission to return photons to the ray. Instead of (15.4.8), the optical depth will then be

$$
\tau = \int_{t_1}^{t_0} \left[1 - \exp\left(-\frac{h\nu_1 R(t_1)}{kT(t)R(t)} \right) \right] \Lambda\left(\nu_1 \frac{R(t_1)}{R(t)}, t \right) dt
$$

$$
+ \int_{t_1}^{t_0} \Sigma\left(\nu_1 \frac{R(t_1)}{R(t)}, t \right) dt \tag{15.4.11}
$$

where $\Sigma(\nu, t)$ is the scattering rate for a photon of frequency ν at time t. It is much more difficult to take into account the effects of photon scattering on the isotropic background, because each scattering contributes a photon to the background for every one it takes away. The one case that is easily dealt with is Thomson scattering, in which both $h\nu$ and kT are much less than the charged particle mass. In such scatterings, there is no change in the photon frequency, so the scattering simply has no effect on the isotropic background. In our calculations of the isotropic background, we shall have to assume that scattering, except for Thomson scattering, is much rarer than absorption. (However, note that resonant scattering must be counted as absorption, if the mean lifetime of the resonant state is long compared with the mean free time of the particles in the medium.)

Let us now apply this formalism to the problem of detecting the "missing mass" in intergalactic space. If the intergalactic medium consists of a rarefied gas of neutral atoms, such as hydrogen, then it will absorb radiation strongly at various discrete frequencies, corresponding to various transitions between atomic states. For simplicity, let us suppose that all absorption takes place within a small frequency interval centered on a single absorption frequency v_a. The absorption rate is then of the form

$$\Lambda(v, t) = n(t)\sigma_a(v)$$

where $n(t)$ is the number density of atoms at cosmic time t, and $\sigma_a(v)$ is the absorption cross-section, assumed to be negligible except within a sharp peak at v_a. According to (15.4.8), essentially all absorption of a ray of light, which leaves a source at time t_1 with frequency $v_1 > v_a$ and arrives here at time t_0 with frequency $v_0 < v_a$, will take place at a time t_a such that

$$R(t_a) = \frac{v_1 R(t_1)}{v_a} = \frac{v_0 R(t_0)}{v_a} \tag{15.4.12}$$

so that the optical depth is

$$\tau \simeq n(t_a)\left[1 - \exp\left(-\frac{hv_a}{kT(t_a)}\right)\right]\int \sigma_a\left(v_1 \frac{R(t_1)}{R(t)}\right)dt$$

By changing variables from t to $v \equiv v_1 R(t_1)/R(t)$, we find

$$\tau \simeq n(t_a)\left[1 - \exp\left(-\frac{hv_a}{kT(t_a)}\right)\right]\left[\frac{R(t_a)}{\dot{R}(t_a)}\right]I_a \tag{15.4.13}$$

where

$$I_a \equiv \frac{1}{v_a}\int \sigma(v)\,dv \tag{15.4.14}$$

the integral being taken over a range of frequencies just large enough to include the whole absorption line. The choice of a particular cosmological model is necessary here only in order to determine the Hubble "constant" $\dot{R}/R$ at time t_a; according to Eqs. (15.3.3), (15.4.12), and (15.3.22), this is

$$\frac{\dot{R}(t_a)}{R(t_a)} = \frac{R(t_0)}{R(t_a)} H_0\left[1 - 2q_0 + 2q_0\frac{R(t_0)}{R(t_a)}\right]^{1/2}$$

$$= \left(\frac{v_a}{v_0}\right) H_0\left[1 - 2q_0 + 2q_0\frac{v_a}{v_0}\right]^{1/2} \tag{15.4.15}$$

Using (15.4.15) in (15.4.13), we see that the optical depth at a *received* frequency v_0 is

$$\tau(v_0) = \frac{v_0 n(t_a)I_a}{v_a H_0}\left[1 - \exp\left(-\frac{hv_a}{kT(t_a)}\right)\right]\left[1 - 2q_0 + 2q_0\frac{v_a}{v_0}\right]^{-1/2} \tag{15.4.16}$$

Of course, this result applies only when (15.4.12) can be satisfied along the light path, that is, for received frequencies in the range

$$\frac{v_a}{(1+z)} \le v_0 \le v_a \tag{15.4.17}$$

where $z \equiv v_1/v_0 - 1$ is the red shift of the *source*. Therefore we expect an *absorption trough* in the received signal for v_0 within this range. The optical depth, caused by a single absorption line at frequency v_a, vanishes for $v_0 < v_a/(1+z)$, jumps up steeply at $v_a/(1+z)$ to a value

$$\tau\left(\frac{v_a}{(1+z)}+\right) = \frac{n(t_1)I_a}{H_0(1+z)}\left[1 - \exp\left(-\frac{hv_a}{kT(t_1)}\right)\right][1 + 2q_0 z]^{-1/2} \tag{15.4.18}$$

then varies more or less smoothly until just below v_a, where it takes the value

$$\tau(v_a-) = H_0^{-1}n(t_0)I_a\left[1 - \exp\left(-\frac{hv_a}{kT(t_0)}\right)\right] \tag{15.4.19}$$

and finally drops down sharply to zero for $v_0 > v_a$.

   At the same time that the intergalactic medium is absorbing light signals, it will also be emitting isotropic background radiation. If the medium has an absorption line at frequency v_a, then it will have a single emission line at the same frequency, and the emitted radiation will be observed at red-shifted frequencies $v_0 < v_a$. Using (15.4.10), and following the same reasoning that led to (15.4.16), we find that the present number density of background photons per unit frequency interval at a received frequency v_0 is

$$\mathcal{N}(v_0, t_0) = \frac{8\pi v_0^3 n(t_a)I_a}{v_a H_0}\exp\left(-\frac{hv_a}{kT(t_a)}\right)\left[1 - 2q_0 + 2q_0\frac{v_a}{v_0}\right]^{-1/2} \tag{15.4.20}$$

with I_a and t_a given by (15.4.14) and (15.4.12), respectively. The background density n varies with v_0 more or less smoothly up to just below the frequency v_a, where

$$\mathcal{N}(v_a-, t_0) = 8\pi v_a^2 H_0^{-1}n(t_0)I_a\exp\left(-\frac{hv_a}{kT(t_0)}\right) \tag{15.4.21}$$

and then drops down sharply to zero for $v_0 > v_a$.

   The detailed v_0-dependence of the optical depth (15.4.16) and the background density (15.4.20) depends on the history of the number density $n(t)$ and the temperature $T(t)$. If the atoms that absorb and emit the line at v_a are neither created nor destroyed during the interval from t_a to t_0, then

$$n(t_a) = n(t_0)\left[\frac{R(t_0)}{R(t_a)}\right]^3 = n(t_0)\left[\frac{v_a}{v_0}\right]^3 \tag{15.4.22}$$

as in Eq. (14.2.17). In particular, for $v_a = v_0(1 + z)$, we have $t_a = t_1$, so

$$n(t_1) = n(t_0)[1 + z]^3 \qquad (15.4.23)$$

If the "missing mass" consists of intergalactic neutral hydrogen atoms, then for $q_0 \simeq 1$ and $H_0 \simeq 75$ km/sec/Mpc, these atoms must have a mass density $\rho_0 \simeq 2 \times 10^{-29}$ g/cm^3 (see Section 15.2) and hence a number density

$$n(t_0) = \frac{\rho_0}{m_H} \simeq 1.2 \times 10^{-5} \text{ cm}^{-3} \qquad (15.4.24)$$

Whether or not we believe this particular estimate, a number density of order 10^{-5} cm^{-3} can serve as a specific target at which to aim in attempts to detect the intergalactic medium.

The most prominent radio-frequency absorption line in atomic hydrogen is the 21-cm hyperfine transition, produced by a flip in the proton and electron spins in the $1s$ state, from total spin zero to total spin unity. The frequency of this line is $v_a = 1420$ MHz, corresponding to a temperature $hv_a/k = 0.068°$K, which almost certainly is much less than the "spin temperature" of whatever intergalactic hydrogen may exist. Hence the correction factor owing to stimulated emission can be approximated here as

$$1 - \exp\left(-\frac{hv_a}{kT}\right) \simeq \frac{hv_a}{kT} = \frac{0.068°\text{K}}{T} \qquad (15.4.25)$$

The absorption coefficient (15.4.14) has the value

$$I_{21 \text{ cm}} = 2.73 \times 10^{-23} \text{ cm}^2 \qquad (15.4.26)$$

In 1959 an ingenious method for detecting weak absorption effects near 21 cm was suggested by Field,[28] and used by him to search for such effects in the spectrum of the radio galaxy Cygnus A. This source has red shift $z = 0.056$, and so according to (15.4.17) an absorption trough should occur in the range of observed frequencies from 1342 to 1420 MHz. This range is sufficiently narrow so that the optical depth throughout the absorption trough should be well approximated by the value (15.4.19). Together with (15.4.25) and (15.4.26), this gives a density-temperature ratio (c.g.s. units)

$$\frac{n_H(t_0)}{T(t_0)} \simeq \frac{kH_0\tau}{hv_a I_a c} \simeq 4.4 \times 10^{-5}\tau \text{ cm}^{-3} \text{ deg}^{-1}\left[\frac{H_0}{75 \text{ km/sec/Mpc}}\right] \qquad (15.4.27)$$

Field[28] could detect no absorption trough, and estimated that $\tau < 0.0075$, which with $H_0 = 75$ km/sec/Mpc implies an upper limit $n_H(t_0)/T(t_0) < 3.3 \times 10^{-7}$ cm^{-3} deg^{-1} for the present density-temperature ratio of neutral hydrogen atoms. This experiment has since been repeated by Field[29] and others,[30] but no absorption trough has yet been definitely established in this frequency range. A recent measurement by Penzias and Scott[30] gives $\tau < 5 \times 10^{-4}$, which with $H_0 = 75$ km/sec/Mpc implies

$$\frac{n_H(t_0)}{T(t_0)} < 2.3 \times 10^{-8} \text{ cm}^{-3} \text{ deg}^{-1} \tag{15.4.28}$$

It seems reasonable to suppose[31] that the effective spin temperature of intergalactic hydrogen should be about equal to $2.7°K$, the temperature of the background microwave radiation (see Section 15.5). In this case, (15.4.28) imposes an upper limit on the intergalactic hydrogen number density

$$n_H(t_0) < 6 \times 10^{-8} \text{ cm}^{-3} \tag{15.4.29}$$

which is 200 times smaller than the expected value (15.4.24). If intergalactic hydrogen really makes up the missing mass, then (15.4.28) requires its temperature to be over $500°K$.

Efforts have also been made to detect red-shifted 21-cm absorption effects in the spectra of the quasi-stellar objects 3C191, PKS 1116+12, and 3C287. No such effects were seen.[32]

One way to set an upper limit on the density of neutral hydrogen in intergalactic space, which would not depend on any assumed upper limit for the spin temperature, is to search for the red-shifted 21-cm radiation that would be emitted. There is in any case a microwave radiation background, so the additional background caused by 21-cm emission would have to be detected by looking for a *step* in the photon number density at $v_0 = v_a = 1420$ MHz. According to Eq. (15.4.21), the number density per unit frequency interval $\mathcal{N}$ should be larger just below than just above v_a by the amount (in c.g.s. units)

$$\Delta\mathcal{N} = 8\pi v_a^2 H_0^{-1} c^{-2} I_a n_H(t_0) \tag{15.4.30}$$

[It is assumed here that $T \gg hv_a/k = 0.068°K$. If this is not the case, then the absence of an absorption trough below 21 cm sets an even smaller upper limit on $n_H(t_0)$ than (15.4.28) or (15.4.29).] Usually measurements of background radiation are reported in terms of an equivalent "antenna temperature" T_A, defined by the Rayleigh-Jeans relation

$$\mathcal{N} \equiv 8\pi v_a k T_A h^{-1} c^{-3} \tag{15.4.31}$$

so (15.4.30) gives

$$n_H(t_0) = \frac{k\,\Delta T_A H_0}{hv_a I_a c} = 4.4 \times 10^{-5} \text{ cm}^{-3} \left(\frac{\Delta T_A}{1°K}\right)\left(\frac{H_0}{75 \text{ km/sec/Mpc}}\right) \tag{15.4.32}$$

where ΔT_A is the step in antenna temperature at 1420 MHz. Penzias and Wilson[33] report that $\Delta T_A < 0.08°K$, so that if $H_0 = 75$ km/sec/Mpc, then

$$n_H(t_0) < 3 \times 10^{-6} \text{ cm}^{-3} \tag{15.4.33}$$

This upper limit is only four times smaller than the expected value (15.4.24), so it is not yet entirely ruled out that the missing mass may consist of hot intergalactic atomic hydrogen.

One other prominent absorption line that has been used in the search for intergalactic hydrogen is the Lyman α line of hydrogen, produced by an electronic transition from the $1s$ to the $2p$ state. This line has a wavelength $\lambda = 1215$ Å, which lies in the ultraviolet, so normally Lyman α would not penetrate the earth's atmosphere. However, a photon that has $\lambda = 1215$ Å when $1.5 < z < 6$ will be shifted into the visible "window" between 3000 Å and 7000 Å when it reaches the earth, and can therefore be detected by ground-based astronomers. Thus intergalactic hydrogen atoms might be detected by observing absorption effects in the spectra of quasi-stellar objects with $z > 1.5$ at emitted frequencies above Lyman α.

There are several reasons why Lyman α absorption provides a more sensitive test for the presence of intergalactic hydrogen atoms than does 21-cm absorption. First, the absorption coefficient (15.4.14) is much larger here:

$$I_{Ly\alpha} = 4.5 \times 10^{-18} \text{ cm}^2 \tag{15.4.34}$$

Also, the frequency ν_a is 2.47×10^{15} Hz, corresponding to a temperature $h\nu_a/k = 118{,}000°$K, and since we are now assuming that the ionization is small, we necessarily have

$$\frac{h\nu_a}{kT} \gg 1 \tag{15.4.35}$$

The factors $[1 - \exp(-h\nu_a/kT)]$ in Eqs. (15.4.16), (15.4.18), and (15.4.19), which represent the suppression of absorption by stimulated emission, can then be set equal to unity. Finally, quasi-stellar object spectra often show Lyman α as an *emission* line, so if there is any appreciable neutral hydrogen nearby, then the blue wing of this line ought to be conspicuously suppressed by a factor $e^{-\tau}$, with τ given by (15.4.18), (15.4.34), and (15.4.35), in c.g.s. units, as

$$\tau\left(\frac{\nu_a}{1+z}+\right) = \frac{n_H(t_1)cI_a}{H_0(1+z)(1+2q_0z)^{1/2}}$$

$$= \frac{5.5 \times 10^{10}}{(1+z)(1+2q_0z)^{1/2}}\left(\frac{n_H(t_1)}{\text{cm}^{-3}}\right)\left(\frac{75 \text{ km/sec/Mpc}}{H_0}\right) \tag{15.4.36}$$

Note that the suppression of the blue wing of Lyman α measures the neutral hydrogen density near the time of emission, not the present. (Also, if quasi-stellar objects are local phenomena, then no suppression is to be expected.)

Attempts to detect Lyman α absorption effects have centered on the quasi-stellar object 3C9, with $z = 2.012$. The first measurements were made in 1965 by Gunn and Peterson;[34] they found a 40% depression in the blue wing of the Lyman α emission line, which would give $\tau\left(\frac{\nu_a}{1+z}+\right) \simeq 0.5$. Taking $q_0 = \frac{1}{2}$ and $H_0^{-1} = 10^{10}$ years, they concluded that the number density of neutral hydrogen atoms at $z \simeq 2$ is about 6×10^{-11} cm^{-3}. A subsequent photoelectric measurement by Oke[35] showed no depression in the blue wing of the 3C9 Lyman α emission line, and was interpreted by the Burbidges[36] as showing that $\tau < 0.05$. With $q_0 = 1$ and $H_0 = 75$ km/sec/Mpc, this gives

$$n_H(z \simeq 2) < 6 \times 10^{-12} \text{ cm}^{-3} \qquad (15.4.37)$$

If (15.4.23) is to be believed, then the "expected" value of n_H at $z = 2$ is 27 times larger than (15.4.24), so the observed upper limit (15.4.37) is 8 orders of magnitude smaller than expected!

It is conceivable that the lack of neutral hydrogen near 3C9 is due to ionizing radiation produced by 3C9 itself. For this reason, it is important to look for an absorption trough extending from the Lyman α emission line toward shorter wavelengths, which would be due to Lyman α absorption of light at great distances from 3C9. [See Eqs. (15.4.16) and (15.4.17).] No such trough was found by Oke,[35] and the observations of Wampler[37] show only a slight depression, with $\tau(v_1)$ no greater than about 0.3.

Other attempts have been made to detect intergalactic absorption at ultraviolet wavelengths in the spectra of quasi-stellar objects, with no better success. Field, Solomon, and Wampler[38] have looked for absorption effects due to intergalactic molecular hydrogen in the spectrum of 3C9, and concluded that the intergalactic mass density of molecular hydrogen is less than about 10^{-32} g/cm^3. There is also a possibility that the intergalactic hydrogen is concentrated in clouds, in which case the absorption of Lyman α should show up in quasi-stellar source spectra as a set of more or less broad lines, one for each cloud along the line of sight. No such effects have been found in analyses of quasi-stellar object spectra by Bahcall and Salpeter,[39] Wagoner,[40] and Peebles,[41] and Peebles concludes that the overall density of neutral hydrogen atoms, even if concentrated in clouds, must be less than a few percent of the expected value (15.4.24). Recently three or four quasi-stellar objects have been found with multiple absorption red shifts very much smaller than the red shift of the corresponding emission line,[42] as if the absorption occurred along the line of sight far from the source. However, this phenomenon is rare, and could well be explained by processes occurring within the quasi-stellar source itself.[43]

If the missing mass is not to be found in the form of neutral hydrogen atoms or molecules, then perhaps it consists of an ionized intergalactic hydrogen plasma, possibly with small admixtures of heavier ions. The high degree of ionization indicated by the absence of Lyman α absorption in quasi-stellar object spectra could be explained in terms of a balance between collisional ionization and radiative recombination, provided[43a] that the temperature at $z \simeq 2$ is above $10^{6\circ}$K.

Such a hot gas would produce X-rays through the familiar bremsstrahlung associated with thermal electron-ion collisions, at a rate per unit volume and per unit frequency interval given (in c.g.s. units) by the formula

$$\Gamma(v) = \left[\frac{32\pi g e^6 Z^3 {n_i}^2}{3hvc^3} \right] \left[\frac{2\pi}{3kT{m_e}^3} \right]^{1/2} \exp \left[\frac{-hv}{kT} \right]$$

where n_i is the ion number density, Z^3 is the mean cubed atomic number, and g is a "Gaunt" correction factor, estimated[43b] to lie between $\frac{1}{2}$ and 2 near the peak of the photon spectrum. Field and Henry[43c] have calculated the resulting cosmic X-ray background, under the assumption that the missing mass consists of H and He^4 (10% by number) which is suddenly heated to an initial temperature T_0 (between $10^{4\circ}$K and $10^{10\circ}$K) at R between $\frac{1}{2}R_0$ and $\frac{1}{10}R_0$, and then cools adiabatically, with $T \propto R^{-2}$. The spectrum falls off rapidly for $h\nu > kT_0$, while the interstellar medium within our galaxy is opaque to soft X-rays with $h\nu < 0.1$ keV, so an intergalactic medium should produce an observable X-ray background only if its initial temperature T_0 is above $10^{6\circ}$K.

In fact, rocket observations (recently summarized by Brecher and Burbidge[44]) do reveal the existence of a diffuse X-ray and γ-ray background extending at least from 250 eV to 100 MeV. This background is highly isotropic,[44a] suggesting that it is at least in part of extragalactic origin. However, until recently the X-ray background was not generally interpreted as providing evidence that the missing mass consists of ionized intergalactic hydrogen. One reason is that estimates of the X-ray intensity were lower than at present, while Field and Henry[43c] had assumed a rather large value for the Hubble constant and hence a large value for the missing mass density, so that it was difficult to construct any thermal history for the intergalactic medium, with a temperature high enough to be consistent with the Lyman α and 21 cm absorption and 21 cm emission results discussed above, and yet low enough not to produce more soft X-rays than observed. Also, after the discovery of the cosmic microwave background, it appeared that the X-ray background might be explained as due to the inverse Compton scattering process discussed at the end of the next section.

The origin of the cosmic X-ray background has been reconsidered very recently by Cowsik and Kobetich.[44b] They find that the X-ray spectrum below 1 keV can be accounted for by the inverse Compton effect, while above 100 keV the spectrum is consistent with that expected from the production of γ-rays by white dwarfs. However, between 1 keV and 100 keV there is an excess "kink" in the X-ray spectrum, which can be roughly fit with the flux per unit energy interval:

$$\Phi_{\text{excess}}(E) \simeq 3 \text{ keV cm}^{-2} \text{ ster}^{-1} \text{ sec}^{-1} \text{ keV}^{-1}$$

$$\times \exp\left(-\frac{E}{30 \text{ keV}}\right)$$

This spectrum is just what we should expect for thermal bremsstrahlung from intergalactic hydrogen, with an effective temperature of $3.3 \times 10^{8\circ}$K and an integrated squared ion density $\int n_i{}^2 \, ds$ of order 10^{17} cm^{-5}. Such a medium could furnish the missing mass, especially if H_0 has a rather low value, near 50 km/sec/ Mpc. However, Field[43a] points out that the excess X-ray background could also be produced by "clumped" matter, such as ionized gas within clusters of galaxies[8a] (see Section 15.3) in which case the required mean mass density is reduced by the ratio of the mean and r.m.s. densities, and falls below the critical density ρ_c.

These considerations have prompted a number of recent studies[45] of the thermal history of the intergalactic medium. According to one interesting suggestion of Rees,[46] the intergalactic medium is supposed to have become ionized at a time corresponding to a critical red-shift z_c between 2 and 3. In this case, the absorption of light by neutral hydrogen at the Lyman $\alpha, \beta, \ldots$ lines and in the Lyman continuum would reduce the luminosity of quasi-stellar objects with $z > z_c$, particularly at the blue end of the spectrum, so that the notable lack of quasi-stellar objects with $z > 3$ could be explained as a selection effect, quasistellar objects being usually identified by their blue appearance in Palomar Sky Survey plates. If the rapid increase of quasi-stellar object density with z found by Schmidt[17] (see Section 15.3) really continues beyond $z = 2$, then these sources may well provide the energy that ionizes the intergalactic hydrogen at $z = z_c$. Alternatively, it may be that the quasi-stellar objects are *formed* at $z = z_c$, and that it is this formation process that ionizes the intergalactic medium. Either way, it seems likely that something peculiar happened on a cosmic scale at $z \simeq 3$.

The effects of ionized intergalactic hydrogen on the propagation of light signals can be readily calculated without detailed assumptions about the plasma temperature. As long as $h\nu$ and kT are much less than 1 MeV, the chief effect of the plasma on light signals is an isotropic elastic scattering, with cross-section per electron given by the Thomson value, $\sigma_T = 0.6652 \times 10^{-24}$ cm^2. The optical depth can then be calculated from (15.4.11) neglecting the first term, and setting the scattering rate equal to

$$\sum (\nu, t) = \sigma_T n_e(t) \qquad (15.4.38)$$

where n_e is the number density of electrons, equal to the number density of protons, Suppose that the whole missing mass consists of ionized hydrogen; then (15.1.22). (15.2.6), and (15.2.3) give

$$n_e(t) \simeq \frac{\rho(t)}{m_H} = \frac{3 q_0 H_0{}^2}{4\pi G m_H} \left(\frac{R(t_0)}{R(t)} \right)^3 \qquad (15.4.39)$$

Also, (15.3.3) gives

$$dt = \frac{dR}{\dot{R}} = \frac{dR(t)}{R(t_0) H_0} \left(1 - 2q_0 + 2q_0 \left(\frac{R(t_0)}{R(t)} \right) \right)^{-1/2} \qquad (15.4.40)$$

Using (15.4.38), (15.4.39), and (15.4.40) in (15.4.11), we find the optical depth to be

$$\tau = \frac{3 q_0 H_0 \sigma_T R^2(t_0)}{4\pi G m_H} \int_{R(t_1)}^{R(t_0)} R^{-3} \left[1 - 2q_0 + 2q_0 \left(\frac{R(t_0)}{R} \right) \right]^{-1/2} dR$$

For a source with red shift $z \equiv R(t_0)/R(t_1) - 1$, the optical depth is then[47]

$$\tau(z) = \frac{\tau_c}{q_0} \left[(3q_0 + q_0 z - 1)(1 + 2q_0 z)^{1/2} + 1 - 3q_0 \right] \qquad (15.4.41)$$

where (c.g.s. units)

$$\tau_c = \frac{H_0 \sigma_T c}{4\pi G m_H} = 0.035 \left(\frac{H_0}{75\text{km/sec/Mpc}} \right) \tag{15.4.42}$$

The quasi-stellar objects with $z = 2$ do not seem particularly faint, so presumably $\tau(2)$ is less than about unity; with $H_0 = 75$ km/sec/Mpc, this gives $q_0 < 10$. With $q_0 = 1$ and $H_0 = 75$ km/sec/Mpc, the optical depth is less than unity out to $z = 6$, so Thomson scattering probably does not play an important role in studies of the quasi-stellar objects.

An intergalactic medium of ionized hydrogen would not only scatter radio signals; it would also *delay*[48] them. The group velocity of an electromagnetic wave of frequency v in an ionized gas with electron number density n_e is given by[49]

$$\beta = \left(1 - \frac{v_p{}^2}{v^2} \right)^{1/2} \tag{15.4.43}$$

where v_p is the *plasma frequency*

$$v_p \equiv \left(\frac{e^2 n_e}{m_e \pi} \right)^{1/2} = 8.97 \times 10^3 \text{ Hz } (n_e[\text{cm}^{-3}])^{1/2} \tag{15.4.44}$$

(Again, this is valid only if hv and kT are both much less than the electron rest-energy.) In a locally inertial coordinate system we have $|d\mathbf{x}| = \beta \, dt$, so the invariant proper time is

$$d\tau^2 = (1 - \beta^2) \, dt^2$$

Equating this to the Robertson-Walker line element with $d\theta = d\phi = 0$, we have

$$(1 - \beta^2) \, dt^2 = dt^2 - \frac{R^2(t) \, dr^2}{1 - kr^2}$$

or, more simply,

$$\beta \frac{dt}{R} = \pm \frac{dr}{\sqrt{1 - kr^2}}$$

For a radio signal that leaves a source with comoving radial coordinate r_1 at time t_1, the time of arrival is now delayed by a time Δt, given by

$$\int_{t_1}^{t_0 + \Delta t} \beta \frac{dt}{R} = \int_0^{r_1} \frac{dr}{\sqrt{1 - kr^2}} \tag{15.4.45}$$

where t_0 is the time the signal would have arrived in the absence of any dispersion, that is,

$$\int_{t_1}^{t_0} \frac{dt}{R} = \int_0^{r_1} \frac{dr}{\sqrt{1 - kr^2}} \tag{15.4.46}$$

In all cases of practical importance, v_p is much less than v, so β is very close to unity,

$$\beta \simeq 1 - \frac{v_p^2(t)}{2v^2(t)} = 1 - \frac{v_p^2(t)R^2(t)}{2v_0^2R^2(t_0)} \tag{15.4.47}$$

where v_0 is the frequency observed at time t_0. Subtracting Eq. (15.4.46) from Eq. (15.4.45) and expanding to first order in Δt and $1 - \beta$, we have then

$$\frac{\Delta t}{R(t_0)} = \int_{t_1}^{t_0} [1 - \beta] \frac{dt}{R} \tag{15.4.48}$$

or, using (15.4.47),

$$\Delta t = \frac{1}{2v_0^2 R(t_0)} \int_{t_1}^{t_0} v_p^2(t) R(t) \, dt \tag{15.4.49}$$

This total time delay is not just the integral of $v_p^2/2v^2$ over time, as might be thought. The extra factor of $R(t_0)/R(t)$ appears in Eq. (15.4.48) because the time delay that has already occurred when a photon reaches any given point along its path causes a slight additional increase in the distance it still has to go.

It is convenient in evaluating Δt to change variables from t to

$$z' \equiv \frac{R(t_0)}{R(t)} - 1$$

Equation (15.3.3) then gives

$$dt = -H_0^{-1}[1 + 2q_0 z']^{-1/2}(1 + z')^{-2} \, dz'$$

Also, if free electrons neither appear nor disappear, then

$$v_p^2(t) = v_{p0}^2 \left(\frac{R(t_0)}{R(t)} \right)^3 = v_{p0}^2(1 + z')^3$$

Now Eq. (15.4.49) gives

$$\Delta t = \frac{v_{p0}^2}{2v_0^2 H_0} \int_0^z [1 + 2q_0 z']^{-1/2} \, dz'$$

and therefore

$$\Delta t = \frac{v_{p0}^2}{2q_0 v_0^2 H_0} \{[1 + 2q_0 z]^{1/2} - 1\} \tag{15.4.50}$$

For instance, suppose that $q_0 \simeq 1$ and $H_0 \simeq 75$ km/sec/Mpc. We then expect a present electron number density $n_{e0} \simeq 1.2 \times 10^{-5}$ cm^{-3} [see Eq. (15.4.24)]. in which case the present plasma frequency (15.4.44) is $v_{p0} \simeq 31$ Hz. In contrast, the frequencies at which quasi-stellar sources are observed[49a] to fluctuate are of order 10,000 MHz, which is greater than v_{p0} by about seven orders of magnitude, so the possible time delays are generally quite short. A sharp fluctuation in a quasi-stellar source at $z \simeq 2$ will appear to us to occur later at $v_0 = 10,000$ MHz than at very high frequencies by a time delay $\Delta t \simeq 2.5$ sec. Unfortunately, although quasi-stellar sources do exhibit fluctuations, there do not seem to be any fluctuations at radio frequencies that have time scales as short as a few days.[50] Also, even if such fluctuations did occur, the intergalactic time delay might be obscured by dispersion within the source itself. However, if these difficulties could be surmounted, then both v_{p0} and q_0 could in principle be determined by measuring time delays for various red shifts and comparing with Eq. (15.4.50).

A more modest and perhaps more practicable program is to measure the intergalactic electron number density near our galaxy by observing the frequency-dependent delay of radio signals from a pulsar in some relatively near galaxy. (This is just an extension of the method actually used to determine the distance of pulsars within our own galaxy, where the electron densities are reasonably well known.) Pulsars are believed to be remnants of supernovae, so they might be found in other galaxies by searching for very rapid radio or optical pulses at the sites of recent supernovae, such as the one in the galaxy M101, at a distance $d \simeq 4$ Mpc. (See Figure 15.2.) At such short distances, we should replace z in Eq. (15.4.50) by

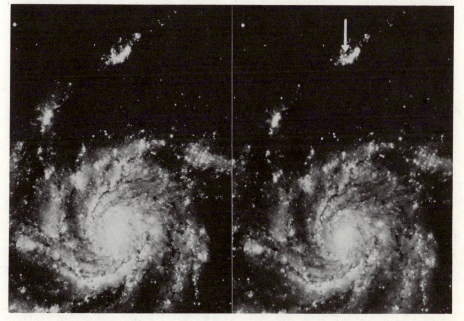

JUNE 9, 1950   FEB. 7, 1951

Figure 15.2 Recent Supernova in the galaxy NGC5457 (M101); photographed with the 200-in. telescope at Mt. Palomar. (Courtesy Mt. Wilson and Mt. Palomar observatories.)

the small quantity $H_0 d$. Also, newborn pulsars would probably emit about 10^4 pulses/sec, so we are here really interested in the difference of the time delay at neighboring frequencies v_0 and $v_0 + dv_0$:

$$-\left(\frac{d\,\Delta t}{dv_0}\right)\delta v_0 = v_{p0}^2 v_0^{-3}\,\delta v_0\,d$$

For instance, if $v_{p0} = 31$ Hz, the difference in arrival times of pulses from a pulsar in M101 at frequencies 1000 MHz and 1001 MHz would be 4×10^{-4} sec, comparable with the expected pulsar period. By working at 100 MHz rather than 1000 MHz, it would be possible to detect electron densities as low as about 10^{-9} cm^{-3}. The problem will be to find a pulsar in some other galaxy.

Other effects of an ionized intergalactic medium on light signals include scintillation,[50a] free-free absorption,[50b] and perhaps Faraday rotation.[50c] Only scintillation now seems promising as a probe for the missing mass.

5   The Cosmic Microwave Radiation Background

The Einstein field equations require that the scale factor $R(t)$ must have been extremely small at some finite time in the past (see Section 15.1). At this early epoch, matter and radiation were presumably in thermal equilibrium, with a very high temperature. As the universe subsequently expanded, both radiation and matter cooled. Eventually, when the temperature had dropped to about 4000°K, the free electrons joined atoms, so that the opacity dropped sharply, breaking the thermal contact between matter and radiation. Whatever radiation existed at that time has since been enormously red-shifted, but it still fills the space around us.

It is widely, though not unanimously, believed, that the microwave radiation background discovered in 1965 is just this left-over radiation, red-shifted by a factor of approximately 1500 since the universe became transparent. If so, then the microwave background provides information of unparalleled value as to the history of the universe, not only back to the time when electrons became bound but, as we shall see, back much further, to the first few seconds of cosmic history.

First, let us consider what sort of background radiation spectrum we would expect on purely theoretical grounds. The proper energy density of the leftover photons, with frequency at the present time t_0 between v and $v + dv$, is given by Eq. (15.4.10) as

$$\rho_{\gamma 0}(v)\,dv = hv \times 8\pi v^2\,dv \int_0^{t_0} \exp\left(-\frac{hvR_0}{kT(t)R(t)}\right)$$

$$\times \Lambda\left(\frac{vR_0}{R(t)},\,t\right)P(t_0, t; v)\,dt \qquad (15.5.1)$$

where h is Planck's constant; k is Boltzmann's constant; R_0 is an abbreviation for $R(t_0)$; $T(t)$ is the temperature of the *matter* (as opposed to the radiation) at time t;

$\Lambda(v, t)$ is the absorption rate for a photon of frequency v at time t; and $P(t_0, t; v)$ is the probability, taking account of stimulated emission, that a photon of frequency $vR_0/R(t)$ present at time t will survive until the present:

$$P(t_0, t; v) \equiv \exp\left\{-\int_t^{t_0}\left[1 - \exp\left(-\frac{hvR_0}{kT(t')R(t')}\right)\right]\Lambda\left(\frac{vR_0}{R(t')}, t'\right)dt'\right\} \quad (15.5.2)$$

The lower limit on the integral (15.5.1) can be chosen as any time t_1 for which $P(t_0, t_1; v)$ is negligible; surely the choice $t_1 = 0$ meets this requirement.

The formula (15.5.1) can usefully be rewritten in the form

$$\rho_{\gamma 0}(v)\, dv = 8\pi hv^3\, dv \int_0^{t_0}\left[\exp\left(\frac{hvR_0}{kT(t)R(t)}\right) - 1\right]^{-1}$$

$$\times \frac{d}{dt} P(t_0, t; v)\, dt \quad (15.5.3)$$

The survival probability P rises from $P = 0$ at $t = 0$ to $P = 1$ at $t = t_0$, so this is just a *weighted average of Planck black-body distributions*. If the opacity drops very steeply at some time t_R, then P is nearly a step function at $t = t_R$, and (15.5.3) gives

$$\rho_{\gamma 0}(v)\, dv \simeq \frac{8\pi hv^3\, dv}{[\exp(hv/kT_{\gamma 0}) - 1]} \quad (15.5.4)$$

where

$$T_{\gamma 0} \equiv \frac{T(t_R)R(t_R)}{R_0} \quad (15.5.5)$$

Thus, *under the assumption of a sharp drop in opacity, the present radiation background should have a black-body spectrum*, with temperature $T_{\gamma 0}$.

It is common to report measurements of the radiation background in terms of a radiation flux $\phi_{\gamma 0}(v)$, the energy received per unit time, per unit receiving area, per unit solid angle, and per unit frequency interval. The flux can be calculated in c.g.s. units from the above formulas for $\rho_{\gamma 0}(v)$ by using

$$\phi_{\gamma 0}(v) = \frac{\rho_{\gamma 0}(v)c}{4\pi}$$

The background measurements are also frequently reported in terms of an *equivalent black-body temperature* $T_{\gamma 0}(v)$, which is defined to be that temperature for which back-body radiation would have the observed energy density or flux at frequency v. That is,

$$\rho_{\gamma 0}(v)\, dv \equiv \frac{8\pi hv^3\, dv}{[\exp(hv/kT_{\gamma 0}(v)) - 1]} \quad (15.5.6)$$

A black-body spectrum is then simply characterized as having $T_{\gamma 0}(v)$ independent of v. Finally, it is occasionally convenient to report background measurements in terms of an *antenna temperature* $T_A(v)$, which is defined to be that temperature for which the low-frequency Rayleigh-Jeans approximation to (15.5.4) would give the observed density or flux at frequency v:

$$\rho_{\gamma 0}(v)\, dv \equiv 8\pi k T_A(v) v^2\, dv \tag{15.5.7}$$

Wherever possible, our discussion here will refer to the black-body temperature $T_{\gamma 0}(v)$.

If we make no assumptions about the thermal history of matter before the drop in opacity, then all we can say is that the radiation background should have roughly a black-body spectrum, with a temperature that tells us the value of $R(t)/R_0$ at the time the universe became transparent. The theoretical situation is tremendously improved if we can assume that, during the time that matter and radiation were in thermal contact, the matter temperature relaxed according to the formula

$$T(t) = \frac{A}{R(t)} \tag{15.5.8}$$

with A a constant. In this case, the first factor in the integrand of (15.5.3) can be taken outside the integral, so we get the black-body formula (15.5.4), *no matter how gradual is the transition from an opaque to a transparent universe*. Further, by taking t_0 in (15.5.3) to be an arbitrary time t, we see now that ρ_γ is given by a black-body formula

$$\rho_\gamma(v,\, t)\, dv = \frac{8\pi h v^3\, dv}{[\exp{(hv/kT_\gamma(t))} - 1]} \tag{15.5.9}$$

with

$$T_\gamma(t) = \frac{A}{R(t)} \tag{15.5.10}$$

at *all* times—after, during, and before the drop in opacity. It is of course not surprising that the radiation would be described by the black-body formula (15.5.9) during the time that matter and radiation were in equilibrium, and naturally its temperature (15.5.10) equaled the matter temperature (15.5.8) during that time. The noteworthy thing here is that the radiation goes on obeying the black-body formula (15.5.9), with temperature given by (15.5.10), throughout the period of transition from high to low opacity, and thereafter until the present. The constant A can be determined by setting $t = t_0$ in (15.5.8), so the radiation temperature at all times is

$$T_\gamma(t) = T_{\gamma 0}\left[\frac{R_0}{R(t)}\right] \tag{15.5.11}$$

and the matter temperature during equilibrium is the same:

$$T(t) \doteq T_{\gamma 0} \left[\frac{R_0}{R(t)} \right] \tag{15.5.12}$$

Thus the present radiation temperature $T_{\gamma 0}$ determines the thermal history of the early universe during the whole period when TR was constant.

In order to see when TR is likely to be constant, let us consider the model of an ideal gas in equilibrium with black-body radiation. The energy density of black-body radiation is given by integrating (15.5.9) over ν:

$$\rho_{\gamma}(t) = aT_{\gamma}^{4}(t)$$

where, in c.g.s. units,

$$a \equiv \frac{8\pi^{5}k^{4}}{15h^{3}c^{3}} = 7.5641 \times 10^{-15} \text{ erg cm}^{-3} \text{ deg}^{-4}$$

Thus the total pressure and energy density in this model are

$$p = nkT + \tfrac{1}{3}aT^{4}$$
$$\rho = nm + (\gamma - 1)^{-1}nkT + aT^{4}$$

where n is the number density of gas particles, m is their mass, and γ is the specific heat ratio of the gas, equal to 5/3 for a monatomic gas like atomic hydrogen. The equation of particle conservation can be written

$$nR^{3} = n_{0}R_{0}^{3} \tag{15.5.13}$$

whereas the equation (15.1.21) of energy conservation reads

$$\frac{d}{dR}[nmR^{3} + (\gamma - 1)^{-1}nkTR^{3} + aT^{4}R^{3}] = -3nkTR^{2} - aT^{4}R^{2}$$

Using (15.5.13) and rearranging terms, this gives

$$\frac{R}{T}\frac{dT}{dR} = -\left[\frac{\sigma + 1}{\sigma + \tfrac{1}{3}(\gamma - 1)^{-1}} \right] \tag{15.5.14}$$

where σk is the *photon entropy per gas particle*:

$$\sigma \equiv \frac{4aT^{3}}{3nk} = 74.0 \frac{[T(\text{deg})]^{3}}{n(\text{cm}^{-3})} \tag{15.5.15}$$

For $\sigma \ll 1$, Eq. (15.5.14) gives

$$T \propto R^{-3(\gamma - 1)} \tag{15.5.16}$$

which is just the usual temperature-volume relation for the adiabatic expansion of an ideal gas. On the other hand, for $\sigma \gg 1$, Eq. (15.5.14) gives

$$T \propto R^{-1} \tag{15.5.17}$$

Even for large σ, as matter goes out of equilibrium with radiation, its temperature curve ultimately shifts from (15.5.17) to (15.5.16). However, if σ is extremely large, then as long as there is any significant thermal contact between matter and radiation, the radiation will continue to overpower the matter, and the matter temperature will have the hoped-for behavior (15.5.8). In this case, (15.5.12), (15.5.13), and (15.5.15) give σ *constant*:

$$\sigma = \frac{4aT_{\gamma 0}{}^3}{3n_0 k} \tag{15.5.18}$$

Hence, if σ is ever very large, then it stays very large. We then say that we are dealing with a *hot universe*. In a hot universe, the background radiation approximately satisfies (15.5.9) and (15.5.11) at all times, and the matter temperature obeys (15.5.12) until the opacity becomes extremely small. Note that the number density of photons in black-body radiation is the integral of $\rho_\gamma(\nu)/h\nu$ over ν, or

$$n_\gamma = \frac{30\zeta(3)}{\pi^4} \frac{aT_\gamma{}^3}{k} = 3.7 \frac{aT_\gamma{}^3}{k}$$

80

$$\sigma = 0.37 \frac{n_{\gamma 0}}{n_0}$$

and the condition for a hot universe can be expressed as a requirement that there are many photons for each proton or neutron in the present universe. None of these considerations gives any clue as to the actual value of $T_{\gamma 0}$, or even as to whether this *is* a hot universe.

The first theoretical estimate of the radiation temperature was based on a theory of element synthesis worked out in the late 1940's by George Gamow and his collaborators.[51] (This subject will be discussed in greater detail in Section 15.7.) At the time when the temperature was $10^9 °K$, corresponding to the dissociation temperature of deuterium, the number density of nucleons must have been roughly 10^{18} cm^{-3}, in order that a fraction of order 10 to 50% of the neutrons and protons could fuse into heavier elements. The specific photon entropy (15.5.15) at that time was then $\sigma \approx 10^{11}$, so in this model the universe is indubitably hot, and RT_γ would therefore have remained constant, both while the universe remained opaque, and thereafter until the present. With a present baryon number density 10^{-6} cm^{-3}, the scale factor at present must be larger than when $T \approx 10^9 °K$ by a factor $(10^{18}/10^{-6})^{1/3}$, or 10^8, so the present radiation temperature should be 10^{-8} times $10^9 °K$, or roughly $10 °K$. A somewhat more detailed analysis along these lines, carried out in 1950 by Alpher and Herman,[52] gave $T_{\gamma 0} \approx 5 °K$. Unfortunately, Alpher and Herman went on to express doubts as to whether this radiation would have survived until the present. It is of course true that the individual photons extant at $T \approx 10^9 °K$ would have been absorbed long before now. However, because $\sigma \gg 1$, the matter temperature must relax like R^{-1}, so that the photons emitted just as the universe is becoming transparent must have had the same value of TR as during the time of element synthesis. Nevertheless, the remarkable prediction of a 5°K black-body radiation background was allowed to slip into obscurity.

The problem of determining $T_{\gamma 0}$ was taken up again in 1965 by Dicke, Peebles, Roll, and Wilkinson.[53] They argued that the universe must once have been hotter than 10^{10}°K, because it either has expanded from a singularity with $R = 0$, or, if it undergoes a cyclic oscillation between finite values of R, it must get hot enough to dissociate the heavy elements left over from the previous cycle. This argument does not fix a value for the present radiation temperature, but Dicke et al. reasoned that the energy density of cosmic black-body radiation should not be large enough to give $q_0 \gg 1$ (see Section 15.2), so that $T_{\gamma 0} \lesssim 40$°K. The really important feature of their work, however, was not this estimate, but rather the fact that at last the black-body radiation background was being taken seriously, with an experiment to measure $T_{\gamma 0}$ being prepared by Roll and Wilkinson.

The difficult part of measuring a radiation temperature less than 40°K is of course that the receiver circuits are at a much higher temperature, so that the signal must be hundreds of times weaker than the receiver noise. In order to pick out the signal, Roll and Wilkinson planned to use a radiometer invented by Dicke in 1945. In this device the radio receiver is switched back and forth a hundred times a second between one horn pointing at the sky and another looking into a bath of liquid helium. The receiver output is filtered to separate just that part that varies with a frequency of 100 Hz, and the strength of this filtered output then measures the difference between the radiation received from the liquid helium and the sky.

Before Roll and Wilkinson could complete a measurement of $T_{\gamma 0}$, they learned that Penzias and Wilson[54] had observed a weak background signal at a radio wavelength $\lambda = 7.35$ cm in the large horn antenna at Holmdel, New Jersey, built to observe the Echo satellite. The antenna temperature could be fit to the curve

$$T_A(\theta) = 4.4°\text{K} + 2.3°\text{K} \sec \theta$$

where θ is the angle between the antenna axis and the zenith. The thickness of atmosphere (taken as a flat slab) through which the antenna beam passes is proportioned to $\sec \theta$, so the second term could be ascribed to radiation from our atmosphere. An additional 0.9°K was estimated as the contribution of ohmic losses in the antenna and radiation from the earth into the antenna side lobes, leaving for the cosmic microwave background a net antenna temperature 3.5°K $\pm$ 1°K. Since $kT_A \gg h\nu$, this is also the equivalent black-body temperature

$$T_{\gamma 0}(7.35 \text{ cm}) = 3.5°\text{K} \pm 1°\text{K}$$

This observation, probably the most important to cosmology since Hubble's discovery of the relation between red shift and distance, was published[54] in 1965 under the modest title "A Measurement of Excess Antenna Temperature at

Table 15.1   Summary of Measurements of the Background Radiation Flux at Microwave and Far-Infrared Wavelengths.

(The temperatures listed are those for which black-body radiation would give the observed flux at the indicated wavelength.)

λ (cm)	Method	Reference	$T_\gamma(\lambda)$ (°K)
73.5	Ground-based radiometer	a	3.7 ± 1.2
49.2	Ground-based radiometer	a	3.7 ± 1.2
21.0	Ground-based radiometer	b	3.2 ± 1.0
20.7	Ground-based radiometer	c	2.8 ± 0.6
7.35	Ground-based radiometer	d	3.5 ± 1.0
3.2	Ground-based radiometer	e	3.0 ± 0.5
3.2	Ground-based radiometer	f	$2.69 \begin{cases} +0.16 \\ -0.21 \end{cases}$
1.58	Ground-based radiometer	f	$2.78 \begin{cases} +0.12 \\ -0.17 \end{cases}$
1.50	Ground-based radiometer	g	2.0 ± 0.8
0.924	Ground-based radiometer	h	3.16 ± 0.26
0.856	Ground-based radiometer	i	$2.56 \begin{cases} +0.17 \\ -0.22 \end{cases}$
0.82	Ground-based radiometer	j	2.9 ± 0.7
0.358	Ground-based radiometer	j'	2.4 ± 0.7
0.33	Ground-based radiometer	k	$2.46 \begin{cases} +0.40 \\ -0.44 \end{cases}$
0.33	Ground-based radiometer	k'	2.61 ± 0.25
0.263	CN $(J = 1/J = 0)$	l	≈ 2.3
0.263	CN $(J = 1/J = 0)$	m	$\begin{cases} 3.22 \pm 0.15 \ \zeta\ \text{Oph} \\ 3.0 \pm 0.6 \ \ \zeta\ \text{Per} \end{cases}$
0.263	CN $(J = 1/J = 0)$	n	3.75 ± 0.50
0.263	CN $(J = 1/J = 0)$	o	$\leqq 2.82$
0.132	CN $(J = 2/J = 1)$	n	< 7.0
0.132	CN $(J = 2/J = 1)$	o	< 4.74
0.0559	CH	n	< 6.6
0.0559	CH	o	< 5.43
0.0359	CH^+	o	< 8.11
0.04–0.13	Rocket-borne IR telescope	p	$8.3 \begin{cases} +2.2 \\ -1.3 \end{cases}$
> 0.05	Balloon-borne IR radiometer	q	$\approx 3.6, 5.5, 7.0$
0.6–0.008	Rocket-borne IR radiometer	r	$3.1 \begin{cases} +0.5 \\ -2.0 \end{cases}$
0.18–1.0	Balloon-borne IR radiometer	s	$2.7 \begin{cases} +0.4 \\ -0.2 \end{cases}$
0.13–1.0	Balloon-borne IR radiometer	s	2.8 ± 0.2
0.09–1.0	Balloon-borne IR radiometer	s	$\leqq 2.7$
0.054–1.0	Balloon-borne IR radiometer	s	$\leqq 3.4$

a T. F. Howell and J. R. Shakeshaft, Nature, **216**, 753 (1967).

b A. A. Penzias and R. W. Wilson, Astron. J., **72**, 315 (1967).

c T. F. Howell and J. R. Shakeshaft, Nature, **210**, 1318 (1966).

d A. A. Penzias and R. W. Wilson, Ap. J., **142**, 419 (1965).

e P. G. Roll and D. T. Wilkinson, Phys. Rev. Letters, **16**, 405 (1966).

f R. A. Stokes, R. B. Partridge, and D. T. Wilkinson, Phys. Rev. Letters, **19**, 1199 (1967).

g W. J. Welch, S. Keachie, D. D. Thornton, and G. Wrixon, Phys. Rev. Letters, **18**, 1068 (1967).

h M. S. Ewing, B. F. Burke, and D. H. Staelin, Phys. Rev. Letters, **19**, 1251 (1967).

i D. T. Wilkinson, Phys. Rev. Letters, **19**, 1195 (1967).

j V. I. Puzanov, A. E. Salomonovich, K. S. Starkovich, Astron. Zh. (USSR), **44**, 1128 (1967) [trans. Soviet A. J., **11**, 905 (1968)].

j' A. G. Kislyakov, V. I. Chernyshev, Yu. V. Lebskii, V. A. Mal'tsev, and N. V. Serov, Ast. Zh., **48**, 39 (1971) [transl. Sov. Astron.—AJ, **15**, 29 (1971)].

k P. E. Boynton, R. A. Stokes, and D. T. Wilkinson, Phys. Rev. Letters, **21**, 462 (1968).

k' M. F. Millea, M. McColl, R. J. Pederson, and F. L. Vernon, Jr., Phys. Rev. Letters, **26**, 919 (1971).

l A. McKellar, Publs. Dominion Astrophys. Observatory (Victoria, B.C.), **7**, 251 (1941).

m G. B. Field and J. L. Hitchcock, Phys. Rev. Letters, **16**, 817 (1966).

n P. Thaddeus and J. F. Clauser, Phys. Rev. Letters, **16**, 819 (1966).

o V. J. Bortolot, J. F. Clauser, and P. Thaddeus, Phys. Rev. Letters, **22**, 307 (1969).

p K. Shivanandan, J. R. Houck, and M. O. Harwit, Phys. Rev. Letters, **21**, 1460 (1968); J. R. Houck and M. Harwit, Ap. J., **157**, L45 (1969). For a recalibration of this data and further data see M. Harwit, J. R. Houck, and R. V. Wagoner, Nature, **228**, 451 (1970); J. L. Pipher, J. R. Houck, B. W. Jones, and M. Harwit, Nature, **231**, 375 (1971).

q D. Muehlner and R. Weiss, Phys. Rev. Letters, **24**, 742 (1970).

r A. G. Blair, J. G. Beery, F. Edeskuty, R. D. Hiebert, J. P. Shipley, and K. D. Williamson, Jr., Phys. Rev. Letters, **27**, 1154 (1971).

s D. Muehlner and R. Weiss, to be published (1972).

4080 MHz" with the paper by Dicke, Peebles, Roll, and Wilkinson[53] appearing as a companion article to explain the fundamental significance of this measurement.

Although Penzias and Wilson reported their result as an "excess antenna temperature," it is important to realize that they had only measured a radiation flux at a *single* wavelength. It remained to verify the Planck form (15.5.4) of the radiation frequency distribution. In Table 15.1 I have listed the measurements of the equivalent black-body temperature of the background radiation that have been carried out at various microwave and far-infrared wavelengths.

At wavelengths above 100 cm, the cosmic background is swamped by the VHF radiation emitted by our galaxy. In the range from 75 to 0.3 cm, the background radiation can be measured with a ground-based microwave radiometer, like that employed by Penzias and Wilson and Roll and Wilkinson. However, below $\lambda = 3$ cm the emission from our atmosphere becomes extremely troublesome, and it is necessary to make observations at mountain altitudes, and at wavelengths, such as 0.9 cm and 0.3 cm, where "windows" appear in the atmosphere. Below $\lambda = 0.3$ cm there are no more useful windows, and the measuring equipment must be carried on a balloon or a rocket. In addition, it is possible to infer a background temperature at certain wavelengths from the absorption of light by molecules in interstellar space. For instance, cyanogen has a visible absorption line at 3874 Å, corresponding to transitions from the ground electronic configuration to an excited electronic configuration. (See Figure 15.3.) Both electronic configurations are split into rotational energy levels, distinguished by the rotational angular momentum J, so this absorption line splits into a number of components,[55] of which the most important are $R(0)$ [$J = 0 \rightarrow J = 1$; $\lambda = 3874.608$Å], $R(1)$ [$J = 1 \rightarrow J = 2$; $\lambda = 3873.998$Å], $P(1)$ [$J = 1 \rightarrow J = 0$; $\lambda = 3875.763$Å], and $R(2)$ [$J = 2 \rightarrow J = 3$; $\lambda = 3873.369$Å]. (These transitions are governed by a dipole selection rule, $\Delta J = \pm 1$.) In 1941 McKellar[56] discovered that cyanogen radicals in an interstellar cloud between us and the star ζ Ophiuchi were absorbing light from that star, not only in the $R(0)$ transition from the $J = 0$ ground state, but also in the $R(1)$ transition from the first excited rotational state, which is at an excitation energy corresponding to a wavelength 2.64 mm. From the relative strength of the two absorption lines, a population for the $J = 1$ state could be inferred, corresponding to a temperature 2.3°K. McKellar could not be sure that there was not some special excitation mechanism at work, so the only conclusion that could be reached was that the radiation background at $\lambda = 2.64$ mm has an equivalent black-body temperature less than about 2.3°K. After the discovery of 3.5°K radiation at 7.35 cm by Penzias and Wilson,[54] Field,[57] Woolf,[58] and Shklovsky[58a]

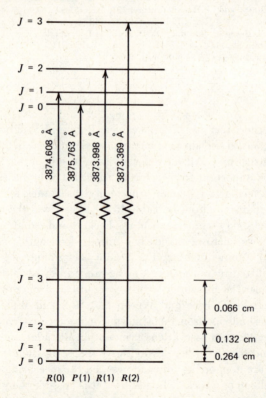

Figure 15.3 Transitions in the cyanogen absorptive spectrum used to set limits on the cosmic microwave radiation background.

independently realized that McKellar's old observations of ζ Ophiuchi might actually have measured the radiation background temperature, and not just set an upper bound on it. This was confirmed by theoretical analyses[57,59] that rejected all other rotational excitation mechanisms, and the measurements were repeated, now including data[60] on the $P(1)$ absorption line, and from a number of other stars. No precise radiation temperature has emerged from these measurements, but it appears pretty certain that T_γ at 2.64 mm is between 2.7 and 3.7°K. There has also been an unsuccessful search[60] for the $R(2)$ absorption line in CN and various absorption lines from excited rotational states in CH and CH$^+$, which allows upper limits to be set on T_γ at wavelengths 1.32 mm, 0.559 mm, and 0.359 mm.

Inspection of Table 15.1 shows that with the exception of the rocket and balloon infrared measurements, all observations are consistent with a 2.7°K black-body distribution. But before we conclude that a black-body distribution has been definitely established, we have to ask how significant this agreement is, and we have to worry about the high-altitude infrared measurements. All of the data at wavelengths above 1 cm unfortunately lie in that part of a 2.7°K Planck distribution that is very well approximated by the Rayleigh-Jeans law

$$\rho_{\gamma 0}(v) \simeq 8\pi k T_{\gamma 0} v^2 \, dv \tag{15.5.19}$$

which can be obtained by letting $v \to 0$ in Eq. (15.5.4). For instance, at $\lambda = 1.5$ cm, the flux for a 2.7°K black body is only 15% below what would be given by the Rayleigh-Jeans formula (15.5.19), and even at $\lambda = 0.856$ cm, the Planck flux is only 35% below the Rayleigh-Jeans flux. (See Figure 15.4.) This is a serious deficiency, because one can imagine a number of models that would give a Rayleigh-Jeans curve (15.5.19) down to wavelengths well below the point where the Planck law begins to drop below the Rayleigh-Jeans law. For instance, suppose that the observed microwave background was emitted at a time t_R when the photon absorption probability $1 - P$ dropped sharply, not from 1 to 0, but from some value $\alpha < 1$ to 0. Then instead of a black-body law, Eq. (15.5.3) would give a *gray-body law,*

$$\rho_{\gamma 0}(v) \, dv \simeq \frac{8\pi\alpha h v^3 \, dv}{[\exp{(hv/kT_\alpha)} - 1]} \tag{15.5.20}$$

with $0 < \alpha < 1$ and $T_\alpha = T(t_R)R(t_R)/R(t_0)$. Then, to account for the data at $\lambda > 1$ cm, it would be necessary to take

$$T_\alpha \simeq \frac{2.7°K}{\alpha}$$

and the flux would then be given by the Rayleigh-Jeans law down to a wavelength $\lambda \approx \alpha$ cm. The total radiant energy density must be less than 10^{-7} erg/cm^3 (see Section 15.2), so α could in principle be as small as 0.08. In order to rule out this sort of theory, we must use data for $\lambda < 1$ cm, and preferably for $\lambda < 0.2$ cm, where a 2.7°K Planck distribution has its maximum. Unfortunately, these are

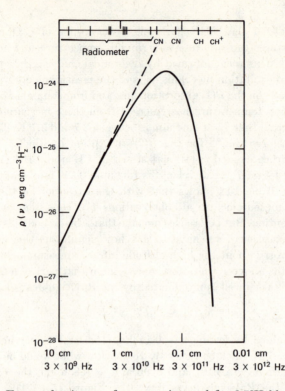

Figure 15.4 Energy density per frequency interval for 2.7°K black-body radiation. The solid curve gives the Planck spectrum (15.5.4); the dashed line gives the Rayleigh-Jeans spectrum (15.5.7) for an antenna temperature of 2.7°K. The short vertical lines mark frequencies at which the black-body temperature has been measured or bounded by radiometer or interstellar absorption observations.

just the wavelengths where the atmosphere begins to interfere with radiometer measurements. The whole case for a black-body distribution, rather than a gray-body distribution, therefore rests on the radiometer measurements[61] at mountain altitude, which give a flux at ≈ 3 mm three times less than expected for the Rayleigh-Jeans law (Eq. (15.5.19) with $T_{\gamma 0} = 2.7°$K), plus the absorption spectra of interstellar molecules, which give upper limits[62] on the flux, at 2.63 mm, 1.32 mm, 0.559 mm, and 0.359 mm, less than Rayleigh-Jeans by factors 2.9, 2.2, 12, and 9.3, respectively. This evidence points strongly to a distribution that does not keep going up like the Rayleigh-Jeans law, but bends over steeply around 0.2 cm, as expected for black-body radiation.

However, this simple picture is contradicted by some of the data taken in the far-infrared by rocket- and balloon-borne equipment. These measurements are essentially bolometric; what is observed is the total power per unit area and solid-angle received by detectors having various complicated spectral-response functions. Initially the rocket observations at Cornell[63] and the balloon observations at M.I.T.[64] both indicated a flux many times larger than expected at these wavelengths for a 2.7°K black-body background. Indeed, taken together with the interstellar absorption measurements, these data were not consistent with *any* smooth spectral distribution, let alone a Planck or Rayleigh-Jeans distribution. The Cornell measurements have since been re-calibrated[64a] and repeated,[64b] and now indicate a much smaller flux, but the flux is still two orders of magnitude greater than expected for a 2.7°K Planck distribution. However, other rocket observations[64c] and new balloon observations by the M.I.T. group[64d] give results consistent with a 2.7°K background. These discrepancies might perhaps arise from a number of strong lines superimposed on a 2.7°K background, or might be due to unexpected sources of atmospheric radiation at high altitudes. These uncertainties will probably be with us until far-infrared measurements can be made with cryogenic equipment carried by artificial satellites.

In checking the agreement of the observed flux of background radiation at various wavelengths with the Planck formula, it is useful to keep in mind the departures from this formula that may be expected on theoretical grounds, even if the observed microwave radiation represents a cosmic background left over from the early universe. With a black-body temperature $T_{\gamma 0} = 2.7°K$, the specific photon entropy (15.5.15) is $\sigma = 1.35 \times 10^8$ for a present mass density $n_0 m_N = 1.8 \times 10^{-29}$ g/cm^3, or $\sigma = 5.4 \times 10^9$ for $n_0 m_N = 4.5 \times 10^{-31}$ g/cm^3. As we have seen, these high σ-values lead us to expect that the matter temperature T would have followed the radiation temperature $T_\gamma \propto R^{-1}$ as long as there was any appreciable thermal contact between matter and radiation. This expectation is borne out by detailed calculations by Peebles of recombination in a universe filled with ionized hydrogen.[65] For a present density $n_0 m_N = 1.8 \times 10^{-29}$ g/cm^3, the fractional ionization dropped sharply from 99.8% to $T_\gamma = 5000°K$ to 0.98% at $T_\gamma = 3000°K$, and then to 0.0053% at $T_\gamma = 1500°K$. However, even though the mean free path of photons at these low ionization levels was very long, the matter temperature when $T_\gamma = 2000°K$ was $T = 1920°K$, and at $T_\gamma = 1500°K$, $T = 1280°K$. For smaller values of the present mass density, T followed T_γ even more closely. In consequence, the departures from a Planck distribution should be quite small. According to Peebles, the largest effects are the excess of photons left over from the $2p \to 1s$ Lyman α transition and the $2s \to 1s$ two-photon transition, by which the recombined hydrogen atoms reached their ground states. These photons now show up red-shifted by a factor of order $\gtrsim 1000$ from λ (Lyman α) = 1215Å and $\lambda(2\gamma) \approx 2500$Å, and thus produce departures from a Planck distribution at wavelengths shorter than 0.015 mm. Unfortunately, at these short wavelengths the cosmic radiation background is much less intense than the radiation from interstellar dust[66] and gas[67] in our galaxy, so it is unlikely that we shall be able to observe these departures from a black-body spectral distribution.

There is one other important possible source of departures from the Planck spectrum. The calculations of Peebles[65] show that by the time the radiation temperature dropped to 200°K, the residual hydrogen ionization was extremely small, of order 10^{-4} to 10^{-5}. However, the experiment on Lyman α absorption discussed in the last section shows that there cannot have been any appreciable amount of *neutral* hydrogen since a time when $T_\gamma \simeq 8°K$, corresponding to $z \simeq 2$. If there really is a good deal of intergalactic hydrogen gas, as indicated by measurements of q_0 (see Section 15.2), then somehow or other this hydrogen must have been reionized at a time when T_γ was between 4000°K and 8°K. If the reionization was very early, then thermal contact would have been reestablished between matter and radiation, and the Planck spectrum would have been distorted by an increase in the individual photon energies. According to Sunyaev,[68] the agreement between the observed background radiation spectrum with the Planck formula already shows that the reionization could not have occurred until T_γ dropped to about 800°K.

There is also a great deal to be learned from the distribution of the microwave background in angle. If this radiation really is left over from an earlier period when matter and radiation were in thermal equilibrium, then we should expect the radiation flux to be isotropic. However, there might be anisotropies of small angular scale, owing to inhomogeneities in the primordial plasma, possibly associated with the presence of nascent galaxies.[66] (See Section 15.8.) There might also be anisotropies of larger angular scale, owing to a departure of the universe as a whole or our local gravitational field[69] from perfect isotropy, and there certainly is a small anisotropy with a 360° angular scale owing to the motion of the solar system relative to the radiation background. If the radiation background does not come from an earlier period of thermal equilibrium, then its angular distribution may reveal its source; for instance, if the radiation comes from a large number of discrete sources, then we should find large anisotropies of very small angular scale, whereas if it comes from our own galaxy, then we should expect a large-scale anisotropy correlated with galactic latitude.

In looking for *anisotropies of small angular scale*, a large antenna pointed at a fixed angle relative to the earth is swept across the sky by the rotation of the earth. If no special care is taken to maintain a stable calibration, then the measured antenna temperature will show a gradual drift in time, which does not concern us here. There will also be a small fluctuation around this general drift, characterized by an r.m.s. fluctuation value $(\Delta T_A)_{\text{obs}}$. If there really is an intrinsic fluctuation ΔT_A with an angular scale θ comparable with the beam width B, then $(\Delta T_A)^2_{\text{obs}}$ will be given by $(\Delta T_A)^2$ plus a term arising from noise in the receiver, so that

$$\Delta T_A \lesssim (\Delta T_A)_{\text{obs}} \qquad \text{for } \theta \approx B \qquad (15.5.21)$$

On the other hand, if the intrinsic fluctuation scale θ is much less than the beam width B, then the beam can be regarded as split into N patches of angular diameter θ, where

$$N \approx \left(\frac{B}{\theta}\right)^2$$

The fluctuation ΔP in the power P from each patch is given by (15.5.7) as

$$\frac{\Delta P}{P} \approx \frac{\Delta T_A}{T_A}$$

The total power received is NP, but the fluctuations have random sign, and so the r.m.s. fluctuation in the total power is $N^{1/2}\Delta P$. Taking account of receiver noise, the observed relative fluctuation in received power will be greater than $N^{1/2}\Delta P/NP$ so

$$\frac{(\Delta T_A)_{\text{obs}}}{T_A} \gtrsim N^{-1/2} \frac{\Delta P}{P} \approx N^{-1/2} \frac{\Delta T_A}{T_A}$$

and therefore

$$\Delta T_A \lesssim N^{1/2}(\Delta T_A)_{\text{obs}} \approx \left(\frac{B}{\theta}\right)(\Delta T_A)_{\text{obs}} \qquad \text{for } \theta \ll B \qquad (15.5.22)$$

A more detailed analysis[70] shows that for fluctuations of arbitrary angular scale θ

$$\Delta T_A \leq \left[1 + \frac{B^2}{\theta^2}\right]^{1/2} (\Delta T_A)_{\text{obs}}$$

in agreement with (15.5.21) and (15.5.22). For a very strong intrinsic fluctuation with $\Delta T_A \approx T_A$, Eq. (15.5.22) sets an upper limit on the angular scale

$$\theta_{\text{max}} \approx \frac{B(\Delta T_A)_{\text{obs}}}{T_A} \qquad (15.5.23)$$

Measurements of $(\Delta T_A)_{\text{obs}}$ at various wavelengths and beam widths are listed in Table 15.2. The anisotropy is evidently less than a few percent on any angular scale larger than a few seconds of arc.

In searching for *anisotropies of large angular scale*, it is not necessary to use a large antenna, but care must be taken to maintain a stable receiver calibration as the antenna beam is swept across the sky by the rotation of the earth. In the work of Partridge and Wilkinson,[71] this is managed by aiming the horn so that it points near the celestial equator, and then for 15 min in each half-hour, inserting a vertical reflector that aims the beam toward the north celestial pole. With and without the reflector, the angle between the antenna beam and the vertical is the same (48°), so the effects of heating of the atmosphere, as well as of the apparatus, should be the same. However, when the reflector is absent, the beam is scanned across the celestial equator as the earth turns, whereas with the reflector inserted, the beam points toward a more or less fixed point on the celestial sphere. Hence any change with time in the *difference* between the radiation flux received with and without the reflector should be a measure of an intrinsic variation of the flux with right ascension (i.e., azimuth) near the celestial equator. This variation must have a

Table 15.2   Summary of Measurements of Fluctuations in the Microwave Background of Small Angular Scale

λ (cm)	T_A (°K)	B	$\Delta T_{A \text{ obs}}$ (°K)	θ_{max}	Reference
7.35	2.56	40′	0.006	5″	a
3.95	2.50	1.4′ × 20′	0.0007	0.1″	a′
2.80	2.45	1°	0.051	75″	b
		6°	0.036	—	
2.80	2.45	10′	0.0061	1.5″	c
		2°	0.0017	—	
0.35	1.14	$\approx 75″$	0.024	1.6″	d
0.35	1.14	80″–100″	0.008	0.7″	d′
0.34	1.11	12.5′	0.2	—	e

Here λ is the wavelength, T_A is the antenna temperature for 2.70°K black-body radiation, B is the beam width, $\Delta T_{A \text{ obs}}$ is the observed r.m.s. fluctuation in antenna temperature, and θ_{max} is the largest angular scale at which the observations would allow gross anisotropies [see Eq. (15.5.23)]. The "beam widths" of 1°, 2°, and 6° were synthesized by integration of data obtained with a 10′ beam width. The measurement of ΔT_A at 0.34 cm really is a measure of the change in slope of $T_A(\theta)$ over an angular interval 12.5″.

[a] A. A. Penzias and R. W. Wilson, Ap. J., **142**, 419 (1965).
[a′] Yu. N. Pariskii and T. B. Pyatunina, Astron. Zh., **47**, 1337 (1970) [transl. Sov. Astron.—AJ, **14**, 1067 (1971)].
[b] E. K. Conklin and R. N. Bracewell, Phys. Rev. Letters, **18**, 614 (1967).
[c] E. K. Conklin and R. N. Bracewell, Nature, **216**, 777 (1967).
[d] A. A. Penzias, J. Schraml, and R. W. Wilson, Ap. J., **157**, L49 (1969).
[d′] P. Boynton and R. B. Partridge, private communication.
[e] E. E. Epstein, Ap. J., **148**, L157 (1967).

24-hr sidereal period, so it can be Fourier-analyzed into components with periods $24/n$ hr, where n is any integer.

Measurements of the anisotropy are summarized in Table 15.3. There is evidently no statistically significant anisotropy observed, and the maximum change in $T_{\gamma 0}$ around the sky is probably less than 1%.

The upper limits in the 24-hr component of the anisotropy $\Delta T_\gamma / T_{\gamma 0}$ are particularly interesting, because they set stringent upper limits on the velocity of the solar system relative to the rest of the universe. Suppose that there is a fundamental reference frame in which the background radiation is perfectly isotropic, with a Planck spectrum, and assume that the earth moves with a velocity $\mathbf{v}_\oplus$ with respect to this fundamental frame. In the fundamental frame,

Table 15.3 Summary of Measurements of Anisotropies of Large Angular Scale in the Microwave Background

(The data used in reference e include those used in reference d.) All values of $\Delta T_\gamma / T_{\gamma 0}$ are based on an assumed value $T_{\gamma 0} = 2.7°\text{K}$.

λ (cm)	Type	$\Delta T_\gamma / T_{\gamma 0}$ (%)	Reference
7.35	r.m.s.	≤ 10	a
7.35	r.m.s.	≤ 3.7	b
3.75	24 hr	0.06 ± 0.03	c
3.2	12 hr	0.18 ± 0.08	d
	24 hr	0.03 ± 0.08	
3.2	12 hr	0.06 ± 0.06	e
	24 hr	0.04 ± 0.06	
0.8	12 hr	0.20 ± 0.24	f
	24 hr	0.28 ± 0.43	

a A. A. Penzias and R. W. Wilson, Ap. J., **142**, 419 (1965).
b R. W. Wilson and A. A. Penzias, Science, **156**, 1100 (1967).
c E. K. Conklin, Nature, **222**, 971 (1969).
d R. B. Partridge and D. T. Wilkinson, Phys. Rev. Letters, **18**, 557 (1967).
e D. T. Wilkinson and R. B. Partridge, quoted by R. B. Partridge, American Scientist, **57**, 37 (1969).
f S. P. Boughn, D. M. Fram, and R. B. Partridge, Ap. J., **165**, 439 (1971).

the photons within a solid angle $\sin \theta \, d\theta \, d\varphi$ and a frequency interval dv contribute to the energy-momentum tensor an amount

$$dT^{\mu\nu} = \left(\frac{p^\mu p^\nu}{h^2 v^2}\right)\left(\frac{\sin \theta \, d\theta \, d\varphi}{4\pi}\right) \rho_{\gamma 0}(v) \, dv$$

$$= 2p^\mu p^\nu h^{-1}[e^{hv/kT_{\gamma 0}} - 1]^{-1} \sin \theta \, d\theta \, d\varphi \, v \, dv$$

where p^μ is the photon momentum four-vector:

$$p^\mu = hv(\sin \theta \cos \varphi, \sin \theta \sin \varphi, \cos \theta, 1)$$

(It follows from (2.8.4) that $dT^{\mu\nu}$ is proportional to $p^\mu p^\nu$; the coefficient of $p^\mu p^\nu$ is determined so that the integral of dT^{00} over θ and φ should be $\rho_{\gamma 0} \, dv$.) In the earth frame these photons have an energy-momentum tensor given by the tensor transformation rule

$$dT'^{\mu\nu} = \Lambda^\mu{}_\rho \Lambda^\nu{}_\sigma \, dT^{\rho\sigma}$$

where Λ is the Lorentz transformation defined by (2.1.17)–(2.1.21), taking $\mathbf{v} = -\mathbf{v}_{\oplus}$. In order to express $dT'^{\mu\nu}$ in terms of earth-frame quantities, we note that

$$p'^{\mu} = \Lambda^{\mu}{}_{\nu}p^{\nu}$$

or, taking the z-axis in the direction of the earth's velocity,

$$v' = \frac{v[1 - v_{\oplus}\cos\theta]}{[1 - v_{\oplus}{}^2]^{1/2}}$$

$$\cos\theta' = \frac{[-v_{\oplus} + \cos\theta]}{[1 - v_{\oplus}\cos\theta]} \qquad \varphi' = \varphi$$

where θ is now the angle between the velocities of the earth and photon. The solid angle then has the transformation rule

$$\sin\theta' \, d\theta' \, d\varphi' = \left(\frac{v}{v'}\right)^2 \sin\theta \, d\theta \, d\varphi$$

and so the differential energy momentum tensor in the earth frame is

$$dT'^{\mu\nu} = 2p'^{\mu}p'^{\nu}h^{-1}[e^{h\nu/kT_{\gamma0}} - 1]^{-1} \sin\theta \, d\theta \, d\varphi\nu \, d\nu$$

$$= 2p'^{\mu}p'^{\nu}h^{-1}[e^{h\nu'/kT_{\gamma0}} - 1]^{-1} \sin\theta' \, d\theta' \, d\varphi'\nu' \, d\nu'$$

where

$$T'_{\gamma0} \equiv \left(\frac{v'}{v}\right)T_{\gamma0} = [1 - v_{\oplus}{}^2]^{-1/2}[1 - v_{\oplus}\cos\theta]T_{\gamma0} \qquad (15.5.24)$$

We see that $dT'_{\mu\nu}$ has the same form as $dT_{\mu\nu}$, so that the *background radiation in the earth frame has a Planck spectrum, but with an angle-dependent temperature* $T'_{\gamma0}$. For $v_{\oplus} \ll 1$, the departure of the measured temperature from the "true" black-body temperature $T_{\gamma0}$ is

$$\Delta T_{\gamma0} \simeq -v_{\oplus}\cos\theta T_{\gamma0} \qquad (15.5.25)$$

In the experiments of Partridge and Wilkinson and of Conklin, the antenna beam scans a circle on the celestial sphere of fixed declination δ once a day, so $\Delta T_{\gamma0}$ should have a 24-hr period, with maximum value given by

$$\frac{(\Delta T_{\gamma0})_{\max}}{T_{\gamma0}} \simeq \frac{v_{\oplus}(\delta)}{c} \qquad (15.5.26)$$

where $v_{\oplus}(\delta)$ is the component of the earth's velocity (in c.g.s. units) along the cone of declination δ. This maximum is attained when the antenna points in the azimuthal direction toward which the earth is moving. The combined data of Partridge and Wilkinson[71] give as a most likely velocity $v_{\oplus}(0°) \simeq 120$ km/sec with a direction toward 0 hr right ascension and a *vector* error of magnitude

180 km/sec. Conklin[72] gives as a most likely velocity $v_\oplus(32°\text{N}) \simeq 160$ km/sec with a direction toward 13 hr right ascension (just the opposite to Partridge and Wilkinson!) and a vector error of magnitude 85 km/sec. It is reasonable to conclude from these two results that

$$|v_\oplus| \lesssim 300 \text{ km/sec} \qquad (15.5.27)$$

This upper limit is already of the same order of magnitude as the velocity of the solar system in the local group of galaxies (owing mostly to the rotation of our own galaxy) which is estimated[73] as 315 km/sec toward 22 hr right ascension. Clearly neither the earth, nor the whole local group of galaxies, is moving at great velocity relative to the radiation background. It will be of very great interest to learn how fast we *are* moving, and in what direction.

Apart from the effects of the earth's motion or local gravitational fields, the microwave background might also exhibit anisotropies owing to a cosmic inhomogeneity at the time t_R that the radiation was last emitted or scattered. If there has been no scattering of the background radiation since the recombination of hydrogen at about 4000°K, then the time t_R corresponds to a red shift z_R given by

$$1 + z_R \equiv \frac{R_0}{R(t_R)} = \frac{T_\gamma(t_R)}{T_{\gamma 0}} \approx \frac{4000°\text{K}}{2.7°\text{K}} = 1500$$

On the other hand, if there is an intergalactic free electron gas with number density $1.2 \times 10^{-5}/\text{cm}^3$, then, as remarked in the last section, the time of last scattering would correspond to a red shift $z_R \approx 6$. It would be very interesting to use the observed isotropy or anisotropy of the present microwave background to determine the scales of distances at which the universe is homogeneous or inhomogeneous at the time t_R.

To this end, consider two photons that leave comoving sources A and B at time t_R, and arrive at the earth at time t_0, traveling along paths separated at the earth by an angle θ. With the earth taken as the origin, Eq. (14.3.1) gives the radial coordinates of the sources A and B as

$$r_A = r_B = r_1 \qquad (15.5.28)$$

where

$$\int_0^{r_1} \frac{dr}{\sqrt{1 - kr^2}} = \int_{t_R}^{t_0} \frac{dt}{R(t)} \qquad (15.5.29)$$

Since photons travel to the earth on trajectories with constant direction $\mathbf{x}/r$, the sources will be separated in the Robertson-Walker coordinate system by exactly the observed angle θ that separates the light rays arriving at the earth. That is,

$$\frac{\mathbf{x}_A \cdot \mathbf{x}_B}{r_1^2} = \cos \theta \qquad (15.5.30)$$

with the scalar product defined in terms of the Robertson-Walker coordinates x^i as if these coordinates were Cartesian:

$$\mathbf{x}_A \cdot \mathbf{x}_B \equiv x_A{}^1 x_B{}^1 + x_A{}^2 x_B{}^2 + x_A{}^3 x_B{}^3$$

$$= r_A r_B [\sin \theta_A \sin \theta_B \cos (\varphi_A - \varphi_B) + \cos \theta_A \cos \theta_B] \qquad (15.5.31)$$

Our problem is to determine the proper distance along a geodesic from A to B at time t_R as a function of θ, for various assumed values of z_R ranging from 6 to 1500.

According to Eq. (14.4.3), the geodesic from A to B can be chosen (setting $\mathbf{x}_1$ equal to a vector $a\mathbf{e}$ normal to $\mathbf{n}$) to have the form

$$\mathbf{x}(\rho) = \mathbf{n}\rho + a\mathbf{e}(1 - k\rho^2)^{1/2} \qquad (15.5.32)$$

where a is a constant, ρ is a variable parameter, and $\mathbf{n}$ and $\mathbf{e}$ are orthogonal unit vectors,

$$\mathbf{n} \cdot \mathbf{e} = 0 \qquad \mathbf{n}^2 = \mathbf{e}^2 = 1 \qquad (15.5.33)$$

the scalar products being defined as in Eq. (15.5.31). The initial and final values of ρ are $-\rho_1$ and $+\rho_1$, with ρ_1 determined by the condition (15.5.28), that is,

$$r_1{}^2 = |\mathbf{x}(\pm \rho_1)|^2 = \rho_1{}^2 + a^2(1 - k\rho_1{}^2)$$

In addition, the condition (15.5.30) gives

$$\cos \theta = \frac{\mathbf{x}(+\rho_1) \cdot \mathbf{x}(-\rho_1)}{r_1{}^2} = \frac{[-\rho_1{}^2 + a^2(1 - k\rho_1{}^2)]}{r_1{}^2}$$

Both ρ_1 and a can thus be expressed in terms of r_1 and θ:

$$\rho_1 = r_1 \sin \frac{\theta}{2}$$

$$a = r_1 \cos \frac{\theta}{2} \left[1 - kr_1{}^2 \sin^2 \frac{\theta}{2} \right]^{-1/2}$$

The proper distance from A to B can now be calculated by integrating the Robertson-Walker line element from $-\rho_1$ to $+\rho_1$:

$$d(\theta) = R(t_R) \int_{-\rho_1}^{\rho_1} \left(\left(\frac{d\mathbf{x}(\rho)}{d\rho} \right)^2 + \frac{k(\mathbf{x}(\rho) \cdot d\mathbf{x}(\rho)/d\rho)^2}{1 - k\mathbf{x}^2(\rho)} \right)^{1/2} d\rho$$

and thus

$$d(\theta) = \frac{2R_0}{1 + z_R} \int_0^{r_1 \sin(\theta/2)} \frac{d\rho}{\sqrt{1 - k\rho^2}} \qquad (15.5.34)$$

If the time t_R of last scattering or emission occurs after the start of the matter-dominated era, then (15.2.5) and (15.3.23) can be used to express R_0 and r_1 in terms of H_0, q_0, and z_R, and we find

$$d(\theta) = \frac{2}{H_0(1 + z_R)\sqrt{2q_0 - 1}}$$

$$\times \sin^{-1}\left\{\frac{\sqrt{2q_0 - 1}\,[z_R q_0 + (q_0 - 1)(-1 + \sqrt{2q_0 z_R + 1})]}{q_0{}^2(1 + z_R)} \sin\frac{\theta}{2}\right\}$$

$$\text{for } q_0 > \tfrac{1}{2},\, k = +1 \qquad (15.5.35)$$

$$d(\theta) = \frac{4}{H_0(1 + z_R)}\{1 - (1 + z_R)^{-1/2}\}\sin\frac{\theta}{2} \qquad \text{for } q_0 = \tfrac{1}{2},\, k = 0 \qquad (15.5.36)$$

$$d(\theta) = \frac{2}{H_0(1 + z_R)\sqrt{1 - 2q_0}}$$

$$\times \sinh^{-1}\left(\frac{\sqrt{1 - 2q_0}\,[z_R q_0 + (q_0 - 1)(-1 + \sqrt{2q_0 z_R + 1})]}{q_0{}^2(1 + z_R)} \sin\frac{\theta}{2}\right)$$

$$\text{for } q_0 < \tfrac{1}{2},\, k = -1 \qquad (15.5.37)$$

In particular, for $\theta \to 0$, Eqs. (15.5.35)–(15.5.37) give

$$d(\theta) \to \frac{[z_R q_0 + (q_0 - 1)(-1 + \sqrt{2q_0 z_R + 1})]\theta}{q_0{}^2(1 + z_R)^2 H_0} \qquad \text{for } \theta \to 0$$

If the homogeneity of the universe is achieved by the physical transport of energy and momentum from one place to another at velocities less than that of light, then we should expect[74] the universe at time t_R to be inhomogeneous over distances larger than twice the "particle horizon" (15.3.32), because no homogenizing signal could travel from any point to a pair of comoving particles separated by a proper distance greater than $2d_H(t_R)$ by time t_R. If this is correct, then the microwave background ought to exhibit large anisotropies for angular scales greater than an angle θ_H, which can be calculated by equating $2d_H(t_R)$, given by Eq. (15.3.33), to $d(\theta_H)$, given by Eqs. (15.5.35)–(15.5.37):

$$\sin\frac{\theta_H}{2} = \frac{q_0\sqrt{2q_0 z_R + 1}}{z_R q_0 + (q_0 - 1)(-1 + \sqrt{2q_0 z_R + 1})} \qquad (15.5.38)$$

If $z_R \simeq 1500$, then we can use the approximation

$$\theta_H \simeq 2\left(\frac{2q_0}{z_R}\right)^{1/2} \simeq 4.2°\sqrt{q_0} \qquad (15.5.39)$$

(This result would not be very much changed if the matter-dominated era began somewhat after the recombination of hydrogen.) If $z_R \simeq 6$ and $q_0 = \tfrac{1}{2}$ or $q_0 = 1$, then $\theta_H \simeq 75°$. However, there is no sign of any appreciable anisotropy in the microwave background at such angular scales—on the contrary, the microwave radiation appears to be highly isotropic on all angular scales greater than 1°. In the light of the above analysis, it is difficult to understand how such a high degree of isotropy could be produced by any physical process occurring at any time since the initial singularity.

The observed distributions of the radiation background in frequency and angle certainly suggest that this is isotropic black-body radiation left over from an earlier period when matter and radiation were in thermal equilibrium. However, other possibilities are not yet excluded by the data. The energy density of starlight within our galaxy is of the order of 5×10^{-13} erg/cm^3, just about the same as the energy density of 2.7°K black-body radiation. For this reason, Hoyle, Narlikar, and Wickramsinghe[75] have suggested that a large fraction of the optical-frequency starlight in our own and other galaxies may be absorbed by interstellar grains, which are heated to a few degrees, and reemit the energy at microwave frequencies, either as a continuum or in discrete lines. It would not be impossible for this re-emitted radiation to be isotropic and to imitate a Planck spectrum, but this seems artificial. Another possibility that has been widely considered is that the microwave background may arise from a large number of discrete sources.[76] Here again, a Planck spectrum would not be impossible over the accessible range of wavelengths, but there is no special reason to expect it. Also, in this case the observed isotropy does put severe limits on any discrete source theory. For example, if the microwave background comes from discrete sources at an average distance of order H_0^{-1}, then we should expect gross anisotropies in the microwave background at an angular scale θ such that the volume $H_0^{-3}\theta^2$ contains about one source, that is, for

$$H_0^{-3}\theta^2 d^{-3} \approx 1 \tag{15.5.40}$$

where d is the mean separation of the sources. The limit $\theta \lesssim 1$ sec given in Table 15.2 thus sets an upper limit $d \lesssim 1$ Mpc, about the same as the mean separation of galaxies. Detailed analyses[77] of the data in specific models show a density even greater than that of galaxies, which would seem to rule out such theories.

The most interesting effects of the cosmic radiation background occur at early times, when the temperature was much greater than at present. These effects will be the subject of the next six sections. However, even at present, the radiation background can have some interesting effects:

(A) A relativistic electron of energy $\gamma_e m_e$ will undergo inverse Compton scattering on the microwave photons, producing recoil photons of average energy[78]

$$\bar{E} = 3.6\gamma_e^2 kT_{\gamma 0} = 8.4 \times 10^{-4}\gamma_e^2 \text{ eV}\left(\frac{T_{\gamma 0}}{2.7°\text{K}}\right) \tag{15.5.41}$$

Hoyle[79] proposed that inverse Compton scattering of cosmic ray electrons within our galaxy is responsible for the diffuse background[13] of cosmic X-rays, but it was pointed out by Gould[80] that the intensity from this mechanism is several hundred times smaller than the observed X-ray background. Soon after, Felton[81] showed that the inverse Compton scattering of cosmic ray electrons in intergalactic space could produce X-rays of the observed intensity. This model has received support from the remark of Brecher and Morrison,[82] that an observed kink in the cosmic ray electron spectrum at $\gamma_e \approx 7 \times 10^3$ would, according to (15.5.41), produce a kink in the spectrum of the diffuse X-ray background at about 40 KeV, just where a kink *is* observed.[13] However, the more recent calculations discussed in the last section indicate that this kink is due to thermal bremsstrahlung in hot intergalactic hydrogen, with inverse Compton scattering important only below 1 keV. The range of high-energy electrons in a 2.7°K background drops sharply for $\gamma_e \gtrsim 10^4$, so if cosmic ray electrons really come to us across intergalactic space, the observed electron energy spectrum should be cut off sharply at energies above 10 GeV.

(B) It has been observed[83] that the very strong radio galaxy Centaurus A emits X-rays in a frequency range 1 to 10 KeV with a total power $L_x = (11 \pm 4) \times 10^{40}$ erg/sec. By using the theory of synchroton emission to account for the observed radio flux, it is estimated[84] that Centaurus A contains about 1.7×10^{59} ergs in cosmic ray electrons, typically with $\gamma_e \simeq 2.5 \times 10^3$. The inverse Compton scattering of these electrons on a 2.7°K radiation background would produce X-rays, at an average energy given by (15.5.41) as 5 keV, with a total power $L_x \simeq 5 \times 10^{40}$ erg/sec, in agreement with the observed value. The most important aspect of this result is that the predicted X-ray power is most sensitive to the background radiation flux at short wavelengths, so that if the early rocket[63] and balloon[64] observations really gave the correct temperature at these wavelengths, the X-ray power from Centaurus A would be more than an order of magnitude larger than observed. However, this interpretation of the Cen A X-ray source is still in doubt.

(C) When a particle of mass m and momentum p strikes a photon of energy w at an angle θ, the total energy in the center-of-mass system is

$$E_c{}^2 = (w + (p^2 + m^2)^{1/2})^2 - (p^2 + 2pw \cos \theta + w^2)$$
$$= 2w[(p^2 + m^2)^{1/2} - p \cos \theta] + m^2 \qquad (15.5.42)$$

In order for a nucleon to have a cross-section on a photon that is of first rather than second order in $\alpha = 1/137$, it is necessary for E_c to be greater than the threshold $m_N + m_\pi$ for the process $\gamma + N \rightarrow \pi + N$:

$$(p^2 + m^2)^{1/2} - p \cos \theta \geq \frac{m_\pi^2 + 2m_N m_\pi}{2w} \simeq \frac{m_N m_\pi}{w}$$

Thus we expect a sharp cutoff[85] in the cosmic ray proton energy spectrum at

$$E_{p,\,\text{max}} \simeq \frac{m_N m_\pi}{k T_{\gamma 0}} \simeq 3 \times 10^{20} \text{ eV}$$

which is just about at the upper limit of current cosmic ray observations. Similarly, for cosmic ray photons, the pair production process $\gamma + \gamma \to e^+ + e^-$ gives a sharp drop[86] in γ-ray range for $\langle E_c \rangle \geq 2m_e$, that is, at an energy

$$E_{e,\,max} \simeq \frac{2m_e{}^2}{kT_{\gamma 0}} \simeq 10^{15} \text{ eV}$$

These upper limits apply only if we assume that the high-energy cosmic ray photons and protons arise outside our own galaxy.

It is not yet certain that the observed microwave background really is black-body radiation left over from an earlier era. However, the case for this view is certainly good enough to warrant a thorough examination of its implications for the early universe. We now turn to a consideration of these consequences.

6   Thermal History of the Early Universe

The energy density of the present 2.7°K microwave background is

$$\rho_{\gamma 0} = a T_{\gamma 0}^4 = 3.97 \times 10^{-13} \text{ erg/cm}^3 = 4.40 \times 10^{-34} \text{ g/cm}^3$$

$$(15.6.1)$$

As already remarked in Section 15.2, this is considerably less than the present nucleonic rest-mass density, so that we presently are in a matter-dominated era, which has lasted throughout most of the history of the universe. This era was discussed in detail in Section 15.3.

We now turn our attention back to an earlier period, when radiation and relativistic particles were more important than ordinary matter. In order to avoid losing the thread of our story in the details of our calculations, it may help to outline first what is now commonly pictured to be the early history of the universe, and then go into the detailed calculations that support this picture. The outline of universal history is currently believed to be something as follows (see Figure 15.5):

(A) At very early times, when the temperature T was above 10^{12}°K, the universe contained a great variety of particles in thermal equilibrium, including photons, leptons, mesons, and nucleons and their antiparticles. The strong interactions among mesons and nucleons make this era very difficult to study; it will be discussed briefly in Section 15.11.

(B) At the time when $T \approx 10^{12}$°K, the universe contained photons, muons, antimuons, electrons, positrons, neutrinos, and antineutrinos. In addition, there was a very small nucleonic contamination, with neutrons and protons in equal numbers. All of these particles were in thermal equilibrium.

(C) As the temperature dropped below 10^{12}°K, the μ^+ and μ^- began to annihilate. After almost all muons were gone, at $T \simeq 1.3 \times 10^{11}$°K, the neutrinos

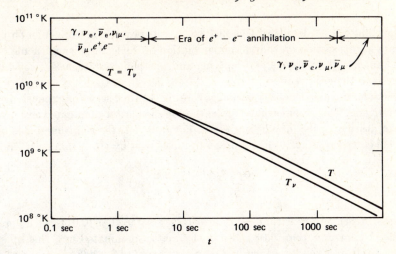

Figure 15.5 Thermal history of the early universe. Here T is the temperature of the $\gamma - e^+ - e^-$ plasma, and T_ν is the temperature of the decoupled ν_e, $\bar{\nu}_e$, ν_μ, and $\bar{\nu}_\mu$.

and antineutrinos decoupled from the other particles, leaving $e^\pm$, γ, and a few nucleons in thermal equilibrium, with $T \propto R^{-1}$. (The electron-type neutrinos may have remained in equilibrium with the other particles a little longer than the muon-type neutrinos, but this makes no difference.)

(D) As the temperature dropped below $10^{11}°$K ($t \simeq 0.01$ sec), the neutron-proton mass difference began to shift the small nucleonic contamination toward more protons and fewer neutrons.

(E) As the temperature dropped below $5 \times 10^9°$K ($t \simeq 4$ sec), the electron-positron pairs began to annihilate, leaving as the dominant constituents of the universe only photons, neutrinos, and antineutrinos in essentially free expansion, with the photon temperature 40.1% higher than the neutrino temperature. At the same time, the cooling of the neutrinos and antineutrinos, and the disappearance of the electrons and positrons, froze the neutron-proton ratio at about 1:5.

(F) At a temperature of about $10^9°$K ($t \simeq 180$ sec), the neutrons rapidly began to fuse with protons into heavier nuclei, leaving an ionized gas of hydrogen and He⁴, with about 27% helium by weight, and a trace of d, He³, and other elements.

(G) The free expansion of the photons, neutrinos, and antineutrinos continued, with $T_\gamma = 1.401 T_\nu \propto R^{-1}$. The ionized gas temperature remained locked to the photon temperature until the hydrogen recombined at $T \approx 4000°$K.

(H) At some temperature between $10^3°$K and $10^5°$K, the energy density of the photons, neutrinos, and antineutrinos dropped below the rest-mass density of hydrogen and helium, and we entered upon the matter-dominated era.

In filling in the details of this history, it will prove very convenient to concentrate in this section on the thermal evolution of the leading constituents of the early universe, the photons and leptons, and postpone our discussion of nucleosynthesis to the next section.

First, let us consider the equation governing the time scale for expansion of the early universe. This is somewhat simpler than in the matter-dominated era, because the curvature of space may be neglected. For $k = \pm 1$, the right-hand side of the Einstein equation (15.1.20) has a present value given by (15.2.5) and (15.2.6):

$$\frac{8\pi G \rho_0 R_0{}^2}{3} = \frac{2q_0}{|2q_0 - 1|}$$

We saw in Section 15.2 that q_0 is probably greater than 0.014, so at present $8\pi G \rho R^2/3$ is greater than 0.03. During the matter-dominated era, this quantity varies as $1/R \propto T$, so it was greater than 10 when T_γ was $1000°K$, and was even larger at earlier times. Hence, during the whole early history of the universe, k was much less than the right-hand side of Eq. (15.1.20), and this equation therefore simplifies to

$$\dot{R}^2 = \frac{8\pi G \rho R^2}{3} \qquad (15.6.2)$$

It will make no difference in our discussion of the early universe whether space is open or closed.

Now we must consider what were the contents of the early universe. At any given time, we can expect to find some particles in thermal equilibrium with each other, other particles in free expansion, and perhaps some particles that are just passing from one condition to the other. In the ideal-gas approximation, the number density $n_i(q) \, dq$ of particles of type i with momentum between q and $q + dq$ is given in thermal equilibrium by a Fermi or a Bose distribution[87]:

$$n_i(q) = 4\pi h^{-3} g_i q^2 \, dq \left[\exp\left(\frac{E_i(q) - \mu_i}{kT}\right) \pm 1 \right]^{-1} \qquad (15.6.3)$$

where $E_i(q) \equiv (m_i{}^2 + q^2)^{1/2}$ is the particle energy, μ_i is the chemical potential, the sign ± 1 is $+1$ for fermions and -1 for bosons, and g_i is the number of spin states, with $g = 1$ for neutrinos and antineutrinos and $g = 2$ for photons, electrons, muons, nucleons, and their antiparticles.

The chemical potentials must be determined from a consideration of the conservation laws obeyed by the various possible reactions. The basic rule is that μ_i is additively conserved in all reactions.[88] In particular:

(A) Photons can be emitted or absorbed in an arbitrary reaction in any number, so $\mu_\gamma = 0$. (Equation (15.6.3) then reduces to the Planck distribution (15.5.9), with $n_\gamma = \rho_\gamma/h\nu$ and $q = E = h\nu$.)

(B) Particle-antiparticle pairs can annihilate into photons, so the chemical potentials of a particle and its antiparticle are equal and opposite.

(C) Electrons and muons can be converted into their associated neutrinos ν_e and ν_μ by collisions with each other or with nucleons, in such reactions as

$$ e^- + \mu^+ \to \nu_e + \bar{\nu}_\mu \qquad e^- + p \to \nu_e + n \qquad \mu^- + p \to \nu_\mu + n, \qquad \text{etc.} $$

The chemical potentials are therefore related by

$$ \mu_{e^-} - \mu_{\nu_e} = \mu_{\mu^-} - \mu_{\nu_\mu} = \mu_n - \mu_p \qquad (15.6.4) $$

Altogether there are just four independent conserved intrinsic quantum numbers: charge, baryon number (nucleons and hyperons minus antinucleons and anti-hyperons), electron-lepton number (e^- and ν_e minus e^+ and $\bar{\nu}_e$), and muon-lepton number[89] (μ^- and ν_μ minus μ^+ and $\bar{\nu}_\mu$). Hence there are just four independent chemical potentials, which can be taken as $\mu_p, \mu_{e^-}, \mu_{\nu_e}, \mu_{\nu_\mu}$. These four independent chemical potentials are to be determined by the values for the charge density N_Q, the baryon number density N_B, the electron-lepton number density N_E, and the muon-lepton number density N_M, all of which simply vary as R^{-3}. The problem of determining the chemical potentials thus leads us to the question: What are the values of the four densities N_Q, N_B, N_E, and N_M?

We know that the average charge density N_Q is zero, or at least very small.[90] We also know that the baryon number density N_B is much less than the number density n_γ of photons, because at present $N_B \simeq n_p + n_n - n_{\bar{p}} - n_{\bar{n}}$ is 8 to 10 orders of magnitude less than n_γ, whereas at earlier times $N_B R^3$ was strictly constant and $n_\gamma R^3 \propto (T_\gamma R)^3$ was roughly constant. Unfortunately we know very little about the present number density of neutrinos, so we cannot estimate the value of $N_E = n_{e^-} + n_{\nu_e} - n_{e^+} - n_{\bar{\nu}_e}$ or $N_M = n_{\mu^-} + n_{\nu_\mu} - n_{\mu^+} - n_{\bar{\nu}_\mu}$. However, since N_B is 8 to 10 orders of magnitude less than n_γ, it is at least a reasonable guess that N_E and N_M are also much less than n_γ. If so, then it is a good approximation to set all the conserved quantum numbers equal to zero:

$$ N_Q = N_B = N_E = N_M = 0 \qquad (15.6.5) $$

Of course, N_B is not really zero, and we shall have to put the baryons back into the calculation in the next section when we consider the synthesis of the elements, but N_B can be ignored in calculating the gross thermal history of the early universe. The question of whether N_E and N_M can also be ignored will be taken up at the end of this section.

The problem of determining the chemical potentials is now very easy. The chemical potentials of particles and antiparticles are equal and opposite, so the four densities N_Q, N_B, N_E, and N_M are *odd* functions of the four independent chemical potentials $\mu_p, \mu_{e^-}, \mu_{\nu_e}, \mu_{\nu_\mu}$. Hence the values of the μ_i determined by (15.6.5) are simply

$$ \mu_i = 0 \qquad (15.6.6) $$

This approximation allows us to deal with energy conservation in a very convenient manner. The total energy density and pressure of all the particles in thermal equilibrium are now evidently just functions of the temperature alone:

$$\rho_{eq}(T) \equiv \sum_{i\,(eq)} \int E_i(q) n_i(q;\,T)\,dq \tag{15.6.7}$$

$$p_{eq}(T) \equiv \sum_{i\,(eq)} \int \left(\frac{q^2}{3E_i(q)}\right) n_i(q;\,T)\,dq \tag{15.6.8}$$

[see Eqs. (2.10.21) and (2.10.22).] According to the second law of thermodynamics, the entropy of the particles in equilibrium at temperature T within a volume V is a function $S(V,\,T)$ with

$$dS(V,\,T) = \frac{1}{T}\,\{d(\rho_{eq}(T)V) + p_{eq}(T)\,dV\} \tag{15.6.9}$$

so that

$$\frac{\partial S(V,\,T)}{\partial V} = \frac{1}{T}\,\{\rho_{eq}(T) + p_{eq}(T)\}$$

$$\frac{\partial S(V,\,T)}{\partial T} = \frac{V}{T}\,\frac{d\rho_{eq}(T)}{dT}$$

The energy density and pressure must then satisfy the integrability condition

$$\frac{\partial}{\partial T}\left[\frac{1}{T}\,\{\rho_{eq}(T) + p_{eq}(T)\}\right] = \frac{\partial}{\partial V}\left[\frac{V}{T}\,\frac{d\rho_{eq}(T)}{dT}\right]$$

or, after a little rearrangement,

$$\frac{dp_{eq}(T)}{dT} = \frac{1}{T}\,\{\rho_{eq}(T) + p_{eq}(T)\} \tag{15.6.10}$$

[This may also be derived directly from Eqs. (15.6.7) and (15.6.8).] As long as the particles in thermal equilibrium interact only with each other, their total energy and pressure must separately satisfy the energy conservation equation (14.2.19):

$$R^3 \frac{dp_{eq}}{dt} = \frac{d}{dt}\,[R^3\{\rho_{eq} + p_{eq}\}] \tag{15.6.11}$$

Using (15.6.10), this may now be written

$$\frac{d}{dt}\left[\frac{R^3}{T}\,\{\rho_{eq}(T) + p_{eq}(T)\}\right] = 0 \tag{15.6.12}$$

This conservation law has a simple interpretation in terms of the entropy. Using (15.6.10) in (15.6.9) gives

$$dS(V,\,T) = \frac{1}{T}\,d[\{\rho_{eq}(T) + p_{eq}(T)\}V] - \frac{V}{T^2}\,\{\rho_{eq}(T) + p_{eq}(T)\}\,dT$$

so, except for a possible additive constant,

$$S(V, T) = \frac{V}{T} \{\rho_{\text{eq}}(T) + p_{\text{eq}}(T)\} \tag{15.6.13}$$

The result (15.6.12) thus simply states the constancy of the entropy in a volume $R^3(t)$:

$$s \equiv S(R^3, T) = \frac{R^3}{T} \{\rho_{\text{eq}}(T) + p_{\text{eq}}(T)\} \tag{15.6.14}$$

In particular, when all the particles in equilibrium are highly relativistic, we can set $E = q$ in (15.6.7) and (15.6.8), so that

$$p_{\text{eq}}(T) = \tfrac{1}{3}\rho_{\text{eq}}(T) \tag{15.6.15}$$

Then (15.6.10) gives

$$\rho_{\text{eq}}(T) \propto T^4 \tag{15.6.16}$$

with a "constant" of proportionality that depends on just which particle types are abundant in equilibrium at these temperatures. [This result can also be obtained directly from (15.6.7) and (15.6.8).] Using (15.6.15) and (15.6.16) in (15.6.12) then gives a temperature decrease

$$T \propto \frac{1}{R} \tag{15.6.17}$$

We shall see that this holds through most, but not all, of the early history of the universe.

Our next task is to decide which particles were in thermal equilibrium at various times. One of the simplifications brought about by our neglect of the chemical potentials is that the only particles that can be present in thermal equilibrium with appreciable number densities (15.6.3) are those with mass $m < kT$. For $kT < m_\pi$, or $T < 1.5 \times 10^{12}{}^\circ$K, these are the $\mu^\pm$, $e^\pm$, ν_μ, $\bar{\nu}_\mu$, ν_e, $\bar{\nu}_e$, and γ. (Gravitons are ignored here, for reasons discussed in Section 15.11.) Throughout the early history of the universe, the processes of pair production and annihilation and Compton scattering kept any extant charged particles in thermal equilibrium with the photons. Hence the photons were described by the Planck law (15.5.9), and the $e^\pm$ and $\mu^\pm$ were described by a Fermi distribution with zero chemical potential:

$$n_{e^-}(q)\, dq = n_{e^+}(q)\, dq = 8\pi h^{-3} q^2\, dq \left[\exp\!\left(\frac{\sqrt{q^2 + m_e{}^2}}{kT}\right) + 1 \right]^{-1} \tag{15.6.18}$$

$$n_{\mu^-}(q)\, dq = n_{\mu^+}(q)\, dq = 8\pi h^{-3} q^2\, dq \left[\exp\!\left(\frac{\sqrt{q^2 + m_\mu{}^2}}{kT}\right) + 1 \right]^{-1} \tag{15.6.19}$$

What about the neutrinos and antineutrinos? We know that they can be produced, destroyed, and scattered in reactions such as

$$
\begin{array}{cc}
e^- + \mu^+ \leftrightarrow \nu_e + \bar{\nu}_\mu & e^+ + \mu^- \leftrightarrow \bar{\nu}_e + \nu_\mu \\
\nu_e + \mu^- \leftrightarrow \nu_\mu + e^- & \bar{\nu}_e + \mu^+ \leftrightarrow \bar{\nu}_\mu + e^+ \\
\nu_\mu + \mu^+ \leftrightarrow \nu_e + e^+ & \bar{\nu}_\mu + \mu^- \leftrightarrow \bar{\nu}_e + e^-
\end{array}
\tag{15.6.20}
$$

As long as $kT < m_\mu$, the cross-sections for all of these reactions will be roughly of order

$$
\sigma_{wk} \approx g_{wk}^2 \hbar^{-4} (kT)^2
\tag{15.6.21}
$$

where $g_{wk} = 1.4 \times 10^{-49}$ erg–cm^3 is the weak coupling constant, known from the observed rate for the muon decay process $\mu^+ \rightarrow e^+ + \nu_e + \bar{\nu}_\mu$. At these temperatures, all particle velocities are of order unity, and (15.6.18) and (15.6.19) give the densities of the charged leptons $e^\pm$ and $\mu^\pm$ as

$$
n_l \approx \left(\frac{kT}{\hbar} \right)^3
\tag{15.6.22}
$$

Hence the rate at which a single neutrino is scattered, and the rate of neutrino production per charged lepton, are both of order

$$
\sigma_{wk} n_l \approx g_{wk}^2 \hbar^{-7} (kT)^5
\tag{15.6.23}
$$

The total energy density is roughly of order

$$
\rho \approx kT \left(\frac{kT}{\hbar} \right)^3
\tag{15.6.24}
$$

so according to (15.6.2), the expansion rate is of order

$$
H \equiv \frac{\dot{R}}{R} \approx (G\rho)^{1/2} \approx G^{1/2} \hbar^{-3/2} (kT)^2
\tag{15.6.25}
$$

Hence, as long as $kT > m_\mu$, or $T > 10^{12}°$K, the ratio of the reaction rate σn_l to the expansion rate H is (now using c.g.s. units)

$$
\frac{\sigma n_l}{H} \approx G^{-1/2} \hbar^{-11/2} c^{-7/2} g_{wk}^2 (kT)^3 \approx \left(\frac{T}{10^{10}°\text{K}} \right)^3
\tag{15.6.26}
$$

However, all of the reactions (15.6.20) require either the presence of a μ^- or μ^+, or enough energy to make a μ^- or μ^+. When $kT < m_\mu$, the number density of muons, and the number densities of other particles with energy $E > m_\mu$, are reduced by factors of order $\exp(-m_\mu/kT)$, and in consequence the ratio of the reaction rate to the expansion rate is of order

$$
\frac{\sigma n_l}{H} \approx \left(\frac{T}{10^{10}°\text{K}} \right)^3 \exp\left(-\frac{10^{12}°\text{K}}{T} \right)
\tag{15.6.27}
$$

The neutrinos and antineutrinos drop out of thermal equilibrium with the other particles when this ratio falls below unity, that is, at about $T \approx 1.3 \times 10^{11} {}^\circ\text{K}$.

Actually, it may be that the ν_e and $\bar{\nu}_e$ remain in thermal equilibrium a little longer than the ν_μ and $\bar{\nu}_\mu$. According to present theories,[91] the weak interactions arise from the coupling of a "weak current" to itself, either directly, or through the agency of a charged spin-1 particle, the "intermediate vector meson." If this is so, then there are additional reactions involving ν_e and $\bar{\nu}_e$,

$$e^- + e^+ \leftrightarrow \nu_e + \bar{\nu}_e \qquad e^\pm + \nu_e \rightarrow e^\pm + \nu_e \qquad e^\pm + \bar{\nu}_e \rightarrow e^\pm + \bar{\nu}_e$$

$$(15.6.28)$$

whose cross-sections are of order (15.6.21) for $kT > m_e$. These reactions do not involve $\mu^\pm$, so the ratio of the ν_e and $\bar{\nu}_e$ reaction rates to the expansion rate H would be given by Eq. (15.6.26) for $kT > m_e$, that is, down to a temperature $T \simeq 5 \times 10^9 {}^\circ\text{K}$. The reactions (15.6.28) could then keep ν_e and $\bar{\nu}_e$ in thermal equilibrium with γ and $e^\pm$ down to a temperature $T \simeq 10^{10} {}^\circ\text{K}$, where the ratio (15.6.26) drops to unity. The same may even be true[173] for ν_μ and $\bar{\nu}_\mu$.

We are now in a position to work out the thermal history of the early universe. Let us start at a temperature between $10^{12} {}^\circ\text{K}$ and $1.3 \times 10^{11} {}^\circ\text{K}$, when the μ^+ and μ^- were rare enough so that their contribution to ρ_{eq} and p_{eq} could be neglected, and yet abundant enough to keep the neutrinos and antineutrinos in thermal equilibrium with the other particles. The important constituents of the universe then were $e^\pm$, γ, ν_e, $\bar{\nu}_e$, ν_μ, and $\bar{\nu}_\mu$, all in thermal equilibrium. The photons had a Planck distribution, the $e^\pm$ had the Fermi distribution (15.6.18), and the neutrinos and antineutrinos had the Fermi distribution

$$n_{\nu_e}(q)\,dq = n_{\bar{\nu}_e}(q)\,dq = n_{\nu_\mu}(q)\,dq = n_{\bar{\nu}_\mu}(q)\,dq$$

$$= 4\pi h^{-3} q^2\,dq \left[\exp\left(\frac{q}{kT}\right) + 1 \right]^{-1} \qquad (15.6.29)$$

Since all these particles were highly relativistic, the temperature was falling in obedience to Eq. (15.6.17), that is, $T \propto R^{-1}$. When T dropped to about $1.3 \times 10^{11} {}^\circ\text{K}$, the ν_μ and $\bar{\nu}_\mu$, and possibly also the ν_e and $\bar{\nu}_e$, decoupled from the particles in equilibrium and began a free expansion. *However, this decoupling had no effect on any of the distribution functions.* The particles remaining in equilibrium still constituted a highly relativistic gas, so their temperature continued to drop like $1/R$. In addition, the number density of the free neutrinos and antineutrinos fell like $1/R^3$ and their momenta were red-shifted by a factor $1/R$ (just as for photons), so that the form of the distribution (15.6.29) was preserved, with a neutrino temperature T_ν proportional to $1/R$. Since T_ν equaled T before decoupling, and T_ν and T both decreased like $1/R$ thereafter, the neutrinos and antineutrinos continued to be described by the Fermi distribution (15.6.29) with $T_\nu = T$, just as if they had remained in thermal equilibrium with the other particles. There may have been a second decoupling, of ν_e and $\bar{\nu}_e$ at $T \simeq 10^{10} {}^\circ\text{K}$, but again this made no difference to the neutrino and antineutrino distribution function, provided that the ν_e and $\bar{\nu}_e$ mostly decoupled while the $e^{\pm}$ were still relativistic. Thus, during the whole of the era $10^{12°}$K $> T > 5 \times 10^{9°}$K, the neutrinos and antineutrinos behaved as if they were in thermal equilibrium, and all particles, γ, $e^{\pm}$, ν_{μ}, $\bar{\nu}_{\mu}$, ν_e, and $\bar{\nu}_e$, were described by Planck or Fermi distributions with the same temperature T, falling like $1/R$. The energy densities of the neutrinos and antineutrinos were thus

$$\rho_{\nu_e} = \rho_{\bar{\nu}_e} = \rho_{\nu_{\mu}} = \rho_{\bar{\nu}_{\mu}} \equiv \rho_{\nu} \tag{15.6.30}$$

where

$$\rho_{\nu} = 4\pi h^{-3} \int_0^{\infty} q^3 \, dq \left[\exp\left(\frac{q}{kT}\right) + 1 \right]^{-1}$$

$$= \frac{7\pi^5}{30h^3} (kT)^4 = \tfrac{7}{16} a T^4 \tag{15.6.31}$$

Also, for $kT > m_e$ the $e^{\pm}$ were relativistic, so

$$\rho_{e^-} = \rho_{e^+} = 2\rho_{\nu} = \tfrac{7}{8} a T^4 \tag{15.6.32}$$

(The densities $\rho_{e^{\pm}}$ are twice ρ_{ν}, because the e^- and e^+ each have two spin states.) The total energy density of the universe during the era from $T < 10^{12°}$K to $T \simeq 10^{10°}$K was thus

$$\rho = \rho_{\nu_e} + \rho_{\bar{\nu}_e} + \rho_{\nu_{\mu}} + \rho_{\bar{\nu}_{\mu}} + \rho_{e^-} + \rho_{e^+} + \rho_{\gamma} = \tfrac{9}{2} a T^4 \tag{15.6.33}$$

The story now becomes a little more complicated. Below $10^{10°}$K, the only important particles left in thermal equilibrium were the $e^{\pm}$ and γ. Their entropy per volume R^3 is given by (15.6.14), (15.6.7), (15.6.8), and (15.6.18):

$$s = \frac{R^3}{T} \{ \rho_{e^-} + \rho_{e^+} + \rho_{\gamma} + p_{e^-} + p_{e^+} + p_{\gamma} \} \tag{15.6.34}$$

For $T > 5 \times 10^{9°}$K, the electrons and positrons were relativistic, so (15.6.15) and (15.6.32) apply, and (15.6.34) gives

$$s = \frac{4R^3}{3T} \{ \rho_{e^-} + \rho_{e^+} + \rho_{\gamma} \} = \tfrac{11}{3} a(RT)^3 \tag{15.6.35}$$

As T dropped below $5 \times 10^{9°}$K, the e^+ and e^- annihilated, eventually leaving just photons, with

$$s = \frac{4R^3}{3T} \rho_{\gamma} = \tfrac{4}{3} a(RT)^3 \tag{15.6.36}$$

But s is constant, so the effect of the disappearance of e^- and e^+ was to increase RT by a factor[92]

$$\frac{(RT)_{T < 10^{9°}\text{K}}}{(RT)_{T > 5 \times 10^{9°}\text{K}}} = \left(\frac{11}{4}\right)^{1/3} \tag{15.6.37}$$

The neutrinos and antineutrinos did not get heated by the electron-positron annihilation, so their temperature just continued to fall like R^{-1}. Hence, for $T < 5 \times 10^9 {}^\circ \text{K}$, we have to distinguish between the temperature T_ν of the neutrinos and antineutrinos, and the temperature T of the photons plus any remaining charged particles. Since RT_ν is constant and RT jumped by the factor $(11/4)^{1/3}$, the photon temperature was eventually greater than the neutrino temperature by just this factor:

$$\left(\frac{T}{T_\nu}\right)_{T < 10^9 {}^\circ \text{K}} = \left(\frac{11}{4}\right)^{1/3} = 1.401 \qquad (15.6.38)$$

In order to determine the behavior of RT or T/T_ν between $5 \times 10^9 {}^\circ \text{K}$ and $10^9 {}^\circ \text{K}$, we have to use the expression (15.6.34), or

$$s = \tfrac{4}{3} a (RT)^3 \mathscr{S}\left(\frac{m_e}{kT}\right) \qquad (15.6.39)$$

where

$$\mathscr{S}(x) \equiv 1 + \frac{45}{2\pi^4} \int_0^\infty y^2 \, dy \left[\sqrt{x^2 + y^2} + \frac{y^2}{3\sqrt{x^2 + y^2}} \right]$$
$$\times \left[\exp\left(\sqrt{x^2 + y^2}\right) + 1 \right]^{-1} \qquad (15.6.40)$$

The constant s can be expressed in terms of the constant RT_ν by replacing T with T_ν in (15.6.35), so that

$$T_\nu = \left(\frac{4}{11}\right)^{1/3} T \left[\mathscr{S}\left(\frac{m_e}{kT}\right) \right]^{1/3} \qquad (15.6.41)$$

A numerical calculation[93] of the function $\mathscr{S}$ shows that T/T_ν had risen only to 1.001 by the time the temperature dropped to $3 \times 10^9 {}^\circ \text{K}$, and T/T_ν did not reach 1.4 until T fell below $10^9 {}^\circ \text{K}$. (See Table 15.4.)

For $T < 10^9 {}^\circ \text{K}$, the only particles in thermal equilibrium with the photons were the small number of nucleons and electrons left over after all the $e^- e^+$ pairs annihilated. Both T_ν and T continued to fall like $1/R$, with a ratio fixed at the value (15.6.37). We saw in the last section that the photon temperature T_γ began to differ from the matter temperature T after T dropped below $4000 {}^\circ \text{K}$, but the photon temperature continued thereafter to drop like $1/R$. Thus there should now be a cosmic "black-body" neutrino and antineutrino background described by Eq. (15.6.29), with temperature

$$T_{\nu 0} = (\tfrac{4}{11})^{1/3} T_{\gamma 0} = 1.9 {}^\circ \text{K}$$

From the time when $T \simeq 10^9 {}^\circ \text{K}$ until the present, the energy density of the photons, neutrinos, and antineutrinos has been

$$\rho_R \equiv \rho_\gamma + \rho_{\nu_e} + \rho_{\bar{\nu}_e} + \rho_{\nu_\mu} + \rho_{\bar{\nu}_\mu}$$
$$= aT_\gamma^4 + \tfrac{7}{4} aT_\nu^4$$
$$= [1 + \tfrac{7}{4}(\tfrac{4}{11})^{4/3}] aT_\gamma^4 = 1.45 aT_\gamma^4 \qquad (15.6.42)$$

This may be compared with the energy density $m_N n_N$ of nonrelativistic matter, which scales as R^{-3}, or $T_\gamma{}^3$:

$$n_N = n_{NO} \left(\frac{T_\gamma}{T_{\gamma 0}} \right)^3$$

Hence the critical temperature T_c, at which $m_N n_N$ equaled ρ_R, is

$$T_c = \frac{m_N n_{NO}}{1.45 a T_{\gamma 0}^3} = 4200°\text{K} \left[\frac{m_N n_{NO}}{10^{-30} \text{g/cm}^3} \right] \tag{15.6.43}$$

For $m_N n_{NO}$ in the range $2 \times 10^{-29} \text{g/cm}^3$ to $3 \times 10^{-31} \text{g/cm}^3$, this temperature lies in the range 84,000°K to 1200°K. It may be noted that the temperature $T_R \simeq 4000°\text{K}$, at which ionized hydrogen recombined, lies within this range, so we are not certain whether the energy density of radiation was greater or less than that of matter at the time when matter and radiation lost thermal contact. This uncertainty did not affect our discussion of the microwave background in the last section; the important point there was that the *number* density of photons is and was much greater than that of baryons.

How long does all this take? During the era when the temperature was between about $10^{12}°\text{K}$ and $5 \times 10^{9}°\text{K}$, and also after it dropped below about $10^{9}°\text{K}$, the only particles present in large numbers were all highly relativistic, so that $p \simeq \rho/3$. According to (15.1.23), the energy density ρ varied as

$$\rho \propto R^{-4}$$

During these periods, the dynamical equation (15.6.2) may be written

$$\frac{\dot\rho}{\rho} = -\frac{4\dot R}{R} = -4 \left(\frac{8\pi G\rho}{3} \right)^{1/2}$$

The solution is

$$t = \left(\frac{3}{32\pi G\rho} \right)^{1/2} + \text{constant} \tag{15.6.44}$$

During the period when $10^{12}°\text{K} > T > 5 \times 10^{9}°\text{K}$, the energy density was given by (15.6.33), so that (using c.g.s. units)

$$t = \left(\frac{c^2}{48\pi GaT^4} \right)^{1/2} + \text{constant}$$

$$= 1.09 \text{ sec} \left[\frac{T}{10^{10}°\text{K}} \right]^{-2} + \text{constant}$$

Starting at $T = 10^{12}°\text{K}$, it took 0.0107 sec for the temperature to relax to $10^{11}°\text{K}$, and another 1.07 sec for the temperature to drop to $10^{10}°\text{K}$.

During the period when $10^9{}^\circ\text{K} > T > T_c$, the energy density was given by (15.6.42), so that

$$t = \left(\frac{c^2}{15.5\pi G a T_\gamma{}^4}\right)^{1/2} + \text{constant}$$

$$= 1.92 \sec\left[\frac{T_\gamma}{10^{10}{}^\circ\text{K}}\right]^{-2} + \text{constant}$$

The time required for the temperature to drop from $10^9{}^\circ\text{K}$ to $10^8{}^\circ\text{K}$ was thus about 5.3 hr. If radiation continued to dominate over matter until the hydrogen recombined at $T = 4000{}^\circ\text{K}$, then the age of the universe at the time of recombination was 4×10^5 years.

Unfortunately, if we want to describe the behaviour of $T(t)$ and $R(t)$ throughout the whole early history of the universe, we have to do a numerical calculation to get through the era of electron-positron annihilation. In order to express R in terms of T, we use the fact that (15.6.39) was constant from $T < 10^{12}{}^\circ\text{K}$ until the present, provided that after T dropped to $4000{}^\circ\text{K}$, we replace T with T_γ. Thus

$$s = \tfrac{4}{3}a(R_0 T_{\gamma 0})^3 \tag{15.6.45}$$

and (15.6.39) can therefore be written

$$\frac{R}{R_0} = \left(\frac{T}{T_{\gamma 0}}\right)^{-1} \mathscr{S}^{-1/3}\left(\frac{m_e}{kT}\right) \tag{15.6.46}$$

The energy density ρ is a function of T, which, for T less than $10^{12}{}^\circ\text{K}$ and greater than both T_c and $4000{}^\circ\text{K}$, may be written

$$\rho = \rho_\gamma + \rho_{\nu_e} + \rho_{\bar\nu_e} + \rho_{\nu_\mu} + \rho_{\bar\nu_\mu} + \rho_{e^+} + \rho_{e^-}$$

$$= aT^4 + \tfrac{7}{4}aT_\nu{}^4 + 16\pi h^{-3}\int_0^\infty E_e(q)q^2\,dq\left[\exp\left(\frac{E_e(q)}{kT}\right) + 1\right]^{-1}$$

Using (15.6.41) for T_ν, this is

$$\rho = aT^4 \mathscr{E}\left(\frac{m_e}{kT}\right) \tag{15.6.47}$$

where

$$\mathscr{E}(x) = 1 + \tfrac{7}{4}(\tfrac{4}{11})^{4/3}\mathscr{S}^{4/3}(x)$$

$$+ \frac{30}{\pi^4}\int_0^\infty \sqrt{x^2 + y^2}\,y^2\,dy\,[\exp(\sqrt{x^2 + y^2}) + 1]^{-1} \tag{15.6.48}$$

Equations (15.6.46) and (15.6.47) can be used in the dynamical equation (15.6.2),

$$dt = \left(\frac{8\pi\rho G}{3}\right)^{-1/2}\frac{dR}{R}$$

and we find a formula for the time as a function of the temperature:

$$t = -\int \left(\tfrac{8}{3}\pi G a T^4 \mathscr{E}\left(\frac{m_e}{kT}\right) \right)^{-1/2} \left(\frac{dT}{T} + \frac{d\mathscr{S}(m_e/kT)}{3\mathscr{S}(m_e/kT)} \right) \qquad (15.6.49)$$

Results[93] for t, R/R_0, and T/T_ν as functions of T are given in Table 15.4.

The only really arbitrary assumption so far has been the conjecture that the lepton number densities N_E and N_M are zero, or at least much less than n_γ. Let us now consider what would be the effect of giving up this assumption. Once the temperature had dropped below $10^{12\circ}$K, the only abundant charged particles were the electrons and positrons, so charge neutrality required that $N_Q = n_{e^+} - n_{e^-}$ vanished. The chemical potential of the electron must then have vanished, so the only particles that might then have had nonvanishing chemical potentials were

Table 15.4 Thermal History of the Universe, from the Annihilation of $\mu^+\mu^-$ Pairs Until the Decoupling of Matter and Radiation[a]

T (°K)	R/R_0	T/T_ν	t (sec)
10^{12}	1.9×10^{-12}	1.000	0
6×10^{11}	3.2×10^{-12}	1.000	1.94×10^{-4}
3×10^{11}	6.4×10^{-12}	1.000	1.129×10^{-3}
2×10^{11}	9.6×10^{-12}	1.000	2.61×10^{-3}
10^{11}	1.9×10^{-11}	1.000	1.078×10^{-2}
6×10^{10}	3.2×10^{-11}	1.000	3.01×10^{-2}
3×10^{10}	6.4×10^{-11}	1.001	0.1209
2×10^{10}	9.6×10^{-11}	1.002	0.273
10^{10}	1.9×10^{-10}	1.008	1.103
6×10^9	3.1×10^{-10}	1.022	3.14
3×10^9	5.9×10^{-10}	1.081	13.83
2×10^9	8.3×10^{-10}	1.159	35.2
10^9	2.6×10^{-9}	1.346	1.82×10^2
3×10^8	9.0×10^{-9}	1.401	2.08×10^3
10^8	2.7×10^{-8}	1.401	1.92×10^4
10^7	2.7×10^{-7}	1.401	1.92×10^6
10^6	2.7×10^{-6}	1.401	1.92×10^8
10^5	2.7×10^{-5}	1.401	1.92×10^{10}
10^4	2.7×10^{-4}	1.401	1.92×10^{12}
4×10^3	6.3×10^{-4}	1.401	1.20×10^{13}

[a] The values for R/R_0 are derived assuming a present radiation temperature $T_{\gamma 0} = 2.7°$K. The last few values for t are derived under the assumption that the energy density of matter was still negligible compared with that of photons and neutrinos. Values of T/T_ν and t for $T > 10^{8\circ}$K are taken from P. J. E. Peebles, Ap. J., **146**, 542 (1966).

the neutrinos and antineutrinos. For $T > 1.3 \times 10^{11}$°K, these particles were in equilibrium with γ, e^+, and e^-, so they were described by the Fermi distributions

$$n_{v_e}(q) \, dq = 4\pi h^{-3} q^2 \, dq \left[\exp\left(\frac{q - \mu_{v_e}}{kT} \right) + 1 \right]^{-1} \tag{15.6.50}$$

$$n_{\bar{v}_e}(q) \, dq = 4\pi h^{-3} q^2 \, dq \left[\exp\left(\frac{q + \mu_{v_e}}{kT} \right) + 1 \right]^{-1} \tag{15.6.51}$$

and likewise for v_μ and $\bar{v}_\mu$. The lepton number densities were then

$$N_E = \int [n_{v_e}(q) - n_{\bar{v}_e}(q)] \, dq = 4\pi \left(\frac{kT}{h} \right)^3 \mathcal{N}\left(\frac{\mu_{v_e}}{kT} \right) \tag{15.6.52}$$

$$N_M = \int [n_{v_\mu}(q) - n_{\bar{v}_\mu}(q)] \, dq = 4\pi \left(\frac{kT}{h} \right)^3 \mathcal{N}\left(\frac{\mu_{v_\mu}}{kT} \right) \tag{15.6.53}$$

where

$$\mathcal{N}(x) \equiv \int_0^\infty \{ [e^{y-x} + 1]^{-1} - [e^{y+x} + 1]^{-1} \} y^2 \, dy \tag{15.6.54}$$

Since electron-lepton number and muon-lepton number are believed to be conserved,[89] the densities N_E and N_M must always vary as R^{-3}. However, we have seen that during the era when 10^{12}°K $> T > 5 \times 10^9$°K, T varies as $1/R$. Hence μ_{v_e}/kT and μ_{v_μ}/kT must have been constant, from the annihilation of the μ^+ and μ^- until the decoupling of the neutrinos and antineutrinos.

After decoupling, the neutrinos and antineutrinos have expanded freely, with number densities dropping as $1/R^3$ and momenta red-shifted by a factor $1/R$. This free expansion preserved the form of the distributions (15.6.50) and (15.6.51) but red-shifted the temperature and the chemical potentials by a factor $1/R$. Hence the neutrino distributions, during the whole period from $T < 10^{12}$°K until the present, are given by

$$n_{v_e}(q) \, dq = 4\pi h^{-3} q^2 \, dq \left[\exp\left(\frac{q - \mu_{v_e}}{kT_v} \right) + 1 \right]^{-1} \tag{15.6.55}$$

$$n_{\bar{v}_e}(q) \, dq = 4\pi h^{-3} q^2 \, dq \left[\exp\left(\frac{q + \mu_{v_e}}{kT_v} \right) + 1 \right]^{-1} \tag{15.6.56}$$

$$n_{v_\mu}(q) \, dq = 4\pi h^{-3} q^2 \, dq \left[\exp\left(\frac{q - \mu_{v_\mu}}{kT_v} \right) + 1 \right]^{-1} \tag{15.6.57}$$

$$n_{\bar{v}_\mu}(q) \, dq = 4\pi h^{-3} q^2 \, dq \left[\exp\left(\frac{q + \mu_{v_\mu}}{kT_v} \right) + 1 \right]^{-1} \tag{15.6.58}$$

where T_v, μ_{v_e}, and μ_{v_μ} all vary as $1/R$, with $T_v = T$ before the electrons and positrons annihilate. The $e^+ - e^-$ annihilation was not affected by the neutrino and antineutrino distributions, so all the previous results for T_ν and R as functions of T still apply.

If N_E and N_M are much less than the photon number density $n_\gamma \simeq (kT/h)^3$, then (15.6.52) and (15.6.53) give

$$|\mu_{\nu_e}| \ll kT_\nu \qquad |\mu_{\nu_\mu}| \ll kT_\nu \tag{15.6.59}$$

and the distributions (15.6.55)–(15.6.58) all reduce to the previously used distribution (15.6.29).

On the other hand, if N_E or N_M is comparable with or greater than n_γ, then the constants $|\mu_{\nu_e}/kT_\nu|$ or $|\mu_{\nu_\mu}/kT_\nu|$ will be of order unity or larger, and the distribution functions (15.6.55)–(15.6.58) will be appreciably different from (15.6.29). In the limit when, say, $\mu_{\nu_e}/kT_\nu \gg 1$, the distribution functions (15.6.55) and (15.6.56) become

$$n_{\nu_e}(q)\, dq \simeq \begin{cases} 4\pi h^{-3} q^2 \, dq & q < \mu_{\nu_e} \\ 0 & q > \mu_{\nu_e} \end{cases} \tag{15.6.60}$$

$$n_{\bar{\nu}_e}(q)\, dq \simeq 0 \tag{15.6.61}$$

This is the case of *complete neutrino degeneracy*. Of course, if $\mu_{\nu_e}/kT_\nu \ll -1$, then the role of the neutrinos and antineutrinos is reversed in (15.6.60) and (15.6.61), and we have complete antineutrino degeneracy. The possibility of complete neutrino degeneracy was suggested[94] several years before the discovery of the microwave background, when it seemed reasonable to suppose that the universe has always been cold enough so that $kT_\nu \ll |\mu_{\nu_e}|$.

The only effect that partial or complete degeneracy would have on the calculations of this section is that it would shorten the time scale. The total energy density of neutrinos and antineutrinos is given by

$$\rho_{\nu+\bar{\nu}} = \int [n_{\nu_e}(q) + n_{\bar{\nu}_e}(q) + n_{\nu_\mu}(q) + n_{\bar{\nu}_\mu}(q)] q\, dq$$

$$= 4\pi h^{-3}(kT_\nu)^4 \left[\mathscr{F}\left(\frac{\mu_{\nu_e}}{kT_\nu}\right) + \mathscr{F}\left(\frac{\mu_{\nu_\mu}}{kT_\nu}\right) \right] \tag{15.6.62}$$

where

$$\mathscr{F}(x) \equiv \int_0^\infty \{[e^{y-x} + 1]^{-1} + [e^{y+x} + 1]^{-1}\} y^3 \, dy$$

This is always *greater* than the energy density $(7/4)aT_\nu^4$ for zero chemical potential [see (15.6.30) and (15.6.31)], so the expansion rate (15.6.2) is increased by degeneracy. In the limit where $|\mu_{\nu_e}/kT_\nu| \gg 1$ or $|\mu_{\nu_\mu}/kT_\nu| \gg 1$ or both, we have

$$\rho \simeq \rho_{\nu+\bar{\nu}} \simeq \pi h^{-3}[\mu_{\nu_e}{}^4 + \mu_{\nu_\mu}{}^4] \tag{15.6.63}$$

The degenerate neutrinos or antineutrinos then dominate the energy density and the expansion rate.

It is interesting to ask whether we could detect a cosmic background of neutrinos and antineutrinos. The most stringent upper limit on $|\mu_{\nu_e}|$ and $|\mu_{\nu_\mu}|$ comes from measurements of the deacceleration parameter q_0. Since q_0 is not much larger than unity, the total energy density cannot be much greater than about $10^{-29} \mathrm{g/cm^3}$ (see Section 15.2), and therefore, according to (15.6.63),

$$[\mu_{\nu_e 0}^4 + \mu_{\nu_\mu 0}^4]^{1/4} \lesssim 0.0075 \, \mathrm{eV} \tag{15.6.64}$$

As we have seen, the present neutrino temperature $T_{\nu 0}$ is about 1.9°K, so $kT_{\nu 0} = 1.7 \times 10^{-4} \mathrm{eV}$. Thus the upper limit on the chemical potential may be written

$$\frac{|\mu_{\nu_e}|}{kT_\nu} \lesssim 45 \qquad \frac{|\mu_{\nu_\mu}|}{kT_\nu} \lesssim 45 \tag{15.6.65}$$

Measurements of q_0 therefore do not rule out nearly complete degeneracy.

We can also try to measure the chemical potentials directly. In allowed β^- decays, such as $\mathrm{H}^3 \rightarrow \mathrm{He}^3 + e^- + \bar{\nu}_e$, we normally expect the number of events for an electron energy between E_e and $E_e + dE_e$ to be given by the Fermi function

$$N_F(E_e) \, dE_e = a p_e E_e (W_0 - E_e)^2 F(E_e) \, dE_e$$

where a is a constant, p_e is the electron momentum, W_0 is the maximum electron energy, and $F(E_e)$ is a known function that corrects for the Coulomb interaction in the final state. However, in the presence of an antineutrino background (15.6.56), the Pauli exclusion principle reduces the β^- decay rate by a factor equal to the fraction of antineutrino states at energy $W_0 - E_e$ that are unfilled:

$$N(E_e) \, dE_e = \left[1 - \frac{h^3 n_{\bar{\nu}_e}(W_0 - E_e)}{4\pi(W_0 - E_e)^2} \right] N_F(E_e) \, dE_e$$

or, explicitly,[94]

$$N(E_e) \, dE_e = \left[1 + \exp\left(\frac{E_e - W_0 + \mu_{\nu_e 0}}{kT_{\nu 0}} \right) \right]^{-1} a p_e E_e (W_0 - E_e)^2 F(E_e) \, dE_e \tag{15.6.66}$$

Since W_0 is much greater than $|\mu_{\nu_e 0}|$ and $kT_{\nu 0}$ for all known beta decays, this correction has little effect over most of the electron spectrum. However, if $\mu_{\nu_e 0} < -kT_{\nu 0}$, the function $N(E_e)$ will show an anomalous depression over the range $W_0 > E_e \gtrsim W_0 - |\mu_{\nu_e 0}|$, very much as if the antineutrino had a mass $|\mu_{\nu_e 0}|$. If $\mu_{\nu_e 0} > 0$, there will not be much of a depression for E_e below W_0, but there will be events with $E_e > W_0$, caused by absorption of cosmic neutrinos in reactions such as $\nu_e + \mathrm{H}^3 \rightarrow e^- + \mathrm{He}^3$. The rate for these events is given[94] by the same formula (15.6.66) as for antineutrino emission, except that now $\mu_{\nu_e 0}$ is replaced with $-\mu_{\nu_e 0}$, and of course $E_e > W_0$. Thus, for $\mu_{\nu_e 0} > kT_{\nu 0}$, the β^- spectrum will rise beyond the endpoint W_0 up to an energy $W_0 + \mu_{\nu_e 0}$, giving the appearance of a violation of the conservation of energy.

By far the best data on the electron spectrum near the endpoint in β^- decay come from studies of the low-energy decay $H^3 \rightarrow He^3 + e^- + \bar{v}_e$, with endpoint $W_0 = 18.7$ keV. In a recent experiment,[95] there were found no anomalous depressions extending more than about 60 eV below the endpoint, and no anomalous events more than about 60 eV above the endpoint. We can conclude that

$$|\mu_{v_e 0}| \lesssim 60 \text{ eV} \tag{15.6.67}$$

for a chemical potential of either sign.

It is also possible to get indirect information about the cosmic neutrino and antineutrino background from the survival of cosmic ray protons. A neutrino or antineutrino with energy q that is struck at an angle θ by a relativistic proton of energy γm_p will appear *in the proton rest-frame* to have energy

$$E \simeq \gamma q(1 - \cos\theta) \qquad \text{for } \gamma \gg 1$$

The total cross-section for pv or $p\bar{v}$ reactions at a "laboratory" energy E is roughly

$$\sigma(E) \simeq AE^2$$

where, in c.g.s. units,

$$A \approx \frac{g_{wk}^2}{\hbar^4 c^4} \approx 10^{-56} \text{ cm}^2/\text{eV}^2$$

The reaction rate for a relativistic proton of energy γm_p in the degenerate v_e (or $\bar{v}_e$) background (15.6.60) is then

$$\Gamma = \int_0^{|\mu_{v_e 0}|} \sigma(\gamma q[1 - \cos\theta]) h^{-3} q^2 \, dq \, \sin\theta \, d\theta \, d\phi$$

or, in c.g.s. units,

$$\Gamma \simeq \frac{4\pi\gamma^2 A |\mu_{v_e 0}|^5}{15 h^3 c^2} \approx 3 \times 10^{-34} \gamma^2 |\mu_{v_e 0}(\text{eV})|^5 \text{ sec}^{-1} \tag{15.6.68}$$

and similarly for degenerate v_μ's or $\bar{v}_\mu$'s. Bernstein, Ruderman, and Feinberg[96] have remarked that, since the cosmic ray protons observed with $\gamma > 10^6$ have certainly been traveling for more than 10^6 sec, both $|\mu_{v_e 0}|$ and $|\mu_{v_\mu 0}|$ must be less than 10^3 eV. Cowsik, Pal, and Tandon[97] assume that protons with $\gamma \approx 10^9$ could not scatter more than about 14 times during a flight time of order 5×10^7 years, and conclude that $|\mu_{v_e 0}|$ and $|\mu_{v_\mu 0}|$ are both less than about 2 eV.

We can also look for kinks in the cosmic ray proton spectrum at the thresholds for various vp or $\bar{v}p$ reactions. For instance, the threshold for the reaction $p + \bar{v}_e \rightarrow n + e^+$ is at $m_e + m_n - m_p = 1.8$ MeV, so if $\mu_{v_e 0}$ is less than $-kT_{v0}$, there should be a downward kink in the cosmic ray proton spectrum at

$$\gamma \approx \frac{1.8 \text{ MeV}}{|\mu_{v_e 0}|} \tag{15.6.69}$$

Konstantinov, Kocharov, and Starbunov[98] note the existence of a kink at $\gamma \approx 2 \times 10^6$, and suggest that this may be due to a degenerate antineutrino background with

$$\mu_{v_e 0} \simeq -0.8 \text{ eV} \tag{15.6.70}$$

This estimate is very much larger in absolute value than the upper limit (15.6.64) allowed by measurements of q_0.

7 Helium Synthesis

The relative abundance of the chemical elements has been under careful study by geologists and astronomers since the pioneering work of Frank Wigglesworth Clarke[99] in the last century. Gradually these studies have revealed a "cosmic" distribution of abundances,[100] with hydrogen and then helium by far the most abundant elements, followed by the group C—N—O—Ne, and with the group Li—Be—B and all elements heavier than nickel scarce. The problem of explaining these abundances has long been considered one of the major challenges facing theoretical astrophysics.

One possible explanation lay in the nuclear reactions that provide energy to the stars. Rutherford's demonstration of nuclear transmutation in the laboratory led Eddington[101] in 1920 to suggest that the sun might derive its energy from the fusion of hydrogen into helium. If so, then perhaps the stars (or at least the first generation of stars) were formed from pure hydrogen, and have gradually produced helium and heavier elements as ashes of their internal fires. The detailed reactions by which the stars burn hydrogen to helium were laid out in 1939 by Hans Bethe,[102] and the subsequent reactions in which helium fuses into heavier elements were explored in the 1950's in a series of papers by Salpeter,[103] the Burbidges, Fowler, and Hoyle,[104] Cameron,[105] and others. Most recently, Clayton and Arnett[106] have emphasized the importance of stellar explosions as an agent in nucleosynthesis.

There is one other competing theory of nucleosynthesis, worked out in the late 1940's by George Gamow and his collaborators.[107] Gamow reasoned that, although the early hot dense period of cosmic expansion was much briefer than the lifetime of a star, there was a large number of free neutrons present at that time, so that the heavy elements could be built up quickly by successive neutron captures, starting with $n + p \rightarrow d + \gamma$. The abundances of the elements would then be correlated with their neutron capture cross-sections, in rough agreement with observation. We have already noted in Section 15.5 that the necessity of avoiding too much helium production in this theory required the presence of black-body radiation, with a present temperature that was estimated[52] as 5°K.

Both the stellar and the cosmological theories of nucleosynthesis have their limitations. There are no stable nuclei with atomic weights $A = 5$ or $A = 8$, so it is difficult to build up elements heavier than helium by p–α, n–α, or α–α collisions. In stars that have converted all hydrogen to helium at their cores, it is possible to bridge the gaps at $A = 5$ and $A = 8$ by the production of small amounts of the unstable nuclide Be^8 in α–α collisions, followed by the production of C^{12} in α-Be^8 collisions.[103] However, the density of the expanding universe at the temperature $t \approx 10^{9\circ}K$ is too low to allow much helium burning to occur. It is generally accepted today that all elements heavier than helium were synthesized in stars.

On the other hand, several authors[108] have noted that the cosmic abundance of helium is too large to be easily explained in terms of nucleosynthesis in stars. The luminosity-to-mass ratio L/M of our galaxy is about one-tenth the solar ratio $L_\odot/M_\odot$, or 0.2 erg/gm sec. If the luminosity of the galaxy has remained constant during the last 10^{10} years, then about 0.06 Mev per nucleon would have been produced. In contrast, the fusion of hydrogen into helium releases about 6 MeV per nucleon, so not more than about 1% of the nucleons in our galaxy could have been fused into helium (or heavier nuclei) by ordinary stellar processes. As we shall see, estimates of the present helium abundance vary, but there is wide agreement that the cosmic abundance of helium by mass is considerably greater than 1%. It is of course possible that the helium could have been synthesized in an earlier, more luminous, epoch of our galaxy; as already remarked in Section 15.5, the released energy could, if thermalized, account for the present 2.7°K microwave background. However, it is more interesting and more natural to assume that the large cosmic helium abundance was produced during the early history of the universe, with the energy of fusion mostly lost in the subsequent red shift.

Let us now calculate the cosmologically produced abundance of helium. It is very convenient to divide this calculation into two parts. First, we calculate the neutron-proton abundance ratio as a function of time, taking account only of the weak interaction processes

$$n + \nu \leftrightarrow p + e^- \qquad n + e^+ \leftrightarrow p + \bar{\nu} \qquad n \leftrightarrow p + e^- + \bar{\nu}$$

$$(15.7.1)$$

(Here ν will mean ν_e.) In the second part of this calculation, we put in the nuclear reactions that lead to helium synthesis.

The number densities of ν, $\bar{\nu}$, e^-, and e^- are given here by the Fermi distributions (15.6.3), with zero chemical potential and with different temperatures T or T_ν for $e^\pm$ (and γ) or ν and $\bar{\nu}$:

$$n_{e^-}(p)\,dp = n_{e^+}(p)\,dp = 8\pi h^{-3}p^2\,dp\left[\exp\left(\frac{E_e(p)}{kT}\right) + 1\right]^{-1}$$

$$n_\nu(p)\,dp = n_{\bar{\nu}}(p)\,dp = 4\pi h^{-3}p^2\,dp\left[\exp\left(\frac{E_\nu(p)}{kT_\nu}\right) + 1\right]^{-1}$$

where

$$E_e(p) = (p^2 + m_e{}^2)^{1/2} \qquad E_\nu(p) = p$$

The rates for the various reactions (15.7.1) are given by the "*V–A*" theory of weak interactions,[109] except that the Pauli exclusion principle supresses these rates by a factor equal to the fraction of all states that are unfilled:

$$1 - \left[\exp\left(\frac{E_e}{kT}\right) + 1\right]^{-1} = \left[1 + \exp\left(\frac{-E_e}{kT}\right)\right]^{-1}$$

$$1 - \left[\exp\left(\frac{E_\nu}{kT}\right) + 1\right]^{-1} = \left[1 + \exp\left(\frac{-E_\nu}{kT}\right)\right]^{-1}$$

The rates (per nucleon) of the processes (15.7.1) are then

$$\lambda(n + \nu \to p + e^-) = A \int v_e E_e^2 p_\nu^2 \, dp_\nu [e^{E_\nu/kT_\nu} + 1]^{-1} [1 + e^{-E_e/kT}]^{-1}$$

$$(15.7.2)$$

$$\lambda(n + e^+ \to p + \bar\nu) = A \int E_\nu^2 p_e^2 \, dp_e [e^{E_e/kT} + 1]^{-1} [1 + e^{-E_\nu/kT_\nu}]^{-1}$$

$$(15.7.3)$$

$$\lambda(n \to p + e^- + \bar\nu) = A \int v_e E_\nu^2 E_e^2 \, dp_\nu [1 + e^{-E_\nu/kT_\nu}]^{-1} [1 + e^{-E_e/kT}]^{-1}$$

$$(15.7.4)$$

$$\lambda(p + e^- \to n + \nu) = A \int E_\nu^2 p_e^2 \, dp_e [e^{E_e/kT} + 1]^{-1} [1 + e^{-E_\nu/kT_\nu}]^{-1}$$

$$(15.7.5)$$

$$\lambda(p + \bar\nu \to n + e^+) = A \int v_e E_e^2 p_\nu^2 \, dp_\nu [e^{E_\nu/kT_\nu} + 1]^{-1} [1 + e^{-E_e/kT}]^{-1}$$

$$(15.7.6)$$

$$\lambda(p + e^- + \bar\nu \to n) = A \int v_e E_e^2 p_\nu^2 \, dp_\nu [e^{E_e/kT} + 1]^{-1} [e^{E_\nu/kT_\nu} + 1]^{-1}$$

$$(15.7.7)$$

Here A is the constant

$$A = \frac{g_V^2 + 3g_A^2}{2\pi^3\hbar^7} \qquad (15.7.8)$$

where g_V and g_A are the vector and axial vector coupling constants of the nucleon, taken here to have the values

$$g_V = 1.418 \times 10^{-49} \text{ erg cm}^3 \qquad g_A = 1.18g_V \qquad (15.7.9)$$

Also, E_e and E_ν are related by

$$E_e - E_\nu = Q \qquad \text{for } n + \nu \leftrightarrow p + e^- \tag{15.7.10}$$

$$E_\nu - E_e = Q \qquad \text{for } n + e^+ \leftrightarrow p + \bar{\nu} \tag{15.7.11}$$

$$E_\nu + E_e = Q \qquad \text{for } n \leftrightarrow p + e^- + \bar{\nu} \tag{15.7.12}$$

where

$$Q \equiv m_n - m_p = 1.293 \text{ MeV} \tag{15.7.13}$$

The integrals (15.7.2)–(15.7.7) are taken over the *positive* values of p_ν and p_e allowed by these relations. It is very convenient to write all integrals in terms of a variable of integration q, taken as E_ν in Eqs. (15.7.2), (15.7.4), and (15.7.5), and as $-E_\nu$ in Eqs. (15.7.3), (15.7.6), and (15.7.7). Replacing $p_e^2 \, dp_e$ with $\nu_e E_e^2 \, dE_e$, the total $n \to p$ and $p \to n$ transition rates are then

$$\lambda(n \to p) \equiv \lambda(n + \nu \to p + e^-) + \lambda(n + e^+ \to p + \bar{\nu}) + \lambda(n \to p + e^- + \bar{\nu})$$

$$= A \int \left(1 - \frac{m_e^2}{(Q+q)^2}\right)^{1/2} (Q+q)^2 q^2 \, dq (1 + e^{q/kT_\nu})^{-1}$$

$$\times (1 + e^{-(Q+q)/kT})^{-1} \tag{15.7.14}$$

and

$$\lambda(p \to n) \equiv \lambda(p + e^- \to n + \nu) + \lambda(p + \bar{\nu} \to n + e^+) + \lambda(p + e^- + \bar{\nu} \to n)$$

$$= A \int \left(1 - \frac{m_e^2}{(Q+q)^2}\right)^{1/2} (Q+q)^2 q^2 \, dq (1 + e^{-q/kT_\nu})^{-1}$$

$$\times (1 + e^{(Q+q)/kT})^{-1} \tag{15.7.15}$$

The integrals now run from $-\infty$ to $+\infty$, leaving out a gap from $-Q - m_e$ to $-Q + m_e$. The differential equation for the ratio X_n of neutrons to all nucleons is

$$-\frac{dX_n}{dt} = \lambda(n \to p)X_n - \lambda(p \to n)(1 - X_n) \tag{15.7.16}$$

The solution of this equation has been calculated by Peebles,[93] and is presented here in Table 15.5. Although the quantitative behavior of $X_n(t)$ can only be obtained by a numerical integration, it is possible to appreciate the main features of this solution through a few qualitative observations:

(A) For $kT \gg Q$, we can set $T = T_\nu$, and put Q and m_e equal to zero in Eqs. (15.7.14) and (15.7.15). The transition rates are then

$$\lambda(n \to p) \simeq \lambda(p \to n) \simeq A \int_{-\infty}^{\infty} q^4 \, dq (1 + e^{-q/kT})^{-1}(1 + e^{q/kT})^{-1}$$

$$= \tfrac{7}{15}\pi^4 A(kT)^5 = 0.361 \text{ sec}^{-1} \left(\frac{T}{10^{10}\text{°K}}\right)^5 \tag{15.7.17}$$

Table 15.5   Neutron Fraction X_n as a Function of Temperature or Time, with Neglect of the Formation of Complex Nuclei[a]

T (deg)	t (sec)	X_n
10^{12}	0	0.496
3×10^{11}	0.001129	0.488
10^{11}	0.01078	0.462
3×10^{10}	0.1209	0.380
10^{10}	1.103	0.241
3×10^{9}	13.83	0.170
1.3×10^{9}	98*	0.150
1.2×10^{9}	119*	0.147
1.1×10^{9}	146*	0.143
1.0×10^{9}	182.0	0.137
9×10^{8}	226*	0.131
8×10^{8}	290*	0.123
7×10^{8}	383*	0.112
3×10^{8}	2080	0.021
10^{8}	18700	10^{-8}

[a] Values of t are taken from the calculations of P. J. E. Peebles, Astron. J., **146**, 542 (1966), except that values marked with an asterisk are interpolated from Peebles' results. Values of X_n for $T \geq 1.0 \times 10^{9}$°K are taken from Peebles, *op. cit.*, Table 4. Values of X_n for $T < 10^{9}$°K are calculated from the value at 10^{9}°K, under the assumption that X_n decreases exponentially at the rate $(1013 \text{ sec})^{-1}$ of free neutron decay.

This may be compared with the "age" t, given by Eq. (15.6.44) and (15.6.33):

$$t = 1.09 \text{ sec} \left(\frac{T}{10^{10}\text{°K}} \right)^{-2} \tag{15.7.18}$$

We see that the product λt is larger than 10 for $T \gtrsim 3 \times 10^{10}$°K, so at these temperatures the neutron fraction X_n should be given by the *equilibrium* solution of Eq. (15.7.16), which is

$$X_n \simeq \frac{\lambda(p \to n)}{\lambda(p \to n) + \lambda(n \to p)} \tag{15.7.19}$$

Note that Eq. (15.7.17) will not be quantitatively correct when T drops to near 3×10^{10}°K, because kT is then not very much larger than Q. However, even though the rates $\lambda(p \to n)$ and $\lambda(n \to p)$ may differ somewhat from Eq. (15.7.17) and from each other, they are still large enough for $T \gtrsim 3 \times 10^{10}$°K to justify the use of the equilibrium solution (15.7.19).

(B) As long as $T_\nu \simeq T$ (that is, for $T > 10^{10\circ}$K) the rates (15.7.14) and (15.7.15) have the ratio

$$\frac{\lambda(p \to n)}{\lambda(n \to p)} = \exp\left(\frac{-Q}{kT}\right) \qquad (15.7.20)$$

Thus Eq. (15.7.19) gives the neutron abundance for $T \gtrsim 3 \times 10^{10\circ}$K as

$$X_n \simeq [1 + e^{Q/kT}]^{-1} \qquad (15.7.21)$$

The neutron abundance starts at $X_n \simeq \frac{1}{2}$ at very early times, and drops slowly as the temperature falls, reaching $X_n \simeq 0.38$ for $T = 3 \times 10^{10\circ}$K. It is a profoundly important fact that the initial condition for Eq. (15.7.16) does not have to be chosen arbitrarily, and does not depend on any detailed model of the very early universe, but follows directly from the singular behavior of the rates λ as $t \to 0^{109a}$.

(C) Once T drops to about $1.3 \times 10^{9\circ}$K, the rates of the two- and three-body reactions $n + \nu \leftrightarrow p + e^-$, $n + e^+ \leftrightarrow p + \bar{\nu}$, and $p + e^- + \bar{\nu} \to n$ become negligible. The only remaining reaction is the "one-body" process $n \to p + e^- + \bar{\nu}$, which at these low temperatures proceeds at the rate of free neutron decay, taken here to have the value used by Peebles[93]:

$$\lambda^{-1}(n \to p + e^- + \nu) = 1013 \text{ sec} \qquad (15.7.22)$$

Thus the neutron abundance from the time when $T \simeq 1.3 \times 10^{9\circ}$K to the start of nucleosynthesis is given by

$$X_n(t) = N \exp\left[-\frac{t(\text{sec})}{1013}\right] \qquad (15.7.23)$$

The only part of the theory of helium synthesis in which detailed numerical calculations are really needed is the evaluation of the constant N. In carrying out this calculation, it is convenient first to ignore neutron decay as well as nucleosynthesis, in which case the neutron abundance is a function $X_n^{(0)}(t)$ that approaches a finite limit as $t \to \infty$. (This is the quantity called X_n in Peebles' Table 1[93]. Peebles also ignores the process $p + e^- + \bar{\nu} \to n$, which is in fact negligible over the whole period of interest.) Neutron decay has a negligible effect until $t \simeq 20$ sec, whereas after that time the temperature is below $3 \times 10^{9\circ}$K, so that the rate $\lambda(p \to n)$ is negligible compared with $\lambda(n \to p)$, and lepton degeneracy has little effect on the rate of neutron decay. It follows that the whole effect of neutron decay is to multiply $X_n^{(0)}(t)$ with an exponential decay factor:

$$X_n(t) \simeq X_n^{(0)}(t) \exp\left[-\frac{t(\text{sec})}{1013}\right] \qquad (15.7.24)$$

Peebles[93] finds that $X_n^{(0)}$ approaches the value 0.1640 as $t \to \infty$, so comparing (15.7.23) with (15.7.24), we have

$$N \simeq X_n^{(0)}(\infty) = 0.1640 \tag{15.7.25}$$

Now we can proceed to the second part of our calculation, and put in the nuclear reactions that lead to the synthesis of complex nucleii. At early times, when $T \gg 10^{10\circ}$K, the various nuclei would be in thermal equilibrium, with the number density n_i of nuclide i given by Eq. (15.6.3). Since the nuclei are highly nonrelativistic and nondegenerate during the whole period that concerns us here, we can use the Maxwell-Boltzmann approximation to Eq. (15.6.3), and write for the total number density of species i:

$$n_i = 4\pi g_i h^{-3} \exp\left\{\frac{\mu_i - m_i}{kT}\right\} \int_0^\infty q^2 \, dq \, \exp\left\{-\frac{q^2}{2m_i kT}\right\}$$

$$= g_i \left(\frac{2\pi m_i kT}{h^2}\right)^{3/2} \exp\left\{\frac{\mu_i - m_i}{kT}\right\} \tag{15.7.26}$$

Of course, we are not given the chemical potentials μ_i, but we know that they are conserved in all reactions. Hence, *if* nuclear reactions can rapidly build up a nucleus i out of Z_i protons and $A_i - Z_i$ neutrons, then μ_i is given by

$$\mu_i = Z_i \mu_p + (A_i - Z_i)\mu_n \tag{15.7.27}$$

It is convenient to write (15.7.26) as a relation between the fractions by weight of nuclide i, *free* neutrons, and *free* protons:

$$X_i \equiv \frac{n_i A_i}{n_N} \qquad X_n \equiv \frac{n_n}{n_N} \qquad X_p \equiv \frac{n_p}{n_N}$$

where n_N is the total number density of nucleons, *bound or free*:

$$n_N = n_{N0}\left(\frac{R_0}{R}\right)^3 = \frac{\rho_{N0}}{m_N}\left(\frac{R_0}{R}\right)^3$$

Using (15.7.27), and approximating $m_p = m_n = m_N$ and $m_i = A_i m_N$ in the 3/2-power in (15.7.26), we have then

$$X_i = \tfrac{1}{2} X_p^{Z_i} X_n^{A_i - Z_i} g_i A_i^{1/2} \varepsilon^{A_i - 1} \exp\left(\frac{B_i}{kT}\right) \tag{15.7.28}$$

where B_i is the binding energy

$$-B_i \equiv m_i - Z_i m_p - (A_i - Z_i)m_n \tag{15.7.29}$$

and ε is the dimensionless quantity

$$\varepsilon \equiv \tfrac{1}{2}h^3 n_N (2\pi m_N kT)^{-3/2}$$

$$= 1.61 \times 10^{-12} \left(\frac{\rho_{N0}}{10^{-30}\ \mathrm{g/cm^3}}\right)\left(\frac{R}{10^{-10}R_0}\right)^{-3}\left(\frac{T}{10^{10\circ}\mathrm{K}}\right)^{-3/2}$$

$$(15.7.30)$$

Since ε is very small in the period of interest, the abundance of a given complex nuclide i will be very small until T drops to the value

$$T_i \simeq \frac{B_i}{k(A_i - 1)\ |\ln \varepsilon|} \qquad (15.7.31)$$

Values of T_i are given for various nuclides and various values of the present density ρ_{N0} in Table 15.6. Note that T_i depends only very weakly on the present density ρ_{N0}, because ρ_{N0} enters only in the quantity $|\ln \varepsilon|$, and this quantity has a value between 25 and 35 over the whole range of relevant temperatures and densities.

Table 15.6   Values for the Temperature T_i Defined by Eq. (15.7.31), for Various Nuclides and Various Values of the Present Density ρ_{N0}[a]

| Nuclide | $\dfrac{B}{k(A-1)}$ $(10^{9\circ}\mathrm{K})$ | $T_i\ (10^{9\circ}\mathrm{K})$ | | |
		$\rho_{N0} = 10^{-29}$ g/cm^3	$\rho_{N0} = 10^{-30}$ g/cm^3	$\rho_{N0} = 10^{-31}$ g/cm^3
H^2	25.8	0.83	0.77	0.72
H^3	49.3	1.6	1.5	1.4
He^3	44.6	1.4	1.3	1.2
He^4, etc.	109	3.9	3.6	3.3

[a] Heavier nuclides have about the same values of T_i as He^4. In thermal equilibrium, T_i is the maximum temperature at which a nuclide i could be abundant.

If nuclear abundances really were governed by the conditions of thermal equilibrium down to temperatures of order $10^{9\circ}$K, then according to Table 15.6, we should expect He^4 and heavier nuclides to appear first, followed by He^3 and H^3, and finally by H^2. However, this is not at all what happens, because thermal equilibrium is *not* maintained down to $10^{9\circ}$K. The number densities at all but the very earliest times are too low to allow nuclei to be built up directly in many-body collisions like $2n + 2p \to \mathrm{He}^4$. Complex nuclei must instead be built up in sequences of *two-body* reactions, such as

$$p + n \leftrightarrow d + \gamma$$

$$d + d \leftrightarrow \mathrm{He}^3 + n \leftrightarrow \mathrm{H}^3 + p$$

$$\mathrm{H}^3 + d \leftrightarrow \mathrm{He}^4 + n$$

etc. $\qquad\qquad$ (15.7.32)

There is no problem with the first step; the rate of deuterium production per free neutron is

$$\lambda_d = [4.55 \times 10^{-20} \ \mathrm{cm}^3/\mathrm{sec}]n_p$$

$$= 27.4 \ \mathrm{sec}^{-1} \left(\frac{R}{10^{-9}R_0}\right)^{-3} \left(\frac{\rho_{N0}}{10^{-30} \ \mathrm{g/cm}^3}\right) X_p \qquad (15.7.33)$$

and this is so much faster than the expansion rate $1/t$ [see Eq. (15.7.18)] that deuterons will appear with the equilibrium abundance (15.7.28):

$$X_d = \frac{3}{\sqrt{2}} X_p X_n \varepsilon \exp\left(\frac{B_d}{kT}\right) \qquad (15.7.34)$$

However, no appreciable quantity of H^3, He^3, He^4, or heavier nuclei can be formed until this equilibrium deuterium abundance is high enough to allow d–d, d–p, or d–n reactions to proceed at an adequate rate. According to Table 15.6, the equilibrium deuterium abundance (15.7.34) is very small for T greater than about 0.8×10^{9}°K. The low binding energy of deuterium thus serves as a "bottleneck," which delays the formation of complex nuclei until T drops to near 0.8×10^{9}°K, or a little earlier in models with a relatively high baryon number density.

Once nucleosynthesis begins, it proceeds very rapidly, because, according to Table 15.6, any temperature less than 1.2×10^{9}°K is low enough to permit high equilibrium concentrations of nuclei heavier than deuterons. However, it is not in fact possible to produce appreciable quantities of elements heavier than helium because, as mentioned above, the lack of stable nuclides with $A = 5$ or $A = 8$ impedes nucleosynthesis via n–α, p–α, or α–α collisions, whereas the Coulomb barrier in the reactions $\mathrm{He}^4 + \mathrm{H}^3 \to \mathrm{Li}^7 + \gamma$ and $\mathrm{He}^4 + \mathrm{He}^3 \to \mathrm{Be}^7 + \gamma$ prevents them from competing effectively with $p + \mathrm{H}^3 \to \mathrm{He}^4 + \gamma$ or $n + \mathrm{He}^3 \to \mathrm{He}^4 + \gamma$. Thus the effect of the nuclear reactions (15.7.32) is very rapidly to incorporate all available neutrons into He^4 nuclei, which have by far the highest binding energies of all nuclei with $A < 5$.

The nucleosynthesis process can only be followed in detail by a numerical integration of a large number of rate equations. This has been done by Peebles[93] for the reactions (15.7.32), and by Wagoner, Fowler, and Hoyle[110] for the reactions (15.7.32) plus the radiative processes

$$p + d \leftrightarrow \mathrm{He}^3 + \gamma \qquad n + d \leftrightarrow \mathrm{H}^3 + \gamma \qquad p + \mathrm{H}^3 \leftrightarrow \mathrm{He}^4 + \gamma$$

$$n + \mathrm{He}^3 \leftrightarrow \mathrm{He}^4 + \gamma \qquad d + d \leftrightarrow \mathrm{He}^4 + \gamma \qquad (15.7.35)$$

plus a large number of other processes leading (weakly) up to nuclei as heavy as Mg^{24}. Fortunately, none of these complications are relevant to our basic problem, that of understanding the helium abundance. All processes proceeding by strong and electromagnetic interactions, such as the reactions (15.7.32) and (15.7.35), will conserve the total numbers of protons and neutrons. The only effect of nucleosynthesis on the neutron-proton ratio is that, by "turning off" the decay of free neutrons, it freezes this ratio at the value it had just before the onset of nucleosynthesis. Before nucleosynthesis begins, the ratio of neutrons to all nucleons is simply the quantity X_n given by Eq. (15.7.23). After nucleosynthesis is over, we have essentially nothing left but free protons and He^4 nuclei, so the fraction of neutrons to all nucleons is one-half the fraction of all nucleons that are bound in helium, or one-half the abundance by weight of helium. Thus the abundance by weight of cosmologically produced helium is simply given by

$$Y \equiv X_{He^4} \text{ (after nucleosynthesis)} = 2X_n \text{ (just before nucleosynthesis)}$$

(15.7.36)

According to the detailed calculations of Peebles, nucleosynthesis begins abruptly at a temperature $0.9 \times 10^{9}\,°K$ for $\rho_{N0} = 7 \times 10^{-31}$ g/cm^3 or at $1.1 \times 10^{9}\,°K$ for $\rho_{N0} = 1.8 \times 10^{-29}$ g/cm^3, just about as we should expect from our qualitative considerations. Using Eq. (15.7.36), we can read off from Eq. (15.7.23) or Table 15.5 that for these two values of the present density, the helium abundance by weight should be 26.2% or 28.6%, respectively. (Peebles[93] actually gives 25.8% and 28.2% in these two cases. This very slight discrepancy is simply due to the small number of free neutrons that decay during the brief duration of nucleosynthesis.) It is safe to say that in the class of cosmological models considered here, a helium abundance by weight of about 27% would be produced cosmologically for any reasonable value of the present density. The reason the helium abundance is so insensitive to the baryon number density is that the neutron-proton ratio before nucleosynthesis is determined by the interaction of the nucleons with the huge number of leptons, not with each other, while the onset of nucleosynthesis is essentially determined by the temperature, not the nucleon density.

Wagoner, Fowler, and Hoyle[110] have calculated the cosmologically produced abundances, not only of the isotopes of hydrogen and helium, but also of Li^7 and heavier elements. Their results are given in Table 15.7. Note that the abundances of all nuclides except H^1 and He^4 are extremely small, so that production or destruction of these nuclides in stars could have a serious effect on their observed "cosmic" abundances. For this reason it is primarily the cosmic abundance of He^4 that serves as a check on models of the early universe. (However, Geiss and Reeves[110a] argue that the H^2 and He^3 observed in the solar system does in fact arise from the early universe. If this is correct, then the cosmic density must be rather low, with a present value of order 3×10^{-31} g/cm^3, to prevent the nuclear reactions, which build up H^2 and He^3 into He^4, from proceeding to completion.)

Table 15.7 Cosmologically Produced Abundances (by Weight) of Various Nuclides, for Various Values of the Present Density ρ_{N0}[a]

	ρ_{N0} (g/cm^3)							
	10^{-31}	3.1×10^{-31}	10^{-30}	3.1×10^{-30}	10^{-29}	3.1×10^{-29}	10^{-28}	3.1×10^{-28}
H^1	0.763	0.748	0.737	0.728	0.719	0.709	0.701	0.691
H^2	6.2×10^{-4}	8.9×10^{-5}	2.3×10^{-5}	2.7×10^{-7}	2.5×10^{-12}	$<10^{-12}$	$<10^{-12}$	$<10^{-12}$
He^3	6.3×10^{-5}	3.8×10^{-5}	2.1×10^{-5}	9.9×10^{-6}	5.6×10^{-6}	4.4×10^{-6}	3.5×10^{-6}	2.4×10^{-6}
He^4	0.236	0.252	0.263	0.272	0.281	0.291	0.299	0.309
Li^7	5.2×10^{-10}	2.1×10^{-10}	4.4×10^{-9}	2.1×10^{-8}	4.3×10^{-8}	1.1×10^{-7}	2.9×10^{-7}	6.8×10^{-7}
Other	$<10^{-12}$	$<10^{-12}$	$<10^{-12}$	$<10^{-12}$	$<10^{-12}$	6×10^{-12}	1.0×10^{-10}	1.9×10^{-9}

[a] Results are taken from Tables 3A and 3B of R. V. Wagoner, W. A. Fowler, and F. Hoyle, Ap. J., **148**, 3 (1967). (A value of 3°K is assumed for the present black-body radiation temperature.)

There are a number of different methods by which the helium abundance can be measured in different parts of the universe.

(A) **Stellar Masses and Luminosity.** The theory[111] of stellar structure and evolution allows us in principle (and even in practice) to calculate a star's luminosity L as a function of time if we are given its mass M and initial chemical composition. The chemical composition is usually specified by three numbers, X, Y, Z, defined as the fraction by weight respectively of H^1, of He^4, and of everything else, with

$$X + Y + Z = 1$$

(The heavy-element abundance Z, though usually small, is an important parameter for any star in radiative equilibrium, such as the sun, because it determines the opacity of the star at a given density and temperature. The helium abundance Y is important because it governs the mean molecular weight appearing in the ideal-gas law.) If we can guess Z and the age of a given star, then comparison of theory with measured values of M and L allows us to compute Y.

The best-studied star is, of course, the sun. Its mass and luminosity are known quite accurately, and its age is believed to be close to the age of the earth, or about 4.5×10^9 years. From the absorption lines of hydrogen and heavy elements it has been estimated[112] that Z/X is about 0.026 to 0.027 in the solar photosphere, though a more recent study[113] gives $Z/X \simeq 0.019$. (Unfortunately, helium lines are much too weak for Y/X to be measured in the sun by this method.) Usually solar evolution calculations are carried out for values of Z in the range 0.01 to 0.04. At the time of the discovery of the cosmic microwave radiation the best solar models[114] gave an initial helium abundance $Y = 0.27$ for $Z = 0.02$ (or $Y = 0.32$ for $Z = 0.04$), so it was regarded as a great victory for the "big-bang" cosmology that, with $T_{\gamma 0} \simeq 3°K$, it gives a primordial helium abundance $Y \simeq 0.27$.

Unfortunately, this happy state lasted only until the advent of neutrino astronomy. The same solar models that are used to calculate Y also can be used to predict the flux of neutrinos from various nuclear reactions in the sun. The sun derives its energy from the fusion of hydrogen into helium in a proton-proton cycle, starting with the reactions

$$H^1 + H^1 \rightarrow H^2 + e^+ + \nu \qquad (\bar{E}_\nu = 0.263 \text{ MeV})$$

$$H^1 + H^1 + e^- \rightarrow H^2 + \nu \qquad (E_\nu = 1.4 \text{ MeV})$$

$$H^2 + H^1 \rightarrow He^3 + \gamma$$

The cycle can then terminate with the "PP I" branch

$$He^3 + He^3 \rightarrow He^4 + 2H^1$$

or it can produce Be^7 by the reaction

$$He^3 + He^4 \rightarrow Be^7 + \gamma$$

In the latter case, one Be^7 nucleus and one proton are converted into two He^4 nuclei by either the "PP II" branch

$$Be^7 + e^- \rightarrow Li^7 + \nu \qquad (\bar{E}_\nu = 0.80 \text{ MeV})$$

$$Li^7 + H^1 \rightarrow He^4 + He^4$$

or the "PP III" branch

$$Be^7 + H^1 \rightarrow B^8 + \gamma$$

$$B^8 \rightarrow Be^8 + e^+ + \nu \qquad (\bar{E}_\nu = 7.2 \text{ MeV})$$

$$Be^8 \rightarrow 2 He^4$$

(Mean neutrino energies are given in parentheses.) Pontecorvo[115] and Alvarez[116] suggested that the neutrinos could be detected in Cl^{37} through the endothermic reaction

$$\nu + Cl^{37} \rightarrow e^- + Ar^{37} \tag{15.7.37}$$

The Ar^{37} decays by electron capture with a convenient half-life of 35 days, so it can be detected by its radioactivity after chemical separation. As pointed out by Bahcall,[117] the energetic neutrinos from B^8 beta decay are particularly effective in the reaction (15.7.37), because they can induce superallowed transitions to an excited state of Ar^{37}. Hence, even though the PP III branch is much less important than the PP II branch, about 90% of the neutrino absorption events in Cl^{37} would be expected to arise from B^8 neutrinos, and about 10% from Be^7 neutrinos. Using the extant solar models[114] with $Y = 0.27$, Bahcall[117] calculated a neutrino capture rate on the earth of $(4 \pm 2) \times 10^{-35} \text{ sec}^{-1}$ per Cl^{37} atom, and Davis[118] set out to measure this rate, using 100,000 gallons of perchlorethylene (C_2Cl_4), a common cleaning fluid, in the Homestake gold mine at Lead, South Dakota. In 1968 Davis et al.[119] announced that they had failed to detect any solar neutrinos and could set an upper limit of $0.3 \times 10^{-35} \text{ sec}^{-1}$ on the absorption rate per Cl^{37} atom, about an order of magnitude less than what had originally been expected! This discrepancy between theory and observation, in the first experiment that ever looked directly into the solar interior, has shaken the general faith in accepted solar models, and in the values they yield for the initial helium abundance of the sun. Needless to say, a great deal of work has been put into recalculation of the expected neutrino fluxes, using improved values for the opacity and for various nuclear reaction rates. In a companion paper to the letter of Davis et al.,[119] Bahcall et al.[120] estimated an absorption rate for $Z = 0.015$ of $(0.75 \pm 0.3) \times 10^{-35} \text{ sec}^{-1}$ per Cl^{37} atom, still too large by a factor of two, and calculations using the Berkeley stellar structure code gave slightly larger rates.[121] Iben[122] has noted that both Y and the neutrino absorption rates are increasing functions of Z, and that the minimum possible absorption rate, attained by $Z = 0$ and $Y \simeq 0.17$, is just about the same as Davis et al.'s upper limit. The latest calculation of

Bahcall and Uhlrich[122a] gives a counting rate of $(0.9 \pm 0.5) \times 10^{-35} \text{ sec}^{-1}$ per Cl^{37} atom.

Meanwhile, Davis' group continued their observations, and have recently announced a counting rate of $(0.15 \pm 0.1) \times 10^{-35} \text{ sec}^{-1}$ per Cl^{37} atom,[122b] about six times less than expected. In view of this discrepancy, the question of the initial solar helium abundance must for the present be regarded as unsettled.

The masses and luminosities are also known for a number of nearby Population I stars that happen to belong to binary systems. Comparison of these M and L values with the theoretical Y-dependent M–L relation yields Y values[123] for these stars lying generally in the range 0.25 to 0.35. It would be very interesting to carry out this analysis for stars of Population II, because they represent an earlier stellar generation, and also because the Davis neutrino experiment has shaken our faith in the theory of stars of Population I. Unfortunately, there are very few stars of Population II near the sun. One of them, μ Cassiopeiae A, belongs to a binary system whose separation has recently been measured by a very ingenious method.[124] The resulting mass value, together with the observed values for L and Z/X, does not agree with theory[125] for *any* value of Y, but fits best a low helium abundance, with $Y \lesssim 0.05$. However, the validity of this mass determination has since been put into doubt.[125a]

(B) **Direct Solar Measurements.** There are a number of different methods for estimating the present helium abundance of the sun, which do not rest on any detailed theory of solar structure and evolution. Measurements of the ratio Y/Z in solar cosmic rays,[126] together with the spectroscopic determination of Z/X in the solar photosphere mentioned above, suggest a helium abundance[127] $Y \simeq$ 0.20 to 0.26. During periods of quiet sun, the He^4/H ratio in the solar wind suggests a value[128] Y about equal to 0.15, but the helium content of the solar wind roughly doubles during magnetic storms.[129] Unfortunately, the solar surface is too cool to allow a spectroscopic determination of Y, but a value of $Y \simeq 0.38$ is suggested by observation of solar prominences.[130]

(C) **Globular Clusters: Theory.** As already mentioned in Section 15.3, the comparison of the numbers of stars in different regions of the Hertzsprung-Russell diagrams of globular clusters with theory yields results for both the age and the initial helium abundance of these clusters. Iben[131] derives values of Y in the range of 0.24 to 0.33, corresponding to ages 18×10^9 years to 9×10^9 years. Comparison of the stellar pulsation theory of Christy[132] with the location of variable stars in the Hertzsprung-Russell diagrams of the globular clusters M3, M15, M92 yields $Y \simeq 0.26$ to 0.32 for these clusters.[133] These studies should be given particular weight, because the globular clusters are believed to be among the first objects to condense out of the primordial gas of hydrogen and helium.

(D) **Stellar Spectra.** Helium lines are visible in the photospheres of a large number of hot stars of both populations. In general, helium abundances appear to be high,[123] with Y/X of order 0.4, and some stars are apparently superabundant in helium. There are several classes of old stars that show anomalously weak helium lines, such as the blue Population II stars on the horizontal branch of the Hertzsprung-Russell diagram of globular clusters.[134] One peculiar star, 3 Centauri A, has a low abundance of helium, most of which is in the form of the isotope He^3! Planetary nebulae[135] and novae[123] generally appear to be superabundant in helium.

(E) **Spectroscopy of Interstellar Matter.** Optical frequency emission lines from H II regions in our galaxy yield helium-hydrogen number ratios[123] that are consistently in the range 0.10–0.14, corresponding to a helium abundance by weight $Y \simeq 0.27$–0.36. It is also possible to observe the recombination of ionized helium at *radio* frequencies;[136] the radiation emitted in a transition $n + 1 \to n$ has a wavelength proportional to n^3 for $n \gg 1$, so that transitions with $n \simeq 100$ have wavelengths of the order of centimeters. The helium-hydrogen number ratios[123] deduced from radio observations of interstellar matter range from 0.06 to 0.16, corresponding to $Y \simeq 0.14$ to $Y \simeq 0.40$.

(F) **Extragalactic Measurements.** The emission lines of helium observed[123] in H II regions in galaxies within and without our local group indicate a helium abundance similar to that of the H II regions in our own galaxy. On the other hand, quasi-stellar sources show remarkably weak helium lines.[123]

There is clearly a good deal of evidence for a universal helium abundance by weight not too different from the predicted value of 27%. Unfortunately, there is also a large body of evidence for a much smaller helium abundance. The clarification of this problem would be of the highest importance for cosmology, because the cosmologically produced helium, together with the 2.7°K radiation background, may be the only relics of the primordial fireball that can serve as clues to the early history of the universe.

In order to keep an open mind about the synthesis of the elements in the early universe, it is useful to consider the possible modifications in physical or astrophysical theory that could affect the production of helium:

(A) **Cool Models.** If the observed microwave background proves not to be black-body radiation left over from the early universe, then we would have to face the possibility that the true present black-body temperature $T_{\gamma 0}$ is very much less than 2.7°K. In this case, the baryon number density at any given past temperature could be much greater than assumed above, with a consequent increase in the rate of nuclear reactions and in the abundance of complex nuclei produced in the early universe. Indeed, it was the high helium abundance produced in such cool models that led Gamow and his collaborators[51] to suggest the presence of a hot radiation background.

(B) **Fast or Slow Models.** Various mechanisms might increase or decrease the expansion rate. In particular, if the universe contained a thermal distribution of additional massless quanta, such as gravitons, Brans-Dicke scalar particles, or new kinds of neutrinos, then the energy density at a given temperature would be greater, and so, according to Eq. (15.6.44), the time required to reach that temperature would be shorter. The rate (15.7.33) of deuteron production per free neutron is normally larger than the expansion rate at $T = 10^9 °K$ by a factor 10 to 10^3 (for a present density of 10^{-31} g/cm^3 to 10^{-29} g/cm^3), so for a moderate shortening of the time scale there would still be plenty of time for nucleosynthesis to occur at $T \simeq 10^9 °K$. In this case, the only effect of the faster expansion would be to cut down the time available for the conversion of neutrons into protons, so that the neutron fraction at $10^9 °K$ would be closer to its initial value of $\frac{1}{2}$, and more helium would be produced. However, if the time scale were extremely short, there would not be time for complex nuclei to be formed before the density (and, for He^3 and He^4 formation, the temperature) falls too low. The detailed calculations of Peebles[137] show that for $T_{\gamma 0} = 3 °K$ and a present density of 7×10^{-31} g/cm^3 to 1.8×10^{-29} g/cm^3, the He^4 abundance rises as the time scale is shortened until it reaches a maximum of 60% to 80% (by weight) for a time scale shortened by a factor 10^{-1} to 10^{-2}, and then falls off again. The deuteron abundance continues to rise with shortening time scale, reaching a maximum of about 9% (by weight) when the time scale is shortened by a factor 3×10^{-3} to 3×10^{-4}, and then falls off again. On the other hand, if the expansion time scale were somehow *lengthened*, the only effect would be that more neutrons would decay into protons before nucleosynthesis occurs, so that less helium would be produced.

(C) **Neutrino-Electron Interactions.**   The thermal history of the early universe was worked out in the last section under the assumption that both electron- and muon-type neutrinos lose thermal contact with the $e^+ - e^- - \gamma$ plasma before the onset of $e^+ - e^-$ annihilation. This assumption is probably valid if the neutrino-electron scattering is produced by the usual Fermi weak interaction with the same strength as in nuclear beta decay or muon decay. However, the neutrino-electron interaction has not yet been measured experimentally, and it could be somewhat stronger than expected.[138] In this case, the ν_e and $\bar{\nu}_e$ (and possibly also the ν_μ and $\bar{\nu}_\mu$) might remain in thermal equilibrium with the $e^+ - e^- - \gamma$ plasma until nearly all $e^+ - e^-$ pairs have annihilated. The effect would be to increase the energy density at any given temperature, and also to eliminate the distinction between T_ν and T in the rates $\lambda(n \to p)$ and $\lambda(p \to n)$. Detailed calculations[139] show that if electron-type neutrinos remain in thermal equilibrium until helium synthesis, then the abundance of cosmologically produced helium would be about 29% instead of 27%.

(D) **Neutrino or Antineutrino Degeneracy.**   It is also interesting to consider the effect of a ν_e or $\bar{\nu}_e$ degeneracy on the helium abundance. One effect is that the increased density would speed up the expansion. In addition, the imbalance between neutrinos and antineutrinos would affect the relative abundance of protons and neutrons. The difference between the neutron and proton chemical potentials in thermal equilibrium is given by Eq. (15.6.4):

$$\mu_n - \mu_p = \mu_{e^-} - \mu_{\nu_e}$$

We saw in the last section that, during the period of interest, μ_{e-} is required to be negligible to maintain charge neutrality, whereas μ_{ν_e}/kT is a constant ν (with $|\nu| \lesssim 45$):

$$\mu_{e-} \simeq 0 \qquad \mu_{\nu_e} \simeq \nu kT$$

The equilibrium neutron fraction is then given by Eq. (15.7.19) as

$$X_n \equiv \frac{n_n}{n_p + n_n} = \left[1 + \exp\left(\nu + \frac{Q}{kT}\right)\right]^{-1}$$

where $Q \equiv m_n - m_p$. Thus, if ν were large and positive, the neutron fraction would start small and remain small, so that very little nucleosynthesis would occur. If ν were moderately negative, say $\nu \approx -1$, then the initial neutron fraction would be high, so that after some neutrons were converted to protons, the neutron fraction at the onset of nucleosynthesis could be close to the optimum value of 50%, and essentially all the matter of the universe could be converted to helium. If ν were large and negative, then the initial neutron abundance would be extremely high, and no nucleosynthesis could occur until some neutrons could decay into protons, at which time the nucleon density would have been too low to allow much synthesis of complex nuclei. Detailed calculations of the abundance of H^2, He^3, He^4, and Li^7 as functions of ν have been carried out by Wagoner, Fowler, and Hoyle,[140] taking into account the effects of the neutrino or antineutrino degeneracy on the rates (15.7.2)–(15.7.7). These calculations show that if the "missing mass" consists of degenerate neutrinos or antineutrinos with $|\nu| \simeq 30$, then the cosmologically produced abundance (by weight) of helium would be considerably less than 1%. On the other hand, if the lepton number density N_E of the universe is of the same order as the baryon number density N_B, then (15.6.52) shows that $|\nu|$ is of order $1/\sigma$, or about 10^{-9}. [See Eq. (15.5.15).] In this case the slight excess of neutrinos or antineutrinos has no appreciable effect on the synthesis of helium.

One final warning: Even if a large cosmic abundance of helium is definitely established, it would not necessarily follow that this helium was formed in the early universe. Geoffrey Burbidge[141] has particularly emphasized the possibility that helium could have been synthesized in an earlier, more luminous phase of our galaxy's history, perhaps in massive galactic objects. A good part of the calculations discussed in this chapter would also apply to nucleosynthesis in the collapse of massive stars.[142]

8 The Formation of Galaxies

In the last two sections we have considered two constituents of the present universe—helium and the microwave background—which may be relics of an earlier era of cosmic history. Looking at the night sky, we see one other possible relic—the clumping of stars into clusters, galaxies, and clusters of galaxies. It is natural to interpret this clumping as the effect of gravitational attraction acting on initially uniform diffuse matter, as first suggested by Newton in a famous letter to Dr. Richard Bentley.[143] Unfortunately, we still do not have even a tentative quantitative theory of the formation of galaxies, anywhere near so complete and plausible as our theories of the origin of the cosmic abundance of helium or the microwave background.

The first serious theory of galaxy formation was proposed by Sir James Jeans early in this century.[144] Jeans supposed the universe to be filled with a non-relativistic fluid, with mass density ρ, pressure p, velocity $\mathbf{v}$, and gravitational field $\mathbf{g}$, governed by the equation of continuity

$$\frac{\partial \rho}{\partial t} + \mathbf{\nabla} \cdot (\rho \mathbf{v}) = 0 \tag{15.8.1}$$

the Euler equation

$$\frac{\partial \mathbf{v}}{\partial t} + (\mathbf{v} \cdot \mathbf{\nabla})\mathbf{v} = -\frac{1}{\rho} \nabla p + \mathbf{g} \tag{15.8.2}$$

and the gravitational field equations

$$\mathbf{\nabla} \times \mathbf{g} = 0 \tag{15.8.3}$$

$$\mathbf{\nabla} \cdot \mathbf{g} = -4\pi G \rho \tag{15.8.4}$$

The effects of gravitation were ignored in the unperturbed "solution," taken to be that for a static uniform fluid:

$$\rho = \text{constant} \qquad p = \text{constant} \qquad \mathbf{v} = 0$$

If we add small perturbations ρ_1, p_1, $\mathbf{v}_1$, $\mathbf{g}_1$, then to first order Eqs. (15.8.1)–(15.8.4) become

$$\frac{\partial \rho_1}{\partial t} + \rho \mathbf{\nabla} \cdot \mathbf{v}_1 = 0$$

$$\frac{\partial \mathbf{v}_1}{\partial t} = -\frac{v_s^2}{\rho} \nabla \rho_1 + \mathbf{g}_1$$

$$\mathbf{\nabla} \times \mathbf{g}_1 = 0$$

$$\mathbf{\nabla} \cdot \mathbf{g}_1 = -4\pi G \rho_1$$

where v_s^2 is the speed of sound,

$$v_s^2 = \frac{p_1}{\rho_1} = \left(\frac{\partial p}{\partial \rho}\right)_{\text{adiabatic}}$$

and all quantities that do not carry a subscript "1" are now understood to refer to the unperturbed "solution." Combining these equations gives a differential equation for ρ_1:

$$\frac{\partial^2 \rho_1}{\partial t^2} = v_s^2 \mathbf{V}^2 \rho_1 + 4\pi G \rho \rho_1$$

The solution takes the form

$$\rho_1 \propto \exp\{i\mathbf{k} \cdot \mathbf{x} - i\omega t\} \tag{15.8.5}$$

with ω and $\mathbf{k}$ related by the "dispersion relation"

$$\omega^2 = \mathbf{k}^2 v_s^2 - 4\pi G \rho \tag{15.8.6}$$

This result bears a very close resemblance to the dispersion relation for longitudinal electrostatic oscillations in a plasma,[49]

$$\omega^2 = \mathbf{k}^2 v_s^2 + \frac{4\pi n_e e^2}{m_e} \tag{15.8.7}$$

where e, m_e, and n_e are the (unrationalized) charge, mass, and number density of electrons. The difference between (15.8.6) and (15.8.7) is that in (15.8.6), n_e is replaced with the number density ρ/m, m_e is replaced with m, e^2 is replaced with the Newtonian "coupling constant" Gm^2, and an extra minus sign is inserted to take account of the attractive nature of gravitation. Because of the minus sign in (15.8.6), the "gravitostatic" waves exhibit an instability that is not present in plasma waves: ω is imaginary for wave numbers below the critical value

$$k_J = \left(\frac{4\pi G \rho}{v_s^2}\right)^{1/2} \tag{15.8.8}$$

so that ρ_1 can grow (or decay) exponentially, with an e-folding rate

$$\text{Im } \omega = v_s(k_J^2 - \mathbf{k}^2)^{1/2} \qquad \text{for } \mathbf{k}^2 < k_J^2 \tag{15.8.9}$$

Unfortunately, Jeans's theory is not applicable to the formation of galaxies in an expanding universe, because Jeans assumed a static medium, whereas the rate of expansion of the universe is given by Eq. (15.1.20) in all cases of interest as

$$\frac{\dot{R}}{R} \simeq \left(\frac{8\pi\rho G}{3}\right)^{1/2} = (\tfrac{2}{3})^{1/2} k_J v_s \tag{15.8.10}$$

which is of the same order as the *maximum* value of the growth rate (15.8.9). The first satisfactory theory of the instabilities of an expanding universe was given in 1946 by Lifshitz.[145] He showed that disturbances at wave numbers below k_J grow, not exponentially, but like a power of t or of $R(t)$. This result will be derived and discussed in detail below, using both the nonrelativistic treatment suggested in 1957 by Bonner[146] (Section 15.9) and a simplified version of Lifshitz's relativistic theory (Section 15.10).

Although we are not yet in a position to determine the rates with which disturbances actually grow, we can rather easily decide which disturbances can grow and which cannot. For sufficiently large wave numbers, the waves described by Jeans's theory become ordinary sound waves, with

$$\omega^2 = \mathbf{k}^2 v_s{}^2 \tag{15.8.11}$$

What are the conditions for this simple dispersion relation to be valid? Gravitational forces will be negligible if the gravitational energy of a sphere of radius $|\mathbf{k}|^{-1}$ is much smaller than its thermal energy:

$$\frac{G(\rho|\mathbf{k}|^{-3})^2}{|\mathbf{k}|^{-1}} \ll \rho v_s{}^2\, |\mathbf{k}|^{-3}$$

Also, the expansion of the universe will have negligible effect if the expansion rate is much less than the frequency:

$$\sqrt{G\rho} \ll |\omega|$$

Both of these conditions will be satisfied by the relation (15.8.11), as long as the wave number satisfies the condition

$$|\mathbf{k}| \gg k_J$$

just as in Jeans's theory. Thus, even when the expansion of the universe is taken into account, we expect there to be a critical wave number, of order k_J, above which disturbances cannot grow, but only oscillate like sound waves.

Since the expansion of the universe causes $\mathbf{k}$ to decrease as $1/R$, it is convenient to characterize disturbances by a constant, the rest-mass within a sphere of radius $2\pi/|\mathbf{k}|$:

$$M = \frac{4\pi n m_H}{3} \left(\frac{2\pi}{|\mathbf{k}|}\right)^3 \tag{15.8.12}$$

where n is the hydrogen number density. According to the above analysis, the only growing disturbances are those with wave number less than k_J, and hence with mass M greater than the *Jeans mass*

$$M_J \equiv \frac{4\pi n m_H}{3} \left(\frac{2\pi}{k_J}\right)^3 = \frac{4\pi n m_H}{3} \left(\frac{\pi v_s{}^2}{G[\rho + p]}\right)^{3/2} \tag{15.8.13}$$

(It proves convenient here to replace ρ with $\rho + p$; this is permissible, because M_J is used only in order-of-magnitude arguments, and p is never more than $\rho/3$.)

We can gain a good deal of insight into the history of a protogalactic fluctuation by following the variations in M_J caused by the expansion of the universe. (See Figure 15.6.)

From the time of $e^+ - e^-$ annihilation ($T \simeq 10^{10}°$K) until the time of recombination of hydrogen ($T \simeq 4000°$K), it is a good approximation to take the contents of the universe as nonrelativistic ionized hydrogen plus black-body electromagnetic radiation, both in thermal equilibrium at temperature T. Also, since the photon entropy σk is so large, we may neglect the pressure, thermal energy, and entropy of the matter. The total energy density, pressure, and specific entropy are then (aside from the uncoupled neutrinos)

$$\rho = nm_H + aT^4 \tag{15.8.14}$$

$$p = \tfrac{1}{3}aT^4 \tag{15.8.15}$$

$$\sigma = \frac{4aT^3}{3nk} \tag{15.8.16}$$

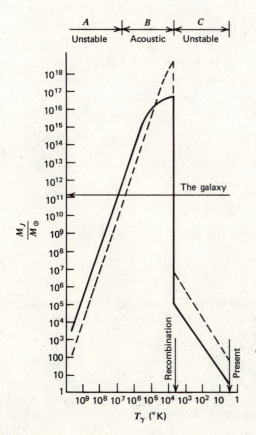

Figure 15.6 Jeans mass as a function of radiation temperature. Solid line is for $\sigma = 0.8 \times 10^8$, corresponding to $T_{\gamma 0} = 2.7°$K, $\rho_0 = 3 \times 10^{-29}$ g/cm³. Dashed line is for $\sigma = 2.4 \times 10^9$, corresponding to $T_{\gamma 0} = 2.7°$K, $\rho_0 = 10^{-30}$ g/cm³. The drop in Jeans mass at recombination is somewhat more gradual than shown here.

In an adiabatic disturbance σ is constant, so n varies as T^3, and hence

$$\delta\rho = [3nm_H + 4aT^4]\frac{\delta T}{T}$$

$$\delta p = [\tfrac{4}{3}aT^4]\frac{\delta T}{T}$$

The speed of sound is therefore

$$v_s{}^2 = \left(\frac{\delta p}{\delta\rho}\right)_{\text{adiabatic}} = \frac{1}{3}\left(\frac{kT\sigma}{m_H + kT\sigma}\right) \tag{15.8.17}$$

and the Jeans mass (15.8.13) has the value

$$M_J = \frac{2\pi^{5/2}k^2\sigma^2}{9a^{1/2}m_H^2 G^{3/2}(1 + \sigma kT/m_H)^3} \tag{15.8.18}$$

or, in terms of the solar mass,

$$M_J = 9.06 M_\odot \sigma^2 \left(1 + \frac{\sigma kT}{m_H}\right)^{-3} \tag{15.8.19}$$

Once the hydrogen recombines at $T_R \simeq 4000°\mathrm{K}$, the radiation pressure becomes ineffective, and the equations of state are those for a monatomic ideal gas with $\gamma = 5/3$:

$$\rho = nm_H + \tfrac{3}{2}nkT \tag{15.8.20}$$

$$p = nkT \tag{15.8.21}$$

The speed of sound here has the familiar value

$$v_s{}^2 = \frac{5}{3}\frac{kT}{m_H} \tag{15.8.22}$$

and the Jeans mass (15.8.13) has the value

$$M_J = 4\left(\frac{\pi}{3}\right)^{5/2}\left(\frac{5kT}{G}\right)^{3/2} n^{-1/2} m_H^{-2} \tag{15.8.23}$$

Just after recombination, the matter temperature T is the same as the radiation temperature, so T can be expressed in terms of n and the specific photon entropy (15.8.16); this gives

$$M_J = \frac{2\pi^{5/2}5^{3/2}k^2\sigma^{1/2}}{9a^{1/2}m_H^2 G^{3/2}} = 102 M_\odot \sigma^{1/2} \tag{15.8.24}$$

As long as no additional heat is put into the gas, its temperature will drop as R^{-2} [see Eq. (15.5.16)], and since n drops as R^{-3}, the Jeans mass (15.8.23) will decrease as $R^{-3/2}$.

We can now see how profound an effect the black-body radiation has on the growth of fluctuations.[147] As emphasized in Section 15.5, the present microwave temperature 2.7°K indicates that σ is very large, of order 10^8 to 10^9. In consequence, the Jeans mass (15.8.19) starts at $T = 10^9$°K at a very low value $10^{13}M_\odot/\sigma$, of order $10^5M_\odot$ to $10^4M_\odot$; then rises like T^{-3} until T reaches a temperature $m_H/k\sigma$, of order 10^5°K to 10^4°K; then levels off at a very high value $9\sigma^2M_\odot$, of order $10^{17}M_\odot$ to $10^{19}M_\odot$ until the recombination of hydrogen; following this, M_J drops precipitously to the value (15.8.24), of order $10^6M_\odot$ to $3 \times 10^6M_\odot$; and it continues to drop as $R^{-3/2}$ thereafter. If we fix our attention on a particular fluctuation whose mass M has the value $M_G \simeq 10^{11}M_\odot$ of a typical good-sized present galaxy, we may distinguish three distinct phases in its growth:

(A) The Jeans mass (15.8.19) will be less than the galactic mass until the temperature drops to the value

$$T_A = \left(\frac{9M_\odot}{\sigma M_G}\right)^{1/3} \frac{m_H}{k} \simeq 10^7°\text{K} \tag{15.8.25}$$

During this period, the amplitude of the fluctuation will have a chance to grow under the influence of its own gravitation. Since the total energy density is dominated in this early phase by radiation, this is a relativistic problem, and the growth rate must be calculated in a general-relativistic formalism. It is shown in Section 15.10 that the fastest-growing normal modes have a density contrast $\delta\rho/\rho$ that grows as t.

(B) From the time when T drops below the value (15.8.25) until the recombination of hydrogen at $T \simeq 4000$°K, the Jeans mass will be larger than the galactic mass, so the protogalactic disturbance will behave like a packet of ordinary sound waves. No appreciable growth is possible during this phase. For a relatively *high* present density, say of order 3×10^{-29} g/cm^3, there is a long period before recombination when $\sigma kT < m_H$, so that the total energy density is dominated by the hydrogen rest mass, and the protogalactic sound waves can be treated by Newtonian mechanics (see Section 15.9). For a relatively *low* present density, say of order 10^{-30} g/cm^3, we have $\sigma kT > m_H$ throughout virtually all of Phase B, so that a relativistic treatment is required. (See Section 15.10.)

(C) From the time of recombination until the present, the Jeans mass will be much less than the galactic mass, so the fluctuation amplitude can again grow. The total energy density in this phase is dominated by hydrogen rest-mass, so this is a nonrelativistic problem, and the growth rate can be calculated by Newtonian methods. The density contrast $\delta\rho/\rho$ is shown in Section 15.9 to grow roughly as $t^{2/3}$.

There is one disappointing aspect of this general picture: Nothing so far gives any clue to the reason for the observed galactic mass distribution. (The Jeans mass just before recombination is very much larger than any galactic mass, whereas the Jeans mass just after recombination appears to be related to the mass of a globular cluster[147a] rather than a galaxy.) Such a clue has recently appeared in calculations[148] of the damping of protogalactic fluctuations while they are undergoing essentially acoustic oscillations in Phase B. Dissipation will be important whenever some particle's mean free time is too long to maintain perfect thermal equilibrium. For photons, the chief collision mechanism during Phase B is scattering by nonrelativistic electrons, and the mean free time here is

$$\tau_\gamma = \frac{1}{n\sigma_T} \tag{15.8.26}$$

where σ_T is the Thomson cross-section

$$\sigma_T \equiv \frac{8\pi e^4}{3m_e{}^2} = 0.6652 \times 10^{-24} \text{ cm}^2$$

In contrast, the mean free time of an electron or a proton, taking into account only Coulomb collisions, will be of order

$$\tau_e \simeq \left[n \left(\frac{kT}{m_e} \right)^{1/2} \frac{e^4}{(kT)^2} \right]^{-1}$$

which is shorter than τ_γ by a factor of order $(kT/m_e)^{3/2}$. Thus the dominant dissipative effects in Phase B will arise from the failure of perfect thermal equilibrium between matter and radiation, rather than from any dissipation in the matter itself. Also, for any reasonable value of the present baryon number density, a fluctuation of mass $10^{11}M_\odot$ will have a radius $2\pi/|\mathbf{k}|$ much larger than the photon mean free path τ_γ throughout virtually all of Phase B, so the interaction between matter and radiation can be treated to first order in τ_γ. In this approximation, the medium of protons, electrons, and photons behaves like an imperfect fluid (see Section 2.11), with coefficients of shear viscosity, bulk viscosity, and heat conduction given by[149]

$$\eta = \tfrac{4}{15} aT^4\tau_\gamma \tag{15.8.27}$$

$$\zeta = 4aT^4\tau_\gamma \left[\frac{1}{3} - \left(\frac{\partial p}{\partial \rho} \right)_n \right]^2 \tag{15.8.28}$$

$$\chi = \tfrac{4}{3}aT^3\tau_\gamma \tag{15.8.29}$$

In general, a sound wave in an imperfect fluid will be damped at a rate[149]

$$\Gamma = \frac{k^2}{2(\rho + p)} \left\{ \zeta + \tfrac{4}{3}\eta + \chi \left(\frac{\partial \rho}{\partial T} \right)_n^{-1} \right.$$

$$\left. \times \left[\rho + p - 2T \left(\frac{\partial p}{\partial T} \right)_n + v_s{}^2 T \left(\frac{\partial \rho}{\partial T} \right)_n - \frac{n}{v_s{}^2} \left(\frac{\partial p}{\partial n} \right)_T \right] \right\} \tag{15.8.30}$$

Using Eqs. (15.8.14), (15.8.15), and (15.8.26)–(15.8.29) in Eq. (15.8.30) gives the damping rate here as

$$\Gamma = \frac{\mathbf{k}^2 a T^4}{6n\sigma_T[nm_H + \frac{4}{3}aT^4]} \left\{ \frac{16}{15} + \frac{n^2 m_H{}^2}{aT^4[nm_H + \frac{4}{3}aT^4]} \right\} \qquad (15.8.31)$$

the two terms in the brackets representing the effects of shear viscosity and heat conduction, respectively. (The bulk viscosity (15.8.28) vanishes here, because with the matter pressure and thermal energy neglected, $(\partial p/\partial \rho)_n$ is just $\frac{1}{3}$.) Since $\mathbf{k}^2 \propto M^{-2/3}$, the protogalactic sound wave will be damped in amplitude by a factor of form

$$D \equiv \exp\left\{ -\int_{t_A}^{t_R} \Gamma\, dt \right\} = \exp\left[-\left(\frac{M_c}{M} \right)^{2/3} \right] \qquad (15.8.32)$$

where M_c is some critical mass (and a subscript R denotes the time of hydrogen recombination). For a relatively *high* present density, there was a long period before recombination when the energy density was dominated by hydrogen rest-mass, so that

$$t \simeq (6\pi n m_H G)^{-1/2}$$

$$\Gamma \simeq \frac{\mathbf{k}^2}{6\sigma_T n} \propto t^{2/3}$$

and the critical mass in Eq. (15.8.32) is

$$M_c \simeq \frac{32\pi^4}{3} \left(\frac{m_H}{10\sigma_T} \right)^{3/2} (6\pi G)^{-3/4} (n_R m_H)^{-5/4} \qquad (15.8.33)$$

For instance, for a present mass density 3×10^{-29} g/cm^3, the critical mass is $5 \times 10^{12} M_\odot$. For a relatively *low* present density, the energy density during virtually all of Phase B was dominated by radiation (including neutrinos), so that

$$t \simeq (15.5\pi a T^4 G)^{-1/2}$$

$$\Gamma \simeq \frac{2\mathbf{k}^2}{15\sigma_T n} \propto t^{1/2}$$

and the critical mass in Eq. (15.8.32) is

$$M_c \simeq \frac{32\pi^4}{3} \left(\frac{4m_H}{45\sigma_T} \right)^{3/2} (15.5\pi a T^4 G)^{-3/4} (n_R m_H)^{-1/2} \qquad (15.8.34)$$

For instance, for a present mass density 10^{-30} g/cm^3, the critical mass is $2 \times 10^{14} M_\odot$. A fluctuation will presumably be unable to survive the damping in Phase B if the exponent $(M_c/M)^{2/3}$ in Eq. (15.8.32) is greater than about 10, so we can conclude that the fluctuations surviving at the time of recombination have a minimum mass between $1.6 \times 10^{11} M_\odot$ and $6 \times 10^{12} M_\odot$, just about the mass of a large galaxy.

So far, we have learned that any small fluctuation of mass between about $10^{11} M_\odot$ and $10^{17} M_\odot$ will grow in Phase A, undergo damped oscillations in Phase B, but will survive to trigger new growth in Phase C. One may still wonder why there is a fairly definite upper limit on galactic mass, of order $10^{12} M_\odot$ to $10^{13} M_\odot$, rather than a smooth distribution of masses extending from $10^{11} M_\odot$ up to much higher values. One possible answer lies in the well-known property of nonlinear effects,[150] of transferring energy from long wavelengths to the shortest wavelengths that can survive the effects of dissipation. Unfortunately, the application of the theory of turbulence to problems of galactic growth has just barely begun.[151]

The theory of the origins of galaxies is of more than academic interest, because the value $(\delta n/n)_R$ of the fractional change in density at the time of recombination may become accessible to observation in the near future.[152] For fluctuations that are approximately adiabatic, the number density is proportional to the cube of the temperature, so the temperature fluctuations at the onset of recombination will be given by

$$\left(\frac{\delta T_\gamma}{T_\gamma}\right)_R = \frac{1}{3}\left(\frac{\delta n}{n}\right)_R \tag{15.8.35}$$

If the universe has remained optically thin since this time, then these fluctuations will show up in the cosmic microwave background, as fluctuations of the observed cosmic radiation temperature with angle. (Note, however, that Thomson scattering could wash out such inhomogeneities[152a] without affecting the Planck shape of the distribution function; see Section 15.4.) According to Eqs. (15.5.35)–(15.5.37) and (15.8.12), a fluctuation of mass M will appear to have angular scale

$$\theta/2 \simeq q_0 H_0 (1 + z_R)\left(\frac{2\pi}{|\mathbf{k}|_R}\right) \simeq q_0 H_0 (1 + z_R)\left(\frac{3M}{4\pi n_R m_H}\right)^{1/3}$$

or, since $n \propto R^{-3}$,

$$\theta/2 \simeq q_0 H_0 \left(\frac{3M}{4\pi n_0 m_H}\right)^{1/3} \tag{15.8.36}$$

For instance, for $q_0 = \frac{1}{2}$, $H_0^{-1} = 13 \times 10^9$ years, and a present density $n_0 m_H = 1.1 \times 10^{-29}$ g/cm^3, the fluctuation corresponding to a nascent galaxy of mass $10^{11} M_\odot$ should have an angular scale $\theta = 30''$. As indicated in Section 15.5, the measurement of even small fluctuations with this angular scale is well within the reach of present techniques. It is therefore a matter of some importance to calculate how strong a fluctuation has to be at the time of recombination to grow into a galaxy by the present time. This problem will be addressed in the next section.

9   Newtonian Theory of Small Fluctuations

We here calculate the behavior of small fluctuations, using the Newtonian equations (15.8.1)–(15.8.4), but now taking into account the expansion of the universe. As remarked at the end of Section 15.1, we can safely employ Newtonian mechanics to deal with astronomical problems in which the energy density is dominated by nonrelativistic particles, so that $p \ll \rho$, and in which the linear scales involved are small compared with the characteristic scale of the universe.[153]

As shown in Section 15.1, there is a simple spatially uniform solution of Eqs. (15.8.1)–(15.8.4), with

$$\rho = \rho_0 \left[\frac{R_0}{R(t)} \right]^3 \tag{15.9.1}$$

$$\mathbf{v} = \mathbf{r} \left[\frac{\dot{R}(t)}{R(t)} \right] \tag{15.9.2}$$

$$\mathbf{g} = -\mathbf{r} \left[\frac{4\pi G\rho}{3} \right] \tag{15.9.3}$$

where $R(t)$ is a scale factor satisfying the differential equation

$$\dot{R}^2 + k = \frac{8\pi\rho G R^2}{3} \tag{15.9.4}$$

or equivalently

$$\frac{\ddot{R}}{R} = -\frac{4\pi\rho G}{3} \tag{15.9.5}$$

We now seek a perturbed solution, by adding to the "zero-order" solution (15.9.1)–(15.9.3) the small perturbations ρ_1, $\mathbf{v}_1$, and $\mathbf{g}_1$. The hydrodynamic equations (15.8.1)–(15.8.4) then give, to first order in these perturbations,

$$\dot{\rho}_1 + 3\frac{\dot{R}}{R}\rho_1 + \frac{\dot{R}}{R}(\mathbf{r}\cdot\nabla)\rho_1 + \rho\nabla\cdot\mathbf{v}_1 = 0 \tag{15.9.6}$$

$$\dot{\mathbf{v}}_1 + \frac{\dot{R}}{R}\mathbf{v}_1 + \frac{\dot{R}}{R}(\mathbf{r}\cdot\nabla)\mathbf{v}_1 = -\frac{1}{\rho}\nabla p_1 + \mathbf{g}_1 \tag{15.9.7}$$

$$\nabla \times \mathbf{g}_1 = 0 \tag{15.9.8}$$

$$\nabla \cdot \mathbf{g}_1 = -4\pi G\rho_1 \tag{15.9.9}$$

Also, since these are for the moment supposed to be adiabatic fluctuations, the pressure perturbation is given by

$$p_1 = v_s^2 \rho_1 \tag{15.9.10}$$

where v_s is the speed of sound. (Here and below, ρ denotes the mass density (15.9.1) in the unperturbed solution.)

The equations (15.9.6)–(15.9.10) are spatially homogeneous, so we expect to find plane-wave solutions. Indeed, solutions can be found with the spatial dependence

$$\rho_1(r, t) = \rho_1(t) \exp \left\{ \frac{i\mathbf{r} \cdot \mathbf{q}}{R(t)} \right\} \tag{15.9.11}$$

and likewise for $\mathbf{v}_1$ and $\mathbf{g}_1$. (The appearance of the factor $1/R$ in the exponential means that the wavelength of these modes is stretched out by the expansion of the universe, as anticipated in the last section.) Equations (15.9.6)–(15.9.10) now become coupled ordinary differential equations:

$$\dot{\rho}_1 + \frac{3\dot{R}}{R} \rho_1 + iR^{-1}\rho\mathbf{q} \cdot \mathbf{v}_1 = 0 \tag{15.9.12}$$

$$\dot{\mathbf{v}}_1 + \frac{\dot{R}}{R} \mathbf{v}_1 = -\frac{iv_s^2}{\rho R} \mathbf{q}\rho_1 + \mathbf{g}_1 \tag{15.9.13}$$

$$\mathbf{q} \times \mathbf{g}_1 = 0 \tag{15.9.14}$$

$$i\mathbf{q} \cdot \mathbf{g}_1 = -4\pi GR\rho_1 \tag{15.9.15}$$

The "field equations" (15.9.14) and (15.9.15) have the obvious solution

$$\mathbf{g}_1 = \frac{4\pi iG\rho_1 R\mathbf{q}}{\mathbf{q}^2} \tag{15.9.16}$$

To solve the equations of motion, it is convenient to decompose $\mathbf{v}_1$ into parts perpendicular and parallel to $\mathbf{q}$,

$$\mathbf{v}_1(t) = \mathbf{v}_{1\perp}(t) + i\mathbf{q}\varepsilon(t) \tag{15.9.17}$$

where

$$\mathbf{q} \cdot \mathbf{v}_{1\perp} = 0$$

$$\varepsilon \equiv -\frac{i\mathbf{q} \cdot \mathbf{v}_1}{\mathbf{q}^2}$$

It is also convenient to express ρ_1 in terms of a fractional change in density δ:

$$\rho_1(t) = \rho(t)\delta(t) = \rho_0 \left[\frac{R_0}{R(t)} \right]^3 \delta(t) \tag{15.9.18}$$

Then (15.9.13) splits into two uncoupled equations,

$$\dot{\mathbf{v}}_{1\perp} + \frac{\dot{R}}{R} \mathbf{v}_{1\perp} = 0 \tag{15.9.19}$$

$$\dot{\varepsilon} + \frac{\dot{R}}{R} \varepsilon = \left(-\frac{v_s^2}{R} + \frac{4\pi G\rho R}{\mathbf{q}^2} \right) \delta \tag{15.9.20}$$

and (15.9.12) simplifies to

$$\mathring{\delta} = \frac{\mathbf{q}^2}{R} \varepsilon \tag{15.9.21}$$

Inspection of Eqs. (15.9.19)–(15.9.21) shows that there are two quite different types of normal mode here. The *rotational* modes, described by $\mathbf{v}_{1\perp}$, simply decay as $1/R$:

$$\mathbf{v}_{1\perp}(t) \propto R^{-1}(t) \tag{15.9.22}$$

On the other hand, the *compressional* modes, described by ε and δ, have a more interesting time dependence. Using Eq. (15.9.21) to eliminate ε in Eq. (15.9.20), we find

$$\ddot{\delta} + \frac{2\dot{R}}{R} \dot{\delta} + \left(\frac{v_s^2 \mathbf{q}^2}{R^2} - 4\pi G\rho \right) \delta = 0 \tag{15.9.23}$$

Note that this goes over to the Jeans dispersion relation (15.8.6) if we set R constant and define the wave number $\mathbf{k}$ as $\mathbf{q}/R$. Equation (15.9.23) is the fundamental differential equation that governs the growth or decay of gravitational condensations in an expanding universe.

The above Newtonian theory becomes applicable at the onset of the matter-dominated era, when the energy density of radiation drops below the rest-mass density, so that $p \ll \rho$. Unfortunately, Eq. (15.9.23) is a little too complicated to allow a solution in closed form that would be valid throughout the whole of the matter-dominated era. We can, however, answer the most interesting questions about the behavior of $\delta(t)$ by solving Eq. (15.9.23) in a number of special cases.

(A) Zero Pressure Solutions

According to the general picture outlined in the last section, a galaxy is supposed to grow out of the small fluctuations present at the time of hydrogen recombination, which are left over from the previous phase of damped acoustic oscillation. The most important problem facing us is how much such a proto-galactic fluctuation could have grown from the time of recombination until the present. Or, to put this in terms relevant to observations, how large would a fluctuation have to be at the time of recombination to have a chance of growing into a galaxy by the present time?

To answer these questions we can simplify Eq. (15.9.23) by neglecting the pressure term $v_s^2 \mathbf{q}^2/R^2$. This term will be negligible compared with the gravitational term $4\pi G\rho$ if the wave number $|\mathbf{k}| \equiv |\mathbf{q}|/R$ is much less than the Jeans wave number (15.8.8) or, equivalently, if the fluctuation mass (15.8.12) is much greater than the Jeans mass (15.8.13). We saw in the last section that the Jeans mass is of order $10^6 M_\odot$ to $3 \times 10^6 M_\odot$ immediately after recombination, and drops as $R^{-3/2}$ thereafter, so a galactic mass of order $10^{11} M_\odot$ will certainly be much greater than the Jeans mass once the hydrogen recombination is over.

In order to carry the solution of Eq. (15.9.23) forward to the present time, it will be necessary to use the parametric formulas for $R(t)$ and $\rho(t)$ derived in Section 15.3. For positive, zero, and negative curvature, these are:

$k = +1$

$$\frac{R(t)}{R_0} = q_0(2q_0 - 1)^{-1}(1 - \cos \theta)$$

$$H_0 t = q_0(2q_0 - 1)^{-3/2}(\theta - \sin \theta)$$

$$\rho = \frac{3H_0{}^2(2q_0 - 1)^3}{4\pi G q_0{}^2(1 - \cos \theta)^3}$$

$k = 0$

$$\frac{R(t)}{R_0} = \left(\frac{3H_0 t}{2}\right)^{2/3}$$

$$\rho = (6\pi G t^2)^{-1}$$

$k = -1$

$$\frac{R(t)}{R_0} = q_0(1 - 2q_0)^{-1}(\cosh \Psi - 1)$$

$$H_0 t = q_0(1 - 2q_0)^{-3/2}(\sinh \Psi - \Psi)$$

$$\rho = \frac{3H_0{}^2(1 - 2q_0)^3}{4\pi G q_0{}^2(\cosh \Psi - 1)^3}$$

With the pressure term neglected, the differential equation (15.9.23) now takes the following forms.

$k = +1$

$$(1 - \cos \theta)\frac{d^2\delta}{d\theta^2} + \sin \theta \frac{d\delta}{d\theta} - 3\delta = 0 \tag{15.9.24}$$

$k = 0$

$$\ddot{\delta} + \frac{4}{3t}\dot{\delta} - \frac{2}{3t^2}\delta = 0 \tag{15.9.25}$$

$k = -1$

$$(\cosh \Psi - 1)\frac{d^2\delta}{d\Psi^2} + \sinh \Psi \frac{d\delta}{d\Psi} - 3\delta = 0 \tag{15.9.26}$$

In each case there are two independent solutions, which we shall call δ_+ and δ_-:

$k = +1$

$$\delta_+ \propto -\frac{3\theta \sin \theta}{(1 - \cos \theta)^2} + \frac{5 + \cos \theta}{1 - \cos \theta} \tag{15.9.27}$$

$$\delta_- \propto \frac{\sin \theta}{(1 - \cos \theta)^2} \tag{15.9.28}$$

k = 0

$$\delta_+ \propto t^{2/3} \tag{15.9.29}$$

$$\delta_- \propto t^{-1} \tag{15.9.30}$$

k = −1

$$\delta_+ \propto -\frac{3\Psi \sinh \Psi}{(\cosh \Psi - 1)^2} + \frac{5 + \cosh \Psi}{\cosh \Psi - 1} \tag{15.9.31}$$

$$\delta_- \propto \frac{\sinh \Psi}{(\cosh \Psi - 1)^2} \tag{15.9.32}$$

In all three cases, the solutions δ_+ and δ_- go over to $t^{2/3}$ and t^{-1}, respectively, for $R(t) \ll R_0$, so that if a disturbance starts with comparable amplitudes for the δ_+ and δ_- modes at recombination, then it will soon be almost purely in the δ_+ mode. We therefore concentrate entirely on the δ_+ mode from now on.

The disturbance is supposed here to start at a time t_R corresponding to the large red shift

$$1 + z_R \equiv \frac{R(t_0)}{R(t_R)} \simeq \frac{4000°\mathrm{K}}{2.7°\mathrm{K}} \simeq 1500$$

The initial values of the parameters θ, t, or Ψ for $k = +1$, $k = 0$, or $k = -1$ are then

$$\theta_R \simeq \left[\frac{2(1 - \cos\theta_0)}{1 + z_R} \right]^{1/2}$$

$$t_R \simeq t_0 (1 + z_R)^{-3/2}$$

$$\Psi_R \simeq \left[\frac{2(\cosh \Psi_0 - 1)}{1 + z_R} \right]^{1/2}$$

Hence the density contrast δ could at most grow by an amplification factor

$$A_0 \equiv \frac{\delta_+(t_0)}{\delta_+(t_R)} \tag{15.9.33}$$

given by

$$A_0 = \begin{cases} \dfrac{5(1 + z_R)}{(1 - \cos\theta_0)^3} \{-3\theta_0 \sin\theta_0 + (1 - \cos\theta_0)(5 + \cos\theta_0)\} & (k = +1) \\[3mm] (1 + z_R) & (k = 0) \\[3mm] \dfrac{5(1 + z_R)}{(\cosh\Psi_0 - 1)^3} \{-3\Psi_0 \sinh\Psi_0 + (\cosh\Psi_0 - 1)(\cosh\Psi_0 + 5)\} & \\[2mm] & (k = -1) \end{cases} \tag{15.9.34}$$

The parameters θ_0 and Ψ_0 can be expressed in terms of q_0 by using Eq. (15.3.10) or (15.3.19). We then find that A_0 is a monotonically increasing function of q_0 for $q_0 > 0$, rising from $A_0 \simeq 5q_0(1 + z_R)$ for $q_0 \ll \frac{1}{2}$, to $A_0 = 1 + z_R$ at $q_0 = \frac{1}{2}$, to $A_0 = 1.45(1 + z_R)$ at $q_0 = 1$, and approaching the upper bound $A_0 = 5(1 + z_R)$ for $q_0 \gg 1$. With q_0 between 0.014 and 2 and $1 + z_R = 1500$, a small fluctuation would have grown from the time of hydrogen recombination until the present by a factor A_0 between 100 and 3000.

The condensations observed in the present universe cannot be considered "small disturbances." For instance, the mass density in a typical cluster of galaxies is of order 10^{-28} g/cm^3, an order of magnitude greater than the maximum likely value for the universe as a whole, and the mass density within a galaxy is of course even greater. The simple linear instability theory described above is therefore not applicable to the whole history of inhomogeneities up to the present moment. However, it seems reasonable to suppose that the present strong condensations grew out of small disturbances, so that a necessary (if perhaps not sufficient) condition for their formation is that the perturbation $\delta_+(t)$ calculated in linear stability theory should have become of order unity at some time before the present. This then sets a lower limit on the magnitude of the initial disturbance at the time of recombination,

$$|\delta_+(t_R)| \gtrsim \frac{1}{A_0} \qquad (15.9.35)$$

so that $|\delta_+(t_R)|$ should be greater than about 10^{-2} to 3×10^{-4}, depending on the value of q_0. In order to say how much greater, it would be necessary to know the time at which the disturbance reached the beginning of the nonlinear regime, with $|\delta_+(t)| \simeq 1$. According to Weymann,[154] the observed binding energies of galaxies indicates that this must have occurred after a time of order 7×10^7 years; if δ_+ reached unity this early, then it must have been already quite large at the time of recombination. On the other hand, if the concentration of quasi-stellar object red shifts at $z \approx 2$ marks the onset of galaxy formation, then (15.9.35) provides a reasonably good estimate of the actual value of $|\delta_+(t_R)|$. As remarked at the end of the last section, the protogalactic fluctuations at the time of recombination would produce fractional fluctuations in the observed microwave background temperature equal to $|\delta_+(t_R)|/3$ over angles of order 30″, *provided* that Thomson scattering during or after recombination does not smooth out the fluctuations.[152a] Even if the nonlinear regime was reached rather recently, $\Delta T_\gamma/T_\gamma$ would be of order 3×10^{-3} to 10^{-4}, which should be observable.

(B) Zero Curvature Solutions

It is also interesting to consider the behavior of the solutions when the pressure term $v_s{}^2\mathbf{q}^2/R^2$ in Eq. (15.9.23) is *not* neglected, particularly in order to locate the precise dividing line between stable and unstable fluctuations. In order to obtain **exact** solutions, it is necessary now to restrict our attention to early times, when

$R(t) \ll R_0$, so that the terms $\dot{R}^2$ and $8\pi\rho G R^2/3$ in Eq. (15.9.4) are much larger than unity, and even for $k = \pm 1$ we can use the $k = 0$ formulas for R and ρ:

$$R \propto t^{2/3} \tag{15.9.36}$$

$$\rho = (6\pi G t^2)^{-1} \tag{15.9.37}$$

(This is not much of a limitation, because for recent times, when these formulas may not be valid, the Jeans mass is so small that its precise value is of little interest.) For a general specific heat ratio γ, the pressure varies as ρ^γ, and the speed of sound is

$$v_s = \left(\frac{\gamma p}{\rho}\right)^{1/2} \propto \rho^{1/2(\gamma-1)} \propto t^{1-\gamma} \tag{15.9.38}$$

Hence Eq. (15.9.23) reads here

$$\ddot{\delta} + \frac{4}{3t}\dot{\delta} + \left(\frac{\Lambda^2}{t^{2\gamma-2/3}} - \frac{2}{3t^2}\right)\delta = 0 \tag{15.9.39}$$

where Λ^2 is the constant

$$\Lambda^2 \equiv \frac{t^{2\gamma-2/3}v_s^2 \mathbf{q}^2}{R^2} \tag{15.9.40}$$

The solutions of Eq. (15.9.39) for $\gamma > 4/3$ are

$$\delta_\pm \propto t^{-1/6} J_{\mp 5/6\nu}\left(\frac{\Lambda t^{-\nu}}{\nu}\right) \tag{15.9.41}$$

where J is the usual Bessel function, and

$$\nu \equiv \gamma - \frac{4}{3} > 0 \tag{15.9.42}$$

The Bessel functions oscillate for $t \ll \Lambda^{1/\nu}$, whereas for $t \gg \Lambda^{1/\nu}$ the solutions behave like

$$\delta_\pm \propto t^{-1/6 \pm 5/6} \tag{15.9.43}$$

By using Eqs. (15.9.37) and (15.9.40), the condition $t > \Lambda^{1/\nu}$ for growth in the δ_+ mode can be written

$$\frac{v_s^2 \mathbf{q}^2}{R^2} \gtrsim 6\pi G\rho$$

which is substantially the same as the Jeans condition $v_s^2 k^2 > 4\pi G\rho$.

The solutions (15.9.41) will apply after recombination, with $\gamma \simeq 5/3$. Also, for a relatively high present density, there is an appreciable period before recombination when the total energy density is dominated by hydrogen rest-mass, but the pressure is dominated by radiation. The cosmic medium then behaves like a nonrelativistic fluid with $\gamma = 4/3$, so that Eq. (15.9.39) reads

$$\ddot{\delta} + \frac{4}{3t}\dot{\delta} + [\Lambda^2 - \tfrac{2}{3}]\frac{\delta}{t^2} = 0 \qquad (15.9.44)$$

with

$$\Lambda^2 \equiv \frac{t^2 v_s^2 \mathbf{q}^2}{R^2} = \frac{v_s^2 \mathbf{q}^2}{6\pi G\rho R^2} \qquad (15.9.45)$$

The solutions of Eq. (15.9.44) are quite simple:

$$\delta_{\pm} \propto t^{\alpha} \qquad \alpha = -\tfrac{1}{6} \pm (\tfrac{25}{36} - \Lambda^2)^{1/2} \qquad (15.9.46)$$

Both solutions undergo a gently damped oscillation for $\Lambda > 5/6$, and decay for $5/6 > \Lambda > \sqrt{2/3}$, whereas for $\Lambda < \sqrt{2/3}$, δ_+ grows and δ_- decays. The condition for growth in the δ_+ mode is then

$$\frac{v_s^2 \mathbf{q}^2}{6\pi G\rho R^2} < \frac{2}{3}$$

which is precisely the same as Jeans's condition $v_s^2 \mathbf{k}^2 < 4\pi G\rho$.

It is not difficult to incorporate dissipative effects in this formalism. The most interesting case here is the damping owing to a finite photon mean free path during the matter-dominated part of "Phase B," that is, during the period prior to recombination when the radiation density aT^4 is much less than the matter density nm_H and the Jeans mass is much greater than a galactic mass. According to Eqs. (15.8.27)–(15.8.30), the effects of viscosity are negligible here compared with the effects of heat conduction, so dissipative effects can be taken into account[155] by using the equation of state to express the pressure perturbation p_1 in Eq. (15.9.7) in terms of the temperature and mass-density perturbations T_1 and ρ_1, and supplementing Eqs. (15.9.6)–(15.9.9) with the usual nonrelativistic equation of radiative heat transfer. We need not go into this further here, since heat conduction as well as viscosity will be incorporated into the general-relativistic theory described in the next section.

10   General-Relativistic Theory of Small Fluctuations

The nonrelativistic analysis presented in the last section is adequate for the study of compressional and rotational perturbations during the matter-dominated era, when $p \ll \rho$. However, a relativistic treatment is needed to deal with the radiation or lepton-dominated eras, when p is of the same order as ρ, or to treat the propagation of gravitational radiation in any era.

The relativistic theory of small disturbances in an expanding universe is rather complicated, so the theory presented here will deal only with the simplest case, when the unperturbed Robertson-Walker metric has curvature $k = 0$. This is not too severe a restriction, because the results for $k = +1$ or $k = -1$ are essentially the same as for $k = 0$, provided that we keep to the early universe when $\dot{R}^2 \gg |k|$, and provided that we keep to perturbations of wavelength much less than R. These are the most interesting cases in any event, particularly since the growth of condensations in the recent past can be treated for any value of k by the nonrelativistic theory of the last section.

It proves convenient here to include the effects of dissipation from the beginning. The medium will be characterized by a coefficient of shear viscosity η and a heat conduction coefficient χ; as mentioned in Section 15.8, the bulk viscosity ζ will be negligible once the pressure and kinetic energy density of matter drop below those of radiation. The effects of dissipation can be taken into account by adding suitable terms to the energy-momentum tensor. These terms were calculated in Section 2.11 for a relativistic imperfect fluid in the absence of gravitation; the correct energy-momentum tensor in the presence of gravitation can be immediately obtained by writing Eq. (2.11.21) in a generally covariant form:

$$T^{\mu\nu} = pg^{\mu\nu} + (p + \rho)U^\mu U^\nu - \eta H^{\mu\rho}H^{\nu\sigma}W_{\rho\sigma}$$
$$- \chi(H^{\mu\rho}U^\nu + H^{\nu\rho}U^\mu)Q_\rho \qquad (15.10.1)$$

where

$$W_{\mu\nu} \equiv U_{\mu;\nu} + U_{\nu;\mu} - \tfrac{2}{3}g_{\mu\nu}U^\gamma{}_{;\gamma} \qquad (15.10.2)$$

$$Q_\mu \equiv T_{;\mu} + TU_{\mu;\nu}U^\nu \qquad (15.10.3)$$

$$H_{\mu\nu} \equiv g_{\mu\nu} + U_\mu U_\nu \qquad (15.10.4)$$

It is easy to check that the dissipative terms in $T^{\mu\nu}$ vanish for a Robertson-Walker metric (with any k), so the Friedmann solutions still provide our starting point. In particular, for $k = 0$ the Einstein equations have the familiar unperturbed solution

$$g_{tt} = -1 \qquad g_{ti} = 0 \qquad g_{ij} = R^2(t)\delta_{ij} \qquad (15.10.5)$$

$$U^t = 1 \qquad U^i = 0 \qquad (15.10.6)$$

where x^i (with $i = 1, 2, 3$) are a set of quasi-Euclidean comoving coordinates, and

$$\dot{R}^2 = \frac{8\pi\rho GR^2}{3} \qquad (15.10.7)$$

The only independent nonvanishing components of the unperturbed affine connection are then given by Eqs. (15.1.3)–(15.1.5) as

$$\Gamma^t_{ij} = R\dot{R}\delta_{ij} \qquad \Gamma^i_{tj} = \frac{\dot{R}}{R}\delta_{ij} \qquad (15.10.8)$$

We shall consider a disturbance, in which the metric is $g_{\mu\nu} + h_{\mu\nu}$, with $h_{\mu\nu}$ small. Before we write down the field equations for $h_{\mu\nu}$, it is useful to recall the remark of Section 10.9, that by performing a coordinate transformation (10.9.6), we can convert the solution $h_{\mu\nu}$ into an equivalent solution

$$h_{\mu\nu}^* = h_{\mu\nu} + \varepsilon_{\mu;\nu} + \varepsilon_{\nu;\mu}$$

with ε_μ an arbitrary small vector field. Using Eq. (15.10.8), this new solution takes the form

$$h_{ij}^* = h_{ij} + \frac{\partial \varepsilon_i}{\partial x^j} + \frac{\partial \varepsilon_j}{\partial x^i} - 2R\dot{R}\delta_{ij}\varepsilon_t \tag{15.10.9}$$

$$h_{it}^* = h_{it} + \frac{\partial \varepsilon_i}{\partial t} + \frac{\partial \varepsilon_t}{\partial x^i} - 2\frac{\dot{R}}{R}\varepsilon_i \tag{15.10.10}$$

$$h_{tt}^* = h_{tt} + 2\frac{\partial \varepsilon_t}{\partial t} \tag{15.10.11}$$

It is extremely convenient to choose ε_μ so that

$$h_{it}^* = h_{tt}^* = 0$$

thus maintaining to the greatest possible extent the form of the unperturbed metric (15.10.5). This can be accomplished by constructing ε_μ according to the prescriptions

$$\varepsilon_t = -\tfrac{1}{2}\int h_{tt}\, dt$$

$$\varepsilon_i = -R^2 \int \left[h_{it} + \frac{\partial \varepsilon_t}{\partial x^i} \right] R^{-2}\, dt$$

Dropping the asterisk, it will simply be assumed from now on that the coordinate system is chosen so that

$$h_{it} = h_{tt} = 0 \tag{15.10.12}$$

The perturbation $\delta g_{\mu\nu} = h_{\mu\nu}$ produces in the affine connection a perturbation (10.9.1), which here has the components

$$\delta\Gamma_{jk}^i = \frac{1}{2R^2}\left[h_{ij;k} + h_{ik;j} - h_{jk;i} \right]$$

$$= \frac{1}{2R^2}\left[\frac{\partial h_{ij}}{\partial x^k} + \frac{\partial h_{ik}}{\partial x^j} - \frac{\partial h_{jk}}{\partial x^i} \right] \tag{15.10.13}$$

$$\delta\Gamma_{jk}^t = -\tfrac{1}{2}[h_{tj;k} + h_{tk;j} - h_{jk;t}]$$

$$= \frac{1}{2}\frac{\partial h_{jk}}{\partial t} \tag{15.10.14}$$

$$\delta\Gamma^i_{tj} = \frac{1}{2R^2}\left[h_{it;j} + h_{ij;t} - h_{tj;i}\right]$$

$$= \frac{1}{2R^2}\left[\frac{\partial h_{ij}}{\partial t} - \frac{2\dot{R}}{R}h_{ij}\right] \tag{15.10.15}$$

$$\delta\Gamma^t_{ti} = \delta\Gamma^i_{tt} = \delta\Gamma^t_{tt} = 0 \tag{15.10.16}$$

Contracting, we find also that

$$\delta\Gamma_\mu \equiv \delta\Gamma^\nu_{\nu\mu} = \delta\Gamma^i_{i\mu} = \frac{\partial}{\partial x^\mu}\left(\frac{h_{kk}}{2R^2}\right) \tag{15.10.17}$$

it being understood that repeated Latin indices are summed over the values 1, 2, 3. The perturbation in the Ricci tensor is then given by

$$\delta R_{ij} = (\delta\Gamma_i)_{;j} - (\delta\Gamma^\mu_{ij})_{;\mu}$$

$$= \frac{\partial\delta\Gamma_i}{\partial x^j} - \frac{\partial\delta\Gamma^k_{ij}}{\partial x^k} - \frac{\partial\delta\Gamma^t_{ij}}{\partial t}$$

$$- R\dot{R}\delta_{ij}\delta\Gamma_t - \frac{\dot{R}}{R}\delta\Gamma^t_{ij} + R\dot{R}[\delta\Gamma^i_{tj} + \delta\Gamma^j_{ti}]$$

$$\delta R_{ti} = (\delta\Gamma_t)_{;i} - (\delta\Gamma^\mu_{ti})_{;\mu}$$

$$= \frac{\partial\delta\Gamma_t}{\partial x^i} - \frac{\partial\delta\Gamma^j_{ti}}{\partial x^j} - \frac{\dot{R}}{R}\delta\Gamma_i + \frac{\dot{R}}{R}\delta\Gamma^j_{ji}$$

$$\delta R_{tt} = (\delta\Gamma_t)_{;t} - (\delta\Gamma^\mu_{tt})_{;\mu}$$

$$= \frac{\partial\delta\Gamma_t}{\partial t} + \frac{2\dot{R}}{R}\delta\Gamma^i_{it}$$

or, more explicitly,

$$\delta R_{ij} = \frac{1}{2R^2}\left[\mathbf{V}^2 h_{ij} - \frac{\partial^2 h_{ik}}{\partial x^j\,\partial x^k} - \frac{\partial^2 h_{jk}}{\partial x^i\,\partial x^k} + \frac{\partial^2 h_{kk}}{\partial x^i\,\partial x^j}\right]$$

$$- \frac{1}{2}\frac{\partial^2 h_{ij}}{\partial t^2} + \frac{\dot{R}}{2R}[\dot{h}_{ij} - \delta_{ij}\dot{h}_{kk}]$$

$$+ \frac{\dot{R}^2}{R^2}[-2h_{ij} + \delta_{ij}h_{kk}] \tag{15.10.18}$$

$$\delta R_{ti} = \frac{1}{2}\frac{\partial}{\partial t}\left[R^{-2}\left(\frac{\partial h_{kk}}{\partial x^i} - \frac{\partial h_{ki}}{\partial x^k}\right)\right] \tag{15.10.19}$$

$$\delta R_{tt} = \frac{1}{2R^2}\left[\ddot{h}_{kk} - 2\frac{\dot{R}}{R}\dot{h}_{kk} + 2\left(\frac{\dot{R}^2}{R^2} - \frac{\ddot{R}}{R}\right)h_{kk}\right] \tag{15.10.20}$$

According to Eq. (15.10.1), the source term on the right-hand side of Einstein's field equations is

$$S_{\mu\nu} \equiv T_{\mu\nu} - \tfrac{1}{2}g_{\mu\nu}T^\lambda{}_\lambda = \tfrac{1}{2}(\rho - p)g_{\mu\nu} + (\rho + p)U_\mu U_\nu$$

$$- \eta H_{\mu\rho}H_{\nu\sigma}W^{\rho\sigma} - \chi(H_{\mu\rho}U_\nu + H_{\nu\rho}U_\mu)Q^\rho \qquad (15.10.21)$$

In order to preserve the normalization of the velocity U, we must have

$$0 = \delta(g_{\mu\nu}U^\mu U^\nu) = -2U_1{}^t$$

The perturbations h_{ij}, U_{1i}, ρ_1, p_1, and T_1 then produce in $S_{\mu\nu}$ the perturbations

$$\delta S_{ij} = \tfrac{1}{2}(\rho - p)h_{ij} + \frac{R^2}{2}\delta_{ij}(\rho_1 - p_1) - \eta R^4 \delta W^{ij} \qquad (15.10.22)$$

$$\delta S_{it} = -R^2(\rho + p)U_1{}^i - \chi \dot{T}\delta H_{it} + \chi R^2 \delta Q^i \qquad (15.10.23)$$

$$\delta S_{tt} = \tfrac{1}{2}(\rho_1 + 3p_1) - 2\chi \dot{T}\delta H_{tt} \qquad (15.10.24)$$

where

$$\delta H_{it} = -R^2 U_1{}^i \qquad \delta H_{tt} = 0 \qquad (15.10.25)$$

$$\delta W^{ij} = R^{-2}\left[\frac{\partial U_1^i}{\partial x^j} + \frac{\partial U_1{}^j}{\partial x^i} - \tfrac{2}{3}\delta_{ij}\nabla \cdot \mathbf{U_1}\right]$$

$$+ R^{-2}\frac{\partial}{\partial t}\{R^{-2}[h_{ij} - \tfrac{1}{3}\delta_{ij}h_{kk}]\} \qquad (15.10.26)$$

$$\delta Q^i = R^{-2}\left(\frac{\partial T_1}{\partial x^i} + T\frac{\partial}{\partial t}(R^2 U_1{}^i)\right) \qquad (15.10.27)$$

Finally, the perturbed Einstein field equations here take the form

$$\delta R_{\mu\nu} = -8\pi G \delta S_{\mu\nu} \qquad (15.10.28)$$

Putting together Eqs. (15.10.18)–(15.10.20) and (15.10.22)–(15.10.28), we find

$$\nabla^2 h_{ij} = \frac{\partial^2 h_{ik}}{\partial x^j \partial x^k} - \frac{\partial^2 h_{jk}}{\partial x^i \partial x^k} + \frac{\partial^2 h_{kk}}{\partial x^i \partial x^j} - R^2 \ddot{h}_{ij}$$

$$+ R\dot{R}[\dot{h}_{ij} - \delta_{ij}\dot{h}_{kk}] + 2\dot{R}^2[-2h_{ij} + \delta_{ij}h_{kk}]$$

$$= -8\pi G(\rho - p)R^2 h_{ij} - 8\pi G R^4 \delta_{ij}(\rho_1 - p_1)$$

$$+ 16\pi G\eta R^4\left(\frac{\partial U_1{}^i}{\partial x^j} + \frac{\partial U_1{}^j}{\partial x^i} - \tfrac{2}{3}\delta_{ij}\nabla \cdot \mathbf{U_1}\right)$$

$$+ 16\pi G\eta R^4 \frac{\partial}{\partial t}\left\{\frac{1}{R^2}[h_{ij} - \tfrac{1}{3}\delta_{ij}h_{kk}]\right\} \qquad (15.10.29)$$

$$\frac{\partial}{\partial t}\left\{\frac{1}{R^2}\left[\frac{\partial h_{kk}}{\partial x^i} - \frac{\partial h_{ki}}{\partial x^k}\right]\right\} = 16\pi G R^2 (\rho + p) U_1{}^i$$

$$- 16\pi G\chi \dot{T} R^2 U_1{}^i - 16\pi G\chi \left(\frac{\partial T_1}{\partial x^i} + T\frac{\partial}{\partial t}(R^2 U_1{}^i)\right) \quad (15.10.30)$$

$$\ddot{h}_{kk} - \frac{2\dot{R}}{R} \dot{h}_{kk} + 2\left(\frac{\dot{R}^2}{R^2} - \frac{\ddot{R}}{R}\right) h_{kk} = -8\pi G(\rho_1 + 3p_1)R^2 \quad (15.10.31)$$

The equations of motion for the fluid can be obtained either from the conservation equations $T^{\mu\nu}_{;\mu} = 0$, or directly from the field equations. By applying the operators $\partial/\partial x^i$ and $\partial/\partial t + 3\dot{R}/R$ to Eqs. (15.10.29) and (15.10.30), and using (15.10.30), (15.10.31), and the trace of (15.10.29) to simplify the result, we obtain the momentum conservation equation

$$\left[\frac{\partial}{\partial t} + 16\pi G\eta\right]\left[R^5 U_1{}^i\{\rho + p - \chi\dot{T}\} - \chi R^3 \left\{\frac{\partial T_1}{\partial x^i} + T\frac{\partial}{\partial t}(R^2 U_1{}^i)\right\}\right]$$

$$= -R^3\frac{\partial p_1}{\partial x^i} + \eta R^3\left\{\nabla^2 U_1{}^i + \frac{1}{3}\frac{\partial}{\partial x^i}\nabla\cdot\mathbf{U}_1 + \frac{2}{3}\frac{\partial}{\partial t}\left(R^{-2}\frac{\partial h_{kk}}{\partial x^i}\right)\right\} \quad (15.10.32)$$

(The vector $\mathbf{U}_1$ is understood to have components $U_1{}^i$, not U_{1i}.) Also, by taking the divergence of Eq. (15.10.30) and using Eqs. (15.10.31) and the trace of Eq. (15.10.29), we obtain the energy conservation equation

$$\dot{\rho}_1 + \frac{3\dot{R}}{R}(\rho_1 + p_1) = -(\rho + p)\left\{\frac{\partial}{\partial t}\left(\frac{h_{kk}}{2R^2}\right) + \nabla\cdot\mathbf{U}_1\right\}$$

$$+ \chi\left\{\dot{T}\nabla\cdot\mathbf{U}_1 + \frac{1}{R^2}\nabla^2 T_1 + \frac{T}{R^2}\frac{\partial}{\partial t}(R^2\nabla\cdot\mathbf{U}_1)\right\}$$

$$(15.10.33)$$

As usual in dealing with dissipative processes, we must also make use of the conservation law for the particle current nU^μ:

$$0 = (nU^\mu)_{;\mu} = U^\mu\frac{\partial n}{\partial x^\mu} + nU^\mu{}_{;\mu}$$

(Strictly speaking, n should be taken as the density of baryon number or lepton numbers.) For the unperturbed solution, this gives the familiar result

$$n \propto R^{-3}$$

and to first order in the perturbations n_1, $\mathbf{U}_1$, and h_{ij}, we have

$$0 = \frac{\partial n_1}{\partial t} + \frac{3\dot{R}}{R} n_1 + n\{\nabla\cdot\mathbf{U}_1 + \delta\Gamma^\nu_{t\nu}\}$$

or, using (15.10,17),

$$\frac{\partial}{\partial t}\left(\frac{n_1}{n}\right) = -\mathbf{V}\cdot\mathbf{U}_1 - \frac{\partial}{\partial t}\left(\frac{h_{kk}}{2R^2}\right) \tag{15.10.34}$$

Equations (15.10.29)–(15.10.34) provide a convenient set of fundamental equations, but it should be kept in mind that these equations are not all independent, as shown by the way they were derived.

One solution of these equations can be found immediately:

$$h_{ij}(\mathbf{x}, t) = R^2(t)\left\{\frac{\partial f_i(\mathbf{x})}{\partial x^j} + \frac{\partial f_j(\mathbf{x})}{\partial x^i}\right\}$$

$$\rho_1 = p_1 = \mathbf{U}_1 = n_1 = T_1 = 0 \tag{15.10.35}$$

with f an arbitrary function of position. [To verify Eq. (15.10.29), use Eq. (15.1.20).] However, reference to Eqs. (15.10.9)–(15.10.11) shows that this is *not* a physical disturbance, but represents the effects of an infinitesimal coordinate transformation of the form (10.9.6):

$$x^\mu \rightarrow x^\mu - \varepsilon^\mu(x)$$

$$\varepsilon^t = 0 \qquad \varepsilon(\mathbf{x}, t) = R^2(t)\mathbf{f}(\mathbf{x}) \tag{15.10.36}$$

whose structure is such as to preserve the vanishing of h_{it} and h_{tt}. We are interested only in physical disturbances, whose form necessarily differs from (15.10.35).

The manifest spatial homogeneity of equations (15.10.29)–(15.10.34) allows us to find solutions with the spatial dependence

$$h_{ij}, \rho_1, p_1, \mathbf{U}_1, n_1, T_1 \propto \exp{(i\mathbf{q}\cdot\mathbf{x})} \tag{15.10.37}$$

with q a constant wave number. Just as in the nonrelativistic case, it is convenient to analyze the general solution into normal modes. Now there are modes of *three* different kinds:

Radiative Modes

There is a simple class of solutions with

$$0 = h_{kk} = q_i h_{ij} = \rho_1 = p_1 = U_{1i} = n_1 = T_1 \tag{15.10.38}$$

Equations (15.10.30)–(15.10.34) are here trivially satisfied, and Eq. (15.10.29), together with Eq. (15.1.20), gives

$$\ddot{h}_{ij} + \left[-\frac{\dot{R}}{R} + 16\pi G\eta\right]\dot{h}_{ij} + \left[\frac{\mathbf{q}^2}{R^2} - \frac{2\ddot{R}}{R} - \frac{32\pi G\eta\dot{R}}{R}\right]h_{ij} = 0 \tag{15.10.39}$$

For very large wave numbers, we can find a general second-order WKB solution

$$h_{ij} \propto R \exp \left\{ \int \left[\frac{\pm i|\mathbf{q}|}{R} - 8\pi G\eta \right] dt \right\} \tag{15.10.40}$$

With R and η slowly varying, this result may be converted from the comoving spatial coordinate system to a nearly Minkowskian system by multiplying h_{ij} by a scale factor $1/R^2$. Thus (15.10.40) corresponds to a plane gravitational wave of the form (10.2.1), with

$$e_{\mu\nu} \propto \frac{1}{R} \exp\{-8\pi G \int \eta\, dt\}$$

$$\mathbf{k} = \frac{\mathbf{q}}{R}$$

According to Eq. (10.3.7), the energy density τ_g^{00} of these gravitational waves decreases as

$$\tau_g^{00} \propto R^{-4} \exp\{-16\pi G \int \eta\, dt\} \tag{15.10.41}$$

The factor R^{-4} is just what we should expect for the *free* expansion of any wave representing a massless particle. [Compare Eq. (15.1.23).] The extra factor in (15.10.41) tells us that gravitational waves in a viscous medium are *absorbed* at a rate:[156]

$$\Gamma_g = 16\pi G\eta \tag{15.10.42}$$

Generally, η will be of the order of the thermal energy density times some typical mean free time τ, so that Γ_g is at most of order $\dot{R}^2\tau/R^2$. Hence as long as the collision rate τ^{-1} is much greater than the expansion rate $\dot{R}/R$, the damping rate Γ_g will be much less than the expansion rate $\dot{R}/R$, and viscosity will have little effect on the wave propagation. With viscosity neglected, and with $R(t)$ assumed to have a power-law time dependence

$$R(t) \propto t^n \tag{15.10.43}$$

we can find a solution of Eq. (15.10.39) valid for all wave numbers,

$$h_{ij} \propto t^{1/2(n+1)} J_{\pm\nu}\left(\frac{|\mathbf{q}|t}{(1-n)R}\right) \tag{15.10.44}$$

where $J_{\pm\nu}$ is the usual Bessel function of order $\pm\nu$, and

$$\nu = \frac{3n-1}{2-2n} \tag{15.10.45}$$

(In the matter-dominated or radiation-dominated eras, Eq. (15.10.43) holds with $n = 2/3$ or $n = 1/2$, respectively.) Unlike the case of electromagnetic waves in a plasma, there is no sharp lower limit to the frequencies at which gravitational waves can propagate; instead, the solutions start at $|q|t \ll R$ with the behavior

$$h_{ij} \propto t^{2n} \quad \text{or} \quad t^{1-n} \tag{15.10.46}$$

and gradually go over for $|\mathbf{q}|t \gg R$ to the wavelike solutions (15.10.40).

Rotational Modes

There is also a simple class of solutions with

$$0 = h_{kk} = \mathbf{q} \cdot \mathbf{U}_1 = \rho_1 = p_1 = n_1 = T_1 \tag{15.10.47}$$

Here Eqs. (15.10.31), (15.10.33), and (15.10.34) are trivially satisfied, and Eq. (15.10.32) becomes an equation for the transverse part of $\mathbf{U}_1$:

$$\left[\frac{\partial}{\partial t} + 16\pi G\eta\right][R^5\{\rho + p - \chi\dot{T}\}\mathbf{U}_1] = -\eta R^3 \mathbf{q}^2 \mathbf{U}_1 \tag{15.10.48}$$

Equations (15.10.29) and (15.10.30) then dictate the gravitational field produced by the rotations represented by $\mathbf{U}_1$; these field equations are automatically consistent with (15.10.48), because the equations of motion from which (15.10.48) was derived were themselves derived from the field equations. With dissipation neglected, Eq. (15.10.48) simply tells us that $\mathbf{U}_1$ has a time dependence inversely proportional to $R^5(\rho + p)$,

$$\mathbf{U}_1 \propto \frac{1}{R^5(\rho + p)} \tag{15.10.49}$$

which may be regarded as the relativistic generalization (in comoving coordinates) of the Newtonian result (15.9.22).

Compressional Modes

Once again, the richest time dependence is displayed by the compressional modes, in which the quantities h_{kk}, $\mathbf{q} \cdot \mathbf{U}_1$, ρ_1, p_1, T_1, and n_1 are *not* constrained to vanish. Equations (15.10.31), (15.10.33), (15.10.34), and the divergence of Eq. (15.10.32) here provide a set of coupled equations for these quantities:

$$\ddot{h}_{kk} - \frac{2\dot{R}}{R}\dot{h}_{kk} + 2\left(\frac{\dot{R}^2}{R^2} - \frac{\ddot{R}}{R}\right)h_{kk} = -8\pi G R^2(\rho_1 + 3p_1) \tag{15.10.50}$$

$$\dot{\rho}_1 + \frac{3\dot{R}}{R}(\rho_1 + p_1) = -(\rho + p)\left\{\frac{\partial}{\partial t}\left(\frac{h_{kk}}{2R^2}\right) + i\mathbf{q} \cdot \mathbf{U}_1\right\}$$

$$+ \chi\left\{i\dot{T}\mathbf{q} \cdot \mathbf{U}_1 + \frac{iT}{R^2}\frac{\partial}{\partial t}[R^2\mathbf{q} \cdot \mathbf{U}_1] - \frac{\mathbf{q}^2}{R^2}T_1\right\}$$

$$\tag{15.10.51}$$

$$\frac{\partial}{\partial t}\left(\frac{n_1}{n}\right) = -i\mathbf{q}\cdot\mathbf{U}_1 - \frac{\partial}{\partial t}\left(\frac{h_{kk}}{2R^2}\right) \tag{15.10.52}$$

$$\left[\frac{\partial}{\partial t} + 16\pi G\eta\right]\left[i\mathbf{q}\cdot\mathbf{U}_1 R^5\{\rho + p - \chi\dot{T}\} + \chi R^3\left\{\mathbf{q}^2 T_1 - i\frac{\partial}{\partial t}(R^2\mathbf{q}\cdot\mathbf{U}_1)\right\}\right]$$

$$= R^3\mathbf{q}^2 p_1 - \eta R^3\mathbf{q}^2\left[\frac{4i}{3}\mathbf{q}\cdot\mathbf{U}_1 + \frac{2}{3}\frac{\partial}{\partial t}\left(\frac{h_{kk}}{R^2}\right)\right] \tag{15.10.53}$$

If we use the equation of state to express p_1 and ρ_1 in terms of n_1 and T_1, these can be regarded as four equations for the four unknowns h_{kk}, $\mathbf{q}\cdot\mathbf{U}_1$, n_1, and T_1. The reader may check that these equations yield the damping rate (15.8.30) for fluctuation wave numbers that are much larger than the Jeans limit, for which gravitation and the expansion of the universe may be neglected. Also, in the nonrelativistic limit with damping neglected, these equations reduce to the previously derived Newtonian equations (15.9.20) and (15.9.21), provided that we identify

$$\delta \equiv \frac{\rho_1}{p} \qquad \varepsilon \equiv -\frac{R}{\mathbf{q}^2}\left\{i\mathbf{q}\cdot\mathbf{U}_1 + \frac{1}{2}\frac{\partial}{\partial t}\left(\frac{h_{kk}}{R^2}\right)\right\}$$

For a thorough discussion of the normal modes described by Eqs. (15.10.50)–(15.10.53), the reader is directed to the review of Field.[157] For our present purposes, it will be sufficient to consider only the limiting case of very small wave numbers. In the limit $\mathbf{q} \to 0$, all dissipative effects vanish; indeed, by eliminating h_{kk} in Eqs. (15.10.51) and (15.10.52), we can show that the perturbations keep the entropy constant, so that

$$p_1 = v_s^2\rho_1 \tag{15.10.54}$$

Also, it is convenient to use Eq. (15.1.21) to write Eq. (15.10.51) as

$$-\frac{1}{2}\frac{\partial}{\partial t}\left(\frac{h_{kk}}{R^2}\right) = \frac{1}{\rho + p}\left\{\dot{\rho}_1 - \frac{\dot{\rho}(1 + v_s^2)\rho_1}{\rho + p}\right\}$$

$$= \frac{1}{\rho + p}\left\{\dot{\rho}_1 - \frac{(\dot{\rho} + \dot{p})\rho_1}{\rho + p}\right\}$$

$$= \frac{\partial}{\partial t}\left(\frac{\rho_1}{\rho + p}\right)$$

The addition of a time-independent term to h_{kk}/R^2 would correspond to a mere coordinate transformation of the form (15.10.36), so the solution is essentially unique,

$$h_{kk} = -2R^2\delta \tag{15.10.55}$$

with δ now defined by

$$\rho_1 = (\rho + p)\delta \tag{15.10.56}$$

Using Eqs. (15.10.54)–(15.10.56) in (15.10.50) yields the second-order differential equation

$$\ddot{\delta} + \frac{2\dot{R}}{R}\dot{\delta} - 4\pi G(\rho + p)(1 + 3v_s{}^2)\delta = 0 \qquad (15.10.57)$$

We are now at last able to calculate the growth rate in "Phase A," that is, in the early period when the Jeans mass is very small and the energy density is dominated by radiation (and neutrinos). In this case, we have

$$R \propto t^{1/2} \qquad \rho = \frac{3}{32\pi G t^2} \qquad p = \frac{\rho}{3} \qquad v_s = \frac{1}{\sqrt{3}}$$

and Eq. (15.10.57) reads

$$\ddot{\delta} + t^{-1}\dot{\delta} - t^{-2}\delta = 0 \qquad (15.10.58)$$

Again, there is a growing solution δ_+ and a decaying solution δ_- :

$$\delta_+ \propto t \qquad \delta_- \propto t^{-1} \qquad (15.10.59)$$

but no exponential growth.

11   The Very Early Universe

The thermal history of the universe was traced in Section 15.6 back to an era when the temperature was about 10^{12}°K. At this early time, the universe was filled with particles—photons, leptons, and antileptons—whose interactions are hopefully weak enough to allow this medium to be treated as a more or less ideal gas. However, if we look back a little further, into the first 0.0001 sec of cosmic history when the temperature was above 10^{12}°K, we encounter theoretical problems of a difficulty beyond the range of modern statistical mechanics. At such temperatures, there will be present in thermal equilibrium copious numbers of strongly interacting particles—mesons, baryons, and antibaryons—with a mean interparticle distance less than a typical Compton wavelength. These particles will be in a state of continual mutual interaction, and cannot reasonably be expected to obey any simple equation of state.

However, the temptation to try to construct some sort of model of the very early universe is irresistible. There are in fact two extremely different simple models that have been widely considered in recent years, and that reflect two divergent views of the nature of the strongly interacting particles. Although neither model can be taken seriously in detail, the hope is that one or the other of these models may come close enough to reality to lead to useful insights about the very early universe.

The first of these two pictures may be called the *elementary particle model*. It is supposed that all particles are made up of a small number of elementary particles—say, photons, leptons, "quarks," and their antiparticles. It is further supposed here that at very high temperatures the forces that bind the elementary particles become negligible, just as the neutron-proton force becomes cosmologically unimportant at temperatures above the dissociation temperature of deuterium. Let there be $\mathcal{N}$ different kinds of elementary particles, counting spin states and antiparticles separately, and counting fermions as $\frac{7}{8}$ of a particle. (See Eq. (15.6.32). For instance, if we include only the familiar photons, leptons, and antileptons, plus three kinds of spin $\frac{1}{2}$ quarks and antiquarks, we have $\mathcal{N} = 26$.) Then for kT above the mass of the heaviest elementary particle, the contents of the universe will behave as if they consisted of $\mathcal{N}/2$ different kinds of black-body radiation, with pressure, energy density, and specific entropy given by

$$3p \simeq \rho \simeq \tfrac{1}{2}\mathcal{N}aT^4 \tag{15.11.1}$$

$$\sigma \simeq \frac{(\rho + p)}{n_B kT} \simeq \frac{2aT^3}{3n_B k}\mathcal{N} \tag{15.11.2}$$

where n_B is the number density of baryons minus antibaryons. (The extra factor $\frac{1}{2}$ enters here to cancel the factor 2 in the Stefan-Boltzmann constant arising from the two photon polarization states.) For an adiabatic expansion σ is constant, and since σ at present has the value (15.5.18), the temperature in the very early universe is given by

$$\frac{T}{T_{\gamma 0}} = \left(\frac{2n_B}{\mathcal{N}n_{B0}}\right)^{1/3} = \left(\frac{2}{\mathcal{N}}\right)^{1/3}\left(\frac{R_0}{R}\right) \tag{15.11.3}$$

The relation between the energy density ρ and the time t here is the same as (15.6.44), and so

$$t \simeq \left(\frac{32\pi G\rho}{3}\right)^{-1/2} \simeq \left(\frac{16\pi G\mathcal{N}aT^4}{3}\right)^{-1/2}$$

$$\simeq \left(\frac{32\pi GaT_{\gamma 0}^4}{3}\right)\left(\frac{\mathcal{N}}{2}\right)^{1/6}\left(\frac{R}{R_0}\right)^2 \tag{15.11.4}$$

In contrast, in a *composite particle model*, it is supposed that there are no true elementary strongly interacting particles but instead that all such "hadrons" must be regarded as composites of one another. We then face a question of principle, whether thermodynamic calculations can be carried out including as "particles" only the one absolutely stable hadron, the proton, or also slowly decaying hadrons, such as the neutron and pi-mesons, or perhaps *all* resonant hadron states, including such rapidly decaying resonances as the rho-meson and the "3–3" π–N resonance. It is an attractive conjecture, that if all resonances were included in our thermodynamic calculations, then to a first approximation, it might not be necessary to take any further account of the particles' mutual interactions. (See Section 11.9.) If so, then the contents of the early universe could be treated as consisting of a great many ideal gases, with $\mathcal{N}(m)\,dm$ types between mass m and $m + dm$. But what is the function $\mathcal{N}(m)$? The greatest possible contrast with the elementary-particle model is achieved for a distribution that grows as fast as possible, that is,

$$\mathcal{N}(m) \to Am^{-B} \exp\left(\frac{m}{kT_M}\right) \qquad \text{for } m \to \infty \qquad (15.11.5)$$

where A, B, and T_M are unknown constants. Thermodynamic quantities will generally involve integrals of $\mathcal{N}(m)\,dm$, with weighting factors that behave like $e^{-m/kT}$ for $m \to \infty$, so these quantities would not converge for a distribution function that grew faster than (15.11.5), and, even for the distribution function (15.11.5), would not converge for $T > T_M$. An ideal-gas model with a number of species given by Eq. (15.11.5) is thus characterized by a *maximum temperature* T_M. The analysis of secondary particle emission in very high-energy reactions[158] and the recent Veneziano model of hadron interactions[159] both independently suggest a number of hadron species given by Eq. (15.11.5), with B of order 2 to 4, and with T_M of order $1.7 \times 10^{12\circ}$K. Leaving aside mesons, leptons, and photons for the moment, the total energy density, pressure, and baryon number density are given by the usual Fermi distribution as

$$\rho = h^{-3} \int \mathcal{N}(m)\,dm \int \{[e^{(E-\mu)/kT} + 1]^{-1} + [e^{(E+\mu)kT} + 1]^{-1}\}E\,d^3p \qquad (15.11.6)$$

$$p = \tfrac{1}{3}h^{-3} \int \mathcal{N}(m)\,dm \int \{[e^{(E-\mu)/kT} + 1]^{-1}$$
$$+ [e^{(E+\mu)/kT} + 1]^{-1}\}E^{-1}\mathbf{p}^2\,d^3p \qquad (15.11.7)$$

$$n = h^{-3} \int \mathcal{N}(m)\,dm \int \{[e^{(E-\mu)/kT} + 1]^{-1} - [e^{(E+\mu)/kT} + 1]^{-1}\}\,d^3p \qquad (15.11.8)$$

where E is the particle energy $(\mathbf{p}^2 + m^2)^{1/2}$, and μ is the chemical potential associated with baryon number. The dimensionless entropy per baryon σ is defined by the second law of thermodynamics as the integral of the perfect differential

$$d\sigma = \frac{1}{kT}\left\{d\left(\frac{\rho}{n}\right) + p\,d\left(\frac{1}{n}\right)\right\}$$

and a straightforward integration gives

$$\sigma = \frac{(\rho + p - \mu n)}{nkT} \qquad (15.11.9)$$

In an adiabatic expansion, ρ and p drop from presumably infinite initial values, while σ must stay constant. The only way that ρ and p can approach infinity while

σ remains fixed is for μ to become infinite while T approaches a finite value T_1 *less* than T_M. In this limit, the integrals (15.11.6)–(15.11.8) approach the values[160]

$$\rho \to A' e^{\mu/kT_M} \mu^{5/2-B} kT_1 \csc\left(\frac{\pi T_1}{T_M}\right)$$

$$\times \left\{ 1 + \left(\frac{\pi kT_1}{\mu}\right)(B - \tfrac{5}{2})\cot\left(\frac{\pi T_1}{T_M}\right) + \left(\frac{3kT_M}{2\mu}\right)(B - \tfrac{1}{4}) + 0\left(\frac{1}{\mu^2}\right) \right\}$$

$$(15.11.10)$$

$$p \to A' e^{\mu/kT_M} \mu^{5/2-B} kT_1 \csc\left(\frac{\pi T_1}{T_M}\right)\left\{\left(\frac{kT_M}{\mu}\right) + 0\left(\frac{1}{\mu^2}\right)\right\} \qquad (15.11.11)$$

$$\mu n \to A' e^{\mu/kT_M} \mu^{5/2-B} kT_1 \csc\left(\frac{\pi T_1}{T_M}\right)$$

$$\times \left\{ 1 + \left(\frac{\pi kT_1}{\mu}\right)(B - \tfrac{3}{2})\cot\left(\frac{\pi T_1}{T_M}\right) + \left(\frac{3kT_M}{2\mu}\right)(B - \tfrac{1}{4}) + 0\left(\frac{1}{\mu^2}\right) \right\}$$

$$(15.11.12)$$

where $A' = (kT_M)^{3/2}\hbar^{-3}(8\pi)^{-1/2}A$. The entropy (15.11.9) then takes the value

$$\sigma = \left(\frac{T_M}{T_1}\right) - \pi \cot\left(\frac{\pi T_1}{T_M}\right) \qquad (15.11.13)$$

Since σ is very large, the initial temperature T_1 is very close to the maximum temperature T_M:

$$\frac{T_1}{T_M} \simeq 1 - \frac{1}{\sigma} + 0\left(\frac{1}{\sigma^2}\right) \qquad (15.11.14)$$

With a finite initial temperature, the energy density and pressure of mesons, leptons, and photons is negligible in the limit $t \to 0$ in comparison with the baryonic contributions (15.11.10) and (15.11.11), justifying the neglect of all particles but baryons in the above calculations. The baryon number density must vary as R^{-3}, so Eqs. (15.11.12) and (15.11.10) give for $R \to 0$

$$\mu \to 3kT_M |\ln R| \qquad (15.11.15)$$

and

$$\rho \to \mu n \propto R^{-3} |\ln R| \qquad (15.11.16)$$

The Einstein field equation (15.1.20) then has a solution (with k neglected) of the form[160]

$$R \propto t^{2/3} |\ln t|^{1/2} \qquad (15.11.17)$$

in contrast with the behavior $R \propto t^{1/2}$ expected in an elementary-particle model.

How can we distinguish among models of the very early universe? As pointed out in Section 15.6, most of the constituents of the universe were in thermal equilibrium at temperatures above $10^{12}°$K, so that the present contents of the universe for the most part depend only on the entropy per baryon, and perhaps on the ratio of the lepton numbers to the baryon number, in the hot early universe. In order to learn something about the behavior of the universe before the temperature dropped to $10^{12}°$K, we need to look for fossils, particles that might have escaped thermal equilibrium before the temperature dropped below $10^{12}°$K.

One such possible fossil particle is the quark, the hypothetical fundamental particle of the strong interactions. Zeldovich[161] has estimated, on the basis of what we have here called an elementary-particle model, that if quarks really can exist as free particles, the density of leftover quarks, which escaped fusion into nucleons in the early universe, would be about the same as the observed present density of gold atoms. Efforts to find quarks in nature have so far failed, so one can conclude either that free quarks do not exist, or that the thermal history of the very early universe was very different from (15.11.4).

One other, less hypothetical, fossil particle is the graviton. As shown by Eq. (15.10.42), a graviton in an imperfect fluid with shear viscosity η will have a mean free time[156]

$$\tau_g = (16\pi G\eta)^{-1} \tag{15.11.18}$$

If τ_g is not much longer than the expansion time t, then the transport of momentum by gravitons will give the medium a viscosity

$$\eta = \tfrac{4}{15}aT^4\tau_g \tag{15.11.19}$$

Eliminating η from these two equations then gives[149]

$$\tau_g = (\tfrac{64}{15}\pi GaT^4)^{-1/2} \tag{15.11.20}$$

In an elementary-particle model, Eqs. (15.11.20) and (15.11.4) give the ratio of the graviton mean free time to the expansion time as

$$\frac{\tau_g}{t} = \left(\frac{4\mathcal{N}}{5}\right)^{1/2} \tag{15.11.21}$$

If $\mathcal{N}$ is not too large, τ_g will be not much greater than t, so that (15.11.19) and (15.11.20) will be roughly correct, and τ_g will thus vanish for $t \to 0$ like t. In an elementary-particle model, then, the "optical depth" $\int \tau_g^{-1}\, dt$ of the very early universe for gravitational radiation diverges logarithmically for $t \to 0$, and the present universe should thus contain left-over black-body gravitational radiation,[162] with a temperature

$$T_{g0} = \frac{(TR)_{t\to 0}}{R_0} = \left(\frac{\mathcal{N}}{2}\right)^{-1/3} T_{\gamma 0} \tag{15.11.22}$$

For instance, if $\mathcal{N} = 26$ and $T_{\gamma 0} = 2.7°$K, then the present gravitational radiation background temperature is about $0.9°$K. In contrast, in a composite-particle model, the product RT vanishes for $t \to 0$, so even if the universe is "optically thick" to gravitational radiation, the present graviton background temperature would be very much less than the value (15.11.22). Thus the presence or the absence of a gravitational radiation background with temperature of order 1°K would provide clear evidence for the behavior of matter in the very early universe. Unfortunately, there does not seem to be any way to detect a 1°K gravitational radiation background directly.[162] Its most important effect would be to shorten somewhat the expansion time scale during the radiation-dominated era, very slightly increasing the cosmic production of helium.

It may be that the heat represented by the huge entropy per baryon in the microwave background provides the most useful clue to the very early history of the universe. Of course, it is possible that this heat was put in at the initial singularity, in which case we must regard σ as a dimensionless fundamental constant, like the fine structure constant. However, it is more attractive to suppose that the present entropy per baryon was generated by physical dissipative processes acting in the early, or the very early, universe.

One such nonadiabatic mechanism for entropy generation is provided by the phenomenon of *bulk viscosity*. We saw in Section 15.10 that the shear viscosity and heat conduction can play no role in a Robertson-Walker model. The only dissipative effect that can enter in the energy-momentum tensor for an isotropic homogeneous expansion is the term in Eq. (2.11.21) proportional to the bulk viscosity, ζ, which in general coordinates takes the form

$$\Delta T^{\mu\nu} = -\zeta(g^{\mu\nu} + U^\mu U^\nu)U^\lambda{}_{;\lambda}$$

where U^μ is the fluid velocity four-vector. In a Robertson-Walker model, we have $U^\lambda{}_{;\lambda} = 3\dot{R}/R$, so the total energy-momentum tensor here is

$$T^{\mu\nu} \equiv \rho U^\mu U^\nu + \left(p - 3\zeta\,\frac{\dot{R}}{R}\right)(g^{\mu\nu} + U^\mu U^\nu)$$

The whole effect of the bulk viscosity is thus to replace the pressure p with

$$p^* = p - 3\zeta\,\frac{\dot{R}}{R} \tag{15.11.23}$$

The bulk viscosity therefore has no effect in the formula, Eq. (15.1.20), for $\dot{R}$ in terms of ρ. However, it does appear in the energy-conservation equation, which now, in place of (15.1.21), reads

$$\frac{d}{dR}(\rho R^3) = -3p^*R^2 = -3pR^2 + 9\zeta\dot{R}R \tag{15.11.24}$$

Since $n \propto R^{-3}$ the specific entropy will in general increase at a rate

$$\dot{\sigma} \equiv \frac{1}{kT}\left[\frac{d}{dt}\left(\frac{\rho}{n}\right) + p\,\frac{d}{dt}\left(\frac{1}{n}\right)\right]$$

$$= \frac{\dot{R}}{nR^2kT}\left[\frac{d}{dR}(\rho R^3) + p\,\frac{d}{dR}\,(R^3)\right]$$

and (15.11.24) gives this as[149]

$$\dot{\sigma} = \frac{9\zeta\dot{R}^2}{nkTR^2} \tag{15.11.25}$$

For instance, for a fluid consisting of material particles with a very short mean free time plus photons with mean free time τ, the bulk viscosity is[149]

$$\zeta = 4aT^4\tau\left[\frac{1}{3} - \left(\frac{\partial p}{\partial \rho}\right)_n\right]^2 \tag{15.11.26}$$

and Eq. (15.11.25) gives an entropy production rate

$$\frac{\dot{\sigma}}{\sigma} = \frac{3\tau\dot{R}^2}{R^2}\left[1 - 3\left(\frac{\partial p}{\partial \rho}\right)_n\right]^2\left(\frac{4aT^3}{3nk\sigma}\right) \tag{15.11.27}$$

(For neutrinos, just multiply by a factor of $\frac{7}{8}$.) The production of entropy here can be understood as due to the fact that the frequency of a free photon will vary as $1/R$ between collisions, whereas the temperature of the material medium will not vary as $1/R$ unless $(\partial p/\partial \rho)_n$ takes the value $\frac{1}{3}$, so that the expansion of the universe is continually pulling radiation and matter out of thermal equilibrium with each other.[163] However, in an elementary-particle model $(\partial p/\partial \rho)_n$ will be close to $\frac{1}{3}$ in the very early universe, whereas in a composite-particle model, the photons (or neutrinos) have only a small share of the total entropy, so that the last factor in (15.11.27) is small. Estimates of $\dot{\sigma}/\sigma$ do not indicate that bulk viscosity can account for the high entropy of the present universe.[149]

If the present entropy of the universe is not due to bulk viscosity, then perhaps it is produced by the effects of shear viscosity or heat conduction in an initially anisotropic or inhomogeneous expansion. Indeed, it may be just these dissipative processes that are responsible for smoothing out initial anisotropies, and hence producing the high degree of isotropy observed in the cosmic microwave radiation background. Misner[164] has shown that neutrino viscosity, acting before the temperature dropped to 2×10^{10}°K, would have reduced the present anisotropy of the black-body radiation, produced in an initially homogeneous but anisotropic expansion, to less than 0.03%. (However, see pp. 525–6.)

Another possible explanation for the observed high entropy per baryon is that the mean baryon number density really vanishes, as in the theories of Klein[165] and Alfvén.[166] When (and if) the temperature was above 10^{13}°K, the number density of nucleons plus antinucleons would have been of order

$$n_N + n_{\bar{N}} \sim \frac{aT^3}{k} \sim \sigma n_B \sim \sigma(n_N - n_{\bar{N}})$$

so if σ has remained constant, the fractional excess of nucleons over antinucleons in the very early universe would have had the very small value

$$\frac{n_N - n_{\bar{N}}}{n_N + n_{\bar{N}}} \sim \frac{1}{\sigma} \sim 10^{-8} \quad \text{to} \quad 10^{-9}$$

In the symmetric cosmology of Klein and Alfvén this small nucleon excess is interpreted as a purely local phenomenon—it is supposed that in other parts of the universe there was a small antinucleon excess, giving rise at present to galaxies of antimatter. The detailed calculations of Omnes[166a] show that reasonable physical processes in a symmetric plasma of matter and antimatter could have produced the required small separation of matter and antimatter. Unfortunately, observational astronomy has not yet provided any definite information as to whether distant galaxies consist of matter or antimatter. Only the gamma-ray spectrum provides a hint[166b] that antimatter may exist on a cosmological scale.

As long as we have the temerity to speculate about the very early universe, we may as well carry our speculations back to the very beginning. The Friedmann solutions, taken literally, indicate that R vanishes as $t \to 0$, as $t^{1/2}$ in the elementary-particle models and as $t^{2/3}|\ln t|^{1/2}$ in a composite-particle model. For all we know, this singularity actually occurs, but it is natural to wonder whether it can be avoided.

One way to avoid a singularity in the very early universe is for the energy density ρ to vanish, owing perhaps to some very short-range attractive force that overbalances the particles' rest masses. If ρ vanishes at some critical value R_c of the scale factor $R(t)$, then $\dot{R}$ vanishes at R_c (or, for finite curvature, near R_c) so that R might have decreased to R_c before beginning its present increasing phase.

Even if the energy density is always positive, it is still conceivable that the universe could escape a general singularity, through anisotropies or inhomogeneities that invalidate the simple Friedmann solutions. Penrose[167] and Hawking[168] have proved a number of powerful theorems that show that a singularity is inescapable under very general conditions. For instance, one of Hawking's theorems states that a singularity is unavoidable, provided that general relativity is valid, that each point of space-time has a small neighborhood that no curve with timelike or null tangents cuts more than once, that the energy-momentum tensor satisfies the positivity condition

$$[T_{\mu\nu} - \tfrac{1}{2}g_{\mu\nu}T^{\lambda}{}_{\lambda}]W^{\mu}W^{\nu} \geq 0$$

for all vectors W^{μ} with $W^{\mu}W_{\mu} < 0$, and that there is a point p such that all the past-directed timelike geodesics through p start converging again within a compact region in the past of p. The last condition is satisfied if there is enough matter to make the world lines through p converge in the past, and Hawking and Ellis[169] have shown that the cosmic microwave background does provide enough energy density in the past to satisfy this condition. However, it is important to note that the Penrose-Hawking theorems do not say that there is a singularity in the past that involves all space, as in the Friedmann solutions, but only that there is some singularity somewhere. The singularity might consist merely of one or more isolated points, which behave like time-reversed collapsing stars.

Finally, it may be that classical general relativity itself breaks down in the very early universe. One simple way for this to happen is through the effects of a cosmological constant, to be discussed in the next chapter. A more intriguing idea is that quantum effects might become important, invalidating any purely classical field theory of gravitation. For a system, consisting of point particles with a mean particle energy E, the relative importance of gravitational "radiative corrections" will be described by a "gravitational fine structure constant"

$$\alpha_g \equiv \frac{GE^2}{\hbar}$$

analogous to the usual electromagnetic fine structure constant $\frac{1}{137}$. Quantum effects will become important when α_g is of order unity, or when E reaches a critical value, given in c.g.s. units by

$$E_c = \left(\frac{\hbar c^5}{G}\right)^{1/2} = 1.22 \times 10^{28} \text{ eV} \qquad (15.11.28)$$

corresponding to a temperature 1.4×10^{32}°K. The temperature in a composite particle model never gets anywhere near so high, but kT would have been greater than E_c at the very beginning of a Friedmann universe composed of a finite number $\mathcal{N}$ of species of elementary particles. In fact, a great many other quantum effects become important at that time.[170] For instance, the oscillation rate of a typical particle wave function at temperature T is $kT/\hbar$, while the expansion rate of the universe at this time is given by Eq. (15.11.4) as

$$\frac{\dot{R}}{R} = \frac{1}{2t} = \left(\frac{4\pi G \mathcal{N} a T^4}{3}\right)^{1/2}$$

Recalling that $a = \pi^2 k^4 / 15 \hbar^3$, the ratio of these rates is

$$\frac{kT/\hbar}{\dot{R}/R} = \left(\frac{45 \hbar}{4\pi^3 \mathcal{N} G(kT)^2}\right)^{1/2} = \left(\frac{45}{4\pi^3 \mathcal{N}}\right)^{1/2} \left(\frac{E_c}{kT}\right)$$

Hence for temperatures above the critical temperature E_c/k, the wave functions of typical particles have an oscillation rate slower than the expansion rate of the universe, so that no classical or semiclassical description can be applied to the particles in thermal equilibrium at that time.

The consideration of first things naturally leads to speculation about last things.[171] We saw in Section 15.1 that a Friedmann universe with $k = +1$ will eventually cease its expansion and begin to contract. Taken literally, such models require that a singularity with $R = 0$ will be reached at some finite time in the future, of order 75×10^9 years for $H_0 = 75$ km sec^{-1} Mpc^{-1} and $q_0 = 1$. However, if it is possible to escape a general singularity in the past, either through negative energy densities, anisotropies, or quantum effects, then presumably it ought to be possible to escape the general singularity in our future. In this case, we might suppose that the universe undergoes an oscillation, with periods of contraction and expansion succeeding one another eternally.

Could this oscillation be periodic? That is, can we restore a steady state picture of the universe, by viewing cosmic history on a sufficiently grand time scale? One obvious objection is that entropy is presumably created, not destroyed, in each cycle. It has been suggested that entropy might be destroyed during contracting phases,[172] because it is the expansion of the universe that, by providing a heat sink, sets the direction of time's arrow in thermodynamic processes. However, there is no detailed model that describes how this can come about. In particular, it is hard to see how time's arrow could be reversed just at the moment when $R(t)$ reaches its maximum value, at which time the background radiation temperature is so low, of order $1°K$, that it can hardly affect terrestrial processes. If somehow or other the second law of thermodynamics can be evaded, and the universe really does expand and contract periodically, then any particles that are not brought into thermal equilibrium during the contracted phase, such (perhaps) as gravitons or neutrinos, would have to be present in large numbers: If N particles are produced in a given comoving volume during each cycle, and the probability that one of these particles is absorbed in one cycle is P, then there would have to be a mean number N/P of particles in this volume in order to maintain a more or less constant population. It is therefore not out of the question that some day we may detect remnants of previous cycles of the history of the universe. For the present, however, such matters remain at the furthest bounds of cosmological speculation.

I5 BIBLIOGRAPHY

See the bibliographies for Chapters 11 and 14. Sources for the special topics covered in this chapter are listed below.

Cosmic Mass Density

☐ G. O. Abell, "Clustering of Galaxies," *Annual Review of Astronomy and Astrophysics*, Vol. 3, ed. by L. Goldberg (Annual Reviews, Inc., Palo Alto, Cal., 1965), p. 1.

☐ G. Burbidge and W. L. W. Sargent, "The Case of the Missing Mass," Comments Astrophys. and Space Phys., **1**, 220 (1969).

☐ R. A. Sunyaev and Ya. B. Zeldovich, "An Open Universe?," Comments Astrophys. and Space Phys., **1**, 159 (1969).

The Intergalactic Medium

☐ G. R. Burbidge, "Intergalactic Matter and Radiation," I. A. U. Symposium No. 44, Uppsala, Sweden, August 1970, to be published.

☐ G. B. Field, "The Physics of the Interstellar and Intergalactic Medium," in *Astrophysics and General Relativity* (1968 Brandeis University Summer Institute in Theoretical Physics), Vol. 1, ed. by M. Chretien, S. Deser, and J. Goldstein (Gordon and Breach Science Publishers, New York, 1969), p. 59.

☐ G. B. Field, "Intergalactic Matter," to be published in *Annual Review of Astronomy and Astrophysics* (Annual Reviews, Inc., Palo Alto, Cal., 1972).

☐ R. J. Gould, "Intergalactic Matter and Radiation," in *Annual Review of Astronomy and Astrophysics*, Vol. 6, ed. by L. Goldberg (Annual Reviews, Inc., Palo Alto, Cal., 1968), p. 195.

The Cosmic Microwave Background

☐ R. B. Partridge, American Scientist, **57**, 3 (1969).

☐ P. J. E. Peebles, "Cosmology and Infrared Astronomy: Closing the Gap between Theory and Practice," Comments Astrophys. and Space Phys., **3**, 20 (1971).

☐ R. A. Sunyaev and Ya B. Zeldovich, "The Spectrum of Primordial Radiation, its Distortions and their Significance," Comments Astrophys. and Space Phys., **2**, 66 (1970).

The Early Universe

☐ E. R. Harrison, "Comments on the Big-bang," Nature, **228**, 258 (1970).

☐ I. D. Novikov and Ya. B. Zeldovich, "Cosmology," in *Annual Review of Astronomy and Astrophysics*, Vol. 5, ed. by L. Goldberg (Annual Reviews, Inc., Palo Alto, Cal., 1967), p. 627.

☐ Ya. B. Zeldovich, "The Universe as a Hot Laboratory for the Nuclear and Particle Physicist," Comments Astrophys. and Space Phys., **2**, 12 (1970).

☐ Ya. B. Zeldovich, "The 'Hot' Model of the Universe," Usp. Fiz. Nauk, **89**, 647 (1966) [trans. Soviet Phys. Usp., **9**, 602 (1967)].

The Origin and Abundance of the Elements

☐ L. H. Aller, *Abundance of the Elements* (Interscience Publishers, New York, 1961).

☐ L. H. Aller, "The Abundance of Elements in the Solar Atmosphere," in *Advances in Astronomy and Astrophysics*, Vol. 3, ed. by Z. Kopal (Academic Press, New York, 1965), p. 1.

☐ G. R. Burbidge, "Cosmic Helium," Comments Astrophys. and Space Phys., **1**, 101 (1969).

☐ A. G. W. Cameron, "Processes of Nucleosynthesis," Comments Astrophys. and Space Phys., **2**, 153 (1970).

☐ D. D. Clayton, *Principles of Stellar Evolution and Nucleosynthesis* (McGraw-Hill, New York, 1968).

☐ I. J. Danziger, "The Cosmic Abundance of Helium," in *Annual Review of Astronomy and Astrophysics*, Vol. 8, ed. by L. Goldberg (Annual Reviews, Inc., Palo Alto, Cal., 1970), p. 161.

☐ W. A. Fowler, "How Now, No Cosmological Helium?," Comments Astrophys. and Space Phys., **2**, 134 (1970).

☐ W. A. Fowler and W. E. Stephens, "Resource Letter OE-1 on Origin of Elements," Am. J. Phys., **36**, 1 (1968).

☐ R. J. Tayler, "The Origin of the Elements," reprint in *Astrophysics* (W. A. Benjamin, New York, 1969).

Fluctuations and Galaxy Formation

☐ G. B. Field, "The Formation and Early Dynamical History of Galaxies," in *Stars and Stellar Systems, Vol. IX: Galaxies and the Universe*, ed. by A. and M. Sandage, to be published.

☐ E. R. Harrison, "Normal Modes of Vibration of the Universe," Rev. Mod. Phys., **39**, 862 (1967).

☐ D. Layzer, "Cosmogonic Processes," in *Astrophysics and General Relativity* (1968 Brandeis University Summer Institute in Theoretical Physics), Vol. 2, ed. by M. Chretien, S. Deser, and J. Goldstein (Gordon and Breach Science Publishers, New York, 1969).

☐ D. Layzer, "A Unified Approach to Cosmology," in *Relativity Theory and Astrophysics 1. Relativity and Cosmology*, ed. by J. Ehlers (American Mathematical Society, Providence, R. I., 1967), p. 237.

☐ J. H. Oort, "Galaxies and the Universe," Science, **170**, 1363 (1970).

☐ M. J. Rees and D. W. Sciama, "The Evolution of Density Fluctuations in the Universe," Comments Astrophys. and Space Phys., **1**, 140, 153 (1969).

15 REFERENCES

1. A. Friedmann, Z. Phys., **10**, 377 (1922); *ibid.*, **21**, 326 (1924).

2. W. H. McCrea and E. A. Milne, Quart. J. Math. (Oxford), **5**, 73 (1934). E. A. Milne, Quart. J. Math. (Oxford), **5**, 64 (1934).

3. Recent surveys are given by G. O. Abell, Ann. Rev. Astron. Astrophys., **3**, (1965); G. R. Burbidge and W. L. W. Sargent, Comments Astrophys. and Space Phys., **1**, 220 (1969). Also see T. Kiang, Mon. Not. Roy. Astron. Soc., **122**, 263 (1961).

4. F. Zwicky, Helv. Phys. Acta, **6**, 110 (1933); E. M. Burbidge, G. R. Burbidge, and R. A. Fish, Ap. J., **133**, 393 (1961).

5. J. H. Oort, in *La Structure et l'Evolution de l'Univers* (Institut International de Physique Solvay, R. Stoops, Brussels, 1958), p. 163.

6. S. van den Bergh, Z. f. Astrophys., **66**, 567 (1961).

7. T. W. Noonan, Pub. Astron. Soc. Pac., **83**, 31 (1971).

7a. S. L. Shapiro, Astron. J., **76**, 291 (1971).

8. G. R. Burbidge and E. M. Burbidge, Ap. J., **130**, 629 (1959); J. D. Karachentsev, Astrofizica, **2**, 81 (1966); S. van den Bergh, Z. f. Ap., **53**, 219 (1961); and refs. 3, 5, 6.

8a. H. Gursky, E. Kellogg, S. Murray, C. Leong, H. Tanenbaum, and R. Giacconi, Ap. J., **167**, L 81 (1971); J. F. Meekins, G. Fritz, T. A. Chubb, H. Friedman, and R. C. Henry, Nature, **231**, 107 (1971). In this connection, also see T. F. Felton, R. J. Gould, W. A. Stein, and N. J. Woolf, Ap. J., **146**, 955 (1966); N. J. Woolf, Ap. J., **148**, 287 (1967); B. Turnrose and H. J. Rood, Ap. J., **159**, 773 (1970); P. D. Noerdlinger, Nature, **232**, 393 (1971); J. R. Gott III and J. E. Gunn, Ap. J, **169**, L 13 (1971); G. A. Welch and G. N. Sastry, Ap. J., **169**, L 3 (1971); D. Goldsmith and J. Silk, Ap. J., to be published (1972); G. B Field, Ann. Rev. Astron. and Astrophy., to be published (1972).

9. V. A. Ambartsumian, Isv. Akad. Nauk. Arm. S. S. R., Ser. Fiz.—Mat., **11**, 9 (1958).

9a. P. J. E. Peebles and R. B. Partridge, Ap. J., **148**, 713 (1967). The observational data on the night sky brightness are given by F. E. Roach and L. L. Smith, Geophys. J. Roy. Astron Soc., **15**, 227 (1968).

10. This is obtained by integrating the number-count distribution N(S) given in the survey of M. Ryle, Ann. Rev. Astron. and Astrophys., **6**, 256 (1968).

11. See the survey of W. Davidson and J. V. Narlikar, in *Astrophysics* (W. A. Benjamin, New York, 1969), Sections 2.5 and 2.6.

12. J. E. Felten, Ap. J., **144**, 241 (1966). Also see G. de Vaucouleurs, Ann. Astrophys., **12**, 162 (1949), and ref. 11.

13. For a survey, see K. Brecher and G. Burbidge, Comments Astrophys. and Space Phys., **2**, 75 (1970).

14. V. L. Ginzburg and S. I. Syrovatskii, *The Origin of Cosmic Rays* (Macmillan, New York, 1964).

14a. Galactic production of gravitational radiation is considered by G. B. Field, M. J. Rees, and D. W. Sciama, Comments on Astrophys. and Space Phys., **1**, 187 (1969); D. W. Sciama, G. B. Field, and M. J. Rees, Phys. Rev. Lett., **23**, 241 (1969). Primordial production of long-wavelength gravitation radiation is considered by M. J. Rees, Mon. Not. Roy. Astron. Soc., **154**, 187 (1971).

15. M. S. Longair, Mon. Not. Roy. Astron. Soc., **133**, 421 (1966).

16. A. G. Doroshkevich, M. S. Longair, and Ya. B. Zeldovich, Mon. Not. Roy. Astron. Soc., **147**, 139 (1970).

17. M. Schmidt, Ap. J., **151**, 393 (1968); also Ann. Rev. Astron. and Astrophys., **7**, 527 (1969); Ap. J., **162**, 371 (1970).

18. E. Rutherford, Nature, **123**, 313 (1929).

19. C. C. Patterson, Geochim. et Cosmochim. Acta, **10**, 230 (1956). A value $(4.53 \pm 0.03) \times 10^9$ years is given by R. G. Ostic, R. D. Russell, and D. H. Reynolds, Nature, **199**, 1150 (1963).

20. E. M. Burbidge, G. R. Burbidge, W. A. Fowler, and F. Hoyle, Rev. Mod. Phys., **29**, 547 (1957).

21. W. A. Fowler and F. Hoyle, Ann. Phys., **10**, 280 (1960). A value 1.89 ± 0.36 is given by P. A. Seeger and D. N. Schramm, Ap. J., **160**, L157 (1970).

22. D. D. Clayton, Ap. J., **139**, 637 (1964).

23. R. H. Dicke, Nature, **194**, 329 (1962); Ap. J., **155**, 123 (1969).

24. I. Iben, Jr., Scientific American, July 1970, p. 27. Also see I. Iben, Jr., and R. T. Rood, Nature, **223**, 933 (1969); Ap. J., **161**, 587 (1970).

25. W. Rindler, Mon. Not. Roy. Ast. Soc., **116**, 663 (1956).

26. See, for example, L. I. Schiff, *Quantum Mechanics* (3rd ed., McGraw-Hill, New York, 1968), p. 531.

27. A. Einstein, Phys. Z., **18**, 121 (1917).

28. G. B. Field, Ap. J., **129**, 525 (1959).

29. G. B. Field, Ap. J., **135**, 684 (1962).

30. J. A. Koehler and B. J. Robinson, Ap. J., **146**, 488 (1966); S. Goldstein, Ap. J., **138**, 978 (1963); R. Allen, Astron. and Astrophys., **3**, 316, 382 (1969); A. A. Penzias and E. H. Scott, III, Ap. J., **153**, L7 (1968); R. D. Davies and R. C. Jennison, Mon. Not. Roy. Astron. Soc., **128**, 123 (1964); R. J. Allen, Astron. and Astrophys., **3**, 316, 382 (1969).

31. J. A. Koehler, Ap. J., **146**, 504 (1966).

32. C. Heiles and G. K. Miley, Ap. J., **160**, L83 (1970).

33. A. A. Penzias and R. W. Wilson, Ap. J., **156**, 799 (1969).

34. J. E. Gunn and B. A. Peterson, Ap. J., **142**, 1633 (1965); also see P. A. G. Scheuer, Nature, **207**, 963 (1965).

35. J. B. Oke, Ap. J., **145**, 668 (1966).

36. G. and M. E. Burbidge, *Quasi-Stellar Objects* (W. H. Freeman and Co., San Francisco, 1967), p. 146.

37. E. J. Wampler, Ap. J., **147**, 1 (1967).

38. G. B. Field, P. M. Solomon, and E. J. Wampler, Ap. J., **145**, 351 (1966).

39. J. N. Bahcall and E. E. Salpeter, Ap. J., **142**, 1677 (1965).

40. R. V. Wagoner, Ap. J., **149**, 465 (1967).

41. P. J. E. Peebles, Ap. J., **157**, 45 (1969).

42. G. R. Burbidge and M. Burbidge, Nature, **222**, 735 (1969); G. R. Burbidge, Ann. Rev. Astron. and Astrophys., **8**, 309 (1970).

43. I. S. Shklovsky, 1969, to be published; M. Rees, Ap. J., **160**, L29 (1970); G. R. Burbidge, invited paper presented at I. A. U. Symposium No. 44, "External Galaxies and Quasi-Stellar Objects," Uppsala, Sweden, August 1970; J. N. Bahcall, Comments Astrophys. and Space Phys., **2**, 221 (1970); and so on.

43a. G. B. Field, ref. 8a.

43b. W. J. Karzas and R. Latter, Ap. J. Suppl., **6**, 167 (1961).

43c. G. B. Field and R. C. Henry, Ap. J., **140**, 1002 (1964).

44. K. Brecher and G. Burbidge, Comments Astrophys. and Space Phys., **2**, 75 (1970).

44a. D. A. Schwartz, Ap. J., **162**, 439 (1970).

44b. R. Cowsik, talk presented at the 12th International Conference on Cosmic Rays (Hobart, Tasmania, August, 1971); R. Cowsik and E. J. Kobetich, to be published (1972).

45. V. L. Ginzburg and L. M. Ozernoi, Astr. Zh., **42**, 943 (1965); R. J. Gould and W. Ramsay, Ap. J., **144**, 587 (1966); R. Weymann, Ap. J., **145**, 560 (1966); *ibid.*, **147**, 887 (1967); J. Arons and R. McCray, Astrophys. Letters, **5**, 123 (1969); J. Bergeron, Astron. and Astrophys., **3**, 364 (1969); P. D. Noerdlinger, Ap. J., **156**, 841 (1969); J. E. Felten and J. Bergeron, Astrophys. Letters, **4**, 155 (1969); and others.

46. M. J. Rees, Astrophys. Letters, **4**, 61 (1969).

47. J. N. Bahcall and E. E. Salpeter, Ap. J., **142**, 1677 (1965).

48. F. T. Haddock and D. W. Sciama, Phys. Rev. Letters, **14**, 1007 (1965).

49. See, for example, L. Spitzer, Jr., *Physics of Fully Ionized Gases* (Interscience Publishers, New York, 1956), Chapter 4.

49a. K. I. Kellerman and I. I. K. Pauliny-Toth, Ann. Rev. Astron. and Astrophys., **6**, 417 (1968).

50. See, for example, ref. 36, Chapter 6.

50a. J. N. Bahcall and E. E. Salpeter, Ap. J., **142**, 1677 (1965). Also see G. B. Field, ref. 8a.

50b. D. D. Noerdlinger, Ap. J., **157**, 495 (1969). Also see G. B. Field, ref. 8a.

50c. Y. Sofue, M. Fujimoto, and K. Kawabata, Pub. Astron. Soc. Japan, **20**, 368 (1969); K. Kawabata, M. Fujimoto, Y. Sofue, and M. Fukui, ibid., **21**, 293 (1969); M. Reinhardt and M. Thiel, Astrophys. Lett., **7**, 101 (1970); H. Arp, Nature, **232**, 463 (1971).

51. G. Gamow, Phys. Rev., **70**, 572 (1946); R. A. Alpher, H. Bethe, and G. Gamow, Phys. Rev., **73**, 803 (1948); G. Gamow, Phys. Rev., **74**, 505 (1948); R. A. Alpher and R. C. Herman, Nature, **162**, 774 (1948); R. A. Alpher, R. C. Herman, and G. Gamow, Phys. Rev., **74**, 1198 (1948); *ibid.*, **75**, 332A (1949); *ibid.*, **75**, 701 (1949); G. Gamow, Rev. Mod. Phys., **21**, 367 (1949); R. A. Alpher, Phys. Rev., **74**, 1577 (1948); R. A. Alpher and R. C. Herman, Phys. Rev., **75**, 1089 (1949).

52. R. A. Alpher and R. C. Herman, Rev. Mod. Phys., **22**, 153 (1950).

53. R. H. Dicke, P. J. E. Peebles, P. G. Roll, and D. T. Wilkinson, Ap. J., **142**, 414 (1965).

54. A. A. Penzias and R. W. Wilson, Ap. J., **142**, 419 (1965).

55. F. A. Jenkins and D. E. Wooldridge, Phys. Rev., **53**, 137 (1938).

56. A. McKellar, Publs. Dominion Astrophys. Observatory (Victoria, B. C.), **7**, 251 (1941).

57. G. B. Field and J. L. Hitchcock, Phys. Rev. Letters, **16**, 817 (1966); Ap. J. **146**, 1 (1966). The significance of the CN absorbtion data was first stated in 1965 by Field, G. H. Herbig, and Hitchcock; for an abstract, see Astron. J. **71**, 161 (1966).

58. N. J. Woolf, quoted in ref. 59.

58a. I. S. Shklovsky, Astronomicheskii Tsircular, No. 364, 1966.

59. P. Thaddeus and J. F. Clauser, Phys. Rev. Letters, **16**, 819 (1966).

60. V. J. Bortolot, Jr., J. F. Clauser, and P. Thaddeus, Phys. Rev. Letters, **22**, 307 (1969).

61. P. E. Boynton, R. A. Stokes, and D. T. Wilkinson, Phys. Rev. Letters, **21**, 462 (1968), M. F. Millea, M. McColl, R. J. Pedersen, and F. L. Vernon, Jr , Phys. Rev. Letters, **26**, 919 (1971); A. G. Kislyakov, V. I. Chernyshev, Yu. V. Lebskii, V. A. Mal'tsev, and N. V. Serov, Ast. Zh., **48**, 39 (1971) [transl. Sov. Astron.—AJ., **15**, 29 (1971)].

62. V. J. Bortolot, J. F. Clauser, and P. Thaddeus, Phys. Rev. Letters, **22**, 307 (1969).

63. K. Shivanandan, J. R. Houck, and M. O. Harwit, Phys. Rev. Letters, **21**, 1460 (1968); J. R. Houck and M. Harwit, Ap. J., **157**, L45 (1969).

64. D. Muehlner and R. Weiss, Phys. Rev. Letters, **24**, 742 (1970).

64a M. Harwit, J. R. Houck, and R. V. Wagoner, Nature, **228**, 451 (1970).

64b. J. L. Pipher, J. R Houck, B. W. Jones, and M. Harwit, Nature, **231**, 375 (1971).

64c. A. G. Blair, J. G. Beery, F. Edeskuty, R. D. Hiebert, J. P. Shipley, and K. D. Williamson, Jr., Phys. Rev. Letters, **27**, 1154 (1971).

64d D. Muehlner and R. Weiss, to be published (1972).

65. P. J. E. Peebles, Ap. J., **153**, 1 (1968). Also see R. A. Sunyaev, Dokl. Akad. Nauk. U.S.S.R., **179**, 45 (1968) [trans. Sov. Phys.—Dokl., **13**, 183 (1968)]; Ya. B. Zeldovich, V. G. Kurt, and R. A. Sunyaev, Zh. Eksp. Teor. Fiz., **55**, 278 (1968) [trans. Sov. Phys.—JETP, **28**, 146 (1969)].

66. R. B. Partridge and P. J. E. Peebles, Ap. J., **148**, 377 (1967).

67. V. Petrosian, J. N. Bahcall, and E. E. Salpeter, Ap. J., **155**, L57 (1969).

68. R. A. Sunyaev, remarks at the Fifth International Conference on Gravity and General Relativity, Tiflis, U.S.S.R., September 1968. Also see Ya. B. Zeldovich and R. A. Sunyaev, Astrophys. and Space Sci., **4**, 285 (1969); Ya. B. Zeldovich, Comments Astrophys. and Space Sci., **2**, 66 (1970).

69. M. J. Rees and D. W. Sciama, Nature, **217**, 511 (1968); R. K. Sachs and A. M. Wolfe, Ap. J., **147**, 73 (1967); A. M. Wolfe, Ap. J., **156**, 803 (1969).

70. E. K. Conklin and R. N. Bracewell, Nature, **216**, 777 (1967).

71. R. B. Partridge and D. T. Wilkinson, Phys. Rev. Letters, **18**, 557 (1967); D. T. Wilkinson and R. B. Partridge, quoted by R. B. Partridge, American Scientist, **57**, 37 (1969).

72. E. K. Conklin, Nature, **222**, 971 (1969).

73. G. de Vaucouleurs and W. L. Peters, Nature, **220**, 868 (1968).

74. C. Misner, private communication.

75. F. Hoyle and N. C. Wickramsinghe, Nature, **214**, 969 (1967); J. V. Narlikar and N. C. Wickramsinghe, Nature, **216**, 43 (1967); *ibid.*, **217**, 1235 (1968). Also see D. Layzer, Astrophys. Letters, **1**, 49 (1968).

76. T. Gold and F. Pacini, Ap. J., **152**, L115 (1968); A. M. Wolfe and G. R. Burbidge, Ap. J., **156**, 345 (1969); R. Wagoner, Nature, **224**, 481 (1969).

77. A. A. Penzias, J. Schraml, and R. W. Wilson, Ap. J., **157**, L49 (1969); C. Hazard and E. E. Salpeter, Ap. J., **157**, L87 (1969).

78. J. E. Felten and P. Morrison, Ap. J., **146**, 686 (1966).

79. F. Hoyle, Phys. Rev. Letters, **15**, 131 (1965).

80. R. J. Gould, Phys. Rev. Letters, **15**, 511 (1965).

81. J. E. Felten, Phys. Rev. Letters, **15**, 1003 (1965), K. Brecher and P. Morrison, Ap. J., **150**, L61 (1967).

82. K. Brecher and P. Morrison, Phys. Rev. Letters, **23**, 802 (1969).

83. E. T. Byram, T. A. Chubb, and H. Friedman, Science, **169**, 366 (1970).

84. Ref. 14, p. 114.

85. K. Greisen, Phys. Rev. Letters, **16**, 748 (1966); F. W. Stecker, Phys. Rev. Letters, **21**, 1016 (1968); G. T. Zatsepin and V. A. Kuz'man, Pis'ma Zh. Eksp. Teor. Fiz., **4**, 114 (1966) [trans. Sov. Phys.—JETP Letters, **4**, 78 (1966)]. For recent data on photonucleon cross-sections, see W. P. Hesse, D. O. Caldwell, V. W. Elings, R. J. Morrison, F. V. Murphy, B. W. Worster, and D. E. Yount, Phys. Rev. Letters, **25**, 613 (1970).

86. R. J. Gould and G. P. Schreder, Phys. Rev. Letters, **16**, 252 (1966); J. V. Jelley, Phys. Rev. Letters, **16**, 479 (1966).

87. See, for example, L. D. Landau and E. M. Lifshitz, *Statistical Physics* (Pergamon Press, London, 1958), Sections 52 and 53.

88. *Ibid.*, Eq. (100.2).

89. For a discussion of the separate conservation of electron-lepton number and muon-lepton number, see, for example, R. E. Marshak, Riazuddin, and C. P. Ryan, *Theory of Weak Interactions in Particle Physics* (Wiley-Interscience, New York, 1969), Sections 1.2 and 3.4.

90. For speculations on a possible nonvanishing cosmic charge density, see R. A. Lyttleton and H. Bondi, Proc. Roy. Soc. (London), **A252**, 313 (1959). Also see V. W. Hughes, in *Gravitation and Relativity*, ed. by H-Y Chiu and W. F. Hoffmann (W. A. Benjamin, New York, 1964), p. 259.

91. See, for example, Marshak, Riazuddin, and Ryan, *op. cit.*, Section 3.3.

92. R. A. Alpher, J. W. Follin, Jr., and R. C. Herman, Phys. Rev., **92**, 1347 (1953).

93. P. J. E. Peebles, Ast. J., **146**, 542 (1966). Also see ref. 92.

94. S. Weinberg, Phys. Rev., **128**, 1457 (1962).

95. K. E. Bergkvist, *Topical Conference on Weak Interactions* (CERN, Geneva, 1962), p. 91. Also see L. M. Langer and R. J. D. Moffat, Phys. Rev., **88**, 689 (1952).

96. J. Bernstein, M. Ruderman, and G. Feinberg, Phys. Rev., **132**, 1227 (1963).

97. R. Cowsik, Y. Pal, and S. N. Tandon, Phys. Letters, **13**, 265 (1964).

98. B. P. Konstantinov and G. E. Kocharov, Soviet Physics J.E.T.P., **19**, 992 (1964); B. P. Konstantinov, G. E. Kocharov, and Yu. N. Starbunov, Izvestiya Akad. Nauk. U.S.S.R., Ser. Fiz., **32**, 1841 (1968).

99. F. W. Clarke, Bull. Phil. Soc. (Washington), **11**, 131 (1889).

100. H. E. Suess and H. C. Urey, Rev. Mod. Phys., **28**, 53 (1956); L. H. Aller, *Abundance of the Elements* (Interscience Publishers, New York, 1961); S. Bashkin, in *Stellar Structure*, ed. by L. H. Aller and D. B. McLaughlin (University of Chicago Press, Chicago, 1965), p. 1.

101. A. S. Eddington, *Report of the Eighty-Eighth Meeting of the British Association for the Advancement of Science*, 1920, p. 34.

102. H. A. Bethe, Phys. Rev., **55**, 434 (1939).

103. E. E. Salpeter, Ap. J., **115**, 326 (1952).

104. Ref. 20. Also see F. Hoyle, Ap. J. Suppl., Ser. 1, No. 5, p. 121, 1954.

105. A. G. W. Cameron, Publ. Astron. Soc. Pacific, **69**, 201 (1957), etc.

106. W. D. Arnett and D. D. Clayton, Nature, **227**, 780 (1970).

107. See refs. 51, 52, 92.

108. G. Burbidge, Pub. Ast. Soc. Pacific, **70**, 83 (1958); F. Hoyle and R. J. Tayler, Nature, **203**, 1108 (1964). Also see J. W. Truran, C. J. Hansen, and A. G. W. Cameron, Can. J. Phys., **43**, 1616 (1965).

109. See, for example, Marshak, Riazuddin, and Ryan, *op. cit.*, p. 29.

109a. C. Hayashi, Prog. Theo. Phys. (Japan), **5**, 224 (1950).

110. R. V. Wagoner, W. A. Fowler, and F. Hoyle, Ap. J., **148**, 3 (1967).

110a. J. Geiss and H. Reeves, Astron. and Ap., to be published (1971).

111. See, for example, M. Schwarzschild, *Structure and Evolution of the Stars* (Princeton University Press, Princeton, 1958); B. Strömgren, in *Stellar Structure*, ed. by L. H. Aller and D. B. McLaughlin (University of Chicago Press, Chicago, 1965), Chapter 4; D. D. Clayton, *Principles of Stellar Evolution and Nucleosynthesis* (McGraw-Hill, New York, 1968).

112. L. Goldberg, E. A. Muller, and L. H. Aller, Ap. J. Suppl., **5**, 1 (1960).

113. D. L. Lambert, Nature, **215**, 43 (1967); Mon. Not. Roy. Ast. Soc., **138**, 143 (1967); Observatory, **87**, 228 (1968). D. L. Lambert and B. Warner, Mon. Not. Roy. Ast. Soc., **138**, 181, 213 (1968).

114. J. N. Bahcall, W. A. Fowler, I. Iben, Jr., and R. L. Sears, Ap. J., **137**, 344 (1963); R. L. Sears, Ap. J., **140**, 477 (1964); P. R. Demarque and J. R. Percy, Ap. J., **140**, 541 (1964).

115. B. Pontecorvo, National Research Council of Canada Report No. P.D. 205, 1946 (unpublished).

116. L. W. Alvarez, University of California Radiation Laboratory Report No. UCRL-328, 1949 (unpublished).

117. J. N. Bahcall, Phys. Rev. Letters, **12**, 300 (1964); Phys. Rev., **135**, B137 (1964).

118. R. Davis, Jr., Phys. Rev. Letters, **12,** 303 (1964).

119. R. Davis, Jr., D. S. Harmer, and K. C. Hoffmann, Phys. Rev. Letters, **20,** 1205 (1968).

120. J. N. Bahcall, N. A. Bahcall, and G. Shaviv, Phys. Rev. Letters, **20,** 1209 (1968); also see J. N. Bahcall and G. Shaviv, Ap. J., **153,** 113 (1968). J. N. Bahcall, N. A. Bahcall, and R. K. Uhlrich, Ap. J., **156,** 559 (1969).

121. S. Torres-Peimbert, E. Simpson, and R. K. Uhlrich, Ap. J., **155,** 957 (1969).

122. I. Iben, Jr., Annals of Physics, **54,** 164 (1960).

122a. J. N. Bahcall and R. K. Uhlrich, Ap. J., to be published, 1971.

122b. R. Davis, Jr., L. C. Rogers, and V. Radeha, paper presented at the April 1971 meeting of the American Physical Society (unpublished).

123. For detailed references, see I. J. Danziger, Ann. Rev. Astron. and Astrophys., **8,** 161 (1970).

124. D. Hegyi and D. Curott, Phys. Rev. Letters, **24,** 415 (1970).

125. J. Faulkner, Ap. J., **147,** 617 (1967).

125a. J. Faulkner, Phys. Rev. Letters, **27,** 206 (1971).

126. S. Biswas and C. E. Fichtel, Ap. J., **139,** 941 (1964).

127. J. E. Gaustad, Ap. J., **139,** 406 (1964), and ref. 113.

128. M. Neugebauer and C. W. Synder, Jr., J. Geophys. Res., **71,** 4469 (1966); A. J. Hundhausen, J. R. Ashbridge, S. J. Bame, H. E. Gilbert, and I. B. Strong, J. Geophys. Res., **72,** 87 (1967); K. W. Ogilvie, L. F. Burlaga, and T. D. Wilkerson, quoted by I. Iben, ref. 122.

129. K. W. Ogilvie et al., ref. 128.

130. A. O. J. Unsöld, Science, **163,** 1015 (1969).

131. I. Iben, Jr., Scientific American, July 1970, p. 27. Also see I. Iben, Jr., and R. T. Rood, Nature, **223,** 933 (1969); Ap. J., **161,** 587 (1970).

132. R. F. Christy et al., Ap. J., **144,** 108 (1966).

133. A. R. Sandage, Ap. J., **157,** 515 (1969); I. Iben, Jr., and J. Huchra, Ap. J., **162,** L43 (1970).

134. W. L. W. Sargent and L. Searle, Ap. J., **145,** 652 (1966).

135. For a summary, see L. H. Aller and W. Liller, *Nebulae and Interstellar Matter*, ed. by B. M. Middlehurst and L. H. Aller (University of Chicago Press, Chicago, 1968).

136. For a review, see A. K. Dupree and L. Goldberg, Ann. Rev. Aston. and Astrophys., **8,** 231 (1970).

137. Ref. 93, Figures 1 and 2. Also see V. F. Shvartsman, Zh. E. T. F. Pis. Red., **9,** 315 (1969).

138. See, for example, R. B. Stothers, Phys. Rev. Letters, **24,** 538 (1970).

139. H. Hecht, Ap. J., **170,** 401 (1971).

140. Ref. 110, figures 5a and 5b. (The quantity Φ_v/kT here is just v.)

141. G. Burbidge, Comments Astrophys. and Space Phys., **1,** 101 (1969).

142. Ref. 110, Section VII.

143. Letters from Sir Isaac Newton to the Reverend Dr. Bentley, Letter I, p. 203 ff, quoted by A. Koyré, *From the Closed World to the Infinite Universe* (Harper and Row, New York, 1958), p. 185.

144. J. Jeans, Phil. Trans. Roy. Soc., **199A,** 49 (1902), and *Astronomy and Cosmogony* (2nd ed., first published by Cambridge University Press in 1928; reprinted by Dover Publications, New York, 1961), pp. 345–350.

145. E. Lifshitz, J. Phys. U.S.S.R., **10,** 116 (1946).

146. W. B. Bonner, Z. Astrophys., **39,** 143 (1956). Also see *Relativity Theory and Astrophysics. 1. Relativity and Cosmology*, ed. by J. Ehlers (American Mathematical Society, Providence, R. I., 1967), p. 263.

147. P. J. E. Peebles, Ap. J., **142,** 1317 (1965); Ya. B. Zeldovich, Usp. Fiz. Nauk, **89,** 647 (1966) [trans. Sov. Phys. Uspekhi, **9,** 602 (1967)]; R. A. Alpher, G. Gamow, and R. Herman, Proc. Nat. Acad. Sci., **58,** 2179 (1967).

147a. R. H. Dicke and P. J. E. Peebles, Ap. J., **154,** 891 (1968).

148. J. Silk, Nature, **215,** 1155 (1967); Ap. J., **151,** 459 (1968); A. G. Doroshkovich, Ya. B. Zeldovich, and I. D. Novikov, Soviet Astron.—AJ, **11,** 233 (1967); P. J. E. Peebles and J. T. Yu, Ap. J., **162,** 815 (1970); G. B. Field, Ap. J., **165,** 29 (1971); K. Tomita, H. Nariai, H. Sato, T. Matsuda, and H. Takeda, Prog. Theor. Phys., **43,** 1511 (1970); H. Sato, Prog. Theor. Phys., **45,** 370 (1971). Also see ref. 149.

149. S. Weinberg, Ap. J., **168,** 175 (1971).

150. See, for example, G. K. Batchelor, *The Theory of Homogeneous Turbulence* (Cambridge University Press, 1959).

151. C. F. von Weizsäcker, Ap. J., **114,** 165 (1951); G. Gamow, Phys. Rev., **86,** 251 (1952); D. Layzer, Ann. Rev. Astron. and Astrophys., **2,** 341 (1964); Ya. B. Zeldovich and I. D. Novikov, Soviet Phys. Usp, **8,** 522 (1966); L. M. Ozernoi and A. D. Chernin, Astron. Zh., **44,** 1131 (1967) [trans. Soviet Astron.—AJ, **11,** 907 (1968)]; *ibid.,* **45,** 1137 (1968) [trans. *ibid.,* **12,** 901 (1969)]; L. Ozernoi, Zh. E. T. F. Pis. Red., **10,** 394 (1969) [trans. JETP Letters, **10,** 251 (1969)]; L. M. Ozernoi and G. V. Chibisov, Astron. Zh., **47,** 769 (1969) [trans. Soviet Astron.—AJ, **14,** 615 (1971)]; H. Sato, T. Matsuda, and H. Takeda, Prog. Theor. Phys., **43,** 1115 (1970); T. Matsuda, H. Sato, and H. Takeda, Publ. Astr. Soc. Japan, **23,** 1 (1971); J. Silk, to be published (1972).

152. R. K. Sachs and A. M. Wolfe, Ap. J., **147,** 73 (1967); J. Silk, Ap. J., **151,** 459 (1967); Nature, **215,** 1155 (1967). Also see ref. 66.

152a. M. S. Longair and R. A. Sunyaev, Nature, **223,** 719 (1969); also see ref. 68.

153. In this connection, see W. H. McCrea, Astron. J., **60,** 271 (1955); C. Callan, R. H. Dicke, and P. J. E. Peebles, Am. J. Phys., **33,** 105 (1965); W. M. Irvine, Ann. Phys., **32,** 322 (1965).

154. Quoted by G. B. Field, in *Stars and Stellar Systems, Vol. IX: Galaxies and the Universe*, ed. by A. and M. Sandage, to be published.

155. J. Silk, Nature, **215,** 1155 (1967).

156. S. W. Hawking, Ap. J., **145,** 544 (1966).

157. G. B. Field, in *Stars and Stellar Systems, Vol. IX: Galaxies and the Universe,* ed. by A. and M. Sandage, to be published.

158. R. Hagedorn, Nuovo Cimento Suppl., **3,** 147 (1965); *ibid.,* **6,** 311 (1968); Nuovo Cimento, **52A,** 1336 (1967); *ibid.,* **56A,** 1027 (1968); R. Hagedorn and J. Ranft, Nuovo Cimento Suppl., **6,** 169 (1968). Also see Yu. B. Rumer, Zh. Eksp. Teor. Fiz., **38,** 1899 (1960) [trans. Soviet Phys.— JETP, **11,** 1365 (1960)].

159. G. Veneziano, Nuovo Cimento, **57A,** 190 (1968); S. Fubini and G. Veneziano, Nuovo Cimento, **64A,** 811 (1969); K. Bardakci and S. Mandelstam, Phys. Rev., **184,** 1640 (1969); S. Fubini, D. Gordon, and G. Veneziano, Phys. Letters, **29B,** 679 (1969).

160. K. Huang and S. Weinberg, Phys. Rev. Letters, **25,** 895 (1970). The case of zero baryon number density and $B = 5/2$ was treated by R. Hagedorn, Astron. and Astrophys., **5,** 184 (1970). Also see J. N. Bahcall and S. Frautschi, Ap. J. Letters, to be published (1971).

161. Ya. B. Zeldovich, Comments Astrophys. and Space Sci., **11,** 12 (1970).

162. J. A. Wheeler, in *La Structure et l'Evolution de l'Univers* (Institut International de Physique Solvay, Brussels, 1958), p. 96; Ya. B. Zeldovich, in *Advances in Astronomy and Astrophysics* (Academic Press, New York, 1965), p. 319; F. Winterberg, Nuovo Cimento, **53B,** 264 (1968); S. Weinberg, in *Contemporary Physics,* Vol. I (International Atomic Energy Agency, Vienna, 1969), p. 559; R. A. Matzner, Ap. J., **154,** 1123 (1968); V. de Sabbata, *Fifth International Conference on Gravitation and the Theory of Relativity,* Tbilisi, 1968, to be published.

163. In this connection, see E. L. Schucking and E. A. Spiegel, Comments Astrophys. and Space Phys., **2,** 121 (1970); J. M. Stewart, M. A. H. MacCallum, and D. W. Sciama, *ibid.,* **2,** 206 (1970).

164. C. Misner, Nature, **214,** 40 (1967); Ap. J., **151,** 431 (1967). Also see A. G. Doroshkovich, Ya. B. Zeldovich, and I. D. Novikov, Sov. Phys.— JETP, **26,** 408 (1968); S. W. Hawking, Mon. Not. Roy. Astron. Soc., **142,** 129 (1969); J. M. Stewart, Astrophys. Letters, **2,** 133 (1969).

165. O. Klein, Soc. Roy. Sci. Liege, Sec. 4, **13,** 42 (1953); in *La Structure et l'Evolution de l'Univers* (Institut International de Physique Solvay, Brussels, 1958), p. 33; in *Werner Heisenberg und die Physik unserer Zeit* (Vieweg and Sohn, Braunschweig, Germany, 1961), p. 58; in *Recent Developments in General Relativity* (Pergamon Press, New York, 1962), p. 293; Astrophys. Norvegica, **9,** 161 (1964); in *Preludes in Theoretical Physics* (North-Holland, Amsterdam, 1966), p. 23; Nature, **211,** 1337 (1966); Ark. Fysik, **39,** 157 (1969); Science, **171,** 339 (1971).

166. H. Alfvén and O. Klein, Arkiv. för Fysik, **23,** 187 (1962); H. Alfvén, Rev. Mod. Phys., **17,** 652 (1965); Physics Today, Feb. 1971, 28; Nature, **229,** 184 (1971).

166a. R. Omnes, Astron. and Astrophysics, **10,** 228 (1971); *ibid.,* **11,** 450 (1971); Phys. Rev. Letters, **23,** 38 (1969).

166b. F. W. Stecker, D L. Morgan, Jr., and J. Bredekamp, Phys. Rev. Letters, **27,** 1469 (1971).

167. R. Penrose, Phys. Rev. Letters, **14,** 57 (1965); in *Contemporary Physics,* Vol. I (International Atomic Energy Agency, Vienna, 1969), p. 545.

168. S. W. Hawking, Phys. Rev. Letters, **15,** 689 (1965); Proc. Roy. Soc., **A294,** 511 (1966); *ibid.,* **295,** 490 (1966); *ibid.,* **300,** 187 (1967). Also see L. C. Shepley, Proc. Nat. Acad. Sci., **52,** 1403 (1965); R. P. Geroch, Phys. Rev. Letters, **17,** 445 (1966); S. W. Hawking and D. W. Sciama, Comments Astrophys. and Space Phys., **1,** 1 (1969).

169. S. W. Hawking and G. F. R. Ellis, Ap. J., **152,** 25 (1968).

170. See, for example, V. L. Ginzburg, Comments Astrophys. and Space Phys., **3,** 7 (1971); C. W. Misner, Phys. Rev., **186,** 1319 (1969); M. P. Ryan, Jr., to be published; K. C. Jacobs, C. W. Misner, and H. S. Zapolsky, to be published; K. C. Jacobs and L. P. Hughston, to be published.

171. See, for example, M. J. Rees, Observatory, **89,** 193 (1969).

172. T. Gold, in *La Structure et l'Evolution de l'Univers* (Institut International de Physique Solvay, Brussels, 1958), p. 81. Also see articles by Gold, and others in *The Nature of Time,* ed. by T. Gold and D. L. Schumacher (Cornell Univ. Press, Ithaca, N.Y., 1967); and B. Gal-Or, Science **176,** 11 (1972).

173. In particular, see S. Weinberg, Phys. Rev. Letters **19,** 1264 (1967); ibid. **27,** 1688 (1971); H. H. Chen and B. W. Lee, to be published (1972); G.'t Hooft, Physics Letters **37B,** 197 (1971).

"And as for certain truth,
no man has seen it, nor will
there ever be a man who
knows about the gods and
about all the things I
mention. For if he succeeds
to the full in saying what is
completely true, he himself
is unaware of it; and
Opinion is fixed by fate upon
all things." *Xenophanes of
Colophon*

16 COSMOLOGY: OTHER MODELS

The big-bang Friedmann model discussed in the last chapter does not at any point come into clear conflict with observation. However, it also cannot yet be said to have been definitely confirmed by observation. Therefore in this chapter we shall take a brief look at some of the other cosmological models that still compete with the "standard" theory.

1 Naive Models: The Olbers Paradox

Throughout the eighteenth and nineteenth centuries, perhaps a majority of astronomers would have subscribed to a simple cosmological picture, in which the universe is supposed to be infinite, eternal, and Euclidean, and the stars are more or less at rest, with constant average luminosity per unit volume. Such naive models would seem to be ruled out by the discovery of the general red shift of distant galaxies, but it is still of some interest to note an argument against the naive cosmologies, which was offered in 1744 by the Swiss astronomer J. P. L. de Cheseaux,[1] and, independently in 1826, by Heinrich Wilhelm Matthias Olbers[2] (1758–1840). Their argument was based on the most ancient of all astronomical observations, that the sky grows dark when the sun goes down.

To see the significance of this observation, note that if absorption is neglected, the apparent luminosity of a star of absolute luminosity L at a distance r in a naive cosmological model will be $L/4\pi r^2$. If the number density of such stars is a constant n, then the number of stars at distances between r and $r + dr$ is $4\pi n r^2\, dr$, so the total radiant energy density due to all stars is

$$\rho_s = \int_0^\infty \left(\frac{L}{4\pi r^2}\right) 4\pi n r^2\, dr = Ln \int_0^\infty dr \qquad (16.1.1)$$

The integral diverges, leading to an infinite energy density of starlight!

In order to avoid this paradox, both de Cheseaux and Olbers postulated the existence of an interstellar medium that absorbs the light from the very distant stars responsible for the divergence of the integral (16.1.1). However, this resolution of the paradox is unsatisfactory,[3] because in an eternal universe the temperature of the interstellar medium would have to rise until the medium was in thermal equilibrium with the starlight, in which case it would be emitting as much energy as it absorbs, and hence could not reduce the average radiant energy density. The stars themselves are of course opaque, and totally block out the light from sufficiently distant sources, but if *this* is the resolution of the Olbers paradox, then every line of sight must terminate at the surface of a star, so the whole sky should have a temperature equal to that at the surface of a typical star.

To see how modern cosmological models avoid the Olbers paradox, we note that according to Eq. (14.4.12), the apparent luminosity of a star of absolute luminosity L at a comoving coordinate r_1 is (now neglecting absorption)

$$l = \frac{L R^2(t_1)}{4\pi R^4(t_0) r_1{}^2}$$

where t_0 is the time the star is observed and t_1 is the time the light was emitted. Also, according to Eq. (14.7.4), the number of stars of luminosity between L and $L + dL$, whose light, observed at time t_0, was emitted between times $t_1 - dt_1$ and dt_1, is

$$dN = 4\pi R^2(t_1) r_1{}^2 n(t_1, L)\, dt_1\, dL$$

where $n(t_1, L)\, dL$ is the number density of stars at time t_1 with luminosity between L and $L + dL$. The total energy density of starlight is therefore

$$\rho_{s0} = \iint l\, dN = \int_{-\infty}^{t_0} \mathscr{L}(t_1) \left[\frac{R(t_1)}{R(t_0)}\right]^4 dt_1 \qquad (16.1.2)$$

where $\mathscr{L}$ is the proper luminosity density

$$\mathscr{L}(t_1) \equiv \int n(t_1, L) L\, dL$$

In a "big-bang" cosmology, there is obviously no paradox, since the integral (16.1.2) is effectively cut off at a lower limit $t_1 = 0$, and the integrand vanishes at $t_1 = 0$, roughly like $R(t_1)$. The question of an Olbers paradox arises only in models, such as the steady state cosmology, in which the universe is supposed to have existed for an infinitely long time. In such models, a necessary condition for avoidance of the Olbers paradox is that

$$t_1 R^4(t_1)\mathscr{L}(t_1) \to 0 \qquad \text{for } t_1 \to -\infty \tag{16.1.3}$$

For neutrinos there is a slightly stronger condition,[4] with $R^3(t_1)$ in place of $R^4(t_1)$, because one of the factors of $R(t_1)/R(t_0)$ in Eq. (16.1.2) arose from the loss of energy by individual red-shifted photons, and for neutrinos the *number* density as well as the *energy* density is in principle observable. The only popular cosmology in which (16.1.3) is not satisfied is the oscillating model discussed in Section 15.11. In this case, absorption is needed to avoid an Olbers paradox, but the absorption occurs during the highly contracted era, and the red shift during the subsequent expansion saves us from an intolerably bright night sky. From this point of view, the 2.7°K microwave background appears as the pale image of the fiery furnace with which we were threatened by de Cheseaux and Olbers.

2   Models with a Cosmological Constant

When Einstein formulated the general theory of relativity in 1916, the universe was generally believed to be static. According to Eqs. (15.1.18) and (15.1.19), the scale factor $R(t)$ can only be constant if

$$\rho = -3p = 3k/8\pi G R^2$$

However, this requires that either the energy density ρ or the pressure p should be *negative*. In order to avoid this unphysical result, Einstein in 1917 modified his equations to read[5]

$$R_{\mu\nu} - \tfrac{1}{2}g_{\mu\nu}R_\rho{}^\rho - \lambda g_{\mu\nu} = -8\pi G T_{\mu\nu} \tag{16.2.1}$$

where λ is a new fundamental constant, the so-called *cosmological constant*.

We have already noted at the end of Section 7.1 that Eq. (16.2.1) is the most general modification of Einstein's equations that preserves the feature that $T_{\mu\nu}$ is set equal to a tensor that is constructed from $g_{\mu\nu}$ and its first and second derivatives, and is linear in the second derivatives of $g_{\mu\nu}$. However, for our present purposes it is more convenient to move $\lambda g_{\mu\nu}$ to the right-hand side of the equations, writing

$$R_{\mu\nu} - \tfrac{1}{2}g_{\mu\nu}R^\rho{}_\rho = -8\pi G \tilde{T}_{\mu\nu} \tag{16.2.2}$$

where $\tilde{T}_{\mu\nu}$ is a modified energy-momentum tensor:

$$\tilde{T}_{\mu\nu} \equiv T_{\mu\nu} - \frac{\lambda}{8\pi G}\, g_{\mu\nu} \tag{16.2.3}$$

If $T_{\mu\nu}$ has the perfect-fluid form (15.1.12), then so does $\tilde{T}_{\mu\nu}$:

$$\tilde{T}_{\mu\nu} = \tilde{p}g_{\mu\nu} + (\tilde{p} + \tilde{\rho})U_\mu U_\nu \tag{16.2.4}$$

with a modified density and pressure

$$\tilde{p} = p - \frac{\lambda}{8\pi G} \qquad \tilde{\rho} = \rho + \frac{\lambda}{8\pi G} \tag{16.2.5}$$

All of the results obtained in Section 15.1 still apply to theories with a cosmological constant, provided that we replace the quantities p and ρ with the modified density and pressure (16.2.5).

In particular, the conditions for a static universe now read

$$\tilde{\rho} = -3\tilde{p} = \frac{3k}{8\pi G R^2} \tag{16.2.6}$$

For a "dust"-filled universe with $p = 0$, this gives

$$\frac{k}{R^2} = \lambda \tag{16.2.7}$$

$$\rho = \frac{\lambda}{4\pi G} \tag{16.2.8}$$

In order to have ρ positive, Eq. (16.2.8) requires that λ should be positive, and Eq. (16.2.7) then tells us that

$$k = +1 \tag{16.2.9}$$

and

$$R = \frac{1}{\sqrt{\lambda}} \tag{16.2.10}$$

The static Einstein universe is therefore *finite* (though of course unbounded), with a positive curvature and a density that are fixed by the fundamental constants λ and G.

Of course, the discovery during the 1920's of a systematic relation between red shift and distance removed any interest in the static Einstein universe as a realistic cosmological model. Nevertheless, the existence of a cosmological constant remains a logical possibility, and cosmologists have thoroughly explored the dynamics of expanding universes with a cosmological constant.[6] We shall restrict our attention here to models with zero pressure, so that Eq. (15.1.21) gives ρR^3 constant. It is convenient to express this constant in terms of the value it would have in a static Einstein model:

$$\rho R^3 = \frac{\alpha}{4\pi G \sqrt{|\lambda|}} \tag{16.2.11}$$

The dynamical equation (15.1.20), with ρ replaced by the modified density $\tilde{\rho}$ defined in Eq. (16.2.5), now reads

$$\dot{R}^2 = \frac{1}{R} \left\{ \frac{\lambda R^3}{3} - kR + \frac{2\alpha}{3\sqrt{|\lambda|}} \right\} \tag{16.2.12}$$

The qualitative behavior of $R(t)$ depends on the pattern of zeros, maxima, and minima of the cubic on the right-hand side. There are three special cases of particular interest, associated with the names of de Sitter, Lemaître, and Eddington and Lemaître.

In the *de Sitter model*,[7] space is essentially empty and flat, so that $k = \alpha = 0$, and λ is positive. Equation (16.2.12) then has the simple solution

$$R \propto e^{Ht} \tag{16.2.13}$$

$$H = \left(\frac{\lambda}{3}\right)^{1/2} \tag{16.2.14}$$

The metric here is the same as in the steady state model discussed in Section 14.8, with the difference that instead of matter being continuously created, there is no matter at all! As discussed in Section 13.3, this metric has a ten-parameter group of isometries, which is just the group of "rotations" in five-dimensions that leave invariant a diagonal matrix with elements $+1$, $+1$, $+1$, $+1$, -1. This group is therefore often called the *de Sitter group*. Although the absence of matter in the de Sitter model removes it from consideration as a serious model of the universe, it should be noted that any model with $\lambda > 0$ goes over to a de Sitter model for $R \to \infty$.

In what is known as the *Lemaître model*,[8] space is positively curved, λ is positive, and more matter is present than in a static Einstein model, so that $k = +1$, and $\alpha > 1$. According to Eq. (16.2.12), the scale factor R starts expanding at $t = 0$ like $t^{2/3}$, but the expansion then slows down, reaching a minimum rate at $R = \alpha^{1/3}/\sqrt{\lambda}$, after which it speeds up again, ultimately approaching the de Sitter result (16.2.13). The most remarkable feature of this model is the existence of a "coasting period" during which $R(t)$ remains close to the value $R = \alpha^{1/3}/\sqrt{\lambda}$ at which $\dot{R}$ has its minimum. During this period, the differential equation (16.2.12) with $k = +1$ takes the approximate form

$$\dot{R}^2 \simeq \alpha^{2/3} - 1 + (\sqrt{\lambda}\, R - \alpha^{1/3})^2$$

The solution is

$$R = \frac{\alpha^{1/3}}{\sqrt{\lambda}} [1 + (1 - \alpha^{-2/3})^{1/2} \sinh (\sqrt{\lambda}\, (t - t_m))]$$

where t_m is the time at which $\dot{R}$ reaches its minimum. If α is very close to unity, then R will remain close to the static Einstein value for a long time, of order

$$\Delta t = \lambda^{-1/2} |\ln (1 - \alpha^{-2/3})| \tag{16.2.15}$$

The *Eddington-Lemaître model* is a limiting case of the Lemaître models, given particular prominence through the work of Eddington.[9] It has the same curvature and mass as the static Einstein model; that is, $k = +1$ and $\alpha = 1$, and behaves like a Lemaître model with an infinitely long "coasting period."

Thus, if we start with $R = 0$ at $t = 0$, then R asymptotically approaches the Einstein value $1/\sqrt{\lambda}$ for $t \to \infty$. On the other hand, if we start with $R = 1/\sqrt{\lambda}$ for $t = 0$, then R grows monotonically, ultimately approaching the de Sitter exponential growth (16.2.13). This incidentally shows that the Einstein model is *unstable*, because if it is subjected to an infinitesimal expansion or contraction, then R must go on expanding or contracting, with a time dependence given by the Eddington-Lemaître model.

The observed concentration of the red shifts of quasi-stellar objects around $z \simeq 2$ (see Sections 11.6, 14.8) has revived interest in the Lemaître models,[10] since it suggests that an unusually large number of QSO's were present at a particular value of the scale factor, $R \simeq R_0/3$, as would be expected in a Lemaître model for which the "coasting" radius $\alpha^{1/3}/\sqrt{\lambda}$ had this particular value. By taking α close to unity, we can make the "coasting period" as long as we want, so that the predominance of the particular red shift $z \simeq 2$ can be made as pronounced as may be needed to account for the QSO observations. With this new motivation, there have recently been carried out studies of the propagation of light signals around the universe,[11] of the radio number counts,[12] and of the formation of galaxies,[13] in Lemaître models. There is no definite evidence against the Lemaître models, but they seem an artificial way to account for what may be merely a detailed feature of the evolution of the quasi-stellar objects.

3   The Steady State Model Revisited

If the universe is not only isotropic and homogeneous in space, but also homogeneous in *time*, then, as shown in Section 14.8, its metric must have the Robertson-Walker form, with

$$k = 0 \qquad R(t) \propto e^{Ht} \tag{16.3.1}$$

where H is the Hubble constant, here truly a constant of nature. Also, all scalars, such as ρ and p, must be time- as well as position-independent:

$$\dot{\rho} = \dot{p} = 0 \tag{16.3.2}$$

The field equations underlying the steady state model were left unspecified in Section 14.8, but it is clear that Einstein's theory would need to be modified to be used here. The Einstein field equations are only consistent with the Bianchi identities if the energy-momentum tensor is conserved, but a constant pressure would violate the energy-conservation equation (14.2.19) unless $\rho = -p$, which would require either the energy-density or the pressure to be negative.

It is therefore necessary to modify the Einstein equations by adding a correction term[14] $C_{\mu\nu}$:

$$R_{\mu\nu} - \tfrac{1}{2}g_{\mu\nu}R^\lambda{}_\lambda + C_{\mu\nu} = -8\pi G T_{\mu\nu} \tag{16.3.3}$$

A straightforward calculation, using Eq. (16.3.1) in (15.1.6), (15.1.7), and (15.1.11), gives

$$R_{\mu\nu} - \tfrac{1}{2}g_{\mu\nu}R^{\lambda}{}_{\lambda} = 3H^2 g_{\mu\nu}$$

so the form of the correction term required by the steady state model is

$$C_{\mu\nu} = -(8\pi G p + 3H^2)g_{\mu\nu} - 8\pi G(\rho + p)U_{\mu}U_{\nu} \qquad (16.3.4)$$

where U^{μ} is the velocity four-vector, with $U^t = 1$ and $U^i = 0$.

In order to learn anything from Eq. (16.3.4), we need to impose some *a priori* ideas as to the form of the tensor $C_{\mu\nu}$. Hoyle[14] suggests that in general

$$C_{\mu\nu} = C_{;\mu;\nu} \qquad (16.3.5)$$

where C is a scalar, called the *C-field*. Hoyle further suggests that in the absence of all inhomogeneities or anisotropies, C is simply proportional to the cosmic time coordinate used in the Robertson-Walker coordinate system:

$$C = At \qquad A \text{ constant} \qquad (16.3.6)$$

It is easy to calculate the second covariant derivative:

$$C_{;\mu;\nu} = -AH(g_{\mu\nu} + U_{\mu}U_{\nu}) \qquad (16.3.7)$$

Comparison of (16.3.7) with (16.3.4) shows that the density must take the value

$$\rho = \frac{3H^2}{8\pi G} \qquad (16.3.8)$$

and the constant of proportionality in C is

$$A = \frac{8\pi G(\rho + p)}{H} \qquad (16.3.9)$$

The pressure can take any value.

The predicted density (16.3.8) is the same as given by a Friedmann model with vanishing curvature. [See Eq. (15.2.1).] Thus the verification of Eq. (16.3.8) would not really serve to confirm the steady state model. Also, the steady state cosmology does not require that the $C_{\mu\nu}$ tensor take the form (16.3.5), (16.3.6), so we would not be forced to give up the steady state metric if the density were found to be different from (16.3.8).

The clearest evidence against the steady state model is the observed cosmic microwave background, which seems to be a remnant of a quite different earlier stage of the universe. (See Section 15.5.) However, it is not out of the question for a microwave background to be created along with the baryons in a steady state model. According to Eq. (15.4.9), the number density of photons per unit frequency interval in a steady state model is

$$
n_\gamma(v) = 8\pi v^2 \int_{-\infty}^{t_0} \exp\left(-\int_t^{t_0} \{\Lambda(ve^{H(t_0-t')}) - \Omega(ve^{H(t_0-t')})\}\, dt'\right)
$$
$$
\times\ \Omega(ve^{H(t_0-t)})\, dt
$$

where $\Lambda(v)$ is the absorption rate of a photon of frequency v, and $8\pi v^2 \Omega(v)\, dv$ is the emission rate per unit volume of photons between frequency v and $v + dv$. By a simple change of variables this can be written in the t_0-independent form

$$
n_\gamma(v) = 8\pi v^2 \int_v^\infty \frac{dv'}{Hv'}\, \Omega(v') \exp\left(-\int_v^{v'} \frac{dv''}{Hv''}\, [\Lambda(v'') - \Omega(v'')]\right) \quad (16.3.10)
$$

Differentiating with respect to v yields a differential equation for $n_\gamma(v)$, which can also be written as a formula for $\Omega(v)$ in terms of $\Lambda(v)$, $n_\gamma(v)$ and $n_\gamma'(v)$:

$$
\Omega(v) = \frac{[\Lambda(v) + 2H]n_\gamma(v) - Hvn_\gamma'(v)}{8\pi v^2 + n_\gamma(v)} \quad (16.3.11)
$$

Thus, by a suitable choice of the photon emission rate, we can arrange to get any background distribution function $n_\gamma(v)$ we want. For instance, if we demand the observed Rayleigh-Jeans low-frequency behavior $n_\gamma(v) \propto v$ [see Eq. (15.5.19)] then (16.3.11) yields in the limit $v \to 0$:

$$
\Omega(0) = \Lambda(0) + H \quad (16.3.12)
$$

The term H represents a purely cosmological continuous creation of photons, unrelated to any absorption processes. We can also obtain a Planck distribution function

$$
n_\gamma(v) = \frac{8\pi v^2}{[\exp(hv/kT) - 1]}
$$

by choosing $\Omega(v)$ as

$$
\Omega(v) = e^{-hv/kT}\Lambda(v) + \frac{Hhv/kT}{[\exp(hv/kT) - 1]} \quad (16.3.13)
$$

The first term represents simply the usual emission processes that will always accompany any absorption [compare Eq. (15.4.7)], while the second term represents a continuous creation of photons. However, there is no *a priori* reason why the rate of continuous creation of photons should have the particular frequency dependence shown in Eq. (16.3.13), so from the standpoint of a steady state model, the Planck distribution law is possible but quite artificial. Indeed, there is no particular reason why low-frequency photons should be continuously created at precisely the rate $8\pi Hv^2\, dv$ required by Eq. (16.3.12), so even the Rayleigh-Jeans low-frequency behavior is somewhat unnatural.

Some support for the steady state model comes from quite a different quarter. From time to time attempts have been made to formulate electrodynamics and other field theories in terms of a direct action at a distance.[15] Such attempts generally foundered, because the electromagnetic effects of charged particles were found to correspond to an equal mixture of advanced and retarded solutions of the Maxwell equations, rather than the usual retarded solution. In 1945 Wheeler and Feynman[16] showed that this difficulty could be overcome by taking into account the electromagnetic interaction at a distance of the accelerated and test charges with all the other charges in the universe. However, they considered a static cosmological model, and so could obtain a net electromagnetic interaction corresponding to either a pure retarded or a pure advanced solution. Hogarth[17] later suggested that this ambiguity could be removed by considering a more realistic model, taking into account the expansion of the universe. According to Hoyle and Narlikar,[18] only a purely retarded solution is possible for a steady state model, while only a purely advanced solution is possible for a Friedmann model with $k \leq 0$. Hoyle and Narlikar subsequently extended these considerations to the C-field,[19] to the theory of gravitation,[20] and to quantum electrodynamics.[21] This line of development certainly represents an intriguing approach to the old problem of relating the physics of the microcosm to the properties of the universe as a whole, a problem to which we shall return in the next section. However, it is too early to conclude that a steady state universe is in any sense required by considerations of microphysics, because there is no reason to suppose that electrodynamics and other field theories *need* to be formulated in terms of an action at a distance.

4 Models with a Varying Constant of Gravitation

Gravitational forces are remarkably weak by the standards of atomic or nuclear physics. For instance, the ratio of the gravitational to the electric force between the electron and the proton has the value

$$Gm_p \frac{m_e}{e^2} = 4.4 \times 10^{-40} \qquad (16.4.1)$$

Despite many attempts,[22] there have been no convincing explanations of why such a tiny dimensionless number should appear in the fundamental laws of physics. However, there is one clue, which suggests that numbers like (16.4.1) are not determined solely by considerations of microphysics, but in part by the influence of the whole universe. This clue is simply the fact that, from the quantities $G, \hbar, c$, and the Hubble constant H_0, it is possible to construct a mass, which is not too different from the mass of a typical elementary particle, such as the pion:

$$\left(\frac{\hbar^2 H_0}{Gc}\right)^{1/3} \approx m_\pi \qquad (16.4.2)$$

(For $H_0{}^{-1} = 10^{10}$ years, the left-hand side has the value 60 MeV$/c^2$, while the pion mass is 140 MeV$/c^2$. If e^2/c were used in place of $\hbar$, the left-hand side of (16.4.2) would become of the same order of magnitude as the electron mass.) Of course, one is perfectly free to regard (16.4.2) as a meaningless numerical coincidence, but it should be noted that the particular combination of $\hbar$, H_0, G, and c appearing in (16.4.2) is very much closer to a typical elementary particle mass than other random combinations of these quantities; for instance, from $\hbar$, G, and c alone one can form a single quantity $(\hbar c/G)^{1/2}$ with the dimensions of a mass, but this has the value 1.22×10^{22} MeV$/c^2$, more than a typical particle mass by about 20 orders of magnitude!

In considering the possible interpretations of Eq. (16.4.2), one should be careful to distinguish it from other numerical "coincidences" such as the rough relation among G, H_0, m_p, and the present cosmic baryon number density n_0:

$$Gn_0m_p \approx H_0{}^2 \tag{16.4.3}$$

This is a relation between *two* cosmological parameters, n_0 and H_0, and is in fact required by various cosmological models, such as the Friedmann models (unless $q_0 \ll 1$ or $q_0 \gg 1$) and Hoyle's version of the steady state model. [See Eqs. (15.2.6) and (16.3.8).] In contrast, Eq. (16.4.2) relates a single cosmological parameter, H_0, to the fundamental constants $\hbar$, G, c and m_π, and is so far unexplained.

There are other numerical coincidences that have been noticed from time to time, but most of these are combinations of (16.4.2) and (16.4.3), sometimes with $e^2/c = 137\hbar$ in place of $\hbar$ and with other masses in place of m_π. For instance, it has often been remarked that the ratio of the atomic time unit e^2/m_ec^3 to the Hubble time $H_0{}^{-1}$ is of the same order of magnitude, about 10^{-40}, as the ratio (16.4.1) of gravitational and electric forces in atoms, but this is equivalent to (16.4.2), with e^2/c in place of $\hbar$ and $m_e{}^{2/3}m_p{}^{1/3}$ in place of m_π.

If we choose to regard the numerical relation (16.4.2) as having a real though mysterious significance, then we must face the problem that in most cosmologies H_0 is not a constant, but a function of the age of the universe. One way of dealing with this problem is to replace H_0 with a quantity of comparable magnitude that *is* a constant; for instance, in a closed Friedmann model we can use the reciprocal of the time it takes for the universe to expand to its maximum extent, while in a steady state model, the Hubble constant itself will do. The only trouble with this approach is that it does not lead anywhere, and in particular, it leaves us with fundamental dimensionless constants, such as (16.4.1) or $Gm^2/\hbar c$, which are inexplicably tiny.

In 1937 a very different approach was suggested by Dirac.[23] He proposed that relations like (16.4.2) are fundamental though as yet unexplained truths, which remain valid, with a constant factor of proportionality, even though the Hubble "constant" $\dot{R}/R$ varies with the age of the universe. It follows then that one or more of the "constants" $\hbar$, G, c, and m_π must vary over cosmic time scales.

In order to avoid having to reformulate the whole of atomic and nuclear physics, Dirac chose G as the "constant" that varies with time, and in order to preserve (16.4.2), he proposed that

$$G \propto \frac{\dot{R}}{R} \tag{16.4.4}$$

In addition, Dirac suggested that relations like (16.4.3) also remain true, with a constant factor of proportionality, as the universe expands. Since $n \propto R^{-3}$, it follows that

$$GR^{-3} \propto \frac{\dot{R}^2}{R^2} \tag{16.4.5}$$

Eliminating $G(t)$ from (16.4.4) and (16.4.5) then yields a differential equation for $R(t)$:

$$\dot{R} \propto R^{-2}$$

with solution

$$R \propto t^{1/3} \tag{16.4.6}$$

Either (16.4.4) or (16.4.5) then gives the gravitational constant a time dependence

$$G \propto t^{-1} \tag{16.4.7}$$

Thus in Dirac's cosmology, there is no fundamental significance to very small dimensionless numbers like 10^{-40}; the reason that (16.4.1) is this small is simply that the universe is old.

For $k = \pm 1$, there are still significant cosmological parameters that are constant and grossly different from unity, such as the number nR^3 of particles within a sphere whose radius is of the order of the radius of curvature of space. To avoid this, Dirac also proposed that space is flat, with $k = 0$, so that the absolute value of the Robertson-Walker scale factor $R(t)$ and pure numbers like nR^3 should be of no physical significance.

If the gravitational constant varies, then general relativity needs to be replaced with some other field theory of gravitation. Dirac did not specify what this field theory would be like, so his cosmological model remained incomplete. Nevertheless, it made a number of definite predictions. First, Eq. (16.4.6) gives a relation between the present Hubble constant H_0 and the present age of the universe t_0:

$$t_0 = \tfrac{1}{3}H_0^{-1} \tag{16.4.8}$$

Even for H_0^{-1} as large as 13×10^9 years, this gives an age of only 4.3×10^9 years, less than the age of the earth and moon determined by radioactive dating (which does not involve assumptions about G). Thus the Dirac theory appears already to conflict with observation. Equation (16.4.6) gives a deacceleration parameter $q_0 = 2$, which cannot be ruled out with present data. (See Section

14.6.) Finally, Eq. (16.4.7) gives a present rate of decrease of the "constant" of gravitation

$$\left(\frac{\dot{G}}{G}\right)_0 = -t_0^{-1} = -3H_0 \qquad (16.4.9)$$

The observable implications of a decreasing gravitational constant are discussed at the end of this section.

Dirac's theory inspired a number of attempts to formulate a field theory of gravitation in which the effective "constant" of gravitation is some function of a scalar field. Jordan[24] proposed one theory, which involved a nonconserved energy-momentum tensor, and was severely criticized on these and other grounds by Fierz[25] and Bondi.[26] A subsequent reformulation[27] removed most of these objections, but Jordan's theory still did not successfully incorporate nonrelativistic matter. The most interesting and complete scalar-tensor theory of gravitation is that proposed by Brans and Dicke[28] in 1961, which we have already discussed in some detail in Sections 7.3 and 9.9. In this theory, the gravitational constant G is replaced with the reciprocal of a scalar field ϕ. In order to incorporate relations such as (16.4.3) into the theory, ϕ is assumed to obey a field equation

$$\Box^2\phi \equiv (\phi^{;\mu})_{;\mu} = \frac{8\pi}{3+2\omega} T^\mu{}_\mu \qquad (16.4.10)$$

where $T_{\mu\nu}$ is the energy-momentum tensor of matter (not including ϕ) and ω is a dimensionless coupling parameter. In order not to interfere with the successes of the Principle of Equivalence, ϕ is assumed not to enter into the equations of motion of ordinary matter and radiation, so that $T^{\mu\nu}$ obeys the familiar conservation law

$$T^{\mu\nu}{}_{;\nu} = 0 \qquad (16.4.11)$$

The Bianchi identities then require the gravitational field equations to take the form (7.3.14), or equivalently,

$$R_{\mu\nu} = -\frac{8\pi}{\phi}\left[T_{\mu\nu} - \left(\frac{1+\omega}{3+2\omega}\right)g_{\mu\nu}T^\lambda{}_\lambda\right]$$

$$-\frac{\omega}{\phi^2}\phi_{;\mu}\phi_{;\nu} - \frac{1}{\phi}\phi_{;\mu;\nu} \qquad (16.4.12)$$

This theory becomes equivalent to that of Jordan[27] in the special case of an energy-momentum tensor with vanishing trace.

In applying the Brans-Dicke theory to cosmology, we again consider the universe to be smeared out into a homogeneous isotropic continuum, as in Chapters 14 and 15. The metric then has the Robertson-Walker form (14.2.1); the energy-momentum tensor has the perfect fluid form (14.2.12), and the scalar field ϕ is a function of time alone. A straightforward calculation using Eqs. (15.1.6), (15.1.7), (15.1.11), and (15.1.13) gives the time-time component of Eq. (16.4.12) as

$$\frac{3\ddot{R}}{R} = -\frac{8\pi}{(3 + 2\omega)\phi} \{(2 + \omega)\rho + 3(1 + \omega)p\} - \frac{\omega\dot{\phi}^2}{\phi^2} - \frac{\ddot{\phi}}{\phi} \qquad (16.4.13)$$

while the space-space components of Eq. (16.4.12) give

$$-\frac{\ddot{R}}{R} - \frac{2\dot{R}^2}{R^2} - \frac{2k}{R^2} = -\frac{8\pi}{(3 + 2\omega)\phi} \{(1 + \omega)\rho - \omega p\} + \frac{\dot{\phi}\dot{R}}{\phi R} \qquad (16.4.14)$$

and the time-space components simply say that zero equals zero. The field equation (16.4.10) for ϕ here reads

$$\frac{d}{dt}(\dot{\phi}R^3) = \frac{8\pi}{(3 + 2\omega)}(\rho - 3p)R^3 \qquad (16.4.15)$$

and the conservation laws (16.4.11) give, as in Chapter 14,

$$\dot{\rho} = -\frac{3\dot{R}}{R}(\rho + p) \qquad (16.4.16)$$

By eliminating $\ddot{R}$ from Eqs. (16.4.13) and (16.4.14), and using Eq. (16.4.15) to eliminate $\ddot{\phi}$, we can derive a first-order equation analogous to (15.1.20):

$$\frac{\dot{R}^2}{R^2} + \frac{k}{R^2} = \frac{8\pi\rho}{3\phi} - \frac{\dot{\phi}\dot{R}}{\phi R} + \frac{\omega\dot{\phi}^2}{6\phi^2} \qquad (16.4.17)$$

We can recover (16.4.13) and (16.4.14) from the derivative of (16.4.17), so the fundamental equations of the Brans-Dicke cosmology may be taken as Eqs. (16.4.15)–(16.4.17), plus an equation of state giving p as a function of ρ. In addition, Eq. (9.9.11) shows that the gravitational "constant," measured by the observation of slowly moving particles or in time-dilation experiments, is

$$G = \left(\frac{2\omega + 4}{2\omega + 3}\right)\phi^{-1} \qquad (16.4.18)$$

For any given equation of state $p = p(\rho)$, Eqs. (16.4.15)–(16.4.17) may be regarded as a second-order differential equation plus two first-order differential equations for the three variables R, ϕ, and ρ. It follows that these equations uniquely determine $R(t)$, $\phi(t)$, and $\rho(t)$ for all t, provided we are given the present value of *four* variables, say R_0, $\dot{R}_0$, ϕ_0, and ρ_0, as well as the constants ω and k. This is rather surprising, because in a Friedmann model we only need to be given the initial values of *three* quantities, say R_0, $\dot{R}_0$, and of course G, in order to be able to calculate $R(t)$ and $\rho(t)$ for all t. [See Eq. (15.2.1).]

Originally, Brans and Dicke[28] eliminated this extra degree of freedom by imposing one additional constraint, that $\dot{\phi}R^3$ vanishes at the initial singularity where $R = 0$:

$$\dot{\phi}R^3 \to 0 \qquad \text{for } R \to 0 \tag{16.4.19}$$

With this initial condition, and with a given ω, k, and equation of state, we can obtain a complete solution for $R(t)$, $\rho(t)$, and $\phi(t)$ by specifying only three present parameters, such as R_0, $\dot{R}_0$, and ϕ_0, or, more conveniently, H_0, G_0, and q_0 (or ρ_0).

A few years later, Dicke[29] suggested that the relevant solutions may in fact be those that do *not* satisfy the constraint (16.4.19). In general, all solutions have a singularity with $R = 0$ at a finite time, which as usual we define to be $t = 0$. Equation (16.4.15) then has the solution:

$$\dot{\phi}(t)R^3(t) = \frac{8\pi}{2\omega + 3} \int_0^t [\rho(t') - 3p(t')]R^3(t') \, dt' - C \tag{16.4.20}$$

where C is an integration constant, which may be positive, negative, or zero. For $C = 0$, we obtain the three-parameter family of models satisfying the initial condition (16.4.19). For $C \neq 0$, we obtain a four-parameter family of solutions, the extra parameter being needed to fix the value of C.

The properties of these various solutions are sufficiently subtle to make it worth our while to study in some detail the one case where (16.4.15)–(16.4.17) can be solved analytically, the case of zero pressure and zero curvature:

$$p = 0 \qquad k = 0$$

Here (16.4.16) gives

$$\rho \propto R^{-3} \tag{16.4.21}$$

so Eq. (16.4.20) gives immediately

$$\dot{\phi} = \frac{8\pi\rho}{2\omega + 3}(t - t_c) \tag{16.4.22}$$

where

$$t_c \equiv \frac{(2\omega + 3)C}{8\pi\rho R^3} \tag{16.4.23}$$

It proves extremely convenient to introduce a new dependent variable

$$u \equiv (t - t_c)\frac{\dot{\phi}}{\phi} = \frac{8\pi\rho(t - t_c)^2}{(2\omega + 3)\phi} > 0 \tag{16.4.24}$$

By expressing ρ and $\dot{\phi}/\phi$ in Eq. (16.4.17) in terms of u, and setting $k = 0$, we can immediately solve for $\dot{R}/R$:

$$\frac{2(t - t_c)\dot{R}}{R} = -u \pm \left(\frac{3 + 2\omega}{3}\right)^{1/2}(u^2 + 4u)^{1/2} \tag{16.4.25}$$

Also, Eq. (16.4.21) and the logarithmic derivative of (16.4.22) give

$$\frac{\dot{u}}{u} = -\frac{3\dot{R}}{R} + 2(t - t_c)^{-1} - \frac{\dot{\phi}}{\phi}$$

or, using (16.4.24) and (16.4.25),

$$(t - t_c)\dot{u} = \tfrac{1}{2}u\left\{u + 4 \mp 3\left(\frac{3 + 2\omega}{3}\right)^{1/2}(u^2 + 4u)^{1/2}\right\} \quad (16.4.26)$$

This first-order equation must be integrated to find $u(t)$, following which (16.4.24) and (16.4.25) can be integrated to determine $\phi(t)$ and $R(t)$.

One obvious class of solutions to Eq. (16.4.26) are those with u a constant, equal to one of the zeros of the expression on the right-hand side of Eq. (16.4.26). In order to have such a zero with $u > 0$, we must take the upper sign of the square root in Eqs. (16.4.26) and (16.4.25), and the solution is

$$u = \frac{2}{3\omega + 4} \quad (16.4.27)$$

For this solution, we must take $t_c = 0$, because otherwise Eq. (16.4.25) would give $R = 0$ only at the time $t = t_c$, and we have agreed to set out clocks so that this singularity occurs at $t = 0$. With $t_c = 0$, Eqs. (16.4.24) and (16.4.25) yield the solutions[28]

$$\phi \propto t^{2/(4+3\omega)} \quad (16.4.28)$$

$$R \propto t^{(2\omega+2)/(3\omega+4)} \quad (16.4.29)$$

$$\frac{4\pi\rho t^2}{\phi} = \frac{(2\omega + 3)}{(3\omega + 4)} \quad (16.4.30)$$

For $t_c \neq 0$, it is necessary to analyze how u in Eq. (16.4.26) moves between the singular points $u = 0$, $u = 2/(3\omega + 4)$, and $u = \infty$. The results depend critically on whether t_c is positive or negative.

$t_c > 0$. Here u drops monotonically from $u = \infty$ at $t = 0$ to $u = 0$ at $t = t_c$, and then rises monotonically to the value (16.4.27) as $t \to \infty$. The sign of the square root in Eqs. (16.4.25) and (16.4.26) switches at t_c from the bottom sign for $t < t_c$ to the top sign for $t > t_c$. The solutions of (16.4.26) are thus given by

$$\ln\left(1 - \frac{t}{t_c}\right) = -2\int_0^u \frac{du}{u\{u + 4 + 3(1 + 2\omega/3)^{1/2}(u^2 + 4u)^{1/2}\}}$$

$$\text{for } t < t_c$$

$$\ln\left(\frac{t}{t_c} - 1\right) = 2\int_0^u \frac{du}{u\{u + 4 - 3(1 + 2\omega/3)^{1/2}(u^2 + 4u)^{1/2}\}}$$

$$\text{for } t > t_c$$

These integrals can be done in closed form, but it is more interesting to look at the behavior of the solutions at very early and very late times. For $t \ll t_c$, we find

$$u \to \frac{(3[1 + 2\omega/3]^{1/2} - 1)}{(4 + 3\omega)(t/t_c)}$$

so Eqs. (16.4.24) and (16.4.25) have the solutions[29]

$$\phi \propto t^{(1 - 3[1 + 2\omega/3]^{1/2})/(4 + 3\omega)} \tag{16.4.31}$$

$$R \propto t^{(1 + \omega + [1 + 2\omega/3]^{1/2})/(4 + 3\omega)} \tag{16.4.32}$$

For $t \gg t_c$, u approaches the value (16.4.27), and the solutions of (16.4.24) and (16.4.25) go over to the forms (16.4.28) and (16.4.29).

$t_c < 0$.   Here u drops monotonically from $u = \infty$ at $t = 0$ to the value (16.4.27) as $t \to \infty$. The square roots in (16.4.25) and (16.4.26) keep the upper sign, so (16.4.26) has the solution

$$\ln\left(1 + \frac{t}{|t_c|}\right) = 2 \int_u^\infty \frac{du}{u\{3(1 + 2\omega/3)^{1/2}(u^2 + 4u)^{1/2} - u - 4\}}$$

For $t \ll |t_c|$, we find

$$u \to \frac{(3[1 + 2\omega/3]^{1/2} + 1)}{(4 + 3\omega)(t/|t_c|)}$$

so Eqs. (16.4.24) and (16.4.25) have the solutions[29]

$$\phi \propto t^{(3[1 + 2\omega/3]^{1/2} + 1)/(4 + 3\omega)} \tag{16.4.33}$$

$$R \propto t^{(1 + \omega - [1 + 2\omega/3]^{1/2}/(4 + 3\omega)} \tag{16.4.34}$$

For $t \gg |t_c|$, u approaches the value (16.4.27), and the solutions of (16.4.24) and (16.4.25) again go over to the forms (16.4.28) and (16.4.29).

Thus there are three kinds of solution, all of which behave alike for $t \gg |t_c|$, but which differ radically for $t \lesssim |t_c|$. Only the simple solution with $t_c = 0$ goes over smoothly to the zero curvature Friedmann solution (ϕ constant, $R \propto t^{2/3}$) in the limit of large ω; the solutions with $t_c > 0$ or $t_c < 0$ have $\phi \to \infty$ or $\phi \to 0$, respectively, as $t \to 0$ for any finite ω.

Although these solutions were derived under the assumption of zero pressure and zero curvature, they exhibit many of the properties of the much more complicated general solutions. In general, the solutions may be classified according to whether the integration constant C in (16.4.20) is zero, or positive, or negative. For sufficiently large t, the integral in Eq. (16.4.20) is dominated by the matter-dominated era, in which $\rho \propto R^{-3}$, so the integral grows like t, and thus the integration constant eventually becomes negligible. In this limit, we have

$$\dot{\phi} = \frac{8\pi\rho t}{2\omega + 3} \qquad (16.4.35)$$

and all solutions converge to the $C = 0$ solution, which unfortunately must be calculated by a numerical solution of (16.4.17) and (16.4.35) with $\rho \propto R^{-3}$. On the other hand, for t sufficiently small, the integration constant will dominate in (16.4.20), provided of course that $C \neq 0$. In this case, the curvature and density terms in (16.4.17) become negligible for $t \to 0$, and the solutions go over to the previously derived forms

$$\phi \propto t^{(1 \mp 3[1 + 2\omega/3]^{1/2})/(4 + 3\omega)} \qquad (16.4.36)$$

$$R \propto t^{(1 + \omega \pm [1 + 2\omega/3]^{1/2})/(4 + 3\omega)} \qquad (16.4.37)$$

with upper sign for $C > 0$ and lower sign for $C < 0$. The $C = 0$ solutions go over smoothly to the Friedmann-model solutions for large ω, but the $C \neq 0$ solutions behave peculiarly at $t = 0$ for any ω.

The Brans-Dicke theory does not offer a satisfactory solution to the numerical relations discussed at the beginning of this section. Generally $\dot{\phi}/\phi$ and $1/t$ will be of the order of the Hubble "constant" H, and ϕ is of the order of $1/G$, so once the integration constant C becomes negligible, Eq. (16.4.35) will become more or less the same as the relation (16.4.3). However, Eq. (16.4.3) is not even approximately valid at very early times when C is not negligible. More important, the mysterious relation (16.4.2) is not explained at all by the Brans-Dicke theory. Indeed, in the simplest case of zero pressure, zero curvature, and zero integration constant t_c, Eqs. (16.4.29) and (16.4.28) show that $H \propto 1/t$ while $G \propto t^{-2/(4 + 3\omega)}$, so the mass $(\hbar^2 H/Gc)^{1/3}$ decreases with time, and the relation (16.4.2) can only be valid for a brief period in the history of the universe.

Now let us turn to the observational implications of this theory. Neither the gravitational field nor the Brans-Dicke field have any direct effect on the nuclear processes that are believed to produce helium in the early universe, but they do affect the rate of expansion of the universe, which in turn governs the amount of helium that can be produced. (See Section 15.7.) For solutions with $C = 0$, the numerical integration[29] of Eqs. (16.4.17) and (16.4.20) shows that in the case $\omega = 5$, $k = 0$, $H_0^{-1} = 9.5 \times 10^9$ years, $\rho_0 = 2 \times 10^{-29}$ g/cm^3, the effect of the Brans-Dicke field is to shorten the time required for the temperature to drop to 10^9°K by a factor of 0.45, so that more neutrons are left when nucleosynthesis begins, and the cosmologically produced abundance of helium is about 42% by weight, rather than 27%. For $k = -1$ models with a smaller present density, the difference between the Friedmann and Brans-Dicke models is considerably less.[30] On the other hand, with a nonvanishing integration constant in (16.4.20), we can make the expansion rate in the early universe essentially anything we like. As remarked in Section 15.7, for a moderate speed-up of the expansion the helium production is enhanced, but if the expansion is speeded up too much, then there will not be enough time for the reaction $n + p \to d + \gamma$ to produce enough deuterium to initiate nucleosynthesis, and very little helium will be produced.

During the more recent epoch within the range of optical telescopes, the integration constant C has presumably (though not certainly) been negligible, so that for large ω the relations among curvature, density, age, Hubble constant, and deacceleration parameter are pretty much the same as for the Friedmann models. For instance, for a zero-pressure zero-curvature model in the limit $t \gg |t_c|$, Eqs. (16.4.28)–(16.4.30) give the relations

$$H_0 t_0 = \frac{(2 + 2\omega)}{(4 + 3\omega)} \tag{16.4.38}$$

$$q_0 = \frac{\omega + 2}{2\omega + 2} \tag{16.4.39}$$

$$\frac{4\pi G \rho_0}{H_0{}^2} = \frac{(4 + 3\omega)(4 + 2\omega)}{(2 + 2\omega)^2} \tag{16.4.40}$$

For $\omega = 6$ these three quantities have the values 0.64, 0.57, 1.80, while the corresponding values for a Friedmann model with $k = 0$ are 0.67, 0.50, and 1.50.

Certainly the most distinctive observable feature of both the Dirac and the Brans-Dicke theories is the decrease of the gravitational constant G with time. In the Brans-Dicke theory the present rate of change of G is given by (16.4.35) as

$$\left(\frac{\dot{G}}{G}\right)_0 = -\left(\frac{\dot{\phi}}{\phi}\right)_0 = -\frac{8\pi\rho_0 t_0}{(2\omega + 3)\phi_0} = -\frac{8\pi G_0 \rho_0 t_0}{(2\omega + 4)} \tag{16.4.41}$$

In general, in order to express ρ_0 and t_0 in terms of G_0, H_0, and q_0, it would be necessary to resort to a numerical solution of the differential equations (16.4.35) and (16.4.17). However, if ω is reasonably large (say $\omega \gtrsim 5$), then the rate of decrease of G may be calculated to a sufficient degree of accuracy by using for ρ_0 and t_0 in (16.4.41) the values calculated using the Einstein equations in Sections 15.2. and 15.3. The general Friedmann-model result for ρ_0 is

$$\frac{4\pi G_0 \rho_0}{3 H_0{}^2} = q_0 \tag{16.4.42}$$

Values for $H_0 t_0$ and the resulting values for the rate (16.4.41) are given in Table 16.1.

In the special case $k = 0$, where we have an analytic solution, Eqs. (16.4.18), (16.4.28), and (16.4.29) yield the "exact" result

$$\left(\frac{\dot{G}}{G}\right)_0 = -\frac{H_0}{(1 + \omega)}$$

so the estimate of this rate given in Table 16.1 is in this case about 12% too low for $\omega = 6$. The estimates in Table 16.1 are in good agreement with the "exact" results that have been computed[30] for $k = -1$ and $\omega = 5$ or $\omega = 10$.

Table 16.1 Rate of Decrease of G in Various Brans-Dicke Models and in the Dirac Model[a]

Model	q_0	$t_0 H_0$ $(\omega = \infty)$	$(\dot{G}/G)_0$
Brans-Dicke	$\ll 1$	1	$-\dfrac{3q_0 H_0}{\omega + 2}$
Brans-Dicke	$\dfrac{1}{2}$	$\dfrac{2}{3}$	$-\dfrac{H_0}{\omega + 2}$
Brans-Dicke	1	$\dfrac{\pi}{2} - 1$	$-\dfrac{1.71 H_0}{\omega + 2}$
Brans-Dicke	$\gg 1$	$\dfrac{\pi}{2\sqrt{2q_0}}$	$-\dfrac{3.34 H_0 \sqrt{q_0}}{(\omega + 2)}$
Dirac	2	$\dfrac{1}{3}$	$-3H_0$

[a]The values of $(\dot{G}/G)_0$ in the Brans-Dicke models are estimated from Eq. (16.4.41), using for t_0 and ρ_0 the Friedmann model (i.e., $\omega = \infty$) results given in the third column and in Eq. (16.4.42), respectively.

For $H_0^{-1} = 10^{10}$ years, q_0 between 0.01 and 1.0, and $\omega = 6$, Table 16.1 gives a rate of decrease in G between 4×10^{-13} parts per year and 2×10^{-11} parts per year. In contrast, the Dirac model predicts a much more rapidly decreasing gravitational "constant"; for $H_0^{-1} = 10^{10}$ years, the rate of decrease of G is 3×10^{-10} parts per year.

The best experimental upper limit on the present rate of change of G comes from the analysis of radar observations of Mercury and Venus.[31] For a planet in a circular orbit with radius r and velocity v, we have $M_\odot G = v^2 r$, so if the orbital angular momentum mrv stays fixed while G changes, then r and v will vary as

$$r \propto \frac{1}{v} \propto \frac{1}{G} \tag{16.4.43}$$

and the orbital period $2\pi r/v$ will vary as

$$2\pi \frac{r}{v} \propto \frac{1}{G^2} \tag{16.4.44}$$

By repeated comparison of the orbital periods of the inner planets over the period 1966–1969, with time as told by an atomic clock (which does not depend on G), Shapiro et al.[31] have set an upper limit

$$\left|\frac{\dot{G}}{G}\right|_0 \lesssim 4 \times 10^{-10}/\text{year}$$

This is almost good enough to rule out the Dirac theory, but is not yet sufficiently stringent to put a useful limit on the Brans-Dicke coupling parameter ω. However, the error in these measurements of $(\dot{G}/G)_0$ is expected to decrease approximately with the 5/2-power of the time span of the observations, so another five years of observation should reduce the upper limit on $\dot{G}/G$ to the value expected in a Brans-Dicke model with q_0 of order unity and $\omega = 6$.

There is also a prospect of setting an upper limit on the rate of change of G from analysis of the flight time of laser signals, sent from the earth to the moon, and reflected back to earth by the corner reflectors placed on the moon's surface by the Apollo expeditions.[32] However, the analysis of these observations is seriously complicated by tidal effects, which play a major role in the dynamics of the earth-moon system. (Fortunately, such tidal effects do not seriously affect the planetary motions that were studied by Shapiro et al.)

Variations in G over the last few millenia can perhaps be determined from the study of ancient eclipse records.[33] A total eclipse of the sun occurs only over a very small portion of the earth's surface, so the knowledge that a particular total eclipse was seen at some particular place provides precise information on the ratio of the length of the day, which does not strongly depend on G, to the length of the year and the lunar month, which vary like $1/G^2$. The analysis by Curott[34] and Dicke[35] of five eclipses, which occurred between 1062 B.C. and 71 A.D., gives an average rate of decrease in the earth's rotation rate relative to planetary periods of $(15.9 \pm 0.7) \times 10^{-11}$ parts per year. The earth's rotation rate is subject to a number of known influences,[36] in particular a tidal deacceleration, between 23.5×10^{-11} and 25.6×10^{-11} parts per year, and an acceleration owing to the rise in sea level and the isostatic recovery of the geoid, between 0.5×10^{-11} and 3.0×10^{-11} parts per year. This leaves a residual unexplained *acceleration* of the earth's rotation between 4×10^{-11} and 10×10^{-11} parts per year. Since the eclipse data measure the rate of the earth's rotation relative to planetary periods, this apparent residual acceleration could be explained in terms of a *deacceleration* of planetary motions, owing to a decrease of G at a rate between 2×10^{-11} and 5×10^{-11} parts per year. [See Eq. (16.4.44).] However, the eclipse data are somewhat ambiguous. (Was Archilochus on Paros or Thasos during the eclipse of 648 B.C.?) More important, there are many uncertainties in the complicated dynamics of the earth-moon system that could account for the small residual apparent acceleration of the earth's rotation, without appealing to a decrease in G.

It may be possible to measure changes over the past 350 million years in the number of days in a lunar month or a year, by counting monthly or annual growth bands and daily growth ridges on fossil corals.[37] However, this approach has not yet yielded results that are precise enough to be of use to cosmologists.

A secular decrease in the constant of gravitation over billions of years would have interesting effects on the evolution of the earth and stars, but unfortunately, none of these effects would yield unambiguous information about whether G really does decrease. With decreasing G, the radius of the earth would increase roughly as $G^{-0.1}$, causing complicated damage to the earth's crust.[38] If G were larger in the past, then stars would have run through their thermonuclear evolution more rapidly;[39] for G decreasing at a rate of $(1-2) \times 10^{-11}$ parts per year, a star whose true age is 6 to 8 billion years would appear to us to be 15 to 25 billions years old.[40] Finally, if G were greater in the past, then the sun's luminosity $L_\odot$ would have been greater by a factor[41] roughly proportional to G^8, and the radius $r_\oplus$ of the earth's orbit would have been smaller by a factor proportional to G^{-1}, so the surface temperature of the earth, which varies more or less as $(L_\odot/r_\oplus{}^2)^{1/4}$, would have been greater by a factor proportional to $G^{2.5}$. If G decreases as $t^{-0.09}$, as would be expected according to Eq. (16.4.28) for a Brans-Dicke model with $k = 0$ and $\omega = 6$, and if the age of the universe is 8×10^9 years, then the temperature of the earth's surface 2×10^9 years ago would have been only about 20°C higher than at present, which need not have had any drastic effect on biological evolution. On the other hand, if G has decreased as $1/t$ as expected in Dirac's cosmology, then the temperature of the earth's surface 10^9 years ago would have been above the boiling point of water, unless the earth's albedo was very much higher than at present.[42] Thus, too large an early value of the constant of gravitation could have prevented the evolution of life forms capable of curiosity about the universe.

16 BIBLIOGRAPHY

See the bibliographies to Chapters 14 and 15, and the following:

☐ V. Petrosian and E. E. Salpeter, "Lemaître Models and the Cosmological Constant," Comments Astrophys. and Space Phys., **2**, 109 (1970).

☐ R. H. Dicke, "Gravitation an Enigma," 27th Joseph Henry Lecture of the Philosophical Society of Washington, J. Wash. Acad. Sci., **48**, 213 (1958).

16 REFERENCES

1. J. P. L. de Cheseaux, *Traité de la Comète* (Lausanne, 1744), pp. 223 ff; reprinted in *The Bowl of Night*, by F. P. Dickson (M.I.T. Press, Cambridge, Mass., 1968), Appendix II.

2. H. W. M. Olbers, *Bode's Jahrbuch*, 111 (1826); reprinted by Dickson, *op. cit.*, Appendix I. Interest in Olbers' paradox has been recently revived by Bondi; see H. Bondi, *Cosmology* (2nd ed., Cambridge University Press, 1960), Chapter III.

3. See Bondi, *op. cit.*, p. 21.

4. S. Weinberg, Nuovo Cimento, Series X, **25,** 15 (1962).

5. A. Einstein, Sitz. Preuss. Akad. Wiss., 142, (1917). For an English translation, see *The Principle of Relativity* (Methuen, 1923, reprinted by Dover Publications), p. 35.

6. See, for example, H. Bondi, *Cosmology* (2nd ed., Cambridge University Press, 1960), Chapter IX.

7. W. de Sitter, Proc. Roy. Acad. Sci. (Amsterdam), **19,** 1217 (1917); **20,** 229 (1917); **20,** 1309 (1917); Mon. Not. Roy. Astron. Soc., **78,** 3 (1917).

8. G. Lemaître, Ann. Soc. Sci. Brux, **A47,** 49 (1927); Mon. Not. Roy. Astron. Soc., **91,** 483 (1931).

9. A. S. Eddington, Mon. Not. Roy. Astron. Soc., **90,** 668 (1930).

10. V. Petrosian, E. E. Salpeter, and P. Szekeres, Ap. J., **147,** 1222 (1967); I. Shklovsky, Ap. J., **150,** L1 (1967); M. Rowan-Robinson, Mon. Not. Roy. Astron. Soc., **141,** 445 (1968).

11. V. Petrosian and E. E. Salpeter, Ap. J., **151,** 411 (1968).

12. N. S. Kardashev, Ap. J., **150,** L135 (1967); G. C. McVittie and R. Stabell, Ap. J., **150,** L141 (1967); V. Petrosian, Ap. J., **155,** 1029 (1969).

13. K. Brecher and J. Silk, Ap. J., **158,** 91 (1969).

14. F. Hoyle, Mon. Not. Roy. Astron. Soc., **108,** 372 (1948); Mon. Not. Roy. Astron. Soc., **109,** 365 (1949).

15. K. Schwarzschild, Nachr. Ges. Wiss. Göttingen, **128,** 132 (1903); H. Tetrode, Z. Phys., **10,** 317 (1922); A. D. Fokker, Z. Phys., **58,** 386 (1929); Physica, **9,** 33 (1929); *ibid.*, **12,** 145 (1932).

16. J. A. Wheeler and R. P. Feynman, Rev. Mod. Phys., **17,** 157 (1945); *ibid.*, **21,** 425 (1949).

17. J. E. Hogarth, Proc. Roy. Soc., **A267,** 365 (1962).

18. F. Hoyle and J. V. Narlikar, Proc. Roy. Soc., **A277,** 1 (1964).

19. F. Hoyle and J. V. Narlikar, Proc. Roy. Soc., **A282,** 178 (1964).

20. F. Hoyle and J. V. Narlikar, Proc. Roy. Soc., **A282,** 184, 191 (1964). Also see S. Deser and F. A. E. Pirani, Proc. Roy. Soc., **A288,** 133 (1965).

21. F. Hoyle and J. V. Narlikar, Ann. Phys., **54,** 207 (1969).

22. See, for example, A. S. Eddington, *Fundamental Theory* (Cambridge University Press, 1946).

23. P. A. M. Dirac, Nature, **139,** 323 (1937); Proc. Roy. Soc., **A165,** 199 (1938).

24. P. Jordan, Nature, **164,** 637 (1949); *Schwerkraft und Weltfall* (2nd ed., Vieweg und Sohn, Braunschweig, Germany, 1955).

25. M. Fierz, Helv. Phys. Acta, **29,** 128 (1956).

26. H. Bondi, *Cosmology* (2nd ed., Cambridge University Press, 1960), p. 163.

27. P. Jordan, Z. Phys., **157,** 112 (1959).

28. C. Brans and R. H. Dicke, Phys. Rev., **124,** 925 (1961).

29. R. H. Dicke, Ap. J., **152,** 1 (1968). This paper uses a modified formulation of the Brans-Dicke theory, given by R. H. Dicke, Phys. Rev., **125,** 2163 (1962). For a treatment closer to that presented here, see G. S. Greenstein, to be published.

30. G. S. Greenstein, Astrophys. Letters, **1,** 139 (1968).

31. I. I. Shapiro, W. B. Smith, M. E. Ash, R. P. Ingalls, and G. H. Pettengill, Phys. Rev. Letters, **26,** 27 (1971).

32. C. O. Alley, P. Bender, R. H. Dicke, J. Faller, P. Franken, H. Plotkin, D. T. Wilkinson, J. Geophys. Res., **70,** 2267 (1965); C. O. Alley et al., Science, **167,** 458 (1970).

33. J. Fotheringham, Mon. Not. Roy. Ast. Soc., **81,** 104 (1920).

34. D. R. Curott, Astron. J., **71,** 264 (1966).

35. R. H. Dicke, in *The Earth-Moon System*, ed. by B. G. Marsden and A. G. W. Cameron (Plenum Press, New York, 1966), p. 93; Physics Today, **20,** 55 (1967).

36. W. H. Munk and G. J. F. MacDonald, *The Rotation of the Earth* (Cambridge University Press, 1960).

37. C. T. Scrutton, Paleontology, **7,** 552 (1965); J. W. Wells, in *The Earth-Moon System*, ed. by B. G. Marsden and A. G. W. Cameron (Plenum Press, New York, 1966), p. 70.

38. R. H. Dicke, Science, **138,** 653 (1962).

39. R. H. Dicke, Rev. Mod. Phys., **34,** 110 (1962).

40. R. H. Dicke, in *Stellar Evolution*, ed. by R. F. Stein and A. G. W. Cameron (Plenum Press, New York, 1966), p. 319; Physics Today, **20,** 55 (1967).

41. R. H. Dicke, Rev. Mod. Phys., **29,** 355 (1957); **34,** 110 (1962).

42. E. Teller, Phys. Rev., **73,** 801 (1948); also see ref. 27.

APPENDIX
SOME USEFUL NUMBERS*

Numerical Constants

$\pi = 3.1415927$ $1'' = 4.8481 \times 10^{-6}$ radians

$e = 2.7182818$ $\ln 10 = 2.3025851$

Physical Constants

Speed of light	$c = 2.9979250(10) \times 10^{10}$ cm sec^{-1}
Gravitational constant	$G = 6.6732(31) \times 10^{-8}$ dyn cm^2 g^{-2}
	$G/c^2 = 7.425 \times 10^{-29}$ cm g^{-1}
Planck's constant	$\hbar = 6.582183(22) \times 10^{-16}$ eV sec
	$= 1.0545919(80) \times 10^{-27}$ erg sec
	$h = 2\pi\hbar = 6.625 \times 10^{-27}$ erg sec
Electron volt	1 eV $= 1.6021917(70) \times 10^{-12}$ erg
Electronic charge (unrat.)	$e = 4.803250(21) \times 10^{-10}$ esu

* Taken from "Review of Particle Properties," Particle Data Group, Rev. Mod.-Phys. **43**, No. 2, Part II, (1971) and *Astrophysical Quantities*, by C. W. Allen (Athlone Press, London, 1955). Where figures are given in parentheses, they indicate the one standard deviation uncertainty in the last digits of the main numbers.

Fine structure constant $\quad \alpha = e^2/\hbar c = 1/137.03602(21)$

Electron mass $\qquad\qquad m_e = 9.109558(54) \times 10^{-28}$ g

$$m_e c^2 = 0.5110041(16) \text{ MeV}$$

Proton mass $\qquad\qquad m_p = 1.67 \times 10^{-24}$ g

$$m_p c^2 = 938.2592(52) \text{ MeV}$$

Neutron mass $\qquad\qquad m_n c^2 = 939.5527(52) \text{ MeV}$

Rydberg $\qquad\qquad m_e e^4/2\hbar^2 = 13.605826(45) \text{ eV}$

Thomson cross-section $\quad 8\pi e^4/3m_e^2 c^4 = 0.6652453(61) \times 10^{-24} \text{ cm}^2$

Weak coupling constant $\qquad g_v c/\hbar^3 = 1.02 \times 10^{-5} m_p^{-2}$

Boltzmann constant $\qquad\qquad k = 1.380622(59) \times 10^{-16} \text{ erg } °\text{K}^{-1}$

$$k^{-1} = 11604.85(49)°\text{K/eV}$$

Black-body constant

$$a \equiv \pi^2 k^4/15 c^3 \hbar^3 = 7.5641 \times 10^{-15} \text{ erg cm}^{-3}°\text{K}^{-4}$$

Typical stellar mass $\quad (\hbar c/G)^{3/2} m_p^{-2} = 3.77 \times 10^{33}$ g

General Astronomical Constants

Sidereal year (1900) $\qquad$ 1 year $= 3.1558149984 \times 10^7$ sec

Light year $\qquad\qquad$ 1 light year $= 9.4605 \times 10^{17}$ cm

Mean earth-sun distance $\qquad$ 1 a.u. $= 1.495985(5) \times 10^{13}$ cm

Parsec $\qquad\qquad\qquad$ 1 pc $= 3.0856(1) \times 10^{18}$ cm

$$= 3.2615 \text{ light year}$$

Hubble time for a Hubble constant of 100 km sec^{-1} Mpc^{-1}

$$[100 \text{ km sec}^{-1} \text{ Mpc}^{-1}]^{-1} = 9.78 \times 10^9 \text{ years}$$

Solar mass $\qquad\qquad M_\odot = 1.989(2) \times 10^{33}$ g

$$M_\odot G/c^2 = 1.475 \text{ km}$$

Solar radius $\qquad\qquad R_\odot = 6.9598(7) \times 10^5 \text{ km}$

Dimensionless solar surface potential

$$M_\odot G/R_\odot c^2 = 2.12 \times 10^{-6}$$

Solar luminosity $\qquad\qquad L_\odot = 3.90(4) \times 10^{33} \text{ erg sec}$

Earth mass $\qquad\qquad M_\oplus = 5.977(4) \times 10^{27}$ g

$$M_\oplus G/c^2 = 0.443 \text{ cm}$$

Earth equatorial radius $\qquad R_{\oplus} = 6.37817(4) \times 10^3$ km

Dimensionless earth surface potential

$$M_{\oplus} G / R_{\oplus} c^2 = 6.95 \times 10^{-10}$$

Acceleration due to gravity at earth's surface

$$g = 980.665 \text{ cm sec}^{-2}$$

Velocity of earth satellite in low orbit

$$v_s = 7.9 \text{ km sec}^{-1}$$

Mean orbital velocity of earth $\qquad v_{\oplus} = 29.78 \text{ km sec}^{-1}$

Lunar mass $\qquad\qquad M_{\mathbb{C}} = 7.35 \times 10^{25}$ g

$$M_{\mathbb{C}} G / c^2 = 5.45 \times 10^{-3} \text{ cm}$$

Lunar radius $\qquad\qquad R_{\mathbb{C}} = 1738$ km

Dimensionless lunar surface potential

$$M_{\mathbb{C}} G / R_{\mathbb{C}} c^2 = 3.14 \times 10^{-11}$$

Mean earth-moon distance $\qquad r_{\mathbb{C}} = 3.84 \times 10^5$ km

Apparent luminosity of star with apparent bolometric magnitude m

$$l = 2.52 \times 10^{-5} \text{ erg cm}^{-2} \text{ sec}^{-1} \times 10^{-2m/5}$$

Absolute luminosity of star with absolute bolometric magnitude M

$$L = 3.02 \times 10^{35} \text{ erg sec}^{-1} \times 10^{-2M/5}$$

Elements of Planetary Orbits

Planet	Symbol	Period T (trop year)	Semilatus Rectum $L(10^6$ km)	Eccentricity e (in 1900)
Icarus		1.12	51.0	0.827
Mercury	☿	0.24085	55.46	0.205615
Venus	♀	0.61521	108.20	0.006820
Earth	⊕	1.00004	149.54	0.016750
Mars	♂	1.88089	225.95	0.093312
Jupiter	♃	11.86223	776.5	0.048332
Saturn	♄	29.45772	1423	0.055890
Uranus	♅	84.013	2863	0.0471
Neptune	♆	164.79	4498	0.0085
Pluto	♇	248.4	5500	0.2494

Selected Galaxies*

Local Group	Type	Distance (Mpc)	m_{pg}	cz (obs.) (km/sec)
Galaxy and Nearest Neighbors				
Galaxy	Sb or Sc	—	—	—
LMC	Ir or SBc	0.049	0.86	+280
SMC	Ir	0.058	2.86	+167
Ursa Minor	dE	0.077	?	?
Draco	dE	0.08	?	?
Sculptor	dE	0.09	10.5	?
Fornax	dE	0.13	9.1	+ 40
Leo I	dE	0.23	11.27	?
Leo II	dE	0.23	12.85	?
NGC6822	Ir	0.52	9.21	− 40
Other Members of Local Group				
NGC224(M31)	Sb	0.65	4.33	−270
NGC205	E6p	0.65	8.89	−240
NGC221(M32)	E2	0.65	9.06	−210
NGC147	dE4	0.65	10.57	?
NGC185	E0	0.65	10.29	−340
NGC598(M33)	Sc	0.74	6.19	−210
IC1613	Ir	0.74	10.00	−240
Maffei 1	E	~1	~5.8(vis)	?
Miscellaneous Bright Galaxies				
NGC3031(M81)	Sb	2.0	7.85	+ 80
NGC3034(M82)	Scp	2.0	9.20	+400
NGC5236(M83)	Sc	2.4	7.0	+320
NGC4826(M64)	?	3.7	9.27	+360
NGC5128(Cen A)	EOp(R)	~4.0	7.87	+260
NGC4736(M94)	Sbp	4.3	8.91	+350
NGC5055(M63)	Sb	4.3	9.26	+2600
NGC5194(M51)	Sb	4.3	9.26	+550
NGC5457(M101)	Sc	4.3	8.20	+400

* Distances, types, and magnitudes are mostly taken from the compilation of S. van den Bergh, *Observors Handbook of the Royal Canadian Astronomical Society*, 1971. Under "Type," E denotes "elliptical," with E0, E1, ... increasingly flat; S denotes "spiral," with SO, Sa, Sb, Sc, ... increasingly open; SB denotes "barred spiral," with SBO, SBa, SBb, SBc, ... increasingly open; Ir denotes "Irregular"; p denotes "peculiar"; d denotes "dwarf"; R denotes "strong radio source." For Maffei 1, see H. Spinrad et al., *Ap. J.*, **163**, L25 (1971).

Messier Galaxies in the Virgo Cluster (Visual Magnitudes)

	NGC4472(M49)	E4	15 ± 5	8.9	
	NGC4579(M58)	SBb	15 ± 5	9.9	
?	NGC4621(M59)	E5	15 ± 5	10.3	
	NGC4649(M60)	E2	15 ± 5	9.3	
	NGC4303(M61)	Sc	15 ± 5	9.7	
	NGC4374(M84)	E ?	15 ± 5	9.8	
	NGC4382(M85)	SO	15 ± 5	9.5	
	NGC4486(M87)	EOp(R)	15 ± 5	9.3	$+1220$
	NGC4501(M88)	Sb	15 ± 5	9.7	
	NGC4552(M89)	EO	15 ± 5	10.3	
	NGC4569(M90)	Sb	15 ± 5	9.7	
?	NGC4192(M98)	Sb	15 ± 5	10.4	
	NGC4254(M99)	Sc	15 ± 5	9.9	
	NGC4321(M100)	Sc	15 ± 5	9.6	
	NGC4594(M104)	Sb	15 ± 5	8.1	$+1020$

Selected Clusters of Galaxies

Cluster	Est. No. Galaxies	cz(km/sec)
Virgo	2500	1150
Pegasus I	100	3800
Pisces	100	5000
Cancer	150	4800
Perseus	500	5400
Coma	1000	6700
Hercules		10300
Pegasus II		12800
Cluster A	400	15800
Ursa Major I	300	15400
Leo	300	19500
Gemini	200	23300
Cor. Bor.	400	21600
Boötes	150	39400
Ursa Major II	200	41000
Hydra		60600

INDEX

Where page numbers are given in italics, they refer to publications cited in bibliographies, lists of references, and tables.